FLORA ZAMBESIACA

Flora terrarum Zambesii aquis conjunctarum

VOLUME NINE: PART FOUR

FLORA ZAMBESIACA

MOZAMBIQUE

MALAWI, ZAMBIA, ZIMBABWE

BOTSWANA

VOLUME NINE: PART FOUR

Edited by
G. V. POPE

on behalf of the Editorial Board:

Published by the Royal Botanic Gardens, Kew,
for the Flora Zambesiaca Managing Committee
1996

Typeset at the Royal Botanic Gardens, Kew, by Christine Beard

Printed in Great Britain by
Whitstable Litho Printers Ltd., Whitstable, Kent

ISBN 1 900347 04 0

CONTENTS

FAMILY INCLUDED IN VOLUME IX, PART 4

153. Euphorbiaceae

Subfamilies: Phyllanthoideae
Oldfieldioideae
Acalyphoideae
Crotonoideae
Euphorbioideae tribe Hippomaneae

LIST OF NEW NAMES PUBLISHED IN THIS WORK

	Page
Bridelia cathartica var. *melanthesoides* (Baill.) Radcl.-Sm. comb. nov.	16
Bridelia cathartica forma *melanthesoides* (Baill.) Radcl.-Sm. comb. nov.	17

Acknowledgements

The Flora Zambesiaca Managing Committee thanks M.A. Diniz and E. Martins of the Centro de Botânica, Lisbon, for their valuable help in reading and commenting on the text.

153. EUPHORBIACEAE

By A. Radcliffe-Smith*

Prostrate, erect or scandent, annual biennial or perennial, dioecious or monoecious herbs, shrubs or trees, succulent or not, spiny or unarmed, sometimes poisonous, sometimes with phylloclades, with or without a milky latex or coloured sap. Indumentum absent or of simple, branched or stellate hairs or peltate scales, the hairs sometimes urticating. Leaves usually alternate, sometimes opposite, rarely whorled, occasionally all three, green or scarious and squamiform, petiolate or sessile, stipulate or exstipulate, simple, lobed or compound, entire or variously toothed, peltate or not, palminerved or penninerved, often with a pair or several pairs of basal glands at or near the petiole insertion, and/or sometimes minutely pellucid-glandular-punctate beneath. Stipules free or connate, sometimes spathaceous, membranaceous, capilliform, glandular or spiny, subpersistent, readily caducous or absent. Inflorescences terminal, or axillary, lateral or leaf-opposed, cymose or paniculate, or racemose or spicate, or with the flowers aggregated into glanduliferous involucrate pseudanthia (cyathia), which may themselves be aggregated into pseudopleiochasial hyperinflorescences, or into involucrate capitula, or else with the flowers fasciculate or solitary, ramiflorous or cauliflorous. Flowers unisexual, usually actinomorphic, small to minute. Calyx in both sexes usually of 3–6 valvate, imbricate or open, equal or unequal lobes, or of free sepals, often dissimilar as between the two sexes, rarely the female calyx spathaceous, sometimes accrescent, or minute or absent. Corolla in one or both sexes of 3–6 free or rarely united subvalvate or imbricate petals, or petals minute or absent. Disk in the male flowers of 5–6 (occasionally more) free extrastaminal and/or interstaminal glands, less often the disk annular or cupular and extrastaminal or lobed, lobulate and intrastaminal with the stamens enfolded by the lobes, or else the disk receptacular, domed and aperturate with the stamens arising through the apertures, or disk absent; in the female flowers hypogynous, annular, flat or cupular, entire or lobed, rarely of separate glands, usually persistent, or absent. Stamens (1)3–100(1000), the filaments free or variously connate, simple or rarely branched, anthers erect or inflexed in bud, 2–4-thecous with the thecae usually parallel and adnate to the connective throughout, sometimes free at the base, occasionally completely free and erect, divaricate, horizontal or pendulous, rarely with the thecae superposed or laterally fused to form an annulus, dehiscing by slits longitudinally, obliquely or laterally, less often by pores (not Africa). Pistillode (non-functional ovary) often present, variously shaped. Ovary superior, sessile or rarely stipitate, 1–4(20)-locular but most commonly 3-locular; placentation axile, with the ovules solitary or paired and collateral in each locule and pendulous from its inner angle, anatropous or hemitropous, rarely orthotropous (*Panda*), crassinucellate, bi-integumental, hilum between the ventral raphe and the epimicropylar obturator, when present; funicle often thickened; styles (1)3(4)(20), free or connate, erect or spreading, entire, bifid, multifid or laciniate, the inner surface usually stigmatic throughout, smooth, granulate, papillose or fimbriate, often reddish- or purplish-tinged. Staminodes absent or occasionally present, usually subulate. Fruit smooth, wrinkled, warty, tubercled, horned or winged, rarely inflated, often schizocarpic, dehiscing septifragally and septicidally into 1–4(20) (but usually 3) bivalved cocci leaving a persistent angled or winged columella, the valves then dehiscing loculicidally, or else fruit loculicidally dehiscent into 3 separate valves, or breaking up irregularly, or else indehiscent, drupaceous or subdrupaceous, with a thin epicarp, with or without a fleshy mesocarp and with a 1–3(more)-locular paleaceous, crustaceous, ligneous or osseous endocarp. Seeds 1–2 per locule, or by abortion 1

* tribe Euphorbieae by Susan Carter and L.C. Leach will constitute volume 9 part 5.

per fruit, caruncutlate or not, often myrmecochorous; sarcotesta sometimes present; sclerotesta thin, crustaceous to thick, osseous; endosperm copious, fleshy, or absent; embryo straight to curved or folded, extending for most of the length of the seed; radicle superior; cotyledons usually broader than the radicle, flat, rarely thick and fleshy or folded.

A large family with over 300 genera and 5000 species, and the sixth largest among the flowering plants after the *Orchidaceae, Compositae, Leguminosae, Gramineae* and *Rubiaceae*; subcosmopolitan, but with the greatest representation in the humid tropics and subtropics.

In the most recent classification of the family, that of G.L. Webster in Ann. Miss. Bot. Gard. **81** (1): 33–144 (1994), 5 subfamilies, 49 tribes, 61 subtribes and 317 genera are recognized. In the Flora Zambesiaca area there are 62 genera and c. 500 species.

Various families have been suggested as having links with the *Euphorbiaceae*, such as the *Flacourtiaceae, Sterculiaceae, Malvaceae, Tiliaceae, Geraniaceae, Icacinaceae, Celastraceae, Sapindaceae, Anacardiaceae, Buxaceae, Ulmaceae* and *Urticaceae*, but the similarities are generally superficial rather than real.

Plagiostyles africana (Müll. Arg.) Prain. There seems to me to be no justification whatsoever for the inclusion of this monotypic genus, which ranges from southern Nigeria to northwestern Zaire, in the "Flora Zambesiaca". The sterile material collected by Milne-Redhead on 15.ii.1938 under his number 4583 from evergreen vegetation by the R. Matonchi in Mwinilunga District, Zambia, and by him confidently thus labelled, prompting J. Léonard to include "Rhodesie du Nord (?)" in the distribution in his account of the species for the *"Euphorbiacées"* for Vol. 8 of the Flora de Congo Belge et du Rwanda-Burundi, albeit sharing with it the glossy, leathery laurocerasiform leaf blades and bipulvinate petioles, nevertheless differs from it vegetatively in a number of significant respects, thus: 1.) *Milne-Redhead* 4583 is heterophyllous, the leaves on the main shoots having rounded bases whilst those on the lateral shoots are cuneate, whereas in *Plagiostyles* the leaves are more or less homophyllous and all narrowly cuneate; 2.) In *Milne-Redhead* 4583 the petioles are mostly short and stout (0.5–1.5 cm long and up to 2 mm thick), whereas in *Plagiostyles* they are 0.5–4(5) cm long and less than 1 mm thick and 3.) the lateral nerves of *Plagiostyles* are all brochidodromous with the loops often closely approaching the margin (i.e. to within 1 mm of it), whereas in *Milne-Redhead* 4583 the lateral nerves are mostly camptodromous, only the upper ones being somewhat brochidodromous and then forming loops well in from the margin (i.e. at least 3 mm from it).

Key to subfamilies (simplified after G.L. Webster, 1994)

1. Locules of ovary each with 2 ovules ············ 2
– Locules of ovary each with one ovule ············ 3
2. Leaves usually alternate, simple; pollen grains not spiny ············ **Phyllanthoideae**
– Leaves often opposite, compound; pollen grains spiny ············ **Oldfieldioideae**
3. Latex absent ············ **Acalyphoideae**
– Latex present ············ 4
4. Latex reddish, yellowish or milky; indumentum often stellate or lepidote ·· **Crotonoideae**
– Latex whitish, often caustic or toxic; indumentum simple or absent ···· **Euphorbioideae**

Key to tribes of subfamily ***Phyllanthoideae*** *in the Flora Zambesiaca area*

1. Sepals valvate ············ Tribe **Bridelieae** (genera **2**, **3**)
– Sepals imbricate or open ············ 2
2. Flowers fasciculate ············ 3
– Flowers capitate, spicate or racemose ············ 5
3. Male disk intrastaminal ············ Tribe **Drypeteae** (genus **10**)
– Male disk extrastaminal or absent ············ 4
4. Petals present ············ Tribe **Wielandieae** (genus **1**)
– Petals absent (except **5. Andrachne**) ············ Tribe **Phyllantheae** (genera **4–9**)
5. Fruits not winged ············ Tribe **Antidesmeae** (genera **11–14**)
– Fruits winged ············ Tribe **Hymenocardieae** (genus **15**)

Subfamily **Oldfieldioideae** is represented only by the tribe **Picrodendreae** in the Flora Zambesiaca area (genera **16**, **17**).

Key to tribes of subfamily ***Acalyphoideae*** *in the Flora Zambesiaca area*

1. Petals present, at least in the male flowers 2
– Petals absent 5
2. Flowers fasciculate; seeds black, shiny 3
– Flowers spicate, racemose or subpaniculate; seeds not as above 4
3. Mostly herbaceous perennials; fruits smooth .... Tribe **Clutieae** (genus **18**)
– Trees or shrubs; fruits tuberculate .... Tribe **Chaetocarpeae** (genus **19**)
4. Trees or shrubs .... Tribe **Agrostistachydeae** (genus **20**)
– Herbs (Flora Zambesiaca area) .... Tribe **Chrozophoreae** (genera **21**, **22**)
5. Plants often scandent or twining; styles unlobed, mostly connate .... Tribe **Plukenetieae** (genera **35–38**)
– Plants rarely scandent; styles mostly free or basally connate 6
6. Male disk present 7
– Male disk absent, or if present then anther cells free but not becoming flexuous-vermiform 8
7. Laminar glands present .... Tribe **Bernardieae** (genera **23**, **24**)
– Laminar glands absent .... Tribe **Pycnocomeae** (genus **25**)
8. Male flowers usually capitate; pollen coarsely reticulate .... Tribe **Epiprineae** (genus **26**)
– Male inflorescences various; pollen finely perforate-tectate to rugulose 9
9. Pollen colpi operculate .... Tribe **Alchorneae** (genus **27**)
– Pollen colpi inoperculate, often reduced .... Tribe **Acalypheae** (genera **28–34**)

Key to the tribes of subfamily ***Crotonoideae*** *in the Flora Zambesiaca area*

1. Indumentum simple or none; petals absent 2
– Indumentum various; petals present, or if absent then indumentum stellate or lepidote 5
2. Plants monoecious; laticifers jointed 3
– Plants mostly dioecious; laticifers not jointed 4
3. Pollen colporate; endosperm oily .... Tribe **Micrandreae** (genus **39**)
– Pollen periporate; endosperm starchy .... Tribe **Manihoteae** (genus **40**)
4. Leaves not pellucid-punctate; inflorescences axillary or terminal .... Tribe **Adenoclineae** (genus **41**)
– Leaves pellucid-punctate; inflorescences leaf-opposed .... Tribe **Gelonieae** (genus **42**)
5. Male sepals fused in bud, splitting valvately into 2–5 lobes .... Tribe **Aleuritideae** (genera **49–54**)
– Male sepals usually imbricate, free, or if connate then not completely covering the petals in bud 6
6. Indumentum simple 7
– Indumentum stellate or lepidote, at least in part 8
7. Inflorescences terminal, dichasial .... Tribe **Jatropheae** (genus **43**)
– Inflorescences mostly axillary, spicate, racemose or paniculate .... Tribe **Codiaeae** (genus **44**)
8. Petals free; fruit usually dehiscent; seeds carunculate .... Tribe **Crotoneae** (genera **45**, **46**)
– Petals coherent; fruit indehiscent; seeds ecarunculate .... Tribe **Ricinodendreae** (genera **47**, **48**)

Key to the tribes of subfamily ***Euphorbioideae*** *in the Flora Zambesiaca area*

Inflorescences racemose or spicate, but if capitate, then not pseudanthial .... Tribe **Hippomaneae** (genera **55–58**)
Inflorescences cyathial, pseudanthial .... Tribe **Euphorbieae** (genera **59–62**, see volume 9 part 5)

Key to the genera of ***Euphorbiaceae*** *in the Flora Zambesiaca area*

1. Male and female flowers much reduced and enclosed within a common cup-like involucre 2
– Not as above 5
2. Cyathia actinomorphic 3
– Cyathia zygomorphic 4
3. Cyathial gland continuous, rim-like .... **60. Synadenium**
– Cyathial glands 2–8, distinct, or single, lateral .... **59. Euphorbia**

4. Cyathial glands exposed ····· **61. Monadenium**
– Cyathial glands enclosed ····· **62. Pedilanthus**
5. Leaves compound ····· 6
– Leaves simple although sometimes deeply lobed ····· 9
6. At least some leaves opposite or subopposite ····· **16. Oldfieldia**
– All leaves alternate ····· 7
7. Petals absent ····· **39. Hevea**
– Petals present ····· 8
8. Petiolules absent; stipules persistent; endocarps smooth ····· **47. Ricinodendron**
– Petiolules distinct; stipules fugacious; endocarp deeply pitted ····· **48. Schinziophyton**
9. Leaves opposite ····· 10
– Leaves alternate ····· 13
10. Plant herbaceous ····· **41. Adenocline**
– Shrubs or trees ····· 11
11. Stipules large, intrapetiolar, completely enclosing the terminal bud ····· **17. Androstachys**
– Stipules small, parapetiolar ····· 12
12. Leaf apex acutely acuminate; stamens numerous ····· **33. Mallotus**
– Leaf apex obtuse or emarginate, mucronulate; stamens few ····· **56. Excoecaria**
13. Petals present, at least in the male flowers ····· 14
– Petals absent ····· 31
14. Indumentum stellate or lepidote ····· 15
– Indumentum simple or absent ····· 18
15. Anthers inflexed in bud ····· **46. Croton**
– Anthers erect in bud ····· 16
16. Fruit massive, indehiscent ····· **49. Aleurites**
– Fruits readily dehiscent ····· 17
17. Monoecious herb ····· **22. Chrozophora**
– Dioecious shrub or tree ····· **45. Mildbraedia**
18. Flowers in axillary fascicles ····· 19
– Flowers in spicate, racemose, paniculate or cymose inflorescences ····· 22
19. Male sepals valvate; fruit generally indehiscent ····· **3. Bridelia**
– Male sepals imbricate; fruit dehiscent ····· 20
20. Disk of male flowers cupular ····· **1. Heywoodia**
– Disk of male flowers of separate glands ····· 21
21. Disk glands in 1–2 series on the receptacle only ····· **5. Andrachne**
– Disk glands in (1)2–3 series on sepals, petals and receptacle ····· **18. Clutia**
22. Male calyx lobes imbricate ····· 23
– Male calyx lobes valvate ····· 25
23. Inflorescences cymose ····· **43. Jatropha**
– Inflorescences racemose ····· 24
24. Stamens 5; ovules 2 per locule ····· **14. Thecacoris**
– Stamens 15–100; ovules 1 per locule ····· **44. Codiaeum**
25. Fruit up to 5 cm in diameter, tardily dehiscent ····· **50. Vernicia**
– Fruit not more than 3 cm in diameter, readily dehiscent ····· 26
26. Stipules fused into an elongate sheath ····· **20. Pseudagrostistachys**
– Stipules small, free or obsolete ····· 27
27. Inflorescences axillary; pistillode (non-functional ovary) present ····· 28
– Inflorescences terminal; pistillode absent ····· 29
28. Shrubs or trees ····· **2. Cleistanthus**
– Herbs ····· **21. Caperonia**
29. Female petals shorter than the calyx ····· **53. Neoholstia**
– Female petals longer than the calyx ····· 30
30. Leaves penninerved; stamens 15–30; fruit 3–5-lobed ····· **51. Cavacoa**
– Leaves palminerved; stamens 7–12; fruit subglobose ····· **52. Tannodia**
31. Male calyx open in bud; milky latex often present ····· 32
– Male calyx closed in bud; milky latex usually absent ····· 35
32. Male flowers in dense ovoid or globose heads ····· **58. Maprounea**
– Male flowers in narrow elongate spikes ····· 33
33. Fruit large, inflated ····· **56. Excoecaria**
– Fruit small, not inflated ····· 34
34. Inflorescences dense at maturity ····· **55. Spirostachys**

– Inflorescences lax at maturity ···· **57. Sapium**
35. Inflorescences involucrate ···· 36
– Inflorescences not involucrate ···· 37
36. Dioecious trees; involucral bracts 5–12 ···· **11. Uapaca**
– Monoecious twiners; involucral bracts paired ···· **38. Dalechampia**
37. Fruit winged ···· 38
– Fruit smooth, pitted, wrinkled, warty or tubercled, but never winged ···· 39
38. Trees or shrubs; leaves entire; fruit wings 2 ···· **15. Hymenocardia**
– Climbers; leaves 3-lobed; fruit wings 4 ···· **35. Pterococcus**
39. Female calyx lobes pinnatifid ···· 40
– Female calyx lobes entire ···· 42
40. Shrubs; indumentum stellate; male flowers capitate ···· **26. Cephalocroton**
– Twiners; indumentum simple, usually urticating; male flowers racemose ···· 41
41. Styles united to form a tube or globose mass ···· **36. Tragiella**
– Styles free above ···· **37. Tragia**
42. Male sepals imbricate ···· 43
– Male sepals valvate ···· 54
43. Leaves usually palmately-lobed to palmatipartite ···· **40. Manihot**
– Leaves simple, entire ···· 44
44. Fruits indehiscent, compressed; endocarp pitted ···· **13. Antidesma**
– Fruits dehiscent, sometimes tardily so; endocarp usually smooth ···· 45
45. Inflorescences leaf-opposed ···· **42. Suregada**
– Inflorescences axillary or cauliflorous ···· 46
46. Inflorescences spicate, racemose or paniculate ···· 47
– Inflorescences fasciculate or glomerulate, or flowers solitary ···· 48
47. Fruits ellipsoid, tardily dehiscent ···· **12. Maesobotrya**
– Fruits 3-lobed, readily dehiscent ···· **14. Thecacoris**
48. Disk absent ···· **9. Breynia**
– Disk present ···· 49
49. Male disk intrastaminal; fruit indehiscent ···· **10. Drypetes**
– Male disk extrastaminal; fruit dehiscent, sometimes tardily ···· 50
50. Pistillode absent ···· 51
– Pistillode present ···· 52
51. Male disk annular; sarcotesta present, bluish ···· **7. Margaritaria**
– Male disk of separate glands (except *P. pinnatus*); sarcotesta absent ···· **8. Phyllanthus**
52. Male disk annular ···· **4. Pseudolachnostylis**
– Male disk of separate glands ···· 53
53. Stamens free; fruit smooth, glabrous ···· **6. Flueggea**
– Stamens connate; fruit tubercled, setose ···· **19. Chaetocarpus**
54. Female bracts usually accrescent; styles laciniate; anthers vermiform ···· **34. Acalypha**
– Not as above ···· 55
55. Leaves palmatilobed, palmatifid or palmatipartite ···· 56
– Not as above ···· 57
56. Robust herbs ···· **28. Ricinus**
– Trees or shrubs ···· **30. Macaranga**
57. Plants often drying purplish; anther thecae erect, separate ···· 58
– Plants not usually drying purplish; anther thecae pendulous, usually contiguous ···· 59
58. Bud scales present, crustaceous to subcoriaceous, persistent ···· **31. Erythrococca**
– Bud scales absent ···· **32. Micrococca**
59. Herbs ···· **29. Leidesia**
– Trees or shrubs ···· 60
60. Anthers 3–4-thecous ···· **30. Macaranga**
– Anthers 2-thecous ···· 61
61. Disk absent; stamens 8 or fewer ···· **27. Alchornea**
– Disk present; stamens 10 or more ···· 62
62. Leaves palminerved; indumentum stellate; inflorescences terminal ···· **54. Neoboutonia**
– Leaves penninerved; indumentum simple; inflorescences axillary ···· 63
63. Leaves stipellate; female sepals accrescent ···· **24. Paranecepsia**
– Leaves exstipellate; female sepals not accrescent ···· 64
64. Petioles often pulvinate and geniculate; styles bifid ···· **23. Necepsia**
– Petioles very short or absent, not as above; styles simple ···· **25. Argomuellera**

1. HEYWOODIA Sim

Heywoodia Sim, For. Fl. Col. Cape Good Hope: 326, t. 140, fig.1 (1907). —Milne-Redhead in Bull. Jard. Bot. État **27**: 329, f. 32 & t. 10 (1957).

Dioecious, completely glabrous evergreen tree. Leaves alternate, petiolate, sometimes peltate in juvenile foliage, stipulate; blades simple, entire, basally symmetrical, coriaceous, penninerved. Flowers axillary, fasciculate, the males densely so, sessile, bracteate, the females in 1–3-flowered cymes, shortly pedicellate; bracts 4, free, imbricate, concave, unequal. Male flowers: sepals 3, free, imbricate, concave, unequal; petals 5, free, imbricate, concave, subequal; disk extrastraminal, lobulate, invaginated amongst the outer stamens, fleshy; stamens (8)10–11(12), in 2 whorls, filaments free, anthers dorsifixed, introrse, thecae parallel, longitudinally dehiscent; pistillode (non-functional ovary) minute, trifid. Female flowers: bracteoles 1–2, sepals and petals similar to those of the male; disk annular, slightly lobulate, shallowly cupular; staminodes 6–8, filiform; ovary 4–5-locular, with 2 ovules per loculus; stigmas 4–5, sessile, 2-lobed. Fruit depressed-subglobose, shallowly 8–10-lobed, tardily septicidally dehiscent, leaving a columella; pericarp thinly fleshy, smooth, becoming rugulose on drying; endocarp bony. Seeds solitary per loculus by abortion, ecarunculate; endosperm papery; embryo minute; cotyledons broad, flat.

A monotypic east and southern African genus.

Heywoodia lucens Sim, For. Fl. Col. Cape Good Hope: 326, t. 140, fig.1 (1907). —Hutchinson in F.C. **5**, 2: 385 (1920). —Engler, Pflanzenw. Afrikas **3**, 2: 36 (1921). —Hutchinson in Bull. Misc. Inform., Kew, **1922**: 115 (1922). —Pax in Engler, Pflanzenr. [IV, fam. 147, xv] **81**: 280 (1922). —Dale & Greenway, Kenya Trees & Shrubs: 204 (1961). —K. Coates Palgrave, Trees Southern Africa, ed. 2, rev.: 395 (1983). —Radcliffe-Smith in F.T.E.A., Euphorb. 1: 86 (1987). —Beentje, Kenya Trees, Shrubs Lianas: 208 (1994). Tab. **1**. Type from South Africa (Eastern Cape Province).

A large evergreen tree up to 25 m tall. Bark grey, irregularly peeling in corky sheets or patches. Twigs pale greyish-brown. Stipules c. 1 × 0.5 mm, elliptic-ovate, subacute. Petioles 1–2 cm long (crown foliage). Leaf blade 6–12 × 5–8 cm (larger on sucker growth), broadly ovate to elliptic-ovate, shortly acuminate, cuneate, rounded or truncate at the base, entire, lateral nerves in 8–12, looped well within the margin, with a second series of loops towards the margin, tertiary nerves loosely reticulate, slightly prominent above, more so beneath. Bracts 1 × 1.5 mm, transversely ovate, erose, chaffy. Male flowers: sepals resembling the bracts, but slightly larger, ciliolate; petals 2–2.5 × 0.75 mm, elliptic-oblong, slightly erose, membranous; disk 0.75 mm in diameter; stamens 3 mm long, anthers 1 mm long; pistillode 1 mm high. Female flowers: pedicels 3–5 mm long, extending to up to 1 cm long in fruit; bracteoles 1.5 × 1 mm, ovate, chaffy; sepals 2–3, resembling the bracteoles; petals 5–6, resembling those of the male; disk 1.5 mm in diameter; staminodes c. 1 mm long; ovary 1.5 × 1.5 mm, ovoid-subglobose, shallowly 4- or 5-lobed; stigmas 0.5 mm long, slightly papillose. Fruit 1 × 1.5 cm, green; endocarp 1 mm thick, with 3–4 vascular traces on each septum; columella 5–7 mm long. Seeds 7–8 × 5–6 × 3–4 mm, smooth, dark brown.

Mozambique. M: Lebombo Mt. Range, south of Estatuene Beacon, fr. ii.1974, *Tinley* 3008 (K; LISC; PRE; SRGH).

Also in Uganda, Kenya, Tanzania, South Africa (KwaZulu-Natal and Eastern Cape Province) and Swaziland. A dominant canopy tree in ravines and an emergent of moist evergreen forest on south facing scree slopes, young trees are common in the deep shade of the adult trees; also in gallery forest.

This species shows a marked discontinuity of distribution: some 2600 km separate the Mozambique locality from the nearest E African one (the Mkomazi Game Reserve in NE Tanzania).

Only fruiting material has been seen from the Flora Zambesiaca area. Therefore non Flora Zambesiaca material had to be referred to in order to build up the floral descriptions.

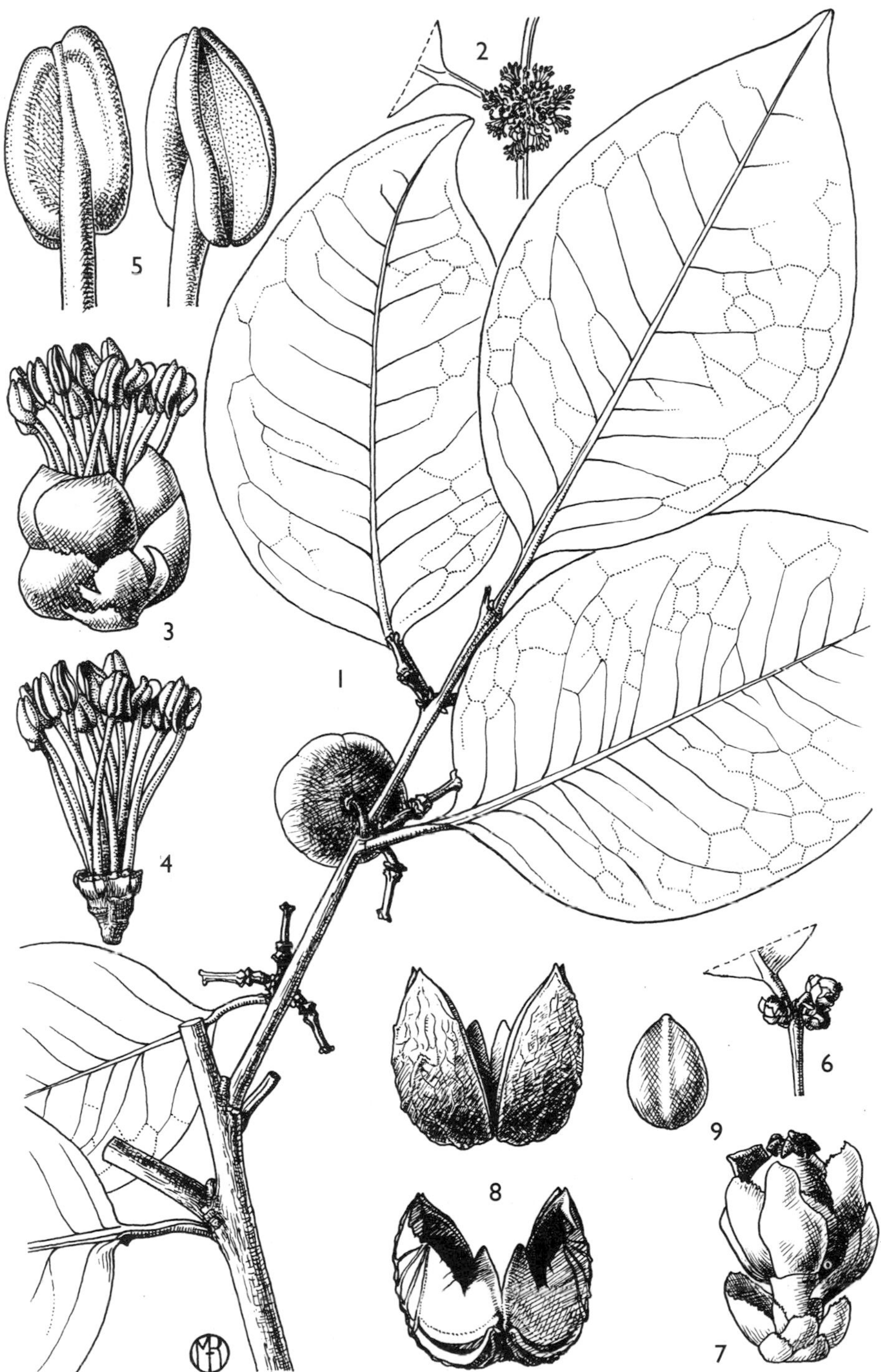

Tab. 1. HEYWOODIA LUCENS. 1, fruiting branch (× 1); 2, male inflorescence (× 1); 3, male flower (× 5); 4, male flower, with sepals removed (× 5); 5, anthers (× 20); 6, female inflorescence (× 1); 7, female flower (× 5); 8, dehisced mericarps (× 2); 9, seed (× 2). Drawn by Olive Milne-Redhead. From F.T.E.A.

2. CLEISTANTHUS Hook.f. ex Planch.

Cleistanthus Hook.f. ex Planch. in Hooker's Icon. Pl. **8**: t. 779 (1848). —Müller Argoviensis in De Candolle, Prodr. **15**, 2: 503 (1866). —Jablonszky in Engler, Pflanzenr. [IV, fam. 147, viii] **65**: 8 (1915).
Kaluhaburunghos Kuntze, Revis. Gen. Pl. **2**: 607 (1891).

Monoecious or more rarely dioecious shrubs or trees. Indumentum simple. Leaves alternate, petiolate, stipulate, simple, entire, penninerved. Inflorescences axillary, rarely terminal, fasciculate or pseudo-racemose, uni- or bisexual, bracteate; bracts soon falling. Flowers proterogynous, pedicellate, fragrant. Male flowers: sepals (4)5(6), valvate; petals (4)5(6), minute; disk ± annular or cupular, ± entire; stamens (4)5(6), filaments united below into a staminal column, free and spreading above, anthers basifixed, introrse, longitudinally dehiscent; pistillode (non-functional ovary) 3-lobed to tripartite. Female flowers: sepals and petals ± as in the male; disk annular, hypogynous or cupuliform, enveloping the ovary; ovary (2)3(4)-locular, ovules 2 per locule; styles (2)3(4), free or united at the base, 1–4 times bifid. Fruit (2)3(4)-lobed, septicidally dehiscent into as many bivalved cocci; endocarp woody, separating from the pericarp; columella persistent, apically dilated. Seeds (1)2 per coccus, ecarunculate; endosperm copious or scanty; cotyledons broad, thin or thick and fleshy, flat or folded.

A palaeotropical genus of approximately 140 species, with 23 in mainland Africa, only 2 of which occur in the Flora Zambesiaca area.

Leaves commonly drying greenish; inflorescence axes less than 1 cm long; indumentum yellowish; female pedicels up to 3 cm long in fruit; male disk and pistillode glabrous; ovary and fruit glabrous or appressed-pubescent; styles once bifid, seeds subglobose · · · · · · · · · 1. *schlechteri*
Leaves commonly drying reddish- or greyish-brown; inflorescence axes up to 4 cm long; indumentum fulvous to ferrugineous; female pedicels less than 1 cm long in fruit; male disk and pistillode pubescent; ovary and fruit pubescent; styles thrice bifid; seeds ovoid · 2. *polystachyus* subsp. *milleri*

1. **Cleistanthus schlechteri** (Pax) Hutch. in F.C. **5**, 2: 382 (1915). —Engler, Pflanzenw. Afrikas **3**, 2: 41 (1921). —J. Léonard in Bull. Jard. Bot. État **30**: 422 (1960). —Drummond in Kirkia **10**: 251 (1975). —K. Coates Palgrave, Trees Southern Africa, ed. 2, rev.: 411 (1983). —Radcliffe-Smith in F.T.E.A., Euphorb. 1: 130 (1987). —Beentje, Kenya Trees, Shrubs Lianas: 189 (1994). Type: Mozambique, Maputo (Lourenço Marques), y. fr. 29.xi.1897, *Schlechter* 11524 (B†, holotype; BR; K; P).
Securinega schlechteri Pax in Bot. Jahrb. Syst. **28**: 18 (1899). —Schinz & Junod in Mém. Herb. Boissier, No. 10: 46 (1900).

A many-stemmed shrub or large tree up to c. 20 m tall with a rounded crown and drooping ultimate branchlets. Bole to 3.5 m high, 4 dm in diameter. Bark on trunk dark grey or reddish-brown to blackish, rough, reticulate-striate, fibrous, paler on the branches. Wood hard. Twigs grey, lenticellate. Young shoots and petioles pubescent to subglabrous. Petioles 2–7 mm long. Stipules 4–5 × 1–2 mm, narrowly lanceolate to falcate-lanceolate, sparingly pubescent, membranaceous, fugacious. Leaf blades 1–9(12) × 0.5–4.5(5.5) cm, elliptic-ovate to elliptic-oblong, obtuse to obtusely acuminate, sometimes emarginate, cuneate-rounded to shallowly cordate at the base, epipetiolar, firmly chartaceous, very sparingly appressed-pubescent along the midrib beneath and otherwise glabrous, more rarely evenly patent-pubescent along the midrib beneath, dark green and shiny above, paler and duller beneath, drying greenish or greyish-green; lateral nerves in 5–8 pairs, brochidodromous, scarcely to slightly prominent above and beneath, bearing acarodomatia in their axils in larger leaves, tertiary nerves reticulate. Inflorescences up to 12-flowered, either all male or male with 1(2) female flowers; axes less than 1 cm long, appressed-pubescent; bracts c. 1 × 1 mm, ± ovate, caducous. Flowers fragrant. Male flowers: pedicels 6–11(15) mm long, sparingly puberulous; buds 3–5 mm long, ellipsoid, rounded at the base; sepals 5–6 × 1–1.5 mm, linear, acute, cucullate, sparingly puberulous without, glabrous within, yellowish-green; petals 1–1.3 mm long, subulate; disk 1.3 mm in diameter, shallowly cupular, slightly pentagonal, glabrous, pink; staminal column 1.5 mm high; anthers 2.5–3 mm long, cream-coloured; pistillode 1 mm high, tripartite, glabrous. Female flowers: pedicels extending to 3 cm long in fruit, pubescent, reddish; buds 5 mm long, ovoid-conic; sepals and petals ± as in the male; disk 2–2.5 mm in diameter, otherwise as in the male; ovary 3 mm in diameter, globose, glabrous or else densely appressed golden-pubescent; styles 3, 2 mm long, united for 0.5 mm, once bifid, glabrous. Fruit 7–10 × 9–15 mm, strongly 3-lobed or rarely 4-lobed, ± smooth to slightly rugulose, glabrous or evenly appressed golden-pubescent, greenish to light brown with 6 reddish lines at first, brownish-black when ripe. Seeds 5 × 4.5 mm, subglobose, smooth, shiny, brownish-pink to dark brown or blackish.

Ovary and fruit glabrous · var. *schlechteri*
Ovary and fruit appressed golden-pubescent · var. *pubescens*

Var. **schlechteri**

Securinega schlechteri Pax in Bot. Jahrb. Syst. **28**: 18 (1899). —Schinz & Junod in Mém. Herb. Boissier No. 10: 46 (1900).

Cleistanthus holtzii Pax in Bot. Jahrb. Syst. **43**: 77 (1909). —Hutchinson in F.T.A. **6**, 1: 622 (1912). —Jablonszky in Engler, Pflanzenr. [IV, fam. 147, viii] **65**: 50, t. 9A–E, 47 (1915). —Engler, Pflanzenw. Afrikas **3**, 2: 41, t. 12 A–E, 40 (1921). —Gardner, Trees Shrubs Kenya Col.: 47 (1936). —Brenan, Check-list For. Trees Shrubs Tang. Terr.: 202 (1949). —Dale & Greenway, Kenya Trees & Shrubs: 188 (1961). —Mogg in Macnae & Kalk, Nat. Hist. Inhaca Isl., Moçamb., rev. ed.: 147 (1969). Type from Tanzania (E. Province).

Cleistanthus johnsonii Hutch. in Bull. Misc. Inform., Kew **1909**: 380 (1909). Type: Mozambique, Manica e Sofala, Sofala, male fl. & fr. 31.x.1906, *Johnson* 26B (K, holotype).

Ovary and fruit glabrous.

Zimbabwe. E: Chipinge Distr., Msilizwe R., fr. (leafless) xi.1962, *Goldsmith* 215/62 (BM; K; LISC; PRE; SRGH). S: Mwenezi Distr., Pombadzi R., o. fr. 19.xi.1957, *Phelps* 202 (K; SRGH). **Malawi**. S: Nsanje (Port Herald) Hills, o. fr. 27.x.1933, *Lawrence* 103 (K). **Mozambique**. N: Cabo Delgado, Pemba, Metoro, 10 km Namatuco–Mecaruma, o. fr. 31.i.1984, *Groenendijk, de Koning, Maite & Dungo* 906 (K; LMU; MO). Z: 500 m, Régulo Ingive–Nante, male fl. & y. fr. 27.ix.1949, *Barbosa & Carvalho* in *Barbosa* 4198 (K; LMA). MS: Búzi Region, L. Ura, male and female fl. 16.x.1964, *Gomes e Sousa* 4840 (FHO; K; LISC; PRE). M: Maputo (Lourenço Marques), Ponta Vermelha, fr. 18.xi.1959, *Lemos & Balsinhas* in *Lemos* 2 (BM; K; LMA; PRE).

Also in Kenya, Tanzania and South Africa (KwaZulu-Natal). On sandy clay soils, rocky hillsides (microphyllous specimens) and on termitaria in deciduous thicket, in dune thickets, miombo and mixed deciduous woodlands with *Combretum, Acacia, Afzelia, Sclerocarya, Pterocarpus, Ostryoderris, Diplorrhynchus, Artabotrys, Dialium, Sapium, Kigelia* and *Pteleopsis*, and in forests with *Parkia, Khaya, Millettia* and *Landolphia*, also in riverine forest and mangrove swamp margins, and in secondary forest and old cultivation with regenerating bush; sea level to 760 m.

Specimens from Maputo District tend to be microphyllous. The leaves of specimens from Malawi are evenly patent-pubescent along the midrib beneath.

The timber is used for railway sleepers, hut poles, roofs and cross-beams because it splits well and easily.

Vernacular names as recorded in specimen data include: "chirre" (Maputo area), "kwekwero", "mueno" (Namagoa area), "m'cua" (Inhamitanga area), "m'tagi" (kiMwani), "muchiche" (Mchengane), "mucua" (ciSena), "muin'hirro" (Zambézia), "macuva" (eMakhuwa), "nacuvali" (Nampula), "nahir" (Mocuba), "sasanjiwa" (Nsanje).

Var. **pubescens** (Hutch.) J. Léonard in Bull. Jard. Bot. État **30**: 423 (1960). —Radcliffe-Smith in F.T.E.A., Euphorb. 1: 133 (1987). Tab. **2**. Type: Mozambique, Manica e Sofala, Sofala, male fl., female fl. and y. fr. 31.x.1906, *Johnson* 26A (K, holotype).

Cleistanthus johnsonii var. *pubescens* Hutch. in Bull. Misc. Inform., Kew **1909**: 380 (1909).

Cleistanthus holtzii var. *pubescens* (Hutch.) Hutch. in F.T.A. **6**, 1: 623 (1912). —Jablonszky in Engler, Pflanzenr. [IV, fam. 147, viii] **65**: 50 (1915).

Ovary and fruit appressed golden-pubescent.

Zimbabwe. E: Chipinge Distr., Umzilizwe (Msilizwe) R., fr. xi.1962, *Goldsmith* 220/62 (BM; K; LISC; SRGH). S: Ndanga Distr., Triangle, fr. xi.1951, *Seward* 85 (K; SRGH). **Mozambique**. N: Angoche (António Enes), male fl. and y. fr. 28.x.1965, *Gomes e Sousa* 4907 (K; PRE). T: 38 km Tete–Chicoa, o. fr. 27.xii.1965, *Torre & Correia* 13850 (LISC). MS: Sofala Prov. (Beira Distr.), Chinizíua R., male fl. and fr. x.1957, *Gomes e Sousa* 4408 (K; LISC; MO; PRE).

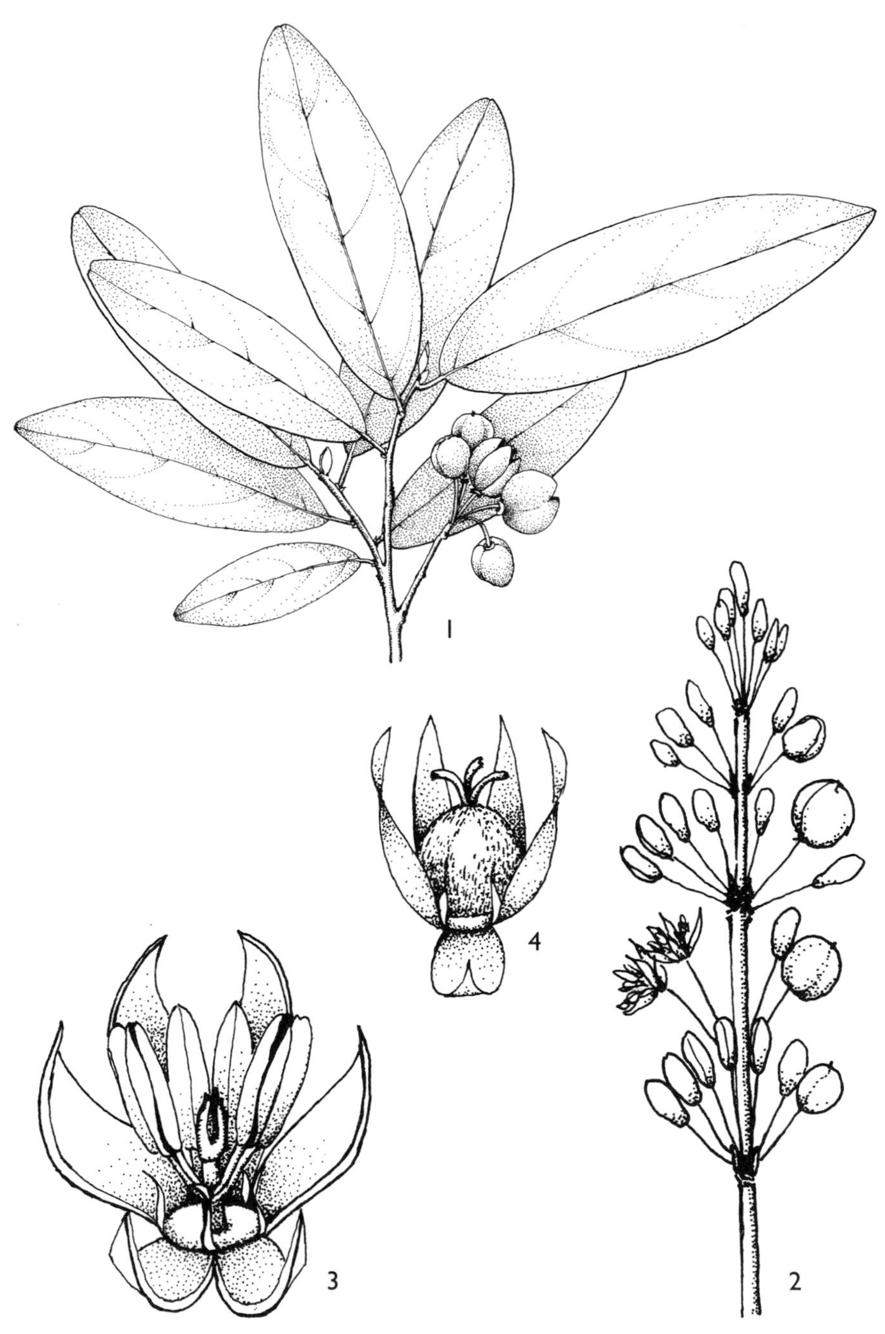

Tab. 2. CLEISTANTHUS SCHLECHTERI var. PUBESCENS. 1, distal part of fruiting branch (× $^{2}/_{3}$) from *Semsei* 3327; 2, inflorescence (× 1$^{1}/_{3}$); 3, male flower (× 4), 2 & 3 from *Faulkner* 2215; 4, female flower (× 4), from *Semsei* 2215. Drawn by G. Papadopoulos. From F.T.E.A.

Also in Kenya and Tanzania. On sandy soil over granite and on rocky hill slopes, in dune woodland and wooded grassland with *Terminalia* and *Combretum* spp., in deciduous woodland with *Pterocarpus* and *Commiphora*, mopane woodland and submontane *Brachystegia spiciformis* and *B. boehmii* forest, also in riverine vegetation and on mangrove swamp margins near the sea; sea level–760 m.

Vernacular names as recorded in specimen data include: “m'taxi” (Cabo Delgado); “mucua” (Cheringoma).

Specimens intermediate between the two varieties are found almost throughout their range. They have the ovary and fruit sparingly pubescent to subglabrous. Examples are:

Zimbabwe. E: Chipinge Distr., 2 km west of Umzilizwe R. near Mt. Selinda–Save (Sabi) road, o. fr. 22.xi.1974, *Müller* 2228 (SRGH). S: Buhera Distr., Nyarushanga R., y. fr. 20.xi.1952, *Jack* s.n. (SRGH). **Mozambique**. N: Ribáuè, M'pipe R., male fl. and y. fr. ix.1941, *Gomes e Sousa* 2267 (COI; FHO; K; LISC; PRE). Z: Mocuba, Namagoa, male fl. and fr. x.1943/5, *Faulkner* P350 (K; PRE; SRGH). MS: Búzi, Mucheve For. Res., fr. 27.x.1963, *Carvalho* 682 (LMU; PRE). M: Boane–Umbelúzi, o. female fl., y. fr. 17.xi.1944, *Torre* 6837 (LISC).

Also in Tanzania. In sandy soil, on rocky hillsides and in limestone valleys, in deciduous and riverine forest, and on silty river banks; 490 m.

Because of the absence of ovaries or fruits the following selection of specimens, in the male or sterile state, are not assignable to variety:

Zimbabwe. E: Chipinge Distr., Makossa, st. 21.v.1956, *Mowbray* 111 (K; SRGH). **Malawi**. S: Lengwe Game Reserve, st. 25.iv.1970, *Hall-Martin* 543 (SRGH). **Mozambique**. N: Eráti, between Namapa and Nivete, male fl. 11.x.1948, *Barbosa* 2353 (LISC). Z: Serra de Morrumbala, near Vila Bocage, st. 10.xii.1971, *Müller & Pope* 1994 (LISC; SRGH). T: 7 km from Tete, male fl. 20.xi.1965, *Neves Rosa* 125 (LISC). MS: Sofala Prov. (Beira Distr.), Gorongosa Nat. Park, st. v.1973, *Tinley* 2811 (LISC; MO; SRGH). GI: Xipembe R., SW of Massangena, male fl. 16.ix.1959, *Goodier* 608 (K; SRGH). M: Inhaca Island, male fl. 18.xii.1956, *Mogg* 26877 (K; PRE).

Vernacular names as recorded in specimen data include: “chire” (Manjacaze), “chrisse” (Inhambane), “muchiti”, “musiti” (chiNdao), “nacuva” (Ribáuè), “nahir” (Maganja).

The leaves are used in the preparation of a cure for burns and scalds. The wood is extremely tough.

2. **Cleistanthus polystachyus** Hook.f. ex Planch. in Hooker's Icon. Pl. **8**: t. 779 (1848). —Müller Argoviensis in De Candolle, Prodr. **15**, 2: 504 (1866). —Hutchinson in F.T.A. **6**, 1: 624 (1912). —Jablonszky in Engler, Pflanzenr. [IV, fam. 147, viii] **65**: 47, t. 9 F, G (1915). —Engler, Pflanzenw. Afrikas **3**, 2: 40 (1921). —Topham, Check List For. Trees Shrubs Nyasaland Prot.: 50 (1958). —J. Léonard in Bull. Jard. Bot. État **30**: 430 (1960); in F.C.B. **8**, 1: 26 (1962). —White, F.F.N.R.: 195 (1962). —Troupin, Fl. Pl. Lign. Rwanda: 251, fig. 87/1 (1982); Fl. Rwanda **2**: 211, fig. 64/1 (1983). —Radcliffe-Smith in F.T.E.A., Euphorb. 1: 133 (1987). —Beentje, Kenya Trees, Shrubs Lianas: 189 (1994). Type from Sierra Leone.

Subsp. **milleri** (Dunkley) Radcl.-Sm. in Kew Bull. **51**, 2: 303 (1996).

Cleistanthus milleri Dunkley in Bull. Misc. Inform., Kew **1937**: 468 (1937). —J. Léonard in Bull. Jard. Bot. État **30**: 437 (1960); in F.C.B. **8**, 1: 24 (1962). Type: Zambia, Mbala Distr., Mwambeshe Stream, female fl. x.1932, *O.B. Miller* D 158 (K, holotype; FHO).

Cleistanthus apetalus S. Moore in J. Linn. Soc., Bot. **40**: 191 (1911). —Hutchinson in F.T.A. **6**, 1: 623 (1912). —Jablonszky in Engler, Pflanzenr. [IV, fam. 147, viii] **65**: 50 (1915). —J. Léonard in Bull. Jard. Bot. État **30**: 435 (1960). —Drummond in Kirkia **10**: 251 (1975). —K. Coates Palgrave, Trees Southern Africa, ed. 2, rev.: 411 (1983). Type: Mozambique, Manica e Sofala, Zona River, Jihu, male and female fl. xi.1906, *Swynnerton* 150 (BM, holotype).

Cleistanthus nyasicus Dunkley on herbarium sheet, based on *Carver* 1 (FHO) from Malawi, ‘W Nyasa Distr.’.

A basally-branched scrambling evergreen shrub, or tree up to 25 m high, sometimes exceeding 12 m, with pendulous branches. Bole to c. 2.5 dm in diameter, trunk fluted. Bark smooth at first, thin, light brown, later becoming roughish, flaking, dark reddish- or greyish-brown. Twigs purplish-grey or -brown, minutely lenticellate. Young shoots and petioles glabrous to fulvous-tomentose. Petioles 3–7 mm long. Stipules 3–8 mm long, linear-lanceolate, acute or obtuse, somewhat striate, puberulous abaxially, glabrous adaxially, ferrugineous-pubescent at the apex, occasionally subpersistent. Leaf blades 3–15 × 1–6 cm, shortly obtusely acuminate, cuneate-rounded to very shallowly cordate at the base, thinly coriaceous, sparingly pubescent towards the base of the midrib and otherwise glabrous or else completely glabrous above and beneath, dark green and glossy above, paler and duller beneath, commonly drying reddish- or greyish-brown; lateral nerves in 5–10 pairs, weakly brochidodromous well within the margin, not prominent above, fairly prominent beneath, tertiary nerves prominently reticulate beneath. Inflorescences up to 40-flowered, usually bisexual; axes 1–4 cm long, densely fulvous- or ferrugineous-tomentose, sometimes bearing small leaves; outer bracts resembling the stipules, soon caducous, inner smaller, more persistent. Flowers fragrant. Male flowers: pedicels 2–4 mm long, tomentose; buds 3–5 mm long, ovoid, truncate at the base, green; sepals 5–6 × 1–2.5 mm, linear to lanceolate, subacute, tomentose without, glabrous within, yellowish-green to creamy white; petals 1–2 mm long, linear-subulate; disk 3 mm in diameter, pulvinate, densely fulvous-pubescent; staminal column 1.5 mm high, greenish-white, free portion of filaments 2.5 mm long; anthers 1.5 mm long, yellow; pistillode 1.25 mm high, tripartite, densely fulvous-pubescent. Female flowers: pedicels 3–7 mm long, not exceeding 1 cm long in fruit, thickened at the apex, tomentose; buds 3 mm long, conical; sepals, petals and disk ± as in the male; ovary 2 mm in diameter, subglobose, densely fulvous- to ferrugineous-pubescent; styles 3, 2 mm long, united for 0.5 mm, thrice bifid, pubescent at the base, otherwise glabrous, brownish-green. Fruit 8–9 × 10–11 mm when dried, 3-lobed, densely verruculose, evenly fulvous- to ferrugineous-pubescent and hirsute, greenish. Seeds 5–6 × 4.5–5 mm, ± ovoid, somewhat chalazally-depressed, smooth, slightly shiny, light to dark brown, often striate.

Zambia. N: c. 48 km south of Ishiba Ngandu (Shiwa Ngandu), female fl. 29.xi.1952, *Angus* 871 (BM; BR; FHO; K). W: Mwinilunga, Lunga R., female fl. 29.xi.1937, *Milne-Redhead* 3428 (BM; BR; K; PRE). C: Kapampa R, Luangwa Valley, o. fr. 19.i.1966, *Astle* 4461 (K). **Zimbabwe.** E: Mutare Distr., Vumba, Burma Valley, Bomponi, y. fr. 4.xii.1961, *Wild & Chase* 5551 (BM; K; LISC; PRE; SRGH). **Malawi**. N: Nkhata Bay Distr., Nkwazi For., 14.5 km south of Nkhata Bay Road, y. fr. 21.xii.1975, *Pawek* 10426 (K; MAL; MO; PRE; SRGH; UC). S: Mulanje Distr., Swazi Estate, st. 5.ix.1970, *Müller* 1698 (K; SRGH). **Mozambique**. N: Maniamba, st. 22.v.1948, *Pedro & Pedrógão,* 3827 (LMA; SRGH). Z: Serra do Gurué, male and female fl. 18.x.1949, *Barbosa & Carvalho* in *Barbosa* 4494 (LMA; PRE). MS: Mt. Spungabera, male and female fl. 21.xi.1960, *Leach & Chase* 10508 (FHO; K; LISC; MO; SRGH).

Also in Zaire and Tanzania. On deep red soil and pink shale outcrops, in miombo and riverine woodlands, in evergreen rainforest margins and the subcanopy, and in fringing and riverine forest with *Anthocleista, Chrysophyllum, Pachystela, Albizia, Myrianthus, Macaranga, Cussonia* and *Aphloia,* also in mushitu swamps; 500–1700 m.

Vernacular names as recorded in specimen data include: "musamvia" (Mambwe in Zambia); "pulipuli" (chiTonga).

Cleistanthus polystachyus subsp. *milleri* has generally been regarded as consisting of two entities in the Flora Zambesiaca area, populations from northern and western Zambia, northern Malawi and northern Mozambique being referred to *Cleistanthus milleri,* and those from southern Malawi, central Mozambique (Z, MS) and eastern Zimbabwe to *Cleistanthus apetalus.* There is no significant difference between them, and the type of *C. apetalus* does have petals! The typical subspecies, larger in all its parts and with less or no indumentum, is a rainforest tree of the Guineo-Congolean region.

3. BRIDELIA Willd.

Bridelia Willd., Sp. Pl. **4**, 2: 978 (1805). —Müller Argoviensis in De Candolle, Prodr. **15**, 2: 493 (1866). —Jablonszky in Engler, Pflanzenr. [IV, fam. 147, viii] **65**: 54 (1915).
Candelabria Hochst. in Flora **26**: 79 (1843).
Pentameria Klotzsch ex Baill., Étud. Gén. Euphorb.: 584 (1858).
Neogoetzea Pax in Bot. Jahrb. Syst. **28**: 419 (1900).
Gentilia Beille in Compt.-Rend. Hebd. Séances Acad. Sci. **114**: 1294 (1907).
Tzellemtinia Chiov. in Ann. Bot. (Rome) **9**: 55 (1911).

Monoecious, or rarely dioecious shrubs or trees. Indumentum simple. Trunk and branches sometimes armed with blunt thorns. Leaves alternate, shortly petiolate, stipulate, simple, entire or more or less so, penninerved; lateral nerves straight, arched or looped, tertiary nerves usually parallel. Flowers small, axillary, glomerulate or fasciculate, sometimes in spikes or panicles of glomeruliform fascicles. Bracts small, scale-like. Male flowers in many-flowered fascicles, sometimes with 1–2 female flowers admixed, sessile, subsessile or shortly pedicellate; sepals (4)5, valvate; petals 5, small, inflexed and not contiguous or imbricate; disk annular or cupuliform, entire or sinuate; stamens 5, filaments united below into a short column, free and spreading above, anthers horizontal, basifixed, the thecae parallel, longitudinally dehiscent; pistillode at the top of the column ampulliform, entire or 2–4-lobed. Female flowers few per cluster or solitary, sometimes pedicellate; sepals and petals perigynous, otherwise ± as in female; outer disk annular, inner encasing the ovary; ovary 2(3)-locular, ovules 2 per locule; styles 2(3), free or connate at the base, bifid or subentire. Fruit drupaceous and indehiscent or else dehiscent; exocarp thin; mesocarp fleshy; endocarp crustaceous, 1- or 2-locular. Seeds solitary per locule by abortion, plano-convex and longitudinally grooved in 2-locular fruits, C-shaped in transverse section and hollowly cylindrical, split down one side in 1-locular fruits; albumen copious, fleshy; embryo curved; cotyledons broad, foliaceous, thin.

A paleotropical genus of 77 species, with 15 in mainland Africa, of which 7 species occur in the Flora Zambesiaca area.

1. Fruits 2-celled · · · · · · · · 2
– Fruits 1-celled · · · · · · · · 3
2. Leaf blades softly pubescent above and beneath; stipules 0.5–1 cm long, falcate-lanceolate, acuminate · · · · · · · · 1. *mollis*
– Leaf blades glabrous above, glabrous or sparingly to evenly but not softly pubescent beneath; stipules not more than 7 mm long, linear-lanceolate, acute · · · · · · 2. *cathartica*
3. Flowers and fruits in terminal leafless panicles · · · · · · · · 4. *brideliifolia*
– Flowers and fruits in axillary glomeruliform fascicles · · · · · · · · 4
4. Lateral nerves in 10–22 pairs, camptodromous; leaves drying dark green or blackish · · · · · · · · 3. *atroviridis*
– Lateral nerves in 6–18 pairs, cheilodromous; leaves drying greenish or brownish · · · · · 5
5. Leaf blades sparingly pubescent to subglabrous along the midrib and main nerves and otherwise minutely sparingly appressed-puberulous to subglabrous beneath · · 5. *micrantha*
– Leaf blades evenly pubescent along the midrib and main nerves and otherwise sparingly so beneath · · · · · · · · 6
6. Stipules 1.5–2 mm wide, lanceolate, acuminate, readily caducous; petioles 5–7 mm long; male glomerules many-flowered · · · · · · · · 6. *ferruginea*
– Stipules 0.5–1 mm wide, subulate-filiform, arcuate, fairly persistent; petioles 3–5 mm long; male glomerules 3 flowered · · · · · · · · 7. *duvigneaudii*

1. **Bridelia mollis** Hutch. in F.T.A. **6**, 1: 612 (1912). —Jablonszky in Engler, Pflanzenr. [IV, fam. 147, viii] **65**: 67 (1915). —Hutchinson in F.C. **5**, 2: 379 (1915). —Eyles in Trans. Roy. Soc. South Africa **5**: 394 (1916). —Engler, Pflanzenw. Afrikas **3**, 2: 43 (1921). —Burtt Davy, Fl. Pl. Ferns Transvaal, part 2: 298 (1932). —Hutchinson, Botanist in Southern Africa: 667 (1946). —O.B. Miller, Check-list For. Trees Shrubs Bech. Prot.: 31 (1948). —White, F.F.N.R.: 195 (1962). —Drummond in Kirkia **10**: 251 (1975). —K. Coates Palgrave, Trees Southern Africa, ed. 2, rev.: 414 (1983). Type: Mozambique, Tete (Tette), ii.1859, *Kirk* Bridelia (1) (K, lectotype, chosen here).

Bridelia stipularis sensu Müll. Arg. in De Candolle, Prodr. **15**, 2: 499 (1866). —Engler, Pflanzenw. Ost-Afrikas **C**: 237 (1895). —Schinz & Junod in Mém. Herb. Boissier, No. 10: 47 (1900). —Sim, For. Fl. Port. E. Afr.: 106 (1909), pp. quoad spec. *Kirk*, non Blume.

Bridelia scandens sensu Eyles in Trans. Roy. Soc. South Africa **5**: 394 (1916), non Willd.

A shrub or small tree up to 9 m in height, branching close to the ground. Bark thick, rough, striated, flaky, dark brownish-grey. Twigs, young shoots and petioles evenly to densely fulvous-tomentose. Petioles 3–5 mm long. Stipules 5–10 × 2–4 mm, falcate-lanceolate, acutely acuminate, densely pubescent, subpersistent. Leaf blades 3–15 × 2–9 cm, broadly elliptic to suborbicular-obovate, obtuse to emarginate, rarely subacute, rounded to shallowly cordulate at the base, firmly chartaceous, softly and usually densely pubescent or almost tomentose above and beneath, light green; lateral nerves in 10–15(20) pairs, closely parallel, arcuate, sometimes reaching the margin or looping, occasionally branched, not prominent above, slightly so beneath, tertiary nerves parallel. Male flowers: pedicels 1.5–2 mm, apically dilated, minutely puberulous; sepals 2.5 × 1.3 mm, elliptic-lanceolate, acute, pubescent without, glabrous within, pale green or yellowish-green; petals 2 × 1.2

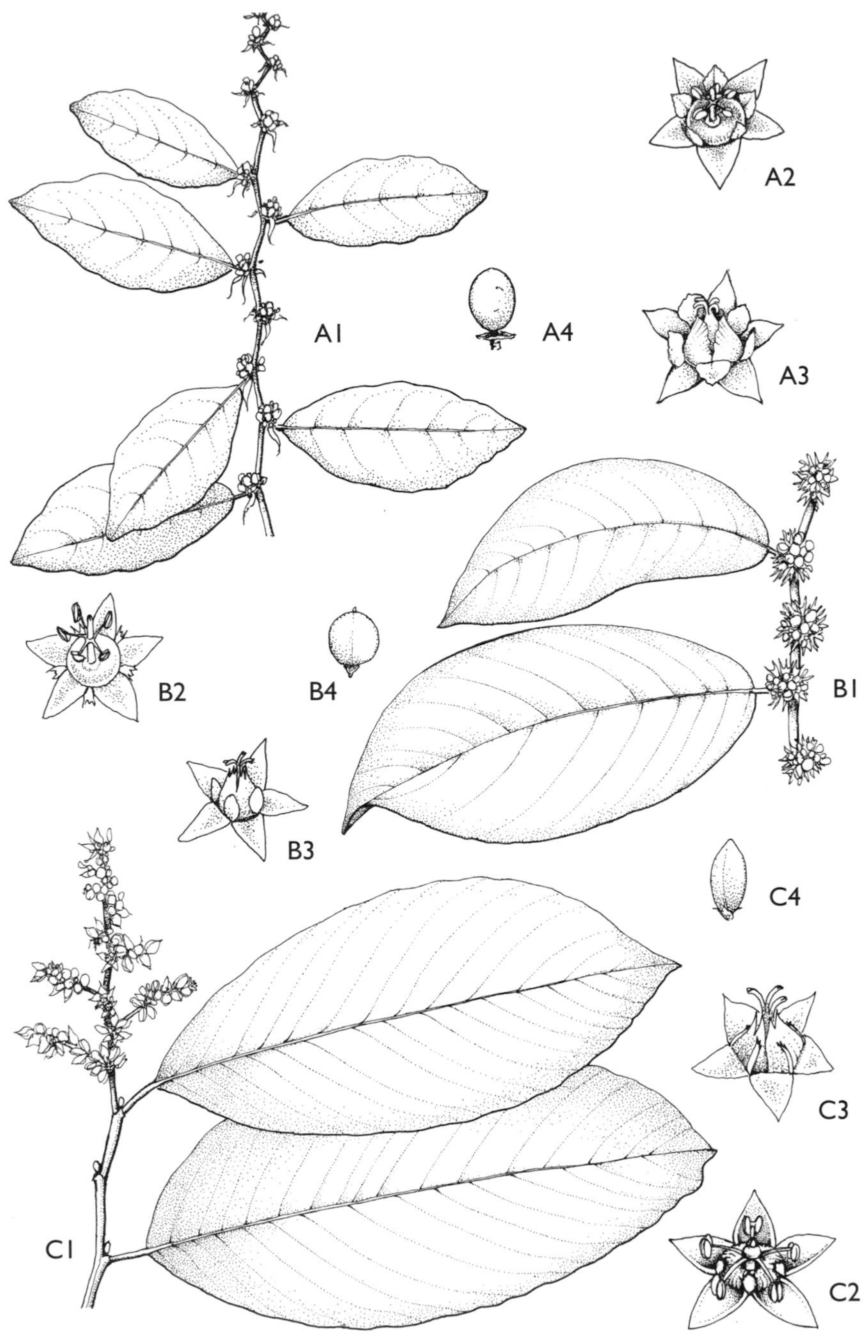

Tab. 3. A —BRIDELIA CATHARTICA. A1, flowering branch (× $^{2}/_{3}$); A2, male flower (× 4), A1 & A2 from *Faulkner* 3793; A3, female flower (× 4), from *Semsei* 3023; A4, fruit (× 1), from *Archbold* 1027. B —BRIDELIA ATROVIRIDIS. B1, portion of flowering branch (× $^{2}/_{3}$); B2, male flower (× 4); B3, female flower (× 4), B1–B3 from *Faulkner* 384; B4, fruit (× 1), from *Faulkner* 1021. C —BRIDELIA BRIDELIIFOLIA. C1, flowering branch (× $^{2}/_{3}$), from *Drummond & Hemsley* 1573; C2, male flower (× 4), from *Semsei* 1084; C3, female flower (× 4), from *Bruce* 976; C4, fruit (× 1), from *Mabberley* 1452. Drawn by G. Papadopoulos. From F.T.E.A.

mm, obovate, slightly erose at apex, glabrous, yellowish; disk 2.5 mm in diameter, annular, flat; staminal column 1.5 mm high, free part of filaments 0.5 mm long, tapering to apex; anthers 0.8 mm long, glabrous; pistillode (non-functional ovary) 1 mm high, conical, verruculose. Female flowers subsessile; sepals 2.5–3 × 1.5 mm, triangular-lanceolate, otherwise as in the male; petals 2 × 0.75 mm, oblong-lanceolate, subentire; outer disk 2 mm in diameter, glabrous; inner disk c. 2 mm high, conical, tunicate, apically lacerate; ovary 1 mm in diameter, glabrous; styles 1.5 mm long. Fruits 0.9–1.1 × 0.7–1 cm when dried, ellipsoid-subglobose, 2-celled, dark brown, black or blue-black when mature, exuding a gummy resin. Seeds c. 7 × 5 × 3 mm, plano-convex, smooth, shiny, chestnut-brown.

Botswana. SE: Palapye Distr., Ratholo, Tswapong Hills, st. 5.iii.1957, *de Beer* T8 (K; SRGH). **Zambia**. E: Chipata (Fort Jameson), fr. 3.vi.1958, *Fanshawe* 4515 (FHO; K; NDO). S: Mazabuka Distr., Gwembe Valley, 17 km from Sinazongwe to Mamba's Village, st. 15.iii.1960, *White* 7789 (FHO). **Zimbabwe**. N: Mutoko Distr., Special Native Area "C", fr. 4.iii.1960, *Cleghorn* 592 (SRGH). W: Bullima-Mangwe (Bulalima Mangwe) Distr., Mtabi's Dam, male fl. 22.x.1972, *Norrgrann* 249 (K; LISC; PRE; SRGH). C: KweKwe (Que Que), Sable Park, o. fr. 1.iv.1976, *Chipunga* 67 (MO; SRGH). E: Chipinge Distr., near the Mozambique border, fr. 13.ii.1957, *Goodier* 124 (K; PRE; SRGH). S: Mwenezi Distr., Rhino Hotel, Runde (Lundi) R., male fl. xii.1955, *Davies* 1739 (K; LISC; PRE; SRGH). **Malawi**. S: Mangochi Distr., Monkey Bay, fr. 29.vi.1981, *Banda, Salubeni & Balaka* 1712 (MAL; MO; SRGH). **Mozambique**. N: Nampula, Mossuril, st. 18.ii.1984, *de Koning, Groenendijk & Dungo* 9723 (LMU; MO). Z: Mopeia, fr. 8.v.1972, *Bowbrick* ST 26a (SRGH). T: Estima–Cahó, fr. 27.i.1972, *Macêdo* 4721 (LISC; LMA; MO). MS: Gorongosa Nat. Park, road from Pungwe R. to Gorongosa (Vila Paiva de Andrada), o. fr. iv.1969, *Tinley* 1814 (LISC; SRGH). M: Delagoa Bay, 1890, *Junod* 227 (Z).

Also in Namibia and South Africa (Transvaal). Common at low and medium altitudes on well drained stony or sandy soils, often among rocks and on granite outcrops (kopjes, dwalas) and low sandstone ridges, in sandveld and dry deciduous bush and thickets, in wooded grassland, deciduous woodland, and riverine vegetation; 200–1525 m.

Vernacular names as recorded in specimen data include: "capsipsa" (Tete area); "mokakole" (setstwana); "murapambare", "musipoia" (chiNdau); "muwowuku" (Masvingo area); "umkumbakumba" (siNdebele).

The fruit is edible.

2. **Bridelia cathartica** G.Bertol. in Mem. Reale Accad. Sci. Ist. Bologna **5**: 476, t. 28 (1854). —Müller Argoviensis in De Candolle, Prodr. **15**, 2: 502 (1866). —Engler, Pflanzenw. Ost-Afrikas **C**: 237 (1895). —Sim, For. Fl. Port. E. Afr.: 106 (1909). —Hutchinson in F.T.A. **6**, 1: 617 (1912). —Jablonszky in Engler, Pflanzenr. [IV, fam. 147, viii] **65**: 61 (1915). —Hutchinson in F.C. **5**, 2: 380 (1915). —Eyles in Trans. Roy. Soc. South Africa **5**: 394 (1916). —Engler, Pflanzenw. Afrikas **3**, 2: 43 (1921). —Burtt Davy, Fl. Pl. Ferns Transvaal, part 2: 298 (1932). —O.B. Miller, Check-list For. Trees Shrubs Bech. Prot.: 31 (1948). —Brenan in Mem. N.Y. Bot. Gard. **9**, 1: 67 (1954). —J. Léonard in Bull. Jard. Bot. État **25**: 360 (1955). —Topham, Check List For. Trees Shrubs Nyasaland Prot.: 49 (1958). —White, F.F.N.R.: 194 (1962). —Mogg in Macnae & Kalk, Nat. Hist. Inhaca Isl., Moçamb., rev. ed.: 147 (1969). —Drummond in Kirkia **10**: 251 (1975). —K. Coates Palgrave, Trees Southern Africa, ed. 2, rev.: 412 (1983). —Radcliffe-Smith in F.T.E.A., Euphorb. 1: 123 (1987). —Beentje, Kenya Trees, Shrubs Lianas: 187 (1994). Tab. **3**, figs. A1–A4. Type: Mozambique, Inhambane, 1842, *Fornasini* s.n. (BOLO†, holotype; B†; P).

A rather variable scrambling single- or many-stemmed much branched shrub or small tree up to 7 m tall with a dense rounded crown and with horizontal or pendent branches. Bark light grey or brownish, smooth or rough, fissured. Twigs brown, lenticellate. Young shoots and petioles evenly to sparingly pubescent, puberulous or subglabrous. Petioles 2–5 mm long. Stipules up to 7 × 1 mm, linear-lanceolate or linear, sparingly pubescent or subglabrous, soon falling. Leaf blades 1–12 × 0.5–7 cm, elliptic-obovate to elliptic-oblong, rounded or obtuse, occasionally subacute, cuneate or rounded, rarely ± truncate at the base, margins sometimes revolute, firmly chartaceous to thinly coriaceous, glabrous above, evenly to sparingly pubescent along the midrib and nerves or else subglabrous to quite glabrous beneath, often bluish-green and shiny above, pale grey-green to glaucous and dull beneath; lateral nerves in 7–14 pairs, cheilodromous or brochidodromous, occasionally camptodromous, not prominent or slightly impressed above, slightly to strongly prominent beneath, tertiary nerves scarcely visible to prominent beneath. Flowers and fruits borne on leafy or wholly or partially leafless shoots. Male flowers sweetly-scented; pedicels very short or 0; sepals 2 × 1–1.5 mm, triangular-ovate, acute, glabrous, pale yellowish-green; petals 1 × 1 mm, flabelliform, erose at apex, pale greenish- or creamy-yellow; disk 1.75 mm in diameter, flat, entire, greenish; staminal column 1.25 mm high, greenish; anthers 0.67 mm long, yellow; pistillode 0.75 mm tall, conic-cylindric, deeply bifid. Female flowers sessile or subsessile; sepals ± as in the male, but somewhat thicker; petals c. 1 × 1 mm, rhombic-obovate, subentire; outer disk 2 mm in diameter, pentagonal, inner of 3 lobes c. 1 mm high, each lobe ± triangular, toothed; ovary c. 1 × 0.75 mm, ovoid, 2-locular; styles 2, 1 mm long, free, shortly bifid, stigmas uneven, greenish. Fruit 6–11 × 7–10 mm when dried, subglobose, 2-locular, green at first, later becoming reddish-purple then blackish. Seeds 7 × 6 × 3 mm, smooth, shiny, chestnut-brown.

1. Leaf blades rarely exceeding 6 cm in length on younger shoots, glaucous beneath and usually drying greyish-brown; lateral nerves in up to 14 pairs, craspedodromous, closely parallel · subsp. *cathartica*
– Leaf blades up to 12 cm in length on younger shoots; lateral nerves in up to 10 pairs, mostly brochidodromous, fairly widely spaced · · · · · · · · · · · · · · · · · · (subsp. *melanthesoides*) 2
2. Venation not prominent beneath, leaf blades usually drying medium green · · · · · · · · · 3
– Venation prominent beneath, leaf blades often drying pale yellowish-green or -brown · · 4
3. Leaves glabrous beneath · var. *melanthesoides* f. *melanthesoides*
– Leaves pubescent beneath · var. *melanthesoides* f. *pubescens*
4. Veins evenly to sparingly pubescent and later glabrescent beneath · var. *lingelsheimii* f. *fischeri*
– Veins completely glabrous beneath · · · · · · · · · · · · · · · · · · · var. *lingelsheimii* f. *niedenzui*

Subsp. **cathartica**

Bridelia schlechteri Hutch. in Bull. Misc. Inform., Kew **1914**: 249 (1914); in F.C. **5**, 2: 380 (1915). —Engler, Pflanzenw. Afrikas **3**, 2: 43 (1921). Type: Mozambique, "Inyamasan", female fl. & y. fr. 21.i.1898, *Schlechter* 12065 (K, holotype; BR).

Leaf blades rarely exceeding 6 cm in length on younger shoots, glaucous beneath and usually drying greyish-brown; lateral nerves in up to 14 pairs, craspedodromous, closely parallel.

Mozambique. MS: Beira, female fl. & y. fr. 25.ii.1912, *Rogers* 4551 (BM; K). GI: Xai-Xai (Vila de João Belo), fr. 9.vi.1960, *Lemos & Balsinhas* in *Lemos* 42 (K; LISC; LMA; PRE). M: Maputo, Manhiça, Vila de Incomáti, male fl. 26.iii.1979, *de Koning* 7372 (BM; K; LMU; MO).

Also in South Africa (KwaZulu-Natal). On sandy soil and dunes, in coastal bush, littoral scrub and dune woodland and forest, also in secondary forest and old cultivations; 2–90(345) m.

Vernacular names as recorded in specimen data include: "catchangate" (Muchopes); "mbalatangati" (Changane); "m'batelo" (Inhambane area); "mun(u)angati" (xiRonga); "munangati", "nunangade", "thlath-langati" (Landim).

Intermediates between this and the following subspecies from Maputo Province are: *Mrs Moss* J28389 (PRE), from Inhaca (in fruit), & J28385 (PRE) from Maputo (in fruit), and *Junod* 510 (LISC; PRE) from Ricatla (female fl.).

Subsp. **melanthesoides** (Baill.) J. Léonard in Bull. Jard. Bot. État **25**: 364 (1955); in F.C.B. **8**, 1: 30 (1962). —Dale & Greenway, Kenya Trees & Shrubs: 187 (1961). —White, F.F.N.R.: 194 (1962). —Drummond in Kirkia **10**: 251 (1975). Type: Mozambique, Inhambane, female fl. no date, *Peters* s.n. (B†, holotype; K; P).

Pentameria melanthesoides Baill., Étud. Gén. Euphorb.: 584 (1858). Type as above.

Bridelia melanthesoides (Baill.) Klotzsch in Peters, Naturw. Reise Mossambique **6**, 1: 103 (1861). —Jablonszky in Engler, Pflanzenr. [IV, fam. 147, viii] **65**: 67 (1915). —Engler, Pflanzenw. Afrikas **3**, 2: 43 (1921). —Brenan, Check-list For. Trees Shrubs Tang. Terr.: 201 (1949).

Leaf blades up to 12 cm in length on younger shoots; lateral nerves in up to 10 pairs, mostly brochidodromous, fairly widely spaced.

Var. **melanthesoides** (Baill.) Radcl.-Sm. comb. nov.

Pentameria melanthesoides Baill., Étud. Gén. Euphorb.: 584 (1858). Type as above.

Venation not prominent beneath; leaf blades usually drying medium green.

forma **melanthesoides** (Baill.) Radcl.-Sm. comb. nov.
Pentameria melanthesoides Baill., Étud. Gén. Euphorb.: 584 (1858). Type as above.

Leaves glabrous beneath.

Caprivi Strip. Schuckmannsburg, fr. 29.x.1970, *Vahrmeijer* 2200 (K; PRE). **Botswana**. N: 40 km Kachikau–Kasane (Gazane), fr. 10.vii.1937, *Erens* 377 in *Pole Evans* 4177 (K; PRE). **Zambia**. B: 9.5 km NW of Sesheke Boma, fr. 9.viii.1947, *Brenan & Keay* 7658 (FHO; K; PRE). N: Chienge Distr., Chipampa R., fr. 1.vi.1933, *Michelmore* 368 (K). C: Mumbwa, fr. 30.v.1961, *Fanshawe* 6632 (K; NDO). E: Petauke Distr., Luangwa R., fr. 5.ix.1947, *Brenan & Greenway* 7800 (FHO; K). S: Livingstone, fr. 28.v.1964, *Lawton* 1150 (K; NDO). **Zimbabwe**. N: Gokwe Distr., Morowa R., fr. 17.vii.1962, *Bingham* 310 (K; LISC; PRE; SRGH). W: Hwange Distr., Victoria Falls, Zambezi Cottages, fr. 5.viii.1976, *Langman* 1 (K; PRE; SRGH). E: Chipinge Distr., Save (Sabi) R., male fl. 6.iv.1959, *Savory* 311 (K; LISC; PRE; SRGH). S: Ndanga Distr., Save (Sabi)-Runde (Lundi) junction, Chitsa's Kraal, fr. 4.vi.1950, *Wild* 3343 (K; LISC; PRE; SRGH). **Malawi**. S: Machinga Distr., Liwonde, fr. 12.v.1988, *Salubeni & Kaunda* 5139 (MAL; MO). **Mozambique**. N: 25 km Palma–Cabo Delgado farol, y. fr. 17.iv.1964, *Torre & Paiva* 12097 (LISC). Z: south of Posto Chire, y. fr. 5.v.1972, *Bowbrick* J236 (LISC; SRGH). T: Chicoa, female fl. 2.iii.1972, *Macêdo* 4965 (LISC; LMA). MS: Marínguè, Sabi R., fr. 24.vi.1950, *Chase* 2531 (BM; SRGH). GI: Macia, Chuale, o. fr. 29.viii.1980, *Nuvunga & Boane* 305 (BM; K; LMU; MO). M: 10 km Namaacha–Boane, male fl. 3.ii.1982, *de Koning* 9162 (K; LMU; MO).

Also in Ethiopia, Somalia, Kenya, Tanzania (including Zanzibar), Namibia, Swaziland and South Africa (Transvaal and KwaZulu-Natal). On sandy soil and floodplain clay and alluvium, in low to medium altitude sandveld, and in riverine vegetation, mixed deciduous woodland and evergreen rainforest; 10–915 m.

Vernacular names as recorded in specimen data include: "msipila" (chiNdau); "mulupanbane" (Nyamenda); "zizuru" (Pambala).

The berries are eaten.

Intermediates between this and forma *pubescens* include *Kirk* s.n. (K) in male fl., from Shupanga, and *Torre & Correia* 18485 (COI; EA; LISC; LMA; PRE) in y. fr. from Chemba, Mozambique.

Intermediates between this and var. *lingelsheimii* formae *fischeri* and *niedenzui* include *Lovemore* 393 (K; LISC; SRGH), in female fl. from the Nganga R., Hurungwe Distr., N Zimbabwe, and *Macêdo* 5259 (LISC; LMA), in fr. from 4.3 km Estima–Candôdo, Tete Province, Mozambique.

Intermediates between this and var. *lingelsheimii* forma *fischeri* include *Torre* 1316 (LISC) both in male and female fl., from Nampula, N Mozambique, and between this and forma *neidenzui* include *Torre* 2743 (LISC) in fr. from between Vilanculos and Mambone, Mozambique.

Intermediates between this and the typical subsp. include *Pawek* 5650 (K) in fr. from Chikale Beach, N Malawi.

forma **pubescens** Radcl.-Sm. in Kew Bull. **51**, 2: 302 (1996). Type: Mozambique, Tete Prov., Chicoa–Mágoè, y. male fl. 13.ii.1970, *Torre & Correia* 17963 (LISC, holotype).

Leaves pubescent beneath.

Zambia. B: Mwandi, fr. 2.ix.1962, *Fanshawe* 7021 (FHO). C: Mumbwa, fr. 30.v.1961, *Fanshawe* 6632 (FHO). **Mozambique**. T: 36 km Chicoa–Mágoè along R. Zambezi, male fl. 17.ii.1970, *Torre & Correia* 18011 (LISC).

Also in Tanzania. On sandy clay soils, in riverine forest; 300–350 m.

Var. **lingelsheimii** (Gehrm.) Radcl.-Sm. in Kew Bull **51**, 2: 302 (1996). Type from Tanzania.
Bridelia lingelsheimii Gehrm. in Bot. Jahrb. Syst. **41**, Beibl. 95: 36 (1908).
Bridelia fischeri var. *lingelsheimii* (Gehrm.) Hutch. in F.T.A. **6**, 1: 616 (1912). —Jablonszky in Engler, Pflanzenr. [IV, fam. 147, viii] **65**: 68 (1915). —Brenan, Check-list For. Trees Shrubs Tang. Terr.: 201 (1949).

Venation prominent beneath. Leaf blades often drying pale yellowish-green or brown.

forma **fischeri** (Pax) Radcl.-Sm. in Kew Bull. **51**, 2: 302 (1996). Type from Tanzania.
Bridelia fischeri Pax in Bot. Jahrb. Syst. **15**: 531 (1893); Pflanzenw. Ost-Afrikas **C**: 237 (1895). —Hutchinson in F.T.A. **6**, 1: 616 (1912). —Jablonszky in Engler, Pflanzenr. [IV, fam. 147, viii] **65**: 68 (1915). —Engler, Pflanzenw. Afrikas **3**, 2: 43 (1921). —O.B. Miller, Check-list For. Trees Shrubs Bech. Prot.: 31 (1948). —Brenan, Check-list For. Trees Shrubs Tang. Terr.: 200 (1949).
Bridelia niedenzui var. *pilosa* Gehrm. in Bot. Jahrb. Syst. **41**, Beibl. 95: 37 (1908). Type from Tanzania.

Bridelia scleroneura sensu R.E. Fries, Wiss. Ergebn. Schwed. Rhod.-Kongo-Exped. **1**, 1: 118 (1914), non Müll. Arg. (1864, 1866).

Veins evenly to sparingly pubescent and later glabrescent beneath.

Botswana. N: Chobe Distr., Kasane Rapids, fr. 2.viii.1950, *Robertson & Elffers* 94 (K; PRE). **Zambia**. N: Luapula Valley, male fl. 10.iv.1961, *Angus* 2802 (FHO; K; LISC). C: Lusaka, male fl. 12.iii.1973, *Chisumpa* 14 (K; NDO). E: Petauke Distr., near Ndefu, fr. 19.iv.1952, *White* 2419 (FHO; K). S: Namwala, v.o. fr. 12.viii.1963, *van Rensburg* 2421 (K). **Zimbabwe**. N: Hurungwe Distr., near Gota Gota Hill, female fl. and y. fr. 21.ii.1956, *Phelps* 125 (K; LISC; SRGH). W: Nyamandhlovu, y. fr. 10.iv.1953, *Plowes* 1588 (K; PRE; SRGH). C: Kadoma (Gatooma), fr. vi.1926, *Herb. Dept. Agr. S.R.* 1234 (K; SRGH). E: Mutare (Umtali), subst. 12.xi.1930, *Fries, Norlindh & Weimarck* 2924 (K; LD). S: Masvingo Distr., Great Zimbabwe, o. fr. vii.1951, *Seward* 57/51 (SRGH). **Malawi**. N: Nkhata Bay–Chikale Beach, o. fr. 7.xii.1975, *Pawek* 10388 (K; MAL; MO; SRGH; UC). C: Dedza Distr., Kirk Range, Ganya, fr. 31.v.1989, *Chikuni, Patel & Nachamba* 141 (MAL; MO). S: Mangochi Distr., Namwera Escarpment, y. fl./st. 15.iii.1955, *Exell, Mendonça & Wild* 892 (BM; K; LISC; SRGH). **Mozambique**. N: 11 km Marrupa–Nungo, o. fr. 5.viii.1981, *Jansen et al.* 66 (MO; WAG). Z: Ile, Errego, 3 km from Mt. Ile, male fl. 3.iii.1966, *Torre & Correia* 14972 (LISC). T: Fíngoè, o. fr. 10.ix.1941, *Torre* 3231 (LISC). MS: Marínguè, st. 26.iv.1973, *Bond* 9B25 (LISC; SRGH).

Also in Zaire (Shaba Province), Somalia, Kenya, Tanzania and Namibia. On Kalahari Sand, sandy loam and clay soils, on rocky outcrops and hill slopes, dambos, floodplains and mushitu margins, in wooded grassland, mopane veld, miombo and mixed deciduous woodland; 470–1370 m.

Vernacular names as recorded in specimen data include: "kalamba bwato" (or "bwatu") (chiBemba); "mulapambare" (chiGowa); "munonyamanzi" (Tok); "mupala-pala" (eMakhuwa); "musamandola" (chiChewa); "sinengwe" (Nyamandhlovu area).

Whereas the MO and SRGH duplicates of *P.A. Smith* 4314 (fr. 14.iv.1983 along the Lesomo–Ngwezumba road in N Botswana) are referable to forma *fischeri*, the PRE duplicate represents the forma *niedenzui*.

forma **niedenzui** (Gehrm.) Radcl.-Sm. in Kew Bull. **51**, 2: 302 (1996). Type from Tanzania.

Bridelia niedenzui Gehrm. in Bot. Jahrb. Syst. **41**, Beibl. 95: 36 (1908). —Hutchinson in F.T.A. **6**, 1: 616 (1912). —Jablonszky in Engler, Pflanzenr. [IV, fam. 147, viii] **65**: 68 (1915). —Eyles in Trans. Roy. Soc. South Africa **5**: 394 (1916). —Engler, Pflanzenw. Afrikas **3**, 2: 43 (1921). —O.B. Miller, Check-list For. Trees Shrubs Bech. Prot.: 31 (1948). —Brenan, Check-list For. Trees Shrubs Tang. Terr.: 201 (1949). —Topham, Check List For. Trees Shrubs Nyasaland Prot.: 50 (1958).

Veins completely glabrous beneath.

Botswana. N: Lesomo–Ngwezumba road, fr. 14.iv.1983, *P.A. Smith* 4314 (PRE). **Zambia**. B: Kaoma (Mankoya) Boma, male fl. 23.ii.1952, *White* 2129 (FHO; K). W: Mpongwe, st. 27.ix.1949, *Hoyle* 1229 (FHO). S: Kafue Gorge, fr. 8.vi.1958, *Angus* 2020 (K; LISC). **Zimbabwe**. N: Mutoko Res., fr. viii.1956, *Davies* 2069 (K; MO; SRGH). W: Hwange Distr., Matetsi, fr. 24.v.1975, *Gonde* 31 (K; SRGH). C: Chegutu Distr., Poole Farm, male & female fl. 3.iv.1946, *Wild* 992 (K; SRGH). E: Nyanga (Inyanga), o. fr. 22.vi.1948, *Chase* 1677 (BM; K; SRGH). S: Masvingo Distr., Mushandike Nat. Park, fr. 28.vi.1972, *Chiparawasha* 495 (K; LISC; SRGH). **Malawi**. N: 67 km west of Karonga, male fl. 16.iv.1976, *Pawek* 11077 (K; MAL; MO; SRGH; UC). C: Ntcheu Distr., Mvai For. Res., o. fr. 30.viii.1986, *Nachamba & Usi* 377 (MAL; MO). S: Mt. Mulanje, Michese Mt., fr. 17.vi.1987, *J.D. Chapman & E.J. Chapman* 8631 (FHO; K; MO; PRE). **Mozambique**. N: Cabo Delgado, Ancuabe, Metoro, Namatuca, y. male fl. 29.i.1984, *Maite, de Koning & Dungo* 156 (K; LMU; MO). Z: Mocuba, female fl. 11.iii.1943, *Torre* 4920 (LISC). T: Moatize, y. fr. 7.v.1948, *Mendonça* 4123 (LISC). MS: Chimoio, foot of Bandula Mt., y. fr. 28.iii.1948, *Garcia* 789 (LISC).

Also in Zaire (Shaba Province), Sudan, Somalia and Tanzania. On Kalahari Sand, quartzitic and granitic soils, heavy black clays and alluvium, on rocky hillsides and granite outcrops, riverbanks and watercourses, in sandveld, grassland, mixed treed savanna, deciduous thicket, miombo and mopane woodlands, riverine vegetation and in periodically inundated ground; 300–1470 m.

Vernacular names as recorded in specimen data include: "duondi", "m'tondi" (Ayaua); "manyambane" (chiShona); "mfura" (Tete area); "mtantanyara", "tantanyerere" (chiNyanja); "m'tundi" (Zomba area); "munonyamanzi" (Tok.); "namazamaza" (Cabo Delgado area).

Whereas the K and SRGH duplicates of *Phipps* 993 (male fl. 1.iii.1958, 65 km north of Mauora, Zimbabwe) are referable to this forma, the LISC duplicate is referable to f. *fischeri*. Likewise the K and NY duplicates of *Brass* 17408 (fr. 24.viii.1946 at Kasungu, Malawi) are of f. *niedenzui*, whilst the BM duplicate is of f. *fischeri*. The K duplicate of *Brummitt* 9295 (male and female fl. 22.iii.1970 at the Dzalanyama For. Res.) is f. *niedenzui*, but the PRE duplicate is intermediate between the 2 formae.

3. **Bridelia atroviridis** Müll. Arg. in J. Bot. **2**: 327 (1864); in De Candolle, Prodr. **15**. 2: 494 (1866). —Hutchinson in F.T.A. **6**, 1: 617 (1912). —Jablonszky in Engler, Pflanzenr. [IV, fam. 147, viii] **65**: 77 (1915). —Eyles in Trans. Roy. Soc. South Africa **5**: 394 (1916). —Engler, Pflanzenw. Afrikas **3**, 2: 44 (1921). —De Wildeman, Pl. Bequaert. **3**, 4: 448 (1926). —Brenan, Check-list For. Trees Shrubs Tang. Terr.: 200 (1949). —Eggeling & Dale, Indig. Trees Uganda, ed. **2**: 117 (1952). —Keay in F.W.T.A., ed. 2. **1**, 2: 370 (1958). —Dale & Greenway, Kenya Trees & Shrubs: 185 (1961). —J. Léonard in F.C.B. **8**, 1: 35 (1962). —Drummond in Kirkia **10**: 251 (1975). —K. Coates Palgrave, Trees Southern Africa, ed. 2, rev.: 412 (1983). —Radcliffe-Smith in F.T.E.A., Euphorb. 1: 125 (1987). —Beentje, Kenya Trees, Shrubs Lianas: 187 (1994). Tab. **3**, figs. B1–B4. Type from Angola.

A forest tree up to 20 m high with a straight trunk up to 45 cm in diameter. Bark pale grey, ± smooth or rough. Heartwood dark. Branches spiny. Twigs brown to dark purplish-brown, sparingly lenticellate. Young shoots and petioles evenly to sparingly puberulous, later glabrescent, or else quite glabrous. Petioles 2–8 mm long. Stipules 3–8 mm long, narrowly lanceolate, acutely acuminate, evenly to sparingly pubescent, soon falling. Leaf blades 2–17 × 1–10 cm, elliptic to oblanceolate, acutely acuminate, rounded-cuneate to subtruncate at the base, membranous, sparingly pubescent along the midrib and otherwise glabrous or else completely glabrous above, evenly to sparingly pubescent along the midrib and veins and sometimes glabrescent beneath, dark green and shiny above, mid-green and dull beneath, almost blackening above in drying; lateral nerves in 10–22 pairs, camptodromous, not or slightly prominent above, somewhat so beneath, tertiary nerves subparallel. Male flowers: pedicels c. 1 mm long, pubescent; sepals c. 2 × 1 mm, triangular-ovate, acute, pubescent without at the base, otherwise glabrous, often pinkish or purplish-tinged; petals 0.75 × 0.75 mm, spathulate, somewhat erose at the apex; disk 1.5 mm in diameter, annular, verruculose, ± entire; staminal column 1 mm high; anthers 0.75 mm long; pistillode 1 mm tall, ampulliform, bifid at the apex. Female flowers subsessile or very shortly pedicellate; sepals ± as in the male; petals 0.5 × 0.5 mm, spathulate, subentire; outer disk as in the male; inner disk 3-lobed, lobes c. 1 × 1 mm, ± triangular, toothed at apex; ovary 1 × 0.75 mm, ovoid, 2-celled, styles 2, c. 1 mm long, ± free, bifid, stigmas ± smooth. Fruit 6–8 × 5–6(7) mm when dried, obovoid-ellipsoid, 1-locular by abortion, green at first, blackish-brown when ripe. Seed 4 mm long, smooth, shiny, chestnut-brown.

Zimbabwe. E: Chipinge Distr., Chirinda For., male fl. i.1966, *Goldsmith* 1/67 (FHO; K; LISC; PRE; SRGH). **Malawi**. C: Dedza Distr., 5 km Linthipe–Chongoni, st. 30.iv.1989, *Radcliffe-Smith, Pope & Goyder* 5805 (K). S: Mt. Mulanje foot, above Power Station at Ruo Gorge entrance, y. fr. 18.ii.1987, *J.D. Chapman & E.J. Chapman* 8340 (FHO; K; MAL; MO). **Mozambique**. MS: Chimoio, Serra de Garuso (Garuzo), fr. 11.iv.1948, *Mendonça* 3904 (LISC).

Widespread in tropical Africa, extending from Sierra Leone eastwards to W Ethiopia and south to Angola, Zimbabwe and Mozambique. Infrequent in evergreen forest, on forest margins and in forest openings, and beside paths in associated woodland; 950–1160 m.

Vernacular name as recorded in specimen data: "mutsangu" (Chirinda Forest area).

4. **Bridelia brideliifolia** (Pax) Fedde in Just's Bot. Jahresber. **36**, 2: 413, in adnot. (1910). —Jablonszky in Engler, Pflanzenr. [IV, fam. 147, viii] **65**: 83 (1915). —Engler, Pflanzenw. Afrikas **3**, 2: 45 (1921) —Robyns & Tournay, Fl. Sperm. Parc Nat. Alb. **1**: 448 (1948). —Brenan, Check-list For. Trees Shrubs Tang. Terr.: 200 (1949). —Eggeling & Dale, Indig. Trees Uganda, ed. 2: 117 (1952). —J. Léonard in F.C.B. **8**, 1: 36 (1962). —Troupin, Fl. Pl. Lign. Rwanda: 250, fig. 86/.3 (1982); Fl. Rwanda **2**: 210, fig. 63/3 (1983). —Radcliffe-Smith in F.T.E.A., Euphorb. 1: 126 (1987). Tab. **3**, figs. C1–C4. Type from Tanzania.

Neogoetzea brideliifolia Pax in Bot. Jahrb. Syst. **28**: 419 (1900).

Bridelia neogoetzea Gehrm. in Bot. Jahrb. Syst. **41**, Beibl. 95: 40 (1908). —Hutchinson in F.T.A. **6**, 1: 619 (1912).

A deciduous tree up to 33 m high with a spreading crown, borne on stilt roots or with fluted buttresses up to 3 m high. Trunk spiny. Bark smooth or scaly, pinkish-brown. Branches ± flattened. Twigs dark purplish-brown or blackish. Young shoots, petioles and inflorescence axes evenly to sparingly pubescent or ± glabrous. Petioles 5–12 mm long. Stipules very fugacious; scars 3 mm long. Leaf blades 4–17 × 2–11 cm, elliptic-ovate to elliptic-oblong, shortly acutely to obtusely acuminate at the apex, cuneate-rounded to truncate or shallowly cordate at the base, firmly chartaceous to thinly coriaceous, glabrous above with the midrib and lateral nerves sparingly pubescent above, more evenly pubescent beneath, dark green above, paler beneath, often drying blackish above and brownish beneath; lateral nerves in 11–20 pairs, cheilodromous, often branched, scarcely prominent above, prominent beneath, tertiary nerves parallel. Flowers in terminal leafless or almost leafless panicles 3–13 cm long. Male flowers: pedicels 1.5 mm long, glabrous; sepals 2.5 × 1 mm, triangular-lanceolate, acute, glabrous, pale greenish; petals 1 × 0.5 mm, spathulate-flabelliform, tridentate, yellowish-green; disk 2 mm in diameter, annular, verruculose, yellow; staminal column 1.5 mm high; anthers 0.8 mm long; pistillode 0.75 mm high, ampulliform, bifid at the apex, the lobes connivent. Female flowers subsessile or very shortly pedicellate; sepals 2 × 1.5 mm, triangular-ovate, otherwise as in male; petals oblanceolate, otherwise as in male; outer disk similar to that of male; inner disk irregularly 3-lobed, lobes c. 1 × 1 mm, ± triangular, irregularly lobulate at apex; ovary 1 × 1 mm, subglobose, 2-celled; styles 2, 1.3 mm long, united at the base, bifid, stigmas ± smooth. Fruit 9–12 × 5–7 mm, ellipsoid to ovoid-ellipsoid, 1-locular by abortion, green at first, later becoming purple or purplish-black. Seed 7 × 5 mm, smooth, brown.

Malawi. N: Viphya Plateau 46.6 km SW of Mzuzu, male and female fls. 9.ii.1976, *Pawek* 10854 (K; MAL; MO; PRE; SRGH; UC). C: Ntchisi Mt. For., fr. 11.v.1984, *Banda & Kaunda* 2165 (K; MAL). S: Goche, Kirk Range, male fls. 30.i.1959, *Robson* 1366 (BM; K; LISC).

Also in Zaire, Rwanda, Burundi, Sudan, Uganda and Tanzania. On edge of dense mixed evergreen relict forest patches; 1600–2000 m.

If Léonard's subsp. *pubescentifolia* (leaf lower surface pubescent to tomentellous, as opposed to puberulous or glabrescent with the midrib and main nerves pubescent or glabrescent) is kept up, then all Malawi material seen is referable to the subsp. *brideliifolia*.

5. **Bridelia micrantha** (Hochst.) Baill., in Adansonia **3**: 164 (1862/3). —Müller Argoviensis in De Candolle, Prodr. **15**, 2: 498 (1866) pro parte. —Engler, Pflanzenw. Ost-Afrikas **C**: 237 (1895). —Schinz & Junod in Mém. Herb. Boissier, No. 10: 47 (1900). —Sim, For. Fl. Port. E. Afr.: 105 (1909). —Hutchinson in F.T.A. **6**, 1: 620 (1912). —R.E. Fries, Wiss. Ergebn. Schwed. Rhod.-Kongo-Exped. **1**, 1: 118 (1914). —Jablonszky in Engler, Pflanzenr. [IV, fam. 147, viii] **65**: 78 (1915). —Hutchinson in F.C. **5**, 2: 381 (1915). —Eyles in Trans. Roy. Soc. South Africa **5**: 394 (1916). —Engler, Pflanzenw. Afrikas **3**, 2: 43 (1921). —De Wildeman, Pl. Bequaert. **3**, 4: 450 (1926). —Burtt Davy, Fl. Pl. Ferns Transvaal **2**: 298 (1932). —Hutchinson, Botanist in Southern Africa: 667 (1946). —Robyns & Tournay, Fl. Sperm. Parc Nat. Alb. **1**: 447 (1948). —Brenan, Check-list For. Trees Shrubs Tang. Terr.: 201 (1949). —Eggeling & Dale, Indig.Trees Uganda, ed. 2: 117 (1952). —Keay in F.W.T.A., ed. 2, **1**, 2: 370 (1958). —Topham, Check List For. Trees Shrubs Nyasaland Prot.: 49 (1958). —Dale & Greenway, Kenya Trees & Shrubs: 187 (1961). —J. Léonard in F.C.B. **8**, 1: 46 (1962). —White, F.F.N.R.: 194 (1962). —Drummond in Kirkia **10**: 251 (1975). —Troupin, Fl. Pl. Lign. Rwanda: 251, fig. 86/1 (1982); Fl. Rwanda **2**: 210, fig. 63/1 (1983). —K. Coates Palgrave, Trees Southern Africa, ed. 2, rev.: 413 (1983). —Radcliffe-Smith in F.T.E.A., Euphorb. 1: 127 (1987). —Beentje, Kenya Trees, Shrubs Lianas: 187 (1994). Type from South Africa.

Candelabria micrantha Hochst. in Flora **26**: 79 (1843).

Bridelia zanzibariensis Vatke & Pax in Bot. Jahrb. Syst. **15**: 530 (1893). —Engler, Pflanzenw. Ost-Afrikas **C**: 237 (1895). —Hutchinson in F.T.A. **6**, 1: 618 (1912). Type from Zanzibar.

Bridelia abyssinica Pax in Bot. Jahrb. Syst. **39**: 630 (1907). —Hutchinson in F.T.A. **6**, 1: 621 (1912). Type from Ethiopia.

Bridelia mildbraedii Gehrm. in Jahres-Ber. Schles. Ges. Bres. **86**, 2b: 29 (1909). —Hutchinson in F.T.A. **6**, 1: 621 (1912). Type from Rwanda.

An evergreen or deciduous, often much-branched, small tree up to 20 m high with a dense spreading crown and ± flattened or pendent branches arising from c. 2 m above ground. Trunk and branches with scattered blunt thorns. Bark smooth and pale grey or pinkish-brown on branches, rough and dark grey or brown at base of trunk. Wood hard, white. Twigs dark grey or brown, lenticellate. Young shoots and petioles evenly to sparingly pubescent or subglabrous. Petioles 5–13 mm long. Stipules 4–7 mm long, linear-lanceolate, acute, puberulous or pubescent. Leaf blades 3–28 × 1.5–12 cm, elliptic to elliptic-oblong, shortly obtusely acuminate, rounded to cuneate at the base, subentire or very shallowly crenate, thinly coriaceous, sparingly pubescent to subglabrous along the midrib and main nerves above and beneath and otherwise ± glabrous above and minutely sparingly appressed-puberulous beneath, carmine-orange in young flush, dark or bright green and shiny above and paler below, with the nerves often pale yellow when mature, often drying greenish-grey above and light brown beneath; lateral nerves in 5–20 pairs, craspedodromous, slightly prominent above and beneath or somewhat more so beneath, tertiary nerves parallel or subparallel, not prominent. Male flowers: pedicels 1 mm long, sparingly appressed-puberulous; sepals 2 × 1 mm, triangular-ovate, acute, appressed-puberulous without, glabrous within, greenish; petals 0.5 × 0.5 mm, obtriangular, apically tridentate, greenish-white; disk 2 mm in diameter, shallowly 5-lobed, ± flat, fleshy, glabrous; staminal column 1 mm high; filaments 0.5 mm long, narrowing apically; anthers 0.75 mm long, yellow; pistillode 0.5 mm tall, ± conical, shallowly lobed at the apex. Female flowers faintly scented, subsessile or shortly stoutly pedicellate; sepals triangular, pale grey-green, otherwise ± as in the male; petals 1 × 0.5 mm, elliptic, subentire; outer disk 1.5 mm in diameter, pentagonal; inner disk 3-lobed, the lobes erose at the apex, closely enfolding the ovary, reddish-brown; ovary c. 0.75 × 0.75 mm, ovoid-subglobose, 2–3-celled; styles 2–3, c. 0.75 mm long, ± free, bifid, stigmas smooth. Fruit 6–8 × 4–5 mm when dried, slightly larger when fresh, ellipsoid or occasionally subglobose, 1-locular by abortion, green at first, black when ripe. Seeds 5 × 3 mm, smooth, slightly shiny, brown.

Zambia. B: Senanga, male fl. 1.viii.1952, *Codd* 7315 (K; PRE). N: Mansa (Fort Rosebery), y. male fl. 16.viii.1952, *Angus* 220 (BM; FHO; K). W: 37 km Solwezi–Mwinilunga, Mutanda R., male fl. 15.ix.1952, *Angus* 456 (BM; FHO; K). C: Mapombo R., SW of Mwambula, y. fr. 5.xi.1972, *Strid* 2468 (C; MO). E: Nsadzu–Chipata (Fort Jameson) road, fr. 25.xi.1958, *Robson & Angus* 700 (BM; K; LISC). S: Choma, y. fr. 9.x.1955, *Bainbridge* 148/55 (FHO; K). **Zimbabwe**. N: Umvukwes, st. 22.ix.1926, *Eyles* 4557 (K; SRGH). C: Rusape, fr. 7.xii.1954, *Munch* 432 (K; LISC; PRE; SRGH). E: Nyanga (Inyanga), y. fr. 7.xi.1930, *Fries, Norlindh & Weimarck* 2654a (BM; K; LD). S: Mberengwa Distr., Mt. Buhwa, male fl. 31.x.1973, *Biegel, Pope & Gosden* 4337 (K; PRE; SRGH). **Malawi**. N: Rumphi Distr., Chelinda R., fr. 17.x.1973, *Pawek* 7408 (K; MAL; MO; PRE; SRGH; UC). C: Mchinji For. Res., Bua R., fr. 21.xi.1983, *Salubeni & Patel* 3477 (MAL; MO). S: Mulanje Distr., Naminjiwa R., fr. 23.x.1983, *Tawakali* 104 (K; MAL; MO). **Mozambique**. N: Chomba, male fl. 2.xi.1960, *Gomes e Sousa* 4587 (K; PRE). Z: Namagoa, fr. xi–xii.1944, *Faulkner* 22 (BM; K; PRE; SRGH). T: Angónia Distr., Ulónguè, near Tchindeque, fr. 3.xii.1980, *Macuácua* 1380 (K; LMA; MO; PRE). MS: Gorongosa Nat. Park, Cheringoma Plateau, female fl. x.1972, *Tinley* 2733 (K; LISC; MO; PRE; SRGH). M: Salamanga, male fl. & fr. 9.x.1947 & i.1948, *Gomes e Sousa* 3621 (K; LISC; PRE).

Throughout tropical Africa from Senegal eastwards to central Ethiopia and south to Angola and South Africa (Eastern Cape Province); also in Réunion. In riverine and gully forests and in evergreen rain and mist forest patches, in miombo and high rainfall woodlands and escarpment woodlands, also in seasonally flooded grassland, dambos, riverine vegetation, swamp forest and mangrove swamp margins, sometimes on granite outcrops and on termitaria; 300–1750 m.

Vernacular names as recorded in specimen data include: "dimualongo" (shiMakonde); "hleha" (xiRonga); "insaba" (Butonga, Inhambane area); "inshepu" (Muchopes area); "massopa" (Cheringoma area); "mecici", "merroco", "murosi", "murroci"; "messunguza" (Manica area); "missambite" (xiRonga); "mpasa" (chiNyanja); "msopa" (chiYao); "mulangale" (Mambwe area); "musewe" ; "mushunguna", "mushungunu", "mutsunguno" (chiNdao); "nsopa" (Zomba area); "saba" (Inharrime-Inhambane area); "sanguso" (Chimoio area); "umhlahlamakwaba".

The pounded bark is used to fill cracks in doors and baskets (Swynnerton). Bark extract is applied to scabies. The wood was used for oxen yokes and is used in furniture making.

The disk of the male flower attracts *Hymenoptera*.

6. **Bridelia ferruginea** Benth. in Hooker, Niger Fl.: 511 (1849). —Hutchinson in F.T.A. **6**, 1: 619 (1912). —R.E. Fries, Wiss. Ergebn. Schwed. Rhod.-Kongo-Exped. **1**, 1: 118 (1914). —Jablonszky in Engler, Pflanzenr. [IV, fam. 147, viii] **65**: 81 (1915). —Hutchinson in F.C. **5**, 2: 381 (1915). —Engler, Pflanzenw. Afrikas **3**, 2: 45 (1921). —De Wildeman, Pl. Bequaert. **3**, 4: 449 (1926). —Keay in F.W.T.A., ed. 2, **1**, 2: 370 (1958). —Topham, Check List For. Trees Shrubs Nyasaland Prot.: 49 (1958). —J. Léonard in F.C.B. **8**, 1: 38 (1962). —White, F.F.N.R.: 194 (1962). Type from N Nigeria.

Bridelia micrantha var. *ferruginea* (Benth.) Müll. Arg. in De Candolle, Prodr. **15**, 2: 498 (1866).

A shrub or small tree up to c. 6 m high with spiny branches. Bark cracked, grey. Twigs dark brown. Young shoots and petioles denseley ferrugineous-tomentose. Petioles 5–7 mm long. Stipules 5–6 × 1.5–2 mm, lanceolate, acuminate, tomentose, readily caducous. Leaf blades 4–10 × 3–5.5 cm, elliptic to elliptic-ovate, subacute to shortly obtusely acuminate, rounded, truncate or sometimes cordulate at the base, thinly coriaceous, evenly pubescent along the midrib and main nerves above and beneath, otherwise sparingly so, glossy green above and paler beneath when fresh, drying dark brown or greyish-brown above, dark reddish-brown beneath; lateral nerves in 7–10 pairs, cheilodromous, scarcely prominent above, prominent beneath, tertiary nerves parallel, fairly prominent beneath, quaternary nerves reticulate. Male flowers: pedicels 1–1.5 mm long, puberulous; sepals 2 × 1 mm, triangular-ovate-lanceolate, acute, puberulous without, glabrous within, yellowish-green; petals 1 × 0.75 mm, obdeltoid-spathulate, incised or tridentate at the apex, cornute, glabrous; disk 1.5–2 mm in diameter, annular, ± smooth; staminal column 1 mm high; anthers 0.7 × 0.5 mm; pistillode 0.5 mm high, conical, notched at apex. Female flowers subsessile; sepals 1.5 × 1.5 mm, triangular, thick, otherwise as in male; petals elliptic-obovate, puberulous without, otherwise as in male; outer disk ± as in male; inner laciniate at the apex, ciliate within, pubescent or subglabrous; ovary 1.5 × 1 mm, ellipsoid, 2-celled; styles 2, c. 1 mm long, ± free, bifid, stigmas ± smooth. Fruit c. 7 × 4–5 mm when dried, ellipsoid to ovoid-ellipsoid, 1-locular by abortion, green at first, then reddening and becoming purplish-black at maturity. Seeds 5 × 3 mm, smooth, brownish.

Zambia. W: Mwinilunga Distr., 1 km south of Matonchi Farm, male fl. 12.x.1937, *Milne-Redhead* 2730 (BM; K; LISC; PRE).

From Guinée and Mali eastwards to the Central African Republic and from Gabon south and east to Angola and extreme NW Zambia. Habitat notes are unavailable for Zambia, but in the Congo Republic it occurs at c. 500 m on laterite in savanna with *Hymenocardia acida*, whilst in Zaire it is found in gallery forest at around 900 m.

The description was completed with help from Angolan and Zairean material.

7. **Bridelia duvigneaudii** J. Léonard in Bull. Jard. Bot. État **25**: 365 (1955); in F.C.B. **8**, 1: 39 (1962). —White, F.F.N.R.: 194 (1962). —Radcliffe-Smith in F.T.E.A., Euphorb. 1: 129 (1987). Type from Zaire (Shaba Province).

Bridelia ferruginea sensu R.E. Fries, Wiss. Ergebn. Schwed. Rhod.-Kongo-Exped. **1**, 1: 118 (1914). —Hutchinson, Botanist in Southern Africa: 533 (1946). —Brenan, Check-list For. Trees Shrubs Tang. Terr.: 200 (1949), non Benth.

Bridelia mollis sensu Brenan, Check-list For. Trees Shrubs Tang. Terr.: 201 (1949), non Hutch.

Bridelia katangensis J. Léonard, name on herbarium sheet only.

Similar to *B. ferruginea*, differing chiefly in having narrower (0.5–1 mm wide), arcuate, more persistent stipules, shorter (3–5 mm long) petioles, generally smaller leaves with a more laxly reticulate venation, fewer (3–5-)flowered glomerules, longer (2–3 mm long) male sepals, female sepals pubescent within towards the apex and ± glabrous inner female disk.

Zambia. B: Zambezi (Balovale), fr. 9.v.1954, *Gilges* 350 (K; PRE). N: Mansa Distr., Samfya, male fl. 30.ii.1959, *Watmough* 202 (K; LISC; PRE; SRGH). W: Chifubwa R. Gorge, 3 km south of Solwezi, o. male fl. 20.iii.1961, *Drummond & Rutherford-Smith* 7124 (FHO; K; LISC; PRE; SRGH). C: 67 km SW of Serenje Corner, o. fr. 25.vii.1930, *Hutchinson & Gillett* 4086 (BM; K; LISC). **Mozambique**. N: Malema, y. fr. 20.iii.1964, *Torre & Paiva* 11309 (LISC). Z: Montes do Ile, fr. 19.vi.1943, *Torre* 5529 (LISC).

Also in Burundi, Zaire, Tanzania and Angola. In high rainfall miombo and Kalahari Sand woodlands and thickets, also in riverine vegetation and occasionally mushitu, sometimes in chipya woodland, often on sandy soil and sometimes amongst rocks and on granite outcrops; 1000–1675 m.

Vernacular names as recorded in specimen data include: "kalambabwato" (chiWemba); "mudemwe" (Sith.); "mufuunji" (chiLunda).

4. PSEUDOLACHNOSTYLIS Pax

Pseudolachnostylis Pax in Bot. Jahrb. Syst. **28**: 19 (1899). —Hutchinson in F.T.A. **6**, 1: 671 (1912). —Pax & K. Hoffmann in Engler, Pflanzenr. [IV, fam. 147, xv] **81**: 206 (1922).

Dioecious or rarely monoecious trees or shrubs. Indumentum simple. Leaves alternate, petiolate, stipulate; blades simple, entire, penninerved. Flowers in axillary pedunculate or subsessile bracteate dense few-(up to 12-)flowered cymes, or females solitary, arising among or below the leaves. Male flowers: buds ovoid; sepals 5(6), imbricate; petals absent; disk extrastaminal, annular, shallowly lobed, the lobes alternating with the sepals; stamens (4)5–7, united for half to two-thirds of their length into a column, anthers basifixed, introrse, thecae parallel, longitudinally dehiscent; pistillode (non-functional ovary) usually trifid. Female flowers: pedicels bibracteolate; buds and sepals as in the male; disk cupular; ovary 3-locular, with 2 ovules per loculus; styles 3, connate at the base, bifid, the stigmas coiled backwards. Fruit globose, subdrupaceous, tardily septicidally dehiscent after the fleshy layer has been eaten by frugivores, and when the sun has dried out the endocarp; exocarp smooth, becoming wrinkled and shiny on drying; mesocarp spongy; endocarp thickly bony or woody, trilocular. Seeds solitary per loculus by abortion, ecarunculate; albumen fleshy; cotyledons broad, flat.

A monotypic east and south tropical African genus in which up to 6 species have been recognized in the past.

Pseudolachnostylis maprouneifolia Pax in Bot. Jahrb. Syst. **28**: 20 (1899). —Hutchinson in F.T.A. **6**, 1: 672 (1912); in F.C. **5**, 2: 407 (1920). —Eyles in Trans. Roy. Soc. South Africa **5**: 392 (1916). —Engler, Pflanzenw. Afrikas **3**, 2: 31 (1921). —Pax in Engler, Pflanzenr. [IV, fam. 147, xv] **81**: 207 (1922). —Burtt Davy, Fl. Pl. Ferns Transvaal: 300 (1932). —Hutchinson, Botanist in Southern Africa: 667 (1946). —O.B. Miller, Check-list For. Trees Shrubs Bech. Prot.: 33 (1948). —Brenan, Check list For. Trees Shrubs Tang. Terr.: 224 (1949). —Suessenguth & Merxmüller, Contrib. Fl. Marandellas Distr. **43**: 85 (1951). —Topham, Check List For. Trees Shrubs Nyasaland Prot.: 53 (1958). —White, F.F.N.R.: 203 (1962). —P.G. Meyer in Merxmüller, Prodr. Fl. SW. Afrika, fam. 67: 40 (1967). —Drummond in Kirkia **10**: 251 (1975). —K. Coates Palgrave, Trees Southern Africa, ed. 2, rev.: 396 (1983). —Radcliffe-Smith in F.T.E.A., Euphorb. 1: 80 (1987). Types from Tanzania (Central Province).

A deciduous tree up to 18 m tall, or rarely a many-stemmed shrub; stem usually unbranched to 3.5 m, up to 25 cm d.b.h.; crown compact and rounded, or ± laxly spreading, the branches sometimes drooping almost to the ground. Bark dark grey or blackish and rough, fissured and flaking in thick flakes, paler grey or whitish and smooth on the branches, cracking and flaking to reveal yellow-grey layer beneath. Twigs brownish, lenticellate. Young shoots, petioles and peduncles glabrous (var. *glabra*), pubescent (var. *maprouneifolia*) or densely fulvous- to ferrugineous-tomentose (vars. *dekindtii* and *polygyna*). Buds perulate (furnished with protective scales); perulae 1 mm long, ovate, dark brown. Stipules 4–5 mm long, lanceolate, pubescent without, glabrous within, soon caducous. Petioles 0.2–1 cm long. Leaf blade 1.3–10(12.5) × 0.9–5.5(6) cm, broadly ovate to elliptic-ovate or sometimes ovate-lanceolate, subacute, obtuse or rounded at the apex, rounded-cuneate to shallowly cordate at the base, entire, chartaceous to thinly coriaceous, glabrous (var. *glabra*) to sparingly or evenly pubescent above and beneath (vars. *dekindtii* and *polygyna*) or else pubescent only along the midrib beneath and otherwise glabrous (var. *maprouneifolia*), pale to deep grey-green above, paler and somewhat glaucescent beneath, reddish when young; lateral nerves in 6–10(12) pairs, somewhat irregularly looped. Flowers faintly musky-odoured, much visited by bees. Male inflorescences 0.5–2 cm long; bracts 3 × 2–3 mm, broadly ovate or elliptic-ovate, acute or subacute, strongly concave, keeled, pubescent without, glabrous within, chaffy. Male flowers sessile; sepals (3)4–5 × (1.5)2–3 mm, ovate or elliptic, obtuse or rounded, glabrous to sparingly pubescent without, glabrous within, yellowish-green, later becoming straw-yellow in colour; disk 2–2.5 mm in diameter, pinkish; staminal column 1.5–2 mm high; anthers 1.5–2 mm long, yellow to brownish; pistillode (non-functional ovary) 0.3–0.5 mm long. Female inflorescences 1–3(7)-flowered, 1–2 cm long; bracts or bracteoles c. 2 × 1 mm, lanceolate, otherwise as in the male. Female flowers: pedicels 1.5–2 mm long; sepals as in the male; disk 2 mm in diameter, finely toothed to coarsely and irregularly-lobed; ovary 1.5 × 1 mm, more or less ovoid, densely pubescent or glabrous; styles 1.5–2 mm long. Fruit 1.3–2 × 1.3–2 cm, ovoid-subglobose to depressed-globose and scarcely 2–4-lobed, glabrous or glabrescent, yellowish-green, yellow or pinkish-green. Seeds 7 × 5 mm, ellipsoid-ovoid, slightly shiny, light brown streaked with darker brown.

1. All vegetative parts quite glabrous ············ var. *glabra*
– All or most vegetative parts pubescent or tomentose ············ 2
2. Female flowers in (1)3(or more)-flowered cymes ············ var. *polygyna*
– Female flowers solitary ············ 3
3. Leaves pubescent only along the midrib beneath; male inflorescences pedunculate; female disk coarsely and irregularly lobed; ovary glabrous ············ var. *maprouneifolia*
– Leaves uniformly pubescent above and beneath; male inflorescences usually sessile; female disk finely toothed; ovary densely pubescent ············ var. *dekindtii*

Var. **maprouneifolia**. Tab. **4**, figs. A1–A4.

Young shoots pubescent; leaves pubescent only along the midrib beneath; male inflorescences pedunculate; female flowers solitary, pedicels not apically thickened; female disk coarsely and irregularly lobed; ovary glabrous.

Botswana. N: Tsessebe, fr. 17.iv.1931, *Pole Evans* 3245 (43) (K; PRE). SE: Sefare (Sefhare), fr. xii.1940, *O.B. Miller* B/243 (PRE). **Zambia**. N: L. Chila, male fl. & fr. 30.ix.1966, *Richards* 21477 (K). W: Kitwe, male & female fl. x.1957, *Fanshawe* 4150 (K; NDO). C: near Mupamadzi R., fr. 9.xi.1966, *Astle* 4041 (K). E: Chadiza–Chipata (Fort Jameson), male fl. 8.x.1958, *Robson & Angus* 30 (BM; K; LISC; PRE). S: Gwembe, st. 7.iv.1952, *White* 2621 (FHO; K). **Zimbabwe**. N: Makonde Distr., Manyame (Hunyani) Mts., y. fr. iv.1920, *Henkel* in *Eyles* 2350 (K; PRE; SRGH). W: Dombodema Mission, Bullima-Mangwe (Bulalima–Mangwe), male fl. 17.x.1972, *Norrgrann* 237 (K; LISC; SRGH). C: L. Chiveru (McIlwaine), st. 19.ix.1954, *Walkerdene* 5 (K; PRE; SRGH). E: 16 km south of Mutare (Umtali), y. fr. 26.xi.1948, *Chase* 1573 (BM; K; SRGH). S: 50 km east of Bikita, st. 21.x.1930, *Fries, Norlindh & Weimarck* 2184 (BM; K; LD). **Malawi**. N: Karonga Distr., fr. 11.ix.1977, *Phillips* 2806 (K; MO; SRGH). C: Chongoni For. Res., male fl. 22.xi.1966, *Jeke* 37 (K; MAL; SRGH). S: Bvumbwe, male fl. 13.x.1985, *la Croix* 3368 (MAL; MO; PRE). **Mozambique**. N: Nampula Distr., Mutuáli, fr. 5.iii.1953, *Gomes e Sousa* 4055 (K; PRE). Z: 67.7 km Alto Ligonha–Alto Molócuè, male fl. 13.x.1949, *Barbosa & Carvalho* in *Barbosa* 4416 (K; LMA). T: Cahora Bassa, fr. 3.v.1972, *Pereira & Correia* 2348 (LISC; PRE; SRGH). MS: Chinizíua, male fl. 17.x.1957, *Gomes e Sousa* 4405 (K; LISC; PRE). M: Magude, 10 km Mapulanguene–Massingir, fr. 1.xii.1944, *Mendonça* 3196 (LISC).

Also in Zaire, Burundi and Tanzania. Common but scattered, in a variety of habitats but usually in mixed deciduous plateau and escarpment woodland, wooded grassland, sandveld, riverine vegetation and on rocky outcrops, sometimes on anthills; 335–1615 m.

Vernacular names as recorded in specimen data include: "messonzoa", "m'sandzua", "muconzoa", "muxonjua" (Manica area); "msolo", "nsolo" (chiYao); "mussouzoa" (Chimoio area); "mutôlo", "mutoto" (Niassa area); "mutoulo", "n'tolo" (Cabo Delgado area); "sangati" (Cheringoma area); "somzoa" (Machado area).

Swynnerton 151 (BM; K) from the Mossurize (lower Umswirizwi) R., in the Manica e Sofala Prov. of Mozambique (male fl. xi.1906), is intermediate between var. *maprouneifolia* and var. *dekindtii*. *Plowes* 3000 (SRGH) from Omega Farm, 56 km SSE of Mutare in Zimbabwe (y. fr. 25.vii.1968), is intermediate between var. *maprouneifolia* and var. *glabra*. *Pawek* 13546 from Karonga in north Malawi (5.i.1978 in y. fr.) is a mixture of the 2 latter varieties (the MO duplicate = var. *glabra*, while the K; MAL; SRGH; UC duplicates = var. *maprouneifolia*).

Var. **dekindtii** (Pax) Radcl.-Sm. in Kew Bull. **33**: 242 (1978); in F.T.E.A., Euphorb. 1: 81 (1987). Tab. **4**, fig. B. Type from Angola (Benguela Province).

Pseudolochnostylis dekindtii Pax in Bot. Jahrb. Syst. **28**: 20 (1899). —Hutchinson in F.T.A. **6**, 1: 673 (1912). —R.E. Fries, Wiss. Ergebn. Schwed. Rhod.-Kongo-Exped. **1**: 120 (1914). —Eyles in Trans. Roy. Soc. South Africa **5**: 392 (1916). —Engler, Pflanzenw. Afrikas **3**, 2: 31 (1921). —Pax & K. Hoffmann in Engler, Pflanzenr. [IV, fam. 147, xv] **81**: 208 (1922). —O.B. Miller, Check-list For. Trees Shrubs Bech. Prot.: 33 (1948). —Brenan, Check-list For. Trees Shrubs Tang. Terr.: 224 (1949). —P.G. Meyer in Merxmüller, Prodr. Fl. SW. Afrika, fam. 67: 40 (1967). Type as above.

Young shoots pubescent or tomentose; leaves uniformly pubescent above and beneath; male inflorescences usually sessile; female flowers solitary, pedicels not or scarcely apically thickened; female disk finely toothed; ovary densely pubescent.

Caprivi Strip. 5 km south of Shamvura, fr. 27.iv.1977, *M. Mueller & Giess* 604 (PRE; WIND); 1.6 km east of Katima Mulilo, fr. 25.ix.1972, *Edwards* 4308 (K; PRE). **Botswana**. N: 4.6 km east of Selinda Spillway, fr. 17.x.1979, *P.A. Smith* 2836 (K; SRGH). **Zambia**. B: Zambezi (Balovale), female fl. 10.ix.1952, *Gilges* 147 (K; PRE). N: Kasama, fr. 14.ii.1961, *Coxe* 187 (K; LISC; SRGH). W: Mufulira, female fl. 7.x.1955, *Fanshawe* 2493 (K; NDO). C: Kabwe (Broken Hill)–Lukanga Swamp, fr. 5.viii.1964, *van Rensburg* 2944 (K; SRGH). S: Livingstone Distr., Natebe, male fl.

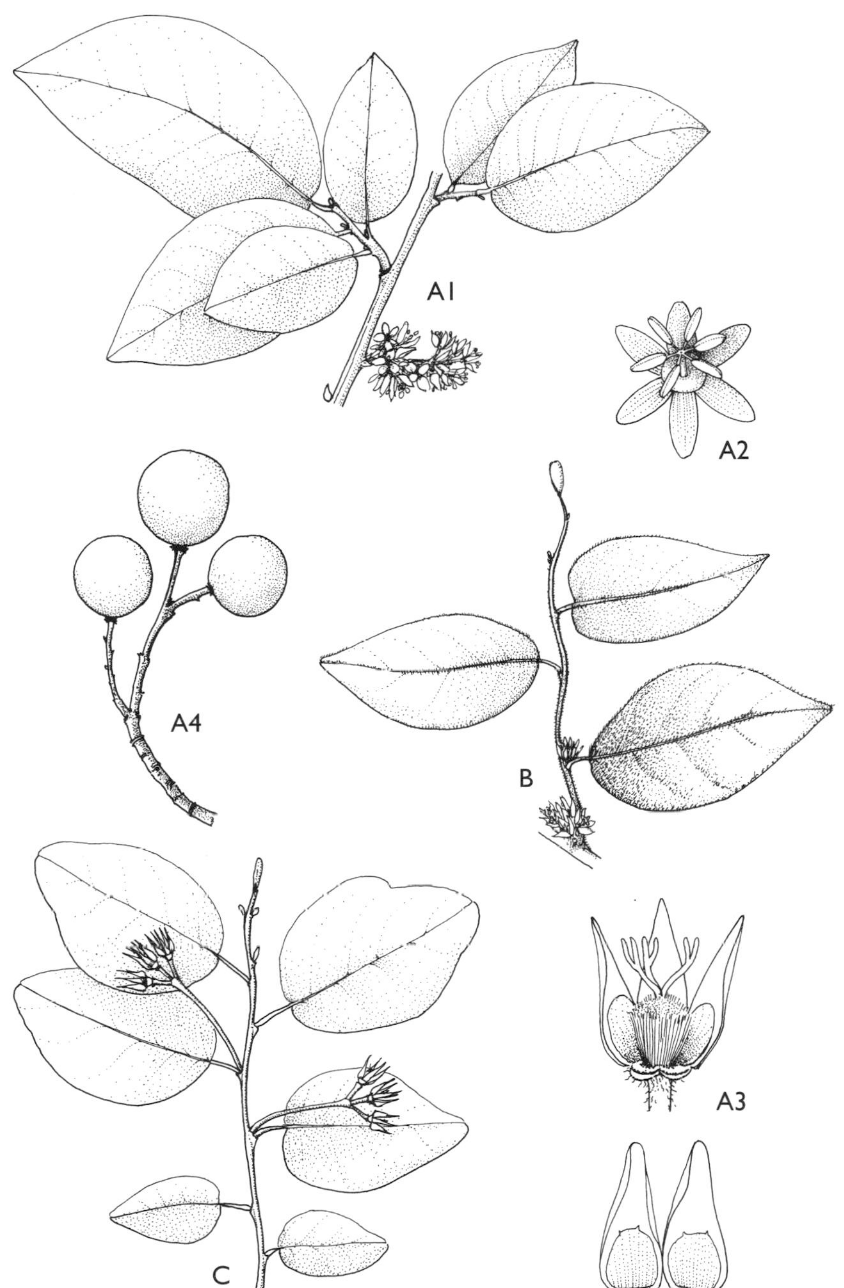

Tab. 4. A —PSEUDOLACHNOSTYLIS MAPROUNEIFOLIA var. MAPROUNEIFOLIA. A1, portion of flowering branch, with male flowers (× $^{2}/_{3}$); A2, male flower (× 2); A3, female flower, with 2 sepals removed (× 6), A1–A3 from *Richards* 26349; A4, infructescence (× $^{2}/_{3}$), from *Greenway & Kanuri* 14645. B —PSEUDOLACHNOSTYLIS MAPROUNEIFOLIA var. DEKINDTII, flowering branch (× $^{2}/_{3}$), from *Semsei* 2453. C —PSEUDOLACHNOSTYLIS MAPROUNEIFOLIA var. POLYGYNA, flowering branch (× $^{2}/_{3}$), from *Stolz* 1754. Drawn by G. Papadopoulos. From F.T.E.A.

9.xi.1955, *Gilges* 423 (K; LISC; SRGH). **Zimbabwe**. N: Mupfure (Umfuli) R., male fl. 12.xii.1950, *Hornby* 3231 (SRGH). W: Hwange (Wankie), male fl. no date, *Rogers* 5991 (BM; K). C: Mtao For. Res., y. fr. 3.iii.1947, *Robertson* 5147 (SRGH). **Malawi**. N: Rumphi Distr., Chilumba (Deep Bay)–Njakwa, fr. 12.v.1952, *White* 2843 (FHO; K). C: Dedza Distr., Mua–Livulezi For. Res., male fl. 9.xii.1953, *Adlard* 15 (MAL; PRE; SRGH). S: Lichenya Hill, fr. 22.iv.1983, *Tawakali & Patel* 77 (K; MAL; MO). **Mozambique**. N: Amaramba, western slopes of Serra Mitucué, fr. 15.ii.1964, *Torre & Paiva* 10583 (LISC). T: Angónia, Vila Mouzinho, male fl. 15.x.1943, *Torre* 6039 (LISC).

Also in Zaire (Shaba), Tanzania, Angola, Namibia and South Africa (Transvaal). In deciduous woodland, forests and thickets on Kalahari Sand, in mixed deciduous plateau and escarpment woodland, mopane woodland, dry evergreen thicket, wooded grassland and riverine vegetation; 580–1615 m.

Vernacular names as recorded in specimen data include: "kabalabala" "kabarabara" (chiLunda); "mbwanyanya" (Dedza area); "messôlo" (eMakhuwa); "mosimba" (Sesheke area); "msolo", "nsolo" (chiYao); "mumwaa" (chiLozi); "musadya" (chindembe); "musalya" (chiLamba); "musangata" (chiBemba); "musangati" (chiWemba); "musolo" (chiKunda).

Var. **polygyna** (Pax & K. Hoffm.) Radcl.-Sm. in Kew Bull. **33**: 242 (1978); in F.T.E.A., Euphorb. 1: 83 (1987). Tab. **4**, fig.C. Types from Tanzania (Southern Highlands Province).

Pseudolachnostylis polygyna Pax & K. Hoffm. in Engler, Pflanzenr. [IV, fam. 147, xv] **81**: 207 (1922); Pflanzenw. Afrikas **3**, 2: 31 (1921). —Brenan, Check-list For. Trees Shrubs Tang. Terr.: 224 (1949). —Topham, Check List For. Trees Shrubs Nyasaland Prot.: 53 (1958). Types as above.

Young shoots pubescent or tomentose; leaves pubescent usually only on the midrib beneath, but sometimes also on the midrib above and on the lamina as well; male inflorescences sessile or pedunculate; female flowers in (1)3–7-flowered cymes, pedicels noticeably apically thickened; female disk finely toothed; ovary densely pubescent.

Zambia. N: Mpika Distr., Lufila R., female fl. 16.x.1957, *Savory* 224 (K; SRGH).

Also in Tanzania (Southern Highlands Province). On riverbanks.

Var. **glabra** (Pax) Brenan in Mem. N.Y. Bot. Gard. **9**, 1: 67 (1954). —Radcliffe-Smith in F.T.E.A., Euphorb. 1: 83 (1987). Type from Angola (Malanje Province).

Cleistanthus glaucus Hiern, Cat. Pl. Afr. Welw. **1**, 4: 955 (1900). Types from Angola (Cuanza Norte Province).

Pseudolachnostylis verdickii De Wild. in Ann. Mus. Congo Belge, Bot. Sér. **4**: 205 (1903). Type from Zaire (Shaba Province).

Pseudolachnostylis dekindtii var. *glabra* Pax. in Bot. Jahrb. Syst. **43**: 75 (1909).

Pseudolachnostylis glauca (Hiern) Hutch. in F.T.A. **6**, 1: 671 (1912). —Engler, Pflanzenw. Afrikas **3**, 2: 31 (1921). —Pax & K. Hoffmann in Engler, Pflanzenr. [IV, fam. 147, xv] **81**: 209 (1922). —O.B. Miller, Check-list For. Trees Shrubs Bech. Prot.: 33 (1948). —Brenan, Check-list For. Trees Shrubs Tang. Terr.: 224 (1949). —Suessenguth & Merxmüller, Contrib. Fl. Marandellas Distr. **43**: 85 (1951).

Pseudolachnostylis bussei Hutch. in F.T.A. **6**, 1: 672 (1912). —Engler, Pflanzenw. Afrikas **3**, 2: 31 (1921). —Pax & Hoffmann in Engler, Pflanzenr. [IV, fam. 147, xv] **81**: 209 (1922). —Brenan, Check-list For. Trees Shrubs Tang. Terr.: 224 (1949). Type from Tanzania (Southern Province).

Young shoots and leaves quite glabrous; male inflorescences usually pedunculate; female flowers solitary, pedicels not or scarcely thickened apically; female disk shallowly lobed; ovary densely pubescent.

Botswana. SE: Palapye, fr. 7.ii.1958, *de Beer* 593 (K; PRE; SRGH). **Zambia**. W: Mwinilunga Distr., 6.5 km NW of Kalene Hill Mission Station, male fl. 21.ix.1952, *Holmes* 900 (FHO; K). C: Lusaka Distr., 6.5 km Luangwa Bridge–Rufunsa, male fl. 6.ix.1947, *Brenan & Greenway* 7820 (FHO; K). E: Great East Road near Kachalolo, male fl. 12.xii.1958, *Robson* 912 (BM; K; LISC; PRE). S: Mazabuka Distr., Sachenga Mica Mine, male fl. 8.x.1930, *Milne-Redhead* 1243 (K; PRE). **Zimbabwe**. N: 8 km north of Banket, fr. 23.iv.1948, *Rodin* 4409 (K; PRE; UC). W: Nyamandhlovu, male fl. 14.x.1929, *Pardy* 4666 (K; SRGH). S: Mwenezi Distr., Shirugwe Hill, 27 km north of Bubye–Limpopo Confl., fr. 12.v.1958, *Loveridge* 85915 (SRGH). **Malawi**. N: Karonga, male fl. 30.xii.1976, *Pawek* 12099 (K; MAL; MO; SRGH; UC). C: Kasungu, female fl. & y. fr. 24.viii.1946, *Brass* 17411 (BM; K; NY; PRE; SRGH). S: Mulanje Massif, Michesi Mt., male fl. 26.ix.1987, *J.D. Chapman & E.J. Chapman* 8890 (FHO; K; MO; MAL). **Mozambique**. N: 12 km Marrupa–Lichinga road, fr. 19.ii.1982, *Jansen & Boane* 7907 (K; MO; WAG). Z: Mocuba, Montes de Metolola, fr. 24.v.1943, *Torre* 5381 (LISC). T: Estima–Inhacapirire, fr. 26.i.1972, *Macêdo* 4695 (LISC; LMA; MO; SRGH). MS: Gorongosa, Parque Nacional de Caça, fr. 9.xi.1963, *Torre & Paiva* 9145 (LISC). GI: Massingir Distr., Chivovo R.–Mazim'chopes, st. 24.vii.1982, *Matos* 5088 (LISC).

Also in Zaire (Shaba), Burundi, Tanzania, Angola, Namibia and South Africa (Transvaal). In dry mixed deciduous woodland, mopane and miombo woodlands, and in riverine and gully forest, usually on sandy soils, Kalahari Sands, on granite outcrops and stony ridges, also on termitaria and disturbed ground at the edge of cultivation; 180–975 m.

Vernacular names as recorded in specimen data include: "kafalafala" (Mwinilunga area); "mazi", "mtolo" (eMakhuwa); "momba" (ciSena); "m'solo" (Zambézia area); "m'sôra", "mussonzoa" (Tete area); "m'toulo", "mutoulo", "ntholo", "ntolo" (Cabo Delgado area); "xene" (Xangaue).

Milne-Redhead 2544 (K; LISC) from Matonchi Farm, near Mwinilunga in W Zambia is intermediate between var. *glabra* and the typical variety.

The leaves are occasionally affected by a mite which causes ramifying excrescentic galls to be produced on the upper surface.

5. ANDRACHNE L.

Andrachne L., Sp. Pl.: 1014 (1753); Gen. Pl., ed. 5: 444 (1754). —Müller Argoviensis in De Candolle, Prodr. **15**, 2: 232 (1866). —Pax & K. Hoffmann in Engler, Pflanzenr. [IV, fam. 147, xv] **81**: 169 (1922).

Monoecious perennial herbs or subshrubs. Indumentum simple, glandular, or absent. Leaves alternate, petiolate, stipulate, simple, entire, penninerved, often small. Flowers axillary, pedicellate, males often fasciculate, females solitary. Male flowers: sepals 5(6), free or almost so, imbricate; petals 5(6), shorter than or equalling the sepals; disk glands 5(6) or 10(12), free, opposite the petals, often 2-lobed, or disk cupular, dentate; stamens 5(6), opposite the sepals, filaments free or connate to halfway, anthers erect and introrse, thecae parallel, distinct, longitudinally dehiscent; pistillode trifid, the arms capitate. Female flowers: sepals larger than in the male; petals smaller, minute or 0; disk glands as in the male; ovary 3-locular, ovules 2 per locule, hemitropous; styles 3, short, bifid or bipartite, stigmas capitate. Fruit globose or subglobose, 3-lobed, dehiscing into 3 bivalved cocci; endocarp thinly woody; columella small, persistent. Seeds 2 per locule, segmentiform, triquetrous or trigonous, smooth, sculptured or ornamented, ecarunculate; albumen fleshy; embryo curved, radicle long, cotyledons broad, flat.

A genus of some 15 species in Central and South America, the West Indies, South Africa, the Mediterranean and SW Asia.

Although the genus *Leptopus* is sundered by Webster (1994) from *Andrachne* to the extent that he places the two genera in separate subtribes of the Tribe *Phyllantheae* primarily on the basis of differences in ovule-attachment (hemitropous in the latter, anatropous in the former), Petra Hoffmann (pers. comm.) points out that in *A. ovalis* an intermediate condition occurs with respect to this character. She feels that on the basis of our present understanding of relationships within the tribe as a whole, *Leptopus* might be best considered merely as deserving of subgeneric rank under *Andrachne*.

Andrachne ovalis (E. Mey. ex Sond.) Müll. Arg. in Linnaea **32**: 78 (1863); in De Candolle, Prodr. **15**, 2: 233 (1866). —Hutchinson in F.C. **5**, 2: 386 (1920). —Burtt Davy, Fl. Pl. Ferns Transvaal: 298 (1932). —Drummond in Kirkia **10**: 250 (1975). —K. Coates Palgrave, Trees Southern Africa, ed. 2, rev.: 394 (1983). Tab. **5**. Type from South Africa (Cape Province).

Phyllanthus ovalis E. Mey. ex Sond. in Linnaea **23**: 135 (1850).

Cluytia ovalis (E. Mey. ex Sond.) Scheele in Linnaea **25**: 583 (1852).

Phyllanthus dregeanus Scheele in Linnaea **25**: 585 (1852). Type from South Africa (Eastern Cape Province).

Andrachne capensis Baill., Adansonia **3**: 163 (1863). Type from South Africa (Eastern Cape Province).

Andrachne dregeana (Scheele) Baill., Adansonia **3**: 164 (1863).

Savia ovalis (E. Mey. ex Sond.) Pax & K. Hoffm. in Engler, Pflanzenr. [IV, fam. 147, xv] **81**: 186 (1922).

A rather lax slender virgate ± glabrous shrub or small tree up to 6 m high, but more commonly 1–3 m high. Bark grey. Young twigs terete, greenish. Petiole 2–5(7) mm long, slender, sparingly pubescent when young, soon glabrescent. Stipules c. 1 mm

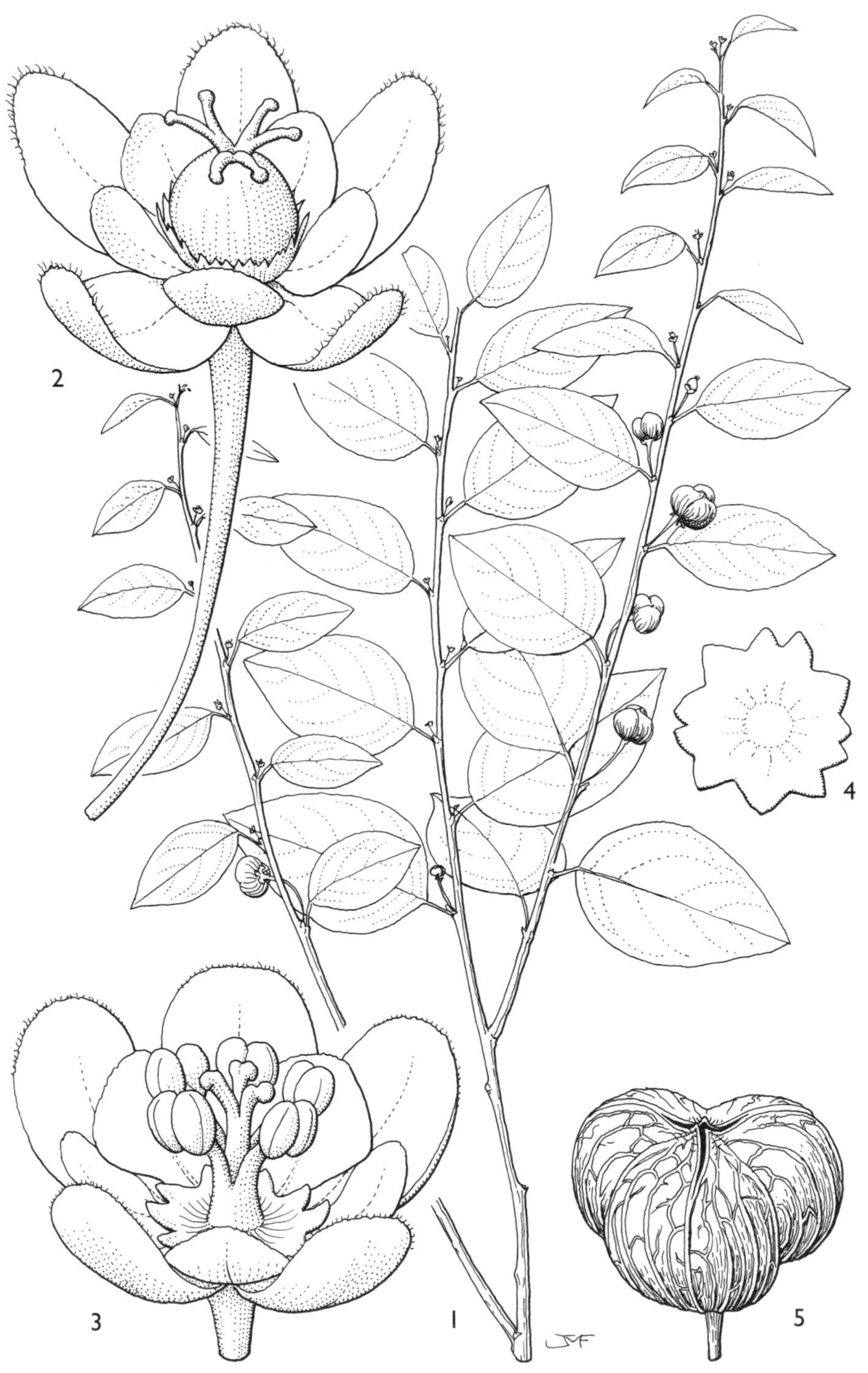

Tab. 5. ANDRACHNE OVALIS. 1, branches with flowers and fruits (× 1); 2, female flower (× 16); 3, male flower (× 16); 4, male disk (× 16), 1–4 from *Meara* 101; 5, fruit (× 4), from *Rogers* 18071. Drawn by J.M. Fothergill.

long, triangular-ovate, sparingly ciliolate. Leaf blades 1–4(6) × 0.5–2(3) cm, ovate to elliptic-ovate, subacute or obtuse at the apex, cuneate or rounded at the base, entire, membranous to chartaceous, dark glossy green above, paler beneath, sometimes paler green above, smelling of HCN when wilted; lateral nerves in 4–6 pairs, scarcely perceptible. Bracts resembling the stipules. Male flowers: pedicels 2–3(5) mm long; sepals 1.5 × 1.3 mm, united at the base, suborbicular-obovate, obtuse, minutely ciliolate, pale green or yellowish-green; petals 1 × 1 mm, suborbicular, white; disk 1.5 mm in diameter, cupular, dentate, the teeth 10, triangular, subacute; stamens 5, 1.5 mm long, connate in the lower third, anthers 0.5 × 0.5 mm, whitish; pistillode (non-functional ovary) 1.5 mm high. Female flowers: pedicels 4–8 mm long, extending to up to 1.5 cm in fruit, apically dilated; sepals 2 × 1.5 mm, ovate, minutely papillose, greenish, becoming hardened and persistent in fruit; petals c. 1 × 1 mm, suborbicular-obovate; disk c. 1–2 mm in diameter, cupular, margin irregularly lobulate and/or denticulate, orange; ovary 1 × 1 mm, globose, sparingly setulose; styles 0.5 mm long. Fruit 5–6 × 8–10 mm, depressed, rounded-3-lobed, strongly reticulate-venose, pubescent at the apex, otherwise glabrous, green. Seeds 3.5 × 3 × 1.5 mm, irregularly compressed-trigonous, dorsally irregularly foveolate-rugulose, ventrally shallowly excavated, dark greyish-brown, hilar region reddish-brown.

Zimbabwe. E. Mutare (Umtali) Heights, female & y. fr. 19.i.1974, *Meara* 101 (K; PRE; SRGH).
Also in South Africa (Cape Province, KwaZulu-Natal, Transvaal) and Swaziland. In ravines, on forest margins, on grassy and stony slopes; 650–1700 m.

Savia laureola Baill., a Madagascan endemic, was indicated as having been collected in Mozambique by J. Forbes, a coastal surveyor who operated in both Madagascar and Mozambique in the first half of the 19th Century, on a specimen at Kew bearing the "Herbarium Hookerianum 1867" stamp. That this is clearly in error for "Madagascar" is evinced from the fact that not only are no other collections of *S. laureola* known from Mozambique, but also from the fact that other known Madagascan endemics collected by him were erroneously signalled as having been obtained in Mozambique as well.

6. FLUEGGEA Willd.

Flueggea Willd., Sp. Pl. **4**, 2: 757 (1805). —G.L. Webster in Allertonia **3**, 4: 273 (1984).
Securinega Juss. sect. *Flueggea* (Willd.) Müll. Arg. in De Candolle, Prodr. **15**, 2: 448 (1866).

Dioecious, or rarely monoecious, usually glabrous trees or shrubs. Leaves alternate, shortly petiolate, stipulate, simple, entire, penninerved, usually deciduous. Flowers axillary, the males numerous and fasciculate, the females few or solitary. Male flowers: pedicels capillary; sepals (4)5(7), unequal, imbricate; petals absent; disk glands (4)5(7), interstaminal, alternating with the sepals; stamens (4)5(7), filaments free, anthers extrorse, longitudinally dehiscent; pistillode usually large, 2–3-partite, the arms often appendiculate or 2-lobed. Female flowers: pedicels and sepals as in the male; petals absent; disk flat, annular, entire or shallowly 5–6-lobed; ovary smooth, glabrous, (2)3(4)-locular, ovules 2 per locule, hemitropous; styles (2)3(4), connate at the base, recurved or spreading, bifid or 2-lobed. Fruit 3-lobed (4-lobed)-subglobose, fleshy and indehiscent or dry and loculicidally dehiscent; exocarp fleshy or thin; endocarp crustaceous, 3-locular or rarely by abortion 2- or 1-locular; columella persistent (not persistent in African taxa). Seeds 2 per locule, or 1 by abortion, trigonous, dorsally convex, hilum often invaginated, ecarunculate; testa smooth or occasionally reticulate or verruculose, thick or thin, crustaceous; albumen scanty or copious; embryo straight or curved; cotyledons ovate, elliptic or obovate, flat, longer than the radicle.

A widespread genus of 14 species with a distribution in the tropics and subtropics of both hemispheres, extending into the warm temperate zone in places. Only one species occurs in the Flora Zambesiaca area.

Flueggea virosa (Roxb. ex Willd.) Voigt, Hort. Suburb. Calcutt.: 152 (1845). —Baillon, Étud. Gén. Euphorb.: 593, tab. 26, figs. 39–43 (1858). —Engler, Pflanzenw. Afrikas **3**, 2: 21 (1921). —Burtt Davy, Fl. Pl. Ferns Transvaal **2**: 300 (1932). —Hutchinson, Botanist in Southern Africa: 667 (1946). —O.B. Miller, Check-list For. Trees Shrubs Bech. Prot.: 32 (1948). —Brenan, Check-list For. Trees Shrubs Tang. Terr.: 215 (1949). —Topham, Check

List For. Trees Shrubs Nyasaland Prot.: 51 (1958). —G.L. Webster in Allertonia **3**, 4: 287 (1984). —Radcliffe-Smith in F.T.E.A., Euphorb. 1: 68 (1987). —Beentje, Kenya Trees, Shrubs Lianas: 207 (1994). Tab. **6**. Type from India.

Phyllanthus virosus Roxb. ex Willd., Sp. Pl. **4**: 578 (1805).

Subsp. **virosa**

Xylophylla obovata Willd., Enum. Pl. Hort. Berol.: 329 (1809). Type cult. in Berlin.

Flueggea microcarpa Blume, Bijdr. Fl. Ned. Ind.: 580 (1826). —Hutchinson in F.T.A. **6**, 1: 736 (1912); in F.C. **5**, 2: 402 (1920). —Eyles in Trans. Roy. Soc. South Africa **5**: 393 (1916). Type from Java.

Securinega abyssinica A. Rich., Tent. Fl. Abyss. **2**: 256 (1850). Types from Ethiopia.

Flueggea abyssinica (A. Rich.) Baill., Étud. Gén. Euphorb.: 593 (1858).

Flueggea senensis Klotzsch in Peters, Naturw. Reise Mossambique, **6**,1: 106 (1861). Type: Mozambique, Manica e Sofala, Sena, *Peters* s.n. (B†, holotype).

Securinega virosa (Roxb. ex Willd.) Baill., Adansonia **6**: 334 (1866). —Pax & Hoffman in Engler & Prantl, Pflanzenfam., ed. 2, **19c**: 60 (1931). —Robyns & Tournay, Fl. Sperm. Parc Nat. Alb. **1**: 439 (1948). —F. W. Andrews, Fl. Pl. Anglo-Egypt. Sudan **2**: 97 (1952). —Keay in F.W.T.A., ed. 2, **1**, 2: 389 (1958). —White, F.F.N.R.: 204 (1962). —P.G. Meyer in Merxmüller, Prodr. Fl. SW. Afrika, fam. 67: 43 (1967). —Mogg in Macnae & Kalk, Nat. Hist. Inhaca Isl., Moçamb.: 147 (1969). —Drummond in Kirkia **10**: 251 (1975). —Biegel, Check List Ornam. Pl. Rhod. Parks & Gard.: 97 (1977). —Troupin, Fl. Pl. Lign. Rwanda: 268, fig. 92/2 (1982); Fl. Rwanda **2**: 240, fig. 73/2 (1983). —K. Coates Palgrave, Trees Southern Africa, ed. 2, rev.: 396 (1983).

Securinega microcarpa (Blume) Müll. Arg. in De Candolle, Prodr. **15**, 2: 434–436 (1866), passim.

Securinega obovata (Willd.) Müll. Arg. in De Candolle, Prodr. **15**, 2: 449 (1866).

Flueggea obovata (Willd.) Wall. ex Fern.-Vill., Noviss. App. Bl. Fl. Fil., ed. 3: 189 (1880), non Baill. (1861). —Engler, Pflanzenw. Ost-Afrikas **C**: 235 (1895). —R. E. Fries, Wiss. Ergebn. Schwed. Rhod.-Kongo-Exped. **1**, 1: 121 (1914). —Schinz & Junod in Mém. Herb. Boissier, No. 10: 46 (1900). —Suessenguth & Merxmüller, Contrib. Fl. Marandellas Distr. **43**: 84 (1951).

Securinega verrucosa sensu Eyles in Trans. Roy. Soc. South Africa **5**: 393 (1916), non (Thunb.) Benth.

A deciduous unarmed, rarely spiny, much-branched sometimes scandent shrub or small tree up to 7.5 m high, branches often weak and pendent. Bark slightly rough, pale to dark grey or greyish-brown, reddish- or purplish-brown on the smaller branches. Twigs virgate, lenticellate. Petioles 2–6 mm long, adaxially grooved, slightly flanged toward the apex, reddish. Stipules 1–2 mm long, triangular-lanceolate, acute, subentire or somewhat fimbriate, chestnut-brown, soon falling. Leaf blades 1–6.5 × 0.5–3.2 cm, suborbicular, obovate or elliptic, obtuse, rounded or occasionally emarginate, less commonly acute or apiculate, cuneate or rounded at the base, thinly chartaceous, yellowish-green above, somewhat glaucous beneath; lateral nerves in 5–9 pairs, camptodromous or irregularly brochidodromous, not prominent above, fairly prominent beneath, tertiary and quaternary nerve networks reticulate, the latter finely so. Bracts resembling the stipules. Male flowers sweetly scented; pedicels 3–7 mm long, buds whitish; sepals dimorphic, the 2 outer 0.7–1 × 0.5 mm, lanceolate, acute, subentire, yellowish-green, the 3 inner sepals 1.3–1.5 × 1–1.3 mm, obovate, rounded-retuse, erose at the apex, pale cream or whitish; disk glands 5, fleshy, knobbly; stamens 5, filaments 2–3.5 mm long, white, anthers 0.5 mm long, yellow; pistillode (non-functional ovary) 2–2.5 mm tall, deeply 3-partite, the divisions appendiculate. Female flowers: pedicels 3–5 mm long, extending to c. 1 cm in fruit; sepals ± as in the male flowers, but slightly smaller, persistent in fruit; disk 0.75–1 mm in diameter, annular or shallowly 5-lobed, thin; ovary 1 mm in diameter, ovoid-subglobose; styles 0.75–1 mm long, yellowish-green. Fruit 3 × 4.5–5 mm (dry), depressed-subglobose, crustaceous and greenish or baccate and translucent ivory- or pearly-white, rarely blackish. Seeds 2.5–3 × 1.5–2 mm, smooth, shiny, chestnut-brown to yellowish-brown.

Caprivi Strip. Kangongo Camp, fr. 12.i.1956, *de Winter & Wiss* 4222 (K; PRE); 3 km east of Katima Mulilo, male fl. & fr. 13.ii.1969, *de Winter* 9161 (K; PRE; SRGH). **Botswana.** N: Mwaku Pan, 23°S, 22°E, fr. 22.iii.1969, *R.C. Brown* s.n. (GAB; K; WIND). SE: Serowe Distr., Pikwe, male fl. xii.1977, *Kerfoot & Falconer* 111 (J; PRE). **Zambia.** B: 11 km SW of Senanga, male fl. 5.viii.1952, *Codd* 7409 (BM; K; PRE; SRGH). N: Mpangara Village, NW of Mafinga Hills, female fl. & fr. 27.xii.1972, *Pawek* 6143 (K; MAL; MO; SRGH; UC). W: Ndola, fr. 8.i.1955, *Fanshawe* 1783 (K; NDO; SRGH). C: Chilanga, Tom Puffet's Farm, fr. 6.i.1966, *Lawton* 1352 (FHO; K;

Tab. 6. FLUEGGEA VIROSA. 1, portion of male flowering branch (× 1); 2, male flower (× 8), 1 & 2 from *Procter* 1711; 3, female flower (× 8); 4, stigma (× 8), 3 & 4 from *Eggeling* 6757; 5, portion of fruiting branch (× 1); 6, fruit (× 4), 5 & 6 from *Richards* 17572. Drawn by G. Papadopoulos. From F.T.E.A.

SRGH). E: Lundazi R., female fl. 19.xi.1958, *Robson & Fanshawe* 666 (BM; K; PRE; LISC; SRGH). S: Mazabuka Distr., Choma, Mapanza, male fl. 30.xii.1957, *Robinson* 2541 (K; PRE; SRGH). **Zimbabwe**. N: Hurungwe Distr., Zwipani Camp, male fl. 20.xi.1957, *Goodier* 395 (FHO; K; PRE; SRGH). W: Victoria Falls, fr. ii.1912, *Rogers* 5562 (BM; K; PRE). C: Chegutu Distr., Munyati (Umniati) R. near Sanyati R. Rest House, female fl. 13.xii.1962, *Müller* 3 (BM; K; LISC; PRE; SRGH). E: Chipinge Distr., fr. 25.i.1957, *Phipps* 171a (K; PRE; SRGH). S: Chibi Distr., Rhino Hotel, Runde (Lundi) R., male fl. xii.1955, *Davies* 1750 (K; MO; PRE; SRGH). **Malawi.** N: Nkhata Bay Distr., Chikale Beach road, fr. 6.i.1976, *Pawek* 10685 (K; MAL; MO; PRE; SRGH; UC). C: Dowa Distr., Chimwere, fr. 24.iii.1970, *Brummitt* 9343 (K; LISC; MAL; PRE; SRGH; UPS). S: Mpondas, Mangochi, male fl. xi.1981, *Salubeni, Balaka & Banda* 3188 (MAL; MO; PRE; SRGH). **Mozambique**. N: Cabo Delgado, Ancuabe, fr. 7.ii.1984, *de Koning, Groenendijk & Dungo* 9537 (K; LMU; MO). Z: 7 km Namacurra–Olinga (Vila Maganja da Costa), fr. 25.i.1966, *Torre & Correia* 14085 (LISC). T: Estima–Cahó, fr. 27.i.1972, *Macêdo* 4726 (LISC; LMA; MO; SRGH). MS: Chibabava, lower Búzi R., fr. 28.xi.1906, *Swynnerton* 1730 (BM; K; SRGH). GI: Caniçado–Papai, fr. 13.xii.1940, *Torre* 2377 (LISC). M: Matutuíne (Bela Vista)–Tinonganine, fr. 11.xii.1961, *Lemos & Balsinhas* in *Lemos* 288 (K; LISC; LMA; PRE; SRGH).

More or less throughout the Old World tropics. Common at medium and low altitudes in mixed deciduous and miombo woodlands, in bush thickets and forest margins, on rocky outcrops and amongst rocks beside dry river beds, often on termitaria, in riverine vegetation and dambo margins, in sand and alluvial soils and disturbed ground beside roadsides and old cultivation; 40–1525 m.

Vernacular names as recorded in specimen data include: “changa” (Omi); “changaume” (Maputo area); “citanyero” (chiTumbuka — ‘broom’); “kapirapira” (chiChewa); “mafoutagomba”, “mouwana”, “m’palo”, “pomboma” (Tete area); “messossoio” (Inhaminga area); “mparapara” (Mzimba area); “mserecheti” (chiYao); “mu-banda” (Kabompo area); “mulianzovu”; “munsosoti”, “musosoti” (chiNdau); “musosia-uria-basimpongo” (chiTonga); “mussossote” (Búzi area); “nameresi” (eMakhuwa); “pombuma” (Manica e Sofala area); “snowberry” (Lilongwe); “umsosoti” (chiNdao).

The fruit is edible and can be made into an alcoholic drink.

Subspecies *melanthesoides* (F. Muell.) G.L. Webster, occurs only in New Guinea and Australia.

Flueggea verrucosa (Thunb.) G.L. Webster is recorded by Eyles in Trans. Roy. Soc. South Africa **5**: 393 (1916) (as *Securinega verrucosa*) from Victoria Falls, but according to G.L. Webster (1984), this species is only found in South Africa from eastern Cape to southern KwaZulu-Natal. I have seen no Flora Zambesiaca area specimens.

7. MARGARITARIA L.f.

Margaritaria L.f., Suppl. Pl.: 66 (1781). —G.L. Webster in J. Arnold Arbor. **60**: 407 (1979).

Phyllanthus sect. *Cicca* subsect. *Margaritaria* (L.f.) Müll. Arg. in De Candolle, Prodr. **15**, 2: 414 (1866).

Dioecious, very rarely monoecious, trees or shrubs, rarely scandent. Indumentum simple. Leaves distichous, shortly petiolate, stipulate, simple, entire, penninerved, usually deciduous. Flowers axillary, usually in the proximal axils of new shoots; the males usually several, fasciculate, often precocious; the females 2–3 per axil or often solitary. Male flowers: pedicels capillary; calyx lobes 4, imbricate; petals absent; disk annular, ± entire; stamens 4, filaments free, anthers extrorse, longitudinally dehiscent; pistillode (non-functional ovary) absent. Female flowers: pedicels stouter than in the male flowers; calyx lobes and disk ± as in the male flowers; ovary smooth, commonly glabrous, (2)3–4(5)-locular, ovules 2 per locule; styles free or variously united, bifid to bipartite. Fruit (2)3–4(5)-lobed, irregularly dehiscent or subindehiscent; exocarp green, usually separating from the paleaceous-crustaceous endocarp. Seeds 2 per locule, sometimes 1 by abortion, hemispherical, ecarunculate; sarcotesta fleshy, bluish; sclerotesta thick, woody or bony, smooth or rugose, chalaza invaginated; endosperm copious, whitish; embryo ± straight; cotyledons thin, flat, much longer than the radicle.

A pantropical genus of 13 species, 4 of which are neotropical, while 4 occur in Madagascar and the Mascarenes, 3 in tropical Asia and 1 in N Australia. One is common and widespread in tropical Africa.

Margaritaria discoidea (Baill.) G.L. Webster in J. Arnold Arbor. **48**: 311 (1967); **60**: 415 (1979). —K. Coates Palgrave, Trees Southern Africa, ed. 2, rev.: 397 (1983). —Radcliffe-Smith in F.T.E.A., Euphorb. 1: 63 (1987). —Beentje, Kenya Trees, Shrubs Lianas: 213 (1994). Type from ‘Senegambia’.

Cicca discoidea Baill., Adansonia **1**: 85 (1860).
Phyllanthus discoideus (Baill.) Müll. Arg. in Linnaea **32**: 51 (1863); in De Candolle, Prodr. **15**, 2: 416 (1866). —Hutchinson in F.T.A. **6**, 1: 707 (1912); in F.C. **5**, 2: 388 (1920). —Engler, Pflanzenw. Afrikas **3**, 2: 24 (1921). —De Wildeman, Pl. Bequaert. **3**, 4: 443 (1926). —Burtt Davy, Fl. Pl. Ferns Transvaal: 299 (1932). —Robyns & Tournay, Fl. Sperm. Parc Nat. Alb. **1**: 441 (1948). —Brenan, Check-list For. Trees Shrubs Tang. Terr.: 223 (1949). —Topham, Check List For. Trees Shrubs Nyasaland Prot.: 52 (1958). —Keay in F.W.T.A., ed. 2, **1**, 2: 387 (1958). —Agnew, Upl. Kenya Wild Fls.: 212 (1974). —Drummond in Kirkia **10**: 251 (1975).

A many-stemmed, densely branched spreading to somewhat sarmentose deciduous shrub or tree up to 25 m high (rarely more) often with a flattened crown. Bole up to 30 cm d.b.h. Bark grey or brownish, flaking in irregular strips. Twigs lenticellate. Petioles 1–9 mm long, terete, grooved or winged, glabrous to densely crisped-puberulous. Stipules 2–13 mm long, linear-lanceolate or oblong, acute or obtuse, entire, membranous, soon falling. Leaf blades 1–11 × 0.5–6 cm, obovate to elliptic-lanceolate, obtuse or rounded to acutely acuminate, rounded to cuneate at the base, chartaceous, glabrous above and beneath or else puberulous at least along the midrib beneath; pale green and dull or dark green and shiny (var. *nitida*) above, paler beneath; lateral nerves in 5–15 pairs, sometimes with 1–2 interstitials between each pair, scarcely prominent above, somewhat so beneath, tertiary nerves reticulate. Inflorescence bracts 2–3 mm long, ovate, chaffy, brownish or blackish, soon falling. Male flowers: pedicels 2–7 mm long, glabrous or sparingly pubescent; sepals 1–3 × 1–1.5 mm, unequal, concave, becoming reflexed, pale yellow-green; disk 0.5–2 mm in diameter, shallowly 4-lobed; stamens 2–3 mm long, anthers 0.5–1.5 mm long. Female flowers: pedicels 2–20 mm long, not or only slightly extending in fruit; sepals ± as in the male flowers, greenish; disk 1.5 mm in diameter, ± entire, thick; ovary 2 mm in diameter, glabrous or sparingly to evenly puberulous; styles 2–4 mm long, stylar column 0–2 mm high, stylar arms 2 mm long, erect, spreading or reflexed. Fruits 5–7 mm long, 1–1.3 cm wide, subglobose to shallowly or deeply (2)3(4)-lobed, smooth or verrucose, often prominently nervose-reticulate, glabrous to sparingly puberulous, golden-brown when ripe. Seeds 5 × 5 × 2.5 mm; exotesta fleshy and bright glossy metallic-blue or purplish-blue when fresh, drying papery and greyish-white; endotesta plano-convex, smooth.

1. Petiole subterete, rarely slightly winged, usually evenly to densely crisped-puberulous, rarely glabrous; stipules obtuse, 2.5 mm long, ciliate; stylar column less than 1 mm high, sometimes 0, style arms erect or suberect; fruit deeply 2-lobed or 3-lobed, rarely 1-lobed by abortion, the lobes subglobose · var. *triplosphaera*
– Petiole adaxially grooved and/or narrowly winged, usually glabrous; stipules acute, up to 13 mm long, glabrous; stylar column up to 2 mm high, style arms spreading or reflexed; fruit subglobose to shallowly 3-lobed or 4-lobed · 2
2. Tall, forest canopy trees up to 30 m in height; leaf blades elliptic-lanceolate, usually acutely acuminate; female pedicels not more than 1 cm long · · · · · · · · · · · · · · · · · var. *fagifolia*
– Small trees or shrubs commonly 1–7 m in height; leaf blades usually obovate and obtuse or rounded; female pedicels up to 2 cm long · var. *nitida*

Var. **fagifolia** (Pax) Radcl.-Sm. in Kew Bull. **36**: 220 (1981); in F.T.E.A., Euphorb. 1: 66 (1987). Type from Tanzania.
Flueggea fagifolia Pax in Engler, Pflanzenw. Ost-Afrikas **C**: 236 (1895). —Hutchinson in F.T.A. **6**, 1: 737 (1912). —Brenan, Check-list For. Trees Shrubs Tang. Terr.: 215 (1949). Type as above.
Margaritaria discoidea subsp. *discoidea* —G.L. Webster in J. Arnold Arbor. **60**: 416 (1979), pro min. parte.
Margaritaria discoidea subsp. *nitida* (Pax) G.L. Webster, J. Arnold Arbor. **60**: 418 (1979), pro min. parte.

Tall, forest canopy tree up to 30 m high; petioles adaxially grooved and/or narrowly winged, glabrous or sparingly pubescent; stipules up to 1.3 cm long, acute, glabrous; leaf blades elliptic-lanceolate, usually acutely acuminate; female pedicels not more than 1 cm long; stylar column up to 2 mm high, style arms spreading or reflexed; fruit subglobose to shallowly 3-lobed.

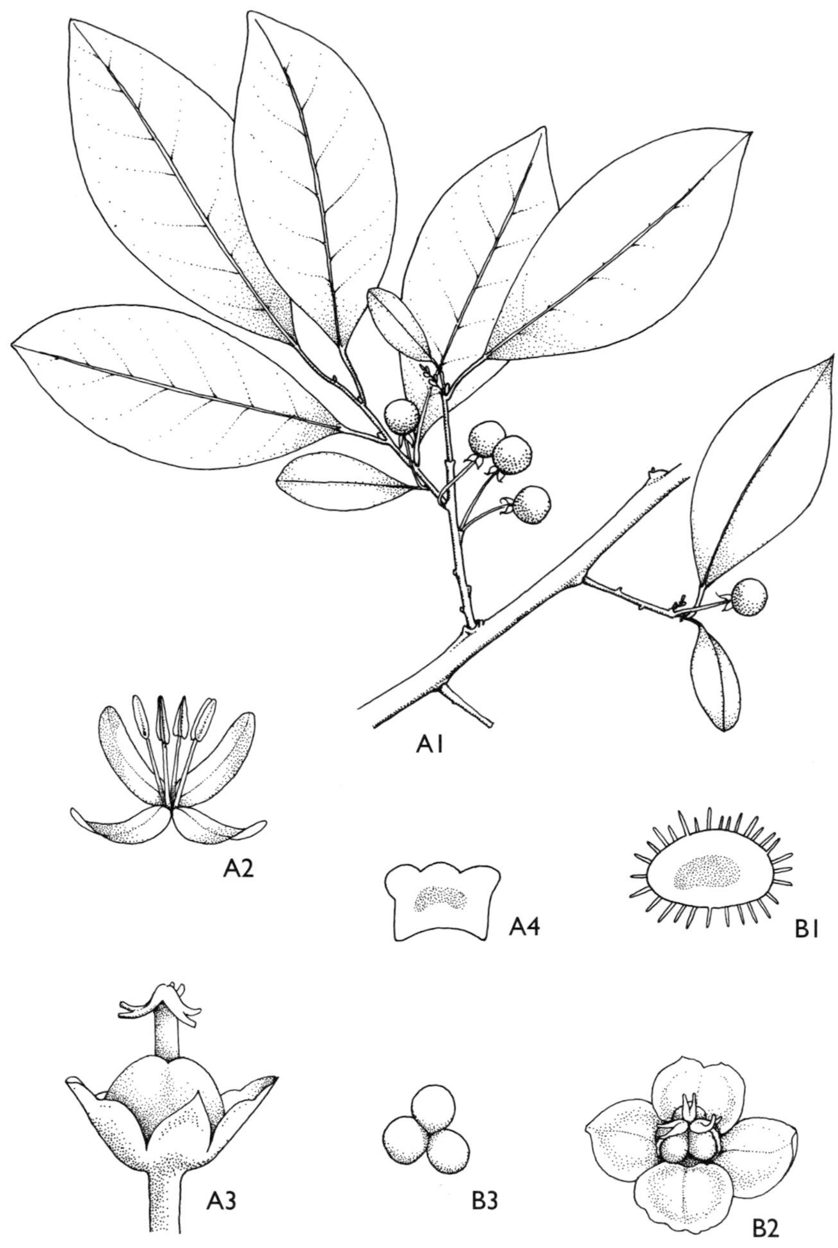

Tab. 7. A —MARGARITARIA DISCOIDEA var. NITIDA. A1, portion of fruiting branchlet (leaves approaching form of var. *fagifolia*) (× $^{2}/_{3}$), from *Shabani* 313; A2, male flower (× 4), from *Semsei* 2966; A3, female flower (× 4), from *Semsei* 3190; A4, t/s of petiole (× 12), from *Shabani* 313. B —MARGARITARIA DISCOIDEA var. TRIPLOSPHAERA. B1, t/s of petiole (× 12), from *Faulkner* 2751; B2, female flower (× 6), from *Vaughan* 1465; B3, fruit (× 1), from *Faulkner* 2751. Drawn by G. Papadopoulos. From F.T.E.A.

Malawi. S: Zomba Distr., Thondwe, Mpita Tea Estate, st. 3.v.1982, *Chapman & Tawakali* 6155 (BR; FHO; K; MAL). **Mozambique**. N: 40 km Malema (Entre Rios) – Ribáuè, Serra Murripa, fr. 15.xii.1967, *Torre & Correia* 16532 (LISC). MS: Mt. Gorongosa, st. xi.1971, *Tinley* 2244 (LISC; MO; PRE; SRGH).

From Guinea-Bissau eastwards to Ethiopia and south to Angola and South Africa (KwaZulu-Natal). In submontane mixed evergreen forest and dense rainforest; 1100–1500 m.

Var. **nitida** (Pax) Radcl.-Sm. in Kew Bull. **36**: 221 (1981); in F.T.E.A., Euphorb. 1: 67 (1987). Tab. **7**, figs. A1–A4. Types: Mozambique, Zambézia Prov., Quelimane (Quilimane), *Stuhlmann* 559 (B†, holotype), and 10 km Nampevo–Mugeba, fr. 31.v.1949, *Barbosa & Carvalho* in *Barbosa* 2950 (K, neotype chosen by G.L. Webster (1979); LMA, isoneotype).

Flueggea nitida Pax in Bot. Jahrb. Syst. **19**: 76 (1894); Pflanzenw. Ost-Afrikas C: 236 (1895). Type as above.

Phyllanthus flacourtioides Hutch. in Bull. Misc. Inform., Kew **1915**: 48 (1915). Synypes: Mozambique, Maputo (Lourenço Marques), st. 4.xii.1897, *Schlechter* 11598 (BOL; K) and 11634 (BOL).

Margaritaria discoidea subsp. *nitida* (Pax) G.L. Webster in J. Arnold Arbor. **60**: 418 (1979), pro max. parte.

Flueggea nov. spec.? Schinz & Junod in Mém. Herb. Boissier, No. 10: 46 (1900).

Small trees or shrubs commonly 1–7 m in height; petioles adaxially grooved and/or narrowly winged, usually glabrous; stipules up to 6 mm long, acute, glabrous; leaf blades usually obovate and obtuse or rounded; female pedicels up to 2 cm long; stylar column up to 2 mm high, style arms spreading or reflexed; fruit subglobose to shallowly 3-lobed or 4-lobed.

Caprivi Strip. Katima Mulilo, fr. 8.xi.1982, *M. Müller* 2950 (PRE; WIND). **Botswana**. N: Kwando R., male fl. 15.xi.1980, *P.A. Smith* 3574 (K; MO; PRE; SRGH). **Zambia**. B: Kabompo R., female fl. 20.xi.1952, *Holmes* 1006 (FHO; K; SRGH). N: Mbala Distr., 1.6 km east of Mpulungu, male fl. 16.xi.1952, *Angus* 771 (FHO; K). W: Ndola, female fl. 17.xi.1954, *Fanshawe* 1664 (K; NDO). E: Chipata Distr., Machinje Hills, st. 15.v.1965, *Mitchell* 2974 (SRGH). S: Mapanza, female fl. 23.xi.1957, *Robinson* 2506 (K; PRE; SRGH). **Zimbabwe**. N: Hurungwe (Urungwe), male fl. 17.xi.1953, *Wild* 4160 (K; LISC; PRE; SRGH). W: Matobo Distr., Quaringa Farm, male fl. xi.1954, *Miller* 2522 (K; LISC; PRE; SRGH). C: Makoni, 8 km from Rusape, male fl. 30.iv.1930, *Fries, Norlindh & Weimarck* 3345 (K; LD). E: Mutare (Umtali), male fl. 1.xii.1953, *Chase* 5150 (BM; K; LISC; PRE; SRGH). S: Mberengwa Distr., Ngesi R., fr. 1.xi.1973, *Pope, Biegel & Gosden* 1157 (K; PRE; SRGH). **Malawi**. N: Mzimba Distr., 1.6 km south of Mzambazi Mission, fr. 28.xii.1975, *Pawek* 10596 (K; MAL; MO; PRE; SRGH; UC). C: Lake Nyasa Hotel, near Salima, fr. 15.ii.1959, *Robson* 1613 (BM; K; LISC; SRGH). S: lower Likulezi Valley, y. fr. 9.xii.1957, *J.D. Chapman* 495 (FHO; K). **Mozambique**. N: Marrupa, fr. 31.i.1981, *Nuvunga* 453 (BM; K; LMU; MO). Z: 29 km Gurué-Ile, male fl. 19.x.1949, *Barbosa & Carvalho* in *Barbosa* 4528 (K; LMA). T: 4.5 km Cahora Bassa Dam to Merocira, fr. 7.ii.1973, *Torre, Carvalho & Ladeira* 19032 (LISC). MS: 20 km south of Muanza, fr. 4.xii.1971, *Müller & Pope* 1860 (K; LISC; SRGH). GI: Nhacoongo, y. fr. 5.iii.1963, *Macêdo & Balsinhas* 1101 (K; LMA; PRE). M: Marracuene, male fl. 1.vi.1959, *Barbosa & Lemos* in *Barbosa* 8544 (K; LISC; LMA; SRGH).

Also in Burundi, Ethiopia, Uganda, Kenya, Tanzania, Namibia, Swaziland and South Africa (Transvaal, KwaZulu-Natal). In mixed dry deciduous woodland, *Uapaca* and miombo woodlands and mutemwa thicket on Kalahari Sand, often on rocky outcrops and hillsides, also on forest margins, in riverine vegetation and dambo margins, and in littoral scrub and coastal thicket; from near sea level up to 1600 m.

Vernacular names as recorded in specimen data include: "kapirapira", "mpichilanyambo" (chiNyanja); "mparapara" (Mzimba area); "mselechete" (chiYao); "muguelexe" (Inhambane area); "mulenkanga" (chiTonga); "namadjia" (Imala); "napala" (eMakhuwa).

Torre 1954, 13.xi.1940 at Polana in Maputo Prov., Mozambique, is unusual in that it is monoecious.

Intermediates between this and the previous variety have been collected from: Ndola in Zambia, *Fanshawe* 731 (K; NDO) – (with leaves of var. *fagifolia* and the fruit of var. *nitida*; from Besna Kobila Farm in Zimbabwe, *O.B. Miller* 2017 (PRE; SRGH); and from Maputo in Mozambique, *R.M. Hornby* 2580 (PRE): and between this and var. *triplosphaera* from Ndola, *Fanshawe* 524 (K; NDO); and *Fanshawe* 1665 (K; NDO).

Var. **triplosphaera** Radcl.-Sm. in Kew Bull. **30**: 680 (1976); in F.T.E.A., Euphorb. 1: 67 (1987). Tab. **7**, figs. B1–B3. Type from Tanzania.

forma **triplosphaera** Radcl.-Sm. in Kew Bull. **39**: 790 (1984).

Flueggea obovata Baill., Adansonia **2**: 41 (1861), non *Xylophylla obovata* Willd. (1809). Type from Tanzania (Zanzibar).

Securinega bailloniana Müll. Arg. in De Candolle, Prodr. **15**, 2: 451 (1866). Type as for *Flueggea obovata.*
Flueggea bailloniana (Müll. Arg.) Pax in Bot. Jahrb. Syst. **19**: 76 (1894), in adnot.; in Pflanzenw. Ost-Afrikas **C**: 236 (1895).
Margaritaria obovata (Baill.) G.L. Webster in J. Arnold Arbor. **60**: 434 (1979).

Shrub 1–4 m in height, usually sarmentose; petiole subterete, rarely slightly winged, evenly to densely crisped-puberulous; stipules 2.5 mm long, obtuse, ciliate; stylar column less than 1 mm high, sometimes absent, style-arms erect or suberect; fruit deeply 2-lobed or 3-lobed, rarely 1-lobed by abortion, the lobes subglobose.

Zambia. N: Kawambwa Boma, male fl. 31.x.1952, *Angus* 687 (BM; FHO; K). W: Luanshya, fr. 24.iii.1955, *Fanshawe* 2207 (K; NDO). **Malawi**. N: Karonga Distr., Vinthukutu For. Res., fr. 2.i.1973, *Pawek* 6316 (K; MAL; MO; SRGH; UC). S: Zomba Distr., Old Naisi Road, female fl. 21.x.1987, *Tawakali & Thera* 1142 (MAL; MO). **Mozambique**. N: Marrupa, R. Massanguezi, fr. 18.ii.1981, *Nuvunga* 597 (BM; K; LMU; MO). Z: 20 km Mopeia Velha–Quelimane, fr. 7.xii.1971, *Müller & Pope* 1939 (K; SRGH). MS: Gorongosa Nat. Park, male fl. x.1971, *Tinley* 2182 (MO; LISC; PRE; SRGH).

Also in Zaire (Shaba), Burundi, Kenya and Tanzania (most provinces, including Zanzibar). In mixed deciduous woodland on sand, also in thickets and forest patches, and in old cultivations; 30–600 m.

Vernacular names as recorded in specimen data include: "mparapara" (Mzimba area); "muselechete" (chiYao).

forma **glabra** Radcl.-Sm. in Kew Bull. **39**: 790 (1984). Type: Mozambique, Niassa Province, 20 km Marrupa–Lichinga, Missor, fr. 16.ii.1981, *Nuvunga* 531 (K, holotype; BM; LMU; MO; SRGH).

As for f. *triplosphaera*, but petioles and stipules glabrous.

Malawi. N: Mzimba Distr., Daulile, Siwonde, fr. 3.ii.1990, *Banda, Kaunda & Chanza* 3771 (MAL; MO). **Mozambique**. N: 7 km Marrupa–Lichinga, fr. 19.ii.1982, *Jansen & Boane* 7868 (K; MO; WAG). Z: Maganja da Costa Distr., 35 km Praia Rarága–Floresta de Gobene, fr. 10.i.1968, *Torre & Correia* 17033 (LISC).

Not known elsewhere. On sandy soil, in vegetation fringing beaches, and in miombo and mixed deciduous woodland; 20–800 m.

Vernacular names as recorded in specimen data include: "ithatjija", "itwonchia" (Niassa area); "kaparapara" (chiTonga).

The specimen *Groenendijk, de Koning & Dungo* 1079 (K; LMU; MO) from Nampula, Monapo, in Northern Mozambique is intermediate between the two formae.

8. PHYLLANTHUS L.

Phyllanthus L., Sp. Pl.: 981 (1753); Gen. Pl., ed. 5: 422 (1754). —Müller Argoviensis in De Candolle, Prodr. **15**, 2: 274 (1866).
Cicca L., Syst. Nat., ed. 12, **2**: 621 (1767).
Kirganelia Juss., Gen. Pl.: 387 (1789).
Macraea Wight, Icon. Pl. Ind. Orient. **5**, 2: 27, t. 1901, 1902 (1852), non Lindl.
Chorisandra Wight, Icon. Pl. Ind. Orient. **6**: 13, t. 1994 (1853), non R.Br.
Chorizonema Jean F. Brunel, Phyllanthus Afr. Intertrop. Mad.: 256 (1987), unpublished thesis.

Monoecious or dioecious herbs, shrubs or trees of various habit, often with the shoots differentiated into 2 or 3 types: long lead shoots of unlimited growth (orthotropic shoots), short lateral shoots of potentially unlimited growth (brachyblasts) and leafy or floriferous lateral shoots of limited growth (plagiotropic shoots) which may resemble pinnate leaves or pseudoracemose inflorescences (see Tab. **8**). Indumentum simple, rarely dendritic (Asia). Leaves often scale-like (cataphylls) on the lead shoots and short shoots, normal (trophophylls) on the lateral leafy shoots and occasionally also on the lead shoots. Foliage leaves alternate, often distichous, shortly petiolate, stipulate, simple, entire, penninerved, the nerves usually looped. Stipules of the scale leaves larger than those of the foliage leaves. Flowers small, axillary; male flowers geminate or fasciculate, usually in the lower axils of the lateral shoots; female flowers solitary in the upper axils, or

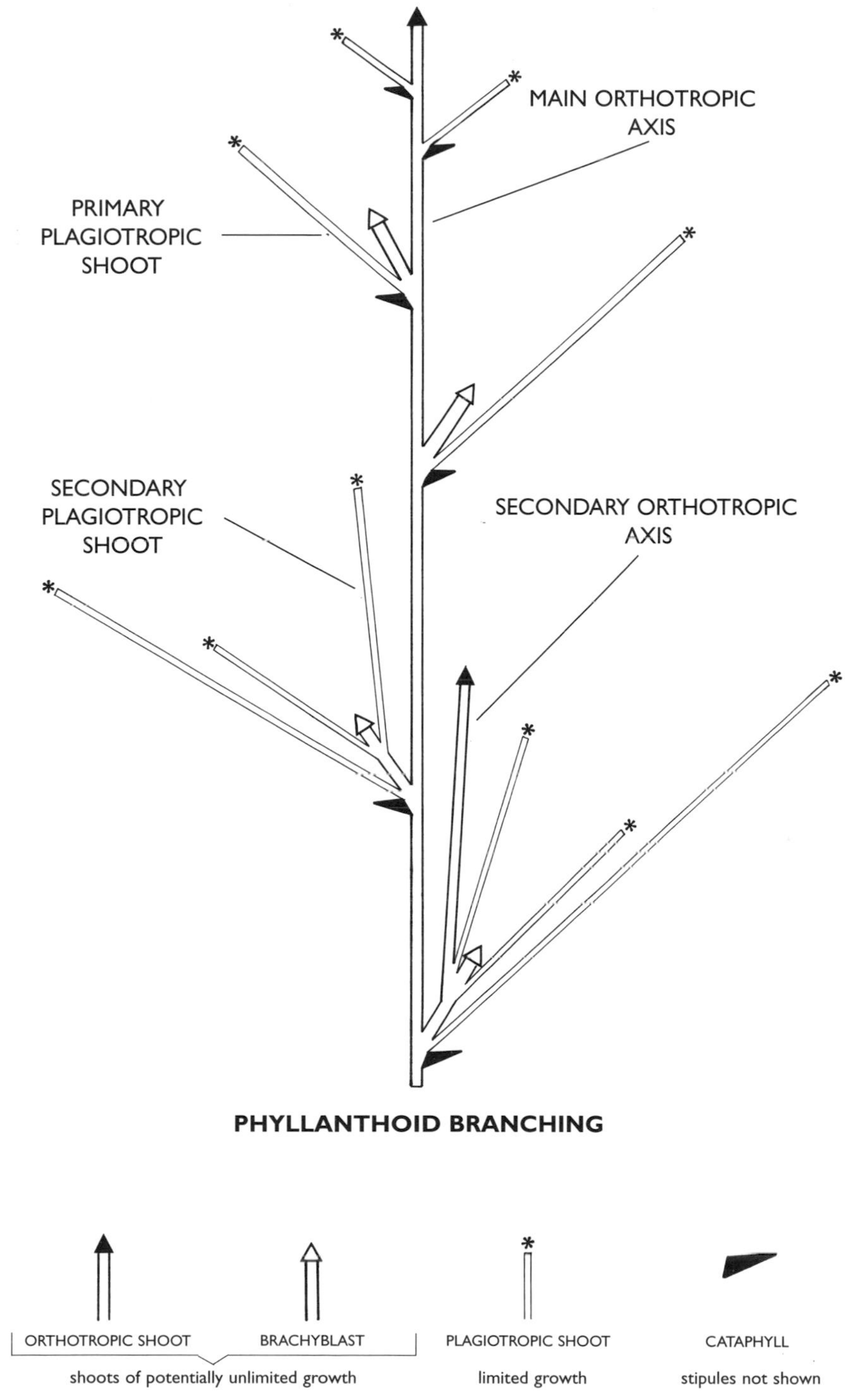
MAIN ORTHOTROPIC AXIS
PRIMARY PLAGIOTROPIC SHOOT
SECONDARY PLAGIOTROPIC SHOOT
SECONDARY ORTHOTROPIC AXIS
PHYLLANTHOID BRANCHING
ORTHOTROPIC SHOOT
BRACHYBLAST
shoots of potentially unlimited growth
PLAGIOTROPIC SHOOT
limited growth
CATAPHYLL
stipules not shown

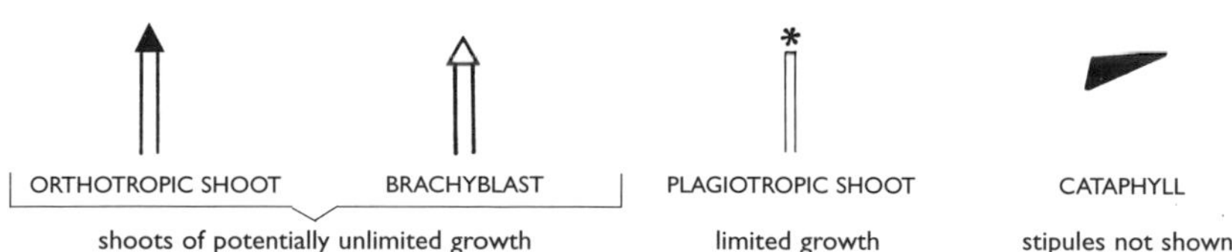

Tab. 8. Phyllanthoid branching patterns. From F.T.E.A.

male and female flowers on leafless lateral shoots, often pendent. Male flowers: pedicels often capillary; sepals (4)5–6, subequal, imbricate; petals absent; disk glands (4)5–6, free, alternisepalous, or rarely disk annular (*P. pinnatus*); stamens 2–6, filaments free or some or all partially or completely united, anthers basifixed, extrorse, variously held and dehiscent, thecae parallel or convergent; pistillode absent. Female flowers: pedicels more robust than in male flowers; sepals larger than but otherwise as in male; petals absent; disk hypogynous, annular, entire or lobed, rarely the glands distinct (e.g. *P. maderaspatensis*); staminodes rarely present; ovary sessile or stipitate, 3(∞)-locular, ovules 2 per locule; styles 3(∞), free or united at the base, variously held, bifid or 2-lobed, rarely simple (*P. ovalifolius*), the stigmas usually recurved. Fruits 3(∞)-celled, dry and septicidally and loculicidally dehiscent or fleshy and subindehiscent; endocarp usually crustaceous. Seeds 2 per locule, usually segmentiform, triquetrous and dorsally convex, rarely ovoid (e.g. *P. inflatus*), tuberculate, ridged, lineate or smooth, ecarunculate; testa usually thinly crustaceous; albumen fleshy; embryo straight or slightly curved; cotyledons flat, straight or rarely flexuous.

A pantropical genus of c. 750 species, of which c. 150 are African, with 67 in the Flora Zambesiaca area. The phyllanthoid branching pattern (see Tab. **8**), with the differentiation of the stems into 2 or 3 distinct types often leading to complex vegetative structures, is found in all except 8 of the species in the Flora Zambesiaca area. Of these, one belongs to the subgenus *Macraea* (Wight) Jean F. Brunel, i.e. *P. glaucophyllus*, 5 to the subgenus *Isocladus* G.L. Webster, i.e. *P. maderaspatensis* and 4 close allies, and 2 to the subgenus *Ceramanthus* (Hassk.) Jean F. Brunel, i.e. *P. welwitschianus* and *P. beillei*.

Key to the species

1. Stems and branches undifferentiated ··· 2
– Stems and branches differentiated into 2 or 3 types (lead, lateral and short shoots), the laterals often resembling pinnate leaves, and usually subtended by scale leaves ··· 9
2. Leaves spirally arranged; female disk of 6 separate glands ··· 3
– Leaves distichous; female disk annular, 6-lobed ··· 7
3. Female disk glands narrower than long; seeds with 12–15(22) rows of rounded tubercles on the dorsal facet, connected by a diamond lattice pattern of ridges ··· 4
– Female disk glands broader than long; seeds with up to 25 rows of rounded or elongate tubercles on the dorsal facet not connected by a lattice ··· 5
4. Leaves variously shaped, flat, obtuse or rounded at the apex ··· 22. *maderaspatensis*
– Leaves narrow, involute, with a spinous apex ··· 23. *incurvus*
5. Leaves oblanceolate; seed tubercles elongate ··· 24. *mendoncae*
– Leaves obovate; seed tubercles rounded ··· 6
6. Stems unbranched; leaves drying greyish; male flowers solitary ··· 25. *serpentinicola*
– Stems branched, the branches hair-like; leaves drying greenish-brown; male flowers in few-flowered clusters ··· 26. *tener*
7. Stamens free, female sepals not venose; female disk thin ··· 10. *glaucophyllus*
– Stamens connate; female sepals venose; female disk thick, stout ··· 8
8. A perennial herb or subshrub up to 90 cm high; leaf blades up to 3.5 × 2.8 cm, usually suborbicular, coriaceous, with 6–18 pairs of lateral nerves ··· 65. *welwitschianus*
– A shrub up to 4 m high; leaf blades up to 7 × 3 cm, elliptic-obovate to oblong-oblanceolate, chartaceous, with (10)15–30 pairs of lateral nerves ··· 66. *beillei*
9. Shrubs or small trees usually with fasciculate branches ··· 10
– Annual or perennial herbs or suffrutices; branches usually not fasciculate ··· 19
10. Short shoots massive, the scale leaves and their stipules stoutly spinous; leaves usually suborbicular, coriaceous, glaucous ··· 7. *engleri*
– Short shoots not as above; if scale leaves spinous, then not stoutly so ··· 11
11. Fruit inflated, 1.5–3 cm in diameter; seeds compressed-ovoid, greyish, brown-mottled ··· 5. *inflatus*
– Fruit not inflated; seeds not as above ··· 12
12. Male flowers densely fasciculate; male disk annular; fruits greenish-yellow and venose when mature; seeds whitish with a dark depression ··· 1. *pinnatus*
– Male flowers in few-flowered cymules; male disk of separate glands ··· 13
13. Scale leaves and their stipules not spinescent ··· 14
– Scale leaves and their stipules spinescent ··· 15

14. Female sepals not accrescent; stamens 3–4(5), free · · · 9. *kaessneri*
– Female sepals accrescent; stamens 3, united · · · 47. *macranthus*
15. Lateral leafy shoots up to 60 cm long; fruit 1.5–2 cm in diameter, fleshy (cultivated) · · · 6. *acidus*
– Lateral leafy shoots usually less than 25 cm long (indigenous) · · · 16
16. Fruit up to 5 cm in diameter, brown and corky at maturity · · · 8. *polyanthus*
– Fruit 3–6 mm in diameter, reddish-purple to bluish-black and fleshy at maturity · · · 17
17. Styles very short, stout; stigmas subsessile; fruit commonly plurilocular · · · 3. *reticulatus*
– Styles up to 1 mm long, slender; stigmas filiform; fruit 3–5-locular · · · 18
18. Foliage leaves 0.7–2.3 × 0.4–1.5 cm, oblong · · · 2. *ovalifolius*
– Foliage leaves 1.5–8 × 1–4.5 cm, ovate · · · 4. *muellerianus*
19. Stamens 5, free or only united at the base · · · 20
– Stamens (2)3, free or united into a column · · · 29
20. Leaf blades linear or linear-lanceolate, up to 4.5 cm long · · · 21
– Leaf blades suborbicular to elliptic-lanceolate, less than 2.5 cm long · · · 22
21. Young lateral shoots usually zigzag at the apex; fruiting pedicels up to 4 mm long · · · 20. *pentandrus*
– Young lateral shoots straight at the apex; fruiting pedicels not more than 1 mm long · · · 21. *mendesii*
22. Young lateral shoots often subtended by foliage leaves · · · 23
– Young lateral shoots subtended by scale leaves · · · 25
23. Much-branched herbs or suffrutices up to 90 cm high, often with fascicled lateral shoots · · · 17. *loandensis*
– Plant tufted up to 35 cm high with several apparently simple stems arising from a woody stock · · · 24
24. Leaves elliptic-ovate to elliptic-lanceolate, acute or subacute, rounded at the base · · · 14. *graminicola*
– Leaves suborbicular, rounded, shallowly cordate at the base · · · 15. *tsetserrae*
25. Fruiting pedicels 5–20 mm long · · · 12. *nummulariifolius*
– Fruiting pedicels 1–8 mm long · · · 26
26. Scale leaves and their stipules 3–4 mm long; seeds with 25 rows of tubercles on the dorsal facet · · · 18. *angolensis*
– Scale leaves and their stipules 1–2 mm long · · · 27
27. Cymes usually bisexual; seeds with 6–8 rows of tubercles on the dorsal facet · · · 19. *parvulus*
– Cymes unisexual · · · 28
28. Annual herb; female flowers often 2 per axil; seeds with 10–14 regular rows of tubercles on the dorsal facet · · · 13. *tenellus*
– Small suffrutex; female flowers solitary; seeds with 19–22 irregular rows of tubercles on the dorsal facet · · · 16. *manicaënsis*
29. Stamens free · · · 11. *martinii*
– Stamens united into a column · · · 30
30. Ovary and fruit sparingly to densely papillose, verruculose or tuberculate · · · 31
– Ovary and fruit smooth · · · 35
31. Female sepals up to 4 mm long in fruit · · · 33. *leucocalyx*
– Female sepals not more than 2 mm long in fruit · · · 32
32. Fruit densely tuberculate · · · 33
– Fruit sparingly to evenly papillose, rarely almost smooth · · · 34
33. Leaves elliptic-lanceolate · · · 32. *taylorianus*
– Leaves linear-lanceolate · · · 35. *tenuis*
34. Leaves narrowly oblong, rounded at apex; lateral nerves visible above; female sepals entire · · · 34. *micromeris*
– Leaves linear-lanceolate to narrowly elliptic-lanceolate, acute or subacute; lateral nerves not visible above; female sepals minutely serrulate · · · 36. *parvus*
35. Sepals of both sexes 5, rarely 6 in some female flowers · · · 36
– Sepals of both sexes 6, rarely 5 in some male flowers · · · 53
36. Leaves markedly heterophyllous · · · 37
– Leaves ± uniform · · · 38
37. Anthers appendiculate · · · 42. *ceratostemon*
– Anthers not appendiculate · · · 63. *zornioides*
38. Female disk collar-like · · · 39
– Female disk flat, undulate or cupular · · · 40

39. Procumbent herb or subshrub; scale leaves swollen and blackened at the base; N Malawi .... 57. *mafingensis*
– Erect herb or subshrub; scale leaves subulate; W Zambia .... 64. *microdendron*
40. Leaf blades mucronulate, mucronate or aristate 41
– Leaf blades not or rarely a few mucronulate 43
41. Plant prostrate or procumbent; leaves suborbicular .... 50. *prostratus*
– Plant erect 42
42. Leaves elliptic-lanceolate, mucronulate, membranaceous to thinly chartaceous; male pedicels 1–2 mm long .... 52. *paxii*
– Leaves broadly elliptic, mucronate-aristate, firmly chartaceous; male pedicels up to 5 mm long .... 53. *caespitosus*
43. Female disk deeply 3–5-lobed 44
– Female disk annular, pentagonal or shallowly 5-lobed 46
44. Female disk 3-lobed; young shoots papillose-puberulous .... 56. *nyikae*
– Female disk 5-lobed; plants completely glabrous 45
45. Female disk lobes triangular; plant of dry often ruderal habitats; nerves not prominent .... 31. *amarus*
– Female disk lobes oblong; plant of moist habitats; nerves somewhat prominent on lower surface of leaves .... 61. *udoricola*
46. Leaves lanceolate 47
– Leaves oblong, elliptic-ovate or elliptic-obovate or suborbicular 48
47. Stipules cream-coloured with a pinkish midrib; Malawi and Zambia .... 36. *parvus*
– Stipules dark brown or black and shiny; Mozambique .... 67. *bernierianus*
48. Plant usually monoecious 49
– Plant usually dioecious 51
49. Leaf margins thickened and slightly recurved; occurring mostly in eastern Zimbabwe and the adjacent area of Mozambique .... 54. *hutchinsonianus*
– Leaf margins flat or ± so 50
50. Lateral shoots often crowded towards the top of the lead shoots; leaves 1–11 mm long; southern Malawi (Mt. Mulanje) only .... 55. *confusus*
– Lateral shoots not crowded; leaves up to 15 mm long; northern Malawi and northern Mozambique .... 62. *boehmii*
51. Leaves up to 28 × 23 mm, glaucous beneath; nerves not thickened beneath; female pedicels 8–14 mm long; female sepals accrescent up to 6 × 4 mm; eastern Zimbabwe and adjacent Mozambique .... 58. *myrtaceus*
– Leaves 3–10 × 2–6 mm, pale green; nerves often thickened beneath; female pedicels 2–3 mm long; female sepals accrescent up to 2.5 × 2 mm 52
52. Lateral nerves arising at an acute angle to the midrib, often unequal, meandering, branching and anastomosing irregularly; plants of widespread occurrence ... 59. *arvensis*
– Lateral nerves very thick, arising perpendicularly to the midrib and closely parallel, or even (the lower pairs) backwardly-directed, looping; plants occurring in Zambia and northern Malawi .... 60. *retinervis*
53. Leaf blades narrowly elliptic-lanceolate to linear-oblong 54
– Leaf blades obovate-suborbicular, ovate, elliptic, oblong or lanceolate 57
54. Subshrub; leaf apex sharply spinescent .... 46. *virgulatus*
– Herb; leaf apex membranous 55
55. Tufted perennial herb with stems arising from a stout woody stock .... 44. *friesii*
– Annual herb 56
56. Plant slender; leaves sharply acute; anthers ellipsoid .... 48. *gossweileri*
– Plant robust; leaves obtuse or rounded; anthers reniform .... 40. *zambicus*
57. Seeds with a collar-like excrescence around the hilum, brown, shiny, 30–40-striate on the back .... 51. *pseudocarunculatus*
– Seeds not as above 58
58. Plant procumbent or semiprostrate 59
– Plant erect or ascending 60
59. Male sepals acute or apiculate; styles 0.1 mm long; Mozambique and southern Zimbabwe .... 41. *delagoensis*
– Male sepals rounded at the apex; styles 0.75 mm long; Zambia, Botswana and central and western Zimbabwe .... 49. *omahakensis*
60. Female disk irregularly divided and laciniate .... 37. *fraternus*
– Female disk annular or regularly 6-lobed 61

61. Female disk not persistent after fruiting · 49. *omahakensis*
– Female disk persistent after fruiting · 62
62. Tufted perennial herb with stems arising from a stout woody stock · · · · · · · · · · · · · · 63
– Annual or perennial herb without a woody stock · 64
63. Leaves ovate-suborbicular or elliptic-ovate, lacking folds when dried; lateral nerves thickened beneath · 43. *crassinervius*
– Leaves ovate-lanceolate to lanceolate, often developing narrow longitudinal folds when dried; lateral nerves inconspicuous · 45. *holostylus*
64. Midrib of female sepals broad, with narrow white margins · 65
– Midrib of female sepals narrow · 66
65. Lateral shoots smooth; female disk annular, thin · · · · · · · · · · · · · · · · · · · 38. *gillettianus*
– Lateral shoots scabrid; female disk shallowly 6-lobed · · · · · · · · · · · · · · · · 39. *asperulatus*
66. Female disk lobes appendiculate · 28. *xiphephorus*
– Female disk lobes not appendiculate · 67
67. Ovary sessile · 29. *pseudoniruri*
– Ovary stipitate · 68
68. Lateral shoots winged; female sepals broadly elliptic, shallowly cordate at the base, broadly white-margined; female disk lobes entire · 27. *leucanthus*
– Lateral shoots angular; female sepals ovate-lanceolate to oblong-lanceolate, rounded at the base, narrowly white-margined; female disk lobes crenulate · · · · · · · · · · 30. *odontadenius*

1. **Phyllanthus pinnatus** (Wight) G.L. Webster in J. Arnold Arbor. **38**: 52 (1957). —Radcliffe-Smith in F.T.E.A., Euphorb. 1: 23 (1987). —Beentje, Kenya Trees, Shrubs Lianas: 218 (1994). Type from India.

Chorisandra pinnata Wight, Icon. Pl. Ind. Orient. **6**: 13, t. 1994 (1853).

Phyllanthus wightianus Müll. Arg. in Linnaea **32**: 6, nec 47 (1863); in De Candolle, Prodr. **15**, 2: 334, nec 425 (1866). Type from India.

Phyllanthus kirkianus Müll. Arg. in Flora **47**: 486 (1864); in De Candolle, Prodr. **15**, 2: 334 (1866). —Engler, Pflanzenw. Ost-Afrikas **C**: 236 (1895). —Hutchinson in F.T.A. **6**, 1: 698 (1912). —Engler, Pflanzenw. Afrikas (Veg. Erde 9) **3**, 2: 22 (1921). —Hutchinson, Botanist in Southern Africa: 667 (1946). —Drummond in Kirkia **10**: 251 (1975). —K. Coates Palgrave, Trees Southern Africa, ed. 2, rev.: 400 (1983). Type: Mozambique, Tete, male fl. & y. fr. xii.1858, *Kirk, Phyllanthus?* (4) (K, holotype).

Phyllanthus senensis Müll. Arg. in De Candolle, Prodr. **15**, 2: 335 (1866). Type: Mozambique, Manica e Sofala Province, Vila de Sena, *Peters* s.n. (B†, holotype).

Cluytiandra schinzii Pax in Bull. Herb. Boissier, Sér. 2, **8**: 635 (1908). Type: Mozambique, Tete Province, Boroma (Boruma), y. fr. x.1890, *Menyharth* 779 (Z, holotype; K).

Chorizonema pinnata (Wight) Jean F. Brunel, Phyllanthus Afr. Intertrop. Mad.: 256 (1987).

An often spindly, unbranched, sometimes decumbent, glabrous shrub or small tree to 4.5 m high; plants dioecious. Bark smooth, mottled, light grey, flaking. Twigs often pinkish-grey. Young lead shoots robust, angular. Lateral leafy shoots up to 12 cm long, sometimes with flowers at the base. Short shoots either giving rise to secondary leafy shoots and lead shoots (in female plants) or else directly floriferous (in male plants). Scale leaves and their stipules 1.5–2 × 0.5–1 mm, narrowly lanceolate, light brown with fimbriate hyaline margins. Foliage leaves distichous; petioles 1.5 mm long; blades up to 4.5 × 3 cm, elliptic, obovate or ± suborbicular, obtuse to rounded at the apex, cuneate to rounded at the base, chartaceous, midrib commonly not running to the apex, bluish-green to yellowish-green above, paler beneath; lateral nerves in 5–7 pairs, not prominent above, scarcely so beneath. Stipules 0.5 mm long, linear-lanceolate, soon falling. Male flowers fragrant, usually in dense fascicles on older leafless twigs; bracts 1 mm long, broadly ovate, erose, chestnut-brown; pedicels 3–4(7) mm long, slender; sepals 6, c. 1 × 0.5–1 mm, obovate-suborbicular, the outer convex, the inner ± flat, fimbriate, brown at the apex, otherwise yellowish; disk 1 mm in diameter, annular, ± entire, thin, flat; stamens (5)6, 2 mm long, filaments free or ± so, anthers 0.3 mm long. Female flowers either in few-flowered fascicles on the older twigs or else solitary in the lowest axils of the leafy shoots; bracts ± as in the male; pedicels 0.5–1 cm long, extending to 2–4 cm long in fruit, slender; outer sepals smaller than inner, otherwise sepals ± as in the male; disk 1.5 mm in diameter, collar-like, shallowly hexagonal; ovary 1.5 mm in diameter, ± sessile, subglobose, smooth; styles 3, 2 mm long, united at the base, ± erect at first, later spreading, slender, bifid, the stigmas filiform, minutely papillose. Fruit 4–5 × 8–10 mm, pendent, rounded, depressed-trigonous, shallowly reticulate-venose, crustaceous, pale green ripening to yellowish. Seeds 2.5–3 × 2–2.5 mm, ovoid, faintly lineate, pale brown to whitish, with a deep, dark, hilar excavation.

Zimbabwe. N: Mutoko Distr., Mkota, o. fr. 19.ix.1951, *Whellan* 557 (SRGH). E: Chipinge Distr., Save (Sabi) R., Dotts Drift, male fl. 16.xi.1959, *Goodier* 653 (K; LISC; PRE; SRGH). S: Gwanda Distr., Marangadzi, o. fr. 10.v.1958, *Drummond* 5751 (K; LISC; PRE; SRGH). **Malawi**. C: Kasungu Distr., Chilanga, st. 18.xi.1983, *Salubeni & Patel* 3444 (MAL). S: Mulanje Distr., Chingozi Hill, Mpinda Village, fr. 6.xi.1983, *Patel & Seyani* 1343 (MAL; MO). **Mozambique**. T: Boroma (Boruma), y. fr. x.1890, *Menyharth* 779 (K; Z). MS: Sofala Prov. (Beira Distr.), Gorongosa Nat. Park, male fl. v.1972, *Tinley* 2611 (K; LISC; SRGH).

Also in Kenya, Tanzania, South Africa (northern Transvaal), India and Sri Lanka. At low altitudes in sandveld and hot dry deciduous and mopane woodlands, along banks of seasonal streams and rivers sometimes with *Androstachys johnsonii*, sometimes among rocks; 100–625 m.

A very distinctive species on account of its dense male inflorescences, annular male disk and excavated seeds.

2. **Phyllanthus ovalifolius** Forssk., Fl. Aegypt.-Arab.: 159 (1775). —Müller Argoviensis in De Candolle, Prodr. **15**, 2: 346 (1866). —Troupin, Fl. Pl. Lign. Rwanda: 267, fig. 90/3 (1982); Fl. Rwanda **2**: 236, ? fig. 71/2 (1983). —Radcliffe-Smith in F.T.E.A., Euphorb. 1: 32 t. 4 (1987). —Beentje, Kenya Trees, Shrubs Lianas: 218 (1994). Type from Yemen.

Phyllanthus lalambensis Schweinf. in Penzig, Atti Congr. Bot. Genova: 360 (1895), nomen; in Bull. Herb. Boissier, **7**, App. 2: 302 (1899). —Hutchinson in F.T.A. **6**, 1: 698 (1912). —Engler, Pflanzenw. Afrikas (Veg. Erde 9) **3**, 2: 24 (1921). —Brenan, Check-list For. Trees Shrubs Tang. Terr.: 222 (1949). Type from Ethiopia (Eritrea).

Phyllanthus guineensis Pax in Bull. Herb. Boissier, **6**: 732 (1898). —Hutchinson in F.T.A. **6**, 1: 699 (1912). —R.E. Fries, Wiss. Ergebn. Schwed. Rhod.-Kongo-Exped. **1**, 1: 120 (1914). —Engler, Pflanzenw. Afrikas (Veg. Erde 9) 3, 2: 23 (1921). —Robyns & Tournay, Fl. Sperm. Parc Nat. Alb. **1**: 441 (1948). —Brenan, Check-list For. Trees Shrubs Tang. Terr.: 222 (1949). —Topham, Check List For. Trees Shrubs Nyasaland Prot.: 52 (1958). —White, F.F.N.R.: 202 (1962). —Agnew, Upl. Kenya Wild Fls.: 212 (1974). —Drummond in Kirkia **10**: 251 (1975). —K. Coates Palgrave, Trees Southern Africa, ed. 2, rev.: 399 (1983). Type from Angola (Huíla Province).

Phyllanthus floribundus sensu Hiern, Cat. Pl. Afr. Welw. **1**, 4: 957 (1900), non Müll. Arg.

Phyllanthus ugandensis Rendle in J. Linn. Soc., Bot. **37**: 210 (1905). Type from Uganda.

A much-branched often dense scrambling or spreading weak-stemmed bush 2–5(10) m high; plants monoecious or dioecious, armed, glabrous; branches long slender horizontal, arching or pendent, blackish. Bark rough, dark brown. Twigs dark grey. Lateral shoots leafy, floriferous or both, not usually more than 10 cm long. Short shoots giving rise to one or more secondary laterals or lead shoots. Scale leaves 1.5–2 mm long, triangular-lanceolate, ciliate, dark brown, their bases becoming hard and spiny; their stipules triangular-ovate, otherwise resembling them. Foliage leaves distichous. Petioles 1 mm long. Stipules 1.3 mm long, oblong-linear to linear-lanceolate, chestnut-brown, ciliate-fimbriate. Leaf blades 0.7–2.3 × 0.4–1.5 cm, obovate-oblong to elliptic-oblong, usually ± parallel sided, obtuse, rounded or truncate, occasionally mucronulate, cuneate or rounded at the base, firmly membranaceous, bright yellow-green above, glaucous beneath; lateral nerves in 6–7(9) pairs, not or scarcely prominent above or beneath, often forming somewhat irregular loops. Flowers in clusters which are all male, or else a few male plus 1 female, or female flowers solitary. Male flowers: pedicels 2–3 mm, capillary; sepals (4)5, the 2 outer 1 × 1 mm, the 3 inner 1.3 × 1.3 mm, obovate-suborbicular, yellowish-green, cream-coloured or occasionally pinkish-, crimson- or purplish-tinged; disk glands (4)5, free, circular, ± smooth, fleshy; stamens (2)4(5), the 2 outer free, the rest united, or all united into a column 1 mm high, anthers 0.3 mm long, vertically held, longitudinally dehiscent. Female flowers: pedicels 1.5–2.5 mm long, slender; sepals ± as in the male; disk 0.67 mm in diameter, annular, crenellate, thick, fleshy; ovary 0.5 mm in diameter, sessile, rhomboid-ovoid, ± smooth; styles 3(4), 1 mm long, connate at the base, ± erect, slender, usually simple, slightly thickened and somewhat recurved at the apex. Fruit 3–4 × 4–5 mm, subglobose, smooth, fleshy, brownish at first later becoming dark reddish-purple, and black when dried. Seeds 1.5 × 1 × 1 mm, triquetrous, ± smooth, shiny, bright reddish-brown, with a round aperture by the hilum.

Zambia. N: Mbala Distr., Saisi Valley, Chitundi Marsh, male & female fl. & fr. 10.x.1970, *Richards & Arasululu* 26255 (K; SRGH). W: Kabompo Distr., Mwinilunga–Kabompo road, male & female fl. 4.x.1952, *Angus* 598 (BM; FHO; K). C: 17.5 km Lusaka–Kabwe, fl. 12.ix.1972, *Strid*

2104 (MO). **Zimbabwe.** E: Mutare Distr., Inyamatshira Mt., male fl. 25.xi.1962, *Chase* 7902 (K; LISC; SRGH). S: Bikita Distr., female fl. & fr. 17.xii.1953, *Wild* 4424 (K; LISC; PRE; SRGH). **Malawi.** N: Chitipa Distr., Misuku Hills, Mughesse Forest, fr. 28.xii.1972, *Pawek* 6181 (K; MAL; MO; SRGH; UC). S: Mt. Mulanje, Likhubula Valley, fr. 29.v.1987, *J.D. Chapman & E.J. Chapman* 8539 (K; MO; PRE). **Mozambique.** N: Malema, Serra Murripa, fr. 15.xii.1967, *Torre & Correia* 16531 (LISC). MS: Dombe, 4 km from Serração de Moribane, y. fr. 4.xii.1965, *Pereira & Marques* 951 (BR; LMU).

Widespread in tropical Africa from southern Nigeria to Ethiopia and south to Angola and Mozambique; also in S Arabia (Yemen). In high rainfall deciduous woodlands, chipya woodland and thickets, and in evergreen rainforest, forest margins and gully forest, also in riverine forest and mushitu, sometimes on well-wooded rocky outcrops and termitaria; 780–1950 m.

Phyllanthus reticulatus var. *glaber* with *P. ovalifolius* produces the hybrid *Phyllanthus* × *collium-misuku* Radcl.-Sm. in Kew Bull. **47**: 680 (1992). It also seems to hybridize with the typical variety.

The flexible stems are used in making fish traps.

3. **Phyllanthus reticulatus** Poir., Encycl. Méth. Bot. **5**: 298 (1804). —Müller Argoviensis in De Candolle, Prodr. **15**, 2: 344 (1866). —Engler, Pflanzenw. Ost-Afrikas **C**: 236 (1895). —Hutchinson in F.T.A. **6**, 1: 700 (1912); in F.C. **5**, 2: 391 (1920). —Eyles in Trans. Roy. Soc. South Africa **5**: 393 (1916). —Engler, Pflanzenw. Afrikas (Veg. Erde 9) 3, 2: 23 (1921). —De Wildeman, Pl. Bequaert. **3**, 4: 444 (1926). —Burtt Davy, Fl. Pl. Ferns Transvaal: 299 (1932). —Bremekamp & Obermeyer in Ann. Transvaal Mus. **16**, 3: 421 (1935). —Gardner, Trees & Shrubs Kenya Col.: 49 (1936). —Hutchinson, Botanist in Southern Africa: 667 (1946). —Brenan, Check-list For. Trees Shrubs Tang. Terr.: 222 (1949). —Eggeling & Dale, Indig. Trees Uganda, ed. 2: 137 (1952). —F.W. Andrews, Fl. Pl. Anglo-Egypt. Sudan **2**: 89 (1952). —Brenan in Mem. N.Y. Bot. Gard. **9**, 1: 68 (1954). —G.L. Webster in J. Arnold Arbor. **38**: 57 (1957). —Keay in F.W.T.A., ed. 2, **1**, 2: 387 (1958). —Dale & Greenway, Kenya Trees & Shrubs: 215 (1961). —White, F.F.N.R.: 202 (1962). —P.G. Meyer in Merxmüller, Prodr. Fl. SW. Afrika, fam. 67: 40 (1967). —Mogg in Macnae & Kalk, Nat. Hist. Inhaca Isl., Moçamb.: 147 (1969). —Agnew, Upl. Kenya Wild Fls.: 212 (1974). —Drummond in Kirkia **10**: 251 (1975). —K. Coates Palgrave, Trees Southern Africa, ed. 2, rev.: 401 (1983). —Radcliffe-Smith in F.T.E.A., Euphorb. 1: 34 (1987). —Beentje, Kenya Trees, Shrubs Lianas: 219 (1994). Type from the 'Indies'.

A much-branched, usually laxly virgate or semiscandent shrub or small tree up to 10 m tall, usually 1–5 m tall; plants coppicing deciduous monoecious; branches pale grey or brownish-white spreading arcuate, almost reaching the ground. Bole up to 25 cm in diameter. Bark light reddish-brown, fissured longitudinally. Twigs grey-brown, roughly lenticellate. Younger twigs greenish. Young lead shoots sometimes glabrous, or else covered with a whitish or ferrugineous indumentum. Lateral leafy shoots up to 25 cm long, floriferous shoots much shorter. Short shoots often bearing fascicles of lateral shoots. Scale leaves 1.5 mm long, lanceolate, reddish-brown, their bases hardening and becoming somewhat spinescent; their stipules broadly triangular, otherwise resembling the scale leaves. Foliage leaves distichous. Petioles 1–4 mm long. Stipules 1–1.5 mm long, linear to narrowly lanceolate, light brown, sparingly fimbriate-ciliate. Leaf blades very variable, 1–5 × 0.5–3 cm, elliptic, ovate-oblong or suborbicular, subacute, obtuse, rounded, truncate or retuse at the apex, rounded-cuneate to truncate at the base, chartaceous to thinly coriaceous, commonly glabrous but occasionally sparingly to evenly crisped-puberulous at least along the midrib and main nerves above and beneath, light green, dull or dark green and pale-veined above, glaucous beneath; lateral nerves in 7–13 pairs, not or scarcely prominent above, slightly so beneath. Flowers fragrant, fasciculate on leafless or leafy lateral shoots, often with 1 female and several male flowers per fascicle, or else female flowers solitary in upper leaf axils; bracts resembling the scale leaf stipules. Male flowers: pedicels 2–4 mm long, capillary; sepals 5(6), 2 × 1–2 mm, elliptic-ovate or obovate-suborbicular, glabrous or puberulous outside, white with a green or yellowish-green median stipe, sometimes tinged pinkish, reddish, brownish or purplish; disk glands 5(6), free, shortly stipitate, angular, fleshy; stamens 5(6), free, or 2(3) inner united at the base and the outer 2(3) free, filaments 1 mm long, anthers 0.5 mm long, basifixed, thecae separated by the connective, longitudinally dehiscent. Female flowers: pedicels shorter and stouter than in the male, often puberulous; sepals ± as in the male; but the innermost petaloid; disk c. 1 mm in diameter, 5-lobed, the lobes thick, ± triangular; ovary 1–1.5 mm in diameter, sessile, subglobose, 3–multilocular, smooth; styles c. 0.25 mm long, 3–∞, free or ± so, incurved, or suberect, crowded, shortly bifid. Fruit 3–5 × 4–6 mm, depressed-subglobose, 3–many lobed, 6–many seeded, smooth, bacciform, with a thinly fleshy pericarp, green at first, later turning reddish-purple or bluish-black. Seeds 2–2.5 × 1.5 × 1 mm, irregularly ovoid-trigonous, minutely reticulate, somewhat shiny, reddish-brown, sometimes 1 per locule by abortion.

1. At least the young flowering shoots, pedicels and outer calyx lobes puberulous · · · · · · · · var. *reticulatus*
– All parts quite glabrous · 2
2. Leaves up to 5 × 3 cm, usually ovate-oblong · · · · · · · · · · var. *glaber*
– Leaves up to 1.8 × 0.8 cm, mostly narrowly elliptic · · · · · · · · var. *orae-solis*

Var. **reticulatus**

At least the young flowering shoots, pedicels and outer calyx lobes puberulous.

Caprivi Strip. Tjaro Island, Okavango R., st. 21.i.1956, *de Winter & Wiss* 4356 (K; PRE). **Botswana**. N: Linyanti R., fr. 6.xi.1974, *P.A. Smith* 1160 (K; MO; PRE; SRGH). **Zambia**. B: Gonye Falls, male fl. 27.vii.1952, *Codd* 7209 (BM; FHO; K; PRE; SRGH). N: Kawambwa, fl. & fr. 28.viii.1957, *Fanshawe* 3685 (K; NDO). C: Mumbwa Distr., Chikumbe Hill, y. fr. 18.viii.1949, *Hoyle* 1146 (BM; FHO; K). E: Luangwa Bridge, fr. 7.x.1958, *Robson & Angus* 1 (BM; K; LISC; SRGH). S: Namwala Distr., Mapanza Mission, fl. and y. fr. 3.viii.1952, *Angus* 148 (BM; FHO; K; PRE). **Zimbabwe**. N: Binga Distr., Sanyati–Zambezi confluence, fr. ix.1955, *Davies* 1524 (K; MO; SRGH). W: Hwange (Wankie), male fl. and fr. 25.vi.1934, *Eyles* 8071 (BM; FHO; GRA; K; SRGH). E: Mutare Distr., Odzi R., Glenshiel, fl. & fr. 18.viii.1957, *Chase* 6699 (K; LISC; PRE; SRGH). S: Chibi Distr., Razi Dam, fl. & fr. 5.v.1970, *Pope* 259 (K; PRE; SRGH). **Malawi**. N: Kilwa, 88 km south of Nkhata Bay, fr. 18.vi.1973, *Pawek* 6901 (K; MAL; MO; SRGH; UC). C: Nkhota Kota Distr., Chia, fl. 7.ix.1946, *Brass* 17563 (BM; FHO; K; NY; SRGH). S: Mangochi Distr., 24 km west of Namwera, fl. 24.viii.1976, *Pawek* 11701 (DAV; K; MAL; MO; SRGH; UC). **Mozambique**. N: Angoche (António Enes), male fl. 16.x.1965, *Mogg* 32232 (K; LISC; PRE). Z: Aguas Quentes, fr. 12.vii.1942, *Torre* 4556 (LISC). T: Boroma Distr., Sisitso Station, fl. & fr. 18.vii.1950, *Chase* 2614 (BM; K; LISC; SRGH). MS: Gorongosa Game Res., fr. 14.vii.1957, *Chase* 6622 (K; LISC; PRE; SRGH). GI: Magul-Macia, fl. & y. fr. 1.vi.1959, *Barbosa & Lemos* in *Barbosa* 8551 (K; LISC; LMA; PRE; SRGH). M: Marracuene, Maxaquene, fr. 22.viii.1958, *Macuácua* 66 (K; LISC; LMA; PRE; SRGH).

Throughout the Old World tropics. Widespread, often forming thickets, occurring in low altitude riverine vegetation and floodplain grassland, in sand dune scrub, in littoral scrub and dune forest and in lowland rainforest, also in mixed deciduous woodlands and scrub, occasionally on termitaria; sea level to 1500 m.

Vernacular names as recorded in specimen data include: “bupsantima”, “busametina” (xiRonga); “cuatima” (Inhambane area); “m’tanta nyerere” (chiTonga); “mugrorgroro” (Mbukushu); “mutantanyere” (Nkhata Bay area); “tandanyelele” (chiNyanja); “tantanyerere” (Nkhota Kota area); “tetenha”, “tetenho” (xiRonga); “zwipera” (chiNdau).

The macerated leaves are placed on burns, and are also taken for stomach disorders; the stems are rubbed on the teeth to whiten them, and the black fruits are used to make ink.

Noel (2467, LMU; SRGH) on herbarium specimen label, reports that the plant emits a strong odour of brewing.

Phyllanthus reticulatus var. *reticulatus* hybridizes with *Phyllanthus muellerianus* to produce the hybrid *P.* × *fluminis-sabi* Radcl.-Sm. in Kew Bull. **51**, 2: 306 (1996). Known from 2 collections in Manica e Sofala Province, Mozambique.

Var. **glaber** (Baill.) Müll. Arg. in Linnaea **32**: 12 (1863); in De Candolle, Prodr. **15**, 2: 345 (1866). —Hutchinson in F.T.A. **6**, 1: 701 (1912); in F.C. **5**, 2: 391 (1920). —Engler, Pflanzenw. Afrikas (Veg. Erde 9) 3, 2: 23 (1921). —G.L. Webster in J. Arnold Arbor. **38**: 59 (1957). —Keay in F.W.T.A., ed. 2, **1**, 2: 387 (1958). —Radcliffe-Smith in F.T.E.A., Euphorb. 1: 36 (1987). Type from Senegal.

Phyllanthus polyspermus Schumach. & Thonn., Beskr. Guin. Pl.: 416 (1827). Type from Ghana.

Kirganelia prieuriana var. *glabra* Baill., Adansonia **1**: 83 (1860).

Phyllanthus prieurianus var. *glaber* (Baill.) Müll. Arg. in Linnaea **32**: 12 (1863).

Kirganelia multiflora var. *glabra* Thwaites, Enum. Pl. Zeyl.: 282 (1861). Type from Sri Lanka.

Phyllanthus nov. spec., Schinz & Junod in Mém. Herb. Boissier, No. 10: 46 (1900).

All parts quite glabrous; leaves up to 5 × 3 cm, usually ovate-oblong.

Zambia. N: Munyamadzi R. (Mwunyamazi), Luangwa Valley, fl. & fr. 6.x.1933, *Michelmore* 644 (K). **Zimbabwe**. W: Victoria Falls, st. 24.vii.1950, *Robertson & Elffers* 40 (K; PRE). C: Kadoma

Distr., Kudu R. Ranch, st. 28.iii.1970, *Burrows* 434 (SRGH). E: Chimanimani Distr., Changadzi R., galled male fls. 11.ix.1949, *Chase* 1757 (BM; K; SRGH). **Malawi**. N: Mzimba Distr., Mzuzu, fl. 7.xi.1969, *Pawek* 2950 (K). S: Mangochi Distr., Namwera, fr. 23.viii.1976, *Pawek* 11683 (DAV; K; MAL; MO). **Mozambique**. N: R. Messala, 68 km Maua–Marrupa R., fl. 14.viii.1981, *Jansen, de Koning & de Wilde* 287 (K; WAG). M: Delagoa Bay, fl. 1890, *Junod* 54 (K; Z).

Throughout the Old World tropics, and introduced into the West Indies. In low altitude riverine vegetation, dry mixed deciduous and miombo woodlands, often beside seasonal rivers and streams, sometimes on termitaria; 610–1370 m.

Vernacular names as recorded in specimen data include: "mufanenele" (eastern Mozambique); "nupanenele" (Lúrio); "uncassiri" (KiMwani).

Intermediates between var. *reticulatus* and var. *glaber*, in which the inflorescence axis is glabrous and the pedicels and calyx lobes pubescent, have been seen from various provinces of Malawi and Mozambique.

Phyllanthus reticulatus var. *glaber* hybridizes with *P. ovalifolius* to produce the hybrid *P.* × *collium-misuku* Radcl.-Sm. in Kew Bull. **47**: 680 (1992). Type: Malawi, Chitipa Distr. Misuku Hills, near Chuwa R., 27.i.1989, *Thompson & Rawlins* 6205 (K, holotype). Known only from this collection.

Var. **orae-solis** Radcl-Sm. in Kew Bull. **51**, 2: 319 (1996). Type: Mozambique, Maputo Prov., between Polana and Costa do Sol, fr. 17.ix.1947, *Pedro & Pedrógão* 1850 (PRE, holotype).

All parts quite glabrous; leaves not exceeding 1.8 × 0.8 cm, mostly narrowly elliptic.

Mozambique. M: Maputo, fl. & fr. 8.x.1940, *Torre* 1730 (LISC).

Not known elsewhere. In dense coastal woodland.

4. **Phyllanthus muellerianus** (Kuntze) Exell, Cat. Vasc. Pl. S. Tomé: 290 (1944). —Brenan, Check-list For. Trees Shrubs Tang. Terr.: 222 (1949). —F.W. Andrews, Fl. Pl. Anglo-Egypt. Sudan **2**: 89 (1952). —Keay in F.W.T.A., ed. 2, **1**, 2: 385 (1958). —White, F.F.N.R.: 202 (1962). —Agnew, Upl. Kenya Wild Fls.: 212 (1974). —Radcliffe-Smith in F.T.E.A., Euphorb. 1: 24 (1987). —Beentje, Kenya Trees, Shrubs Lianas: 218 (1994). Type from Guinée.

Kirganelia floribunda Baill., Adansonia **1**: 83 (1860), non Spreng. (1826). Type as above.

Phyllanthus floribundus (Baill.) Müll. Arg. in Linnaea **32**: 14 (1863); in De Candolle, Prodr. **15**, 2: 343 (1866). —Engler, Pflanzenw. Ost-Afrikas **C**: 236 (1895). —Hutchinson in F.T.A. **6**, 1: 701 (1912). —R.E. Fries, Wiss. Ergebn. Schwed. Rhod.-Kongo-Exped. **1**, 1: 120 (1914). —Eyles in Trans. Roy. Soc. South Africa **5**: 393 (1916). —Engler, Pflanzenw. Afrikas (Veg. Erde 9) 3, 2: 22 (1921). —De Wildeman, Pl. Bequaert. **3**, 4: 443 (1926), non Kunth (1817).

Diasperus muellerianus Kuntze, Revis. Gen. Pl., part 2: 597 (1891).

A scandent shrub with numerous stems from the base, or small tree up to 7.5 m tall, monoecious, evergreen, completely glabrous, spiny; branches arched, pendulous almost to the ground. Bole to 20 cm in diameter. Bark greyish-brown, smooth. Twigs dark brown, with scattered corky lenticels. Lateral leafy shoots to 25 cm long, floriferous shoots rarely more than 10 cm long, the latter borne in fascicles on short shoots co-axillary with the former. Scale leaves 3–4 mm long, triangular-lanceolate, spinescent, decurved, purplish-brown; stipules broadly triangular, otherwise similar to the scale leaves. Foliage leaves distichous; petioles 2–5 mm long; stipules 2 mm long, subulate, chaffy, brownish. Leaf blades 1.5–8 × 1–4.5 cm, broadly ovate to ovate-lanceolate, usually subacute or obtuse, sometimes shortly acuminate, cuneate, rounded or truncate at the base, thinly to firmly chartaceous, dark green and shiny above, paler and duller beneath; lateral nerves in 6–16 pairs, slightly prominent above and beneath. Flowers malodorous, fasciculate on leafless often zigzag lateral shoots, often with 1 female and 1–3 male flowers per fascicle, rarely the female flowers solitary on leafy shoots; bracts minute. Male flowers: pedicels 2 mm long, capillary; sepals 5, c. 1 × 1 mm, subequal, elliptic-ovate to suborbicular, concave, pale greenish-yellow; disk glands 5, free, turbinate, verruculose, fleshy; stamens 5, free, 3(2) short, 2(3) long, the longer less than 1 mm long, anthers minute, ellipsoid, ± vertically-held, longitudinally dehiscent. Female flowers: pedicels shorter and stouter than in the male; sepals ± as in the male, but early caducous; disk glands 5, free or united in pairs, transversely ellipsoid, fleshy, ± smooth; ovary 0.75 mm in diameter, sessile, ellipsoid-subglobose, 3–4(5)-locular, smooth; styles 3–4(5), 0.67 mm long, free, ± erect at first, later spreading or recurved, bifid, the stigmas filiform, velvety. Fruit 2–3 × 3–4 mm, subglobose, ± smooth, bacciform, with a fleshy pericarp, green at first, later ripening through pink to reddish-brown or black. Seeds 1.3 × 1 × 1 mm, triquetrous, minutely reticulate, shiny, reddish-brown.

Zambia. B: Zambezi Distr., Chavuma, y. fl. 14.x.1952, *Gilges* 243 (K). N: Mansa Distr., Lake Bangweulu, Samfya, fr. 30.i.1959, *Watmough* 175 (K; LISC; PRE; SRGH). W: Mwinilunga Distr., 16 km west of Kabompo R., fl. 20.xii.1969, *Simon & Williamson* 1876 (K; MO; SRGH). C: 13 km west of Lusaka, fl. & fr. 29.xii.1969, *Simon & Williamson* 2074 (K; PRE; SRGH). E: Nsadzu to Chipata (Ft. Jameson), fl. & y. fr. 25.xi.1958, *Robson* 699 (BM; K; LISC; SRGH). S: Mazabuka Distr., Mapanza, Choma, fl. 1.xii.1957, *Robinson* 2518 (K; PRE; SRGH). **Malawi**. N: Nkhata Bay Distr., 22.5 km east of Mzuzu, fl. & fr. 6.i.1976, *Pawek* 10687 (K; MAL; MO; SRGH; UC). C: Mchinji Distr., Bua R., fl. 14.xi.1983, *Patel & Salubeni* 1352 (K; MAL; MO). **Mozambique**. T: 3–10 km from Furancungo to Bene, fr. 19.iii.1966, *Pereira, Sarmento & Marques* 1875 (BR; LMU).

Widespread in tropical Africa from Guinée eastwards to the Sudan and south to Angola and Tanzania. In riverine vegetation and fringing forest, and in dry evergreen forest and thicket (mateshi), and high rainfall miombo and chipya woodlands, also in swamp forest (mushitu) and on dambo edges and in tall grassland; 500–1750 m.

Vernacular names as recorded in specimen data include: "mupetwalupe" (chiBemba, chiWemba); "mupetwandupe" (Mambwe area).

The hybrid, *Phyllanthus* × *fluminis-sabi* Radcl-Sm. in Kew Bull. **51**, 2: 306 (1996) (*P. muellerianus* × *P. reticulatus* var. *reticulatus*), is known from Manica e Sofala Province in Mozambique.

5. **Phyllanthus inflatus** Hutch. in Bull. Misc. Inform., Kew **1920**: 334 (1920). —F.W. Andrews, Fl. Pl. Anglo-Egypt. Sudan **2**: 89 (1952). —Agnew, Upl. Kenya Wild Fls.: 212 (1974). —Drummond in Kirkia **10**: 251 (1975). —K. Coates Palgrave, Trees Southern Africa, ed. 2, rev.: 400 (1983). —Radcliffe-Smith in F.T.E.A., Euphorb. 1: 25 (1987). —Beentje, Kenya Trees, Shrubs Lianas: 218 (1994). Type from Sudan.

Phyllanthus polyanthus sensu Eggeling & Dale, Indig. Trees Uganda, ed. 2: 137 (1952), non Pax (q.v.).

A multi-stemmed tree up to 9 m tall with a spreading crown, monoecious, usually glabrous, deciduous, spiny; bole smooth, grey; branches trailing, brown at first, later flaking to become greyish. Lead shoots robust, green at first, soon turning reddish-brown. Lateral leafy shoots appearing after the flowers, up to 45 cm long, strongly resembling pinnate leaves, crowded at the branch ends, green; lateral flowering shoots up to 5 cm long, borne on the old wood from short shoots, in fascicles; rarely the leafy shoots with axillary flowers also. Scale leaves 4 mm long, lanceolate, sometimes fimbriate at the apex, purplish-brown, spinescent; stipules 2–3 mm long otherwise similar to the scale leaves. Foliage leaves distichous, hysteranthous; petioles 1–2 mm long; stipules 2–3 mm long, linear-lanceolate to filiform, usually subentire, pale green and/or light brown. Leaf blades 3–6 × 1–2 cm, lanceolate, the lowest sometimes suborbicular, acute or subacute, the lowest sometimes truncate or retuse and brown-mucronate, rounded to cuneate at the base, membranaceous to thinly chartaceous, usually glabrous above and beneath, pale green; lateral nerves in 6–12 pairs, scarcely prominent above, slightly so beneath, looped well within the margin. Flowers fasciculate, usually on leafless lateral shoots, either with 1 female and several male flowers per fascicle or else fascicles all male, which often arise below the leafy shoots; bracts 1.5 mm long, triangular to spathulate, fimbriate-laciniate, dark brown. Male flowers: pedicels c. 1 mm long; sepals 2–2.5 × 1–1.5 mm, ovate, pale green; disk glands 5, free or ± so, irregularly transversely ovoid, ± smooth, fleshy; stamens 5, free, filaments 0.5 mm long, anthers 0.3 mm long, broadly ovoid, vertically-held, longitudinally dehiscent. Female flowers: pedicels 2 mm long, extending to 1 cm or more in fruit; sepals 5, 2 × 2 mm, suborbicular, erose; disk 1.5 mm in diameter, annular, flat, thin but with a slightly thickened margin, subentire; ovary 2 mm in diameter, ± sessile, 3-lobed to subglobose, smooth; styles 3, c. 1 mm long, very slightly connate at the base, appressed to the top of the ovary, bifid, the stigmas subulate. Fruit 1.5–3 × 1.5–3 cm, 3-lobed to subglobose, smooth, inflated, somewhat fleshy at first, later with a papery pericarp, glabrous, pale green when ripe. Seeds 7 × 6 × 4.5 mm, compressed-ovoid, smooth, shiny, buff, marbled with dark brown; hilum with a small aperture.

Zambia. B: Kabompo, male fl. 23.ix.1964, *Fanshawe* 8926 (K; NDO). **Zimbabwe**. E: Chipinge Distr., Chirinda Forest, male & female fl. viii.1966, *Goldsmith* 63/66 (K; LISC; SRGH). **Malawi**. S: Blantyre Distr., Soche Mt., st. 6.x.1960, *Chapman* 971 (MAL; SRGH). **Mozambique**. MS: Bandula Mt., fr. 6.xi.1957, *Chase* 6737 (K; LISC; SRGH).

From Sudan south to eastern Zimbabwe and adjacent Mozambique. An uncommon understorey tree of evergreen forest, also in evergreen gully forest and swamp forest; 900–1200 m.

6. **Phyllanthus acidus** (L.) Skeels in U.S. Dept. Agr. Bur. Pl. Industr. Bull. **148**: 17 (1909). —Brenan, Check-list For. Trees Shrubs Tang. Terr.: 222 (1949). —G.L. Webster in J. Arnold Arbor. **38**: 66 (1957). —Keay in F.W.T.A., ed. 2, **1**, 2: 388 (1958). —Radcliffe-Smith in F.T.E.A., Euphorb. 1: 36 (1987). Type from 'India', Linnean Herbarium 592/3.
Averrhoa acida L., Sp. Pl.: 428 (1753).
Cicca disticha L., Mant. Pl. Alt.: 124 (1767). Type from India, Linnean Herbarium 1108/1.
Phyllanthus distichus (L.) Müll. Arg. in De Candolle, Prodr. **15**, 2: 413 (1866).

Similar to *Phyllanthus inflatus*, but with larger foliage leaves (up to 9 × 4 cm), floral whorls usually in 4s, staminodes in the female flowers, fruit 1–2-locular, 1–1.5 × 1.5–2 cm, not inflated and more markedly fleshy.

Mozambique. M: Maputo (Lourenço Marques), Jardim Tunduru (Jardim Vasco da Gama), fr. 15.xii.1972, *Balsinhas* 2458 (K; LISC).
Cultivated for it's edible fruits, occasionally escaping. Country of origin unknown.

7. **Phyllanthus engleri** Pax in Engler, Pflanzenw. Ost-Afrikas **C**: 236 (1895). —Hutchinson in F.T.A. **6**, 1: 699 (1912). —Engler, Pflanzenw. Afrikas (Veg. Erde 9) **3**, 2: 24 (1921). —Brenan, Check-list For. Trees Shrubs Tang. Terr.: 222 (1949). —White, F.F.N.R.: 202 (1962). —Drummond in Kirkia **10**: 251 (1975). —K. Coates Palgrave, Trees Southern Africa, ed. 2, rev.: 399 (1983). —Radcliffe-Smith in F.T.E.A., Euphorb. 1: 27 (1987). Type from Tanzania.

A stout, deciduous shrub or small tree up to 8 m tall, rarely taller, dioecious, glabrous, spiny, branched from near the base; bole up to 12 cm d.b.h. Bark grey, irregularly fissured. Branches long, up to 2–3 cm thick, unbranched, greyish. Lead shoots robust, dark reddish-brown. Lateral leafy shoots up to 30 cm long, floriferous or not at the base; lateral non-leafy flowering shoots 2–3 cm long. Scale leaves 4–5 mm long, triangular-lanceolate, dark reddish-brown, soon hardening and thickening into stout spines; their stipules 3–4 mm long, lanceolate, slightly spinescent. Short shoots or spur shoots 2–3(6) cm long, densely spiny, giving rise to secondary lateral shoots. Foliage leaves distichous; petioles 1–2 mm long; stipules 1–4 mm long, lanceolate, fimbriate to subentire, dark or light reddish-brown or greenish, soon deciduous. Leaf blades 1–3 × 0.75–2.5 cm, elliptic, ovate or suborbicular, obtuse or rounded, sometimes mucronulate, rounded to shallowly subcordate at the base, firmly chartaceous to thinly coriaceous, deep dull green to yellowish green above, paler and somewhat glaucous beneath; lateral nerves in c. 5 pairs, not prominent and often scarcely visible above, scarcely prominent beneath. Flowers fasciculate. Bracts resembling the larger stipules. Male flowers: pedicels 1 mm long; sepals (4)5, 1–2 × 1–1.5 mm, strongly imbricate, ovate, obtuse to subacute, pale green; disk glands 5, minute, free, rounded, ± smooth; stamens (4)5, 0.5 mm long, ± free, anthers 0.3 mm long, longitudinally dehiscent. Female flowers not known. Styles c. 1 mm long, ± free, recurved, bifid, stigmas filiform. Fruit 1.5–2 × 2–2.5(3) cm when dried (3–4 cm in diameter when fresh), 3-lobed to subglobose, smooth when fresh, yellowish-green; exocarp papery; mesocarp spongy or corky; endocarp crustaceous. Seeds 6, or fewer by abortion, c. 1 cm long, irregularly ovoid-trigonous, smooth, dull, dark purplish-brown, blotched with reddish- and yellowish-brown.

Zambia. B: Zambezi Distr., Mombezi–Kabompo confluence, fr. 8–10.v.1953, *Holmes* 1090 (FHO; K). C: Chalimbana, 20 km east of Lusaka, male fl. 15.x.1972, *Strid* 2326 (MO). S: Monze, fr. 11.vii.1930, *Hutchinson & Gillett* 3564 (BM; K; LISC). **Zimbabwe**. N: Mazowe Distr., Chipoli, Shamva, male fl. 2.xi.1958, *Moubray* in *GHS* 87779A (K; PRE; SRGH). **Malawi**. S: near Fundi Pass, Tuchila Plain, st. 22.ix.1929, *Burtt Davy* 21954 (FHO). **Mozambique**. N: Gazimbe to Mecanhelas, st. 14.ix.1970, *Trevor-Jones* NP4 (SRGH). MS: Moribane, fr. xi–xii.1911, *Dawe* 499 (K).
Also in Tanzania. In mixed dry deciduous woodland, mopane and *Acacia* woodlands, less often in miombo woodlands, also in *Baikiaea* forest, Kalahari Sand dambos and on alluvial flats, often on termitaria; 900–1250 m.
Vernacular names as recorded in specimen data include: "mpululwe" (chiNyanja); "mufweba-bacazhi" (Lusaka, Namwala areas); "mulia-baushina" (Lusaka area); "mululwe" (T.); "pupwe" (chiWemba).
The description of the styles and seeds was drawn up from Tanzanian material.
Phyllanthus engleri hybridizes with *P. polyanthus* to give rise to the hybrid *P.* × *fluminis-zambesi* Radcl.-Sm. in Kew Bull. **51**, 2: 306 (1996). Known only from the type in western Zambia.

8. **Phyllanthus polyanthus** Pax in Bot. Jahrb. Syst. **28**: 19 (1899). —Hutchinson in F.T.A. **6**: 1: 703 (1912). —Engler, Pflanzenw. Afrikas (Veg. Erde 9) **3**, 2: 23 (1921). —White, F.F.N.R.: 202 (1962). Type from Zaire.

Phyllanthus delpyanus Hutch. in F.T.A. **6**, 1: 1047 (1913). —Engler, Pflanzenw. Afrikas (Veg. Erde 9) **3**, 2: 24 (1921). —Radcliffe-Smith in F.T.E.A., Euphorb. 1: 27 (1987). —Beentje, Kenya Trees, Shrubs Lianas: 217 (1994). Types from Gabon.

Phyllanthus cedrelifolius Verdoorn in Bull. Misc. Inform., Kew **1924**: 259 (1924). —K. Coates Palgrave, Trees Southern Africa, ed. 2, rev.: 398 (1983). Type from South Africa.

Similar to *Phyllanthus engleri*, but differing in having lead shoots much less robust, spur shoots with less markedly spinescent scale leaves; foliage leaves acute or subacute, shiny, up to 6 × 4 cm; fruits up to 5 cm in diameter, depressed-globose and often 4-locular, becoming brown and corky at maturity, and seeds uniformly reddish-brown in colour.

Zambia. N: Samfya, L. Bangweulu, st. 21.iv.1989, *Radcliffe-Smith, Pope & Goyder* 5755 (K). W: 6.5 km north of Kalene Hill Mission, fr. 21.ix.1952, *Angus* 513 (BM; BR; FHO; K; PRE). C: Kafue R., Bolenga, fr. 25.xi.1919, *Shantz* 49606 (BM). S: Ngoma, Kafue Nat. Park, fr. 29.xii.1963, *Bainbridge* 953 (FHO; K; NDO; SRGH). **Malawi**. S: Thondwe, Mpita, fr. 6.v.1982, *Chapman & Tawakali* 6187 (BR; FHO; K; MAL). **Mozambique**. N: Angoche (António Enes), fl. 22.x.1965, *Mogg* 32523 (LISC). Z: Maganja da Costa Distr., Gobene Forest, y. fr. 12.ii.1966, *Torre & Correia* 14546 (LISC). GI: Manjacaze, 30 km Serração de M'crusse-Santos Gil, fr. 18.iii.1948, *Torre* 7513 (LISC).

Also in São Tomé, Cameroon, Gabon, Central African Republic, Congo, Zaire, Kenya, Tanzania, Angola and South Africa. Rare in dry evergreen forest and *Cryptosepalum* evergreen thicket (mavunda) on Kalahari Sand, also in chipya woodland, often beside perennial rivers, rocky streamsides and lakeside sand dunes, also in littoral scrub and coastal forest; 20–1200 m.

Phyllanthus polyanthus hybridizes with *P. engleri* in W Zambia to produce the hybrid *P.* × *fluminis-zambesi* Radcl.-Sm. in Kew Bull. **51**, 2: 306 (1996).

9. **Phyllanthus kaessneri** Hutch. in Bull. Misc. Inform., Kew **1911**: 315 (1911); in F.T.A. **6**, 1: 706 (1912). —Engler, Pflanzenw. Afrikas (Veg. Erde 9) **3**, 2: 29 (1921). —Radcliffe-Smith in F.T.E.A., Euphorb. 1: 38 (1987). —Jean F. Brunel, Phyllanthus Afr. Intertrop. Mad.: 308 (1987). —Beentje, Kenya Trees, Shrubs Lianas: 218 (1994). Type from Kenya (Coast Province).

A shrub or subshrub up to 3 m tall, monoecious or sometimes dioecious; stems several from the base, slender, wiry and often with scandent branches; twigs dark greyish-brown. Lead shoots and lateral shoots and petioles glabrous, or sparingly white-puberulous with multicellular hairs, especially at the nodes. Lateral leafy shoots 2–8 cm long. Scale leaves 1 mm long, linear-oblong, fimbriate, dark brown; stipules triangular-lanceolate, otherwise resembling the scale leaves. Foliage leaves distichous; petioles 0.5 mm long; stipules c. 1 mm long, linear, pinkish-brown; blades 3–17 × 2–10 mm, elliptic-ovate, obovate or suborbicular, sometimes somewhat asymmetric, rounded to obtuse, rounded-cuneate at the base, membranaceous, usually glabrous above and beneath, pale yellow-green above, glaucescent beneath; lateral nerves in 4–7 pairs, not or scarcely prominent above or beneath. Male flowers 1(2) per axil, female flowers solitary. Male flowers: pedicels 4–8 mm long, capillary; sepals 5(6), c. 1 × 1 mm, ovate-suborbicular, whitish with a pale greenish-cream midrib; disk glands 5(6), minute, free, flat, petaloid; stamens 3–4(5), free or united at the base, filaments 0.8–1 mm long, erect, anthers 0.3 mm wide, horizontal, transversely dehiscent, connective purplish. Female flowers: pedicels (2)4–5 mm long, not extending in fruit, slender; sepals resembling those of the male; disk 0.65 mm in diameter, thin, flat, shallowly 5(6)-lobed, the lobes rounded, entire; ovary 1 mm in diameter, sessile, subglobose, smooth; styles 3, 0.5 mm long, free, appressed to the top of the ovary, bifid, stigmas linear. Fruit 2 × 3 mm, rounded-trigonous, smooth, pale green. Seeds 1.5 × 1 mm, triquetrous, faintly lineate, buff.

Young shoots and petioles glabrous · var. *kaessneri*
Young shoots and petioles sparingly white-puberulous, the hairs multicellular · var. *polycytotrichus*

Var. **kaessneri**

Young shoots and petioles glabrous.

Zambia. N: Mbala Distr., Lunzua R., st. 24.vi.1957, *Richards* 10197 (K).
Also in Kenya and Tanzania. In riverine forest in deep shade; 840 m.

Var. **polycytotrichus** Radcl.-Sm. in Kew Bull. **35**: 769 (1981); in F.T.E.A., Euphorb. 1: 39 (1987). —Jean F. Brunel, Phyllanthus Afr. Intertrop. Mad.: 308 (1987). Type from Tanzania.

Young shoots and petioles sparingly white-puberulous, the hairs multicellular.

Zambia. N: 19 km from Kawambwa to Mansa (Fort Rosebery), fl. 30.x.1952, *Angus* 669 (BR; FHO; K). W: Kitwe, fr. 3.xii.1968, *Mutimushi* 2859 (K; NDO).
Also in Zaire, Kenya and Tanzania. In dense evergreen thicket around granite boulders, on edge of evergreen fringing forest and in mushitu.

10. **Phyllanthus glaucophyllus** Sond. in Linnaea **23**: 133 (1850). —Müller Argoviensis in De Candolle, Prodr. **15**, 2: 393 (1866). —Engler, Pflanzenw. Ost-Afrikas **C**: 236 (1895). —Hutchinson in F.C. **5**, 2: 394 (1920). —Engler, Pflanzenw. Afrikas (Veg. Erde 9) **3**, 2: 29 (1921). —Burtt Davy, Fl. Pl. Ferns Transvaal: 299 (1932). —Agnew, Upl. Kenya Wild Fls.: 212 (1974). —Jean F. Brunel, Phyllanthus Afr. Intertrop. Mad.: 299 (1987). —Radcliffe-Smith in F.T.E.A., Euphorb. 1: 19 (1987). Type from South Africa (Transvaal).
Phyllanthus glaucophyllus var. *major* Müll. Arg. in Flora **47**: 514 (1864); in De Candolle, Prodr. **15**, 2: 393 (1866). —Hutchinson in F.T.A. **6**, 1: 713 (1912). Type from South Africa (KwaZulu-Natal).
Phyllanthus alpestris Beille in Bull. Soc. Bot. Fr. **55**, Mém. **8**: 56 (1908). —Hutchinson in F.T.A. **6**, 1: 712 (1912). —Keay in F.W.T.A., ed. 2, **1**, 2: 387 (1958). Type from Guinée.
Phyllanthus glaucophyllus var. *suborbicularis* Hutch. in F.C. **5**, 2: 395 (1920). Type from South Africa (Cape Province).

A semiprostrate to erect tufted perennial herb, or suffrutex up to 45 cm tall, with several simple or sparingly branched flattened slightly winged stems arising from a small woody rootstock, monoecious or dioecious, glabrous. Leaves distichous, sensitive, becoming flattened against the stem when touched. Petioles 0.5–1 mm long. Stipules 1.25 × 0.75 mm, triangular-ovate, acuminate, sparingly fimbriate or denticulate to subentire, reddish-brown. Leaf blades 0.5–2 × 0.4–1.3 cm, elliptic-lanceolate to ovate or sometimes suborbicular, subacute, obtuse or rounded, sometimes mucronulate, cuneate-rounded to subcordate at the base, margins revolute, firmly chartaceous to thinly coriaceous, deep green above, paler and slightly glaucous beneath; lateral nerves in 7–10 pairs, looped well within the margin, slightly prominent above and beneath. Flowers usually solitary. Male flowers: pedicels 5–7 mm long, capillary; sepals 6, c. 1 × 0.75 mm, obovate, pale pink with a reddish midrib; disk glands 6, free, orbicular, ± flat and smooth, pinkish; stamens 3, minute, free, filaments pale yellow-green, anthers horizontal, thecae apically convergent, laterally dehiscent, yellow. Female flowers: pedicels 5–10 mm long, extending to up to 1.5 cm in fruit, horizontal, slender, reddish; sepals 6, c. 1.5 × 1 mm, the outer sepals obovate, the inner oblong, pale greenish, tinged reddish-pink, or brownish; disk annular, thin, flat, shallowly 6-lobed; ovary 1.5 mm in diameter, sessile, 6-lobed, ± smooth, pale green; styles 3, 0.75 mm long, free, deeply bipartite, spreading, yellowish-green. Fruit 2 × 3.5–4 mm, depressed 3-lobed to subglobose, ± smooth, pale glaucous-green. Seeds 1.75 × 1.5 mm, broadly segmentiform, brown, with numerous irregular rows or arcs of minute darker brown tubercles on each facet.

Zambia. N: Mbala Distr., Nkali Dambo, female fl. & fr. 5.i.1955, *Richards* 3902 (K). W: Mwinilunga Distr., Dobeka Bridge, female fl. 23.xii.1937, *Milne-Redhead* 3792 (BM; K; LISC; SRGH). C: Mkushi Distr., Fiwila, fr. 12.i.1958, *Robinson* 2746 (K; SRGH). S: Choma, male fl. & o. fr. 12.i.1952, *White* 1893 (FHO; K). **Zimbabwe**. N: Guruve Distr., Nyamunyeche Estate, Great Dyke, fl. 16.iii.1981, *Nyariri* 898 (PRE; SRGH). C: Harare Distr., Kilworth, o. fr. 2.ii.1946, *Wild* 762 (K; SRGH). E: 5 km north of Nyanga (Inyanga), female fl. & o. fr. 25.xi.1930, *Fries, Norlindh & Weimarck* 3216 (K; LD). **Malawi**. C: Dedza Distr., Kangoli (Kanjoli) Hill, male fl. & fr. 17.i.1967, *Salubeni* 498 (K; LISC; SRGH). S: Likhubula Valley, male fl. & fr. 9.iv.1986, *J.D. Chapman & E.J. Chapman* 7395 (K; MO). **Mozambique**. T: Macanga, Furancungo, o. fr. 15.v.1948, *Mendonça* 4241 (LISC). M: Namaacha, o. fr. ii.1931, *Gomes e Sousa* 429 (K).
Also in West Africa from Senegal to Côte d'Ivoire (*P. alpestris*), and in central and South Africa, from Central African Republic south to South Africa (*P. glaucophyllus* s.str.). Plateau miombo and *Uapaca* woodlands, and in submontane miombo woodland and grassland, also in dambos and at stream sides; 600–1650 m.

11. **Phyllanthus martinii** Radcl.-Sm. in Kew Bull. **51**, 2: 311 (1996). Type: Zambia, Barotseland, without precise locality, fl. & fr. 28.ii.1933, *Martin* 575/33 pro parte (FHO, holotype).

An erect annual herb to c. 30 cm tall, monoecious, completely glabrous. Lead shoots very slightly asperulous; lateral shoots up to 11 cm long, quite smooth, later co-axillary with secondary lead shoots. Scale leaves 0.75 mm long, subulate; stipules 0.75–1 mm long, triangular-lanceolate. Foliage leaves distichous; petioles 0.5–0.75 mm long; stipules variable, those on the primary lateral shoots resembling those of the scale leaves, those on the secondary shoots c. 0.5 mm long, narrowly lanceolate; leaf blades up to 13 × 7 mm, ovate to elliptic-ovate, obtuse, rounded to rounded-cuneate at the base, membranous, medium green above, paler beneath; lateral nerves in 5–6 pairs, not prominent above, slightly so beneath. Male flowers 2–3 per fascicle, accompanied by a female flower, or not; pedicels less than 1 mm long, capillary; sepals 5, 0.4 × 0.5 mm, broadly ovate-suborbicular, pale yellowish, translucent, with a faint midrib; disk glands 5, free, circular, flat, petaloid; stamens 3, free, filaments 0.25 mm long, suberect-spreading, anthers 0.1 mm across, ± horizontally held, transversely dehiscent. Female flowers either solitary or else accompanied by male flowers; pedicels 1 mm long, extending to 2 mm in fruit, stouter than in the male; sepals 5, 0.7 × 0.3 mm, scarcely accrescent in fruit, triangular-ovate, white with a broad reddish-green midrib; disk c. 0.5 mm in diameter, shallowly cupular, roundly pentagonal; ovary 0.5 mm in diameter, sessile, ± globose, smooth; styles 3, 0.2 mm long, ± free, appressed to the top of the ovary, bifid, stigmas slightly decurved. Fruit 1 × 1.75–2 mm, rounded-3-lobed, strongly depressed, smooth. Seeds 0.8 × 0.75 × 0.7 mm, segmentiform, light brown, with 8 rows of tubercles on the dorsal facet, and 6–7 concentric rows of tubercles on each ventral facet.

Zambia. B: Barotseland, without precise locality, fl. & fr. 28.ii.1933, *Martin* 575/33 pro parte (FHO).

Known only from the type. In understorey thicket in *Baikiaea* forest on Kalahari Sand (mutemwa), especially on old lines.

Superficially resembles *P. tenellus*, but the bisexual fascicles and especially the androecium serve to set it apart.

12. **Phyllanthus nummulariifolius** Poir., Encycl. Méth. Bot. **5**: 302 (1804). —Müller Argoviensis in De Candolle, Prodr. **15**, 2: 337 (1866). —Engler, Pflanzenw. Ost-Afrikas **C**: 236 (1895). —Hutchinson in F.T.A. **6**, 1: 710 (1912); in F.C. **5**, 2: 392 (1920). —Eyles in Trans. Roy. Soc. South Africa **5**: 393 (1916). —Engler, Pflanzenw. Afrikas (Veg. Erde 9) **3**, 2: 25 (1921). —Burtt Davy, Fl. Pl. Ferns Transvaal: 299 (1932). —Robyns & Tournay, Fl. Sperm. Parc Nat. Alb. **1**: 442 (1948). —Brenan, Check-list For. Trees Shrubs Tang. Terr.: 223 (1949). —F.W. Andrews, Fl. Pl. Anglo-Egypt. Sudan **2**: 90 (1952). —Brenan in Mem. N.Y. Bot. Gard. **9**, 1: 68 (1954). —Troupin, Fl. Pl. Lign. Rwanda: 267, fig. 90/2 (1982); Fl. Rwanda **2**: 236, fig. 71/3 (1983). —Radcliffe-Smith in F.T.E.A., Euphorb. 1: 28 (1987). Type from Madagascar.

An erect or semiscandent annual or perennial herb, or weak-stemmed sparingly-branched subshrub up to 3 m tall, but commonly much less, monoecious or dioecious; stem often reddish at the base with brownish-green or purplish-brown branches. Lead shoots and lateral shoots glabrous, scaberulous, papillose or evenly to densely whitish-puberulous, the hairs multicellular. Lateral leafy shoots (5)10–15(25) cm long, later co-axillary with short shoots and secondary lead shoots. Scale leaves c. 2 mm long, linear-subulate; stipules c. 2.5 mm long, narrowly lanceolate. Foliage leaves distichous; petioles 0.5–1 mm long; stipules 1–1.5 mm long, linear-lanceolate, reddish-brown with hyaline margins; blades 0.2–2.3 × 0.1–1.4 cm, suborbicular, obovate or elliptic, subacute to rounded or truncate, rounded to cuneate at the base, membranous to chartaceous, usually glabrous above and beneath, rarely somewhat papillose-puberulous along the midrib and main nerves, light yellow-green to bright blue-green or dark grey-green above, paler and somewhat glaucous beneath, margins slightly recurved and sometimes purplish-tinged; lateral nerves in 4–12 pairs, not prominent above, scarcely so beneath. Male flowers in few-flowered fascicles, usually with 1 female flower per fascicle in the more distal axils of the lateral shoots. Male flowers: pedicels 5–6 mm long, capillary; sepals 5, 1 × 1 mm, suborbicular-obovate, cream-coloured or whitish with a green midrib, sometimes pink-tinged; disk glands 5, minute, free, flat, petaloid; stamens 5, free, filaments 0.75 mm long, erect, anthers minute, horizontal, transversely dehiscent, yellow. Female flowers: pedicels (5)7–20 mm long, capillary, red; sepals 5(6), 1 × 0.75 mm, elliptic-ovate, yellowish-green, often reddish- or brownish-tinged; disk 0.75 mm in diameter, annular, flat, entire; ovary 1 mm in diameter, sessile, depressed-subglobose, smooth; styles 3, 0.5 mm long, free, spreading, bifid, stigmas curved. Fruit 1 × 2 mm, depressed 3-lobed to subglobose, smooth, pale green, sometimes reddish-tinged. Seeds 0.8 × 0.6–0.7 mm, segmentiform, light brown, with 10–12 rows of minute tubercles on the dorsal facet, and 9–10 concentric rows of tubercles on each ventral facet.

Young lateral shoots and/or lead shoots glabrous or minutely scaberulous · var. *nummulariifolius*
Young lead shoots and/or lateral shoots evenly to densely whitish-puberulous, the hairs multicellular · var. *capillaris*

Var. **nummulariifolius**

Phyllanthus tanzaniensis Jean F. Brunel, Phyllanthus Afr. Intertrop. Mad.: 305 (1987), unpublished thesis.

Young lateral shoots and/or lead shoots glabrous or minutely scaberulous.

Zambia. B: Mongu, o. fr. 14.iii.1966, *Robinson* 6882 (K; SRGH). N: Mansa Distr., Samfya, L. Bangweulu, fr. 8.ii.1959, *Watmough* 227 (K; LISC; PRE; SRGH). W: Kitwe, fr. 29.iii.1969, *Mutimushi* 3019 (K; NDO; PRE). C: Lusaka, fl. 5.ii.1965, *Fanshawe* 9146 (K; NDO; SRGH). E: Chipata (Fort Jameson), fr. 6.vi.1954, *Robinson* 840 (K; SRGH). S: Victoria Falls, o. fr. 1909, *Monro* 417 (BM). **Zimbabwe**. N: Mazowe (Mazoe), fr. iii.1906, *Eyles* 292 (BM; SRGH). C: Harare (Salisbury), male fl. & o. fr. 13.iii.1984, *Bayliss* in *GHS* 10111 (MO; PRE). E: Mutare Distr., Zimunya's Res., fl. & fr. 8.i.1956, *Chase* 5940 (BM; BR; K; LISC; PRE; SRGH). S: Runde (Lundi) R.–Beitbridge, fl. & o. fr. 15.ii.1955, *Exell, Mendonça & Wild* 371 (LISC; SRGH). **Malawi**. N: Rumphi Distr., Livingstonia Escarpment, fr. 28.xii.1976, *Pawek* 12060 (DAV; K; MAL; MO; SRGH; UC). C: Namitete R., Lilongwe to Chipata (Fort Jameson), o. fr. 5.ii.1959, *Robson* 1460 (BM; K; LISC; PRE; SRGH). S: Mulanje Mt., Luchenya Plateau, fl. & fr. 25.vi.1946, *Brass* 16427 (BM; BR; K; NY; PRE; SRGH). **Mozambique**. N: Eráti, Namapa, R. Lúrio, fr. 29.iii.1961, *Balsinhas & Marrime* 326 (BM; COI; K; LISC; LMA). Z: Morrumbala Mt., fr. 10.xii.1971, *Pope & Müller* 579 (K; LISC; SRGH). T: Macanga, Mt. Furancungo, fr. 15.iii.1966, *Pereira, Sarmento & Marques* 1759 (BR; LMU). MS: Chimanimani Mt., fl. & y. fr. 31.v.1969, *Müller* 1224 (K; LISC; MO; SRGH). GI: Mangorro–Panda, o. fr. 7.iv.1959, *Barbosa & Lemos* in *Barbosa* 8527 (K; LISC; LMA).

Also in the Sudan, Zaire, Rwanda, Burundi, Uganda, Kenya, Tanzania, Angola and South Africa (Transvaal, KwaZulu-Natal), and in Madagascar, the Mascarenes and the Seychelles. Submontane woodland usually on rocky slopes, and montane grassland often in marshy ground beside streams, also on evergreen rainforest floor, in the understorey and on the margins and in clearings, and in dense riverine forest, also in high rainfall miombo usually on hillsides, and in moist dambos and grasslands often appearing after an annual burn; 5–2100 m.

Var. **capillaris** (Schumach. & Thonn.) Radcl.-Sm. in Kew Bull. **51**, 2: 316 (1996). Types from Ghana.

Phyllanthus capillaris Schumach. & Thonn., Beskr. Guin. Pl.: 417 (1827); in Dansk. Vid. Selsk. Skr. **4**: 191 (1829). —Müller Argoviensis in De Candolle, Prodr. **15**, 2: 338 (1866). —Engler, Pflanzenw. Ost-Afrikas **C**: 236 (1895). —Hutchinson in F.T.A. **6**, 1: 709 (1912). —Engler, Pflanzenw. Afrikas (Veg. Erde 9) **3**, 2: 25 (1921). —De Wildeman, Pl. Bequaert. **3**, 4: 442 (1926). —Robyns & Tournay, Fl. Sperm. Parc Nat. Alb. **1**: 442 (1948). —Brenan, Check-list For. Trees Shrubs Tang. Terr.: 222 (1949). —F.W. Andrews, Fl. Pl. Anglo-Egypt. Sudan **2**: 90 (1952). —Keay in F.W.T.A., ed. 2, **1**, 2: 387 (1958). —Agnew, Upl. Kenya Wild Fls.: 212 (1974).

Phyllanthus stuhlmannii Pax in Engler, Pflanzenw. Ost-Afrikas **C**: 236 (1895). Type from Uganda.

Young lead shoots and/or lateral shoots evenly to densely whitish puberulous, the hairs multicellular.

Zambia. W: Solwezi, o. fr. 15.v.1969, *Mutimushi* 3337 (K; NDO). **Zimbabwe**. C: Goromonzi, male fl. & o. fr. 17.iv.1927, *Eyles* 4885 (K; SRGH). S: Runde (Lundi) R.–Beitbridge, fl. & fr. 15.ii.1955, *Exell, Mendonça & Wild* 371 (BM). **Malawi**. N: Chitipa Distr., Misuku Hills, Matipa For., fl. & fr. 27.xii.1977, *Pawek* 13419 (DAV; K; MAL; MO). **Mozambique**. N: 25 km Marrupa–Lichinga, y. fr. 10.viii.1981, *Jansen, de Koning & de Wilde* 166 (K; WAG). Z: Quelimane, Mocuba, Namagoa, fr.

1945, *Faulkner* 371 (K; PRE; SRGH). MS: Sofala Prov. (Beira Distr.), Gorongosa Nat. Park, y. fr. iii.1972, *Tinley* 2473 (K; LISC; SRGH).

From Guinée eastwards to Uganda and south to Angola and Mozambique. Plateau miombo, evergreen forest margins, in marshy ground and tall grasses beside streams, and in dambos; 60–1920 m.

The gatherings from central Zimbabwe and Mozambique cited above are somewhat intermediate in the indumentum character between the two varieties. *Exell, Mendonça & Wild* 371 is a mixture: the LISC and SRGH duplicates of this number are referable to the typical variety, whereas the BM duplicate represents var. *capillaris*.

13. **Phyllanthus tenellus** Roxb., Fl. Ind., ed. 2, **3**: 668 (1832). —Müller Argoviensis in De Candolle, Prodr. **15**, 2: 338 (1866). —G.L. Webster in J. Arnold Arbor. **38**: 52 (1957). —Coode in Fl. Masc., fam. 160, Euphorb.: 26 (1982). Type cultivated in Calcutta, from seed collected in the Mascarenes.

Very like *Phyllanthus nummulariifolius*, but always an annual herb, with leaves elliptic-obovate to elliptic-oblanceolate, never truncate at the apex, the margins flat; male flowers minute (pedicels to 1.5 mm long, sepals 0.5 × 0.5 mm), and female pedicels not more than 8 mm long, often with 2 female flowers per axil.

Mozambique. M: Maputo, Jardim Tunduru (Jardim Vasco da Gama), male fl. & fr. 24.x.1971, *Balsinhas* 2250 (K; LISC; SRGH).

Native of the Mascarenes, and introduced into southern Arabia, India, Mozambique, SE USA, West Indies and Brazil. In alluvial sandy flats by riverbanks, as a weed of cultivation and a roadside ruderal; sea level to 200 m.

Material from Malawi and Zimbabwe has been referred to this taxon, but the material is insufficient for this to be confirmed. *Barbosa* 7631 (LMA), Inhaca, 10.vii.1957, resembles *P. tenellus* but is a suffrutex with bisexual cymes.

14. **Phyllanthus graminicola** Hutch. ex S. Moore in J. Linn. Soc., Bot. **40**: 191 (1911). —Hutchinson in F.T.A. **6**, 1: 708 (1912). —R.E. Fries, Wiss. Ergebn. Schwed. Rhod.-Kongo-Exped. **1**, 1: 120 (1914). —Eyles in Trans. Roy. Soc. South Africa **5**: 393 (1916). —Suessenguth & Merxmüller, Contrib. Fl. Marandellas Distr. **43**: 85 (1951). Type: Zimbabwe, Eastern Province, Chirinda, male fl. 7.ii.1906, *Swynnerton* 261 (K, holotype; SRGH).

Phyllanthus rogersii Hutch. in J. Bot. **57**: 160 (1919). —Burtt Davy, Fl. Pl. Ferns Transvaal: 299 (1932). Type from South Africa (Transvaal).

Phyllanthus sofalaënsis Jean F. Brunel, on herbarium sheets only.

A perennial herb up to 35 cm tall, dioecious or rarely monoecious; stems several, clustered, arising from a vertical thickened woody rootstock. Lead shoots and lateral shoots smooth or minutely asperulous, terete, glabrous. Lateral shoots 4–12(20) cm long, usually subtended by foliage leaves, ± vertically oriented. Foliage leaves of the lateral shoots usually ± distichous; petioles 0.3–0.5 mm long; stipules 1–1.5(2) cm long, linear-lanceolate, subentire, pinkish; blades 0.3–1.5 × 0.1–0.5 cm, elliptic-ovate to -lanceolate, acute or subacute, rarely obtuse, usually rounded at the base, margin somewhat thickened, revolute, chartaceous, glabrous or sometimes minutely scaberulous along the midrib and nerves beneath; lateral nerves in 2–6(7) pairs, looped within the margin, not prominent above, fairly prominent beneath. Male flowers in 1–3-flowered axillary cymules; pedicels 3–3.5 mm long, capillary; sepals 5, 1.5 × 1.25 mm, obovate, white or pinkish with a red midrib; disk glands 5, 0.2–0.3 mm across, free, flat, thin, smooth, transversely ovate, ± truncate; stamens 5, united at the base, c. 1 mm long, the free parts of the filaments c. 0.6 mm long, erect, anthers 0.25 mm across, horizontal, transversely dehiscent. Female flowers solitary; pedicels 2 mm long, extending to 4–5 mm in fruit, slender, expanding apically; sepals 5(6), 1–1.5 × 0.75–1 mm, slightly accrescent in fruit, elliptic, reddish-pink with whitish margins and a green midrib; disk 0.8–1 mm in diameter, annular, thin, undulate; ovary 0.7 mm in diameter, sessile, depressed-subglobose, smooth; styles 3, 0.5 mm long, united at the base, spreading, bifid, stigmas recurved. Fruit c. 2 × 3 mm, depressed 3-lobed, smooth, often reddish-tinged. Seeds 1.3 × 1 × 1 mm, segmentiform, yellowish-brown, with c. 15 rows of minute shiny brown tubercles reticulately connected on the dorsal facet, and c. 12 concentric rows on each ventral facet.

Zimbabwe. N: Guruve Distr., Great Dyke, Nyamunyeche Estate, male fl. 15.xii.1978, *Nyariri* 586 (SRGH). W: Matobo Distr., Besna Kobila Farm, male fl. & fr. xi.1956, *O.B. Miller* 3870 (K;

SRGH). C: 22.5 km Harare to Domboshawa, male & female fl. 26.x.1955, *Drummond* 4880 (K; LISC; PRE; SRGH). E: Chimanimani Distr., Martin For. Res., male & o. fr. 1.x.1966, *Simon* 910 (K; PRE; SRGH). **Mozambique**. MS: Sofala Province, 204 km Mambone–Nova Lusitânia, male fl. 9.x.1965, *Torre & Pereira* 12353 (LISC). GI: Mangola Mts., male fl. no date, *Vasse* 19 (K; PRE).

Also in South Africa (Transvaal). In plateau and wooded grasslands, also in submontane grassland and mist forest boundaries, rare in *Hymenocardia* chipya, often appearing after a fire; (40)1150–2100 m.

Angus 876, *Mutimushi* 829, *Richards* 1991, 21519, 22325 and *Sanane* 273, all from northern Zambia, are intermediate between this species and *P. nummulariifolius* Poir.

The BM duplicate of *Swynnerton* 261 is referable to *P. glaucophyllus* Sond.

15. **Phyllanthus tsetserrae** Jean F. Brunel ex Radcl.-Sm. in Kew Bull. **51**, 2: 325 (1996). Tab. **9**, figs. A1 & A2. Type: Mozambique, Manica e Sofala Prov., Tsetserra, male & female fl. & o. fr.-cal. 9.ii.1955, *Exell, Mendonça & Wild* 325 (SRGH, holotype; BM; LISC).

Phyllanthus tsetserrae Jean F. Brunel, Phyllanthus Afr. Intertrop. Mad.: 310 (1987), with respect to specimen, *Exell, Mendonça & Wild* 325, but not the description (which was based on *Exell, Mendonça & Wild* 290), name invalid.

Like *Phyllanthus graminicola*, but with suborbicular leaves shallowly cordate at the base.

Mozambique. MS: Tsetserra, male, female fl. & o. fr.-cal. 9.ii.1955, *Exell, Mendonça & Wild* 325 (SRGH, holotype; BM; LISC).

Not known elsewhere. In montane grassland; 1830 m.

In his doctoral thesis, Brunel has confused two very distinct new species from the Tsetserra Mountains, both collected by Exell, Mendonça & Wild, on successive days. The epithet '*tsetserrae*' is, however, clearly attached to *Exell, Mendonça & Wild* 325, although the description given for it is of *Exell, Mendonça & Wild* 290. I have rectified the situation in my paper in Kew Bulletin.

The SRGH duplicate of *Exell, Mendonça & Wild* 325 has been chosen as the holotype rather than the LISC one, since it has both male and female flowers.

16. **Phyllanthus manicaënsis** Jean F. Brunel ex Radcl.-Sm. in Kew Bull. **51**, 2: 309 (1996). Tab. **9**, figs. B1–B4. Type: Mozambique, Manica e Sofala Prov., Tsetserra, fl. & fr. 8.ii.1955, *Exell, Mendonça & Wild* 290 (LISC, holotype; BM; SRGH).

Like *Phyllanthus graminicola*, but with scale leaves at each node of the lead shoots, and with densely papillose lateral shoots, broader chaffy brown stipules, and ovate-suborbicular leaf blades with rounded apices and revolute margins.

Mozambique. MS: Serra Zuira, Tsetserra, fl. & fr. 2.iv.1966, *Torre & Correia* 15563 (LISC).

Not known elsewhere. Amongst rocks in mist forest with *Aphloia*, *Rapanea*, *Curtisia* and *Podocarpus*; 1940–2100 m.

The type designated by Brunel (the LISC duplicate of *Exell et al.* 325, loc. idem) for his *Phyllanthus tsetserrae* (unpublished), is substerile, but the description was clearly drawn up using the type of *P. manicaënsis*.

Wild 2970 from the Chimanimani Mts., Zimbabwe, at 1525 m in burnt grassland, is intermediate between this species, *P. tsetserrae* and *P. graminicola*.

17. **Phyllanthus loandensis** Welw. ex Müll. Arg. in J. Bot. **2**: 329 (1864); in De Candolle, Prodr. **15**, 2: 342 (1866). —Hutchinson in F.T.A. **6**, 1: 702 (1912). —Engler, Pflanzenw. Afrikas (Veg. Erde 9) **3**, 2: 22 (1921). Type from Angola (Luanda Province).

Phyllanthus angolensis sensu Radcl.-Sm. in F.T.E.A., Euphorb. 1: 30 (1987), non Müll. Arg.

An erect, much-branched annual or perennial herb, subshrub or shrub up to 90 cm tall, monoecious or dioecious; stems and branches strict, tough wiry. Lead shoots and lateral shoots smooth, angular, glabrous. Lateral shoots 1–7(10) cm long, divaricate, usually subtended by foliage leaves on the young lead shoots, but densely fasciculate on the short shoots of older plants, subtended by scale leaves. Scale leaves 2 mm long, subulate; stipules 2 × 1 mm, triangular-lanceolate, dark purplish-brown. Foliage leaves of the lateral shoots strongly distichous; petioles c. 0.3 mm long; stipules 0.5–1 mm long, subulate-filiform to narrowly linear-lanceolate, reddish-purple with white margins; blades 2–7 × 1–4 mm, elliptic, acute or subacute, often mucronulate, cuneate to rounded at the base, chartaceous, glabrous above and beneath, light green above, paler beneath; lateral nerves in 2–6

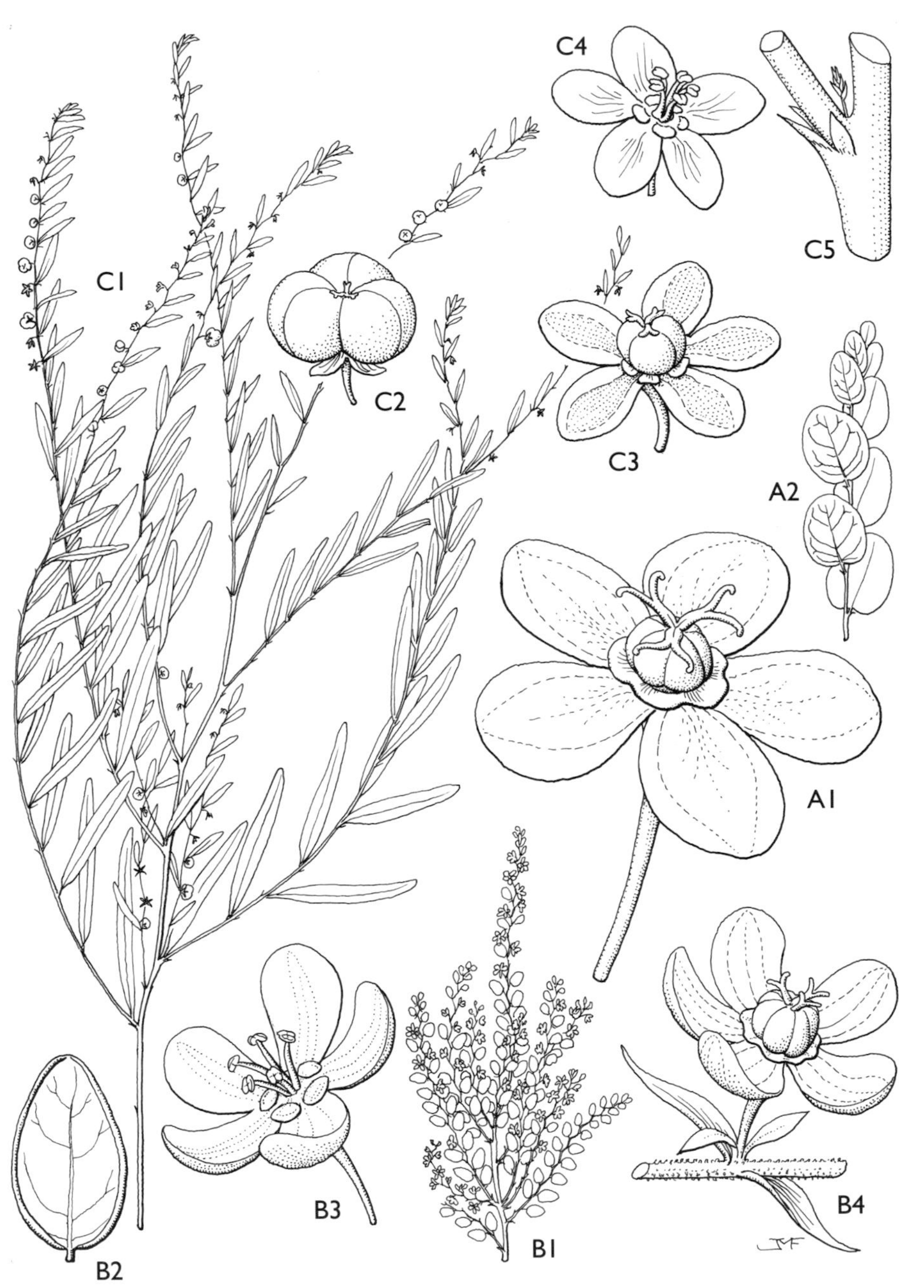

Tab. 9. A —PHYLLANTHUS TSETSERRAE. A1, female flower (× 16); A2, shoot (× 1), A1 & A2 from *Exell et al.* 325. B —PHYLLANTHUS MANICAËNSIS. B1, habit (× 1); B2, underside of leaf (× 6); B3, male flower (× 16); B4, female flower (× 16), B1–B4 from *Exell et al.* 290. C —PHYLLANTHUS MENDESII. C1, habit (× 1), from *Rushworth* 1433; C2, fruit (× 10); C3, female flower (× 16), C2 & C3 from *Robinson* 6794; C4, male flower (× 16), from *White* 2045; C5, cataphyll and stipules (× 20), from *Rushworth* 1433. Drawn by J.M. Fothergill.

pairs, usually scarcely visible, looped within the margin. Male flowers in 1–3-flowered cymules in the median axils of the lateral shoots, female flowers solitary in the distal axils, sometimes with a bisexual cymule in between. Male flowers: pedicels 1.5 mm long, capillary; sepals 5, 1.2 × 0.8 mm, obovate, greenish or whitish with a red midrib; disk glands 5, 0.25 mm across, free, transversely ovate, thin, flat, smooth, ± truncate; stamens (4)5, free, filaments c. 0.75–1 mm long, erect, anthers 0.2 mm across, inclined, obliquely dehiscent, yellow. Female flowers: pedicels 1 mm long, slender, extending to 2 mm in fruit, smooth, glabrous; sepals 5, 1–1.3 × 0.5–0.8 mm, ovate-lanceolate to lanceolate, pinkish-white with a pinkish-green midrib; disk 0.5 mm in diameter, annular, thin, undulate; ovary 0.7 mm in diameter, sessile, globose, smooth, green; styles 3, 0.5 mm long, united at the base, ± erect, bifid, stigmas recurved. Fruit 1.5 × 2.5 mm, depressed 3-lobed, smooth, glaucous-green, olive-green when dried. Seeds 1.3 × 1 × 1 mm, segmentiform, reddish-brown, with c. 20 rows of minute tubercles on the dorsal facet, and c. 17 concentric arcs on each ventral facet.

Zambia. B: Kaoma (Mankoya), Luampa Mission, o. fr. 20.ii.1952, *White* 2101 (FHO; K). N: Chipili, o. fr. 14.vi.1957, *Robinson* 2252 (K; SRGH). W: Mwinilunga Distr., R. Luao, fl. 18.xii.1937, *Milne-Redhead* 3725 (BM; K; LISC). C: Great East Rd., Undaunda to Rufunza, male fl. & fr. 14.i.1972, *Kornaś* 847 (K). **Zimbabwe**. N: Shavanhohwe (Shanawe) R., o. fr. 2.i.1937, *Eyles* 8907 (K; SRGH). C: Kadoma Distr., Umniati R., 40 km from Kwe Kwe (Que Que) to Kadoma (Gatooma), fl. 31.iv.1972, *Grosvenor* 759 (SRGH). E: Nyumquarara Valley, fr. ii.1935, *Gilliland* P1673 (BM; K). **Malawi**. N: 21 km from Mzuzu to Ekwendeni, fr. 3.iii.1977, *Pawek* 12361 (DAV; K; MAL; MO; SRGH; UC). C: Lilongwe Distr., Dzalanyama For. Res., 5 km SE of Choulongwe Falls, fr. 22.iii.1970, *Brummitt* 9318 (K; LISC; MAL; PRE; SRGH; UPS). **Mozambique**. N: 20 km Marrupa to Lichinga, o. fr. 16.ii.1981, *Nuvunga* 526 (K; LMU; MO; SRGH).

Also in Zaire, Burundi, Tanzania, Angola, Namibia and South Africa (Transvaal). Riverine vegetation amongst rocks and on dry rocky hills, in *Brachystegia* (miombo) woodland, short grassland and *Cryptosepalum* woodland on Kalahari Sand; 1067–1870 m.

18. **Phyllanthus angolensis** Müll. Arg. in J. Bot. **2**: 329 (1864); in De Candolle, Prodr. **15**, 2: 339 (1866). —Hutchinson in F.T.A. **6**, 1: 712 (1912). —Engler, Pflanzenw. Afrikas (Veg. Erde 9) **3**, 2: 25 (1921). Type from Angola (Huíla).

An erect or procumbent annual or perennial herb or subshrub up to 60 cm tall, monoecious, later developing a vertical woody rootstock. Lead shoots somewhat scabrulous, subterete; lateral shoots up to 8 cm long, terete, scaberulous-papillose. Scale leaves 3 mm long, linear-filiform; stipules 3.5–4 mm long, narrowly lanceolate, irregularly denticulate, dark purplish-brown in the upper half. Foliage leaves distichous; petioles 0.7 mm long; stipules c. 2–2.5 mm long, linear-filiform, pinkish with white margins, darkened at the tips; blades 3–12 × 1–7 mm, broadly elliptic to elliptic-lanceolate, acute, subacute or obtuse, cuneate to rounded at the base, thinly chartaceous, sparingly to evenly papillose above, more densely so beneath, mid- to pale green above, glaucescent beneath; lateral nerves in 2–5 pairs, often scarcely visible, arcuate. Male flowers single or geminate in the lowest axils of the lateral shoots, female solitary in the distal axils. Male flowers: pedicels c. 1 mm long, capillary; sepals 5, 1–1.2 × 0.8–1 mm, elliptic-obovate, cream-coloured with a pinkish midrib, sometimes pinkish-flecked; disk glands 5, 0.3 mm across, free, transversely ovate, thin, flat, smooth, ± truncate; stamens (4)5, ± free, filaments 0.75 mm long, erect, anthers 0.25 mm across, inclined or horizontal, obliquely or laterally dehiscent. Female flowers: pedicels 1.5 mm long, extending to 2–3 mm in fruit, scaberulous to minutely papillose; sepals 5, 1.2 × 0.8 mm, slightly accrescent in fruit, elliptic to elliptic-oblong, reddish-green with whitish margins; disk 0.6 mm in diameter, annular, thin, undulate; ovary 0.7 mm in diameter, sessile, 3-lobed to subglobose, smooth; styles 3, 0.5 mm long, free, suberect, bifid, stigmas recurved. Fruit 2 × 3–3.5 mm, depressed 3-lobed to subglobose, ± smooth to faintly venose, glabrous, yellowish-green to dark purplish-brown. Columella 1 mm long. Seeds 1.5 × 1.2 × 1.2 mm, segmentiform, dark reddish-brown, with c. 25 rows of minute somewhat darker tubercles on the dorsal facet, and c. 25 concentric arcs on each ventral facet.

Zambia. W: Kitwe, o. fr. 9.iii.1967, *Fanshawe* 9954 (K; NDO; SRGH). S: Mazabuka Distr., Choma West For. Res., fr. 28.i.1960, *White* 6526 bis (FHO; K). **Malawi**. N: Mzimba Distr., Mzuzu,

Marymount, o. fr. 19.iv.1977, *Pawek* 12602 (DAV; K; MAL; MO; PRE; SRGH; UC). C: Dedza Distr., Chongoni Forest Station, fr. 22.ii.1966, *Jeke* 29 (K; MAL; SRGH).

Also in Angola. On laterite outcrops and sandy soil amongst sparse grasses, in wooded grassland, and in miombo and *Uapaca kirkiana* woodlands; 1370 m.

19. **Phyllanthus parvulus** Sond. in Linnaea **23**: 132 (1850). —Hutchinson in F.C. **5**, 2: 391 (1920). —Burtt Davy, Fl. Pl. Ferns Transvaal: 299 (1932). Type from South Africa (Transvaal).

An erect, often much-branched, annual to subperennial herb to c. 50 cm tall, monoecious. Lead shoots ± smooth, lateral shoots smooth or finely scabrid to papillose-puberulous. Lateral leafy shoots up to 27 cm long, but most often c. 5–10 cm long. Scale leaves c. 1 mm long, subulate-filiform; stipules c. 1 mm long, narrowly triangular-lanceolate. Foliage leaves distichous; petioles 0.3–0.5 mm long; stipules c. 1 mm long, linear-lanceolate, buff coloured; blades 3–15 × 1–4 mm, elliptic to elliptic-lanceolate, acute, subacute, obtuse or rounded at the apex, cuneate to rounded at the base, membranaceous, sometimes minutely scaberulous along the midrib beneath, otherwise glabrous, dull green to glaucous; lateral nerves in 3–6 pairs, not conspicuous. Flowers commonly in few-flowered bisexual cymes, with 1 male and 1 female flower per axil. Male flowers: pedicels c. 1 mm long, capillary; sepals 5, 0.6–0.8 × 0.5–0.6 mm, ovate-suborbicular, somewhat mucronulate, cream-coloured with a thin pinkish midrib; disk glands 5, minute, free, flat, smooth, transversely ovate, ± truncate; stamens 5, free, filaments 0.2 mm long, erect, anthers 0.2 mm across, horizontal, transversely dehiscent. Female flowers: pedicels 0.5 mm long, extending to c. 2 mm in fruit, slender, sometimes scaberulous; sepals ± as in the male, but with a broader reddish-green midrib; disk 0.5 mm in diameter, annular, flat, thin, entire; ovary 0.5 mm in diameter, sessile, 3-lobed to subglobose, ± smooth; styles 3, 0.25 mm long, ± free, spreading, bifid, stigmas straight. Fruit 1 × 2 mm, depressed 3-lobed to subglobose, ± smooth or minutely scaberulous-papillose, pale green, often reddish-tinged. Seeds 0.8 × 0.7 × 0.6 mm, segmentiform, light brown, with 6–8 rows of minute slightly darker tubercles connected by a lattice of ridges on the dorsal facet, and 5–6 concentric rows on each ventral facet.

Lateral shoots finely scabrid to papillose-puberulous · var. *parvulus*
Lateral shoots smooth, glabrous · var. *garipensis*

Var. **parvulus**

Phyllanthus tenellus var. *scabrifolius* Müll. Arg. in Linnaea **32**: 7 (1863). Type as above.

Phyllanthus tenellus var. *parvulus* (Sond.) Müll. Arg. in De Candolle, Prodr. **15**, 2: 339 (1866).

Phyllanthus mozambicensis Gandoger in Bull. Soc. Bot. Fr. **66**: 287 (1920). Type: Mozambique, Maputo, Ponta Vermelha, iv.1893, *Quintas* 59 (COI, holotype).

Phyllanthus humilis sensu Hutchinson in F.C. **5**, 2: 392 (1920). —Burtt Davy, Fl. Pl. Ferns Transvaal: 299 (1932), non Pax.

Phyllanthus burchellii sensu Bremekamp & Obermeyer in Ann. Transvaal Mus. **16**, 3: 421 (1935), non Müll. Arg.

Phyllanthus seydelii Jean F. Brunel, Phyllanthus Afr. Intertrop. Mad.: 313 (1987), unpublished thesis.

Lateral shoots finely scabrid to papillose-puberulous.

Botswana. N: Aha Hills, fr. 13.iii.1965, *Wild & Drummond* 6956 (K; LISC; SRGH). SW: Kaotwe, fr. 8.iv.1930, *van Son* in *Tvl. Mus. Herb.* 28836 (BM; K; PRE). SE: Orapa, fr. 16.iv.1971, *Pope* 312 (K; PRE; SRGH). **Zimbabwe**. W: Maitengwe river banks and islands, fr. recd. v.1883, *Holub* s.n. (K). C: Gweru Distr., 29 km SSE of Kwe Kwe (Que Que), fr. 3.ii.1966, *Biegel* 878 (K; MO; SRGH). E: Odzi, o. fr. 28.v.1936, *Eyles* 8625 (K; SRGH). S: Mwenezi Distr., Malangwe R., fr. 6.v.1958, *Drummond* 5615 (K; LISC; PRE; SRGH). **Mozambique**. M: Marracuene, fr. 3.x.1957, *Barbosa & Lemos* in *Barbosa* 7966 (COI; K; LMA).

Also in Zaire, Namibia and South Africa (Transvaal). On Kalahari Sands in low rainfall dry deciduous woodlands and short grasslands with scattered trees and shrubs, and in mopane woodlands on sand, also in pans on dry sandy river banks amongst stones, and on limestone outcrops; 400–1525 m.

Vernacular name as recorded in specimen data: "nkanga" (Diriko).

Var. **garipensis** (E. Mey. ex Drège) Radcl.-Sm. in Kew Bull. **51**, 2: 317 (1996). Type from South Africa (Cape Province).

Phyllanthus garipensis E. Mey. ex Drège, Zwei Pflanzengeogr. Dokum.: 93 (1843), nom. nud.

Phyllanthus burchellii Müll. Arg. in Linnaea **32**: 7 (1863); in De Candolle, Prodr. **15**, 2: 340 (1866). —Hutchinson in F.C. **5**, 2: 394 (1920). —Burtt Davy, Fl. Pl. Ferns Transvaal: 299 (1932). Type from South Africa (Cape Province).

Phyllanthus tenellus var. *natalensis* Müll. Arg. in Linnaea **32**: 7 (1863); in De Candolle, Prodr. **15**, 2: 338 (1866). Type from South Africa (KwaZulu-Natal).

Phyllanthus tenellus var. *garipensis* (E. Mey. ex Drège) Müll. Arg. in Linnaea **32**: 7 (1863); in De Candolle, Prodr. **15**, 2: 339 (1866).

Phyllanthus tenellus var. *exiguus* Müll. Arg. in Linnaea **32**: 7 (1863); in De Candolle, Prodr. **15**, 2: 339 (1866). Type from South Africa (Cape Province).

Lateral shoots smooth, glabrous.

Zambia. S: Katombora, fl. & fr. 3.iv.1956, *Robinson* 1413 (K). **Mozambique**. M: Inhaca, road to Cabo Mponduíne, male fl. & o. fr. 10.vii.1957, *Barbosa* 7631 (LMA; PRE; SRGH).

Also in South Africa (Cape Province, Transvaal, KwaZulu-Natal). In low altitude hot dry scrubland and *Themeda triandra* grassland with scattered trees, also on fixed coastal sand dunes; sea level to 900 m.

20. **Phyllanthus pentandrus** Schumach. & Thonn., Beskr. Guin. Pl.: 419 (1827); in Dansk. Vid. Selsk. Skr. **4**: 193 (1829). —Klotzsch in Peters, Naturw. Reise Mossambique **6**, 1: 104 (1861). —Müller Argoviensis in De Candolle, Prodr. **15**, 2: 336 (1866). —Engler, Pflanzenw. Ost-Afrikas **C**: 236 (1895). —Schinz & Junod in Mém. Herb. Boissier, No. 10: 46 (1910). —Hutchinson in F.T.A. **6**, 1: 710 (1912). —R.E. Fries, Wiss. Ergebn. Schwed. Rhod.-Kongo-Exped. **1**, 1: 120 (1914). —Eyles in Trans. Roy. Soc. South Africa **5**: 393 (1916). —Hutchinson in F.C. **5**, 2: 393 (1920). —Engler, Pflanzenw. Afrikas (Veg. Erde 9) **3**, 2: 25 (1921). —Burtt Davy, Fl. Pl. Ferns Transvaal: 299 (1932). —F.W. Andrews, Fl. Pl. Anglo-Egypt. Sudan **2**: 90 (1952). —Keay in F.W.T.A., ed. 2, **1**, 2: 387 (1958). —P.G. Meyer in Merxmüller, Prodr. Fl. SW. Afrika, fam. 67: 39 (1967). —Radcliffe-Smith in F.T.E.A., Euphorb. 1: 31 (1987). Types from Ghana.

Phyllanthus scoparius Welw., Apont. Phytog.: 591 (1859). Types from Angola.

Phyllanthus deflexus Klotzsch in Peters, Naturw. Reise Mossambique **6**, 1: 104 (1861). Type: Mozambique, Manica e Sofala Province, Sena, *Peters* s.n. (B†, holotype).

Phyllanthus dilatatus Klotzsch in Peters, Naturw. Reise Mossambique **6**, 1: 106 (1861). Type: locality and details as for *Phyllanthus deflexus*.

Phyllanthus tenellus var., sensu Pax in Warburg, Kunene-Sambesi-Exped. Baum: 282 (1903), non Roxb.

A delicate, erect or decumbent glabrous or minutely scaberulous annual or subperennial herb up to 60 cm tall, monoecious; stem little-branched at first but later becoming much-branched and woody at the base. Lead shoots terete, zigzag, pale green. Lateral leafy and flowering shoots 5–20 cm long. Scale leaves 1 mm long, linear-lanceolate; stipules 1 mm long, narrowly triangular-lanceolate, often reddish-tinged. Foliage leaves: petioles c. 1 mm long; stipules resembling those of the scale leaves; blades 0.5–4.5 × 0.1–0.3 cm, linear to linear-lanceolate or occasionally elliptic-lanceolate, acute, subacute or obtuse, usually ± rounded at the base, membranaceous, smooth or minutely scaberulous along the midrib and main nerves beneath, green above, glaucous beneath, occasionally pinkish-tinged; lateral nerves in 4–6 pairs, widely spaced, the lower free, the upper looping, indistinct above, faint and scarcely prominent beneath. Flowers commonly in few-flowered bisexual cymes with 1–3 male flowers and 1 female flower per cyme, or else male flowers in unisexual cymes in the lower axils of the lateral shoots, and female flowers solitary in the upper axils. Male flowers: pedicels 0.5 mm long; sepals 5, 0.67 × 0.67 mm, obovate-suborbicular, apiculate, white, cream-coloured or greenish-white with a red or pinkish midrib; disk glands 5, free, 0.4 mm across, transversely ovate, ± truncate, thin, flat, smooth; stamens 5, 0.5 mm long, usually free, anthers 0.2 mm across, horizontally held, laterally dehiscent, yellow. Female flowers: pedicels 1–1.5 mm long, extending to 2–3(4) mm in fruit, ± smooth; sepals 5, unequal, the 3 outer slightly larger than those of the male flowers, but otherwise similar; disk 1 mm in diameter, annular, shallowly saucer-shaped, thin, smooth; ovary 0.67 mm in diameter, sessile, subglobose, smooth; styles 3, 0.33 mm long, ± free, closely appressed to the top of the ovary, shortly bifid. Fruit 1.2 × 2 mm, depressed-subglobose, smooth, olivaceous. Seeds 0.9–1 × 0.8 × 0.8 mm, segmentiform, pale yellowish-brown or greyish, shiny, with 8–10 rows of darker brown tubercles on the dorsal facet, and 7–9 concentric arcs of such tubercles on each ventral facet, the tubercles connected by a lattice of shallow ridges.

Botswana. N: Qoroqwe Island, Santantadibe R., o. fr. 22.iv.1976, *P.A. Smith* 1716 (K; MO; PRE; SRGH). **Zambia**. B: near Senanga, o. fr. 4.viii.1952, *Codd* 7385 (BM; K; PRE). N: Chemba to Cascalawa, fl. & fr. 16.ii.1960, *Richards* 12478 (K). C: Luangwa Valley Game Reserve South, o. fr. 20.i.1967, *Prince* 71 (K; LISC; SRGH). E: Chadiza, o. fr. 1.xii.1958, *Robson* 803 (BM; K; LISC). S: 72 km SE of Choma, fl. 17.xii.1956, *Robinson* 1994 (K; NDO; SRGH). **Zimbabwe**. N: Hurungwe Distr., Mensa (Mansa) Pan, 17.5 km ESE of Chirundu Bridge, y. fr. 30.i.1958, *Drummond* 5356 (K; LISC; PRE; SRGH). W: Hwange Distr., Safari Area H.Q., fr. 15.xii.1979, *Gonde* 263 (MO; SRGH). C: Chegutu Distr., Makwiro, o. fr. 24.ii.1931, *Fries, Norlindh & Weimarck* 5136 (K; LD). E: Mutare Distr., o. fr. 7.ii.1955, *Chase* 5459 (BM; K; LISC; SRGH). S: Masvingo Distr., Makoholi Exptl. Station, fr. 13.xii.1977, *Senderayi* 117 (K; SRGH). **Malawi**. N: 8 km west of Karonga, Chaminade Secondary School, fr. 15.iv.1976, *Pawek* 11059 (K; MAL; MO; PRE; SRGH; UC). C: Lilongwe Distr., Chitedze, o. fr. 22.iii.1955, *Exell, Mendonça & Wild* 1114 (BM; LISC; SRGH). S: Mangochi Distr., Monkey Bay, fl. 11.i.1980, *Masiye, Tawakali & Salubeni* 224 (MAL; MO; SRGH). **Mozambique**. N: Eráti, Namapa, near R. Lúrio, o. fr. 29.iii.1961, *Balsinhas & Marrime* in *Balsinhas* 324 (K; LISC; LMA). Z: Alto Molócuè , Mamala, fr. 20.xii.1967, *Torre & Correia* 16655 (LISC). T: Estima, near R. Sanângoè, o. fr. 28.iii.1972, *Macêdo* 5112 (K; LISC; LMA). MS: Dombe, fr. 18.xi.1965, *Pereira & Marques* 720 (BR; LMU). GI: Caniçado, Guijá (Chamusca), fr. 19.vi.1947, *Pedrógão* 359 (K; LMA). M: Marracuene–Bobole, o. fr. 31.iii.1959, *Barbosa & Lemos* in *Barbosa* 8411 (K; LISC; LMA; PRE; SRGH).

Widespread in tropical Africa from Senegal eastwards to Ethiopia and south to Namibia and South Africa (Transvaal, KwaZulu-Natal), but apparently absent from Kenya. In sandy localities, sometimes amongst stones and rocks, on sandy banks of rivers and dams and in dry river beds and lakeside dunes, in *Brachystegia* and mopane woodlands on Kalahari Sand and mutemwa on Kalahari Sand, also in miombo woodlands on rocky slopes and dry deciduous sandy woodlands, sandy grasslands and dambos, and often as a weed of cultivated and disturbed ground; 15–1463 m.

Vernacular name as recorded in specimen data: "kauluzi" (chiKunda).

Drummond 5356 from Hurungwe Distr., Zimbabwe is atypical in having most leaves broadly elliptic-ovate and a puberulous-papillose indumentum. As these features represent an extreme condition of tendencies apparent in some other gatherings, no formal recognition is accorded it.

21. **Phyllanthus mendesii** Jean F. Brunel ex Radcl.-Sm. in Kew Bull. **51**, 2: 312 (1996). Tab. **9**, figs. C1–C5. Type from Angola (Cuando-Cubango Province).

Phyllanthus mendesii Jean F. Brunel, Phyllanthus Afr. Intertrop. Mad.: 314 (1987), unpublished thesis.

An erect glabrous annual or subperennial herb to c. 80 cm tall, but more often 20–40 cm tall, monoecious. Lead shoots bearing foliage leaves at the lower 8–10 nodes, with no branches, and scale leaves at the upper nodes, subtending lateral shoots up to 22 cm long, departing from the main axis at a narrow angle. Scale leaves c. 1 mm long, narrowly lanceolate; stipules c. 1 mm long, triangular-lanceolate. Foliage leaves: petioles c. 1 mm long; stipules smaller than those of the scale leaves, but otherwise similar; leaf blades 1–4.5 × 0.5–4 mm, those of the main axis larger than those of the branches, linear, sometimes strongly curved, obtuse or subacute at the apex, attenuate at the base, pale green above, glaucous beneath; lateral nerves in 4–6 pairs, widely spaced, the lower free, the upper looping, not conspicuous above, scarcely so beneath. Flowers commonly in few-flowered bisexual cymes with 1 male and 1 female per axil. Male flowers: pedicels c. 0.33 mm long; sepals 5, 0.5 × 0.5 mm, obovate-suborbicular, white or cream-coloured; disk glands 5, free, minute, transversely ovate, truncate, thin, flat, smooth; stamens 4–5, 0.5 mm long, united at the base, anthers horizontally held, 4-lobed, laterally dehiscent. Female flowers: pedicels 0.5 mm long, extending to c. 1 mm in fruit; sepals 0.7–0.8 × 0.5–0.7 mm, ovate, green edged with white; disk 0.5 mm in diameter, 5-lobed, the lobes alternisepalous, thin, flat; ovary 0.6 mm in diameter, sessile, 3-lobed to subglobose, ± smooth; styles 3, 0.25 mm long, united at the base, closely appressed to the top of the ovary, shortly bifid. Fruit 1.2 × 2 mm, depressed 3-lobed to subglobose, ± smooth, olivaceous. Seeds 0.8 × 0.7 × 0.6 mm, segmentiform, dark reddish-brown, with 12–14 rows of almost contiguous tubercles on the dorsal facet, and c. 12 irregular concentric arcs of contiguous tubercles on each ventral facet.

Caprivi Strip. Okavango R., Shamvura, o. fr. 11.ii.1956, *de Winter & Marais* 4599 (K; PRE); Mpola, 24 km from Katima Mulilo to Ngoma, fr. 5.i.1959, *Killick & Leistner* 3301 (K; PRE; SRGH). **Botswana**. N: Movombe, fl. & fr. 14.ii.1983, *P.A. Smith* 4032 (PRE; SRGH). SW: 93 km from Ghanzi to Lobatse, fl. 31.i.1980, *Skarpe* 387 (K; PRE; SRGH). **Zambia**. B: Sesheke Distr., Sichinga Forest, fr. 29.xii.1952, *Angus* 1077 (FHO; K). **Zimbabwe**. W: Hwange (Wankie) Nat. Park, Dom Road, fr. 23.i.1969, *Rushworth* 1433 (K; LISC; SRGH).

Also in Angola and Namibia. Kalahari Sand habitats including short grasslands, mutemwa thicket, *Baikiaea* forest and open woodlands, often here as a weed of cultivated and disturbed ground; 914–1066 m.

22. **Phyllanthus maderaspatensis** L., Sp. Pl.: 982 (1753). —Müller Argoviensis in De Candolle, Prodr. **15**, 2: 362 (1866). —Engler, Pflanzenw. Ost-Afrikas **C**: 236 (1895). —Schinz & Junod in Mém. Herb. Boissier, No. 10: 46 (1900). —Hutchinson in F.T.A. **6**, 1: 722 (1912). —Eyles in Trans. Roy. Soc. South Africa **5**: 393 (1916). —Hutchinson in F.C. **5**, 2: 395 (1920). —Engler, Pflanzenw. Afrikas (Veg. Erde 9) **3**, 2: 26 (1921). —Burtt Davy, Fl. Pl. Ferns Transvaal: 299 (1932). —Bremekamp & Obermeyer in Ann. Transvaal Mus. **16**, 3: 421 (1935). —Robyns & Tournay, Fl. Sperm. Parc Nat. Alb. **1**: 442 (1948). —Brenan, Check-list For. Trees Shrubs Tang. Terr.: 223 (1949). —F.W. Andrews, Fl. Pl. Anglo-Egypt. Sudan **2**: 93 (1952). —Keay in F.W.T.A., ed. 2, **1**, 2: 388 (1958). —P.G. Meyer in Merxmüller, Prodr. Fl. SW. Afrika, fam. 67: 38 (1967). —Agnew, Upl. Kenya Wild Fls.: 212 (1974). —Troupin, Fl. Rwanda **2**: 234, fig. 71/1 (1983). —Radcliffe-Smith in F.T.E.A., Euphorb. 1: 18 (1987). Type from India.

Phyllanthus vaccinioides Klotzsch in Peters, Naturw. Reise Mossambique **6**, 1: 104 (1861). Type: Mozambique, Manica e Sofala Province, Sena, *Peters* s.n. (B†, holotype).

Phyllanthus gueinzii Müll. Arg. in Linnaea **32**: 18 (1863); in De Candolle, Prodr. **15**, 2: 363 (1866). Type from South Africa (KwaZulu-Natal).

Phyllanthus paxianus Dinter in Fedde, Repert. Spec. Nov. Regni Veg. **22**: 379 (1926). Type from Namibia.

Phyllanthus magudensis Jean F. Brunel, Phyllanthus Afr. Intertrop. Mad.: 323 (1987). Type: Mozambique, Xai Xai (Vila de João Belo), o. fr. 22.iv.1941, *Torre* 2598 (LISC, holotype).

An erect, suberect or decumbent much-branched glabrous annual or perennial herb up to 1 m tall, monoecious. Lead shoots and lateral shoots differing chiefly in leaf size, angular, often reddish- or brownish-tinged. Leaves spirally arranged; petioles (0.5)1(2) mm long; stipules 1–4 mm long, asymmetrically triangular-lanceolate, acutely acuminate, subentire, cordate-auriculate on one side at the base, whitish or pinkish cream with a median brown stripe. Leaf blades 1–5.5(7.5) × 0.2–1.5(1.8) cm on the lead shoots, 0.5–3 × 0.05–0.9 cm on the lateral shoots, linear, lanceolate, oblong, oblanceolate or obovate, acute, subacute, obtuse or rounded, rounded-cuneate to attenuate at the base, chartaceous, smooth or slightly asperulous-papillose beneath and along the marginal nerve, grey-green above, somewhat glaucous beneath, sometimes edged with red; lateral nerves in 4–10 pairs, apically-directed, sometimes looped towards the apex, not prominent above, usually prominent beneath. Flowers present in almost all axils, distal axils with male or bisexual few-flowered fascicles, proximal with solitary female flowers. Male flowers: pedicels 0.5–1 mm long; sepals 6, biseriate, the outer sepals 1.1–1.2 × 0.8 mm, obovate-spathulate, acute, the inner sepals 1.2 × 0.6 mm, oblong, rounded, yellowish-green; disk glands 6, minute, free, rounded, thin, flat, smooth; stamens 3, 1 mm long, filaments united into a short column, anthers 0.6 mm long, vertically held, apically free, thecae bilobed, longitudinally dehiscent, yellow. Female flowers: pedicels c. 2 mm long, not extending in fruit; sepals 6, biseriate, the outer accrescent to 1.8 × 1.2–1.3 mm, suborbicular-obovate, subacute, the inner sepals to 1.8 × 1–1.2 mm, somewhat spathulate, obtuse, dull green, often pinkish- or purplish-tinged, white-margined; disk glands 6, 0.3 mm across, free, squarish, thin, flat, smooth; ovary c. 1 mm in diameter, sessile, rounded-3-lobed, smooth; styles 3, 0.75 mm long, free, erect, shortly bifid. Fruit 2 × 3 mm, depressed rounded-3-lobed, smooth, somewhat shiny, olivaceous, often reddish-tinged. Seeds 1.5 × 1.2 × 1.1 mm, segmentiform, dark brown, slightly shiny, with 12–15(22) rows of darker brown or blackish tubercles on the dorsal facet, and 11–12 concentric arcs of such tubercles on each ventral facet, the tubercles connected by a lattice of shallow lineate ridges.

Botswana. N: Gcoha Hills, o. fr. 19.v.1977, *P.A. Smith* 2050 (MO; PRE; SRGH). SW: Central Kalahari G. Res., Deception Pan, fr. 27.iv.1975, *Owens* 61 (MO; SRGH). SE: Orapa, fr. 16.iv.1971,

Pope 313 (K; PRE; SRGH). **Zambia**. B: Masese, o. fr. 9.i.1961, *Fanshawe* 6099 (FHO; K; NDO). C: Mfuwe, fr. 12.iii.1969, *Astle* 5593 (K); Chilanga Distr., Quien Sabe Farm, fr. 30.viii.1929, *Sandwith* 30 (K). S: Mapanza, Choma, fr. 21.xii.1958, *Robinson* 2951 (K; PRE; SRGH). **Zimbabwe**. N: Makonde Distr., from Mtorashanga to Mpinga, fr. 22.ii.1961, *Rutherford-Smith* 586 (K; SRGH). W: Hwange Distr., Matetsi, fl. & fr. 26.i.1980, *Gonde* 279 (MO; SRGH). C: Gweru Distr., Mlezu Govt. Agric. School Farm, o. fr. 3.xi.1965, *Biegel* 504 (MO; SRGH). E: Mutare (Umtali) Circular Drive, fr. 9.ii.1962, *Chase* 7665 (BM; K; LISC; SRGH). S: Ndanga Distr., Chipinda Pools, o. fr. 30.i.1961, *Goodier* 97 (LISC; PRE; SRGH). **Malawi**. N: 32 km west of Karonga, Stevenson Road, o. fr. 16.iv.1976, *Pawek* 11071 (K; MAL; MO; SRGH; UC). C: Chitala to Kasache road, fl. 12.ii.1959, *Robson* 1577 (BM; K; LISC; SRGH). S: Nsanje Distr., Thangadze and Lilanje Rivers, y. fr. 25.iii.1960, *Phipps* 2687 (K; PRE; SRGH). **Mozambique**. Z: Vicente to Cuácua (Quaqua) R. junction, fr. 22.ii.1888, *Scott* s.n. (K). T: 33 km from Chicoa to Mágoè, fr. 17.ii.1970, *Torre & Correia* 18018 (LISC). MS: Chemba, Chiou, fr. 18.iv.1960, *Lemos & Macuácua* 117 (BM; K; LISC; LMA; SRGH). GI: Xai-Xai (Vila de João Belo), o. fr. 22.iv.1941, *Torre* 2598 (LISC). M: Magude, Chobela, R. Incomáti, y. fr. 14.ii.1953, *Myre & Balsinhas* 1470 (K; LISC; LMA; SRGH).

Very common and widespread throughout the Old World tropics and subtropics. Usually on heavy clay and alluvial soils of low altitude river valleys, on river banks and floodplains, in mopane and *Acacia* woodlands, in short grasslands and in seasonally flooded dambos and pans on clay and sandy soils, also in *Brachystegia allenii* and miombo woodlands and dry deciduous woodlands and scrub on sandy soil; sea level to 1400 m.

Two specimens from near Moamba in southern Mozambique, *de Koning* 7667 (K; LMU; MO) and *Jansen, de Koning, Nuvunga & Macuácua* 7562 (K; WAG), differ from typical *P. maderaspatensis* in having generally smaller leaves, in being dioecious and in having the female flower with minute, rounded-oblong disk glands. Although inclusion of these extends the circumscription of *P. maderaspatensis* they are treated here as belonging to this species.

23. **Phyllanthus incurvus** Thunb., Prodr. Pl. Cap., part 1: 24 (1794); Fl. Cap., ed. 2, 2: 499 (1823). —Müller Argoviensis in De Candolle, Prodr. **15**, 2: 362 (1866). —Hutchinson in F.C. **5**, 2: 396 (1920). —Engler, Pflanzenw. Afrikas (Veg. Erde 9) **3**, 2: 27 (1921). —Burtt Davy, Fl. Pl. Ferns Transvaal: 299 (1932). Type from South Africa (Cape Province).

Phyllanthus genistoides Sond. in Linnaea **23**: 134 (1850); in De Candolle, Prodr. **15**, 2: 360 (1866). Type from South Africa (Transvaal).

Phyllanthus multicaulis Müll. Arg. in Linnaea **32**: 18 (1863); in De Candolle, Prodr. **15**, 2: 360 (1866). Type from South Africa (Cape Province).

Very like *Phyllanthus maderaspatensis* but is distinguished chiefly by its caespitose habit; by its leaves being uniformly narrow and often involute with a curved, spinous apex, and without visible lateral nerves; by the female pedicels being 2–6 mm long; and by the smaller female disk and seeds.

Botswana. N: Ngamiland, Gcwihaba Hills, y. fr. 17.iii.1987, *Long & Rae* 343 (E; K; MO). ?SW: !Kungbushmanland, fl. x.1967, *Whiting* #1b (PRE). **Zimbabwe**. S: Beitbridge Distr., between Chiturupazi and Chikwarakwara, o. fr. 24.ii.1961, *Wild* 5372 (K; PRE; SRGH). **Mozambique**. Z: 13 km from Mocuba to Olinga (Maganja da Costa) crossroad, o. fr. 7.ii.1966, *Torre & Correia* 14451 (LISC). M: Goba Estação, female fl. 10.i.1980, *de Koning* 7930(2) (K; LMU; MO).

Also in Namibia, Swaziland, South Africa (Cape Province, Free State, Transvaal, KwaZulu-Natal). Usually on black basaltic and sandy clay soils of low altitude river valleys, on river banks, in dry river beds and amongst rocks, also on limestone pavement, on serpentine hills, and in open *Brachystegia* woodland; 15–930 m.

24. **Phyllanthus mendoncae** Jean F. Brunel ex Radcl.-Sm. in Kew Bull. **51**, 2: 313 (1996). Type: Mozambique, Inhambane Prov., between Vilanculos and Mapinhane, fl. 31.viii.1942, *Mendonça* 66 (LISC, holotype).

Phyllanthus mendoncae Jean F. Brunel, Phyllanthus Afr. Intertrop. Mad.: 324 (1987), unpublished thesis.

Like *Phyllanthus maderaspatensis* but a suffrutex with small foliage leaves (4–5 × 2 mm) mucronate at the apex, with broader male disk glands and female disk lobes (up to 0.5 mm across), and with styles divergent, bifid, and seeds usually with rows of elongated tubercles.

Mozambique. N: Cabo Delgado Prov., 15 km Pemba (Porto Amélia) to Montepuez, fr. 27.i.1984, *Groenendijk, Maite & Dungo* 831 (K; LMU; MO). MS: Nhamatanda (Vila Machado), between Chiluvo Mts. and Mucuzi, o. fr. 23.iv.1948, *Mendonça* 4041 (LISC). GI: Vilanculos to R. Save, o. fr. 21.x.1967, *Torre & Correia* 15805 (LISC). M: 5 km Boane to Porto Henrique, fr. 29.i.1983, *de Koning & Groenendijk* 9254 (LMU; MO).

Also in Ethiopia, Burundi, Kenya and Tanzania. On sandy clay and black basaltic soils, in open woodland and wooded grassy floodplains; 100 m.

In the material attributable to this species, there is a certain amount of variation in the shape of the female disk lobes and the seed tubercles. Thus in *Mendonça* 4041 (LISC) the female disk lobes are transversely rectangular whereas in *de Koning & Groenendijk* 9254 (LMU; MO) they are obovate. Furthermore, in *Mendonça* 4041 the seed tubercles are almost rounded, whereas in *Groenendijk et al.* 831 (K; LMU; MO) they are elongate.

Intermediates occur between this species and *P. maderaspatensis*. Thus in *Exell, Mendonça & Wild* 1435 (BM; LISC) from Mazabuka in Zambia, and in *Kelly* 530 (LISC; SRGH) from between Chipinda Pools and Chiredzi in Zimbabwe, the female disk lobes and styles are as in *P. maderaspatensis*, whilst the seed tubercles are as in typical *P. mendoncae*. Intermediates also occur between these species in southern Malawi, e.g. *Salubeni* 238, from Kasupe Distr., on the Liwonde to Sitola road.

25. **Phyllanthus serpentinicola** Radcl.-Sm. in Kew Bull. **51**, 2: 320 (1996). Tab. **10**, figs. A1–A4. Type: Zimbabwe, Kadoma Distr., Great Dyke between Battlefields and Ngesi, male, female fl., y. & o. fr. 15.i.1952, *Wild* 5590 (K, holotype; PRE; SRGH).

Very like *Phyllanthus maderaspatensis*, but a suffrutex with small (less than 1.5 × 1 cm) obovate leaves; male flowers solitary; female disk with broad rounded lobes; seeds smaller with more numerous (20–25) rows of tubercles on the dorsal facet, the tubercles not connected by a triangular latticework.

Zimbabwe. N: Makonde Distr., Great Dyke between Mutorashanga (Mtoroshanga) and Mpinga, fl. and o. fr. 22.ii.1961, *Rutherford-Smith* 586 (K; LISC; SRGH); on Mutorashanga to Kildonan road, 12.iv.1959, *Drummond* 6054 (SRGH).

Endemic on the Great Dyke. On serpentine slopes of Great Dyke, with chrome seams; c. 1200 m.

Drummond 6054 (SRGH) resembles the type except for its narrower leaves (as also does *Young* 579, 10.iv.1927 at Waterval Boven, in the Transvaal).

26. **Phyllanthus tener** Radcl.-Sm. in Kew Bull. **51**, 2: 323 (1996). Tab. **10**, fig. B. Type: Zambia, Mazabuka Distr., 3 km (2 miles) from Chirundu Bridge on Lusaka road, male, female fl. and y. fr. 1.ii.1958, *Drummond* 5408 (K, holotype; SRGH).

Differs from *Phyllanthus serpentinicola* by its much more graceful habit, its capilliform branches, its leaves greenish-brown and not greyish on drying; and by the male flowers in few-flowered clusters and not solitary.

Zambia. S: Mazabuka Distr., c. 3 km from Chirundu Bridge on Lusaka road, male, female fl. & y. fr. 1.ii.1958, *Drummond* 5408 (K; SRGH).

Not known elsewhere. Mopane woodland on river valley floor.

27. **Phyllanthus leucanthus** Pax in Bot. Jahrb. Syst. **15**: 524 (1893); Pflanzenw. Ost-Afrikas **C**: 236 (1895). —Hutchinson in F.T.A. **6**, 1: 728 (1912). —Engler, Pflanzenw. Afrikas (Veg. Erde 9) **3**, 2: 26 (1921). —Troupin, Fl. Rwanda **2**: 234, fig. 72/2 (1983). —Radcliffe-Smith in F.T.E.A., Euphorb. 1: 48 (1987). Type from Tanzania (Western Province).

Phyllanthus eylesii S. Moore in J. Bot. **58**: 79 (1920). Type: Zimbabwe, Victoria Falls, fl. iv.1918, *Eyles* 1296 (BM; K; SRGH).

Phyllanthus merripaensis Jean F. Brunel, Phyllanthus Afr. Intertrop. Mad.: 363 (1987), unpublished thesis. Based on: *Torre & Paiva* 10464 (LISC), Mozambique, Serra de Merripa, fl. & fr. 5.ii.1964.

A scaberulous or glabrous erect annual or perennial herb up to 90 cm tall, usually much less, monoecious or rarely dioecious; stems reddish or purplish later becoming wiry. Lead shoots angular. Lateral shoots up to 15 cm long, narrowly 2-winged, the older ones often co-axillary with secondary shoots. Short shoots sometimes developing. Scale leaves c. 1 mm long, narrowly triangular-lanceolate to subulate; stipules triangular-lanceolate, otherwise similar to scale leaves. Foliage leaves distichous; petioles 0.5–1 mm long; stipules c. 1 mm long, narrowly triangular-lanceolate, pallid. Leaf blades 0.5–3.5 × 0.4–1.5 cm, elliptic to oblong, subacute, obtuse or rounded, cuneate or rounded at the base, firmly membranaceous, light to medium green above, paler and somewhat glaucescent or purplish-tinged beneath; lateral nerves in 5–8(10) pairs, usually looped near the margin, not prominent

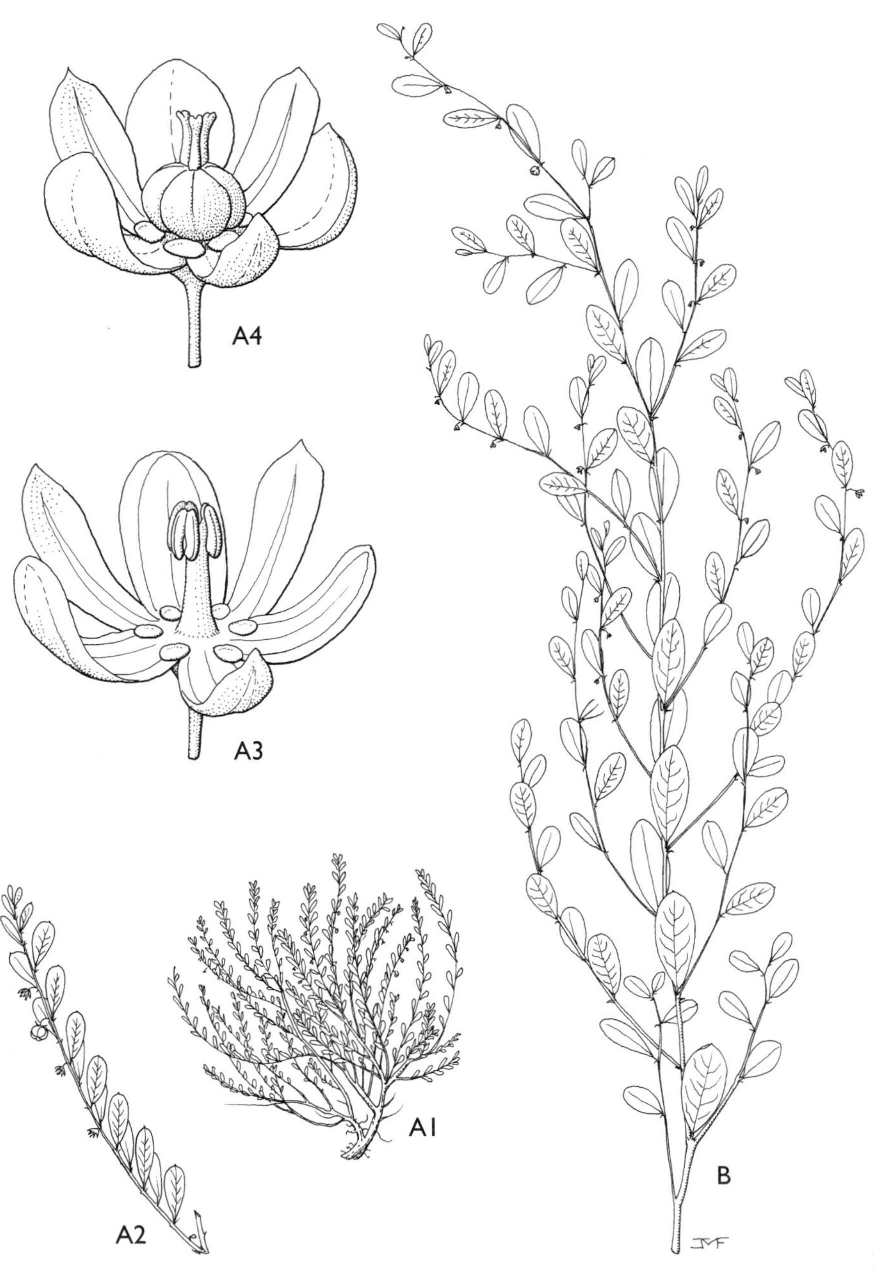

Tab. 10. A —PHYLLANTHUS SERPENTINICOLA. A1, habit (× $^1/_4$); A2, lateral shoot (× 1); A3, male flower (× 16); A4, female flower (× 16), A1–A4 from *Rutherford-Smith* 586. B — PHYLLANTHUS TENER, habit (× 1), from *Drummond* 5408. Drawn by J.M. Fothergill.

above, slightly so beneath. Male flowers in few-flowered clusters in the lower parts of the lateral shoots, female flowers pendulous and solitary in the upper parts of the lateral shoots. Male flowers: pedicels 0.5 mm long; sepals 6, c. 1 × 1 mm, suborbicular-obovate, rounded, cream-coloured with a greenish midrib; disk glands 6, free, 0.3 mm in diameter, flat or somewhat verruculose, minutely lobulate; stamens 3, filaments connate in the lower two-thirds, 0.75 mm long, anthers 0.3 mm long, vertically held, longitudinally dehiscent. Female flowers: pedicels 1 mm long, extending to 2 mm in fruit, somewhat flattened; sepals 6, in 2 whorls of 3, 1.5 × 1 mm, accrescent to 3 × 2 mm in fruit, the outer broadly elliptic and rounded to cordate at the base, the inner obovate and attenuate-cuneate at the base, white with a narrow green midrib; disk 0.75 mm in diameter, shallowly 6-lobed, the lobes entire; ovary 0.5 mm in diameter, shortly stipitate, 6-lobed, subglobose, smooth; styles 3, 0.67 mm long, united at the base, closely appressed to the top of the ovary, deeply bifid, the stigmas slender, recurved. Fruit 1.8 × 2.8 mm, depressed 3-lobed to subglobose, smooth, olivaceous, enclosed by the persistent sepals. Seeds 1.5 × 1.2 × 1 mm, segmentiform, light to dark brown, with 7–9 shallow longitudinal ridges on the dorsal facet, and 6–7 concentric ridges on each ventral facet, with innumerable faint transverse striae between them.

Zambia. N: Ishiba Ngandu (Shiwa Ngandu), o. fr. 5.ii.1955, *Fanshawe* 1994 (K; NDO; SRGH). W: Kitwe, Mindolo, fl. & y. fr. 28.iv.1963, *Mutimushi* 296 (K; NDO; SRGH). C: Mt. Makulu, fr. 18.i.1965, *Lawton* 1193 (K; SRGH). S: Kafue Gorge, o. fr. 14.iv.1956, *Robinson* 1482 (K; SRGH). **Zimbabwe**. N: Hurungwe Distr., Mauora R., fl. 26.ii.1958, *Phipps* 903 (K; PRE; SRGH). W: Shangani Distr., Gwampa For. Res., o. fr. ii.1958, *Goldsmith* 52/58 (K; LISC; SRGH). C: Harare (Salisbury), fl. & y. fr. i.1918, *Eyles* 912 (BM; SRGH). E: Narwatsi Zone, Nyamkwarara (Nyumquarara) Valley, o. fr. ii.1935, *Gilliland* K1387 (K). **Malawi**. N: Nkhata Bay, Chinteche, o. fr. 7.vi.1974, *Pawek* 8688 (K; MAL; MO; SRGH). C: Goche, Kirk Range, o. fr. 30.i.1959, *Robson* 1364 (BM; K; LISC; SRGH). S: Namwera Escarpment, Jalasi, o. fr. 15.iii.1955, *Exell, Mendonça & Wild* 901 (BM; BR; LISC; SRGH). **Mozambique**. N: Marrupa, Massanguezi, o. fr. 19.ii.1981, *Nuvunga* 607 (BR; K; LMU; MO).

Also in Zaire, Rwanda, Ethiopia, Djibouti, Somalia, Uganda, Kenya and Tanzania. Common, usually in shady places in dense riverine and lakeshore vegetation, high rainfall miombo woodland and evergreen forest margins, also in dambos and plateau and submontane grasslands, and in disturbed and cultivated ground; 520–1645 m.

Vernacular name as recorded in specimen data: "kamachanga" (chiKunda).

Angus 2127 (SRGH) from Petauke in eastern Province of Zambia is probably this species, but the material is rather immature.

28. **Phyllanthus xiphephorus** Jean F. Brunel ex Radcl.-Sm. in Kew Bull. **51**, 2: 327 (1996). Tab. **11**, figs. A1–A4. Type: Zambia, Northern Prov., Chilongowelo Farm, fl. & y. fr. 4.iii.1952, *Richards* 884 (BR, holotype; K).

Phyllanthus xiphophorus Jean F. Brunel, Phyllanthus Afr. Intertrop. Mad.: 344 (1987), unpublished thesis.

Very like *Phyllanthus leucanthus*, but differing in having 6 filiform staminode-like appendages arising from the middle of the disk lobes in the female flowers, and in having a sessile ovary and non-winged lateral shoots.

Zambia. N: Chilongowelo Farm, fl. & y. fr. 28.i.1955, *Richards* 4265 (K; SRGH).

Not known elsewhere. In rich soil in farmyard, in grassy and other herbaceous associations; 1463 m.

It is just possible that this is an artificially induced variant of either *P. leucanthus* or *P. pseudoniruri*.

29. **Phyllanthus pseudoniruri** Müll. Arg. in Flora **47**: 539 (1864); in De Candolle, Prodr. **15**, 2: 365 (1866). —Engler, Pflanzenw. Ost-Afrikas **C**: 236 (1895). —Hutchinson in F.T.A. **6**, 1: 726 (1912). —Engler, Pflanzenw. Afrikas (Veg. Erde 9) **3**, 2: 26 (1921). —F.W. Andrews, Fl. Pl. Anglo-Egypt. Sudan **2**: 93 (1952). —Troupin, Fl. Rwanda **2**: 236, fig. 72/4 (1983). —Radcliffe-Smith in F.T.E.A., Euphorb. 1: 48 (1987). Type from Uganda (Western Province).

Phyllanthus odontadenioides Jean F. Brunel, Phyllanthus Afr. Intertrop. Mad.: 342 (1987), unpublished thesis.

Phyllanthus sananei Jean F. Brunel, Phyllanthus Afr. Intertrop. Mad.: 342 (1987), unpublished thesis. Based on *Sanane* 877 (B; K), Zambia, Mbala Distr., Lake Lyapa, 10.ix.1969.

Tab. 11. A —PHYLLANTHUS XIPHEPHORUS. A1, habit (× 1); A2, male flower (× 16); A3, female flower (× 16); A4, female disk glands showing filiform extensions (× 32), A1–A4 from *Richards* 4265. B —PHYLLANTHUS TENUIS. B1, habit (× 1); B2, distal part of flowering branch (× 2); B3, male flower (× 16); B4, stamens and disk glands (× 32); B5, female flower (× 16); B6, fruit (× 16), B1–B6 from *Richards* 15075. C —PHYLLANTHUS ZAMBICUS. C1, flowering branch (× 1); C2, stamens (× 32); C3, male flower (× 16); C4, female flower (× 16), C1–C4 from *Mitchell* 5/38. Drawn by J.M. Fothergill.

Very like *Phyllanthus leucanthus*, but having narrower female sepals and a sessile ovary.

Zambia. N: Mbala, o. fr. 12.ix.1969, *Fanshawe* 10630 (K; NDO; SRGH). W: Kitwe, fr. 30.ix.1970, *Fanshawe* 10921 (K; NDO). **Zimbabwe**. W: Victoria Falls, o. fr. 8.vii.1930, *Hutchinson & Gillett* 3441 (K). E: Nyamkwarara (Nyumquarara) Valley, o. fr. ii.1935, *Gilliland* 1387 (BM). **Malawi**. N: Mafinga Hills, Chisenga, fl. & fr. 24.viii.1962, *Tyrer* 539 (BM). C: Lake Malawi (Nyasa), Uvera (Mvera), fl., recd. 1901, *Kenyon* 4 (K).

Widespread in central and east tropical Africa from Cameroon eastwards to Somalia and south to Zimbabwe.

On black clay and sandy soils, in moist dambos and grassland beside rivers streams and lakes, also in riverine forest and cultivated ground; 610–1525 m.

Taylor 155 from Mfuwe, Luangwa Nat. Park, Zambia, is referred to this species, but its leaves are narrower than usual.

30. **Phyllanthus odontadenius** Müll. Arg. in J. Bot. **2**: 331 (1864); in De Candolle, Prodr. **15**, 2: 365 (1866). —Hutchinson in F.T.A. **6**, 1: 727 (1912). —Engler, Pflanzenw. Afrikas (Veg. Erde 9) **3**, 2: 26 (1921). —F.W. Andrews, Fl. Pl. Anglo-Egypt. Sudan **2**: 95 (1952). —Keay in F.W.T.A., ed. 2, **1**, 2: 388 (1958). —Agnew, Upl. Kenya Wild Fls.: 212 (1974). —Troupin, Fl. Rwanda **2**: 236, fig. 72/1 (1983). —Radcliffe-Smith in F.T.E.A., Euphorb. 1: 47 (1987). Type from Angola (Malanje Province).

Very like *Phyllanthus leucanthus*, but with the lateral shoots angular and denticulate, not winged, the female sepals somewhat narrower and scarcely cordate, and the female disk lobes often minutely crenulate.

Zambia. W: Mwinilunga Distr., south of Matonchi Farm, y. fl. 22.x.1937, *Milne-Redhead* 2894 (K). **Zimbabwe**. S: Ndanga Disr., Chipinda Pools, o. fr. 30.i.1961, *Goodier* 97b (K; SRGH). **Malawi**. N: Chitipa Distr., Mughesse For. Res., fl. 28.i.1989, *Thompson & Rawlins* 6255 (CM; K). C: Ntchisi For. Res., o. fr. 19.vi.1970, *Brummitt* 11570 (K).

Widespread in tropical Africa from Sierra Leone eastwards to Kenya, and south to Angola and Zimbabwe. In high rainfall *Brachystegia-Uapaca* and miombo woodlands, and in evergreen forest shade; 1590–1890 m.

31. **Phyllanthus amarus** Schumach. & Thonn., Beskr. Guin. Pl.: 421 (1827); in Dansk. Vid. Selsk. Skr. **4**: 195 (1829). —Hutchinson in F.T.A. **6**, 1: 717 (1912). —Engler, Pflanzenw. Afrikas (Veg. Erde 9) **3**, 2: 28 (1921). —Robyns & Tournay, Fl. Sperm. Parc Nat. Alb. **1**: 445 (1948). —F.W. Andrews, Fl. Pl. Anglo-Egypt. Sudan **2**: 93 (1952). —G.L. Webster in J. Arnold Arbor. **38**: 313 (1957). —Keay in F.W.T.A., ed. 2, **1**, 2: 387 (1958). —Agnew, Upl. Kenya Wild Fls.: 212 (1974). —Radcliffe-Smith in F.T.E.A., Euphorb. 1: 34 (1987). Type from Ghana.

Phyllanthus niruri sensu Klotzsch in Peters, Naturw. Reise Mossambique **6**, 1: 104 (1861) non L.

Phyllanthus niruri var. *genuinus* sensu Müll. Arg. in De Candolle, Prodr. **15**, 2: 406 (1866), pro parte non L.

Phyllanthus niruri var. *debilis* sensu Müll. Arg. in De Candolle, Prodr. **15**, 2: 407 (1866), pro parte non Klein ex Willd.

A glabrous erect or ascending annual herb up to 75 cm tall, sometimes woody at the base, monoecious or rarely dioecious. Lead shoots terete. Lateral shoots up to c. 15 cm long, the older ones usually co-axillary with secondary lead shoots. Scale leaves 1–1.3 mm long, linear-subulate, blackening at the apex; stipules 1.5 × 1 mm, broadly triangular-lanceolate, asymmetrical at the base, dark brown. Foliage leaves distichous; petioles 0.3–0.5 mm long; stipules 1 mm long, linear-lanceolate, cream-coloured with a brownish midrib. Leaf blades 5–10 × 2–5 mm, mostly oblong, rounded-subtruncate at apex and base, membranaceous, dull green above, paler beneath; lateral nerves in 4–6 pairs, looped at the apex, inconspicuous above, not prominent beneath. Male and female flowers often occurring together in the distal axils, female flowers usually solitary in the proximal axils. Male flowers: pedicels 1 mm long; sepals 5, 0.6 × 0.5 mm, ovate, acute, pale greenish-cream; disk glands 5, free, minute, lobulate; stamens 3, filaments connate into a column c. 0.4 mm high, anthers 0.25 mm across, sessile, vertical, 2 bithecate, reniform, obliquely dehiscent with confluent slits, 1 monothecate. Female flowers: pedicels c. 1 mm long, extending to 1.5 mm in fruit, apically thickened; sepals 5, 0.75–1 × 0.3–0.5 mm, not or slightly accrescent in fruit, ovate-oblong, subacute, greenish with white margins;

disk c. 0.5 mm across, flat, 5-lobed, the lobes triangular, basally lobulate or not; ovary 0.3 mm in diameter, subsessile, subglobose, smooth; styles 3, 0.1 mm long, free, suberect, later ± spreading, 2-lobed. Fruit 1–1.2 × 1.5–2 mm, oblate, smooth, ochreous-olivaceous. Seeds 0.8–1 × 0.6–0.8 × 0.5–0.7 mm, segmentiform, light brown, with 5–7 longitudinal ridges on the dorsal facet, and 4–6 irregularly concentric ridges on each ventral facet, with innumerable faint transverse striae between them.

Malawi. N: Karonga Distr., St. Anne's, 3 km north of Chilumba, fr. 23.iv.1969, *Pawek* 2302 (K; MAL). S: Mangochi Distr., Maldeco Hill, y. fr. 9.ii.1989, *Tawakali & Nachamba* 1340 (MAL; PRE).

Widespread in tropical Africa from Gambia eastwards to Djibouti and south to Angola and Malawi; native to America, but now a pantropical weed.

In grassland under *Acacia*, in mixed woodland and by roadsides; 470 m.

32. **Phyllanthus taylorianus** Jean F. Brunel ex Radcl.-Sm. in Kew Bull. **51**, 2: 322 (1996). Type from Tanzania (Southern Province).

Phyllanthus niruroides sensu Radcl.-Sm. in F.T.E.A., Euphorb. 1: 60 (1987), non Müll. Arg.

Phyllanthus taylorianus Jean F. Brunel, Phyllanthus Afr. Intertrop. Mad.: 355 (1987), unpublished thesis.

A glabrous erect annual herb 10–25(45) cm tall, monoecious; stems reddish or purplish. Lead shoots terete. Lateral shoots up to 8 cm long, terete. Scale leaves 0.5–1 mm long, linear-subulate, soon darkening apically; stipules triangular-lanceolate, otherwise similar to the scale leaves. Foliage leaves distichous; petioles 0.3 mm long; stipules c. 1 mm long, linear-setaceous, cream-coloured with a reddish midrib, blackening at the apex. Leaf blades 3–11 × 1–4 mm, elliptic-lanceolate, acute or subacute, rounded-cuneate, membranaceous, green above, glaucous beneath, margin often reddish-tinged; lateral nerves in 4–6 pairs, inconspicuous. Male flowers in small cymules in the proximal axils of the lateral shoots, females solitary in the distal axils, rarely both in same axil. Male flowers: pedicels 0.5 mm long, capillary; sepals 5, 0.4 × 0.4 mm, suborbicular, greenish-cream; disk glands 5, free, 0.2 mm in diameter, tuberculate-lobulate; stamens (2)3, filaments connate into a column 0.2 mm high in the lower half, free above, anthers 0.2 mm across, rounded, horizontally held, laterally dehiscent. Female flowers: pedicels 0.5–1 mm long, extending to 2 mm in fruit, relatively stout; sepals 5(6), 0.7–0.8 × 0.5 mm, scarcely accrescent in fruit, elliptic-ovate, acute or subacute, whitish with a green midrib; disk c. 0.5 mm in diameter, flat, 15-lobulate, the lobules ± rounded; ovary 0.3 mm in diameter, sessile, shallowly 6-lobed, densely papillose, pale green; styles 3, 0.25 mm long, free, spreading, bifid, stigmas filiform, recurved. Fruit 1 × 1.5–2 mm, somewhat depressed 6-lobed, subglobose, densely tuberculate, greenish. Seeds 0.8–0.9 × 0.7–0.8 × 0.6–0.7 mm, segmentiform, pale to mid-brown, with 9–14 indistinct longitudinal ridges on the dorsal facet, and 7–10 scarcely visible concentric lines on each ventral facet, with innumerable transverse striae between them.

Zambia. N: Mbala, Abercorn Agric. Station, fr. 29.iii.1960, *Angus* 2193 (K; SRGH). W: Kitwe, fl. & fr. 29.iii.1969, *Mutimushi* 3017 (K; NDO). S: Mapanza Mission, fr. 28.ii.1953, *Robinson* 108 (K). **Zimbabwe**. C: Harare Distr., Cranborne, fl. 17.ii.1946, *Wild* 887 (K; SRGH). **Malawi**. N: Mzimba Distr., 9.6 km north of Mzambazi, o. fr. 10.iii.1978, *Pawek* 13964 (DAV; K; MAL; MO).

Also in Cameroon, Central African Republic, Zaire, Burundi, Ethiopia, Uganda and Tanzania.

Usually in moist shady places in miombo woodland, riverine forest and wooded grassland, also as a weed of cultivation; 600–1525 m.

33. **Phyllanthus leucocalyx** Hutch. in Bull. Misc. Inform., Kew **1911**: 316 (1911); in F.T.A. **6**, 1: 716 (1912). —Engler, Pflanzenw. Afrikas (Veg. Erde 9) **3**, 2: 28 (1921). —Brenan, Check-list For. Trees Shrubs Tang. Terr.: 223 (1949). —Agnew, Upl. Kenya Wild Fls.: 211 (1974). —Radcliffe-Smith in F.T.E.A., Euphorb. 1: 61 (1987). Type from Tanzania (Southern Province).

Phyllanthus rotundifolius var. *leucocalyx* sensu Müll. Arg. in De Candolle, Prodr. **15**, 2: 406 (1866), pro parte quoad spec. Kirk, non Müll. Arg. in Linnaea **32**: 43 (1863).

Very like *Phyllanthus taylorianus*, differing chiefly in the rather more robust habit, the female sepals up to 4 mm in length, the female disk annular and indistinctly crenulate, the fruit tuberculate in the upper half only, and the slightly larger seeds with up to 25 parallel lines on the dorsal facet.

Mozambique. N: Cabo Delgado, Macomia, Ingoane, o. fr. 12.ix.1948, *Barbosa* 2066 (LISC). Z: 42 km from Olinga (Maganja da Costa) to beach, fr. 15.ii.1966, *Torre & Correia* 14658 (LISC).
Also in Somalia, Uganda, Kenya and Tanzania. Sandy soil, in herbaceous savanna and with grasses and sedges in pools; 15 m.

34. **Phyllanthus micromeris** Radcl.-Sm. in Kew Bull. **35**: 769 (1981); in F.T.E.A., Euphorb. 1: 59 (1987). Type from Tanzania (Southern Highlands Province).

Very like *Phyllanthus taylorianus*, but differing in the following features: leaves narrowly oblong, lateral nerves up to 10 pairs and usually fairly prominent beneath; female disk ± annular to distinctly 5-lobulate; ovary shallowly and evenly papillose-verruculose to almost smooth; fruit sparingly shallowly papillose-verruculose; seeds slightly smaller, illineate, or only weakly 3–10-lineate or finely 17–18-lineate on the dorsal facet.

1. Lateral nerves prominent beneath; sepals of female flowers less than 1 mm long · · · · · · · · var. *micromeris*
– Lateral nerves not prominent beneath; sepals of female flowers more than 1 mm long · · · 2
2. Annual herb up to 20 cm tall; stipules 1.5–2 mm long; disk of female flowers annular, entire; fruit almost smooth; seeds not or weakly 10-lineate · · · · · · · · · · · · · · · · var. *mughessensis*
– Perennial herb up to 60 cm tall; stipules 0.5–1 mm long; disk of female flowers 5-lobed; fruit evenly papillose; seeds finely but distinctly 17–18-lineate · · · · · · · · · · · · · var. *sesbanioides*

Var. **micromeris**

Annual or perennial herb up to 30 cm tall; lateral nerves of foliage leaves prominent beneath; stipules 1–1.5 mm long; female sepals less than 1 mm long; female disk indistinctly 5–6-lobed; fruit sparingly shallowly papillose; seeds not or weakly 3–4-lineate.

Malawi. N: Mzimba Distr., Viphya Plateau, Champoyo, male, female fl. & fr. 12.viii.1971, *Salubeni* 1685 (K; MO; SRGH).
Also in Tanzania. In submontane grassland, under low scrub; 1980 m.

Var. **mughessensis** Radcl.-Sm. in Kew Bull. **51**, 2: 316 (1996). Type: Malawi, Chitipa Distr., Misuku Hills, Mughesse For. Reserve, male, female fls. & fr. 24.v.1989, *Radcliffe-Smith, Pope & Goyder* 5932A (K, holotype).

Annual herb up to 20 cm tall; lateral nerves of foliage leaves not prominent beneath; stipules 1.5–2 mm long; female sepals more than 1 mm long; female disk annular, entire; fruit almost smooth; seeds not or weakly 10-lineate.

Malawi. N: Chitipa Distr., Misuku Hills, Mughesse For. Res., male, female fls. & fr. 24.v.1989, *Radcliffe-Smith, Pope & Goyder* 5932A (K).
Not known elsewhere. By pathside in moist evergreen forest on steep slopes; 1500 m.

Var. **sesbanioides** Radcl.-Sm. in Kew Bull. **51**, 2: 316 (1996). Type: Malawi, Mangochi Distr., Cape Maclear, Chembe Village, male, female fls. & fr. 11.viii.1987, *Salubeni & Patel* 5082 (MO, holotype; MAL).

Perennial herb up to 60 cm tall; lateral nerves of foliage leaves not prominent beneath; stipules 0.5–1 mm long; female sepals more than 1 mm long; female disk 5-lobed; fruit evenly papillose; seeds finely but distinctly 17–18-lineate.

Malawi. S: Mangochi Distr., Cape Maclear, Chembe Village, male, female fls. & fr. 11.viii.1987, *Salubeni & Patel* 5082 (MO; MAL).
Not known elsewhere. In mango woodland; 475 m.

35. **Phyllanthus tenuis** Radcl.-Sm. in Kew Bull. **51**, 2: 323 (1996). Tab. **11**, figs. B1–B6. Type: Zambia, Mbala Distr., Chikwalala Ridge, fr. v.1955, *Lawton* 189 (K, holotype).

Very like *Phyllanthus taylorianus* but differing in the following features; the much more graceful habit; the lead shoots slender and the lateral shoots capillary; foliage leaf blades 3–6(8) × 1–2 mm, linear-lanceolate, with the lateral nerves generally invisible; female disk annular, somewhat irregularly lobed, the lobes subentire; seeds with only 7–8 longitudinal ridges on the dorsal facet.

Zambia. N: Kawambwa Distr., Muchinga Escarpment, west of Kawambwa, 09°50'S, 28°57'E, fl. & fr. 19.iv.1989, *Radcliffe-Smith, Pope & Goyder* 5727 (K).

Not known elsewhere. On sandy soil amongst rocks with *Xerophyta*, and in wet grassland with scattered small trees and shrubs including *Hymenocardia acida* on top of escarpment; 1800 m.

This species is closely related to *Phyllanthus tukuyuanus* (a species proposed by Brunel in his unpublished thesis) from southern Tanzania which has ovate-lanceolate foliage leaves, 5 sepals in female flowers, an entire disk in the female flowers and 13–14 longitudinal ridges on the dorsal facet of each seed.

36. **Phyllanthus parvus** Hutch. in Bull. Misc. Inform., Kew **1911**: 316 (1911); in F.T.A. **6**, 1: 732 (1912). —Radcliffe-Smith in F.T.E.A., Euphorb. 1: 62 (1987). Type from Tanzania (Western Province).

A glabrous erect annual herb up to 20 cm tall, often reddish- or pinkish-tinged, monoecious. Lead shoots terete; lateral shoots up to 9 cm long, commonly much less. Scale leaves 1 mm long, linear-subulate, soon darkening, sometimes replaced by foliage leaves; stipules 0.6 mm long, triangular. Foliage leaves distichous; petioles 0.5–0.6 mm long; stipules c. 1 mm long, linear-setaceous, subentire, cream-coloured with a pinkish midrib. Leaf blades 0.5–1.5 × 0.1–0.2 cm, linear-lanceolate to narrowly elliptic-lanceolate, acute or subacute, cuneate or attenuate, membranaceous, green above, glaucous and often pinkish-tinged beneath with reddish margins; lateral nerves in 2–4 pairs, upwardly directed, invisible above, scarcely visible beneath. Male flowers in few-flowered cymules in the proximal axils of the lateral shoots, female flowers solitary in the distal axils. Male flowers: pedicels 0.3–0.5 mm long, capillary; sepals 5, 0.5–0.6 × 0.5 mm, suborbicular, minutely apiculate, cream-coloured with a pink midrib; disk glands 5, free, minute, rounded, ± flat, smooth; stamens 2(3), filaments connate into a column 0.2 mm high, anthers 0.2 mm across, free, rounded, horizontally held, laterally dehiscent. Female flowers: pedicels 0.5 mm long, extending to 1.5 mm in fruit; sepals 5, 1.2–1.3 × 0.5–0.7 mm, accrescent to 2 × 1 mm in fruit, elliptic-oblong, subacute, minutely serrulate, white with a reddish midrib; disk 0.6 mm in diameter, annular to shallowly 5-lobed, undulate, thin; ovary 0.3 mm in diameter, sessile, 6-lobed, ± smooth, pinkish; styles 3, c. 0.2 mm long, ± free, divergent, bifid, stigmas recurved. Fruit 1.2 × 2 mm, oblate, very shallowly papillose in the upper half, otherwise ± smooth, greenish-brown. Seeds 1 × 0.8 × 0.7 mm, segmentiform, pale yellowish-brown to dark grey-brown, with 9–10 faint longitudinal ridges on the dorsal facet, and c. 7 scarcely visible concentric lines on each ventral facet, with numerous more strongly marked transverse lines or ridges between them.

Zambia. C: Serenje Distr., Kundalila Falls, y. fl. 29.iii.1984, *Brummitt, Chisumpa & Nshingo* 16962 (K; NDO). **Malawi**. N: Mzimba Distr., Mzuzu, o. fr. 20.iii.1974, *Pawek* 8236 (BR; K; MAL; MO; PRE; SRGH; UC). C: Ntchisi Distr., Ntchisi Mt., fl. & fr. 16.iv.1991, *Radcliffe-Smith* 5991 (K).

Also in Zaire (Shaba Province), Tanzania and Angola. Submontane grassland and amongst rock outcrops in short grass, and in *Brachystegia* and *Uapaca* woodlands on steep slopes, also beside waterfalls and as a roadside weed; 1370–2340 m.

37. **Phyllanthus fraternus** G.L. Webster in Contrib. Gray Herb., No. 176: 53 (1955); in J. Arnold Arbor. **38**: 309 (1957). —Troupin, Fl. Rwanda **2**: 234, fig. 72/3 (1983). —Radcliffe-Smith in F.T.E.A., Euphorb. 1: 49 (1987). Type from India.

Phyllanthus niruri var. *scabrellus* Müll. Arg. in Linnaea **32**: 43 (1863); in De Candolle, Prodr. **15**, 2: 406 (1866). Syntypes from India and W Africa.

Phyllanthus niruri sensu Hook.f., Fl. Brit. Ind. **5**: 298 (1887), pro parte. —Engler, Pflanzenw. Ost-Afrikas **C**: 236 (1895). —Schinz & Junod in Mém. Herb. Boissier, No. 10: 46 (1900). —Hutchinson in F.T.A. **6**, 1: 731 (1912), pro parte. —Eyles in Trans. Roy. Soc. South Africa **5**: 393 (1916). —Hutchinson in F.C. **5**, 2: 399 (1920). —Engler, Pflanzenw. Afrikas (Veg. Erde 9) **3**, 2: 28 (1921). —F.W. Andrews, Fl. Pl. Anglo-Egypt. Sudan **2**: 93 (1952). —Keay in F.W.T.A., ed. 2, **1**, 2: 388 (1958). —P.G. Meyer in Merxmüller, Prodr. Fl. SW. Afrika, fam. 67: 39 (1967), non L.

Phyllanthus fraternus subsp. *togoensis* Jean F. Brunel & Roux in Bull. Soc. Bot. Fr. **122**: 153 (1975). Type from Togo.

An asperulous to subglabrous erect annual herb commonly up to 45 cm tall, rarely taller, monoecious. Lead shoots angular, straw coloured; lateral shoots up to 10 cm long, commonly much less. Scale leaves 1 mm long, linear-subulate; stipules 0.8–1 mm long, triangular-lanceolate. Foliage leaves distichous; petioles 0.5 mm long; stipules 1 mm long, linear-lanceolate, whitish with a brownish midrib. Leaf blades 0.5–1.3 × 0.15–0.5 cm, elliptic-oblong, obtuse or rounded at the apex, cuneate to rounded at the base, membranaceous, glabrous, dark green above, paler and somewhat greyish-green beneath; lateral nerves in 4–7 pairs, scarcely visible above, slightly prominent beneath. Male flowers in few-flowered cymules in the axils of the lowest quarter of the lateral shoots, females solitary in the rest, with occasionally the transitional axil bisexual. Male flowers: pedicels 0.5 mm long, capillary; sepals 6, 0.7 × 0.5 mm, obovate-suborbicular, the 3 outer apiculate, the inner obtuse, translucent with an opaque greenish-yellow midrib; disk glands 6, free, 0.1 mm across, flattened, lobulate and tuberculate; stamens 3, filaments connate into a column 0.25 mm high, anthers 0.2 mm across, reniform, obliquely held, apically dehiscent. Female flowers: pedicels 0.5 mm long, extending to 1 mm in fruit; sepals 6, 1 × 0.5 mm, not or scarcely accrescent, the outer oblong-lanceolate, the inner oblong to oblanceolate, obtuse to rounded, white with an olive-green midrib; disk 0.6 mm across, very irregularly laciniate or divided; ovary 0.5 mm in diameter, subsessile, 3-lobed to subglobose, smooth, greenish; styles 3, 0.25 mm long, divergent, shortly bifid, stigmas recurved. Fruit 1.1–1.2 × 1.6 mm, rounded-3-lobed, smooth, ochreous. Seeds 1 × 0.8 × 0.7 mm, segmentiform, yellowish-brown, with 6–7 longitudinal ridges of close packed darker brown, transversely lineate tubercles on the dorsal facet, and 5–7 concentric ridges on each ventral facet, with innumerable parallel transverse striae between them.

Botswana. N: Xaudum (Chadum) Valley, 16 km west of Nxaunxau (Knauknau), o. fr. 14.iii.1965, *Wild & Drummond* 7018 (K; LISC; SRGH). SE: Mochudi, o. fr. iii.1914, *Rogers* 6538 (K) pro parte, see *P. gillettianus*. **Zambia**. C: Luangwa Valley Game Res. South, 5 km south of Lubi R., fl. 2.ii.1967, *Prince* 112 (K; SRGH). E: Luangwa Valley, Mulila Munkanya, o. fr. 2.iii.1968, *R. Phiri* 80 (K; SRGH). S: Mazabuka, o. fr. 29.iii.1963, *van Rensburg* 1855 (K; SRGH). **Zimbabwe**. N: Hurungwe Distr., Mensa Pan, fr. 29.i.1958, *Drummond* 5324 (K; LISC; PRE; SRGH). W: Victoria Falls, o. fr. vii.1908, *Rogers* 5008 (K). E: Chipinge Distr., fl. & o. fr. ii.1960, *Soane* 279 (SRGH). S: Mwenezi Distr., near Malipate, o. fr. 25.iv.1961, *Simon* 32 (SRGH). **Malawi**. N: Kondowe to Karonga, o. fr. vii.1896, *Whyte* s.n. (K). C: Golomoti to Monkey Bay Road, o. fr. 11.iii.1990, *Salubeni & Tawakali* 5801 (MAL; MO). S: Nsanje Distr., between Thangadzi and Lirangwe (Lilanje) rivers, o. fr. 25.iii.1960, *Phipps* 2696 (K; PRE; SRGH). **Mozambique**. Z: near Chupanga (Shupanga), o. fr. i.1859, *Kirk* s.n. (K). MS: Sofala Prov. (Beira Distr.), Gorongosa Nat. Park, Urema R., o. fr. vii.1970, *Tinley* 1953 (SRGH).

Native of Pakistan and NW India. Introduced into many parts of tropical Africa and Arabia, and also the West Indies.

On sandy or clay soils, often numerous in mopane and mixed *Acacia* woodland and in sand at edge of dry seasonal pans, also on termitaria in floodplains, in tall grassland and on rocks in Kariba Gorge, and a weed of cultivated ground; 90–1830 m.

38. **Phyllanthus gillettianus** Jean F. Brunel ex Radcl.-Sm. in Kew Bull. **51**, 2: 306 (1996). Type from Kenya (Coast Province).

Phyllanthus gillettianus Jean F. Brunel, Phyllanthus Afr. Intertrop. Mad.: 365 (1987), unpublished thesis.

Very like *Phyllanthus fraternus*, but the female sepals have a broad green median band, and the female disk is annular, thin, and closely appressed to the base of the ovary.

Botswana. SE: Mochudi, fr. iii.1914, *Rogers* 6538 (K) pro parte see *P. fraternus*. **Zambia**. S: Mazabuka, o. fr. 17.i.1963, *Angus* 3528 (FHO; K). **Zimbabwe**. S: 8 km south of Bubye R. on Masvingo (Ft. Victoria)–Beitbridge road, o. fr. 17.iii.1967, *Rushworth* 376 (BR; K; LISC; SRGH).

Also in Zaire and Kenya. On heavy black clay in floodplain grasslands, and hot dry deciduous woodlands with short grasses on sandy granitic soils; 15–625 m.

Two unlocalized *Buchanan* gatherings from Malawi, collected in 1895 (128 (BM), and in *Medley-Wood* 7038 (PRE)), may be referable to this species.

39. **Phyllanthus asperulatus** Hutch. in Bull. Misc. Inform., Kew **1920**: 27 (1920); in F.C. **5**, 2: 399 (1920). —Burtt Davy, Fl. Pl. Ferns Transvaal: 300 (1932). Type from South Africa (Transvaal).

Very like *Phyllanthus gillettianus*, but the lateral shoots are distinctly scabrid, and the female disk is more like that of *Phyllanthus leucanthus* and *Phyllanthus pseudoniruri*.

Botswana. SE: Mochudi, Phutodikobo Hill, fr. 14.iii.1967, *Lady Naomi Mitchison* 42 (K). **Zimbabwe**. S: Gwanda Distr., Special Native Area 'G', Weir 9, fl. 15.xii.1956, *Davies* 2310A (K). **Mozambique**. M: Goba–Moamba crossroads, fr. 26.ii.1948, *Torre* 7429 (LISC).

Also in South Africa (Transvaal). In mopane woodland and dry short grasslands, also colonizing bare ground in irrigation areas; 560–1067 m.

Torre 7429, cited above, varies somewhat in the direction of the previous species.

40. **Phyllanthus zambicus** Radcl.-Sm. in Kew Bull. **51**, 2: 328 (1996). Tab. **11**, figs. C1–C4. Type: Zambia, Chunga, Kafue Nat. Park, male, female, y. & o. fr. 29.i.1961, *Mitchell* 5/38 (SRGH, holotype).

This species differs from *Phyllanthus gillettianus* in its foliage leaves which are narrowly lanceolate or elliptic-lanceolate, and about twice as long as those in *P. gillettianus*, and in having its styles closely appressed to the top of the ovary.

Zambia. C: Mfuwe, Luangwa Nat. Park, o. fr. 15.iii.1969, *Taylor* 155 (SRGH).

Known only from these two localities. In floodplain grassland, and in mopane woodland on clay; c. 600 m.

Vernacular names as recorded in specimen data include: "butongi" (cinkoya); "changu" (citoka); "mbuluwe" (cisenga).

41. **Phyllanthus delagoensis** Hutch. in Bull. Misc. Inform., Kew, **1920**: 28 (1920); in F.C. **5**, 2: 400 (1920). —Burtt Davy, Fl. Pl. Ferns Transvaal: 300 (1932). —Mogg in Macnae & Kalk, Nat. Hist. Inhaca Isl., Moçamb.: 147 (1969). Type: Mozambique, Maputo (Lourenço Marques), o. fr. 8.xii.1897, *Schlechter* 11663 (K, holotype; BR; P).

A glabrous to asperulous semiprostrate annual to perennial herb up to 60 cm tall, but often much less, monoecious. Lead shoots angular; lateral shoots 2–8 cm long. Scale leaves 1.2 mm long, linear-lanceolate; stipules 1–1.2 mm long, triangular-lanceolate, sometimes with an abaxial filiform lobule. Foliage leaves distichous; petioles 0.5 mm long; stipules 0.8 mm long, linear, pallid with a light brown midrib. Leaf blades 2–5 × 1–2 mm, ovate to elliptic-oblong, subacute or obtuse, rounded to rounded-cuneate at the base, membranaceous, smooth to minutely papillose above, often scabridulous beneath, uniformly green (Mozambique) or somewhat discolorous (Zimbabwe); lateral nerves in 4–6 pairs, not visible above, visible or not beneath. Male flowers solitary or paired in the proximal axils of the lateral shoots, females solitary in the distal axils. Male flowers: pedicels 0.3–0.4 mm long, slender; sepals 6, very unequal, the outer 1 × 0.5 mm, lanceolate, acute, the inner 0.6 × 0.4 mm, ovate, apiculate and subcucullate, white with a narrow greenish midrib; disk glands 6, free, c. 0.1 mm across, ± rounded, smooth; stamens 3, filaments connate into a column 0.5 mm high, anthers 0.25 mm across, sessile, 4-lobed, horizontally held, transversely dehiscent, yellow. Female flowers: pedicels 0.25 mm long, extending to 2 mm in fruit; sepals 6, somewhat unequal, the outer 0.8 × 0.4 mm, the inner 0.6 × 0.3 mm, accrescent to 1.5 × 0.6 mm and 1.2 × 0.5 mm respectively, lanceolate, subacute, greenish; disk 0.5 mm in diameter, hexagonal, thin, undulate; ovary 0.3 mm in diameter, sessile, 6-lobed, subglobose, smooth; styles 3, 0.1 mm long, free, slightly divergent, shortly 2-lobed, stigmas rounded, subsessile. Fruit 1.3 × 2 mm, oblate, minutely foveolate-reticulate above, ± smooth beneath, pale green. Seeds 1 × 0.9 × 0.8 mm, segmentiform, yellowish-brown, with c. 14–17 fine and somewhat irregularly spaced longitudinal ridges on the dorsal facet, and c. 12 somewhat fractured concentric ridges on each ventral facet, with innumerable parallel transverse striae between them.

Zimbabwe. S: Masvingo Distr., Mutirikwi (Mtilikwe) R., Bangala (Bangara) Falls, o. fr. 13.xii.1953, *Wild* 4370 (K; LISC; PRE; SRGH). **Mozambique**. N: Mafamede Island, fl. & fr. 28.x.1965, *Mogg* 32542 (LISC; PRE). Z: Quelimane, y. fr. 1908, *Sim* 20768 (PRE). MS: Macúti,

y. male fl. 23.iii.1960, *Wild & Leach* 5238 (BM; K; MO; PRE; SRGH). GI: Bazaruto Island, male fl. viii.1937, *Gomes e Sousa* 1971 (K; LISC; SRGH). M: Costa do Sol to Marracuene, fl. 19.ii.1981, *de Koning & Boane* 8647(6) (BM; BR; K; LMU; MO).

Also in South Africa (Transvaal, KwaZulu-Natal). Locally common at low altitudes in sandy soil, in coastal scrub and grassland and in swampy places, also in moist sand and amongst rocks on river banks, and as a ruderal weed; 30–250 m.

42. **Phyllanthus ceratostemon** Brenan in Kew Bull. **21**: 259 (1967). —Jean F. Brunel, Phyllanthus Afr. Intertrop. Mad.: 367 (1987), unpublished thesis. —Radcliffe-Smith in Kew Bull. **44**: 452 (1989). Type: Zambia, Kaputa District, Mweru-Wantipa, road to Bulaya, fl. & fr. 12.iv.1957, *Richards* 9170 (K, holotype; BR).

A slender annual herb up to 45 cm tall, suberect, unbranched at the base, glabrous, monoecious. Lead shoot usually leafy; lateral shoots 2–5(8) cm long. Scale leaves 1–3 mm long, linear-subulate, often not developed; stipules triangular-lanceolate. Foliage leaves dimorphic, those of the lead shoot spirally arranged with petioles c. 0.5 mm long, and blades 0.7–4 cm × 1–2 mm, linear, acute, tapered to the base, lateral nerves in c. 3 pairs, invisible above, scarcely so beneath; those of the lateral shoots distichous, subsessile, 3–7 × 2–4 mm, broadly ovate, subacute, wide-cuneate to rounded or subtruncate at the base, lateral nerves in 2–3 pairs, scarcely visible; all leaves chartaceous. Stipules c. 1 mm long, linear-subulate, soon blackening in the upper half. Male flowers solitary or paired in short multibracteate cymules in the proximal axils of the leafy shoots, females solitary in the distal axils. Male flowers very shortly pedicellate; sepals 5, 1.5–1.75 × 1–1.25 mm, elliptic-obovate, rounded, yellowish-green with white margins; disk glands 5, free, 0.3 mm across, transversely elliptic, flattened, lobulate, papillose; stamens 3, filaments connate into a column 0.75 mm high, anthers 0.3 mm long, subsessile, strongly decurved, vertically held, connective produced into a small cylindric appendage, dehiscence longitudinal. Female flowers: pedicels 1.5–2 mm long, not or scarcely extending in fruit; sepals 5, 3 × 1.5–2 mm, scarcely accrescent in fruit, elliptic, obtuse, green with narrow hyaline margins; disk 0.8 mm in diameter, somewhat irregularly 5–6-lobed; ovary 0.7 mm in diameter, subsessile, 6-lobed, smooth; styles 3, 0.4 mm long, free, slightly divergent, shortly bifid, stigmas papillose. Fruit c. 2 × 2.5 mm, slightly depressed-subglobose, smooth, olivaceous. Seeds 1.4 × 1 × 1 mm, segmentiform, reddish-brown, with 14–15 somewhat irregular longitudinal rows of closely arranged transversely-disposed dark brown tubercles on the dorsal facet, and c. 12 irregular and fractured concentric arcs of such tubercles on each ventral facet, with innumerable parallel transverse striae between them.

Zambia. N: Kaputa Distr., Mweru-Wantipa (Mweru-wa-Ntipa), road to Bulaya, fl. & fr. 12.iv.1957, *Richards* 9170 (BR; K).

Also in Chad, Central African Republic and Zaire. In shallow soil amongst flat ironstone rocks; 1050 m.

43. **Phyllanthus crassinervius** Radcl.-Sm. in Kew Bull. **35**: 766 (1981); in F.T.E.A., Euphorb. 1: 52 (1987). —Jean F. Brunel, Phyllanthus Afr. Intertrop. Mad.: 373 (1987), unpublished thesis. Type from Tanzania (Southern Highlands Province).

A glabrous or minutely asperulous perennial herb or suffrutex up to 40 cm tall, monoecious or dioecious; stem branched from near the base, the leafy lead shoots arising from a stout cylindrical woody rootstock. Lateral shoots (2)5–10 cm long. Scale leaves 1 mm long, subulate, often not formed; stipules 1.4–1.8 × 0.5–0.8 mm, triangular-lanceolate. Foliage leaves spiral or distichous; petioles 0.5–0.75 mm long; stipules on the lateral shoots 1.5 mm long, linear-lanceolate, reddish-brown with cream-coloured margins. Leaf blades 3–13 × 2–8 mm, those of the lead shoots obovate-suborbicular, truncate to obtuse at the apex and rounded at the base, those of the lateral shoots elliptic or elliptic-ovate, acute or subacute and rounded-cuneate, chartaceous; lateral nerves in 6–9 pairs, not or scarcely prominent above, prominent and thickened beneath, looped and sometimes reddish-tinged. Male flowers usually solitary in the proximal axils of the lateral shoots and females in the distal axils, or occasionally the shoots may be either all male or all female. Male flowers: pedicels 1 mm long; sepals 6, 2 × 1 mm, oblong, obtuse, angular, yellowish-green; disk glands 6, free, c. 0.5 mm in diameter, ± circular, tuberculate, the tubercles apically pitted; stamens 3, filaments connate into a column 1 mm high, anthers 0.5 mm long, sessile, reflexed, ovoid, obiquely and apically dehiscent. Female flowers: pedicels 1 mm long, extending to 2 mm in fruit; sepals 6, 2.5–3 × 1.5 mm, elliptic-oblong, rounded, reddish-green with narrow hyaline margins, sometimes rose tinted; disk 1.5 mm in diameter, shallowly 6-lobed, granular-tuberculate; ovary c. 1 mm in diameter, shortly stipitate, 6-lobed, subglobose, smooth; styles 3, free, 0.5 mm long, divaricate, 2-lobed, stigmas rounded, smooth. Fruit 2 × 3 mm, 3-lobed to subglobose, smooth, ochraceous, sometimes purplish-tinged. Seeds 1.4 × 1 × 1 mm, segmentiform, ochraceous to brownish, with 14–15 slightly darker rows of minute transverse tubercles on the dorsal facet, and 11–12 concentric arcs of such tubercles on each ventral facet.

Zambia. W: Solwezi Distr., Kansanshi Mine, fl. & fr. 14.ix.1952, *White* 447A (FHO; K). **Malawi**. N: Mzimba Distr., Viphya Plateau, fl. 25.ix.1962, *Pawek* 5824 (K; MAL; MO; SRGH; UC).

Also in Zaire and Tanzania. In chipya woodland and submontane grassland, usually after annual fire, also in moist dambos; 1700–1740 m.

Pawek 13232 (DAV; MO), also from the Viphya, is somewhat intermediate between this species and *P. caespitosus.*

44. **Phyllanthus friesii** Hutch. in R.E. Fries, Wiss. Ergebn. Schwed. Rhod.-Kongo-Exped. **1**, 1: 121 (1914). —Radcliffe-Smith in Kew Bull. **35**: 767 (1981); in F.T.E.A., Euphorb. 1: 54 (1987). Type: Zambia, Northern Province, near Luapula, fr. 7.ix.1911, *R.E. Fries* 566 (UPS, holotype; K, fragment of holotype).

Phyllanthus angustatus Hutch. in R.E. Fries, Wiss. Ergebn. Schwed. Rhod.-Kongo-Exped. **1**, 1: 121 (1914). Type: Zambia, Northern Prov., 'between Bangweulu (Bangweolo) and Tanganyika', fr. 30.x.1911, *R.E. Fries* 1165 (UPS, holotype).

An erect virgate tufted perennial herb up to 60 cm tall, glabrous, dioecious; stems numerous wiry, arising from a stout carrot-shaped woody rootstock. Lead shoots leafy or not, often reddish-green; lateral shoots 5–15 cm long, vertically disposed. Scale leaves 1 mm long, subulate; stipules 1.25 mm long, triangular, auriculate adaxially, purplish-brown. Foliage leaves spirally disposed; petioles 0.3 mm long or leaves subsessile; stipules 1 mm long, linear-lanceolate, brownish. Leaf blades 4–10 × 0.5–1 mm, narrowly elliptic- to oblong-linear, subacute or acute, sometimes mucronate, margins usually involute, tapered to the base, thinly coriaceous, grey-green or glaucous; nerves invisible. Male flowers: pedicels 1 mm long; sepals 6, 2 × 1 mm, oblong, obtuse to rounded, creamy-white, often pink-tinged; disk glands 6, free, 0.3 mm in diameter, rounded, minutely verruculose; stamens 3, filaments connate into a column 1 mm high, anthers 0.5 mm long, sessile, contiguous, vertically held, soleiform, obliquely dehiscent. Female flowers: pedicels 2 mm long, extending to 8 mm in fruit; sepals accresent to 3.5 × 1.5 mm in fruit, reddish-green with a narrow hyaline margin, otherwise as in the male; disk 1 mm in diameter, annular, colliform, thick, slightly crenulate; ovary 0.6 mm in diameter, sessile, 6-lobed, subglobose, smooth; styles 3, ± free, 1.4 mm long, suberect, bifid, the stigmas slightly recurved. Fruit 3 × 4 mm, subglobose, smooth, greenish, often pinkish-tinged. Seeds 1.8–2 × 1.5 × 1.5 mm, segmentiform, pale greyish-brown, with 20–28 shallow darker longitudinal ridges on the dorsal facet, and 12–17 concentric ridges on each ventral facet, with innumerable parallel transverse striae between them.

Zambia. N: 5 km east of Kasama, fr. 16.ix.1960, *Robinson* 3832 (BR; K; LISC; SRGH). W: Mwinilunga Distr., Sinkabolo Dambo, fl. & y. fr. 9.xii.1937, *Milne-Redhead* 3576 (BR; K; LISC). C: Serenje, fr. 24.ix.1961, *Fanshawe* 6707 (BR; K; SRGH).

Also in Tanzania. Of scattered occurance on Kalahari Sands and dry sandy soils in watershed grasslands, wooded grasslands and dambos; also on termite mounds in swampy grassland and on rocky outcrops; 1200–1350 m.

45. **Phyllanthus holostylus** Milne-Redh. in Bull. Misc. Inform., Kew **1937**: 414 (1937). —Jean F. Brunel, Phyllanthus Afr. Intertrop. Mad. 374 (1987), unpublished thesis. Type: Zambia, Solwezi, male & female fl. 21.ix.1930, *Milne-Redhead* 1164 (K, holotype).

Phyllanthus oxycoccifolius sensu Radcl.-Sm. in F.T.E.A., Euphorb. 1: 23 (1987), pro parte quoad perscriptionem zambicum, non Hutch.

An erect, usually tufted perennial herb up to 15 cm tall, sometimes a suffrutex up to 50 cm tall, glabrous, dioecious; stems several arising at intervals from a woody rhizome. Lead shoots mostly leafy; lateral shoots up to 10 cm long, vertically disposed. Scale leaves c. 1 mm long, linear-lanceolate; stipules up to 2 mm long, triangular-lanceolate, scarious, fimbriate-denticulate, auriculate, reddish-brown. Foliage leaves spirally disposed or subdistichous; petioles 0.25 mm long or leaves subsessile; stipules as for those of the scale leaves. Leaf blades 5–15 × 2–7 mm, ovate-lanceolate to lanceolate, subacute or acute, rounded at the base, margins sometimes slightly thickened, thinly coriaceous, somewhat glaucous, often with narrow longitudinal folds when dried; lateral nerves in 4–7 pairs, often inconspicuous. Male flowers: pedicels 1.5–2 mm long; buds constricted; sepals 6, 3 × 1 mm, elliptic-oblong, obtuse, greenish, whitish or pink-tinged; disk glands 6, c. 0.3 mm in diameter, ± contiguous, flat, closely appressed to the sepals, truncate, minutely verruculose; stamens 3, filaments connate into a stout column 2 mm high, anthers 1 mm long, sessile, ± vertically held, ellipsoid, longitudinally dehiscent. Female flowers: pedicels 1–2 mm long, extending to c. 5 mm long in fruit; sepals 6, 2 × 1.5 mm at first, later accrescent to 3 × 2 mm, broadly ovate, obtuse, reddish-green with hyaline margins; disk 1.1 mm in diameter, annular, thick, scarcely crenulate to subentire; ovary 1 mm in diameter, sessile, 6-lobed, subglobose, smooth, fleshy; styles 3, free, 1.2 mm long, erect, not or scarcely 2-lobed, stigmas broadened, reflexed, rounded, smooth, ± persistent. Fruit 2.5 × 4.5 mm, somewhat depressed-subglobose, smooth, greenish or reddish-tinged. Seeds 2–2.2 × 1.8 × 1.5 mm, segmentiform, pale brown, faintly 18–20-lineate on the dorsal facet and even more faintly 12-arced on each ventral facet.

Zambia. W: Solwezi, fr. 10.ix.1952, *Angus* 399 (FHO; K). **Malawi**. N: Mafinga Mts. above Chisenga, o. fr. 11.xi.1958, *Robson & Fanshawe* 565 (BM; K; LISC; SRGH).

Also in Zaire and Angola. A pyrophyte of watershed grassland and dambos, also in high rainfall miombo woodland and chipya; 1510–1960 m.

46. **Phyllanthus virgulatus** Müll. Arg. in J. Bot. **2**: 330 (1864); in De Candolle, Prodr. **15**, 2: 360 (1866). —Hutchinson in F.T.A. **6**, 1: 705 (1912). —Engler, Pflanzenw. Afrikas (Veg. Erde 9) **3**, 2: 27 (1921). —Jean F. Brunel, Phyllanthus Afr. Intertrop. Mad.: 390 (1987), unpublished thesis. Type from Angola (Malanje Province).

Very like *Phyllanthus holostylus*, but differing chiefly in the more shrubby habit and in the coriaceous linear-lanceolate foliage leaves with sharply spinescent apices and hyaline cartilaginous margins.

Zambia. B: Kaoma (Mankoya), o. fr. 15.x.1964, *Fanshawe* 8956 (BR; K; NDO; SRGH). W: Mwinilunga Distr., Tshikundula Stream, male fl. 4.x.1952, *Angus* 594 (BR; FHO; K).

Also in Zaire and Angola. On Kalahari Sand, in miombo woodland with long grass.

Holmes 1054 (FHO) from c. 15 km south of Chavuma, combines the habit of *P. holostylus* with the leaf apex of *P. virgulatus*.

47. **Phyllanthus macranthus** Pax in Bot. Jahrb. Syst. **19**: 77 (1894); in Engler, Pflanzenw. Ost-Afrikas **C**: 236 (1895). —S. Moore in J. Linn. Soc., Bot. **40**: 192 (1911). —Hutchinson in F.T.A. **6**, 1: 705 (1912). —Engler, Pflanzenw. Afrikas (Veg. Erde 9) **3**, 2: 27 (1921). —Brenan, Check-list For. Trees Shrubs Tang. Terr.: 222 (1949). —Drummond in Kirkia **10**: 251 (1975). —Jean F. Brunel, Phyllanthus Afr. Intertrop. Mad.: 374 (1987), unpublished thesis. —Radcliffe-Smith in F.T.E.A., Euphorb. 1: 43 (1987). Type: Mozambique, without precise locality, 2.i.1889, *Stuhlmann* 850 (B†, holotype); 29 km Liupo–Mogincual, o. fr. 30.iii.1964, *Torre & Paiva* 11468 (LISC, lectotype chosen by Brunel, 1987, but not effectively published).

A bushy shrub up to 1.5 m tall, glabrous, asperulous, papillose or puberulous, monoecious or sometimes dioecious; stems with longitudinally peeling bark. Twigs light brown at first, later becoming greyish. Lateral leafy shoots 1–3(10) cm long. Short shoots developing, giving rise to secondary lateral shoots. Scale leaves c. 1 mm long, linear-lanceolate; stipules c. 1.5 mm long, triangular-lanceolate, purplish-brown. Foliage leaves distichous; petioles 0.5 mm long; stipules 1 mm long, linear-lanceolate, pale brown. Leaf blades 3–17 × 2–9 mm, elliptic, obovate or suborbicular, obtuse, rounded or retuse, sometimes mucronate, cuneate to rounded at the base, thinly chartaceous when dried, slightly asperulous along the midrib and main nerves above and beneath and on the margin or else quite glabrous, fleshy, pale green;

lateral nerves in 3–5 pairs, weakly brochidodromous, scarcely prominent above, slightly so beneath. Flowers solitary. Male flowers: pedicels 1 mm long; sepals 6, 2-seriate, the outer 2 × 0.5 mm, oblong, obtuse, glabrous, the inner 2.5 × 1 mm, lanceolate, subacute, minutely granulate without, white or cream-coloured with a greenish midrib; disk glands 6, in 3 pairs opposite the inner sepals, 0.3 × 0.2 mm, oblong, flat, ± smooth; stamens 3, filaments connate into a slender column 2 mm high, anthers 1 × 0.4 mm, free, divaricate, fusiform, connective broad, longitudinally dehiscent. Female flowers: pedicels 2–4 mm long, extending to 5–6 mm in fruit; sepals 6, 3 × 1.5–2 mm, accrescent to 5 × 4 mm, broadly ovate-suborbicular, subacute to rounded, pale green with a creamy-white margin, venose, completely enclosing the fruit; disk 1.2 mm in diameter, 6-lobed, thick, papillose; ovary 1–1.5 mm in diameter, shortly stipitate, 6-lobed, subglobose, smooth; styles 3, ± free, 1 mm long, spreading, deeply bifid, stigmas filiform, ± straight or coiled. Fruit 2.2 × 3.5 mm, depressed, somewhat 3-lobed to subglobose, smooth, olivaceous. Seeds 2 × 1.5 × 1.3 mm, rounded-trigonous, ochreous, with 6–8 shallow scarcely perceptible longitudinal ridges on the dorsal facet, and 6 concentric such ridges on each ventral facet, with innumerable transverse striae between them.

Plant glabrous, asperulous or papillose, the asperities or papillae unicellular, rarely a few bicellular · var. *macranthus*
Plant puberulous, the hairs tricellular, sometimes multicellular · · · · · · · · · · · · · · · var. *gilletii*

Var. **macranthus**

Plant glabrous, asperulous or papillose, the papillae are unicellular or rarely a few bicellular.

Zimbabwe. N: Mutoko Distr., Mkota Communal Land (Tribal Trust Land), fl. 4.xii.1968, *Müller & Burrows* 946 (K; SRGH). E: Chimanimani Distr., Umvumvumvu R., y. fr. 19.xi.1956, *Chase* 6245 (K; LISC; SRGH). S: Ndanga Distr., 3 km north of Chipinda Pools, y. fl. 17.xi.1959, *Goodier* 729 (K; SRGH). **Mozambique**. N: Nampula, Monapo, fr. 10.ii.1984, *Groenendijk, de Koning & Dungo* 971 (K; LMU; MO). MS: Madanda Forests, fl. 5.xii.1906, *Swynnerton* 1756 (BM; K). GI: Mavume, ± st. no date, *Gomes e Sousa* 2169 (K).

Also in Tanzania and Angola. Sandy soils, often on rocky outcrops with *Brachystegia glaucescens* and in low altitude dry mixed deciduous woodland on sand; 50–1067 m.

Var. **gilletii** (De Wild.) Jean F. Brunel ex Radcl.-Sm. in Kew Bull. **51**, 2: 308 (1996). Type from Zaire (Kinshasa Province).

Phyllanthus gilletii De Wild. in Ann. Mus. Congo Belge, Bot., Sér. 5, **2**: 266 (1908).

Phyllanthus macranthus var. *gilletii* (De Wild.) Jean F. Brunel, Phyllanthus Afr. Intertrop. Mad.: 374 (1987), unpublished thesis.

Plant puberulous, the hairs tricellular or multicellular.

Mozambique. N: Angoche (António Enes), st. 25.vi.1937, *Torre* 1557 (COI; LISC). MS: Madanda Forests, fr. 5.xii.1906, *Swynnerton* 1757 (BM).

Also in Zaire and Angola; doubtfully reported from Zambia. In sandy soils, in dry woodland and in cashew plantation; 30–120 m.

48. **Phyllanthus gossweileri** Hutch. in Bull. Misc. Inform., Kew **1911**: 315 (1911); in F.T.A. **6**, 1: 724 (1912). —Engler, Pflanzenw. Afrikas (Veg. Erde 9) **3**, 2: 26 (1921). —Jean F. Brunel, Phyllanthus Afr. Intertrop. Mad.: 379 (1987), unpublished thesis. Type from Angola (Bié Province).

A slender erect annual herb up to c. 30 cm tall, glabrous, monoecious. Lead shoots terete, reddish; lateral shoots few, erect, up to 25 cm long. Scale leaves 0.75 mm long, subulate; stipules 0.75 mm long, triangular-ovate. Foliage leaves distichous, vertically held against the axis; petioles 0.5 mm long; stipules 1 mm long, subulate, brownish. Leaf blades 6–12 × 1–3.5 mm, narrowly elliptic-lanceolate, sharply acute, rounded-cuneate at the base, firmly chartaceous, lateral nerves invisible, dull. Male flowers 1–3 per axil in short multibracteate cymules in the lower half of the lateral shoots, female flower solitary in the upper axils. Male flowers: pedicels 0.3 mm long; sepals 6, 1–1.2 × 0.6–0.8 mm, elliptic-ovate, rounded, dull yellowish-white; disk glands 6, 0.2 mm in diameter, circular, thin, flat, smooth, widely spaced; stamens 3, filaments connate into a stout column 1 mm high, anthers 0.5 × 0.3 mm, free, parallel, ellipsoid, longitudinally dehiscent. Female flowers: pedicels 0.5 mm long, extending to 2 mm in fruit; sepals 6, 1.5 × 0.6 mm, accrescent to 2 × 1.2 mm, elliptic-oblong, rounded or obtuse, green with narrow white margins; disk 0.5 mm in diameter, shallowly 6-lobed, very thin, flat, smooth, not readily visible in older flowers; ovary 1 mm in diameter, sessile, 6-lobed, subglobose, smooth; styles 3, ± free, 0.4 mm long, spreading and appressed to the top of the ovary, shortly bifid, stigmas slightly recurved. Fruit 2.3 × 3–4 mm, depressed-subglobose, smooth, ochreous. Seeds 2 × 1.3 × 1.2 mm, sharply triquetrous-segmentiform, yellowish-brown to dull greyish-brown, often with c. 8–10 longitudinal stripes of darker brown and 15–20 shallow longitudinal ridges on the dorsal facet, and with 8 concentric bands and 15 concentric ridges on each ventral facet, with innumerable transverse striae between them which exfoliate from older seeds.

Zambia. W: 45 km east of Mwinilunga, fl. & fr. 15.iv.1960, *Robinson* 3557 (K; SRGH).
Also in Zaire and Angola. In dry, barren watershed grassland; 1400 m.

49. **Phyllanthus omahakensis** Dinter & Pax in Bot. Jahrb. Syst. **45**: 234 (1910). —Hutchinson in F.T.A. **6**, 1: 725 (1912). —Engler, Pflanzenw. Afrikas (Veg. Erde 9) **3**, 2: 26 (1921). —P.G. Meyer in Merxmüller, Prodr. Fl. SW. Afrika, fam. 67: 39 (1967). —Jean F. Brunel, Phyllanthus Afr. Intertrop. Mad.: 379 (1987), unpublished thesis. Type from Namibia.
Phyllanthus milanjicus Hutch., ined. (on herbarium sheets only).

An erect to procumbent or semi-prostrate annual herb up to 30 cm tall or 60 cm across, glabrous, asperulous or papillose-pilose, monoecious; stems often much-branched from the base. Lowest nodes of the lead shoots bearing leaves and secondary lead shoots, the upper bearing scale leaves and lateral shoots. Lateral shoots 2–8 cm long. Scale leaves 1 mm long, subulate; stipules 1 mm long, linear-lanceolate. Foliage leaves mostly distichous; petioles 0.3–0.5 mm long; stipules 1–1.25 mm long, linear to filiform, cream-coloured to pale brownish or reddish. Leaf blades 3–15 × 1–4.5 mm, ovate to ovate-lanceolate, acute, subacute or obtuse, sometimes mucronulate, rounded at the base, margins sometimes minutely asperulate-serrulate, chartaceous, glabrous on both surfaces, glaucous; lateral nerves in c. 5 pairs, but usually not visible. Male flowers solitary or paired in the axils of the lower half of the lateral shoots, female flower solitary in the upper axils. Male flowers: pedicels 0.75 mm long; sepals 6, 1.5 × 0.8–1 mm, oblong, rounded, pale greenish-white; disk glands 6, 0.2 mm in diameter, circular, very thin, flat, smooth; stamens 3, filaments connate into a conical column c. 1 mm high, anthers 0.4 × 0.3 mm, free, sessile, divergent, ellipsoid, longitudinally dehiscent. Female flowers: pedicels 1 mm long, extending to 2–3 mm in fruit; sepals 6, 1.2 × 0.75 mm, slightly accrescent to 2 × 1 mm in fruit, elliptic-oblong, rounded, green with white margins; disk 0.5 mm in diameter, shallowly 6-lobed, flat, smooth, only clearly visible in young flowers; ovary 0.6 mm in diameter, sessile, shallowly 6-lobed, subglobose, smooth; styles 3, united in the lower third, 0.75 mm long, erect, bifid, stigmas subglobose. Fruit 2 × 2.2–2.5 mm, subglobose, ± smooth to slightly granulate, often yellowish or reddish-tinged. Seeds 1.2–1.7 × 0.75–1 × 0.7–1.1 mm, rounded trigonous- to sharply triquetrous-segmentiform, yellowish-brown or tawny, with 14–20 shallow longitudinal ridges on the dorsal facet and 10–16 concentric ridges on each ventral facet, with innumerable transverse striae between them.

Botswana. N: 69 km west of Nokaneng, fl. & fr. 12.iii.1965, *Wild & Drummond* 6892 (K; LISC; SRGH). SW: 21 km south by east of Grootkalk, Nossob (Nosop) R., o. fr. 21.iv.1960, *Leistner* 1866 (B; K; PRE). **Zambia**. B: Mongu, fl. & fr. 12.i.1966, *Robinson* 6793 (BR; K; SRGH). C: 25 km from Lusaka to Mumbwa, fr. 20.iii.1965, *Robinson* 6446 (B; K; P; SRGH). S: Machili, fl. 1.i.1961, *Fanshawe* 6071 (K; NDO; SRGH). **Zimbabwe**. W: Shangani Distr., Gwampa For. Res., fl. & fr. 2.ii.1955, *Goldsmith* 42/55 (K; SRGH). C: Hatfield, fr. 14.v.1934, *Gilliland* 132 (BM; K; SRGH).
Also in Angola and Namibia. On Kalahari Sand in woodland and mutemwa thickets, also in sandy soil in seasonally wet grassland, in well drained woodland, and in sand veld; 975–1035 m

50. **Phyllanthus prostratus** Welw. ex Müll. Arg. in J. Bot. **2**: 330 (1864); in De Candolle, Prodr. **15**, 2: 361 (1866). —Hutchinson in F.T.A. **6**, 1: 720 (1912). —Engler, Pflanzenw. Afrikas (Veg. Erde 9) **3**, 2: 27 (1921). —Jean F. Brunel, Phyllanthus Afr. Intertrop. Mad.: 379 (1987), unpublished thesis. Type from Angola (Huíla Province).

Very like *Phyllanthus omahakensis*, differing chiefly in being a small glabrous procumbent or prostrate herb often rooting at the nodes, with suborbicular mucronate foliage leaves.

Zimbabwe. C: Goromonzi, fl. 17.iv.1927, *Eyles* 4918 (K; SRGH).
Also in Angola. In open sandveld; 1585 m.

51. **Phyllanthus pseudocarunculatus** Radcl.-Sm. in Kew Bull. **51**, 2: 318 (1996). Type from Zaire (Shaba Province).
Phyllanthus carunculatus Jean F. Brunel, Phyllanthus Afr. Intertrop. Mad.: 380 (1987), unpublished thesis.

A glabrous annual herb with a short erect terminal lead shoot up to 10 cm high and several decumbent or semi-prostrate secondary lead shoots arising from near the base and spreading to 15–20 cm, monoecious. Stems red. Lateral leafy shoots 2–10 cm long. Scale leaves 1 mm long, subulate; stipules 1 × 0.5 mm, triangular-lanceolate, reddish-brown. Foliage leaves distichous, with a few spiral at the lowest nodes of the lead shoots; petioles 0.3 mm long; stipules 1–1.5 mm long, narrowly linear-lanceolate, creamy-white, sometimes reddish at the apex. Leaf blades 3–8 × 1–3 mm, ovate-lanceolate or oblong-lanceolate, subacute or obtuse, rounded to cuneate at the base, chartaceous, pseudopapillose beneath when dried, glaucous to pale grey-green, often reddish-tinged; lateral nerves in up to 9 pairs, mostly invisible. Male flowers in small, few-flowered multibracteate cymules in the lower axils of the lateral shoots, female flowers solitary in the upper axils. Male flowers: pedicels c. 0.5 mm long, slender; sepals 6, 0.8–1 × 0.5 mm, obovate-oblong, rounded, cream-coloured with hyaline margins; disk glands 6, 0.1 mm in diameter, ovoid, smooth; stamens 3, filaments connate into a squat column 0.5 mm high, anthers 0.25 mm long, free, sessile, erect, 2-lobed, longitudinally dehiscent. Female flowers: pedicels 0.5 mm long, extending to 1.5–2 mm in fruit; sepals 6, 0.75 × 0.7 mm, slightly accrescent to 1.5 × 1 mm in fruit, obovate-suborbicular, rounded, strongly concave, pale glaucous-green with white margins; disk 0.5 mm in diameter, shallowly 6-lobed, flat, smooth, not clearly visible in older flowers; ovary 0.5 mm in diameter, sessile, 6-lobed, subglobose, smooth; styles 3, united in the lower half, 0.2 mm long, erect, 2-lobed, stigmas subglobose, recurved. Fruit 2 × 3 mm, depressed-subglobose, smooth, pale yellow-green, drying brownish. Seeds 1.3 × 1 × 0.9 mm, roundly trigonous-segmentiform, with a collar-like excrescence around the terminal hilum, shiny, warm chestnut-brown, with 30–40 fine longitudinal lines on the dorsal facet, and 20–30 concentric arcs on each ventral facet; transverse striae scarcely visible.

Zambia. N: Lake Mweru, near Nchelenge, fl. & fr. 23.iv.1951, *Bullock* 3825 (BR; K); Cassava (Casawa), Lake Tanganyika, 14.iv.1957, *Richards* 9210 (K); Lake Bangweuleu, Samfya, saw mill, 21.iv.1989, *Radcliffe-Smith et al.* 5742 (K).
Also in Zaire. Locally common on lakeshores, in deep dry sand and on sand dunes; 975–1050 m.

52. **Phyllanthus paxii** Hutch. in Bull. Misc. Inform., Kew **1911**: 316 (1911); in F.T.A. **6**, 1: 718 (1912). —Engler, Pflanzenw. Afrikas (Veg. Erde 9) **3**, 2: 27 (1921). —Jean F. Brunel, Phyllanthus Afr. Intertrop. Mad.: 392 (1987), unpublished thesis. —Radcliffe-Smith in F.T.E.A., Euphorb. 1: 56 (1987). Type: Malawi, Northern Prov., without precise locality or date, *Whyte* s.n. (K, holotype).

A stiffly erect, sometimes tufted, woody herb or small shrub up to 120 cm tall, usually unbranched at base, completely glabrous, monoecious. Lead shoots terete, often purplish-grey; lateral shoots up to 15 cm long. Scale leaves 1.5–2 mm long, triangular-lanceolate, acute; stipules 2.5–3 × 1 mm, ovate-lanceolate, irregularly denticulate to subentire, strongly auriculate adaxially, reddish-brown. Foliage leaves distichous; petioles 0.5–0.75 mm long; stipules 1–1.2 mm long, linear-lanceolate, buff, soon darkening to brown. Leaf blades 0.5–1.8 × 0.1–0.7 cm, elliptic to elliptic-lanceolate, acute, subacute or obtuse, usually mucronulate, rounded-cuneate at the base, margin somewhat thickened, subentire to minutely denticulate, membranaceous to chartaceous, yellowish-green to dark green above, glaucous beneath, closing up when touched; lateral nerves in 7–12 pairs, indistinct to slightly prominent above, fairly prominent beneath. Male flowers in 2–3-flowered cymules in the middle axils of the lateral shoots, female flowers solitary in the most distal axils. Male flowers: pedicels 1–2 mm long, slender; sepals 5, 2 × 1 mm, oblong, rounded, white with a cream-coloured median stripe; disk glands 5, free, 0.25 mm across, transversely ovoid, distinctly tuberculate, creamy-white; stamens 3, filaments connate into a stout column 1.5 mm high, anthers 0.5 × 0.25 mm, ellipsoid, ± free, sessile, vertically held, longitudinally dehiscent, yellow. Female flowers: pedicels 1 mm long, extending to 5 mm in fruit, stouter than in the male and broadening at the apex, dull purplish; sepals 5, 1.5 × 0.5 mm, accrescent to up to 4 × 2 mm in fruit, oblong-lanceolate, obtuse, white with a broad cream-coloured, greenish or reddish median band; disk 0.4 mm in diameter, shallowly cupular, slightly undulate-lobulate on the margin; ovary 0.5 mm in diameter, sessile, 6-lobed, subglobose, smooth, cream-coloured, sometimes reddish-tinged; styles 3, united at the base, 0.4 mm long, erect, 2-lobed, stigmas strongly recurved. Fruit 2–2.5 × 3.5–4 mm, depressed 6-lobed, subglobose, ± smooth, yellow-green to bright red. Seeds 2 × 1.5 × 1.5 mm, segmentiform, brown or buff, with 14–20 fine longitudinal lines on the dorsal facet, 12–14 concentric arcs on each ventral facet, and innumerable finer transverse striae between them.

Zambia. N: Kasama Distr., Luombi R., fr. 31.iii.1955, *Exell, Mendonça & Wild* 1363 (BM; LISC; SRGH). W: Mwinilunga Distr., Kalene Hill, o. fr. 18.v.1989, *Mutimushi* 3200 (K; NDO; SRGH). **Malawi**. N: Nkhata Bay Distr., Chikale Beach, o. fr. 15.iv.1977, *Pawek* 12593 (DAV; K; MAL; MO; PRE; SRGH; UC). C: Kasungu Distr., Chimaliro Forest, o. fr. 5.vii.1976, *Pawek* 11465 (K; MAL; MO; SRGH; UC). **Mozambique**. N: Lichinga Distr., Litunde, o. fr. 1.vii.1934, *Torre* 154 (LISC).

Also in Zaire, Burundi, Tanzania and Angola. On sandy soil and rocky hillsides, in high rainfall *Brachystegia/Uapaca* woodlands, miombo and chipya woodlands, also in shade in riverine forests and lakeshore vegetation, and submontane wooded grassland on slopes; 520–1525 m.

53. **Phyllanthus caespitosus** Brenan in Kew Bull. **21**. 258 (1967). Type: Zambia, Kasama Distr., 33 km ESE of Kasama, fr. 18.ii.1961, *Robinson* 4391 (K, holotype).

Very close to *Phyllanthus paxii*, but with leaf blades more broadly elliptic, mucronulate-aristate and firmly chartaceous; male pedicels up to 5 mm long; disk glands 0.75 mm across and coarsely muricate; female flowers more numerous, occupying all the axils in the upper half of the lateral shoots.

Zambia. N: 33 km ESE of Kasama, fr. 18.ii.1961, *Robinson* 4391 (K). W: Ndola, male fl. 21.xii.1956, *Fanshawe* 2902 (K; NDO).

Not known elsewhere. A pyrophyte of plateau miombo woodlands; 500–1000 m.

In northern Malawi intermediates between this and *P. caespitosus* exist (e.g. 'Migubu' (?Misuku) Hills, fr. 17.iii.1953, *J. Williamson* 230 (BM); Nkhata Bay Distr., Chikale Beach, o. fr. 10.v.1970, *Brummitt* 10580 (K; LISC; MAL; PRE; SRGH); Nkhwadzi For. Res., fr. 13.vi.1987, *la Croix* 4527 (MO)).

The style differences described by Brenan are not significant.

54. **Phyllanthus hutchinsonianus** S. Moore in J. Linn. Soc., Bot. **40**: 192 (1911). —Hutchinson in F.T.A. **6**, 1: 718 (1912), pro parte excl. spec. nyasica. —Eyles in Trans. Roy. Soc. South Africa **5**: 393 (1916). —Engler, Pflanzenw. Afrikas (Veg. Erde 9) **3**, 2: 29 (1921). —Drummond in Kirkia **10**: 251 (1975). —Jean F. Brunel, Phyllanthus Afr. Intertrop. Mad.: 392 (1987), unpublished thesis. —Radcliffe-Smith in F.T.E.A., Euphorb. 1: 56 (1987). Type: Mozambique, Chimanimani Mts., fl. & fr. 26.ix.1906, *Swynnerton* 1524 (K, holotype; BM; SRGH).

An erect virgate shrub or subshrub 0.5–2.5 m tall, often much-branched, glabrous, monoecious or sometimes dioecious. Lead shoots angular. Lateral shoots (2)5–10 cm long, often borne close together at the top of the lead shoots. Scale leaves c. 2 mm long, subulate, dark purplish-brown; stipules 3 × 1 mm, triangular-lanceolate, acutely acuminate, strongly auriculate adaxially, scarious, dark brown. Foliage leaves closely distichous; petioles c. 0.5 mm long; stipules c. 1

mm long, narrowly triangular-lanceolate, brownish. Leaf blades 3–13 × 2–7.5 mm, elliptic-obovate to elliptic-oblong, obtuse or rounded, sometimes apiculate, rounded-cuneate at the base, thickened and slightly recurved at the margins, chartaceous, dark green above, paler beneath, often reddish on the margins; lateral nerves in 3–6 pairs, scarcely visible above, often indistinct beneath. Male flowers in few-flowered bracteate cymules in the middle axils of the leafy shoots, female flowers solitary in the uppermost axils, either on the same or different shoots or plants. Male flowers: pedicels up to 3 mm long, slender; sepals 5(6), 2 × 1.5 mm, obovate-suborbicular, cream-coloured or greenish with white margins; disk glands 5(6), free, c. 0.5 mm in diameter, ± circular, shallowly tuberculate, the tubercles with apical pits; stamens 2(3), filaments united for the most part into a column c. 1 mm high, anthers 0.4 × 0.5 mm, transversely ovoid, free, ± horizontally held, laterally dehiscent, yellow. Female flowers: pedicels 2–5 mm long, more robust than in the male flower and broadening towards the apex; sepals 5, 1.75 × 1.25 mm, accrescent to up to 3.5 × 2.5 mm, obovate-suborbicular, cream-coloured to yellowish, often flushed reddish; disk 0.8 mm in diameter, roundly pentagonal, slightly undulate at the margin; ovary 0.75 mm in diameter, sessile, 6-lobed, subglobose, smooth; styles 3, 0.6 mm long, ± free, suberect at first, later divergent, bifid, the stigmas slightly recurved. Fruit 2–3 × 3–4 mm, depressed-globose, ± smooth, ochreous, enclosed by the persistent sepals. Seeds 1.5–2 × 1.2–1.6 × 1–1.4 mm, segmentiform, pale brown, with 15–20 fine longitudinal lines on the dorsal facet, and c. 12 concentric arcs on each ventral facet, with innumerable transverse striae between them.

Zimbabwe. E: Nyanga Distr., Pungwe Gorge, male & female fl. & fr. 15.vii.1955, *Chase* 5661 (B; BM; BR; COI; K; LISC; PRE; SRGH). **Malawi**. *Burtt Davy* 22100 (FHO) unlocalized. **Mozambique**. Z: Serra do Gurué, Namuli, male fl. 6.xi.1967, *Torre & Correia* 15936 (LISC). MS: Báruè, Serra de Choa, 9 km from Choa to Catandica (Vila Gouveia), male fl. 25.v.1971, *Torre & Correia* 18658 (LISC).

Also in Tanzania. Montane grassland usually in shelter of rocks, on rocky slopes and beside streams, also in evergreen forest margin understorey, on edge of kloof forest and gallery forest; 1200–2500 m.

Burtt Davy 22100 (FHO) from Malawi appears to be this species rather than any of the 3 following related species. However, no locality is given.

55. **Phyllanthus confusus** Brenan in Mem. N.Y. Bot. Gard. **9**, 1: 68 (1954). —Jean F. Brunel, Phyllanthus Afr. Intertrop. Mad.: 396 (1987), unpublished thesis. Type: Malawi, Mulanje Mt., Lichenya Plateau, male, female fl. & fr. 27.vi.1946, *Brass* 16472 (K, holotype; NY).

Phyllanthus rotundifolius var. *leucocalyx* sensu Baker f. in Trans. Linn. Soc. London, Bot. **4**: 38 (1894), non Müll. Arg.

Phyllanthus hutchinsonianus sensu Hutch. in F.T.A. **6**, 1: 718 (1912), pro parte, quoad spec. nyasica, non S. Moore.

Phyllanthus sp. 1, Brenan, in Mem. N.Y. Bot. Gard. **9**, 1: 69 (1954).

Phyllanthus brenanianus Jean F. Brunel, Phyllanthus Afr. Intertrop. Mad.: 392 (1987), unpublished thesis. Based on *Brass* 16571 (BR; K; NY; PRE), Malawi, Mulanje Mt., Lichenya Plateau, fr. 1.vii.1946.

Very like *Phyllanthus hutchinsonianus*, but with shorter internodes on the lateral shoots; smaller leaves with blades 1–11 × 0.5–5 mm, oblong to elliptic-oblong, membranaceous, not or scarcely thickened on the margins, lateral nerves usually not visible above, scarcely so beneath; female pedicels 1–3 mm long; female sepals accrescent to 2–3 × 1 mm, elliptic-oblong, not enfolding the fruit; fruit 2 × 3 mm; and seeds 1.2 × 1.1 × 1 mm.

Malawi. S: Mt. Mulanje, Litchenya Plateau, male fl. 25.vi.1946, *Brass* 16426 (BM; K; NY; SRGH).

Endemic to Mt. Mulanje. Locally abundant on and amongst rocks, on rocky slopes and seepage slopes, in submontane grassland with scattered shrubs, and at stream and forest margins, also in *Pinus patula* plantations and pathsides; 1200–2750 m.

56. **Phyllanthus nyikae** Radcl.-Sm. in Kew Bull. **51**, 2: 317 (1996). Type: Malawi, Nyika Plateau, male fl. 19.viii.1946, *Brass* 17332 (K, holotype; BM; NY; PRE; SRGH).

Phyllanthus sp. 2. —Brenan in Mem. N.Y. Bot. Gard. **9**, 1: 69 (1954).

Very like *Phyllanthus confusus* from Mt. Mulanje but with lead shoots and lower part of the lateral shoots minutely papillose-puberulous with multicellular hairs; scale leaves strongly reflexed, their stipules scarcely or slighly auriculate adaxially; male disk glands shallowly lobulate and tuberculate; and the female disk deeply 3-lobed.

Malawi. N: Rumphi Distr., Nyika Plateau, Nyamkowa, 6.5 km west of Livingstonia, female fl. 23.ii.1978, *Pawek* 13851 (DAV; K; MAL; MO; SRGH; UC).

Endemic to Nyika Plateau. In montane grassland and on the grassy edges of montane forest; 2030–2340 m.

57. **Phyllanthus mafingensis** Radcl.-Sm. in Kew Bull. **51**, 2: 308 (1996). Type: Malawi, Chitipa Distr., Mafinga Mts., north end of ridge, fl. & fr. 2.iii.1982, *Brummitt, Polhill & Banda* 16263 (K, holotype; BR; C; LISC; MAL; MO; SRGH).

Very like *Phyllanthus nyikae*, but the plants procumbent with indumentum composed of simple papillae; scale leaves swollen and blackened at the base, not strongly reflexed; male disk glands not lobulate; female disk collar-like, pentagonal; and the ovary stipitate.

Malawi. N: Chitipa Distr., Mafinga Mts., north end of ridge, male & female fl. & fr. 2.iii.1982, *Brummitt, Polhill & Banda* 16263 (BR; C; K; LISC; MAL; MO; SRGH).

Endemic to the Mafinga Mts. Known only from the type. On rock outcrops in submontane grassland; 2340 m.

58. **Phyllanthus myrtaceus** Sond. in Linnaea **23**: 134 (1850). —Müller Argoviensis in De Candolle, Prodr. **15**, 2: 397 (1866). —S. Moore in J. Linn. Soc., Bot. **40**: 192 (1911). —Hutchinson in F.T.A. **6**, 1: 726 (1912); in F.C. **5**, 2: 397 (1920). —Eyles in Trans. Roy. Soc. South Africa **5**: 393 (1916). —Engler, Pflanzenw. Afrikas (Veg. Erde 9) **3**, 2: 29 (1921). —Drummond in Kirkia **10**: 251 (1975). —Jean F. Brunel, Phyllanthus Afr. Intertrop. Mad.: 392 (1987), unpublished thesis. Type from South Africa (KwaZulu-Natal).

Very like *Phyllanthus hutchinsonianus*, but the plants generally dioecious with scale leaf stipules not as strongly auriculate; foliage leaves not as closely distichous, the blades up to 2.8 × 2.3 cm and often broadly elliptic-ovate, glaucous beneath, occasionally retuse at the apex and truncate at the base and with up to 12 pairs of lateral nerves; female pedicels 8–14 mm long; female sepals accrescent to up to 6 × 4 mm; seeds silky-textured and brown with only 8–9 longitudinal rows of shallow closely-set reddish-brown transversely lineate tubercles on the dorsal facet, and c. 6 concentric arcs of such tubercles per ventral facet, connected by parallel transverse striae.

Zimbabwe. E: Chimanimani Distr., Makurupini area, male & female fl. 28.viii.1969, *T. Wild* 37 (K; LISC; PRE; SRGH). **Mozambique**. MS: Chimanimani Mts., female fl. 28.v.1969, *Müller* 1077 (K; LISC; SRGH).

Also in South Africa (KwaZulu-Natal and Eastern Cape Province). Submontane grassland often at streamsides, also on forest margins and in *Brachystegia* woodland at south eastern foot of Chimanimani Mts; 305–2250 m.

H.Wild 2936 from the Chimanimani Mts., and *Biegel* 277 and *Chase* 2875 from Nyanga, are intermediate between this species and *P. hutchinsonianus.*

59. **Phyllanthus arvensis** Müll. Arg. in J. Bot. **2**: 332 (1864); in De Candolle, Prodr. **15**, 2: 405 (1866). —Hutchinson in F.T.A. **6**, 1: 714 (1912). —Engler, Pflanzenw. Afrikas (Veg. Erde 9) **3**, 2: 29 (1921). —Radcliffe-Smith in F.T.E.A., Euphorb. 1: 57 (1987). —Jean F. Brunel, Phyllanthus Afr. Intertrop. Mad.: 396 (1987), unpublished thesis. Type from Angola (Huíla Province).

An erect, procumbent or prostrate node-rooting mat-forming woody perennial herb, or subshrub up to 2 m tall, but usually much less, glabrous, dioecious or rarely monoecious; stems red, fleshy, arising from a horizontal woody root system. Lateral shoots 2–4(6) cm long. Scale leaves 1.5–2 mm long, linear-subulate; stipules 2.5 × 0.5–0.7 mm, lanceolate, slightly auriculate, dark reddish-brown. Foliage leaves closely distichous; petioles 0.5 mm long; stipules 2 mm long, linear-lanceolate, pale brown. Leaf blades 3–10 × 2–6 mm, ovate-suborbicular to elliptic-obovate, obtuse, rounded, truncate or sometimes retuse or mucronulate, rounded-cuneate to shallowly cordate at the base, margins sometimes slightly revolute and reddish-tinged, membranaceous to chartaceous, pale green; midrib sometimes meandering, lateral nerves in 4–8 pairs, often unequal, meandering and irregularly branching and anastomosing, usually not prominent above, prominent and somewhat thickened beneath. Male flowers in 2–4-flowered bracteate axillary cymules; bracts resembling the stipules; pedicels 1.2–1.5 mm long, broadening at the apex; sepals 5, 1.5–2 × 1.2–1.5 mm, ovate-suborbicular, concave, rounded, creamy-white; disk glands 5, free, 0.3 mm across, lobulate, ± flat; stamens 3, filaments connate into a column c. 1 mm high; anthers 0.4 mm across, strongly 2-lobed, free, horizontally held, minutely granulate, laterally dehiscent. Female flowers: pedicels 2 mm long, increasing to 3 mm in fruit; sepals 5(6), 1.5 × 0.8 mm, accrescent to up to 2.5 × 2 mm in fruit, elliptic or elliptic-ovate, rounded, green with white margins; disk 0.8 mm in diameter, shallowly cupular, 6-lobed to subentire, the lobes ± truncate, becoming flattened in fruit; ovary 0.75 mm in diameter, sessile, 6-lobed, subglobose, smooth; styles 3, 0.5–0.7 mm long, ± free, suberect at first, later divergent, bifid, the stigmas recurved, reddish pigmented. Fruit 2 × 3 mm, depressed-subglobose, smooth, yellowish-green, ochreous-olivaceous when dried, partially enveloped by the persistent sepals. Seeds 1.5 × 1.2 × 1 mm, segmentiform, greyish-brown, ± smooth or with c. 5–6 indistinct longitudinal lines on the dorsal facet, and c. 3–4 imperceptible concentric arcs on each ventral facet, with innumerable transverse refringent striae between them.

Zambia. B: Kabompo, road from Mwinilunga, male fl. 4.x.1952, *Angus* 595 (BR; FHO; K). N: Lake Chila, male fl. 4.i.1952, *Richards* 372A (K). W: Solwezi, male & female fl. 16.x.1953, *Fanshawe* 437 (K; NDO). **Zimbabwe**. C: Harare Distr., Domboshawa, male & female fl. 7.iii.1944, *Wild* 896 (K; SRGH). E: Nyanga Distr., upper Nyamaziwa R., male fl. 14.i.1951, *Chase* 3681 (BM; BR; LISC; SRGH). **Malawi**. N: Nyika Plateau, Chelunduo Stream, male fl. 26.x.1958, *Robson & Angus* 383 (BM; K; LISC; SRGH). C: Kasungu Distr., 16 km south of Champhila (Champira), male fl. & fr. 23.iv.1974, *Pawek* 8476 (K; MAL; MO). S: Zomba Distr., Zomba Plateau, William's Falls, female fl. 10.x.1979, *Salubeni & Banda* 2627 (MAL; MO; SRGH).

Also in Zaire, Tanzania and Angola. Moist localities, usually in peaty soil in submontane grassland beside streams, swamps and in seepage areas, also in moist watershed dambos and Kalahari Sand dambos, and seepage areas on granite outcrops and stream sides at lower altitudes; 1220–2285 m.

60. **Phyllanthus retinervis** Hutch. in F.T.A. **6**, 1: 735 (1912). —R.E. Fries, Wiss. Ergebn. Schwed. Rhod.-Kongo-Exped. **1**, 1: 120 (1914). —Radcliffe-Smith in F.T.E.A., Euphorb. 1: 58 (1987). Type: Zambia, Mbala Distr., Fwambo, female ix.1893, *Carson* s.n. (K, holotype).

Very like *Phyllanthus arvensis*, but with the lateral nerves arising perpendicularly to the midrib and closely parallel to each other, or even backwardly directed, often much-branched towards the margin, and very markedly thickened beneath.

Zambia. N: Lumangwe, male & female fl. 14.xi.1957, *Fanshawe* 4019 (K; NDO; SRGH). W: Chisera Stream, Solwezi to Mwinilunga road, male fl. 16.ix.1952, *Angus* 466 (BM; BR; FHO; K). C: Serenje Distr., Kundalila Falls, fr. 15.x.1967, *Simon & Williamson* 1013 (K; LISC; PRE; SRGH). **Malawi**. N: Mzimba Distr., Katoto, 5 km west of Mzuzu, male fl. 22.ix.1972, *Pawek* 5764 (K; MAL; MO; SRGH).

Also in Zaire and Tanzania. In moist or wet localities, sometimes in water, often in peaty soil, in permanently wet watershed grassland dambos, swamp forest (mushitu) margins, in marshy ground in long grass at edge of evergreen riverine vegetation and dambo, also in swampy areas in submontane grassland, and amongst spray drenched rocks beside waterfalls; 460–1740 m.

61. **Phyllanthus udoricola** Radcl.-Sm. in Kew Bull. **51**, 2: 326 (1996). Tab. **12**, figs. A1–A5. Type: Zambia, Nkali (Kali) Dambo, Kawimbe Mission, fl. & fr. 6.v.1952, *Richards* 1614 (K, holotype).

Phyllanthus pusillus Jean F. Brunel, Phyllanthus Afr. Intertrop. Mad.: 400 (1987), unpublished thesis.

Very like *Phyllanthus arvensis*, but plants monoecious, with smaller (not exceeding 7 × 3 mm) foliage leaves, smooth anthers, and with a deeply 5-lobed female disk, the lobes narrowly oblong.

Tab. 12. A —PHYLLANTHUS UDORICOLA. A1, habit (× 1), from *Astle* 2388; A2, leaf (× 6); A3, male flower (× 16); A4, stamens and glands (× 32); A5, female flower and disk (× 16), A2–A5 from *Exell et al.* 944. B —PHYLLANTHUS ZORNIOIDES. B1, habit (× 1); B2, male flower (× 16); B3, stamens and glands (× 32); B4, female flower and disk (× 16); B5, distal leaf (× 2); B6, lower leaf (× 2), B1–B6 from *Richards* 5318. Drawn by J.M. Fothergill.

Zambia. N: Nkali (Kali) Dambo, Kawimbe Mission, fl. & fr. 6.v.1952, *Richards* 1614 (K). W: Mwinilunga Distr., Kalenda Plain, fr. 12.xii.1937, *Milne-Redhead* 3627 (K). C: Kabwe (Broken Hill), o. fr. v.1909, *Rogers* 8160 (K). S: 17.5 km east of Choma, fr. 28.v.1955, *Robinson* 1276 (K; SRGH). **Zimbabwe**. C: Harare, male fl. & fr. vi.1920, *Eyles* 2265 (K; PRE; SRGH). **Malawi**. S: Ntcheu Distr., lower Kirk Range, Chipusiri, male fl. & y. fr. 17.iii.1955, *Exell, Mendonça & Wild* 944 (BM; BR; LISC; SRGH).

Not known elsewhere. In moist or wet localities, usually in peaty soils of moist dambos often with tall grasses, also in shallow water over laterite; 1220–1555 m.

62. **Phyllanthus boehmii** Pax in Bot. Jahrb. Syst. **15**: 525 (1893). —Engler, Pflanzenw. Ost-Afrikas **C**: 236 (1895). —Hutchinson in F.T.A. **6**, 1: 719 (1912). —Engler, Pflanzenw. Afrikas (Veg. Erde 9) **3**, 2: 26 (1921). —Jean F. Brunel, Phyllanthus Afr. Intertrop. Mad.: 398 (1987), unpublished thesis. —Radcliffe-Smith in F.T.E.A., Euphorb. 1: 54 (1987). Type from Tanzania (Western Province).

Phyllanthus paivanus Jean F. Brunel, Phyllanthus Afr. Intertrop. Mad.: 353 (1987), unpublished thesis.

Very like *Phyllanthus arvensis*, but with foliage leaf blades elliptic-oblong to oblanceolate-oblong and often parallel-sided; lateral nerves much less conspicuous and more symmetrically arranged; and the male disk glands ± smooth.

Malawi. N: Chitipa Distr., Jembya Forest Reserve, 18 km SSE of Chisenga, male fl. & fr. 18.xii.1988, *Thompson & Rawlins* 5613 (CM; K). **Mozambique**. N: Maniamba, Serra Jeci near Malulo, fl. 3.iii.1964, *Torre & Paiva* 10986 (LISC) (type of *P. paivanus*).

Also in Zaire, Ethiopia, Uganda, Kenya and Tanzania. In moist localities, in seasonally flooded grassland beside rivers and streams; 1700–1870 m.

63. **Phyllanthus zornioides** Radcl.-Sm. in Kew Bull. **51**, 2: 328 (1996). Tab. **12**, figs. B1–B6. Type: Zambia, Mbala to Mpulungu road, male & o. fr. 5.iv.1955, *Richards* 5318 (K, holotype).

An erect annual herb up to 65 cm tall, sometimes much-branched, glabrous, monoecious. Lead shoots angular. Lateral shoots up to 20 cm long. Scale leaves c. 1 mm long, linear-lanceolate, light brown at first, later darkening; stipules 1.25 × 0.75 mm, obliquely triangular-lanceolate, not or slightly adaxially auriculate, sometimes replaced by foliage leaves at the lower nodes. Foliage leaves spiral (lead shoots) or distichous and folding upwards (lateral shoots); petioles 0.75 mm long; stipules c. 1 mm long, linear-lanceolate, light brown with tawny margins; leaf blades 0.3–2.8 × 0.2–0.8 mm, those of the lead shoots and lower nodes of the lower lateral shoots usually linear to linear-lanceolate, the rest elliptic-lanceolate to elliptic-obovate, subacute, obtuse or rounded, the lowest mucronulate, cuneate to rounded at the base, margins slightly revolute, membranaceous, bright green above, paler beneath; lateral nerves in 5–7 pairs, running to the margin, indistinct above, visible beneath. Male flowers in few-flowered cymules in the axils of the lowest quarter to one-third of the lateral shoots, female flowers solitary in the axils of the upper two-thirds to three-quarters of the lateral shoots. Male flowers: pedicels c. 1 mm long; sepals 5, 1.25 × 1 mm, obovate-suborbicular; disk glands 5, free, 0.25 mm across, irregularly lobulate, shallowly tuberculate; stamens 3, filaments united into a column 0.5 mm high, anthers 0.3 mm across, broadly ovoid, free, reflexed, apically and obliquely dehiscent. Female flowers pendulous; pedicels 1 mm long, extending to 3 mm in fruit; sepals 5, 1.3–1.5 × 1 mm, accrescent to up to 2.5 × 1.5 mm in fruit, elliptic-ovate or -obovate, slightly fused at the base, pale greenish-yellow with a broad white border; disk 0.6 mm in diameter, very shallowly 5-lobed, flat, thin; ovary 0.5 mm in diameter, sessile, subglobose, smooth; styles 3, 0.2 mm long, ± united at the base, divergent, 2-lobed, stigmas shortly cylindric. Fruit 2 × 2.5–2.8 mm, trigonous-subglobose, ± smooth, fissuring on drying, often tinged reddish-purple at the apex, ± enclosed by the persistent sepals. Seeds 1.5–1.6 × 1.2 × 1.1–1.2 mm, segmentiform, mid- to dark brown, with 17–19 longitudinal ridges on the dorsal facet, and 14–16 concentric ridges on each ventral facet, with innumerable transverse striae between them.

Zambia. N: Mbala (Abercorn)–Mpulungu road, male & o. fr. 5.iv.1955, *Richards* 5318 (K). C: 9.5 km east of Lusaka, female fl. & y. fr. 30.i.1956, *King* 284 (K; SRGH). **Zimbabwe**. N: Makonde Distr., Plateau Farm south of Brian Shaft, o. fr. 16.ii.1969, *Jacobsen* 3686 (PRE). **Malawi**. N: Karonga Distr., Kayelekera, male fl. & o. fr. ix.1989, *Collinson & Davy* s.n. (K). C: Mchinji, o. fr.

7.i.1959, *Robson* 1072 pro parte (BM; K; LISC; SRGH).
Not known elsewhere. In escarpment and plateau miombo and grassland; 860–1280 m.

64. **Phyllanthus microdendron** Welw. ex Müll. Arg. in J. Bot. **2**: 330 (1864); in De Candolle, Prodr. **15**, 2: 359 (1866). —Hutchinson in F.T.A. **6**, 1: 716 (1912). —Engler, Pflanzenw. Afrikas (Veg. Erde 9) **3**, 2: 26 (1921). —Jean F. Brunel, Phyllanthus Afr. Intertrop. Mad.: 402 (1987), unpublished thesis. Type from Angola (Huíla Province).
Phyllanthus antunesii Pax in Bot. Jahrb. Syst. **23**: 519 (1898). Type from Angola (Huíla Province).

Var. **asper** Radcl.-Sm. in Kew Bull. **51**, 2: 315 (1996). Type: Zambia, Kitwe, male, female & fr. 4.ii.1963, *Fanshawe* 7658 (K, holotype; NDO).

A much-branched annual or perennial herb or subshrub up to 60 cm tall, monoecious. Lead shoots slightly asperous. Lateral shoots 2–7 cm long, densely scabrid-papillose. Scale leaves 1 mm long, subulate; stipules 2.5–3 × 1 mm, broadly lanceolate, acutely acuminate, strongly adaxially auriculate, scarious, greyish-brown. Foliage leaves distichous; petioles 0.5–0.75 mm long; stipules 1.5–2 mm long, linear-lanceolate, acuminate, light brown; leaf blades 3–14 × 2–6 mm, elliptic, obtuse or subacute, the youngest on each shoot mucronate, rounded-cuneate and sometimes slightly asymmetrical at the base, margin very slightly thickened, often reddish, thinly chartaceous, minutely scaberulous on both surfaces, glaucous beneath; lateral nerves in 7–10 pairs, running to the margin, scarcely visible above, fairly distinct beneath. Male flowers in few-flowered cymules in the axils of the lowest half to three-quarters of the lateral shoots, female flowers solitary in the axils of the upper quarter to half of the lateral shoots. Male flowers: pedicels 0.5 mm long; sepals 5, 1.5 × 1 mm, obovate-spathulate, rounded, white with a faint midrib; disk glands 5, free, 0.2 mm in diameter, circular, flat, ± smooth; stamens 3, filaments united into a column 1–1.5 mm high; anthers 0.5 × 0.5 mm, free, the cells distinct, separated by a connective, ± vertically held and longitudinally dehiscent. Female flowers: pedicels 1 mm long, extending to 2–3 mm in fruit, expanded apically; sepals 5, 2 × 1 mm, accrescent to 4 × 2–3 mm in fruit, ovate or elliptic, nervose, pale green with a white border; disk c. 1 mm in diameter, 5-lobed, thick and upwardly prolonged to form a collar at the base of the ovary; ovary 0.8 mm in diameter, stipitate, 6-lobed, smooth; styles 3, 0.6 mm long, ± free, divergent, bifid, the stigmas recurved. Fruit 2.5 × 4 mm, depressed-globose, ± smooth, ochreous, enclosed by the persistent sepals. Seeds 1.8–1.9 × 1.5 × 1.4 mm, segmentiform, dark brown, with c. 17 longitudinal ridges on the dorsal facet, and c. 12–14 concentric ridges on each ventral facet, with innumerable often pigmented transverse striae between them.

Zambia. B: 77 km from Mongu on road to Kaoma, o. fr. 30.i.1975, *Brummitt, Chisumpa & Polhill* 14172 (K; NDO). W: Kitwe, male & female fl. & o. fr. 4.ii.1963, *Fanshawe* 7658 (K; NDO).
Not known elsewhere. In miombo woodland and *Guibourtia*, *Baikiaea* woodland on Kalahari Sand; ?1000–1250 m.
The typical variety is from Angola.

65. **Phyllanthus welwitschianus** Müll. Arg. in J. Bot. **2**: 330 (1864); in De Candolle, Prodr. **15**, 2: 351 (1866). —Hutchinson in F.T.A. **6**, 1: 723 (1912). —Engler, Pflanzenw. Afrikas (Veg. Erde 9) **3**, 2: 27 (1921). —Radcliffe-Smith in F.T.E.A., Euphorb. 1: 20 (1987). —Jean F. Brunel, Phyllanthus Afr. Intertrop. Mad.: 412 (1987), unpublished thesis. —Beentje, Kenya Trees, Shrubs Lianas: 220 (1994). Types from Angola (Malanje and Huíla Provinces).

An erect perennial herb or subshrub up to 90 cm tall, dioecious or occasionally monoecious; stems several from a woody rootstock, monomorphic, simple or branched, glabrous or minutely puberulous, reddish-brown. Scale leaves 0. Foliage leaves distichous or ± so, often folded up together; petioles 1–2 mm long; stipules 1–4 mm long, linear-lanceolate to subulate, adaxially auriculate, fimbriate-laciniate at the base, dark reddish-brown; leaf blades 0.5–3.5 × 0.3–2.8 cm, suborbicular to elliptic-oblong, rounded or emarginate, sometimes mucronulate, rounded, truncate or shallowly cordate at the base, firmly chartaceous to thinly coriaceous, glabrous on both surfaces, deep blue-green above, paler and glaucous beneath, often reddening;

lateral nerves in 6–18 pairs, scarcely or slightly prominent above, more so beneath, often somewhat irregularly looped and branched, tertiary nerves reticulate. Male flowers in 2 or more flowered axillary cymules, female flowers solitary or rarely in pairs per axil, rarely accompanied by 1–2 male flowers. Male flowers: pedicels 4–6 mm long, slender; sepals 6, very unequal, the 3 outer 2 × 1.2 mm, ovate-lanceolate, strongly concave, hardened, reddish-tinged, the 3 inner 2.5 × 2 mm, broadly ovate-suborbicular, ± flat, soft, petaloid, white to pale yellowish-green; disk glands 6, in 3 pairs opposite the outer sepals, 0.7 mm long, oblong, thick, rugulose; receptacle conical; stamens 3, filaments connate into a column 1.5 mm high; anthers 1 × 0.5 mm, oblong, strongly reflexed, connective thick, fused to the column, thecae obliquely dehiscent. Female flowers: pedicels 3–6(10) mm long; sepals 6, very unequal, the 3 outer 3.5 × 2 mm, elliptic, somewhat venose, the 3 inner 3 × 2.5–3 mm, ovate-suborbicular, strongly venose, pale yellowish-green with yellow veins, often pink-tinged; disk 2–2.5 mm in diameter, thick, shallowly 6-lobed, the lobes somewhat convoluted and rugulose; ovary 1.5 mm in diameter, sessile, subglobose, smooth; styles 3, 1.2 mm long, thick, fused to c. halfway, somewhat divergent, shortly bifid, the stigmas brownish. Fruit 4 × 7–8 mm, 3-lobed, shallowly rugulose-venose to ± smooth, pale green, often reddish-tinged. Seeds 3.2–3.4 × 2.6–2.8 × 2.4–2.5 mm, somewhat asymmetrically rounded-segmentiform, pale yellowish-brown, with c. 35 longitudinal rows of minute glistening darker reddish-brown tubercles on the dorsal facet, and c. 30 irregular and rather broken arcs of such tubercles on each ventral facet.

Zambia. B: Kaoma (Mankoya) to Luampa, female fl. 21.ii.1952, *White* 2110 (BM; FHO; K; PRE). N: 21 km Chipili–Mansa (Fort Rosebery), o. fr. 15.vi.1960, *Symoens* 7668 (BR; K; P). W: south of Matonchi Farm, o. fr. 3.ii.1938, *Milne-Redhead* 4446 (BM; BR; K; LISC; SRGH). C: Serenje, male & female fl. 23.i.1955, *Fanshawe* 1851 (K; NDO). E: Lundazi Distr., Lukusuzi Nat. Park, female fl. & fr. 19.ii.1971, *Sayer* 944 (SRGH). S: 40 km NE of Livingstone, o. fr. 10.vii.1930, *Hutchinson & Gillett* 3505 (BM; COI; K; LISC; SRGH). **Malawi**. N: Mzimba Distr., 11 km south of Euthini (Eutini), female fl. 31.i.1976, *Pawek* 10790 (K; MAL; MO; SRGH; UC). C: Mlanda (Tamanda) Mission, male & female fl. 8.i.1959, *Robson* 1102 (BM; K; LISC; SRGH). **Mozambique**. N: 1.5 km Marrupa–Nungo, male fl. 30.i.1981, *Nuvunga* 447 (BR; BM; K; LMU; MO).

Also in Zaire (Shaba Province), Tanzania and Angola. In sandy soils and Kalahari Sands, usually in miombo, also in mixed deciduous woodland and wooded grassland and in sandy watershed dambos; 800–1550 m.

Vernacular name as recorded in specimen data: "katandabalva" (chiKaonde).

66. **Phyllanthus beillei** Hutch. in F.T.A. **6**, 1: 733 (1912). —Engler, Pflanzenw. Afrikas (Veg. Erde 9) **3**, 2: 29 (1921). —Keay in F.W.T.A., ed. 2, **1**, 2: 388 (1958). —Jean F. Brunel, Phyllanthus Afr. Intertrop. Mad.: 412 (1987), unpublished thesis. Types from Central African Republic.

Phyllanthus stolzianus Pax & K. Hoffm. in Engler, Pflanzenw. Afrikas (Veg. Erde 9) **3**, 2: 29 (1921). —Brenan, Check-list For. Trees Shrubs Tang. Terr.: 223 (1949). —Dale & Greenway, Kenya Trees & Shrubs: 216 (1961). —Drummond in Kirkia **10**: 251 (1975). Types from Tanzania (Southern Highlands Province).

Phyllanthus nyassae Pax & K. Hoffm. in Notizbl. Bot. Gart. Berlin-Dahlem: **10**: 383 (1928). —Brenan, Check-list For. Trees Shrubs Tang. Terr.: 224 (1949). Type from Tanzania (Southern Highlands Province).

Phyllanthus grahamii Hutch. & M.B. Moss in Gardner, Trees & Shrubs Kenya Col.: 49 (1936); in Bull. Misc. Inform., Kew **1937**: 413 (1937). Type from Kenya (Coast Province).

Phyllanthus welwitschianus var. *beillei* (Hutch.) Radcl.-Sm. in Kew Bull. **35**: 775 (1981); in F.T.E.A., Euphorb. 1: 21 (1987).

Similar to *Phyllanthus welwitschianus*, but a much-branched dense twiggy virgate shrub up to 4 m tall and 2.5 m in diameter; foliage leaves up to 7 × 3 cm, mostly elliptic-obovate to oblong-oblanceolate, subacute, obtuse or rounded at the apex and rounded-cuneate or cuneate at the base, thinly chartaceous with lateral nerves in (10)15–30 pairs; plants never monoecious; male flowers with a taller staminal column (up to 2 mm high); styles bipartite spreading or reflexed.

Zambia. N: Kasama Distr., Chishimba Falls, o. fr. 31.iii.1955, *Exell, Mendonça & Wild* 1356 (BM; LISC; SRGH). W: Kitwe, o. fr. 27.ii.1955, *Fanshawe* 2112 (B; K; NDO). **Zimbabwe**. E: Mutare Distr., road to Mandambiri Mt., male & female fl. 19.i.1958, *Chase* 6798 (BM; K; LISC; PRE; SRGH). S: Bikita Distr., Turgwe R. Gorge, fr. 6.v.1969, *Biegel* 3037 (K; LISC; SRGH). **Malawi**. N: Nkhata Bay Distr., Luwazi R., male fl. 29.iii.1986, *la Croix* 3734 (MO; PRE). S: Mulanje Distr., Ruo–Litchenya junction, o. fr. 8.ix.1970, *Müller* 1580 (K; MAL; SRGH). **Mozambique**. N:

Ribáuè, Serra de Mepáluè, female fl. 23.i.1964, *Torre & Paiva* 10160 (LISC). MS: Dondo, female fl. 23.iii.1960, *Wild & Leach* 5205 (BM; K; MO; SRGH). GI: 10 km Chipenhe–Mainguelane, o. fr. 19.ix.1980, *Nuvunga, Boane & Conjo* 327 (BM; K; LMU; MO).

Also in W Africa from Guinée eastwards to the Cental African Republic, and in Zaire, Kenya, Tanzania and Angola, with disjunct outliers in Thailand and Cambodia. Usually in sand, often locally frequent in shade of stream side and riverbank vegetation, in riverine forest understorey, on mushitu margins, also in high rainfall miombo, *Androstachys* forest, in semi-evergreen forest and mixed evergreen forest; 600–1460 m.

Richards 278 (K), from Chilongowelo in northern Zambia, and *Brummitt and Synge* WC 227 (K), from the lower Mondwe R., Nyika Plateau, northern Malawi, are intermediate between *P. welwitschianus* and *P. beillei* in habit and vegetative features.

67. **Phyllanthus bernierianus** Baill. ex Müll. Arg. in De Candolle, Prodr. **15**, 2: 361 (1866). — Leandri, Cat. Pl. Madag., Euph.: 22 (1935); in Humbert., Fl. Madag., fam. 111: 76 (1958). Type from Madagascar.

Var. **glaber** Radcl.-Sm. in Kew Bull. **51**, 2: 305 (1996). Type: Mozambique, Chimanimani Mts., male & female 17.iv.1960, *Goodier* 1006 (K, holotype; SRGH).

Phyllanthus sp. no. 1. —Drummond in Kirkia **10**: 251 (1975).

An erect twiggy bush up to 90 cm tall, completely glabrous, monoecious or dioecious; stems blackening on drying. Shoots angular. Lateral shoots 5–10 cm long, often borne close together towards the top of the lead shoots. Scale leaves 0.5 mm long, triangular, black; stipules 0.5 × 1 mm, broadly triangular, slightly auriculate, shiny, black. Foliage leaves distichous; petioles 0.5–0.75 mm long; stipules 0.75–1 mm long, triangular-lanceolate, dark brown; leaf blades 4–10 × 1.5–3 mm, elliptic-lanceolate, acute or subacute, rounded-cuneate at the base, revolute on margins, thinly coriaceous, dark green above, glaucous beneath, dark brown on drying; lateral nerves in c. 6–8 pairs, invisible above, scarcely visible beneath. Male flowers in small few-flowered cymules in most axils, female flowers solitary in uppermost axils. Male flowers: pedicels 1.5 mm long; sepals 5, c. 1 × 0.75 mm, obovate, yellowish with a thin red midrib; disk glands 5, free, 0.25 mm in diameter, circular, flat, ± smooth; stamens 2, filaments united into a column 0.5 mm tall; anthers 0.25 × 0.3 mm, transversely ovoid, 2-lobed, free, ± horizontally held, laterally dehiscent. Female flowers: pedicels 3 mm long, dilated at the apex; sepals 5, 2–2.5 × 1 1.5 mm, ovate-oblong, pale green with a white border; disk c. 1 mm in diameter, 5-lobed, slightly undulate or ± flat; ovary 0.8 mm in diameter, sessile, subglobose, smooth; styles 3, 0.5 mm long, ± free, spreading, scarcely 2-lobed, stigmas recurved. Fruit 2 × 3 mm, depressed 3-lobed to subglobose, smooth. Seeds 1.5 × 1.2 × 1.2 mm, segmentiform, pale yellowish-brown, with c. 14 very faint longitudinal lines on the dorsal facet, and 10–12 concentric arcs on each ventral facet, with innumerable transverse striae at right angles to them.

Zimbabwe. E: Chimanimani Distr., Makurupini Falls, o. fr. 27.viii.1969, *T. Wild* A14 (SRGH). **Mozambique**. MS: Mevumoze (Mevumozi) R. tributary, male fl. 6.v.1965, *Whellan* 2230 (K; SRGH).

Not known elsewhere. Eastern slopes of Chimanamani Mountains, beside river and at forest edges, also with shrubs in grassland and amongst rocks beside streams; (430)915–1525 m.

The typical variety is from Madagascar.

Species unknown

Gibbs Russell 2115 from a *Typha* marsh on the edge of L. Chilwa, in southern Malawi, is an aquatic plant with a long fleshy horizontal rootstock, producing erect leafy aerial stems at intervals. Unfortunately the material is completely sterile and therefore its affinities and status are inassessable.

9. BREYNIA J.R. & G. Forst.

Breynia J.R. & G. Forst., Char. Gen. Pl.: 145, t. 73 (1776). —Müller Argoviensis in De Candolle, Prodr. **15**, 2: 438 (1866), nom. conserv.

Monoecious or apparently dioecious shrubs or small trees, with or without a simple indumentum, often blackening on drying. Branching phyllanthoid (see

Tab. **8**). Leaves alternate, shortly petiolate, stipulate, simple, entire, penninerved, borne on plagiotropic shoots (leafy or floriferous lateral shoots of limited growth, see *Phyllanthus*). Flowers axillary, the male fasciculate or solitary, usually in the proximal axils, the female solitary, usually in the distal axils. Male flowers: pedicels often capillary; calyx obconic or turbinate, calyx lobes 6, imbricate, sharply inflexed; petals absent; disk absent; stamens 3, united into a short column, anthers elongate, thecae linear, extrorse, adnate to the column, longitudinally dehiscent; pistillode absent. Female flowers: pedicels sometimes capillary; calyx lobes 6, imbricate, not inflexed, usually larger than in the male flowers, accrescent; disk absent; ovary 3-locular, ovules 2 per locule; styles 3, free, short, erect, simple or bifid. Fruit ± baccate, tardily and often incompletely loculicidally dehiscent; exocarp sometimes somewhat fleshy; endocarp crustaceous. Seeds trigonous, ecarunculate; testa membranous; albumen fleshy; cotyledons broad; radicle long.

An Indo-Pacific genus of 25 species, some of which are widely cultivated ornamentals, one being ± naturalized in parts of Africa.

Breynia stipitata Müll. Arg. from Queensland considered by H.K. Airy Shaw as possibly not distinct from *B. cernua* (Poir.) Müll. Arg. but with distinctly stipitate fruits, is recorded as having once been cultivated in Harare, National Botanic Gardens (*Müller* 1705, 3.i.1971 (K; SRGH)).

Breynia disticha J.R. & G. Forst., Char. Gen. Pl.: 146, t. 73 (1775). —Müller Argoviensis in De Candolle, Prodr. **15**, 2: 439 (1866). Type from Vanuatu, Tana (Tanna) Island.

Var. **disticha**

forma **nivosa** (W. Bull) Croizat in Sargentia **1**: 48 (1942), emend Radcl.-Sm. in Kew Bull. **35**: 498 (1980); **37**: 612 (1983). —Troupin, Fl. Rwanda **2**: 208, fig. 62/2 (1983). —Radcliffe-Smith in F.T.E.A., Euphorb. 1: 104 (1987). Type from Vanuatu, Tana (Tanna) Island.

Phyllanthus nivosus W. Bull, Cat.: 9 (1873). —W.G. Sm., Flor. Mag. n.s. **30**: t. 120 (1874).

Breynia nivosa (W. Bull) Small in Bull. Torr. Bot. Club **37**: 516 (1910). —Biegel, Check List Ornam. Pl. Rhod. Parks & Gard.: 30 (1977).

A glabrous shrub up to 2 m high, densely branched. Twigs purplish or reddish-brown at first, later becoming greyish. Cataphylls 2 mm long, narrowly triangular-lanceolate, acute. Cataphyllary stipules 1–1.5 × 1–1.5 mm, broadly triangular-ovate, acuminate. Plagiotropic shoots (leafy or floriferous lateral shoots of limited growth, see *Phyllanthus*) and their leaves distichous. Petioles 3–4 mm long. Stipules 1.5 mm long, triangular-ovate to lanceolate, acutely acuminate, subentire, green with hyaline margins. Leaf blades 1.5–3.5(6) × 1–3(3.5) cm, ovate-elliptic to suborbicular, rounded or emarginate, truncate, rounded or wide-cuneate at the base, membranaceous, olive-green, grey-green or bluish-green above and paler beneath, sometimes reddish-brown tinged, edged or blotched with creamy-white, white with pale grey-green patches and/or small dark olive flecks, entirely white, white suffused with pink or else entirely reddish-pink; lateral nerves in 5–7 pairs, ascending, camptodromous, scarcely prominent above, slightly so beneath. Male flowers (1)2(3) per axil, 2–3 mm in diameter; pedicels 6–9 mm long, capillary; calyx lobes obtuse, green, tinged reddish; staminal column 1 mm high. Female flowers 0.7–1 cm in diameter; pedicels 3–9 mm long; calyx lobes 1–2 × 1–2 mm, accrescent to 3–4 × 4–5 mm, obovate-obdeltate, obtuse, rounded or retuse, green or grey-green, reddish-tinged, sometimes irregularly edged and/or streaked white or pale pinkish; ovary 1 × 1 mm, 3-lobed to subglobose, smooth; styles minute, ± conical, shortly bifid. Fruit 3 × 5 mm, somewhat depressed 3-lobed, subglobose, ± smooth, greenish. Seeds c. 2.5 × 1.5 mm, ± smooth, greyish.

Zimbabwe. C: Harare, garden ornamental, female fl. 20.xi.1973, *Biegel* 4372 (K; SRGH). **Mozambique**. M: Maputo (Lourenço Marques), male & female fl. 23.xi.1963, *Balsinhas* 679 (K; LMA).

Native to Vanuatu, but widely cultivated throughout tropical Africa both for its ornamental foliage and as a hedge plant. Occasionally a garden escape in the Flora area. The “Snow Bush”.

Breynia disticha f. *disticha* has uniformly concolorous or monochrome dark green leaves.

10. DRYPETES Vahl

Drypetes Vahl, Eclog. Amer., part 3: 49 (1807).
Cyclostemon Blume, Bijdr. Fl. Ned. Ind.: 597 (1826).
Lingelsheimia sensu Hutch. in F.T.A. **6**, 1: 690 ff. (1912), pro parte excl. *L. frutescens*, non Pax (1909).

Dioecious trees or shrubs. Indumentum simple. Buds perulate (furnished with protective scales) or not. Leaves alternate, stipulate, shortly petiolate; blades simple, entire or toothed, usually asymmetrical at the base, often coriaceous, penninerved. Flowers borne in leaf axils or axils of recently fallen leaves, or cauliflorous on branches or trunk, fasciculate, usually pedicellate. Male flowers: sepals 4–5(6), imbricate, broad, concave, often unequal; petals absent; stamens (3)4–30(50), filaments free, anthers usually introrse, thecae parallel, longitudinally dehiscent; disk intrastaminal, sometimes convoluted and enfolding the stamens; pistillode minute or absent. Female flowers: sepals caducous, otherwise as in male flowers; hypogynous disk annular, cupular or absent; ovary 1–2(4)-locular, with 2 ovules per loculus; styles usually very short or almost obsolete, stigmas dilated and variously shaped. Fruit drupaceous; pericarp somewhat fleshy, becoming hardened on drying; endocarp coriaceous, chartaceous or osseous. Seeds solitary per loculus or fruit by abortion, ecarunculate, sometimes with a thin sarcotesta; endosperm copious; cotyledons broad, flat.

A pantropical genus of c. 200 species, of which c. 60 occur in Africa.
Drypetes roxburghii (Wall.) Hurusawa is cultivated in Zimbabwe.

1. Plants cauliflorous; leaf blades usually spinulose-denticulate, cuneate on one side and rounded on the other at the base; stamens (11)18–30; fruits up to 3.3 cm in diameter · · · · · · · · · · 1. *natalensis*
– Plants ramiflorous, flowers borne in axils of leaves or in axils of recently fallen leaves; stamens 4–15; fruits not exceeding 2 cm in diameter · · · · · · · · 2
2. Base of leaf blades often shallowly cordate, or if not cordate, then leaf margins either sharply serrate or else margins entire and blades not strongly reticulate; stamens 5–15 · · · 3
– Base of leaf blades cuneate, rounded or rarely truncate; stamens 4 · · · · · · · · 4
3. Perulae (protective bud scales) striate; leaf margins spinulose-serrate; flowers and fruits in leaf axils; stamens 15; fruiting pedicels usually not exceeding 5 mm in length; fruit subglobose · · · · · · · · 2. *arguta*
– Perulae smooth; leaf margins entire; flowers and fruits in axils of fallen leaves; stamens (5)8(10); fruiting pedicels up to 3.5 cm long; fruit ellipsoid · · · · · · · · 3. *mossambicensis*
4. Leaf margins shallowly and remotely crenate-serrate or crenate-dentate to subentire; tertiary nerves regularly reticulate; stigmas linear · · · · · · · · 4. *gerrardii*
– Leaf margins entire; tertiary nerves strongly and somewhat irregularly reticulate; stigmas obdeltoid, somewhat bifid at the apex · · · · · · · · 5. *reticulata*

1. **Drypetes natalensis** (Harv.) Hutch. in F.C. **5**, 2: 404 (1920). —Engler, Pflanzenw. Afrikas (Veg. Erde 9) **3**, 2: 35 (1921). —Pax in Engler, Pflanzenr. [IV, fam. 147, xv] **81**: 243 (1922). —Burtt Davy, Fl. Pl. Ferns Transvaal: 300 (1932). —Brenan in Mem. N.Y. Bot. Gard. **9**, 1: 70 (1954). —Dale & Greenway, Kenya Trees & Shrubs: 193 (1961). —Mogg in Macnae & Kalk, Nat. Hist. Inhaca Isl., Moçamb.: 147 (1969). —Drummond in Kirkia **10**: 251 (1975). —K. Coates Palgrave, Trees Southern Africa, ed. 2, rev.: 403 (1983). —Radcliffe-Smith in F.T.E.A., Euphorb. 1: 92 (1987). —Beentje, Kenya Trees, Shrubs Lianas: 195 (1994). Syntypes from South Africa (KwaZulu-Natal).

Var. **natalensis**

Cyclostemon natalensis Harv., Thes. Cap. **2**: 64, t. 200 (1863). —Müller Argoviensis in De Candolle, Prodr. **15**, 2: 483 (1866).

Drypetes zombensis Dunkley in Bull. Misc. Inform., Kew **1937**: 468 (1937). —Topham, Check List For. Trees Shrubs Nyasaland Prot.: 50 (1958). Type: Malawi, Zomba Mountain, x.1929, *Clements* 35 (K, holotype).

A large evergreen cauliflorous shrub or tree up to 15 m high. Bole up to c. 30 cm in diameter; branches borne above 4.5 m, upright then spreading to form a dense rounded crown; bark light to dark grey or grey-green, dull, smooth except for the transverse knobby ridges or bosses on which the flowers and fruits are borne; bosses produced from near the base to c. 6 m up the trunk and on the main branches. Young shoots and petioles pubescent at first, becoming glabrescent. Stipules 3–4 mm long, linear to subulate, pubescent, soon falling. Petioles (2)3–9 mm long. Leaf blades 4–21 × 1–8.5 cm, elliptic-lanceolate to oblong-lanceolate, shortly acuminate, usually cuneate on one side at the base and rounded on the other, usually spinulose-denticulate on the margins, rarely subentire, puberulous along the midrib at least beneath when very young, soon becoming completely glabrous, coriaceous, dark green and glossy above, paler and duller beneath; lateral nerves in 7–10(12) pairs, strongly brochidodromous, looped well within the margin and with a second series of loops towards the margin, tertiary nerves reticulate, all fairly prominent above, prominent beneath. Flowers fasciculate on the trunk and lower branch bosses, sometimes with over 100 flowers per boss, often c. 1.5 m above ground level, with a disagreeable odour. Male flowers: pedicels 2–8(15) mm long, very sparingly pubescent; sepals 5–6, subequal, 3.5–4 × 3.5 mm, ovate-suborbicular to orbicular, glabrous, minutely ciliolate, yellowish or greenish-white; stamens (11)12, in 2 whorls, 5–7 mm long, anthers 1.5–2 mm long; disk 11–12-lobed, with the lobes extruded between the bases of the filaments, somewhat irregularly rugulose, but without a central projection, glabrous. Female flowers: pedicels (0.5)1–2 cm long, patent-pubescent at first, later glabrescent; sepals as in the male flowers; disk c. 3 mm in diameter, annular, scarcely lobed, somewhat crenellate, fleshy, glabrous; ovary 2–2.5 × 2.5–3 mm, (2)3-locular, subglobose, densely sericeous-tomentose; styles (2)3, persistent; stigmas 1.5 mm across, subsessile, spreading, broadly obdeltoid, slightly crenellate, smooth, fleshy. Fruit 1.7–2.5 × 2–2.7 cm, up to 3 cm in diameter when fresh, 3-lobed to subglobose, occasionally 2-lobed, smooth or rugulose, glabrous (E Africa only) or pubescent and variously glabrescent, green at first, later becoming orange-yellow. Seeds c. 1.1 × 1 × 0.7 cm, ovoid, orange.

Zimbabwe. E: Mutare Distr., south of Mutare, female fl. 25.x.1953, *Chase* 5115 (BM; K; LISC; PRE; SRGH). **Malawi**. C: Ntchisi Mt., For. Res., o. fr. 23.i.1962, *Chapman* 6102 (FHO). S: Thyolo (Cholo), male fl. 29.ix.1946, *Brass* 17876 (BM; K; NY; SRGH). **Mozambique**. N: Nampula, Mossuril, ? fr. 19.ii.1984, *de Koning et al.* 9744 (K; LMU). Z: Mocuba Distr., Namagoa Plantation, fr. x.1944, *Faulkner* 217 (K; PRE; SRGH). MS: Sofala Prov. (Beira Distr.), Cheringoma, ? fr. v.1973, *Tinley* 2921 (K; LISC; PRE; SRGH). GI: Bazaruto Island, y. fr. 29.x.1958, *Mogg* 28742 (J; LISC; SRGH). M: Marracuene, Muntanhane, fr. 12.xi.1960, *Balsinhas* 243 (BM; K; LISC; LMA; PRE).

Also in Sudan, Somalia (var. *leiogyna* Brenan only), Kenya (both vars.), Tanzania (both vars., var. *leiogyna* only in Zanzibar) and South Africa (KwaZulu-Natal). An understorey tree of evergreen rainforest, often beside rivers in submontane gully forest and in coastal forest and woodland and dune thickets; sea level–1100 m.

Vernacular names as recorded in specimen data include: "kwekewei" (Zambézia); "chucuane" (xiRonga).

This species is frequently confused with two trees in the family Flacourtiaceae; *Casearia gladiiformis* Mart. and *Rawsonia lucida* Harv. & Sond. The former may be distinguished by the characteristic 'dot-and-dash' pattern of pellucid glands in its young leaves, and by its axillary fascicles of bisexual flowers. The latter, like *Drypetes*, also has unisexual flowers, but these are in short axillary, spiciform racemes (not fascicles as in *Drypetes*).

The variation in the ovule and fruit character of *Drypetes natalensis* parallels that of *D. usambarica*, an East African species, as both have glabrous and pubescent states. *Drypetes natalensis* var. *leiogyna* Brenan (from Somalia, Kenya and Zanzibar) has a completely glabrous ovary and fruit, as do the varieties *usambarica*, *mrimae* and *stylosa* of *D. usambarica*, whilst *D. natalensis* var. *natalensis* and the varieties *rugulosa* and *trichogyna* of *D. usambarica* have pubescent ovaries and fruit. Recently material of one of the pubescent varieties of *D. usambarica* in Tanzania has been collected which appears to be intermediate between them and between them and *D. natalensis* var. *natalensis*.

2. **Drypetes arguta** (Müll. Arg.) Hutch. in F.C. **5**, 2: 404 (1920). —Engler, Pflanzenw. Afrikas (Veg. Erde 9) **3**: 2: 35 (1921). —Pax in Engler, Pflanzenr. [IV, fam. 147, xv] **81**: 261 (1922). —Burtt Davy, Fl. Pl. Ferns Transvaal: 300 (1932). —Mogg in Macnae & Kalk, Nat. Hist. Inhaca Isl., Moçamb.: 147 (1969). —Drummond in Kirkia **10**: 251 (1975). —K. Coates Palgrave, Trees Southern Africa, ed. 2, rev.: 402 (1983). —Radcliffe-Smith in F.T.E.A., Euphorb. 1: 89 (1987). Type from South Africa (KwaZulu-Natal).

Cyclostemon argutus Müll. Arg. in De Candolle, Prodr. **15**, 2: 485 (1866). —Sim, For. Fl. Port. E. Afr.: 106 (1909).

A shrub or small tree up to 8 m high. Bark more or less smooth or finely longitudinally fissured, grey; twigs greyish or yellowish-grey. Young shoots and petioles evenly to sparingly yellowish-pubescent or puberulous at first, later glabrescent. Buds perulate (furnished with protective scales); perulae up to 5 mm long, broadly ovate, subscarious, closely longitudinally striate, ciliolate, persistent. Stipules up to 7 mm long, linear-lanceolate, faintly striate, sparingly fimbriate, glabrous, soon falling. Petioles 2–4 mm long. Leaf blades 3–11 × 1–4 cm, obliquely lanceolate to oblong-lanceolate, acuminate, rounded to cordate at the base, spinulose-serrate on the margins, the teeth tipped with subulate glands, thinly coriaceous, quite glabrous when mature, bright green and somewhat shiny above, dull green below, usually drying pale green; lateral nerves in 7–13 pairs, 3–4 of which are crowded at the base, usually somewhat irregularly looped well within the margin, tertiary nerves reticulate, not prominent above, fairly prominent beneath. Flowers axillary, solitary. Male flowers: pedicels 3–5 mm long, pubescent; sepals 4, 3 × 2.5–3 mm, suborbicular, pubescent and puberulous without and within, greenish; stamens 15, in 2 whorls, 2.5 mm long, anthers 1 mm long; disk 2.5 mm in diameter, margin strongly crisped-convolute, the lobes enveloping thc filaments, pubescent, smooth, with 1–2 minute central conical projections. Female flowers: pedicels 0.3–1.2 cm long, sparingly pubescent; calyx lobes 4–5, reflexed then inflexed, 4 × 3 mm, accrescent to up to 7 × 5.5 mm in fruit, suborbicular-ovate, closely longitudinally striate, ciliate; disk 3 mm in diameter, accrescent to 5 mm in fruit, annular, thick, tomentose; ovary 4 × 4 mm, 2-locular, 2-lobed to subglobose, densely fulvous-tomentose; styles 2, united at the base, divaricate, 2 mm long, tubular, tomentose, persistent; stigmas 2.5 mm broad, broadly rhombic to reniform, glabrous, papillose. Fruit 1.5–1.7(2) × 1.7–2(2.5) × 0.7–1 cm, 2-lobed to subglobose, slightly laterally compressed, evenly pubescent and sparingly hirsute, bright orange-red to yellow-orange when mature; endocarp firmly corky. Seeds 1–2 per fruit, 10 × 9 × 8 mm, ellipsoid-subglobose, brownish.

Zimbabwe. E: Chimanimani Distr., Haroni/Makurupini Forest, y. fr. 3.xii.1964, *Wild* 6612 (FHO; K; LISC; SRGH). S: Chibi Distr., Nyoni Mts., fr. 20.iv.1967, *Müller* 603 (K; SRGH). **Mozambique**. Z: Maganja da Costa Distr., Gobene Forest, male & female fl. 10.i.1968, *Torre & Correia* 17020 (LISC). MS: Macúti, fr. 23.iii.1960, *Wild & Leach* 5192 (K; PRE; SRGH). GI: Manjacaze, st. 3.xii.1942, *Mendonça* 1575 (LISC). M: Fonte de Goba, fr. 13.v.1975, *Marques* 2763 (K; LISC; LMU; SRGH).

Also in Tanzania, Swaziland, South Africa (KwaZulu-Natal and Eastern Cape Province). Understorey tree of low altitude evergreen forest and evergreen kloof forest, also in coastal mixed deciduous woodland and scrub; near sea level to 600 m.

3. **Drypetes mossambicensis** Hutch. in F.T.A. **6**, 1: 1046 (1913). —Pax & K. Hoffmann in Engler, Pflanzenr. [IV, fam. 147, xv] **81**: 237 (1921). —White, F.F.N.R.: 197 (1962). —Drummond in Kirkia **10**: 251 (1975). —K. Coates Palgrave, Trees Southern Africa, ed. 2, rev.: 403 (1983). Tab. **13**. Type: Mozambique, Manica e Sofala Province, Murraça (Mourassa) Village, Púngoè (Pungoué) Valley, *Vasse* 319 (P, holotype; K).

A deciduous shrub or tree up to 20 m high, sometimes more or less evergreen; branching at 3–5 m to form a dense broadly conical crown up to 10 m in diameter. Bark smooth and silver-grey at first, later darkening and cracking and flaking rectangularly at the base. Twigs lenticellate. Young shoots and petioles sparing to densely minutely puberulous. Buds perulate (furnished with protective scales); perulae c. 3 mm long, suborbicular, smooth, ciliate, soon falling. Stipules c. 1 mm long, lanceolate, acute, more or less glabrous without, puberulous within, soon falling. Petioles 3–8 mm long. Leaves 3–11 × 1.5–5 cm, narrowly oblong to elliptic-oblong, rounded and sometimes emarginate at the apex, rarely obtuse, obliquely rounded and cordulate to distinctly cordate at the base, entire on the margins; blades chartaceous to thinly coriaceous, sometimes minutely puberulous at the base of the midrib beneath at first, otherwise completely glabrous, usually dark green and glossy above, glaucous beneath; lateral nerves in 6–10 pairs, weakly brochidodromous, tertiary nerves reticulate. Flowers borne in the axils of the scars of the previous season's fallen leaves, below the terminal flush; male fascicles many-flowered; female flowers solitary or paired, rarely ternate. Male flowers: pedicels c. 4 mm long, minutely puberulous; sepals 4(5), somewhat unequal, 2.5–3 × 1–2 mm, usually ± ovate,

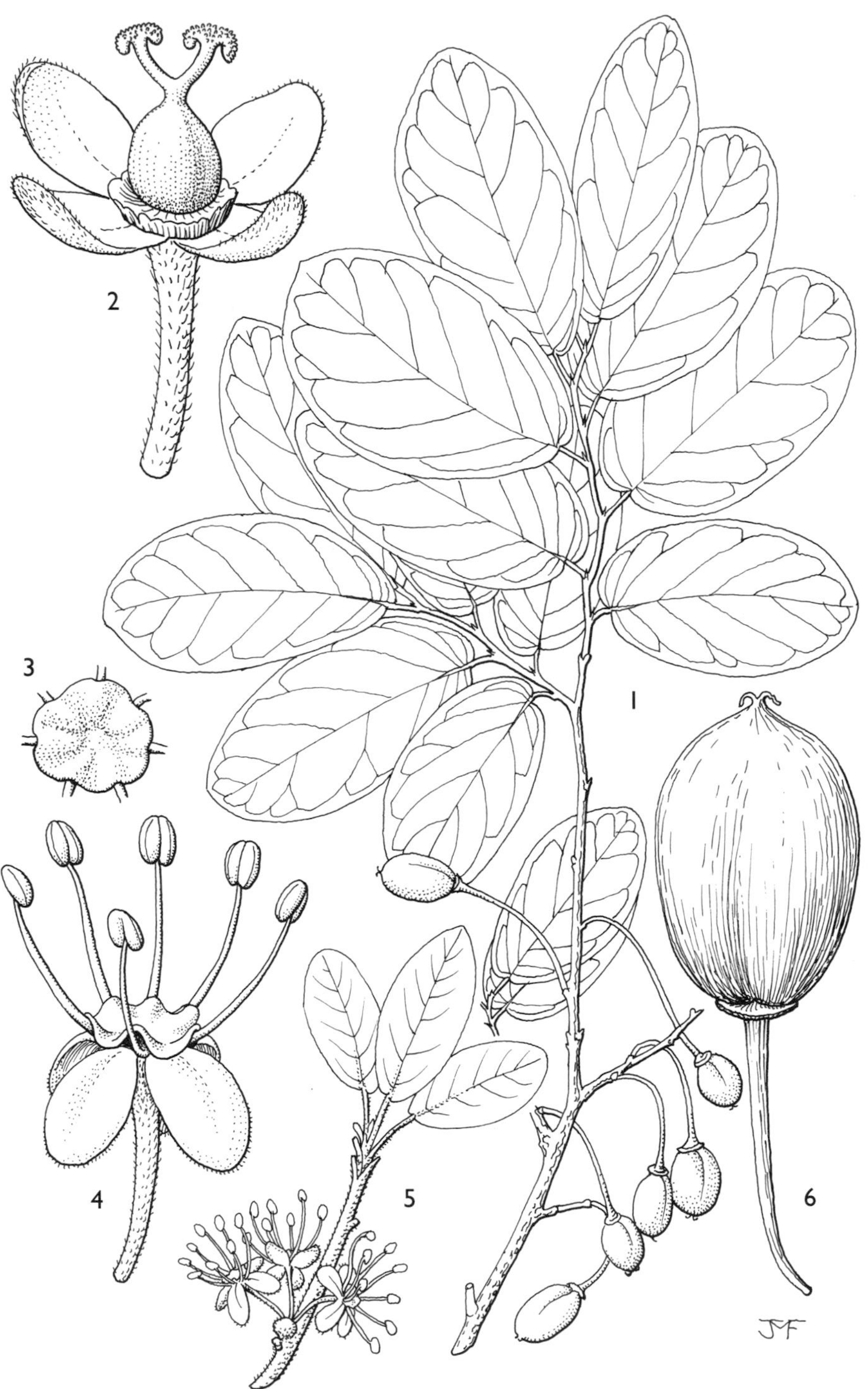

Tab. 13. DRYPETES MOSSAMBICENSIS. 1, distal part of branch with immature fruit (× $^{2}/_{3}$), from *Goldsmith* 227/62; 2, female flower (× 6), from *Fanshawe* 8136; 3, male disk (× 6); 4, male flower (× 6); 5, male inflorescence (× 2), 3–5 from *Goodier* 677; 6, fruit (× 2), from *Chase* 2375. Drawn by J.M. Fothergill.

rounded at the apex, fulvous- or ferrugineous-tomentellous at the apex without, glabrous within, pale lime-green; stamens 6–10, enclosed by the disk folds, 2.5–3 mm long, anthers 1.2 mm long; disk 2.5 mm across, more or less acetabuliform, plicate, with as many marginal folds as stamens, smooth, with no central projection, glabrous. Female flowers: pedicels c. 5 mm long, stouter than in the male, extending to up to 3.5 cm long in fruit; sepals resembling those of the male flowers, but slightly larger, green; disk 3.5–4 mm in diameter, annular, subentire; ovary c. 2 mm in diameter, 2-locular, ovoid-subglobose, glabrous, dark green; styles 2, united at the base, erect, persistent, 1.5 mm long, bifid, stigmas spreading, papillose. Fruit 1.3–1.5(2) × 0.75–1(1.5) cm, ellipsoid, shallowly 2-lobed and 2-celled, or not lobed and 1-celled by abortion, smooth, glabrous, green at first later becoming yellow to yellow-orange and somewhat fleshy. Seeds 0.8–1.4 × 0.6–1 × 0.3–0.7 cm, somewhat compressed-ellipsoid; sarcotesta drying pale brownish or yellowish-brown.

Zambia. C: Katondwe, y. fr. 14.xi.1963, *Fanshawe* 8136 (FHO; K; NDO); Luangwa, Mfuwe, st. 26.iv.1965, *B.L. Mitchell* 2672 (K; SRGH). E: Luangwa Valley, 13°00'S, 32°00'E, fr. 28.xi.1957, *Stewart* 68 (K). S: Gwembe, fr. 15.xi.1955, *Bainbridge* 188/55 (FHO; K). **Zimbabwe**. N: Kariba, y. fr. xii.1959, *Goldsmith* 6159 (K; LISC; PRE; SRGH). E: Chimanimani Distr., Save (Sabi) R. Valley, male fl. 17.ix.1953, *Chase* 5072 (BM; K; LISC; SRGH). S: Save (Sabi) R., fr. 5.vi.1950, *Chase* 2374 (BM; FHO; K; LISC; SRGH). **Malawi**. C: Dedza Distr., Mankhanba Rest House, male fl. 8.xii.1953, *Adlard* 11 (K; MAL; SRGH). S: Blantyre Distr., Mpatamanga Gorge, female fl. & y. fr. 30.ix.1980, *Patel* 737 (MAL; SRGH). **Mozambique**. N: Imala, between Muite and R. Lurio, y. fr. 25.x.1948, *Barbosa* 2565 (LISC). Z: sine loc. cert., st. 28.ix./xi.1972, *Tawse* 22 (LISC; SRGH). T: Estima to Candôdo, fr. 25.i.1972, *Macêdo* 4686 (K; LISC; LMA; SRGH). MS: Gorongosa Nat. Park, y. fr. xii.1970, *Tinley* 2008 (K; LISC; PRE; SRGH). GI: Guijá, estrada da Malvérnia, y. fr. 16.xi.1957, *Barbosa & Lemos in Barbosa* 8173 (K; LISC; LMA). M: Goba, near R. Maivavo, male fl. 5.xi.1960, *Balsinhas* 208 (BM; K; LISC; LMA; SRGH).

Also in South Africa (Transvaal). Usually in riverine woodland or thicket on alluvium of low altitude river valleys, often beside seasonal watercourses, also in floodplain mopane woodland, beside pans and at base of rocky escarpments, sometimes on dambo margins and on termitaria; 90–760 m.

Fruit sought after by large bats, and eaten by monkeys.

Vernacular names as recorded in specimen data include: "munocorre", "mushakwari", "mutenakovari", "muulukwa" (chiTonga).

4. **Drypetes gerrardii** Hutch. in F.C. **5**, 2: 405 (1920). —Pax in Engler, Pflanzenr. [IV, fam. 147, xv] **81**: 253 (1922). —Burtt Davy, Fl. Pl. Ferns Transvaal: 300 (1932). —Brenan in Mem. N.Y. Bot. Gard. **9**, 1: 70 (1954). —Topham, Check List For. Trees Shrubs Nyasaland Prot.: 50 (1958). —Dale & Greenway, Kenya Trees & Shrubs: 193 (1961). —White, F.F.N.R.: 197 (1962). —Drummond in Kirkia **10**: 251 (1975). —Troupin, Fl. Pl. Lign. Rwanda: 257, fig. 89/3 (1982). —K. Coates Palgrave, Trees Southern Africa, ed. 2, rev.: 402 (1983). —Radcliffe-Smith in F.T.E.A., Euphorb. 1: 97 (1987). —Beentje, Kenya Trees, Shrubs Lianas: 194 (1994). Type from South Africa (KwaZulu-Natal).

A large evergreen ramiflorous shrub to c. 3 m high, or tree to 30 m; bole slightly fluted to sharply buttressed up to c. 1 m, (15)23(38) cm d.b.h., unbranched to 10 m, then much branched; branches in one plane from c. 5 m forming a narrow dense crown. Bark smooth, or flaking in oblong or circular patches, pale grey or brown. Twigs greyish. Young shoots and petioles densely yellowish-tomentose to evenly or sparingly pubescent or subglabrous. Stipules c. 1 × 0.5 mm, triangular, soon falling. Petioles 3–10 mm long. Leaves 2–14(17) × 1–7(9) cm, obliquely ovate to rhombic-elliptic or lanceolate, acute to subacute or obtuse, acuminate or not at the apex, weakly to strongly asymmetrically rounded or cuneate at the base, usually shallowly crenate-serrate to subentire on the margins; blades thinly coriaceous, often yellowish-pubescent along the midrib above and beneath and otherwise glabrous, or else completely glabrous, dark green and glossy on the upper surface, paler beneath, those of the upper crown usually brighter yellow-green; lateral nerves in 5–9 pairs, looped well within the margin, and often with a second series of loops towards the margin, tertiary nerves reticulate, parallel near the midrib, scarcely to fairly prominent above, usually prominent beneath. Flowers axillary, male flowers in few-flowered glomerules, the females solitary. Male flowers: pedicels 3–4 mm long, yellowish-pubescent; sepals 4, 3 × 3 mm, suborbicular, densely yellowish-pubescent without, sparingly pubescent within, creamy-yellow; stamens 4, 3 mm long, anthers 1 mm long; disk c. 2 mm in diameter, 4-lobed, the lobes somewhat crenulate, protruding a little between the stamens, smooth, pubescent, with a shallow central boss. Female flowers: pedicels 1–2 mm long, extending to 1(2) cm in fruit; sepals as in the male; disk c. 1.3 mm in diameter, shallowly cupular, subentire, glabrous; ovary 2-locular, 1.5 × 2 mm, subglobose, densely sericeous-tomentose; styles 2, free, reflexed,1–1.5 mm long, persistent; stigmas ± linear. Fruit 1–1.3 × 1.3–1.7 cm, obovoid-subglobose, usually shallowly bilobed, evenly tomentose, apple-green to grey-green when fresh, yellowish-brown when dried. Seeds c. 1 mm long, compressed-ovoid, brownish, streaked whitish.

1. Leaf blades not more than 5 cm long, narrowly elliptic-lanceolate, coarsely irregularly serrate-dentate in the upper half ········ var. *angustifolia*
– Leaf blades (2)5–14(17) cm long, broadly ovate, rhombic-elliptic or lanceolate, shallowly crenate-serrate to subentire ········ 2
2. Young shoots and petioles densely yellowish-tomentose; leaf blades 2–6 × 1–3.5 cm; fruits shallowly bilobed ········ var. *tomentosa*
– Young shoots and petioles evenly or sparingly yellowish-pubescent to subglabrous; leaf blades (3)4–14(17) × 2–7(9) cm; fruits bilobed or not lobed ········ 3
3. Young shoots and petioles evenly to sparingly pubescent; leaf blades (3)4–12 × 2–6 cm; fruits bilobed ········ var. *gerrardii*
– Young shoots and petioles sparingly pubescent to subglabrous; leaf blades 6–14(17) × 3–7(9) cm; fruits not lobed ········ var. *grandifolia*

Var. **gerrardii**

Drypetes subdentata Mildbr. in Siebenlist, Forstwirtsch. Deutsch Ost-Afrika: 93 (1914). — Pax, loc. cit. (1922). —Brenan, Check-list For. Trees Shrubs Tang. Terr.: 207 (1949). Type from Tanzania.

Drypetes battiscombei Hutch. in Bull. Misc. Inform., Kew **1924**: 261 (1924). —Gardner, Trees & Shrubs Kenya Col.: 48 (1936). —Brenan, Check-list For. Trees Shrubs Tang. Terr.: 207 (1949). Type from Kenya.

Young shoots and petioles evenly to sparingly pubescent; leaf blades (3)4–12 × 2–6 cm, broadly ovate, rhombic-elliptic or lanceolate, shallowly crenate-serrate to subentire; fruits 2-lobed.

Zambia. E: Makutus, male fl. 28.x.1972, *Fanshawe* 11604 (K; NDO). **Zimbabwe**. E: Chipinge Distr., Chirinda Forest, female fl. x.1966, *Goldsmith* 96/66 (K; LISC; PRE; SRGH). **Malawi**. N: Chitipa Distr., Misuku Hills, Mughesse, y. fr. 4.i.1974, *Pawek* 7776 (K; MAL; MO; SRGH; UC). C: Ntchisi Forest, st. 6.v.1961, *Chapman* 1282 (FHO; MAL; SRGH). S: Thyolo (Cholo) Mt., fr. 27.ix.1946, *Brass* 17837 (K; NY; SRGH). **Mozambique**. Z: Gurué, fr. 4.i.1968, *Torre & Correia* 16893 (LISC). MS: Mt. Gorongosa, st. 23.vii.1970, *Müller & Gordon* 1390 (K; LISC; SRGH).

Also in Burundi, Uganda, Kenya, Tanzania, Swaziland, South Africa (Transvaal, KwaZulu-Natal and Eastern Cape Province). Mostly a subcanopy tree of submontane mixed evergreen rainforest; 780–2000 m.

Vernacular names as recorded in specimen data include: "mutaga", "ntaga" (Chisukwa).

Chapman & Tawakali 6169 (K; MAL) from Thondwe, southern Malawi, is more or less intermediate between *D. gerrardii* and *D. reticulata.*

Var. **grandifolia** Radcl.-Sm. in Kew Bull. **32**: 476 (1978); in F.T.E.A., Euphorb. 1: 98 (1987). Type from Uganda.

Young shoots and petioles sparingly pubescent to subglabrous; leaf blades 6–14(17) × 3–7(9) cm, otherwise as in typical variety; fruits not lobed.

Malawi. N: Mzimba Distr., Viphya, Uzumara, st. 24.i.1964, *Chapman* 2220 (MAL; SRGH). S: Mulanje Distr., Ruo Gorge, st. 2.ix.1970, *Müller* 1493 (K; SRGH).

Also in Uganda, SW Kenya, NW Tanzania and Angola. Subcanopy tree of submontane mixed evergreen forest; 1000–2000 m.

Vernacular name as recorded in specimen data: "kalembo" (chiNyanja, ciSena).

Müller 1594 (K; SRGH), from Nkhata Bay District, Malawi, is intermediate between this and the typical variety.

Var. **tomentosa** Radcl.-Sm. in Kew Bull. **32**: 477 (1978); in F.T.E.A., Euphorb. 1: 98 (1987). Type from Uganda.

Young shoots and petioles densely yellowish-tomentose; leaf blades 2–6 × 1–3.5 cm, otherwise more or less as in typical variety; fruits shallowly bilobed.

Zambia. W: Ndola, fr. 21.iv.1969, *Fanshawe* 10583 (K; NDO; SRGH). S: Mumbwa Distr., Nambala Mt., y. male fl. 18.ix.1947, *Brenan & Greenway* 7877 (FHO; K; PRE). **Malawi**. S: Machemba Hill, male fl. 28.viii.1984, *Patel & Morris* 1543 (K; MAL; MO).

Also in Rwanda, Sudan (Equatoria), Uganda, S Kenya, NW Tanzania and South Africa (KwaZulu-Natal). Usually in mteshi (dry evergreen thicket), also in mushitu, evergreen forest and riverine fringing vegetation, often on anthills; 1250–1450 m.

Patel & Tawakali 1019 (K; MAL) from Chikala Mt., Liwonde For. Res., of southern Malawi, is intermediate between this and the previous variety, whilst *White* 3197D from Solwezi and *Fanshawe* 2160 from Kitwe, western Zambia, are intermediate between it and the typical variety.

Var. **angustifolia** Radcl.-Sm. in Kew Bull. **47**: 678 (1992). Type: Mozambique, Manica e Sofala, 18.xi.1935, *Kleinschmidt* s.n. (K, holotype; PRE).

Young shoots sparingly pubescent to subglabrous; leaf blades 2–5 × 0.7–1.5 cm, narrowly elliptic-lanceolate, coarsely irregularly serrate-dentate in the upper half; fruits unknown.

Mozambique. MS: male fl. 18.xi.1935, *Kleinschmidt* s.n. (K; PRE).

Not known elsewhere. No habitat data available.

Dr J. Hutchinson wrote to the collector, H. Kleinschmidt, Forest Officer at Penhalonga on the Zimbabwe/Mozambique border, on 17.xii.1935 requesting female flowers, fruits and exact locality details, but, as far as can be ascertained from the Kew Archives, never received any reply.

5. **Drypetes reticulata** Pax in Bot. Jahrb. Syst. **43**: 219 (1909). —Hutchinson in F.T.A. **6**, 1: 682 (1912). —Engler, Pflanzenw. Afrikas (Veg. Erde 9) **3**, 2: 35 (1921). —Pax in Engler, Pflanzenr. [IV, fam. 147, xv] **81**: 253 (1922). —Brenan, Check-list For. Trees Shrubs Tang. Terr.: 207 (1949). —Radcliffe-Smith in F.T.E.A., Euphorb. 1: 99 (1987). —Beentje, Kenya Trees, Shrubs Lianas: 195 (1994). Type from Tanzania.

A several-stemmed shrub or long-branched small tree up to 15 m high, very like *D. gerrardii*, but with the leaf blades quite entire, light green, drying greyish- or blackish-green, shiny on both surfaces, the tertiary nerves more strongly and irregularly reticulate, and with the stigmas obdeltoid, somewhat bifid at the apex, and soon falling.

Zimbabwe. E: Chipinge Distr., western end of Mwangazi Gap, y. male fl. 30.i.1975, *Pope, Biegel & Russell* 1465 (K; SRGH). S: Ndanga Distr., Chipinda Pools, y. fr. 14.x.1951, *Mullins* 113/51 (FHO; K; LISC; SRGH). **Malawi**. S: Mulanje Distr., Machemba Hill, y. male fl. 31.viii.1984, *Patel & Morris* 1546 (K; MAL; MO). **Mozambique**. N: Nampula, Mossuril, Floresta de Cruce, fr. 18.ii.1984, *de Koning, Groenendijk & Dungo* 9685 (K; LMU; MO). Z: sine. loc. cert., st. 27.ix.1972, *Tawse* 11 (LISC; SRGH). MS: Sofala Prov. (Beira Distr.), Gorongosa Nat. Park, Sangarassa Forest, 18.iv.1973, *Tinley* 2740 (K; LISC; PRE; SRGH). M: Moamba, Machatuine, Mangulane, fr. 9.iv.1974, *Balsinhas* 2694 (LISC).

Also in Somalia, Kenya, Tanzania (including Zanzibar) and South Africa (Transvaal, KwaZulu-Natal). In high rainfall savanna woodland, low altitude evergreen kloof forest and riverine fringe vegetation, submontane evergreen rainforest, *Brachystegia glaucescens* woodland on rocky hillsides; 275–800 m.

11. UAPACA Baill.

Uapaca Baill., Étud. Gén. Euphorb.: 595 (1858). —Müller Argoviensis in De Candolle, Prodr. **15**, 2: 489 (1866). —De Wildeman, Contrib. Ét. Esp. Gen. Uapaca (1936).

Dioecious trees or shrubs. Indumentum simple and/or minutely pseudolepidote. Trunks occasionally stilt-rooted. Twigs stout, often with pronounced leaf scars, producing an exudate when cut, which hardens and darkens on drying. Leaves alternate, usually crowded towards the end of the twigs, petiolate or subsessile, stipulate or not; blades simple, often obovate, entire, penninerved. Inflorescences axillary, or borne on older wood, solitary or fasciculate, pedunculate, with a whorl of 5–12 involucrate imbricate tepaloid bracts surrounding the flowers; male inflorescences many-flowered, the flowers in dense globose capitula; female inflorescences 1-flowered. Male flowers sessile; calyx campanulate or turbinate, truncate, dentate or irregularly or regularly lobed, the lobes imbricate; petals absent; disk absent; stamens (4)5(6), free, episepalous, anthers erect, subbasifixed, introrse, thecae parallel, longitudinally dehiscent; pistillode cylindric-obconic, infundibuliform, hypocrateriform, pileiform or sometimes lobed. Female flowers sessile; calyx minute, truncate, sinuate or lobed, disciform; petals absent; disk absent; ovary (2)3(5)-locular, with 2 ovules per loculus; styles (2)3(5), free, thick, recurved, covering the ovary, multipartite to laciniate. Fruit drupaceous, indehiscent; mesocarp spongy; pyrenes (2)3(4), dorsally carinate and bisulcate, indurate, tardily bivalved. Seeds mostly 1 per pyrene, compressed, ecarunculate; endosperm fleshy; embryo straight; cotyledons broad, flat, green.

An Afro-Malagasy genus of 61 species of which 49 are restricted to tropical Africa, the remaining 12 being endemic to Madagascar.

1. Leaf blades elliptic-oblong to elliptic-oblanceolate, not more than 15 × 8 cm, shiny above, usually with conspicuous interstitial nerves alternating with the main lateral ones; petiole slender; fruits 1–1.5 cm in diameter ··· 2
– Leaf blades usually oblanceolate to broadly obovate, up to 50 × 25 cm, dull or shiny above, often without conspicuous interstitial nerves; petioles usually stout; fruits often more than 1.5 cm in diameter ··· 3
2. Young shoots and petioles glabrous; leaf blades cuneate at the base ··· 2. *nitida*
– Young shoots and petioles crisped-ferrugineous-pubescent; leaf blades rounded at the base ··· 3. *rufopilosa*
3. Leaf blades completely glabrous; midrib ± straight; lateral nerves in 4–7(11) pairs; pyrenes neither carinate, sulcate nor lobed; tree with stilt roots, usually in moist habitats ··· 4. *lissopyrena*
– Leaf blades pubescent, tomentose or pilose, at least sparingly so beneath; midrib often distally zigzag; lateral nerves in (5)7–24 pairs; pyrenes carinate, sulcate and lobed; trees usually without stilt roots, usually in fairly dry habitats ··· 4
4. Indumentum of lower leaf surface sparse, often confined to the midrib and main nerves, consisting of straight, whitish hairs, or leaves glabrescent ··· 5
– Indumentum of lower leaf surface sparse to dense, not confined to the midrib and main nerves, and consisting of strongly crisped ferrugineous hairs, detersible (easily detached) only when very mature ··· 6
5. Leaves sessile or petiolate; young leaves pubescent or pilose on both surfaces; lobes of pyrenes irregularly crenellate; shrub or small tree up to 4.5 m high ··· 1. *pilosa*
– Leaves petiolate; young leaves glabrous above, sparingly hirsute-pubescent to subglabrous beneath; lobes of pyrenes smooth, entire; tree up to 25 m high with a clear bole up to 6 m high ··· 5. *sansibarica*
6. Indumentum of lower leaf surface sparse to evenly puberulous, subpersistent or ultimately detersible (easily detached); twigs commonly less than 1 cm in diameter; stipules 5 mm long ··· 6. *kirkiana*
– Indumentum of lower leaf surface dense, often floccose, generally persistent; twigs generally stout, up to 1.5 cm in diameter; stipules 1–2.5 cm long ··· 7. *robynsii*

1. **Uapaca pilosa** Hutch. in F.T.A. **6**, 1: 635 (1912). —R.E. Fries, Wiss. Ergebn. Schwed. Rhod.-Kongo-Exped. **1**, 1: 118 (1914). —Engler, Pflanzenw. Afrikas (Veg. Erde 9) **3**, 2: 37 (1921). —Pax in Engler, Pflanzenr. [IV, fam. 147, xv] **81**: 301 (1922). —De Wildeman, Contrib. Ét. Esp. Gen. Uapaca: 163 (1936). —White, F.F.N.R.: 206 (1962). —Radcliffe-Smith in F.T.E.A., Euphorb. 2: 566 (1988). Type: Zambia/Malawi, Stevenson Road, xi.1894, *Scott-Elliot* 8272 (K, holotype).

Uapaca masuku De Wild., Ann. Soc. Sci. Bruxelles **45**: 311 (1926); op. cit., Sér. B, **56**: 142 (1936). Type from Zaire.

An open-branched shrub or small tree up to 4.5 m high, with stout, brittle branchlets. Bark smooth or rough, longitudinally or quadrangularly-fissured. Young twigs pubescent. Stipules 5–7 mm long, linear to narrowly lanceolate, pubescent, soon falling. Leaves sessile or shortly petiolate, or petioles up to 7 cm long (var. *petiolata*). Leaf blades up to 40 × 25 cm, broadly obovate to obovate-oblanceolate, rounded at the apex, cuneate-attenuate at the base, thinly coriaceous; indumentum pubescent or pilose at first, especially along the midrib and main nerves, later ± glabrescent on leaf upper surface, minutely glandular-lepidote between the hairs on leaf lower surface; lamina glaucous or pale green, turning yellow when old; lateral nerves in up to 15 pairs, the lower camptodromous, the upper weakly to strongly brochidodromous, not prominent above, prominent beneath, tertiary nerves arcuate. Inflorescences usually borne just below the leaves. Male peduncles c. 4–6(9) cm long, pubescent; bracts 10, 10–15 × 3–7 mm, oblong, pubescent without, glabrous within, yellowish-green at first, turning cream-yellow; head 0.7–1 cm in diameter. Male flowers: calyx lobes 5–6, 1.5 mm long, linear-oblong to setaceous, pubescent at the apex; stamens 5, filaments very short, anthers 1 mm long, pale yellow, turning buff; pistillode hypocrateriform, pubescent at the apex. Female peduncles c. 1 cm long, extending to up to 6 cm long in fruit, stouter than the male; bracts more or less as in male inflorescence. Female flower: calyx lobes 1 × 1 mm, broadly triangular, pubescent at the apex; ovary 4–5 × 4–5 mm, subglobose, glabrous, yellowish-green, white-speckled; styles 3(5), 7–8 mm long, somewhat irregularly multipartite, the segments linear-setaceous, pale yellow. Fruits 3 × 4–5 cm, depressed-globose, shallowly (3)4(5)-lobed, smooth, glabrous, apple-green with brownish markings; mesocarp c. 3 mm thick, sticky, yellow; pyrenes 3–5, c. 2 × 1.5 cm, strongly carinate, the lateral lobes irregularly crenellate.

Leaves sessile or subsessile . . . . var. *pilosa*
Leaves markedly petiolate . . . . var. *petiolata*

Var. **pilosa**

Leaves sessile or subsessile.

Zambia. N: Mpika to Ishiba Ngandu (Shiwa Ngandu), fr. 28.vii.1938, *Greenway & Trapnell* 5535 (K). W: Ndola Dist., Butondo R. near Mufulira, male fl. 28.ix.1947, *Brenan, Miller & Greenway* 7973 (FHO; K). C: Kashitu Bridge, st.vii.1909, *Rogers* 8311 (K; SRGH). N: 16 km from Tunduma to Chipita, male fl. 2.xi.1966, *Gillett* 17548 (EA; K). **Malawi**. ?N: sine loco, pyr. & sds., iv.1890, *Johnston* 6 (K).

Also in Cameroon, Zaire (Shaba) and Tanzania. Miombo, and mixed deciduous woodlands with *Bauhinia, Combretum,* other *Uapaca* spp., *Burkea, Erythrophloeum, Parinari, Maprounea* and *Diplorrhynchus*; 1050–1700 m.

Vernacular names as recorded in specimen data include: "imasuku, imapangwa" (fr.) (Bangweulu area); "mompangwe" (chiLamba), "msuku" (K); "mupangwa"; "mupangwu" (chiWemba); "mukonkola" (Kipushi).

Var. **petiolata** P.A. Duvign. in Bull. Inst. Roy. Colon. Belg. **20**, 4: 890 (1949). Type from Zaire (Shaba).

Uapaca macrocephala Pax in Engler, Pflanzenr. [IV. fam. 147. xv] **81**: 305 (1922). Type from Tanzania.

Uapaca sapinii De Wild., Contrib. Ét. Esp. Gen. Uapaca: 174 (1936). Type from Zaire.

Uapaca benguelensis sensu Topham, Check List For. Trees Shrubs Nyasaland Prot.: 53 (1958); White, F.F.N.R.: 206 (1962), pro parte, non Müll. Arg.

Leaves markedly petiolate.

Zambia. N: 64 km from Kawambwa to Mansa (Ft. Rosebery), y. female fl. & o. fr. 29.x.1952, *Angus* 663 (BM; FHO; K). W: R. Matonchi, male fl. 15.x.1937, *Milne-Redhead* 2799 (K; LISC). S: Mumbwa Distr., Kafue Nat. Park, fr. 7.iv.1963, *Mitchell* 19/76 (SRGH). **Malawi**. N: Mzuzu, Marymount, fr. 27.x.1973, *Pawek* 7447 (K; MAL; MO; SRGH; UC).

Also in Zaire and Tanzania. Miombo and *Brachystegia/Uapaca* woodlands, sometimes in pure stands in Kalahari Sand, also with *Diplorrhynchus, Swartzia, Vitex, Vangueriopsis, Burkea, Monotes* and *Marquesia*; 1000–1525 m.

Vernacular names as recorded in specimen data include: "msuku", "masuku" (fr.) (chiTumbuka).

2. **Uapaca nitida** Müll. Arg. in Flora **47**: 517 (1864); in De Candolle, Prodr. **15**, 2: 491 (1866). —Engler, Pflanzenw. Ost-Afrikas **C**: 237 (1895). —Hutchinson in F.T.A. **6**, 1: 639 (1912). —R.E. Fries, Wiss. Ergebn. Schwed. Rhod.-Kongo-Exped. **1**, 1: 119 (1914). —Engler, Pflanzenw. Afrikas (Veg. Erde 9) **3**, 2: 37 (1921). —Pax in Engler, Pflanzenr. [IV, fam. 147, xv] **81**: 307 (1922). —De Wildeman, Contrib. Ét. Esp. Gen. Uapaca: 156 (1936). —Brenan,

Check-list For. Trees Shrubs Tang. Terr.: 228 (1949); in Mem. N.Y. Bot. Gard. **9**, 1: 70 (1954). —Topham, Check List For. Trees Shrubs Nyasaland Prot.: 53 (1958). —White, F.F.N.R.: 206 (1962). —Drummond in Kirkia **10**: 251 (1975). —K. Coates Palgrave, Trees Southern Africa, ed. 2, rev.: 409 (1983). —Radcliffe-Smith in F.T.E.A., Euphorb. 2: 568 (1988). Type: Zambia, Livingstone Distr., Batoka Country highlands, vii-x. 1860, *Kirk* s.n. (K, holotype).

Uapaca microphylla Pax in Bot. Jahrb. Syst. **23**: 523 (1897). Type from Angola (Malanje Province).

A small glabrous evergreen tree up to 12 m high, usually much less, branching from c. 0.5 m above the base; crown open rounded; bole usually less than 1 m in diameter, ± irregular; bark smooth at first, or flaking, later rough, grey, dark brown or black, deeply quadrately fissured to show reddish underbark, and exuding a gummy red "varnish"; young twigs fairly slender. Stipules c. 1 mm long, triangular, soon falling, or not developed. Leaves long-petiolate, the petioles (1)1.5–6 cm long. Leaf blades 4–16 × 1.5–8 cm, elliptic-oblong to elliptic-oblanceolate (narrowly elliptic in var. *longifolia*), rounded or obtuse, sometimes subacute (attenuate in var. *longifolia*) at the apex, cuneate or attenuate at the base, thickly chartaceous, shiny dark green above, paler and duller beneath; lateral nerves in 8–12 pairs, usually with conspicuous interstitials alternating with them, brochidodromous, scarcely prominent, tertiary nerves weakly reticulate. Inflorescences usually borne among the leaves. Male peduncles 0.7–1.5(2) cm long; inflorescence bracts 7–10, 5–10 × 3–7 mm, elliptic-obovate, rounded, yellow-green, cream-coloured or whitish; head c. 5 mm in diameter. Male flowers with a sweet, musky scent; calyx c. 1 mm long, obconic, 5-lobed, the lobes irregularly lobulate, glabrous or pubescent, white; stamens 4–5, filaments 1.5 mm long, creamy-white, anthers 0.7 mm long, pale yellow; pistillode 1 mm high, infundibuliform, irregularly lobed, pubescent without, glabrous within. Female peduncles and bracts ± as in the male. Female flower: calyx 4 mm in diameter, shallowly cupular, 6-lobed, the lobes truncate; staminodes rarely present; ovary 5 × 4 mm, ovoid-subglobose, 3-locular; styles 3, 4–5 mm long, flabelliform, laciniate, flattened, adaxially minutely granulate, abaxially smooth, ochreous. Fruits 1.6–2 × 1.4–1.5 cm when fresh (1.4–1.6 × 1–1.3 cm when dried), ovoid-ellipsoid, shallowly longitudinally 12-ribbed, pale green at first, later becoming reddish-green or brownish. Pyrenes 3, 1–1.5 cm × 7–9 mm, carinate, apiculate, the lateral lobes smooth, entire.

Leaves elliptic-oblong to elliptic-oblanceolate, rounded or obtuse at the apex, sometimes subacute · var. *nitida*
Leaves narrowly elliptic, attenuate at the apex · var. *longifolia*

Var. **nitida**

Leaves elliptic-oblong to elliptic-oblanceolate, rounded or obtuse at the apex, sometimes subacute.

Zambia. N: Mbala (Abercorn) Lake, male fl. 19.vii.1930, *Hutchinson & Gillett* 3873 (BM; K; SRGH). W: 27 km from Mwinilunga to Kabompo, fr. 6.vi.1963, *Loveridge* 814 (K; SRGH). C: 3 km north of Kabwe (Broken Hill), y. male fl. 15.iii.1961, *Drummond & Rutherford-Smith* 6911 (K; SRGH). E: Petauke, male fl. 21.iv.1952, *White* 2430 (FHO; K). S: Mazabuka Distr., 2 km to Gwembe, fr. 14.vii.1952, *White* 3010 (FHO; K; PRE). **Zimbabwe**. N: Gokwe, male fl. 11.iii.1963, *Bingham* 486 (K; LISC; PRE; SRGH). C: Harare Distr., Norton, fr. 15.v.1954, *McGregor* 4/54 (SRGH). **Malawi**. N: Chitipa Distr., Kaseye Mission, fr. 6.x.1977, *Pawek* 13121 (K; MAL; MO; PRE; SRGH; UC). C: Bunda Agric. College, 23 km south of Lilongwe, male fl. 31.iii.1970, *Brummitt* 9557 (K; PRE; SRGH). S: Mulanje Mt., Likhubula (Likabula) Gorge, female fl. & y. fr. 20.vi.1946, *Brass* 16375 (K; NY; PRE; SRGH). **Mozambique**. N: Nampula Prov., 10 km from Mutuali to Malema, y. male fl. 4.iii.1953, *Gomes e Sousa* 4051 (K; LISC; PRE). Z: Pebane, male fl. 4.x.1949, *Barbosa & Carvalho* in *Barbosa* 4294 (K; LMA). T: Z Z: Pebane, male fl. *Hornby* 2755 (K; SRGH). MS: between Inhaminga and Mupa, male fl. 11.vii.1946, *Simão* 778 (LMA; PRE; SRGH). GI: Macovane, male fl. 1.vi.1947, *Hornby* 2720 (K; SRGH (PRE duplicate under this number is a mixture of *U. sansibarica* and *U. kirkiana*)).

Also in Zaire (Kinshasa, Shaba), Burundi, Kenya, Tanzania and Angola. Plateau and escarpment deciduous woodlands, typically *Brachystegia/Uapaca* and miombo woodlands, also in *Parinari/ Terminalia sericea* and *Monotes/Syzygium* savanna, and sometimes on dambo margins and in mushitu, often on rocky hillsides; 40–1830 m.

Vernacular names as recorded in specimen data include: "cagigoura" (Tete area); "cochocorre", "cotho'co'rre", "mucunapa" (Cabo Delgado area); "kasakolowe" (chiChewa, chiNyanja); "kasokolowe" (Ngoni, chiTonga, chiTumbuka, chiYao); "kochokore", "metongolo", "tongolo" (Mataca); "metoto" (Niassa); "molundu" (Choma area); "msechela", "msechera", "mtoto" (chiYao); "msokolowe", "musokolowe" (chiBemba, chiLungu); "mulengu" (chiLunda); "mulundu" (Toki, Sik., chiTonga); "munundu" (chiKaonde, chiTonga); "musokolobe", "musokolole", "musokolowe" (chiShinga, chiWemba, chiWisa); "musorkobani" (chiLienda, chiNyanja); "mutela", "m'tela", "n'tela", "tela" (eMakhuwa); "teela" (Zambézia).
Uses: Soft wood used for hut poles; fruits edible.

Var. **longifolia** (P.A. Duvign.) Radcl.-Sm. in Kew Bull. **48**: 616 (1993). Type from Zaire (Shaba).
Uapaca nitida var. *sokolobe* f. *longifolia* P.A. Duvign. in Bull. Inst. Roy. Colon. Belg. **20**, 4: 890 (1949).

Leaves narrowly elliptic, attenuate.

Zimbabwe. N: Hurungwe Distr., Mwami (Miami), st. 4.x.1946, *Wild* 1275 in *GHS* 15442 (K; SRGH). **Malawi**. N: Nkhata Bay, Chikale Beach, male fl. 9.viii.1977, *Pawek* 12866 (K; MAL; MO; PRE; SRGH; UC). **Mozambique**. N: Cabo Delgado, Palma to Nangade, fr. 21.x.1942, *Mendonça* 1023 (LISC). Z: near Luabo, st. 18.xi.1942, *Mendonça* 1443 (LISC). MS: Dondo, fr. 18.viii.1947, *Pimenta* 34 (SRGH).
Also in Zaire (Shaba). Escarpment miombo, often with *Brachystegia boehmii* and *Uapaca kirkiana* on rocky hillsides; 520–1370 m.
Vernacular names as recorded in specimen data include: "mtototo" (Cabo Delgado area); "mutongaro" (Dondo area).

3. **Uapaca rufopilosa** (De Wild.) P.A. Duvign. in Bull. Inst. Roy. Colon. Belg. **20**: 890 (1949). —Radcliffe-Smith in Kew Bull. **37**: 428 (1982). Type from Zaire (Shaba Province).
Uapaca nitida var. *rufopilosa* De Wild., Contrib. Ét. Esp. Gen. Uapaca: 161 (1936). —White, F.F.N.R.: 206 (1962).

A small deciduous tree up to 7.5 m high. Bark light brown, slightly gnarled. Twigs somewhat thickened. Young shoots densely ferrugineous-floccose-pubescent. Stipules minute. Leaves long-petiolate, the petioles 2–7 cm long. Leaf blades 4–12 × 2–5.5 cm, oblong-elliptic, ± rounded at apex and base, entire, chartaceous, sparingly ferrugineous-pubescent beneath at first, later glabrescent, otherwise glabrous; midrib bifurcate just before the apex; lateral nerves in 9–12 pairs, scarcely prominent above or beneath, the lower camptodromous, the upper brochidodromous or else all brochidodromous, tertiary nerves often indistict. Inflorescences borne among or just below the leaves. Male peduncles 7–10(20) mm long, bracteate; bracts of main inflorescence 11–12, 5–9 × 3–6 mm, obovate-suborbicular, glabrous, pale yellowish-green; head of male flowers 8.5 mm in diameter. Male flowers: calyx lobes 5–6, 1.5 × 1 mm, triangular, acute, glabrous, yellowish-green; stamens 6 with white filaments 1 mm long and pale yellow anthers 1 mm long; pistillode 2 mm high, unequally trilobed, the lobes somewhat foliaceous. Female peduncles as in the male; bracts 7, otherwise resembling the male. Female flower: calyx 4 mm in diameter, shallowly 5(6)-lobed, the lobes minutely crenellate, sparingly ciliate; ovary 6 × 4 mm, ellipsoid, 3(4)-locular, glabrous; styles 3(4), 5 × 3 mm, irregularly 5–6-partite, the segments crenellate. Fruits 1.4–1.5 × 1.2–1.3 cm, ellipsoid-subglobose, smooth, glabrous, 3(4)-seeded. Mesocarp thin. Pyrenes 3(4), c. 1 × 0.7 cm, carinate, apiculate apically and basally.

Zambia. W: 27 km from Mwinilunga to Kabompo, female fl. 6.vi.1963, *Loveridge* 818 (K; LISC; PRE; SRGH); Lisombo R., male fl. 20.v.1969, *Mutimushi* 3221 (K; NDO).
Also in Zaire (Shaba) and Angola. Miombo and Isenga woodlands, often on rocky hillsides; c. 1370 m.

4. **Uapaca lissopyrena** Radcl.-Sm. in Kew Bull. **48**: 612 (1993). Type: Zambia, Mwinilunga, fr. 5.ix.1955, *Holmes* 1164 (K, holotype; NDO).
Uapaca sp. 1 sensu White, F.F.N.R.: 206 (1962). —Drummond in Kirkia **10**: 251 (1975). —K. Coates Palgrave, Trees Southern Africa, ed. 2, rev.: 410 (1983).
Uapaca guineensis sensu Topham, Check List For. Trees Shrubs Nyasaland Prot.: 53 (1958). —Radcliffe-Smith in F.T.E.A., Euphorb. 2: 571 (1988), pro parte, non Müll. Arg.

An evergreen tree up to 30 m high with an open spreading crown. Stem up to

c. 1 m in diameter, unbranched up to 8 m, with stilt roots up to 2.5 m; bark smooth, developing rectangular cracks when old, pale grey with brown lenticels; wood pinkish, soft, fibrous. Young twigs usually fairly slender, glabrous, but sometimes apparently lepidote owing to the minutely fracturing greyish waxy pruina. Stipules absent. Leaves petiolate, the petioles on the crown shoots 1–5 cm long, but up to 10 cm long on sucker shoots. Leaf blades 3–15(30) × 2–10(15) cm, obovate or elliptic-obovate, rounded at the apex (shortly acuminate on sapling, shade and sucker leaves), attenuate or cuneate, rarely rounded or cordulate, at the base, coriaceous, glabrous, dark green and glossy above, pallid beneath, reddish-pink when young; midrib ± straight, lateral nerves in 4–7 pairs (up to 11 on the sucker leaves), widely spaced, camptodromous, often impressed above, somewhat prominent beneath, tertiary nerves often indistinct. Inflorescences and fruits borne among and immediately below the leaves. Male peduncles 1–2.5 cm long, minutely bracteolate; inflorescence bracts 10–11, (5)9–12 × (3)5–6 mm, oblong-oblanceolate, rounded, minutely crenulate to subentire, glabrous, pale yellow; head 7 mm in diameter. Male flowers: calyx lobes 5, c. 1 mm long, acute, subglabrous; tube 0.5 mm long; stamens 5, filaments 1.5 mm long, flattened, anthers 0.5 mm long, cream-coloured; pistillode 1.5 mm high, cylindric-obconic, sparingly pubescent. Female peduncles 0.5–1.5 cm long; bracteoles and bracts fewer than in the male flowers. Female flower delicately scented; calyx 2.5–3 mm in diameter, 6-lobed, the lobes c. 1 mm long, broadly triangular, swollen at the base, sparingly pubescent to glabrous without, with a ring of hairs within; ovary 4–5 × 3–4 mm, ellipsoid-subglobose, 3(4)-locular, smooth, sparingly pubescent towards the apex, pale greenish-yellow; styles 3(4), 3–4 mm long, usually deeply 3-lobed, the lobes laciniate, flattened, smooth, pubescent at the base, pale greenish-yellow. Fruits 1.8–2.3 × 1.5–1.7 cm, ovoid-subglobose, ± smooth, greenish, later turning pale yellow. Pyrenes 3(4), 1.25–1.3 cm × 7.5–8 mm × 3–4 mm, compressed-ellipsoid, slightly apiculate at the apex, rounded at the base, without keel, grooves or lobes, smooth, entire, thin. Seeds 9 × 5 × 2.5 mm, resembling the pyrenes in shape, smooth, greyish.

Zambia. N: Misamfu, 8 km north of Kasama, fr. 17.v.1958, *Angus* 1971 (FHO; K; PRE; SRGH). W: Mwinilunga Distr., 6.5 km north of Kalene Hill, fr. 26.ix.1952, *Angus* 556 (BM; FHO; K; PRE). S: Mazabuka Distr., 2.5 km to Gwembe, fr. 14.vii.1952, *White* 3010 (BM; FHO). **Zimbabwe**. E: Chimanimani Distr., Haroni/Makurupini Forest, fr. 4.xii.1964, *Wild, Goldsmith & Müller* 6636 (FHO; K; LISC; PRE; SRGH). **Malawi**. N: Nkhata Bay Distr., Nkwazi For. Res., st. 13.ix.1970, *Müller* 1615A (SRGH). C: Lilongwe Distr., Dzalanyama., male fl. 7.ii.1963, *Chapman* 1795 (FHO; K; SRGH). **Mozambique**. Z: Serra do Gurué, R. Licungo, fr. 9.xi.1967, *Torre & Correia* 16026 (LISC). MS: Chimanimani Mts., Musapa Gorge, fr. 16.x.1950, *Chase* 2930 (BM; PRE; SRGH).

Not known elsewhere. Dense evergreen swamp forest (mushitu), wet plateau riverine forest and fringing woodland, also in swampy ground in low altitude evergreen forest, less often in drier *Brachystegia/Uapaca* woodland; 400–1650 m.

Vernacular names as recorded in specimen data include: "chitoto", "mtoto" (chiYao); "kasokolowe" (chiTambuka); "mlemba" (chiTonga); "mufemba", "mulemba" (Tok.); "mulengu" (chiLunda); "musokolowe wa mushitu" (chiWemba).

Used for making canoes.

5. **Uapaca sansibarica** Pax in Bot. Jahrb. Syst. **34**: 370 (1904). —S. Moore in J. Linn. Soc., Bot. **40**: 193 (1911). —Hutchinson in F.T.A. **6**, 1: 636 (1912). —R.E. Fries, Wiss. Ergebn. Schwed. Rhod.-Kongo-Exped. **1**, 1: 118 (1914). —Eyles in Trans. Roy. Soc. South Africa **5**: 394 (1916). —Engler, Pflanzenw. Afrikas (Veg. Erde 9) **3**, 2: 37 (1921). —Pax in Engler, Pflanzenr. [IV, fam. 147, xv] **81**: 304 (1922). —De Wildeman, Contrib. Ét. Esp. Gen. Uapaca: 172 (1936). —Brenan, Check-list For. Trees Shrubs Tang. Terr.: 228 (1949). —Eggeling & Dale, Indig. Trees Uganda, ed. 2: 143 (1952). —F.W. Andrews, Fl. Pl. Anglo-Egypt. Sudan **2**: 100 (1952). —Brenan in Mem. N.Y. Bot. Gard. **9**, 1: 71 (1954). —Topham, Check List For. Trees Shrubs Nyasaland Prot.: 54 (1958). —White, F.F.N.R: 206 (1962). —Drummond in Kirkia **10**: 251 (1975). —K. Coates Palgrave, Trees Southern Africa, ed. 2, rev.: 409 (1983). —Radcliffe-Smith in F.T.E.A., Euphorb. 2: 568 (1988). Tab. **14**. Syntypes: Mozambique, Quelimane (Quilimane), *Stuhlmann* I. 577 (B, syntype); Malawi, sine loc. cert., 1891, *Buchanan* 221 (K, syntype); and from Tanzania.

An evergreen tree up to 25 m high, with divaricate branches, a dense rounded crown and a clear bole up to 6 m; bark smooth, striate, or finely quadrangularly fissured and peeling in small flakes, dark grey or blackish; stilt roots rarely produced;

Tab. 14. UAPACA SANSIBARICA. 1, distal portion of flowering branch ($\times \frac{2}{3}$); 2, tip of leaf, undersurface (× 2); 3, male inflorescence (× 1); 4, male flower (× 8), 1–4 from *Vesey-FitzGerald* 5165; 5, female flower, with 2 inflorescence bracts removed (× 4), from *Davis* 176; 6, fruit (× 4), from *Procter* 904. Drawn by Christine Grey-Wilson. From F.T.E.A.

young twigs fairly slender to quite robust, ± glabrous. Stipules absent. Leaves petiolate, the petioles (0.3)1–2(6.5) cm long, usually fairly slender, densely pubescent to glabrous. Leaf blades (3)5–21(25) × (1.5)2.5–11(15) cm, obovate to oblanceolate, rounded at the apex, attenuate to a cuneate or narrowly rounded-cuneate base, thinly coriaceous, glabrous above, sparingly hirsute-pubescent to subglabrous or glabrous beneath, deep green and glossy above, paler and duller or greyish-green beneath; midrib usually distally zigzag, lateral nerves in (5)7–14 pairs, rarely with interstitials, mostly camptodromous, the distal 3–4 pairs sometimes brochidodromous, somewhat impressed above, prominent beneath, tertiary nerves subparallel. Inflorescences usually borne among the leaves, the fruits borne below them. Male peduncles 1–3 cm long, sometimes 1–2-bracteolate; inflorescence bracts 9–10, 0.5–1.2 cm × 3–7 mm, elliptic, rounded, the outer ones often becoming reflexed, pubescent without at the base, otherwise glabrous, yellowish; head 5–6 mm in diameter. Male flowers: calyx lobes 4–5, 1 mm long, truncate, pubescent at the apex; stamens 4, filaments 1 mm long, flattened, anthers 0.5 mm long, yellow; pistillode 1 mm high, cylindric-obconic, densely pubescent. Female peduncles 0.5–1.5 cm long; bracts ± as in the male. Female flower: calyx 4 mm in diameter, shallowly cupular, 5-lobed, the lobes 0.5 mm long, rounded, pubescent at the apex; ovary 4–5 × 3–4 mm, ovoid-subglobose, 3(4)-locular, smooth, glabrous; styles 3(4), 4–6 mm long, flabelliform, laciniate, flattened, smooth, pubescent at the base, greenish-yellow. Fruits 1.7–2 × 1.5 cm, ellipsoid, ± smooth, yellowish-green, later turning yellow or reddish. Pyrenes 3(4), 1–2.5 cm × 8–9 mm × 5–6 mm, shallowly carinate, slightly apiculate at the apex, ± truncate at the base, lateral lobes smooth, entire. Seeds 1 × 0.4 cm, cylindric-compressed, sulcate.

Zambia. N: 72 km from Kasama to Luwingu, male fl. 5.iv.1961, *Angus* 2708 (FHO; K; SRGH). W: c. 3 km NE of Katandana Bridge on old Solwezi–Chingola road, fr. 4.ix.1949, *Hoyle* 1185 (BM; FHO; K). C: Serenje to Mpika, fr. 16.vii.1930, *Pole Evans* 2905 (12) (K; PRE; SRGH). E: 6.5 km north of Lundazi Boma, y. fr. 27.iv.1952, *White* 2485 (FHO; K). S: near Mswebi on the Mumbwa road, o. fr. 16.v.1963, *van Rensburg* 2164 (K; SRGH). **Zimbabwe**. E: Chimanimani Distr., Haroni Gorge, fr. 10.i.1969, *Macdonald* 27 (K; LISC; PRE; SRGH). **Malawi**. N: Mzimba Distr., Kaning'ina For. Res., Mzuzu, fr. 17.viii.1984, *Balaka & Kaunda* 560 (K; MAL; MO). C: Nkhota Kota Distr., Chia, fr. 1.ix.1946, *Brass* 17466 (BM; K; NY; PRE; SRGH). S: Mulanje Distr., Mimosa Res. Station, female fl. 9.iii.1987, *J.D. Chapman & E.J. Chapman* 8377 (FHO; K; MAL; MO; PRE). **Mozambique**. N: 12 km from Quiterajo to Mocímboa da Praia, fr. 11.xi.1953, *Balsinhas* 78 (BM; LMA). Z: 40 km from Munhamade to Mucubi, fr. 25.ix.1949, *Barbosa & Carvalho* in *Barbosa* 4170 (K; LMA). T: Zóbuè, male fl. 18.vi.1941, *Torre* 2902 (LISC). MS: Chimoio Distr., west of Beira/Chicamba Dam road junction, male fl. 11.iii.1962, *Chase* 7650 (K; LISC; MO; SRGH).

Also in Zaire (Shaba), Burundi, Sudan, Uganda, Tanzania and Angola. Usually in miombo and *Brachystegia/Uapaca* woodlands, also in *Marquesia/Uapaca* and high rainfall plateau woodlands, sometimes in riverine woodland, mushitu margins and sparsely treed dambos, also in submontane woodland or scrub on steep lower mountain or valley sides; 500–1830 m.

Vernacular names as recorded in specimen data include: "chitoto cha m'adzi", "mtoto" (chiYao); "coroco'ee" (Gurué area); "kasakolowe", "kasokolowe" (chiNyanja); "nakashunkuri" (eMakhuwa); "mazange" (Chimoio area); "messexera" (Niassa area); "metongoro", "m'tongouro" (Manica area); "msokolowe" (Mzimba area); "m'tangora m'chena" (Zambézia area); "mucurucuro" (Cabo Delgado area); "mulengu" (chiLunda); "musankwa("male")–masuku" (Toka); "musika" (chiBemba); "musokolobe" (wabutu); "mutengere", "mutongoro" (Honde Valley area); "mutongoro" (Chirinda, Cheringoma); "mutu" (Milange area); "m'zanja" (Beira area); "tongolo" (Dondo area).

Bark used for fish poison; bark extract used for diarrhoea; wood used in building construction; fruit edible.

6. **Uapaca kirkiana** Müll. Arg. in Flora **47**: 517 (1864); in De Candolle, Prodr. **15**, 2: 491 (1866). —Engler, Pflanzenw. Ost-Afrikas **C**: 237 (1895). —S. Moore in J. Linn. Soc., Bot. **40**: 194 (1911). —Hutchinson in F.T.A. **6**, 1: 636 (1912). —R.E. Fries, Wiss. Ergebn. Schwed. Rhod.-Kongo-Exped. **1**, 1: 118 (1914). —Eyles in Trans. Roy. Soc. South Africa **5**: 394 (1916). —Engler, Pflanzenw. Afrikas (Veg. Erde 9) **3**, 2: 37 (1921). —Pax in Engler, Pflanzenr. [IV, fam. 147, xv] **81**: 302 (1922). —De Wildeman, Contrib. Ét. Esp. Gen. Uapaca: 135 (1936). —Brenan, Check-list For. Trees Shrubs Tang. Terr.: 228 (1949); in Mem. N.Y. Bot. Gard. **9**, 1: 71 (1954). —Topham, Check List For. Trees Shrubs Nyasaland Prot.: 53 (1958). —White, F.F.N.R.: 207 (1962). —Drummond in Kirkia **10**: 251 (1975). —K. Coates Palgrave, Trees Southern Africa, ed. 2, rev.: 408 (1983). —Radcliffe-Smith in F.T.E.A., Euphorb. 2: 567 (1988); in Kew Bull. **48**: 612 (1993). Type: Malawi, Southern Province, Soche Hill, Manganja Country, male fl. 8.iii.1862, *Kirk* s.n. (K, holotype).

Uapaca benguelensis Müll. Arg. in J. Bot. **2**: 332 (1864); in De Candolle, Prodr. **15**, 2: 491 (1866). —Hutchinson in F.T.A. **6**, 1: 637 (1912). —Engler, Pflanzenw. Afrikas (Veg. Erde 9) **3**, 2: 37 (1921). —Pax in Engler, Pflanzenr. [IV, fam. 147, xv] **81**: 303 (1922). —De Wildeman, Contrib. Ét. Esp. Gen. Uapaca: 95 (1936). Type from Angola (Benguela Province).

Uapaca goetzei Pax in Bot. Jahrb. Syst. **28**: 418 (1900). Type from Tanzania (Southern Highlands Province).

Uapaca kirkiana var. *goetzei* (Pax) Pax in Bot. Jahrb. Syst. **34**: 370 (1904).

Uapaca greenwayi Suesseng. in Trans. Rhod. Sci. Assoc. **43**: 85 (1951). Syntypes: Zimbabwe, Marondera (Marandellas), 22.x.1941, *Dehn* 373; 374 (M, syntypes); 29.i.1942, *Dehn* 666 (M, syntype).

A much-branched evergreen tree up to 12 m high, with a wide spreading rounded crown and a short clear bole (usually up to c. 1–2 m) up to c. 30 cm d.b.h.; bark striated to longitudinally fissured and finely transversely cracked or reticulate, corky, brittle, dark grey or blackish, occasionally exuding a viscous transparent fluid. Wood somewhat pulpy, reddish. Young twigs 0.4–1 cm in diameter, fairly stout, evenly to densely crisped-puberulous or crisped-pubescent. Stipules 5 mm long, filiform, pubescent, soon falling. Leaves subsessile to shortly-petiolate, petioles 0.5–2(4.5) cm long, stout. Leaf blades (7)10–27(30) × (5)7–17 cm, broadly obovate to suborbicular-obovate, saddle-shaped, rounded or retuse at the apex, attenuate to rounded-cuneate at the base, entire or undulate-sinuate, usually thickly coriaceous, glabrous above, usually ferrugineous crisped-puberulous and minutely glandular-lepidote beneath, sometimes glabrescent, dull mid- to dark green above, paler beneath, midrib often pale yellowish-green; lateral nerves in 12–24 pairs, sometimes almost perpendicular to the midrib, not prominent or slightly impressed above, prominent beneath, subcraspedodromous, interstititials rarely present, tertiary nerves subparallel. Inflorescences usually borne for 15–30 cm below the leaves. Male peduncles 0.7–2 cm long, sometimes 1–2 bracteolate; inflorescence bracts (5)7–11, 5–7 × 3–5 mm, broadly elliptic or elliptic-obovate, the outer pubescent without, the inner mostly glabrous, pale yellow or whitish; head (5)7–8 mm in diameter. Male flowers: calyx lobes 4–5, 1 mm long, acute or bifid, glabrous; stamens 4, filaments 1 mm long, flattened, anthers 1 × 1 mm, yellow; pistillode 1.5 mm high, infundibuliform, pubescent. Female peduncles 4–5 mm long, extending to c. 1 cm and thickening in fruit, 2-bracteolate, minutely lepidote, the bracteoles ferrugineous crisped-puberulous; bracts more or less as in the male. Female flower: calyx 3 mm in diameter, shallowly cupular, 5–8-lobed, the lobes 0.5 mm long, broadly triangular; ovary 3–4 × 3–4 mm, subglobose, 4-locular, densely ferrugineous-tomentose; styles 4, 3 mm long, multifid-flabelliform, the segments oblong, rounded, flattened, shallowly rugulose, abaxially puberulous, stigmatic surface glabrous. Fruits 4 × 4 cm (fresh), 3 × 3 cm (dried), subglobose to imperceptibly 4-lobed, smooth, reddish-brown or greenish at first, later becoming orange-yellow. Pyrenes 4, 2 × 1.1–1.2 × 0.8 cm, strongly carinate, bisulcate, apiculate, tricuspidate-emarginate at the base, lateral lobes smooth, entire. Seeds 1.5 × 0.8 × 0.5 cm, resembling the pyrenes in shape, brown.

Zambia. B: Chavuma, o. fr. 5.vii.1954, *Gilges* 396 (K; PRE; SRGH). N: Misamfu, 6.4 km north of Kasama, fr. 5.iv.1961, *Angus* 2676 (FHO; K; SRGH). W: c. 27 km south of Mwinilunga on Kabompo road, fr. 6.vi.1963, *Loveridge* 791 (K; LISC; SRGH). C: north of Kabwe (Broken Hill), male fl. i.1906, *Allen* 446 (K; SRGH). E: west of Mwangozi R., st. iii.1930, *Bush* 65 (K). S: Mazabuka Distr., 67 km WNW of Chirundu, male fl. 28.ii.1960, *Leach* 9806 (K; LISC; PRE; SRGH). **Zimbabwe**. N: near Umvukwe Mts., 6.4 km north of Banket, male fl. & fr. 23.iv.1948, *Rodin* 4410A (K; SRGH; UC). C: Harare, male fl. ii.1920, *Eyles* 2051 (K; PRE; SRGH). C/S: Masvingo (Fort Victoria)–Gweru (Gwelo), fr. 3.vii.1930, *Pole Evans* 2733 (40) (K; PRE; SRGH). E: Chipinge Distr., Gungunyana For. Res., male fl. iii.1967, *Goldsmith* 39/67 (K; LISC; SRGH). **Malawi**. N: Mzimba Distr., Mzuzu, y. male fl. 5.iii.1975, *Pawek* 9117 (K; MAL; MO; PRE; SRGH; UC). C: Ntchisi For. Res., male fl. 25.iii.1970, *Brummitt* 9365 (K; PRE; SRGH). S: Mulanje Mt., Likhubula (Likabula) Gorge, y. fr. 20.vi.1946, *Brass* 16376 (K; NY; SRGH). **Mozambique**. N: 6 km from Marrupa to Nungo, o. fr. 4.viii.1981, *Jansen, de Koning & de Wilde* 23 (K; MO; SRGH; WAG). Z: entre o Alto Ligonha e o Alto Molócuè, 14.6 km do Alto Ligonha, o. fr. 13.x.1949, *Barbosa & Carvalho* in *Barbosa* 4407 (K; LMA). T: Zóbuè male fl. 2.x.1942, *Mendonça* 551 (LISC). MS: Chimoio (Vila Pery) Distr., Mavita–frontier, fr. 20.vi.1942, *Torre* 4388 (LISC).

Also in Zaire (Shaba), Burundi, Tanzania and Angola. Plateau woodlands on well drained soils, often as understorey tree of *Brachystegia* woodlands, sometimes locally dominant on gravelly soils, or co-dominant as *Brachystegia/Uapaca* wooodland, or with *Marquesia* and *Isoberlinia* or in *Brachystegia taxifolia/B. spiciformis* woodlands, often on rocky escarpment and hillsides, and among granite boulders, also on dambo margins; 30–1830 m.

Vernacular names as recorded in specimen data include: "chibaba" (Chisamba area); "mahobohobo" (Kabwe arae); "mapopolo" (Chavuma area); "masugwa" (chiKaonde); "masuko", "masuku", "musuku" (chiChewa, ciSenga, chiTonga, chiLozi, chiBemba); "mesange", "metongoso", "m'jange" (Manica area); "messuco" (Tete); "mobura" (Chilinda area); "mpotopoto" (Ngoni area); "m'suku" (Ayaua, chiNyanja, chiYao); "mukumapa" (Zambézia area); "n'tjunku", "n'chunkuri" (eMakhuwa); "wild loquat".

Fruit edible and much sought after, the wood although soft does not rot quickly and is used for hut poles and planks, and for making spoons.

There is considerable overlap between this species and *Uapaca benguelensis* Müll. Arg. (e.g. in petiole length, numbers of lateral nerves in the leaves and peduncle length and number). The author therefore placed the latter species in synonymy under *Uapaca kirkiana* Müll. Arg. (Kew Bulletin **48**: 612 (1993)). However, having subsequently examined further Zambian material it is considered possible that this may be a separate taxon after all and if, as a result of further investigation, this is recognized as distinct then material such as *Harder, Merello & Nkhoma* 2181 (K; MO) should be referred to *Uapaca benguelensis* Müll. Arg. sensu stricto. The indumentum of the leaf lower surface does not provide a reliable taxonomic character, since crisped ferrugineous hairs, short straight whitish hairs or hairs ± absent may be found in either taxon.

7. **Uapaca robynsii** De Wild. in Ann. Soc. Sci. Brux. **53**, B: 60 (1933); Contrib. Ét. Esp. Gen. Uapaca: 167 (1936). —White, F.F.N.R.: 206 (1962). Type from Zaire (Shaba Province).
Uapaca benguelensis sensu Topham, Check List For. Trees Shrubs Nyasaland Prot.: 53 (1958), non Müll. Arg.

A much-branched spreading, often stunted, shrub or small tree up to 7 m high, with gnarled divaricate branches often arising from near the ground, and with a more or less rounded crown; bark longitudinally or quadrately deeply-fissured, grey to black or brownish-black. Young twigs (0.7)1–1.5 cm in diameter, stout, densely floccose or tomentose. Stipules 1–2.5 cm long, subulate-filiform, densely tomentose, soon falling. Leaves petiolate; petioles 1–8 cm long, fairly stout. Leaf blades 6–23 × 5–17 cm, broadly obovate to suborbicular-obovate, sometimes elliptic-oblong, saddle-shaped, rounded or somewhat truncate at the apex, cuneate to rounded or rarely shallowly cordate at the base, entire or rarely undulate-sinuate on the margins, thickly coriaceous, glabrous above, except on the midrib and lateral nerves when young, densely whitish or ferrugineous tomentose or floccose beneath, dark mid- or dull grey-green on upper surface, with the midrib often bright yellow; midrib sometimes bifurcate near the apex; lateral nerves in 10–18 pairs, often almost perpendicular to the midrib, not prominent or slightly impressed above, not or scarcely prominent beneath, subcraspedodromous, tertiary nerves subparallel, often not clearly visible. Inflorescences usually borne among or just below the leaves. Male peduncles 1–2(3) cm long, 0–1-bracteolate, often glabrous; inflorescence bracts 9–12, variable, 6–9 × 2–6 mm, elliptic-lanceolate to broadly elliptic, glabrous, the outer sepaloid and reddish-brown, the inner petaloid and yellow; head 6–9 mm in diameter. Male flowers: calyx lobes 5, 0.5–1 mm long, filiform, acute or truncate, sparingly pubescent; stamens 5, filaments 2 mm long, flattened, anthers c. 1 × 1 mm, creamy-yellow; pistillode 1.5 mm high, infundibuliform, densely pubescent. Female peduncles 3–4 mm long, extending to 5 mm in fruit, pubescent; bracts as in male. Female flower: calyx 4 mm in diameter, patelliform, scarcely lobed; ovary 5 × 3–4 mm, ellipsoid-subglobose, 3–4-locular, densely whitish or fulvous- to ferrugineous-tomentose; styles 3–4 capping the ovary, 3–3.5 mm long, multifid-flabelliform, the segments linear, terete, almost smooth, abaxially pubescent, adaxially glabrous, yellow at first, later turning dark brown. Fruits 3 × 2.5 cm, ellipsoid, smooth, densely ferrugineous-tomentose at first, later partially glabrescent, yellowish when fresh, reddish-brown when dried. Pyrenes and seeds more or less as in *U. kirkiana.*

Zambia. N: Mpika, male fl. 4.ii.1955, *Fanshawe* 1980 (K; NDO; SRGH). W: Mwinilunga Distr., north of Kalene Hill, fr. 24.ix.1952, *Angus* 538 (BM; FHO; K). C: Serenje Distr., Kundalila Falls, female fl. 12.ii.1973, *Strid* 2796 (FHO). E: 20 km from Chikwanda, male fl. 25.vi.1963, *Symoens* 10460 (BR; K; LSHI). **Malawi**. N: Rumphi Distr., Nyika Plateau, fr. 15.viii.1975, *Pawek* 10061 (K; MAL; MO; SRGH; UC).

Also in Zaire (Shaba). *Brachystegia, Brachystegia/Isoberlinia* and *Brachystegia/Uapaca* woodlands, and wooded grassland, sometimes locally dominant, usually in sandy or rocky, well drained soils on hillsides or escarpment edges, also in watershed wooded grassland and plains on Kalahari Sand, sometimes on dambo margins; 1300–2000 m.

Vernacular name as recorded in specimen data: "mubula" (chiLunda).

12. MAESOBOTRYA Benth.

Maesobotrya Benth. in Hooker's Icon. Pl. **13**: t. 1296 (1879). —J. Léonard in Bull. Jard. Bot. Belg. **63**: 4 (1994).
Pierardia sect. *Isandrion* Baill., Adansonia IV: 140 (1863–4).
Baccaurea sect. *Isandrion* (Baill.) Müll. Arg. in DC., Prodr. **15**, 2: 464 (1866), p.p.
Staphysora Pierre, Bull. Soc. Linn., Paris **2**: 1233 (1896).

Dioecious trees or shrubs with a simple indumentum. Leaves alternate, often long-petiolate, stipulate, simple, entire or toothed, penninerved; petioles bipulvinate; stipules minute and deciduous (or foliaceous and persistent elsewhere). Inflorescences axillary (or cauliflorous elsewhere), solitary (or fasciculate elsewhere), racemose (or subspicate elsewhere); bracts usually 1-flowered; flowers shortly pedicellate. Male flowers: calyx (4)5-lobed, imbricate; petals absent; disk glands (4)5, alternating with the stamens, fleshy, contiguous; stamens (4)5(6), opposite the sepals; filaments free; anthers erect, dorsifixed, introrse, thecae parallel, longitudinally dehiscent; pistillode cylindric, not lobed. Female flowers: calyx as in the male; petals absent; disk hypogynous, cupular, entire; ovary (1)2(4)-locular, with 2 ovules per locule; styles short; stigmas bifid, recurved. Fruit subglobose or ellipsoid, subdrupaceous, tardily loculicidally dehiscent; pericarp thin; endocarp 1-locular by suppression. Seeds solitary by abortion, ellipsoid, ecarunculate; testa thin; albumen copious; cotyledons broad, flat, green.

A tropical African genus of c. 20 species, only one of which reaches the Flora Zambesiaca area.

Maesobotrya floribunda Benth. in Hooker's Icon. Pl. **13**: t. 1296 (1879). —Hutchinson in F.T.A. **6**, 1: 665 (1912). —Engler, Pflanzenw. Afrikas (Veg. Erde 9) **3**, 2: 16 (1921). —Pax in Engler, Pflanzenr. [IV, fam. 147, xv] **81**: 20 (1922). —Keay in F.W.T.A., ed. 2, **1**, 2: 374 (1958). —J. Léonard in Bull. Jard. Bot. Belg. **63**: 46 (1994). Type from Zaire (Orientale Province).

Var. **hirtella** (Pax) Pax & K. Hoffm. in Engler, Pflanzenr. [IV, fam. 147, xv] **81**: 20 (1922). Tab. **15**. Type from Zaire (Equateur Province).
Maesobotrya hirtella Pax in Bot. Jahrb. Syst. **28**: 21 (1900). —Hutchinson in F.T.A. **6**, 1: 666 (1912).

A shrub or small tree up to 12 m high, with drooping branches. Bark greyish-brown, longitudinally grooved. Young twigs angular, sparingly to evenly ± densely yellowish-brown pubescent. Petioles 1–7 cm long, adaxially canaliculate and pubescent in the groove. Leaf blades 4–18 × 2.5–7 cm, elliptic to oblong-oblanceolate, usually shortly acutely acuminate at the apex, cuneate or rounded at the base, distantly shallowly glandular-toothed on the margins in the upper half to subentire, chartaceous, glabrous on the upper surface except for the sparingly pubescent midrib, or upper surface completely glabrous, sparingly pubescent beneath, more densely so along the midrib and main nerves, seldom completely glabrous beneath, often drying dark greenish-brown on the upper surface, paler beneath; lateral nerves in 6–9 pairs, weakly brochidodromous, not prominent above, prominent beneath. Stipules 2.5–4 mm long, linear-lanceolate, evenly pubescent, deciduous. Male inflorescences 3–10 cm long, axillary, solitary, paired or ternate, racemose; axis sparingly puberulous to subglabrous; bracts 1 mm long, triangular. Male flowers: pedicels 1–1.5 mm long, jointed; calyx lobes 1 × 0.7 mm, triangular, creamy-yellow; stamens 1.5 mm long, anthers 0.2 × 0.3 mm; disk glands 0.3 × 0.3 mm, turbinate, truncate; pistillode 0.75 mm high, pubescent at the apex. Female inflorescences shorter but otherwise as in the male inflorescences. Female flowers: pedicels 1 mm long, extending to 2 mm in fruit; calyx lobes c. 1 × 1 mm, ovate, greenish-cream; disk 1.5 mm in diameter; ovary 1.5 × 1.2 mm, ovoid-ellipsoid, 2-locular, densely appressed-setulose; styles 2, 0.5 mm long, united at the base, persistent; stigmas 1 mm long, shallowly minutely papillose. Fruit 5–6 × 4.5–5 mm, ellipsoid, smooth, sparingly appressed-pubescent at the base, otherwise glabrous, green, 1-seeded. Seed 6 × 3.5 mm, ellipsoid, purplish-grey.

Zambia. W: Mwinilunga Distr., Lisombo R., male fl. 10.vi.1963, *Loveridge* 903 (K; LISC; SRGH).

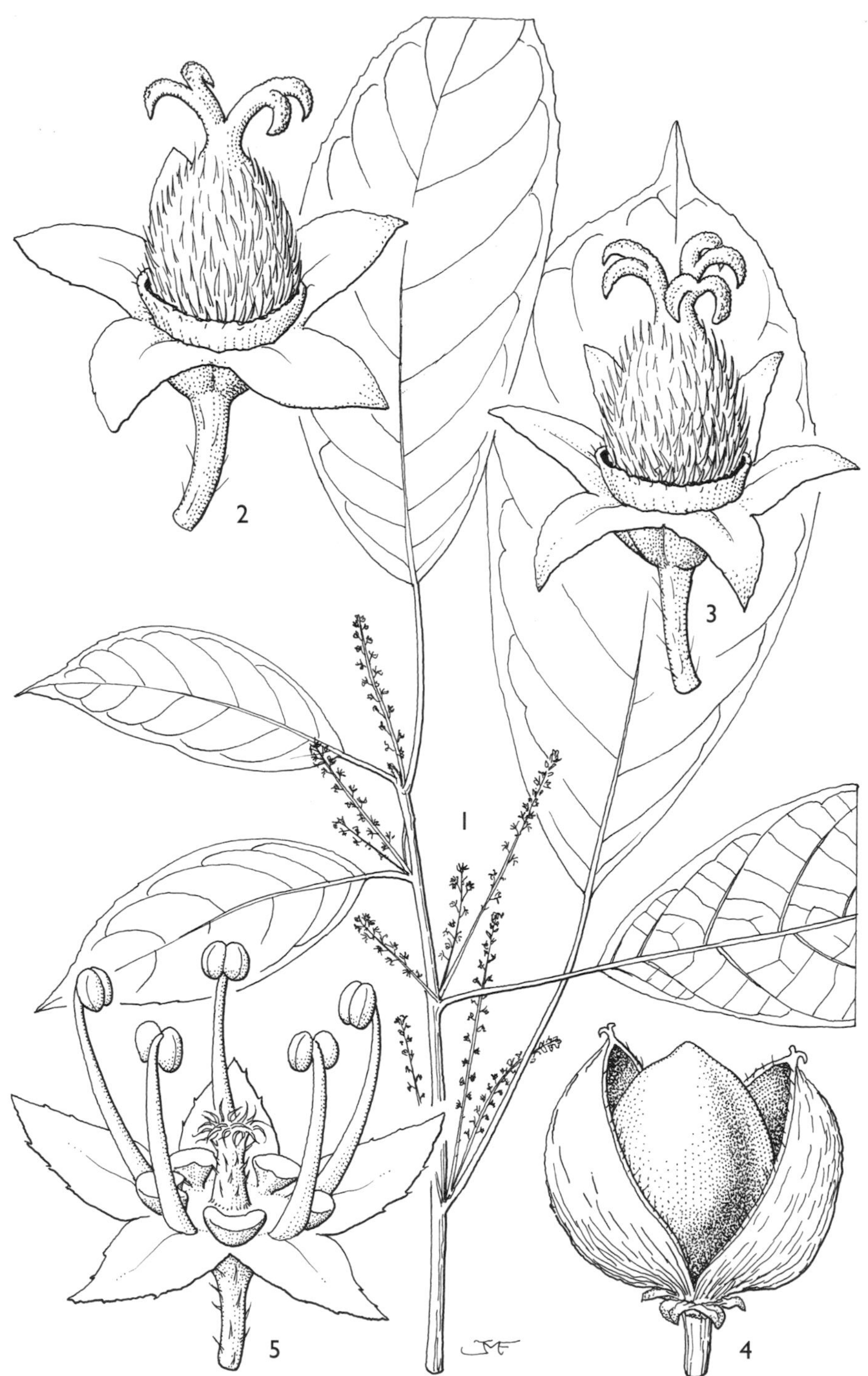

Tab. 15. MAESOBOTRYA FLORIBUNDA var. HIRTELLA. 1, distal portion of flowering branch (× ⅔), from *Edwards* 759; 2, female flower, 5 sepals, 2 styles (× 16); 3, female flower, 6 sepals, 3 styles (× 16), 2 & 3 from *Holmes* 1161; 4, fruit (× 6), from *Léonard* 250 (Zaire); 5, male flower (× 16), from *Loveridge* 903. Drawn by J.M. Fothergill.

Also in Cameroon, Gabon, Congo and Zaire. Mushitu (swamp forest) and riverine forest; c. 900 m.

The typical variety occurs in N Zaire, and may be distinguished by its glabrous ovaries.

The description of the fruit above is based on material from Zaire (Kivu Province), as no Zambian fruiting material has been seen.

13. ANTIDESMA L.

Antidesma L., Sp. Pl.: 1027 (1753); Gen. Pl., ed. 5: 451 (1754). —Müller Argoviensis in De Candolle, Prodr. **15**, 2: 247 (1866). —J. Léonard in Bull. Jard. Bot. Belg. **58**, 1/2: 4 (1988).

Stilago L., Mant. **1**: 16 (1767).

Dioecious trees or shrubs with a simple indumentum. Leaves alternate, shortly petiolate, stipulate, simple, entire, penninerved and usually brochidodromous, acarodomatiiferous, eglandular; stipules entire and deciduous (or laciniate and persistent, but not so in Flora Zambesiaca area). Inflorescences axillary, leaf-opposed, terminal or cauliflorous, solitary or few-fascicled, spicate, racemose or subpaniculate, shortly pedunculate, densely flowered; bracts small, 1-flowered; flowers small, sessile or subsessile. Male flowers: calyx 3–5(8)-lobed, lobes imbricate; petals absent; disk extrastaminal, the glands free or connate; stamens (2)3–5(10), opposite the sepals; filaments free, exserted; anthers inflexed in bud, later erect, apicifixed, connective thick, thecae distinct, divergent, basally and longitudinally dehiscent; pistillode small, often cylindric, sometimes absent. Female flowers: calyx as in the male; petals absent; disk hypogynous, annular or cupular; ovary 1(2)-locular, with 2 pendulous ovules per locule; styles (2)3(5), very short, united at the base, terminal or lateral, usually 2-lobed, often persistent. Fruit drupaceous, unilocular, small, often oblique, laterally compressed, indehiscent, often red or black; mesocarp fleshy; endocarp hardened, reticulate-foveolate. Seeds 1(2) per fruit, ecarunculate; albumen fleshy, not copious; cotyledons broad, flat.

A large palaeotropical genus of c. 170 species, only 7 of which are African. The 4 species of the Flora Zambesiaca area have been greatly confused in the past by collectors and authors.

1. Fruits 3–5 × 2–3.5 mm when dried, not or only slightly compressed; female inflorescences not deformed; female calyx lobed to halfway; disk of male flowers continuous, enfolding the stamens; styles usually terminal; a deciduous tree or shrub characteristic of forest edges ··· 3. *membranaceum*
– Fruits 5–10 × 4–7.5 mm when dried, strongly compressed; female inflorescences deformed or not; female calyx truncate or shallowly lobed; disk of male flowers continuous or composed of separate glands; styles terminal or lateral; scramblers, shrubs or trees ···· 2
2. Leaf blades acutely acuminate at the apex, glabrous or almost so, drying green, reticulation lax by reflected light, dense by transmitted light; female inflorescences unbranched, not deformed; an evergreen shrub or small tree of forest understorey ······· 4. *vogelianum*
– Leaf blades rounded or obtuse, sometimes acuminate, indumentum various, drying greyish or brownish; female inflorescences usually with lateral branches at the base, often deformed ··· 3
3. Disk of male flowers continuous or fragmented, partially or completely surrounding the stamens; styles terminal; leaf blades commonly densely ferrugineous- or fulvous-tomentose or -pubescent and rarely glabrescent beneath, reticulation lax; a deciduous sometimes scrambling shrub or small tree characteristic of open bushland ··········· 1. *venosum*
– Disk of male flowers with glands free and alternating with the sepals, or connate and extrastaminal; styles terminal or lateral; leaf blades glabrous except along the nerves; reticulation dense, the nervation fine and reddening after drying; an evergreen shrub or small tree of riparian habitats ··· 2. *rufescens*

1. **Antidesma venosum** E. Mey. ex Tul. in Ann. Sci. Nat., Bot., sér. 3, **15**: 232 (1851). —Müller Argoviensis in De Candolle, Prodr. **15**, 2: 260 (1866). —Engler, Pflanzenw. Ost-Afrikas **C**: 237 (1895). —Schinz & Junod in Mém. Herb. Boissier, No. 10: 46 (1900). —Sim, For. Fl. Port. E. Afr.: 106 (1909). —Hutchinson in F.T.A. **6**, 1: 646 (1912). —R.E. Fries, Wiss. Ergebn. Schwed. Rhod.-Kongo-Exped. **1**, 1: 119 (1914). —Eyles in Trans. Roy. Soc. South

Africa **5**: 394 (1916). —Hutchinson in F.C. **5**, 2: 406 (1920). —Engler, Pflanzenw. Afrikas (Veg. Erde 9) **3**, 2: 19 (1921). —Pax in Engler, Pflanzenr. [IV, fam. 147, xv] **81**: 139 (1922). —De Wildeman, Pl. Bequaert. **3**: 517 (1926). —Burtt Davy, Fl. Pl. Ferns Transvaal: 300 (1932). —Bremekamp & Obermeyer in Ann. Transvaal Mus. **16**, 3: 421 (1935). —O.B. Miller, Check-list For. Trees Shrubs Bech. Prot.: 31 (1948). —Brenan, Check-list For. Trees Shrubs Tang. Terr.: 200 (1949). —F.W. Andrews, Fl. Pl. Anglo-Egypt. Sudan **2**: 55 (1952). —Brenan in Mem. N.Y. Bot. Gard. **9**, 1: 71 (1954). —Topham, Check List For. Trees Shrubs Nyasaland Prot.: 49 (1958). —White, F.F.N.R.: 193 (1962). —P.G. Meyer in Merxmüller, Prodr. Fl. SW. Afrika, fam. 67: 6 (1967). —Mogg in Macnae & Kalk, Nat. Hist. Inhaca Isl., Moçamb.: 147 (1969). —Drummond in Kirkia **10**: 251 (1975). —K. Coates Palgrave, Trees Southern Africa, ed. 2, rev.: 406 (1983). —Radcliffe-Smith in F.T.E.A., Euphorb. 2: 573 (1988). —J. Léonard in Bull. Jard. Bot. Belg. **58**, 1/2: 39 (1988). —Beentje, Kenya Trees, Shrubs Lianas: 185 (1994). Types from South Africa (KwaZulu-Natal).

Antidesma bifrons Tul. in Ann. Sci. Nat., Bot. sér. 3, **15**: 229 (1851). Type from Sudan.

Antidesma biovinianum Baill., Adansonia **2**: 45 (1861). Type from Zanzibar.

Antidesma membranaceum auctt. non Müll. Arg.

Antidesma membranaceum var. *molle* Müll. Arg. in Linnaea **34**: 68 (1865); in De Candolle, Prodr. **15**, 2: 261 (1866), pro parte. Types from Angola.

A spreading, sometimes scrambling, deciduous shrub or small tree commonly up to 8 m tall, branched from the base or with a clear bole to up to 4.5 m; branches drooping; crown often dense; bark smooth, furrowed or scaly, light yellowish-brown to dark grey. Twigs grey-brown, lenticellate. Young shoots and buds sparingly to densely fulvous- or ferrugineous-tomentose, or pubescent. Leaves stratified. Petioles 0.2–1 cm long, adaxially canaliculate, pubescent at least in the groove. Leaf blades 2–12(20) × 1–6(7) cm, elliptic-obovate to oblong-oblanceolate, rounded or obtuse at the apex, less often shortly acuminate, rounded or cuneate, rarely shallowly cordulate at the base, thickly chartaceous to thinly coriaceous, pubescent along the midrib but otherwise glabrous or almost so on upper surface, sparingly pubescent to densely fulvous- or ferrugineous-tomentose beneath, dark green and glossy above, paler and duller beneath; lateral nerves in 4–8(9) pairs, the lower camptodromous, the upper brochidodromous, somewhat impressed above, prominent beneath, reticulation lax. Stipules 2–8(15) mm, lanceolate to linear-lanceolate, acute, evenly to densely pubescent or tomentose, subpersistent. Male inflorescences terminal or subterminal on short lateral shoots, 4–10(13) cm long, spicate, usually with 1–2 lateral spikes at the base; bracts minute. Male flowers: calyx (3)4(5)-lobed, the lobes 0.5–1 mm long, unequal, subacute or obtuse, pubescent without, ciliate, yellowish-green; disk variable, continuous or fragmented, thick, partially or completely surrounding the (3)4(5) stamens; filaments 2.5 mm long, white, anthers 0.5 mm long, often reddish; pistillode minute, usually slender, sparingly pubescent. Female inflorescences 1.5–9(12) cm long, compound-spicate with 1–4 lateral spikes in the lower half, often galled and then irregularly densely paniculate, especially in the upper half, otherwise as in the male. Female flowers: pedicels 0.5–1 mm long, extending to 2 mm in fruit; calyx 1 mm long, cupular, shallowly 3–4-lobed, otherwise ± as in the male; disk c. 1 mm in diameter, shallowly cupular, crenellate, ovary 1 × 0.75 mm, compressed-ellipsoid, smooth, glabrous; styles 2–4, ± terminal, 0.75 mm long, bifid, strongly recurved, glabrous, whitish. Fruits 5–7(8) × 4–5(6) mm when dried, 6–8(9) × 5–6(7) mm when fresh, ellipsoid-suborbicular, slightly asymmetrical, strongly laterally compressed, irregularly and coarsely reticulate-rugulose when dried, green at first, later becoming whitish, red, purple or black when ripe. Seeds ellipsoid.

Caprivi Strip. Tjaro Island, Okavango R., fr. 21.i.1956, *de Winter* 4354 (K; PRE). **Botswana**. N: Kasane-Chobe R., fr. vii.1930, *van Son* in *Tvl. Mus. Herb.* 28832 (BM; K; PRE). **Zambia**. B: Zambezi (Balovale), fr. xii.1953, *Gilges* 312 (K; PRE; SRGH). N: Mansa Distr., Samfya, L. Bangweulu, female (galled inflorescences), 8.ii.1959, *Watmough* 230 (K; SRGH). W: Mufulira, fr. 6.iii.1955, *Fanshawe* 2126 (K; NDO). C: Lusaka to Kabwe (Broken Hill), fr. 4.v.1958, *Benson* 238 (BM; LISC; SRGH). E: Njumbe–Mchinji ("Jumbe–Machinje") Hills, y. fr. (galled infrs.), 11.x.1958, *Robson & Angus* 39 (BM; K; LISC; SRGH). S: Kafue River Gorge, female 6.x.1957, *Angus* 1730 (FHO; K). **Zimbabwe**. N: Hurungwe (Urungwe) Nat. Res., Nyagugutu R., y. fr. 21.xi.1957, *Goodier* 399 (K; LISC; PRE; SRGH). W: Hwange Distr., Hwange (Wankie) Nat. Park, Robin's Camp, male 19.ii.1973, *Chiparawasha* 630 (K; SRGH). C: Makoni Distr., near Inyazura, y. fr. 9.ix.1956, *Whellan* 1131 (SRGH). E: Chimanimani Distr., Umvumvumvu R., female (galled inflorescences) 7.vi.1965, *Chase* 8299 (K; LISC; PRE; SRGH). S: Mberengwa (Belingwe), near Mnene, fr. 26.ii.1931, *Norlindh & Weimarck* 5163 (BM; PRE; UPS). **Malawi**. N: Chitipa Distr., 9.5 km from crossroads toward

Karonga, female (galled inflorescences) + fr. 20.xii.1972, *Pawek* 6261 (K; MAL; MO; SRGH; UC). C: Kasungu Game Res., female (galled inflorescences) 6.viii.1970, *Hall-Martin* 1644 (PRE; SRGH). S: Machinga Distr., Lifune Stream, male (galled inflorescences), 28.iv.1982, *Patel* 861 (K; MAL). **Mozambique**. N: Marrupa, Massanguezi, fr. 18.ii.1981, *Nuvunga* 601 (BM; K; LMU; MO). Z: 1.2 km from Muaquiua (Macuia), fr. 17.v.1949, *Barbosa & Carvalho* in *Barbosa* 2694 (LMA; SRGH). T: Angónia Distr., Ulónguè-aldeia de Tchindeque, female + fr. 3.xii.1980, *Macuácua* 1379 (K; LMA; MO; PRE). MS: Chimoio, Massequece, female (galled inflorescences) 24.ix.1947, *Pimenta* 18 (LISC; PRE; SRGH). GI: Massinga, male xii.1936, *Gomes e Sousa*, 1940 (K). M: Marracuene female + fr. (galled inflorescences) 10.ii.1960, *Macuácua* 84 (BM; K; LMA; LISC).

Throughout subsaharan Africa, extending southwards to Eastern Cape Province. Common at medium and low altitudes in riverine and lakeshore vegetation, on sandy banks and alluvial soils of permanent and seasonal rivers and lakes, also in mixed deciduous *Acacia* and miombo woodlands and dry evergreen thickets, on dambo margins and on termite mounds in grassy floodplains, sometimes in granite outcrop vegetation, gully forest, coastal forest and dune vegetation; sea level to 1830 m.

Vernacular names as recorded in specimen data include: "dangramunho" (Chimoio area); "kamimba" (chiTumbubza); "lavechovecho" (Niassa area); "mpungalira" (Mwanza area); "mpungulila" (chiNyanja); "muchongo" (Vila Machado area); "mujanje-mutongoro" (Nyumquarara Valley area); "mukenanene" (K); "mukoso", "murombe", "muzaru" (Mbukushu); "munyonyamanzi", "munyonyamenda" (I); "muoiele", "namatjamatja", "neuechuiegé", "n'ruinta" (Macúa); "murgongi", "yongue" (Búzi area); "murungamungu", "murunga-munyu", "marungumunyu", "mushongo" (chiNdao- "stir in the salt"); "musabubamfwa" (Ndola area); "musambarabwabwa" (Gutu area); "musamborabyangu" (Harare area); "mussungano" (Massequece area); "mutomamena" (K); "namasa-masai" (Cabo Delgado area); "nawossowosso" (Marrupa area); "noduli" (Zambézia area); "simai" (Sik.); "tsongue" (Magude area); "txungi" (Inhaca Island).

The fruit is edible.

2. **Antidesma rufescens** Tul. in Ann. Sci. Nat., Bot., Sér. 3, **15**: 231 (1851). —J. Léonard in Bull. Jard. Bot. Belg. **57**: 453 (1987); **58**, 1/2: 9 (1988). —Radcliffe-Smith in F.T.E.A., Euphorb. 2: 574 (1988), in adnot. Type from Senegambia.

Antidesma sassandrae Beille, Bull. Soc. Bot. Fr. **57**, Mém 8C: 123 (1910). Type from Côte d'Ivoire.

Antidesma membranaceum var. *glabrescens* Müll. Arg. in Linnaea **34**: 68 (1865). Type from N Nigeria.

Antidesma membranaceum auctt. non Müll. Arg.

Antidesma venosum auctt. non E. Mey. ex Tul.

Very close to *A. venosum*, but differing chiefly in the leaves being glabrous (except along the nerves), and densely reticulately nerved with the nervation fine and reddening after drying; also differing in the disk glands of male flowers being interstaminal and free or extrastaminal and connate, and in the styles either terminal or lateral.

Caprivi Strip. Katima Mulilo, female & y. fr. 25.xi.1972, *P.A. Smith* 291 (K; LISC; SRGH). **Zambia**. B: near Senanga, male, 30.vii.1952, *Codd* 7265 (BM; K; PRE; SRGH). N: Mbala Distr., Ulungu, Kanda Village, female 27.xi.1948, *Bredo* 6357 (BR; K; PRE). W: Angola border west of Kalene Hill, female 22.ix.1952, *White* 3332 (BM; FHO; K). C: Kafue R., male, 25.x.1951, *Holmes* 588 (FHO). E: Lundazi, o. fr. vii.1977, *Sylvestre* 74 (SRGH). S: Kafue National Park, Chunga Camp, female & y. fr. 19.x.1986, *Linder* 3934 (K; PRE). **Zimbabwe**. E: Chimanimani Distr., Umvumvumvu R., male, xii.1955, *Armitage* 195/55 (SRGH). **Malawi**. N: Mzimba Distr., Lunyangwa Res. Station, male, 19.xii.1986, *la Croix* 4242 (MO; PRE). S: Shire R., male, 12.xii.1865, *Kirk* s.n. (K). **Mozambique**. N: 32 km Liupo-Mogincual, female (galled infl.) 30.iii.1964, *Torre & Paiva* 11481 (LISC). Z: Namacurra, male, 27.i.1966, *Torre & Correia* 14197 (LISC). T: Nhandoa, male, 10.x.1905, *Le Testu* 865 (BM; LISC; P). MS: Sofala Prov. (Beira Distr.), Gorongosa Nat. Park near Dingedinge Plains, male, x.1972, *Tinley* 2732 (K; LISC; PRE; SRGH). M: Matutuíne (Bela Vista), fr. 12.iv.1948, *Torre* 7662 (LISC).

Widespread in West, Central, South and Southeast tropical Africa. Riverine vegetation, often in pure stands and forming thickets, usually on sand banks, also on lagoon margins, swampy scrub of floodplains and coastal dunes; sea level to 1066 m.

Vernacular names as recorded in specimen data include: "chiongui" (Maputo area); "musangula" (chiBemba).

3. **Antidesma membranaceum** Müll. Arg. in Linnaea **34**: 68 (1865/6); in De Candolle, Prodr. **15**, 2: 261 (1866), excl. var. *glabrescens* Müll. Arg. —Engler, Pflanzenw. Ost-Afrikas **C**: 237 (1895). —Hutchinson in F.T.A. **6**, 1: 645 (1912), pro parte. —Eyles in Trans. Roy. Soc. South Africa **5**: 394 (1916). —Engler, Pflanzenw. Afrikas (Veg. Erde 9) **3**, 2: 20 (1921).

—Pax in Engler, Pflanzenr. [IV, fam. 147 xv] **81**: 141 (1922), pro parte. —De Wildeman, Pl. Bequaert. **3**: 516 (1926), pro parte. —Brenan, Check-list For. Trees Shrubs Tang. Terr.: 199 (1949). —F.W. Andrews, Fl. Pl. Anglo-Egypt. Sudan **2**: 55 (1952). —Keay in F.W.T.A., ed. 2, **1**, 2: 375 (1958). —Dale & Greenway, Kenya Trees & Shrubs: 184 (1961). —White, F.F.N.R.: 193 (1962). —Drummond in Kirkia **10**: 251 (1975). —Troupin, Fl. Pl. Lign. Rwanda: 249, fig. 85/2 (1982); Fl. Rwanda **2**: 206, fig. 61/2 (1983). —K. Coates Palgrave, Trees Southern Africa, ed. 2, rev.: 405 (1983). —Radcliffe-Smith in F.T.E.A., Euphorb. 2: 574 (1988). —J. Léonard in Bull. Jard. Bot. Belg. **58**, 1/2: 28 (1988). Type from Angola (Malanje Province).

Antidesma meiocarpum J. Léonard in Bull. Jard. Bot. État **17**: 260, tt. 23, 24 (1945). Types from Zaire (Bas-Congo, Ubangi-Uele).

Antidesma venosum auctt. non E. Mey. ex Tul.

Resembling *A. venosum*, but with leaves more narrowly elliptic-oblong and more obviously acutely acuminate, the leaf lower surface often yellowish and with an indumentum less dense and never really tomentose except on very young leaves; the basal lateral nerves camptodromous, running further up the leaf blade at a smaller angle from the midrib; the female inflorescences never deformed; the calyx of female flowers lobed to the middle; the fruits only 3–5 × 2–3.5 mm when dried and only slightly compressed.

Zambia. N: Kawambwa Distr., Kawambwa-Mambilima Falls (Johnston Falls) Rd., male, 4.xii.1961, *Richards* 15508 (K). W: Ndola, Chichele For. Res., fr. 1951, *Holmes* 682 (FHO; SRGH). **Zimbabwe**. E: Chimanimani Distr., Haroni R., male 4.xii.1964, *Wild, Goldsmith & Müller* 6664 (BM; FHO; K; LISC; SRGH). S: Bikita Distr., Turgwe Gorge, fr. 6.v.1969, *Biegel* 3036 (K; PRE; SRGH). **Malawi**. N: Mzimba Distr., Mzuzu, road to Lunyangwa R. Waterworks, male, 29.xi.1972, *Pawek* 6050 (K; MAL; MO; SRGH; UC). C: Mua-Livulezi For. Res., fr. 21.i.1959, *Robson & Jackson* 1286, pro parte (BM; K; LISC; PRE; SRGH). S: Zomba, Mponda Stream, fr. 28.vi.1954, *Banda* 29 (BM; LISC; SRGH). **Mozambique**. N: Ribáuè, Serra de Mepáluè, male, 23.i.1964, *Torre & Paiva* 10181 (LISC). Z: Alto Molócuè, Nauela, fr. 1959, *Uatola* 8 (LISC). T: Moatize, 5 km Zóbuè-Metengo Balama, fr. 10.i.1966, *Correia* 331 (LISC). MS: Serra Mocuta, o. male, 6.vi.1971, *Müller & Gordon* 1811 (LISC; SRGH). GI: Chongoene-Chibuto, fr. 4.iii.1963, *Macedo & Balsinhas* 1094 (K; LMA; PRE). M: 4 km from Alvor, fr. 17.iii.1981, *Jansen & Macuácua* 7685 (K; MO; PRE).

More or less throughout tropical Africa except for the dry northeast and southwest. Plateau and high rainfall woodlands, and mixed woodland/evergreen forest, often in miombo, submontane *Brachystegia* and *Brachystegia/Uapaca* woodlands, also in riverine vegetation in kloofs and ravines, lakeshore vegetation and coastal forest and woodland, sometimes associated with termite mounds; near sea level to 1830 m.

Vernacular names as recorded in specimen data include: "chidiapumbwa", "chidikafumbwa" (Mwanza); "chongo", "murungamunho" (Manica region); "dagramunho" (Inhamitanga area); "inlyamasolokoto" (Malawi); "ishongi", "ixongi", "shongi" (xiRonga); "mapilakukutu" (chiNyanja); "mpulira" (chiYao); "mpungulila" (C); "munya munya" (Dao); "musambara" (Bgabga); "siriko" (chiTonga); "txuangi", "utxungi" (Shangana).

4. **Antidesma vogelianum** Müll. Arg. in Flora **47**: 529 (1864); in De Candolle, Prodr. **15**, 2: 260 (1866). —Hutchinson in F.T.A. **6**, 1: 645 (1912). —Engler, Pflanzenw. Afrikas (Veg. Erde 9) **3**, 2: 20 (1921). —Pax in Engler, Pflanzenr. [IV, fam. 147, xv] **81**: 119 (1922). —Keay in F.W.T.A., ed. 2, **1**, 2: 375 (1958). —White, F.F.N.R.: 194 (1962). —Drummond in Kirkia **10**: 251 (1975). —K. Coates Palgrave, Trees Southern Africa, ed. 2, rev.: 406 (1983). —Radcliffe-Smith in F.T.E.A., Euphorb. 2: 576 (1988). —J. Léonard in Bull. Jard. Bot. Belg. **58**, 1/2: 22 (1988). Tab. **16**. Type from S Nigeria.

Antidesma staudtii Pax in Bot. Jahrb. Syst. **26**: 327 (1899). Type from Cameroon.

Antidesma membranaceum var. *crassifolium* Pax & K. Hoffm. in Engler, Pflanzenr. [IV, fam. 147, xv] **81**: 141 (1922). —Brenan, Check-list For. Trees Shrubs Tang. Terr.: 200 (1949). Syntypes: Zimbabwe, Chimanimani District, Mt. Pene, male, 28.ix.1906, *Swynnerton* 1111 (BM, syntype; K, isosyntype), and from Tanzania (Southern Highlands Province),

Antidesma venosum f. *glabrescens* De Wild., Ann. Mus. Cong. Belge, Bot., Ser. 4, **1**: 79 (1902). Type from Zaire (Shaba Province).

Antidesma membranaceum auctt. non Müll. Arg.

A spreading, sometimes scrambling, evergreen shrub or tree up to 15 m tall, often with only a few long lax drooping branches; bole up to 50 cm in diameter, aerial roots sometimes formed. Bark smooth, longitudinally furrowed, or reticulate and flaking, pale grey or brownish. Wood hard, yellowish-white. Twigs sparingly lenticellate. Young shoots and buds densely ferrugineous-pubescent at first, soon glabrescent. Petioles 2–8(15) mm long, adaxially canaliculate. Leaf blades 3–24 × 1.5–9 cm,

Tab. 16. ANTIDESMA VOGELIANUM. 1, portion of flowering branch ($\times \frac{2}{3}$), from *Faulkner* 2528; 2, male inflorescence (× 4), from *Faulkner* 3327; 3, female inflorescence (× 4), from *Faulkner* 2528; 4, fruit, two views (× 4), from *Faulkner* 2496. Drawn by Christine Grey-Wilson. From F.T.E.A.

elliptic to elliptic-oblanceolate, usually acutely acuminate, often long-acuminate and mucronate at the apex, cuneate-rounded and sometimes slightly asymmetric at the base, coriaceous, glabrous or sometimes the midrib and main nerves pubescent above and beneath, coppery-red when young, becoming deep glossy green above and paler or ferrugineous-tinged with darker green veins beneath; lateral nerves in 5–11 pairs, brochidodromous, not prominent above, somewhat prominent beneath, reticulation lax by reflected light, dense by transmitted light. Stipules 0.4–1.5(3.5) cm × 1–2(4) mm, linear to narrowly elliptic, acute, pubescent, foliaceous on suckers, deciduous or subpersistent. Male inflorescences 3–13 cm long, terminal or subterminal usually on lateral shoots, spicate, often with a lateral basal spike, pendulous; bracts minute, but up to 5 mm long on galled inflorescences. Male flowers: pedicels very short or absent; calyx 3–4-lobed, the lobes 0.5 mm long, subequal, obtuse, glabrous without, pubescent within, ciliate, pale green; disk continuous, thick, enfolding the 3 stamens; filaments 2–2.5 mm long, anthers 0.3 × 0.5 mm, crimson to dark brown; pistillode 1 mm tall, cylindric, scarcely 2–3-lobed at the apex, glabrous. Female inflorescences 1–8 cm long, extending to up to 17 cm long (much longer in W Africa and Zaire) in fruit, pendulous, unbranched, not galled, otherwise as in male. Female flowers: pedicels 0.5–1 mm long, extending to up to 4 mm long in fruit; calyx 1 × 1 mm, squarely cupular, truncate or very shallowly and broadly 3(4–5)-lobed to one-third, otherwise ± as in the male; disk c. 1 mm in diameter, annular, entire or crenellate; ovary 2 × 1 mm, fusiform-ellipsoid, smooth, glabrous, brown; styles 3–4, usually terminal, 1 mm long, strongly recurved, usually glabrous. Fruits (6)7–9(10) × (4)5–7.5 mm when dried, 8–10(12) × 5–8 mm when fresh, ellipsoid, slightly asymmetrical, strongly compressed, irregularly and coarsely foveolate when dried, glabrous, shiny, olive-green at first, turning through whitish to red, reddish-yellow, brownish-pink, purple or black when ripe, very juicy. Seeds 6 × 3 × 1 mm, ellipsoid, often aborted.

Zambia. N: Ishiba Ngandu (Shiwa Ngandu), fr. 3.viii.1938, *Greenway & Trapnell* 5572 (EA; FHO; K). W: 11 km north of Solwezi Boma, female & fr. 14.ix.1952, *Holmes* 865 (FHO; K; PRE; SRGH). C: Mkushi R., y. fr. 23.vi.1963, *Symoens* 10426 (K; LSHI). **Zimbabwe**. E: Chimanimani Distr., Muchira R., male, xi.1969, *Goldsmith* 83/69 (BM; FHO; K; LISC; PRE; SRGH). **Malawi**. N: Nkhata Bay Distr., 8 km east of Mzuzu, female, 5.xi.1972, *Pawek* 5924 (MO; SRGH; UC). S: Mulanje Mt., Lukulezi R., male, 1.x.1957, *Chapman* 446 (BM; FHO; K). **Mozambique**. N: Ribáuè, Serra de Mepáluè, male, 9.xii.1967, *Torre & Correira* 16428 (LISC). Z: Gurué, near Malema R., male, 6.xi.1967, *Torre & Correia* 15944 (LISC). MS: Chimanimani Mts., Mussapa Gap, 3 km from Zimbabwean Border, fr. 30.i.1962, *Wild* 5627 (FHO; K; PRE; SRGH).

More or less throughout Central, East and South tropical Africa. Understorey tree or shrub of swamp forest (mushitu) and riverine forest (often evergreen), also in montane and submontane mixed evergreen forest and sheltered kloof forest; 20–1980 m.

14. THECACORIS A. Juss.

Thecacoris A. Juss., Euph. Gen. Tent.: 12, t.1, f.1 (1824). —Müller Argoviensis in De Candolle, Prodr. **15**, 2: 245 (1866). —J. Léonard in Bull. Jard. Bot. Belg. **64**: 26(1995).

Cyathogyne Müll. Arg. in Flora **47**: 536 (1864). —Radcliffe-Smith in F.T.E.A., Euphorb. 1: 110 (1987).

Baccaureopsis Pax in Bot. Jahrb. Syst. **43**: 318 (1909).

Dioecious, rarely monoecious, trees shrubs or perennial herbs. Indumentum simple. Leaves alternate or subfasciculate, shortly petiolate, stipulate, simple, entire or subentire, penninerved. Inflorescences axillary, solitary (in Flora Zambesiaca area), geminate or fasciculate, pedunculate, spicate, racemose or rarely paniculate, few- to many-flowered; flowers 1 per bract. Male flowers: sepals 5(6), imbricate; petals 5, small, or absent; disk glands 5, free, interstaminal, alternating with the sepals, thick; stamens 5, opposite the sepals, filaments free, anther-thecae distinct, introrse, parallel and pendulous at first, later extrorse, divaricate and erect, longitudinally dehiscent; pistillode large, obconic, turbinate or cyathiform, truncate and entire or 3–5-lobed. Female flowers: pedicels patent, elongating and deflexing in fruit; sepals somewhat connate at the base, otherwise ± as in the male; petals smaller than in the male, or absent; staminodes sometimes present; disk hypogynous, annular, crenulate; ovary 3-locular, ovules 2 per locule; styles 3, free or connate at the base, recurved, bifid; stigmas ± smooth. Fruit 3-lobed, dehiscent into bivalved cocci; pericarp thin, separating from the thinly woody endocarp; columella persistent. Seeds ovoid-subglobose to obovoid-pyriform, carunculate; testa thin, striate, shiny; endosperm copious, fleshy; cotyledons greenish.

An Afromalagasy genus of 25 species. J. Léonard in Bull. Jard. Bot. Belg. **64**: 13–52 (1995) treats *Cyathogyne* as a separate genus.

Suffrutex; leaves 1–3 × 0.5–1.2 cm, emarginate; male inflorescences 1(2) cm long, few-flowered; female inflorescences spicate, extending to 3 cm long in fruit · · · · · · · · · · 1. *spathulifolia*
Scrambling shrub or small slender tree; leaves 2–21 × 0.5–8 cm, obtuse or acuminate, mucronate; male inflorescences 0.5–4.5 cm long, many-flowered; female inflorescences racemose, extending to 25 cm long in fruit · 2. *trichogyne*

1. **Thecacoris spathulifolia** (Pax) Leandri in Mém. Inst. Sci. Madagascar, Sér. B, Biol. Veg. **8**: 211 (1957); in Fl. Mad. Com. Euph. **1**: 11 (1958). Type from Tanzania (Eastern Province).
Cyathogyne spathulifolia Pax in Bot. Jahrb. Syst. **33**: 281 (1903). —Hutchinson in F.T.A. **6**, 1: 653 (1912). —Brenan, Check-list For. Trees Shrubs Tang. Terr.: 206 (1949).
Cyathogyne bussei Pax in Bot. Jahrb. Syst. **33**: 280 (1903). —Hutchinson in F.T.A. **6**, 1: 653 (1912). —Brenan, Check-list For. Trees Shrubs Tang. Terr.: 206 (1949). —Radcliffe-Smith in F.T.E.A., Euphorb. 1: 110 (1987). —J. Léonard in Bull. Jard. Bot. Belg. **64**: 44 (1995). Type from Tanzania (Tanga Province).
Thecacoris bussei (Pax) Radcl.-Sm. in Kew Bull. **51**, 2: 330 (1996).

A dioecious shrub up to 2 m high, but usually much less. Branches straight, often ± fastigiate; twigs usually pale greyish-brown; young shoots evenly fulvous-pubescent. Petioles 1–3 mm long. Leaf blades (0.5)1–3 × (0.3)0.5–1.2 cm, elliptic-obovate or elliptic-oblanceolate to oblong-spathulate, emarginate at the apex, cuneate-attenuate at the base, entire, slightly fleshy, glabrous on the upper surface, subglabrous or sparingly fulvous-pubescent along the midrib beneath, dark green above, paler beneath; lateral nerves in 2–3 pairs, sometimes scarcely visible, brochidodromous. Stipules 1.5–2 × 0.7–1 mm, triangular-lanceolate, entire or fimbriate in the lower quarter, brown with a tawny midrib. Male inflorescences 1(2) cm long, axis densely fulvous-pubescent. Male flowers: pedicels 1(2) mm long; sepals 1 × 1 mm, obovate, pubescent; greenish; disk glands 0.5 mm long, clavate, pubescent; stamens 2 mm long, filaments broad at the base, narrowing upwards, white, anthers 0.3 mm long, yellow; pistillode 1 mm high, turbinate, truncate. Female inflorescences 1–1.5 cm long, extending to 2.5–3 cm in fruit. Female flowers: pedicels 1 mm long; sepals 1 × 1 mm, oblong-ovate; disk 0.75 mm in diameter; ovary 1 × 1 mm, subglobose, minutely puberulous to subglabrous; styles 1 mm long, sparingly appressed-puberulous, cream-coloured, stigmas red. Fruit 4–5 × 7–8 mm, strongly 3-lobed, subglabrous, green. Seeds 2 × 2 mm, chestnut-brown, hilum white.

Mozambique. N: Messalo (Msalu) R., male fl. 20.iii.1912, *Allen* 147 (K).
Also in Somalia, Kenya, Tanzania and Madagascar. On sandy hills.
Note: As only male material is available from the Flora Zambesiaca area, the above description was drawn up with the help of East African material.

2. **Thecacoris trichogyne** Müll. Arg. in J. Bot. **2**: 328 (1864); in De Candolle, Prodr. **15**, 2: 246 (1866). —Hutchinson in F.T.A. **6**, 1: 660 (1912). —R.E. Fries, Wiss. Ergebn. Schwed. Rhod.-Kongo-Exped. **1**, 1: 119 (1914). —Engler, Pflanzenw. Afrikas (Veg. Erde 9) **3**, 2: 14 (1921). —Pax in Engler, Pflanzenr. [IV, fam. 147, xv] **81**: 11 (1922). —White, F.F.N.R.: 205 (1962). —J. Léonard in Bull. Jard. Bot. Belg. **64**: 30 (1995). Tab. **17**. Type from Angola (Malanje Province).

A small, several-stemmed scrambling shrub or slender tree up to 6 m high; bark smooth, pale grey. Twigs green at first and shiny, later becoming brownish or purplish-grey. Young shoots densely fulvous-pubescent. Petioles 0.5–1 cm long, 1–3 mm thick. Leaf blades 2–21 × 0.5–8 cm, elliptic-oblanceolate to oblong-lanceolate, obtuse or acuminate, mucronate, asymmetrically rounded-cuneate at the base, undulate on the margins, chartaceous to thinly coriaceous, sparingly pubescent along the margin and the midrib and lateral nerves beneath, otherwise glabrous,

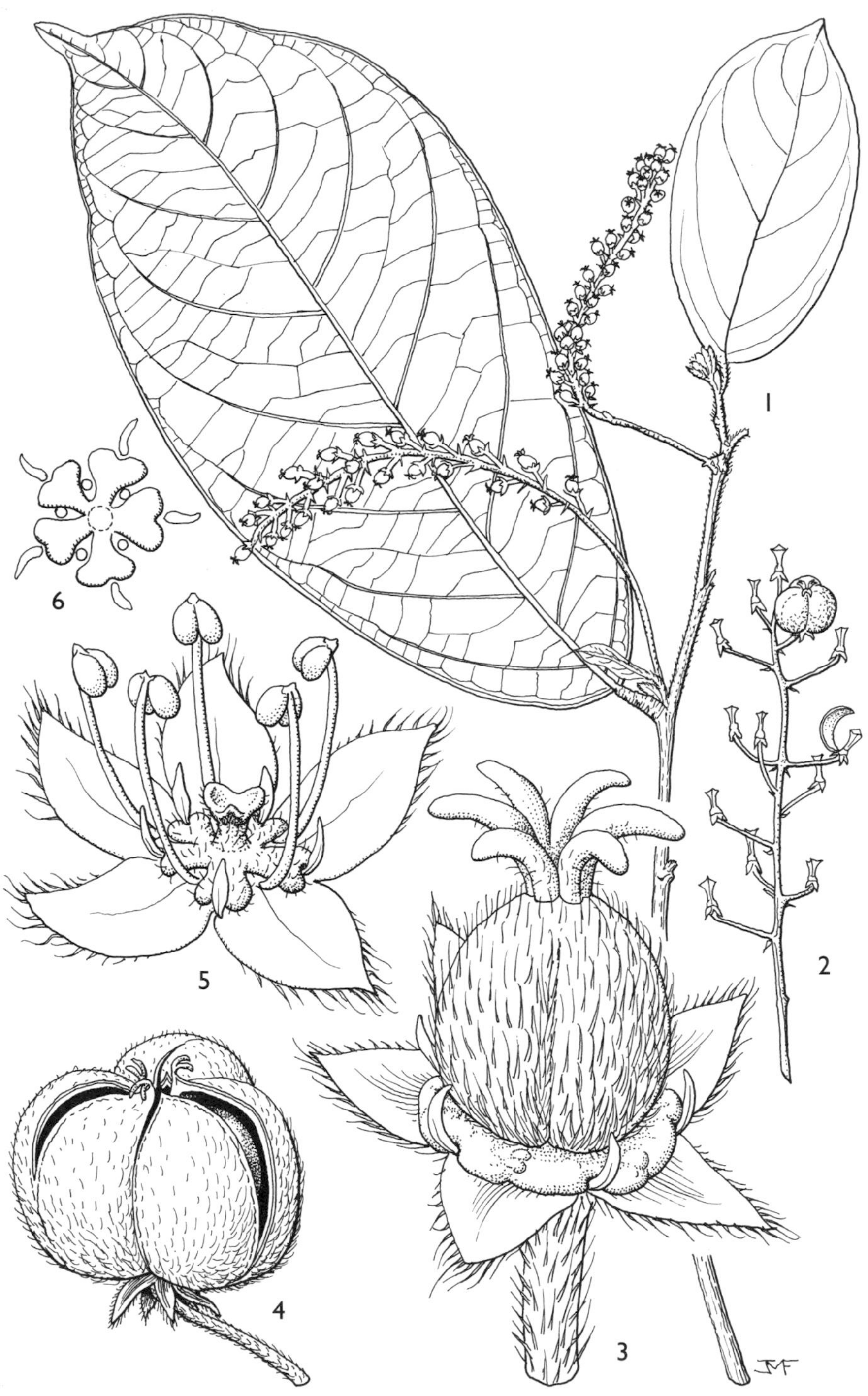

Tab. 17. THECACORIS TRICHOGYNE. 1, distal portion of branch with female flowers (× ²⁄₃), from *White* 3538; 2, infructescence (× ²⁄₃), from *Fanshawe* 4023; 3, female flower (× 12), from *White* 3538; 4, fruit (× 4), from *Fanshawe* 275; 5, male flower (× 12); 6, male disk glands and petals (× 12), 5 & 6 from *Richards* 4623. Drawn by J.M. Fothergill.

deep glossy green above, paler and duller beneath; lateral nerves in 6–12 pairs, impressed above, fairly prominent beneath, strongly brochidodromous, tertiary nerves prominently reticulate above and beneath. Stipules 0.5–1 cm long, lanceolate to setaceous, acute, pubescent, green, subpersistent. Male spikes 0.5–4.5 cm long, amentiform, pendulous, shortly pedunculate; axis densely pubescent. Male flowers: sepals 1.5 × 1 mm, ovate, subacute, pubescent towards the apex without, glabrous within, greenish; petals, when present, 0.5 mm long, subulate; disk glands 2-lobed, contiguous, apically setose; stamens 2 mm long, filaments uniformly slender, anthers 0.5 mm long; pistillode 0.5 mm high, cylindric, pubescent, 3-lobulate at the apex, lobules glabrous. Female racemes 6–17 cm long, extending to 25 cm in fruit. Female flowers: pedicels 1–3 mm long, extending to up to 1.5 cm in fruit and becoming sharply decurved at the tip; sepals ± as in the male; petals, when present, 1 mm long, subulate, acute or linear, 2-lobed; disk 3 mm in diameter, glabrous; ovary 2.5 × 3 mm, 3-lobed to subglobose, densely fulvous-pubescent, grey-green; styles 1.5–2 mm long, sparingly pubescent, greenish, stigmas glabrous, yellowish. Fruit 6–7 × 8–11 mm, depressed, rounded-3-lobed, sparingly appressed-pubescent, pale grey-green, sometimes pinkish-tinged. Seeds 4–5 × 3–4 mm, ovoid-subglobose, smooth, shiny, dark green at first, later becoming dark brown; hilum and caruncle tawny-brownish.

Zambia. N: 19 km south of Kawambwa, male fl. 20.x.1952, *White* 3535 (FHO; K; PRE). W: Mwinilunga Distr., Lisombo R., 16 km SW of Kalene Hill Mission, female fl. & o. fr. 11.vi.1963, *Drummond* 8286 (K; LISC; PRE; SRGH).

Also in Cameroon, Zaire and Angola. Riverine and swamp forest (mushitu) understorey and margins, also in sand of lakeshore fixed dunes; 900 m.

This is possibly not specifically distinct from *T. annobonae* Pax & K. Hoffm., from Cameroon and Annobon Island.

15. HYMENOCARDIA Wall. ex Lindl.

Hymenocardia Wall. ex Lindl., Nat. Syst. Bot., ed. 2, 441 (1836). —Müller Argoviensis in De Candolle, Prodr. **15**, 2: 476 (1866). —J. Léonard in Bull. Soc. Roy. Bot. Belgique **90**: 15 (1957); Fl. Afr. Cent. (Hymenocardiaceae): 2 (1985), nom. conserv.
Samaropyxis Miq., Fl. Ind. Bat. Suppl.: 464 (1860).

Dioecious, deciduous trees or shrubs. Indumentum simple. Leaves alternate, shortly petiolate, stipulate, simple, entire, penninerved, acarodomatiiferous, gland-dotted beneath. Stipules deciduous. Male inflorescences often precocious, axillary, solitary or paucifasciculate, spicate or subpaniculate, densely flowered; bracts minute. Male flowers: calyx cupular, 4–6(8)-lobed, lobes imbricate; petals absent; disk absent; stamens 4–6, opposite the sepals, filaments short, elongating and spreading at anthesis, free or connate at the base, glandular at the apex, anthers 2-locular, dorsifixed, introrse then extrorse (due to bending of filaments), thecae parallel, distinct, longitudinally dehiscent; pistillode cylindric, entire or 2-lobed. Female flowers axillary, solitary or paired, or in few-flowered terminal racemes; pedicels slender; sepals 4–6(9), usually free, open or imbricate, narrow, caducous; petals absent; disk absent; ovary 2-locular, compressed perpendicularly to the septum, with 2 apical pendulous anatropous ovules per locule; styles 2, free, elongate, simple with smooth or papillose stigmas. Fruit stipitate or not, 2-coccous, flattened, winged, septifragally dehiscent or not; pericarp somewhat crustaceous; endocarp thinly membranous; columella persistent. Seeds usually 1 per coccus, flat, winged or not; testa thin, striate, shiny; endosperm sparse; cotyledons broad, flat, thin; radicle elongate.

A small palaeotropical genus of 6–7 species, one of which is Asiatic, the rest African. Only 2 species occur in the Flora Zambesiaca area.

Leaves elliptic-ovate, often subacutely acuminate, chartaceous, sparingly gland-dotted beneath; male inflorescences 1–1.5(3) cm long, lax, often branched; galls commonly absent; fruit indehiscent, suborbicular, surrounded by a membranous wing · · · · · · · · · · · · 1. *ulmoides*

Leaves ± oblong, usually obtuse, subcoriaceous, often densely gland-dotted beneath; male inflorescences 2–7 cm long, dense, not branched; galls often formed; fruit splitting into 2 rhombic indehiscent unilaterally winged cocci leaving a persistent columella · · · · · 2. *acida*

1. **Hymenocardia ulmoides** Oliv. in Hooker's Icon. Pl. **12**: t. 1131 (1873). —Engler, Pflanzenw. Ost-Afrikas **C**: 236 (1895). —Hutchinson in F.T.A. **6**, 1: 648 (1912); in F.C. **5**, 2: 410 (1920). —Engler, Pflanzenw. Afrikas (Veg. Erde 9) **3**, 2: 18, t. 6 (1921). —Pax in Engler, Pflanzenr. [IV, fam. 147, xv] **81**: 73 (1922). — De Wildeman, Pl. Bequaert. **3**: 446 (1926); **5**: 396 (1932). —Brenan, Check-list For. Trees Shrubs Tang. Terr.: 216 (1949). —F.W. Andrews, Fl. Pl. Anglo-Egypt. Sudan **2**: 80 (1952). —Dale & Greenway, Kenya Trees & Shrubs: 206 (1961), excl. spec. cit. —White, F.F.N.R.: 200 (1962). —Mogg in Macnae & Kalk, Nat. Hist. Inhaca Isl., Moçamb.: 147 (1969). —Drummond in Kirkia **10**: 251 (1975). —K. Coates Palgrave, Trees Southern Africa, ed. 2, rev.: 405 (1983). —J. Léonard et Mosango, Fl. Afr. Cent. (Hymenocardiaceae): 3 (1985). —Radcliffe-Smith in F.T.E.A., Euphorb. 2: 577, t. 107 (1988). Tab. **18**. Type from Tanzania (Eastern Province).

Hymenocardia poggei Pax in Bot. Jahrb. Syst. **15**: 528 (1893). Types from Zaire (Kasai Province).

Hymenocardia ulmoides var. *capensis* Pax in Bot. Jahrb. Syst. **28**: 22 (1899). Type: Mozambique, Maputo Province, Matola, male, 10.xii.1897, *Schlechter* 11725 (B†, holotype; BM).

Hymenocardia capensis (Pax) Hutch. in Bull. Misc. Inform., Kew **1920**, 10: 334 (1920).

Hymenocardia ulmoides var. *longistyla* De Wild., Pl. Bequaert. **3**: 447 (1926). Type from Zaire (Oriental Province).

A tree up to 20 m high with a spreading crown, or a shrub branched from the base, sometimes scandent. Bark smooth or longitudinally furrowed, tough, grey or brownish. Young shoots and petioles puberulous or pubescent, older twigs glabrescent. Petioles 2–8 mm long. Leaf blades 1–5.5 × 0.5–4 cm, elliptic-ovate or elliptic-lanceolate, subacute, obtuse or rounded, occasionally slightly acuminate at the apex, rounded or cuneate at the base, firmly chartaceous, pubescent along the midrib but otherwise ± glabrous above and beneath except for the tufts of domatial hairs beneath, sparingly gland-dotted beneath, bright or dark green above, paler beneath; lateral nerves inconspicuous, in 4–5(6) pairs. Stipules 3–5 mm long, linear-lanceolate, pubescent, soon falling. Male inflorescences usually on short lateral shoots, laxly spicate or subpaniculate, 1–1.5(3) cm long, the axis slender; bracts 1 mm long, spathulate. Male flowers: buds reddish, orange or pink; calyx c. 1 mm across, shallowly 5-lobed, the lobes ciliate, whitish; anthers 0.8 mm long, yellowish; pistillode 2-lobed, 1 mm high. Female inflorescences terminal on short lateral shoots, 1.5 mm long, few-flowered, leafy at the base; bracts 2 mm long, lanceolate. Female flowers: pedicels 1.5–3 mm long, extending to 7 mm in fruit, very slender; sepals 5–6, 3–4 mm long, linear, pubescent towards the apex; ovary 1 mm long, obovoid to oblong-ellipsoid, glabrous, with a few scattered gland-dots; styles 2–5(10) mm long, extending to 1.5 cm fruit, slender, divergent, red, caducous. Fruits 1.2–2.3 × 1–2.3 cm, suborbicular to obcordate, completely or almost completely surrounded by the wing, flat, borne on a stipe 1–3 mm long, emarginate at the apex, with the wing-tips sometimes overlapping, rounded or slightly decurrent on to the stipe at the base, glabrous, membranous, brown, red, yellow, whitish or greenish. Seeds 7–9 × 2–4 mm, semicircular-oblong, smooth, purplish-brown or blackish.

Zambia. N: Samfya, fr. 5.v.1958, *Fanshawe* 4398 (K; NDO). W: Mwinilunga, male fl. 17.v.1969, *Mutimushi* 3477 (K; NDO). **Zimbabwe**. E: Chimanimani Distr., Haroni R., fr. 7.i.1969, *Mavi* 817 (K; PRE; SRGH). S: Chivirira (Chiribira) Falls, Save-Runde Junction (Sabi-Lundi Junction), fr. 6.vi.1950, *Wild* 3424 (K; LISC; SRGH). **Malawi**. S: Shire R., Mpatamanga Gorge, fr. 28.ii.1961, *Richards* 14484 (K; SRGH). **Mozambique**. N: Marrupa, fr. 20.ii.1981, *de Koning & Nuvunga* 655 (BM; K; LMU; MO). Z: Mocuba Distr., Namagoa, fr. 12.i.1949, *Faulkner* 371 (K; SRGH). MS: Chinizíua, fr. 12.iv.1957, *Gomes e Sousa* 4354 (FHO; K). GI: Inharrime, fr. 22.vi.1960, *Lemos & Balsinhas* 164 (K; LISC; LMA; SRGH). M: Marracuene, male fl. 23.ii.1961, *Macuácua* 90 (BM; K; LISC; LMA; PRE; SRGH).

Widespread in central and southern Africa from Cameroon east to Sudan, south to Angola and southeast to Tanzania and South Africa (KwaZulu-Natal). Often in sandy soil in riverine vegetation, *Androstachys* thickets and lakeshore dunes and thickets, in high rainfall woodland, gully forests on rocky outcrops, also on mushitu margins and coastal forest and dunes; sea level to 800 m.

Vernacular names as recorded in specimen data include: "chimbua", "chingua" (Cheringoma area); "itatalatani", "mtatalatane", "shatalatani" (Ihonga, xiRonga); "leva", "muleva", "naspalala", "nkulumbina", "ntuagka", "wasipalale" (eMakhuwa); "nacuva", "nakwuva" (Niassa area); "nassipalala" (Mocubela); "nas(s)ivalala" (Cabo Delgado area); "tatalatani", "tchatchalatani" (Inhaca Island).

Tab. 18. HYMENOCARDIA ULMOIDES. 1, portion of male flowering branchlet (× 1); 2, male flower (× 12); 3, stamen, rear view, showing gland (× 20), 1–3 from *Eggeling* 6400; 4, female inflorescence (× 1); 5, female flower (× 6), 4 & 5 from *Eggeling* 6398; 6, fruiting branch (× $^{2}/_{3}$), from *B.D. Burtt* 4633; 7, fruit (× 2), from *Semsei* 548. Drawn by Pat Halliday. From F.T.E.A.

2. **Hymenocardia acida** Tul. in Ann. Sci. Nat., Bot., sér. 3, **15**: 256 (1851). —Müller Argoviensis in De Candolle, Prodr. **15**, 2: 477 (1866). —Engler, Pflanzenw. Ost-Afrikas **C**: 236 (1895). —Hutchinson in F.T.A. **6**, 1: 651 (1912). — R.E. Fries, Wiss. Ergebn. Schwed. Rhod.-Kongo-Exped. **1**, 1: 119 (1914). —Eyles in Trans. Roy. Soc. South Africa **5**: 394 (1916). — Engler, Pflanzenw. Afrikas (Veg. Erde 9) **3**, 2: 18, t. 6 (1921). —Pax in Engler, Pflanzenr. [IV, fam. 147, xv] **81**: 76, t. 9 (1922). —De Wildeman, Pl. Bequaert. **3**: 445 (1926). — Brenan, Check-list For. Trees Shrubs Tang. Terr.: 216 (1949). —F.W. Andrews, Fl. Pl. Anglo-Egypt. Sudan **2**: 80 (1952). —Keay in F.W.T.A., ed. 2, **1**, 2: 377, t. 132 (1958). —Dale & Greenway, Kenya Trees & Shrubs: 204 (1961). —White, F.F.N.R.: 200 (1962). — Drummond in Kirkia **10**: 251 (1975). —Troupin, Fl. Pl. Lign. Rwanda: 264, fig. 90/1 (1982); Fl. Rwanda **2**: 227, fig. 62/3 (1983). —K. Coates Palgrave, Trees Southern Africa, ed. 2, rev.: 404 (1983). —J. Léonard et Mosango, Fl. Afr. Cent. (Hymenocardiaceae): 10 (1985). —Radcliffe-Smith in F.T.E.A., Euphorb. 2: 579 (1988). —Beentje, Kenya Trees, Shrubs Lianas: 209 (1994). Types from Guinée and Gambia.

A shrub or small tree up to 10 m high, often straggling or untidy; crown flat or rounded; upper branches spreading, lower branches drooping; bole 15–30 cm in diameter at breast height, often stunted or contorted; bark smooth, light brown or grey, flaking off to leave a powdery rufous or fulvous underbark. Young shoots and petioles pubescent or puberulous, glabrescent. Petioles 0.2–1.6 cm long. Leaf blades 2.5–9.5 × 1.5–5 cm, elliptic-ovate to oblong-oblanceolate, rounded, obtuse or rarely subacute at the apex, rounded or sometimes cuneate at the base, thinly coriaceous at maturity, dark green above, paler beneath; upper surface sparingly or evenly pubescent to subglabrous; lower surface densely subferruginous- or fulvous-tomentose, or subglabrous except along the midrib and main nerves and the domatial hair tufts, sparingly to evenly gland-dotted; lateral nerves in 5–8(10) pairs, fairly prominent and camptodromous on leaf lower surface, scarcely prominent on upper surface and brochidodromous. Stipules 1–3 mm long, linear-lanceolate to filiform, pubescent, soon falling. Male inflorescences 2–7 cm long, densely spicate, amentiform, the spikes solitary or fasciculate in the axils of fallen leaves; axis pubescent; bracts 1 mm long, spathulate, ciliate. Male flowers: buds open, deep reddish-carmine or brownish; calyx 1.5–2 mm across, shallowly 5-lobed, the lobes obtuse, ciliate, pinkish; anthers 1.2 mm long, dorsally purplish-tinged, pollen pale yellow; pistillode 2-lobed, 1 mm high. Female flowers axillary or solitary, or in few-flowered racemes up to 3 cm long terminating lateral leafy shoots; axis and bracts ± as in the male. Female flowers: pedicels 1 mm long, extending to up to 2 cm long in fruit; sepals 5–9, 1.5–4 mm long, free or sometimes partially united, linear, usually readily caducous, pinkish; ovary 2 × 1 mm, obovoid-oblong to obcordate, 2-winged in the upper half, gland-dotted, glabrous or densely pubescent, glaucous or crimson; styles 2(3), 0.2–2 cm long, the arms 0.25–1 mm thick, filiform, rugose-papillose, reddish-purple. Fruits borne on a stipe 2 mm long, flat, 2–3.5 × 2.3–4 cm, V-shaped, with 2 apical divergent rounded-rhomboid membranous striate wings, rounded or cordate at the base, reticulate, glabrous or pubescent, gland-dotted or not, somewhat shiny, yellow-green at first, turning pink then reddish-brown, dehiscent into 2 subtrapeziform cocci leaving a compressed-fusiform columella. Seeds 1 × 0.5 cm, compressed-semicircular, smooth, shiny, dark purplish-brown streaked with black.

Fruits glabrous; leaves often glabrous on lower surface · var. *acida*
Fruits pubescent; leaves often pubescent on lower surface · · · · · · · · · · · · · · · · · · · var. *mollis*

Var. **acida**

Hymenocardia mollis var. *glabra* Pax in Bot. Jahrb. Syst. **15**: 528 (1893). Type from Zaire.
Carpodiptera minor Sim, For. Fl. Port. E. Afr. 21, t. III c (1909). Type: Mozambique, Maputo, Polana, *Sim* 5555 (GRA, holotype).

Fruits glabrous; leaves often glabrous beneath.

Zambia. B: near Senanga, fr. 1.viii.1952, *Codd* 7319 (BM; K; PRE; SRGH). N: 15.5 km west of Kawambwa, fr. 31.x.1952, *White* 3560 (BM; FHO; K). W: Mwinilunga Distr., Kalene Hill, fr. 15.xii.1963, *Robinson* 6059 (K; PRE; SRGH). C: 20 km east of Lusaka, fr. 15.x.1972, *Strid* 2335 (MO). S: Monze, fr. 11.vii.1930, *Hutchinson & Gillett* 3556 (BM; K; LISC; SRGH). **Zimbabwe**. W: Hwange Distr., Victoria Falls National Park, Zambezi Camp, fr. 26.xi.1978, *Mshasha* 119 (K; MO; PRE; SRGH). C: Rusape Distr., SE Wedza Res., fr. 22.v.1962, *Cleghorn* 699 (SRGH). E: Gungunyana For. Res., male fl. xi.1967, *Goldsmith* 130/67 (K; LISC; PRE; SRGH). **Malawi**. N:

Chitipa Distr., Kaseye Mission, fr. 26.xii.1977, *Pawek* 13378 (K; MAL; MO; SRGH; UC). C: Kasungu Nat. Park, male (galled) 26.xi.1972, *Pawek* 6017 (K; MAL; MO; SRGH; UC). S: foot of Mt. Mulanje, fr. 10.i.1987, *J.D. Chapman & E.J. Chapman* 8323 (K; MO; PRE). **Mozambique**. N: Eráti, male fl. 11.x.1948, *Barbosa* 2360 (LISC). Z: 30 km Nicuadala–Campo, fr. 2.ii.1966, *Torre & Correia* 14352 (LISC). MS: Dombe–Mavita, fr. 19.vi.1942, *Torre* 4367 (LISC).

More or less throughout tropical Africa from Senegal to Ethiopia and south to southern Angola, central Zimbabwe and central Mozambique. Deciduous woodlands (*Brachystegia* and *Cryptosepalum*), watershed grasslands and dambos on Kalahari Sand, and lakeshore sand dunes, also in high rainfall miombo and mixed deciduous woodlands, dambos, riverine fringes and mushitu margins; 640–1740 m.

Vernacular names as recorded in specimen data include: "akapempe" (Lamba, Wemba); "icholwa muyandi" (chiLozi); "kapape", "kapepe" (chiLunda); "kapembe" (chiBemba); "kapempe" (chiKaonde); "mupepe" (Lovale).

Var. **mollis** (Pax) Radcl.-Sm. in Kew Bull. **28**: 324 (1973); in F.T.E.A., Euphorb. 2: 580 (1988). Type from Tanzania (Lake Province).

Hymenocardia mollis Pax in Bot. Jahrb. Syst. **15**: 528 (1893). —Engler, Pflanzenw. Ost-Afrikas **C**: 237 (1895). —Hutchinson in F.T.A. **6**, 1: 651 (1912). —R.E. Fries, Wiss. Ergebn. Schwed. Rhod.-Kongo-Exped. **1**, 1: 119 (1914). —Engler, Pflanzenw. Afrikas (Veg. Erde 9) **3**, 2: 18 (1921). —Pax in Engler, Pflanzenr. [IV, fam. 147, xv] **81**: 75 (1922). —Brenan, Check-list For. Trees Shrubs Tang. Terr.: 216 (1949). —Topham, Check List For. Trees Shrubs Nyasaland Prot.: 51 (1958).

Hymenocardia lasiophylla Pax in Bot. Jahrb. Syst. **19**: 79 (1894). —Engler, Pflanzenw. Ost-Afrikas **C**: 237 (1895). Type from Tanzania (Western Province).

Hymenocardia mollis var. *lasiophylla* (Pax) Pax in Engler, Pflanzenr. [IV, fam. 147, xv] **81**: 76 (1922). —Brenan, Check-list For. Trees Shrubs Tang. Terr.: 216 (1949).

Fruits pubescent; leaves often pubescent beneath.

Zambia. B: Kalabo, fr. 14.xi.1959, *Drummond & Cookson* 6452 (K; LISC; PRE; SRGH). N: L. Bangweulu (Bangweolo) north of Samfya Mission, male fl. & y. fr. 7.x.1947, *Brenan & Greenway* 8062 (BM; FHO; K; PRE). W: Solwezi Distr., Chifubwa R. Gorge, fr. 20.iii.1961, *Drummond & Rutherford-Smith* 7122 (LISC; PRE; SRGH). C: Mkushi, fr. 27.iii.1961, *Angus* 2543 (FHO). E: Petauke–Sasare road, fr. 6.xii.1958, *Robson* 860 (BM; K; LISC; PRE; SRGH). S: Namwala, male fl. 14.x.1963, *van Rensburg* 2532 (K; SRGH). **Zimbabwe**. C: Makoni Distr., Chiduku Res., fr. i.1962, *Davies* 2963 (SRGH). E: 29 km south of Mutare (Umtali), y. fr. 20.xi.1955, *Chase* 5861 (BM; K; PRE; SRGH). **Malawi**. N: Mzimba Distr., 9.5 km Eutini–Vuvumwe Bridge, male fl. 26.xi.1976, *Pawek* 11964 (K; MAL; MO; SRGH; UC). C: Mchinji Distr., Mbonjera, fr. 24.xi.1983, *Salubeni & Patel* 3573 (MAL; MO). S: Kasupe Distr., Liwonde Ferry, fr. 12.ii.1964, *Salubeni* 239 (MAL; SRGH). **Mozambique**. N: Malema, Mutuáli–Cuamba (Nova Freixo), fr. 10.iv.1962, *Lemos & Marrime* in *Lemos* 332 (BM; K; LISC; LMA; SRGH). Z: 43 km Mocuba–Régulo Mataia, fr. 27.v.1949, *Barbosa & Carvalho* in *Barbosa* 2892 (LMA; SRGH). T: Chiputo (Vila Vasco da Gama), Fíngoè, o. fr. 12.viii.1941, *Torre* 3264 (LISC). MS: Gorongosa Nat. Park, Pungwe R. –Gorongosa (Vila Paiva), fr. i.1972, *Tinley* 2371 (LISC; PRE; SRGH).

Also in Rwanda, Zaire and Tanzania. *Brachystegia* woodland, usually on sand, and on rocky outcrops and escarpments, also in riverine vegetation, wooded grassland and dambo, sometimes on lakeshore dunes; 90–1740 m.

Vernacular names as recorded in specimen data include: "chikolowanga", "chipundu" (chiYao); "chimbali" (Manica area); "chitata", "chiteta" (Chimoio area); "chitswati" (Dao); "mkulangondo" (chiNyanja); "mpothakodji", "mputhakwochi" (Niassa area); "pitacodje" (eMakhuwa); "tchimberi" (Machengane); "tero-tero", "terro-terro" (Zambézia area).

16. OLDFIELDIA Benth. & Hook.f.

Oldfieldia Benth. & Hook.f. in Hooker's J. Bot. Kew Gard. Misc. **2**: 184, t. 6 (1850). —Müller Argoviensis in De Candolle, Prodr. **15**, 2: 1259 (1866). —J. Léonard in Bull. Jard. Bot. État **26**: 338 (1956).

Paivaeüsa Benth. & Hook.f., Gen. Pl. **1**, 3: 993 (1867). —Welwitsch in Trans. Linn. Soc. London, **27**: 20, t. 7 (1869).

Cecchia Chiov., Fl. Somala **2**: 397, t. 227 (1932).

Dioecious pachycaul trees or shrubs with a simple indumentum. Twigs with prominent leaf scars. Buds perulate (furnished with protective scales). Leaves alternate, opposite-decussate or verticillate in whorls of 3, long-petiolate, exstipulate, digitately 3–8-foliolate; leaflets subsessile or petiolulate, entire, penninerved, brochidodromous. Inflorescences axillary, solitary or geminate, cymose; male inflorescences subsessile or pedunculate, many-flowered, lax or capitate; female inflorescences pedunculate, the peduncles extending in fruit, 1–3-flowered. Male flowers small, shortly pedicellate; sepals 5–8, unequal, united at the base, imbricate; petals absent; stamens 4–12, free, filaments inserted between the lobes of the disk, unequal, anthers dorsifixed, extrorse, minutely papillose, longitudinally widely dehiscent; disk central, thick, fleshy, lobate-sinuate, pubescent; pistillode filiform, minute or absent. Female flowers: pedicels very short, lengthening slightly in fruit; sepals longer, narrower and stouter than in the male, but otherwise similar, persistent in fruit; disk hypogynous, annular, short, fleshy, crenellate; ovary 2–3-locular, with 2 ovules per locule; styles 2–3, short, connate at the base; stigmas ± reniform, shallowly 2–4-lobulate, caducous. Fruits 2–3-locular, depressed obovoid-subglobose, tardily loculicidally dehiscent into 2–3 valves each with an attached septum; pericarp coriaceous; mesocarp crustaceous; endocarp chartaceous; columella persistent, often with the seeds attached. Seeds 1–2 per locule, usually 3 per fruit, somewhat compressed; exotesta fleshy; endotesta crustaceous; funicle thickened, carunculoid; albumen fleshy; embryo green, cotyledons broad, flat.

A tropical African genus of 4 species ranging from Sierra Leone to Somalia and southwards to Tanzania, Zambia and Angola. Two species occur in the Flora Zambesiaca area.

Leaf scar margins raised to form tubercles; leaflets 3–5(7), unequal, rounded or subacute; male inflorescences shortly pedunculate to subsessile; mature fruits sparingly to evenly pubescent · 1. *dactylophylla*
Leaf scar margins not or scarcely raised; leaflets (3)5–7(8), subequal, shortly acuminate; male inflorescences distinctly pedunculate; mature fruits glabrescent · · · · · · · · · · 2. *somalensis*

1. **Oldfieldia dactylophylla** (Welw. ex Oliv.) J. Léonard in Bull. Jard. Bot. État **26**: 340 (1956). —White, F.F.N.R.: 201 (1962). —Radcliffe-Smith in F.T.E.A., Euphorb. 1: 115 (1987). Tab. **19**. Type from Angola (Huíla Province).

Paivaeüsa dactylophylla Welw. ex Oliv. in F.T.A. **1**: 328 (1868). —Welwitsch in Trans. Linn. Soc. London, **27**: 21, t. 7 (1869). —Prain in F.T.A. **6**, 1: 626 (1912). —R.E. Fries, Wiss. Ergebn. Schwed. Rhod.-Kongo-Exped. **1**, 1: 118 (1914). —Engler, Pflanzenw. Afrikas (Veg. Erde 9) **3**, 2: 38 (1921). —Pax in Engler, Pflanzenr. [IV, fam. 147, xv] **81**: 296 (1922). —Brenan, Check-list For. Trees Shrubs Tang. Terr.: 221 (1949).

A small, spreading often somewhat stunted tree to 10 m high, with a milky latex. Bole short, straight, c. 25 cm in diameter; bark rough, reticulate, c. 2 cm thick, dark grey; wood dull brown. Twigs thick, ferrugineous-pubescent at first, later glabrescent; leaf scar margins raised, tuberculiform. Leaves alternate, subopposite or opposite; petioles 2–11 cm long; leaflets 3–5(7), the median often much larger than the laterals, 6–15 × 2.5–5.5 cm, elliptic-oblong, oblanceolate or obovate, rounded, obtuse or subacute at the apex, attenuate or cuneate at the base, thinly coriaceous, the midrib sunk in a channel, fairly prominent beneath, evenly pubescent along the midrib and otherwise sparingly pubescent above at first, later glabrescent, evenly to densely fulvous- to sericeous-tomentose beneath, shiny and dark green above, duller and paler beneath; lateral nerves in 10–17 pairs, brochidodromous well within the margin, often with interstitials, slightly prominent above; petiolules 1–7 mm long, or leaflets subsessile. Male inflorescences subsessile, or with peduncles 1–7 mm long, densely ferrugineous-tomentose; bracts 1.5 mm long, suborbicular-ovate, pubescent; flowers densely capitate. Female peduncles 3 mm long, extending a little in fruit, commonly 1-flowered; bracts 3 mm long, oblong-lanceolate. Male flowers: pedicels 0.5 mm long; sepals 6–8, 2 mm long, oblong, obtuse, ferrugineous-pubescent without, glabrous within; stamens 7, 5–6 mm long, anthers c. 1 mm long, bright yellow; disk 2 mm in diameter, slightly convex; pistillode minute or absent. Female flowers: pedicels 0.5 mm long, extending to 5 mm in fruit, stout, ferrugineous-pubescent; sepals 2.5–3 × 1.5 mm, triangular-lanceolate, minutely pubescent within at the apex, otherwise more or less as in the male; disk 2 mm in diameter; ovary 2-locular, 1.5–3 × 1.5–3 mm, ovoid-subglobose, densely ferrugineous-tomentose; styles 2, 1–1.5 mm long, densely tomentose, stigmas velvety, greyish-yellow-brown. Fruit 1.5 × 1.6 cm, obovoid-subglobose, 2-lineate, brownish velvety-tomentose, orange when ripe. Seeds 1 × 1 × 0.4 cm, compressed-rhomboid, pale yellowish-brown.

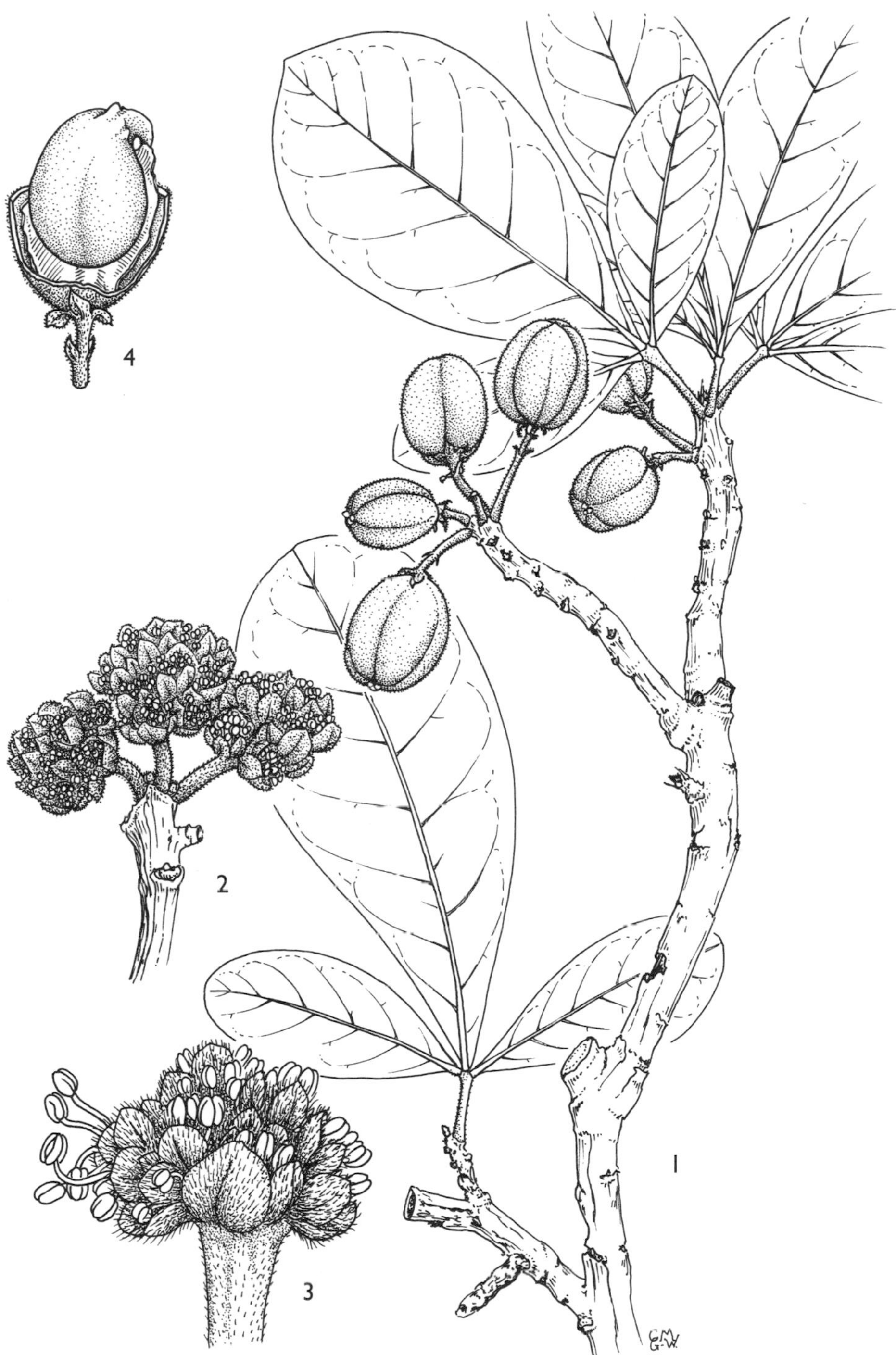

Tab. 19. OLDFIELDIA DACTYLOPHYLLA. 1, distal portion of fruiting branch (× $^2/_3$), from *Greenway & Kanuri* 14762; 2, male inflorescence (× 2); 3, male flowers (× 6), 2 & 3 from *Boaler* 66; 4, fruit, partly cut away to show attachment of seed (× 2), from *Greenway & Kanuri* 14762. Drawn by Christine Grey-Wilson. From F.T.E.A.

Zambia. B: Zambezi (Balovale), y. fr. vii.1933, *Trapnell* 1216 (K). N: Nkole Mfumu (Nkolemfumu), Protected Forest Area, c. 48 km south of Kasama on Mpika road, male fl. 7.v.1958, *Lawton* 366 (FHO; K; NDO). W: 19 km west of Solwezi, st. 18.iii.1961, *Drummond & Rutherford-Smith* 7029 (K; LISC; PRE; SRGH). C: 99 km south of Ndola, male fl. 23.ix.1963, *Angus* 3754 (FHO; K). **Malawi**. N: Chitipa (Fort Hill) Distr., Katutula, fr. no date, *Lewis* 62 (FHO).

Also in Angola, Zaire (Kasai, Shaba) and Tanzania. Deciduous plateau woodland, often on sandy soils, usually as a subcanopy tree or shrub in *Brachystegia* and mixed deciduous woodlands, and woodland on Kalahari Sand, also on rocky hills and dambo margins; 1035–1830 m.

Vernacular names as recorded in specimen data include: "kali" (chiWemba); "kalikali" (chiBemba); "kazonga" (chiLunda); "kazong'we" (Zambezi area); "mutengulu", "munyasha" (Mambwe area); "mutobakuso" (Kaonde).

2. **Oldfieldia somalensis** (Chiov.) Milne-Redh. in Kew Bull. **3**: 456 (1949). —Dale & Greenway, Kenya Trees & Shrubs: 211 (1961). —Radcliffe-Smith, F.T.E.A., Euphorb. 1: 117 (1987). —Beentje, Kenya Trees, Shrubs Lianas: 216 (1994). Type from Somalia.

Cecchia somalensis Chiov., Fl. Somala **2**: 397, t. 227 (1932).

Paivaeüsa orientalis Mildbr. in Notizbl. Bot. Gart. Berlin-Dahlem **12**: 710 (1935). —Brenan, Check-list For. Trees Shrubs Tang. Terr.: 221 (1949). Type from Tanzania (Southern Province).

Oldfieldia sp., Gardner, Trees & Shrubs of Kenya Col.: 47 (1936). —Brenan, Check-list For. Trees Shrubs Tang. Terr.: 221 (1949).

A much-branched evergreen tree up to 20 m high. Bark grey or brownish, deeply fissured. Twigs with the leaf scar margins not or scarcely raised, fulvous- or ferrugineous-pubescent at first, later glabrescent, greyish. Leaves opposite or subopposite; petioles 4–7 cm long; leaflets (3)5–7(8), subequal, 4–10 × 1.5–4 cm, elliptic or elliptic-oblanceolate, shortly acuminate at the apex, attenuate or cuneate at the base, firmly chartaceous, evenly pubescent along the midrib above and beneath, and elsewhere sparingly pubescent or subglabrous at first, later glabrescent above, shiny and dark green above, duller and paler beneath; midrib canaliculate above, prominent beneath, lateral nerves in 10–20 pairs, brochidodromous well within the margin, slightly prominent above and beneath; petiolules 1–5 mm long. Male inflorescences: peduncles 1–1.5 cm long; bracts minute; flowers subcapitate. Female inflorescences: peduncles 5 mm long, extending to 1 cm long in fruit, 1(3)-flowered; bracts 3–6 mm long, oblong, fulvous- or ferrugineous-pubescent. Male flowers: pedicels up to 2 mm long; sepals 6–7, 1 mm long, oblong, obtuse, ferrugineous-pubescent without, glabrous within except towards the apex; stamens 5–7, 3–4 mm long, anthers c. 1 mm long; disk 1 × 1 mm, irregularly lobulate, conical at the apex; pistillode minute or absent. Female flowers: pedicels 5 mm long, extending to c. 1 cm in fruit; sepals 5–7 × 1.5–3 mm, oblong, foliaceous, obtuse, pubescent; disk 3 mm in diameter; ovary 2–3-locular, 4–5 × 3–4 mm, ovoid, fulvous-sericeous; styles 2–3, 2 mm long, densely white-pubescent, stigmas velvety, brownish-grey. Fruit 1.5–2 × 1.5–1.8 cm, subglobose, 2–3-lineate, glabrescent, yellow when ripe. Seeds 1–1.3 × 0.8–1 × 0.3 cm, flattened-ovoid, brownish or blackish.

Mozambique. Z: Maganja da Costa Distr., Gobene Forest, 42 km from Olinga (Vila da Maganja) to the coast, fr. 14.ii.1966, *Torre & Correia* 14603 (LISC).

Also in Somalia, Kenya and Tanzania. In dense mixed coastal forest with abundant lianes, on dunes and sandy soil, with *Brachystegia, Cynometra, Trachylobium, Albizia, Craibia* and *Mimusops*; c. 20 m.

17. ANDROSTACHYS Prain

Androstachys Prain in Bull. Misc. Inform., Kew **1908**: 438 (1908).
Weihea sensu Sim, For. Fl. Port. E. Afr.: 66 (1909), non Spreng.

Dioecious trees with a simple indumentum. Branching monopodial. Leaves opposite, decussate, stipulate, petiolate, sometimes peltate, simple, entire, palminerved. Stipules large, intrapetiolar, connate at first into an elongate flattened sheath enclosing the terminal bud. Flowers axillary, the males in triads, pendent; the females solitary, pedicellate. Male flowers: sepals 2–5, free, spirally arranged or ± whorled, bract-like; petals absent; disk absent; stamens numerous, spirally disposed on an elongate receptacle, filaments short and recurved or absent, anthers dorsifixed, elongate, extrorse, bilocular, thecae free at apex and base, connective setose when young, longitudinally dehiscent; pistillode absent. Female flowers: sepals 5–6, imbricate, caducous; petals absent; disk absent; staminodes absent; ovary 3(5)-locular, with 2 ovules per loculus; stylar column elongate, stigmas 3, simple, reflexed. Fruit 3-lobed, apically depressed, the keels subcarinate, dehiscing septicidally and loculicidally into 3 bivalved cocci or 6 separate valves; endocarp crustaceous. Seeds 6, laterally compressed, shiny, ecarunculate; albumen fleshy; cotyledons broad and flat.

A monotypic genus restricted to South tropical Africa and western Madagascar.

Androstachys johnsonii Prain in Bull. Misc. Inform., Kew **1908**: 439 (1908); **1909**: 201–204 (1909). —Hutchinson in F.T.A. **6**, 1: 741 (1912); 1049 (1913); in F.C. **5**, 2: 377 (1915). —Engler, Pflanzenw. Afrikas (Veg. Erde 9) **3**, 2: 38 (1921). —Pax & Hoffmann in Engler, Pflanzenr. [IV, fam. 147, xv] **81**: 287 (1922); in Engler & Harms, Pflanzenfam., ed. 2, **19c**: 75 (1931). —Burtt Davy, Fl. Pl. Ferns Transvaal: 298 (1932). —Hutchinson, Botanist in Southern Africa: 667 (1946). —White, F.F.N.R.: 193 (1962). —Airy Shaw in Kew Bull. **18**: 251 (1965); in Adansonia, Sér. 2, **10**, 4: 520 (1970). —Drummond in Kirkia 10: 251 (1975). —K. Coates Palgrave, Trees Southern Africa, ed. 2, rev.: 415 (1983). Tab. **20**. Syntypes: Mozambique, Inhambane, female fl. viii. 1883, *O'Neill* s.n. (K); Manica e Sofala, Beira, male fl. & fr. xi. 1908, *Johnson* 283 (K).

Weihea (?) subpeltata Sim, For. Fl. Port. E. Afr.: 66, t. 61A (1909). Type: Mozambique, Maputo, Lebombo Mts., *Sim* 6387 (PRE, holotype; GRA; NU).

Androstachys subpeltatus (Sim) Phillips, in Mem. Bot. Surv. S. Africa **25** (Gen. S. Afr. Fl. Pl., ed. 2): 460 (1951).

A tree up to 20 m high. Bark blackish-grey, longitudinally fissured. Wood hard. Twigs grey. Young growth densely whitish-tomentose. Petioles 0.5–4 cm long. Stipular sheath 1–3.5 × 0.4–0.7 cm, oblong, rounded, detersibly tomentellous (indumentum easily detached). Leaf blades 3–9(13) × 2–7(12) cm, broadly ovate or asymmetrically hemirhombic-ovate, less often elliptic-ovate, obtuse or rounded at the apex, cuneate to rounded or truncate at the base, or peltate, entire, coriaceous, glabrous, shiny and dark green on upper surface, densely whitish-tomentose beneath, 5–7-nerved from the base; lateral nerves in 5–8 pairs. Male inflorescences up to 3 cm long, 3-flowered, with the central flower longer than the laterals. Male flowers: pedicels 5–6 mm long in lateral flowers, or 1–1.3 cm long in central flowers, pubescent; sepals 2–3 in lateral flowers, or 5 in central flowers, 4 × 0.5 mm, linear-oblanceolate, subacute, sericeous-hirsute without, glabrous within; receptacle up to 1.5 cm long in lateral flowers, or 2 cm long in central flowers; stamens up to c. 35 in lateral flowers, or to c. 50 in central flowers, filaments of lowest stamens up to 1 mm long, anthers 3 × 0.5 mm, connective sparingly sericeous-pubescent at first, later glabrescent, thecae yellow. Female flowers: pedicels 1–1.5 cm long, extending to up to 3 cm long in fruit; sepals 7–8 × 2–2.5 mm, ovate-lanceolate, obtusely acuminate or sometimes bifid at the apex, minutely crisped-puberulous on the margins and at the base, otherwise ± glabrous; ovary c. 3 mm in diameter, ovoid, densely tomentose; stylar column with styles c. 7 mm long, puberulous, stigmas minutely papillose. Fruit 8–10 × 12–13 mm, minutely crisped puberulous, bright green at first, yellowish to light brown when mature; indumentum detersible (easily detached). Seeds 6.5–7(8) × 4.5–5(6) × 2–3 mm, laterally compressed-ovoid, shallowly longitudinally striate-ridged, chestnut-brown.

Zimbabwe. E: Chipinge Distr., Mangaze Native Area, 32 km SW of Chirinda, male fl. xi.1965, *Goldsmith* 50/65 (K; LISC; SRGH). S: Masvingo Distr., Bangara Falls, Mtilikwe R., fr. 13.xii.1953, *Wild* 4635 (K; LISC; SRGH). **Mozambique**. N: Nampula, Mossuril, Cruce Forest, o. fr. & seed 18.ii.1984, *de Koning, Groenendijk & Dungo* 9664 (K; LMU; MO). MS: Maringua, 9.5 km north of Save (Sabi) R., fr. 23.vi.1950, *Chase* 2439 (BM; SRGH). GI: Mocodoene–Funhalouro, Xilaule Forest, st. 28.x.1947, *Barbosa* 614 (K; LMA). M: Goba, Libombo Spring, fr. 8.xi.1961, *Lemos & Balsinhas* in *Lemos* 220 (BM; K; LISC; LMA; SRGH).

Also in Swaziland, South Africa (northern and eastern Transvaal) and Madagascar. Locally common below 1000 m, in hot dry localities on well drained soils, usually gregarious on rocky hillsides and along seasonal watercourses, often forming dense shrub thickets, also in mixed deciduous woodland; 100–900m.

The specimen *Dr. Williams* (Wellcome Chemical Research Lab.) 28.x.1925, without precise locality, but said to come from NE Rhodesia, is unlikely to have been collected in Zambia (NE Rhodesia).

Tab. 20. ANDROSTACHYS JOHNSONII. 1, distal portion of fruiting branch (× 1), from *Lemos & Balsinhas* 220; 2, male inflorescence (× 2); 3, male flower (× 6), 2 & 3 from *Gomes e Sousa* 1946; 4, female inflorescence (× 1); 5, female flower (× 6), 4 & 5 from *Groenendijk* 2133; 6, columella (× 2); 7, fruit (× 2), 6 & 7 from *Wild* 4365. Drawn by J.M. Fothergill.

Vernacular names as recorded in specimen data include: "cimbirre", "simbiri" (Manjacaze, Muchopes areas); "cruce", "crusse", "m'crusse", "mecrusse", "mucrusse" (eMakhuwa, Goba area); "makruss" (Inhambane area); "mezimbite", "m'zimbiti", "zimbiti" (Beira area); "msumbiti" (Mewenezi area); "musimbite" (chiNdau); "simbite" (Chipinge area); "uluti" (eMakhuwa).

The timber is resistant to termite attack, and is therefore of value for the construction of bridge and house piles. Although hard and heavy it can be readily sawn into planks for flooring, steps, causeways etc. It is also employed for cabinet-making. The pollen, however, produces poisonous honey.

18. CLUTIA L.

Clutia L., Sp. Pl.: 1042 (1753); Gen. Pl., ed. 5: 464 (1754).
Cluytia Ait., Hort. Kew. **3**: 419 (1789).

Dioecious, occasionally monoecious, small trees, shrubs or woody herbs. Indumentum simple or absent. Leaves alternate, often exstipulate, shortly petiolate, simple, entire, penninerved, often pustulate. Flowers in axillary fascicles, or female flowers solitary. Male flowers: pedicels jointed; sepals 5, united at the base, imbricate; petals 5, free, clawed, open or imbricate; disk glands in 1–3 series: 2–7 per sepal, (0)1–3 per petal, 0–∞ on the receptacle; stamens 5, oppositipetalous, borne on a receptacular column, anthers dorsifixed, introrse, longitudinally dehiscent; pistillode cylindrical, borne on top of the column. Female flowers: sepals and petals ± as in the male, but persistent and becoming hardened in fruit; disk glands usually only in 1 series, on the sepals; ovary trilocular, with 1 ovule per loculus; styles 3, free or united at the base, bifid or bipartite. Fruit ± spherical, dehiscing septicidally into 3 bivalved cocci; endocarp woody; septa thin, free or adhering to the columella. Seeds ovoid-ellipsoid; testa crustaceous, shiny, black; caruncle conical, protuberant; albumen fleshy; cotyledons broad, flat.

A genus of some 70 species from tropical and South Africa and Arabia.

1. Plants monoecious ··· 2
– Plants dioecious ··· 3
2. Leaves sessile or subsessile, oblanceolate-oblong, rounded to emarginate, crowded towards the tips of the branches ··· 7. *conferta*
– Leaves shortly petiolate, narrowly lanceolate, acute or subacute, evenly spaced on the stems ··· 12. *angustifolia*
3. Perennial herbs with usually simple stems arising from a woody stock ··· 4
– Shrubs ··· 6
4. Petals of male flowers each with 1 disciform gland at the base ··· 1. *whytei*
– Petals of male flowers each with 3–5 spherical glands at the base ··· 5
5. Plants completely glabrous; leaves pustulate; pedicels of male flowers up to 17 mm long; fruiting pedicels up to 25 mm long ··· 13. *monticola*
– Plants sparingly to densely pubescent or hirsute; leaves not or rarely pustulate; pedicels of male flowers not exceeding 2 mm; fruiting pedicels not exceeding 5 mm in length ··· 10. *hirsuta*
6. Leaves sessile, cordate ··· 7
– Leaves petiolate, or if sessile, then blades not cordate ··· 8
7. Stems closely beset with longitudinal centrally punctate tubercles; petals of male flowers each with a solitary rhomboid gland ··· 6. *brassii*
– Stems smooth; petals of male flowers each with c. 5 spherical glands ··· 14. *sessilifolia*
8. Petals of male flowers deltate, unguiculate, each with 1 spherical or subspherical gland at the base; fruiting pedicels up to 25 mm long ··· 9
– Petals of male flowers obovate- or rhombic-spathulate, each with 2 or more spherical glands at the base; fruiting pedicels not exceeding 6 mm in length ··· 12
9. Plants completely glabrous; leaves ovate, reticulate, rounded to truncate at the base ··· 5. *punctata*
– Plants glabrous to sparingly or evenly pubescent; leaves suborbicular-ovate to elliptic-lanceolate, not markedly reticulate, attenuate, cuneate or rounded at the base ··· 10
10. Petioles not more than 3 mm long; leaves strongly glaucous ··· 4. *galpinii*
– Petioles 10–35 mm long; leaves green on upper surface, somewhat glaucous beneath ··· 11

11. Leaf blades mostly more than twice as long as broad, ovate to elliptic-lanceolate, narrowed to the apex · 2. *abyssinica*
– Leaf blades mostly less than twice as long as broad, broadly ovate-suborbicular, rounded at the apex · 3. *pulchella*
12. Petals of male flowers with 2–3 glands at the base; fruit glabrous · · · · · · · · · · · · · · · 13
– Petals of male flowers with 3–5 glands at the base; fruit hirsute · · · · · · · · · · · · · · · · 14
13. Leaf blades up to 11 × 3.5 cm, oblanceolate; petals of male flowers usually with only 2 glands at the base · 8. *robusta*
– Leaf blades up to 6 × 2.5 cm, obovate to elliptic-lanceolate; petals of male flowers with 2–3 glands at the base · 9. *paxii*
14. Leaf blades ovate to obovate-oblong, up to 5 cm wide · · · · · · · · · · · · · · · 11. *swynnertonii*
– Leaf blades narrowly lanceolate, not more than 2.5 cm wide · · · · · · · · · · 12. *angustifolia*

1. **Clutia whytei** Hutch. in F.T.A. **6**, 1: 806 (1912) as "*Cluytia*". —Engler, Pflanzenw. Afrikas (Veg. Erde 9) **3**, 2: 124 (1921) as "*Cluytia*". —Brenan, Check-list For. Trees Shrubs Tang. Terr.: 203 (1949). —Topham, Check List For. Trees Shrubs Nyasaland Prot.: 50 (1958). —Radcliffe-Smith in F.T.E.A., Euphorb. 1: 333 (1987). Type: Malawi, Nyika Plateau, male fl. & fr. vi.1896, *Whyte* 175 (K, lectotype, selected by A. Radcliffe-Smith, 1987).

Perennial herb up to 120 cm high, usually pubescent at first, later glabrescent, sometimes glabrous, purplish-green, dioecious. Stems numerous simple, from a woody rootstock. Stipules absent. Petioles 2–5 mm long, or leaves subsessile. Leaf blades 2–5.5(6.5) × 1–3.5 cm, broadly ovate-suborbicular in lower leaves, ovate-lanceolate in upper leaves, obtuse or subacute at the apex, cuneate or rounded, or rarely shallowly cordate at the base, entire, chartaceous to thinly coriaceous, greenish on upper surface, glaucous beneath; lateral nerves in 5–9 pairs. Male fascicles 3–9-flowered; bracts imbricate, c. 1 × 1 mm, ovate, crustaceous, stramineous. Male flowers: pedicels up to 5(10) mm long, pubescent; sepals 2 × 1.5 mm, elliptic-obovate, rounded, each with 3 uncinate glands near the base, pubescent without, glabrous within, greenish-white; petals 2 × 1 mm, spathulate, each with a small disciform gland at the base, white; staminal column c. 2 mm high, free filaments c. 0.5 mm long, anthers 0.67 mm long; pistillode 0.67 mm high, cylindric, peltate-capitate above. Female flowers 1–3(4) per axil; pedicels c. 4 mm long at first, usually extending to c. 13 mm in fruit; sepals 2.25 × 1 mm, elliptic-oblong, each with a flattened truncate or erose gland at the base, otherwise as in the male; petals eglandular, otherwise as in the male; ovary 1.25 mm in diameter, glabrous; styles c. 1 mm long, spreading. Fruits 4.5–5 × 4.5–5 mm, pusticulate, glabrous, olivaceous or brownish. Seeds 3 × 2 × 2 mm, ellipsoid, ± smooth; caruncle 1 × 1 mm, brownish.

Stems pubescent at first, later glabrescent; petioles 2–5 mm long; lamina cuneate or rounded at the base; pedicels of male flowers not more than 5 mm long · · · · · · · · · · · · var. *whytei*
Stems glabrous; petioles 1–2 mm long, or leaves subsessile; lamina rounded or shallowly cordate at the base; pedicels of male flowers up to 10 mm long · · · · · · · · · · · · · var. *monticoloides*

Var. **whytei**

Stems pubescent at first, later glabrescent; petioles 2–5 mm long; lamina cuneate or rounded at the base; pedicels of male flowers not more than 5 mm long.

Zambia. N: Luwingu Distr., Chishinga Ranch, male fl. 11.i.1963, *Astle* 1941 (SRGH). W: Kitwe, fr. 6.iii.1955, *Fanshawe* 2130 (K; NDO). E: Nyika, male fl. 31.xii.1962, *Fanshawe* 7390 (K; NDO). **Malawi**. N: Nkhata Bay Distr., Viphya Plateau, c. 8 km north of Luwawa Dam, fr. 6.vii.1976, *Pawek* 11477 (K; MAL; MO; PRE; SRGH; UC). C: Dedza Mt., male fl. 3.iii.1977, *Grosvenor & Renz* 1011 (K; MAL; SRGH). S: Blantyre Distr., Michiru Mt. For. Res., female fl. 8.vii.1983, *Seyani & Patel* 1105 (MAL).

Also in Tanzania (Western and Southern Highlands Provinces). Submontane grassland often by streams, and in high rainfall *Brachystegia* woodland; 1675–2375 m.

Simon & Williamson 1444 (Zambia, Danger Hill, c. 29 km north of Mpika, male fl. 18.xii.1967, K; LISC; MO; SRGH) is somewhat anomalous in that although the foliage and male floral characters indicate *whytei*, the branching habit conveys upon it a rather different gestalt.

Var. **monticoloides** Radcl.-Sm. in Kew Bull. **47**: 117 (1992). Type: Zambia, Mbala District, Kambole Rd., male fl. 1.i.1955, *Richards* 3825 (K, holotype; PRE, as '*Bock* 107').

Stems glabrous; petioles 1–2 mm long, or leaves subsessile; lamina rounded or shallowly cordate at the base; pedicels of male flowers up to 1 cm long.

Zambia. N: Mbala (Abercorn Township), fr. 18.iv.1961, *Richards* 15082 (K; SRGH).
Known only from this locality. High rainfall plateau grassland, amongst bracken, and in dambos; 1525–1590 m.
Superficially resembles *C. cordata* Bernh. ex Krauss from South Africa (KwaZulu-Natal), but in that species the male flowers are larger and have c. 50 glands within.
Intermediate specimens between the 2 varieties have subsequently been collected in Tanzania (*Bidgood, Mbago & Vollesen* 2537), in the Sumbawanga region.

2. **Clutia abyssinica** Jaub. & Spach, Illustr. Pl. Orient. **5**: 77, t. 468 (1855). —Müller Argoviensis in De Candolle, Prodr. **15**, 2: 1045 (1866). —Pax in Engler, Pflanzenw. Ost-Afrikas **C**: 241 (1895); Pflanzenr. [IV, fam. 147, iii] **47**: 56 (1911) as "*Cluytia*". —Hutchinson in F.T.A. **6**: 1: 807 (1912) as "*Cluytia*". —Engler, Pflanzenw. Afrikas (Veg. Erde 9) **3**, 2: 124 (1921) as "*Cluytia*". —Robyns & Tournay, Fl. Sperm. Parc Nat. Alb. **1**: 471 (1948). —Brenan, Check-list For. Trees Shrubs Tang. Terr.: 202 (1949). —Topham, Check List For. Trees Shrubs Nyasaland Prot.: 50 (1958). —Dale & Greenway, Kenya Trees & Shrubs: 188 (1961). —J. Léonard in F.C.B. **8**, 1: 99 (1962). —White, F.F.N.R.: 195 (1962). —Agnew, Upl. Kenya Wild Fls.: 219 (1974). —Drummond in Kirkia **10**: 252 (1975). —Troupin, Fl. Pl. Lign. Rwanda: 252, fig. 87/2 (1982); Fl. Rwanda **2**: 212, fig. 64/2 (1983). —K. Coates Palgrave, Trees Southern Africa, ed. 2, rev.: 431 (1983). —Radcliffe-Smith in F.T.E.A., Euphorb. 1: 333 (1987). —Beentje, Kenya Trees, Shrubs Lianas: 190 (1994). Type from Ethiopia.
Cluytia glabrescens Knauf ex Pax in Bot. Jahrb. Syst. **30**: 340 (1901). Types from Tanzania.
Cluytia abyssinica var. *calvescens* Pax & K. Hoffm. in Engler, Pflanzenr. [IV, fam. 147, iii] **47**: 57 (1911). —Brenan in Mem. N.Y. Bot. Gard. **9**, 1: 73 (1954). Types from Tanzania.
Cluytia anomala Pax & K. Hoffm. in Engler, Pflanzenr. [IV, fam. 147, vii, Addit. v] **63**: 405 (1914); Pflanzenw. Afrikas (Veg. Erde 9) **3**, 2: 124 (1921). —Brenan, Check-list For. Trees Shrubs Tang. Terr.: 202 (1949). Type from Tanzania.

An erect lax shrub, glabrous to evenly pubescent, dioecious; stems up to 6 m tall, brittle, flagelliform. Stipules absent. Petioles 1–3.5 cm long. Leaf blades 2–16 × 1–7 cm, ovate to elliptic-lanceolate, narrowed to the subacute or obtuse apex, attenuate, cuneate or rounded at the base, membranous, glabrous to sparingly or evenly pubescent on both surfaces, pale green on upper surface, somewhat glaucous beneath, often turning orange before falling; lateral nerves in 5–12(15) pairs, strongly brochidodromous. Male fascicles dense, usually many-flowered; bracts 0.5–1 mm long, broadly deltate, chaffy, stramineous. Male flowers: pedicels c. 5 mm long, jointed; sepals 2.5 × 1.25 mm, elliptic-obovate, rounded, each with 3–4 uncinate clavate glands at the base, pale green; petals 2 × 1.5 mm, deltate, unguiculate, each with a single globose gland at the base, white, green at the base; staminal column 1.25 mm high, free filaments 0.5 mm long, anthers 0.5 mm long, yellow; pistillode 1 mm high, cylindric, slightly broadened apically, yellowish-orange. Female fascicles 1–many-flowered; bracts as in male. Female flowers: pedicels 5 mm long, extending to 25 mm in fruit; sepals 2 × 1 mm, oblong-lanceolate, obtuse, each with a 2–3-lobed pale yellow-green gland at the base, pale green; petals 2 × 1 mm, spathulate, eglandular, white; ovary 1 mm in diameter, usually glabrous; styles 1 mm long, divaricate, pale yellow-green. Fruit 5 × 4.5 mm, pusticulate, glabrous or subglabrous, pale green, the pustules whitish; endocarp 0.5 mm thick. Seeds 3 × 2 × 1.75 mm, ovoid, minutely pitted; caruncle 1 × 1.25 mm, flattened, bifid, stramineous.

Stems and leaves glabrous or sparingly pubescent and soon glabrescent · · · · · · var. *abyssinica*
Stems and leaves evenly pubescent · var. *pedicellaris*

Var. **abyssinica.** Tab. **21.**

Stems and leaves glabrous or sparingly pubescent and soon glabrescent.

Zambia. N: Mbala, Itimbwe Gap (Itembwe Gorge), male fl. 7.i.1968, *Richards* 22866 (K). W: 74 km Mwinilunga–Kabompo, male fl. 25.i.1975, *Brummitt, Chisumpa & Polhill* 14124 (K; SRGH). C: Serenje, fr. 18.ii.1955, *Fanshawe* 2085 (K; NDO; SRGH). E: Lundazi, Nyika Plateau, Kangampande Mt., fr. 2.v.1952, *White* 2552 (FHO; K). **Zimbabwe**. E: Mutasa, Nyakombe R.

Tab. 21. CLUTIA ABYSSINICA var. ABYSSINICA. 1, portion of male flowering branch ($\times \frac{2}{3}$); 2, young male flower (× 6); 3, young male flower, from above, stamens removed ($\times 7\frac{1}{2}$), 1–3 from *Milne-Redhead & Taylor* 10209; 4, portion of female flowering and fruiting branch ($\times \frac{2}{3}$); 5, female flower (× 6); 6, female sepals, outside (× 6); 7, fruit (× 3); 8, seed (× 10), 4–8 from *Milne-Redhead & Taylor* 10210. Drawn by Pat Halliday. From F.T.E.A.

tributary, fr. 16.i.1980, *Pope & Müller* 1747 (MO; PRE; SRGH). **Malawi**. N: Chitipa Distr., 5 km NE of Nganda, fr. 23.viii.1972, *Synge WC* 291 (FHO; K; MAL; SRGH). C: Dedza Distr., Kangoli Hill, Chongoni Forest, male fl. 25.xi.1965, *Banda* 728 (K; SRGH). S: Mchese Mt., Ft. Lister Gap, female fl. & y. fr. 8.viii.1958, *Chapman* 624 (FHO; K; MAL; MO; SRGH). **Mozambique**. N: Ribáuè Mt., fr. 19.vii.1962, *Leach & Schelpe* 11404 (K; LISC; MO; PRE; SRGH). Z: Serra do Gurué, o. fr. 17.ix.1949, *Barbosa & Carvalho* in *Barbosa* 4136 (K; LMA; PRE). T: Zóbuè, male fl. 3.x.1942, *Mendonça* 571 (LISC). MS: c. 28 km west of Dombe, female fl. 25.iv.1974, *Pope & Müller* 1304 (K; LISC; PRE; SRGH).

From Sudan and Somalia south to Angola and South Africa (KwaZulu-Natal). Constituent of submontane evergreen forest margins and clearings and plateau swamp forest (mushitu) margins, also as an understorey tree of high rainfall plateau and submontane *Brachystegia* woodland, of gully and ravine forests and riverine forests; 300–2300 m.

Vernacular name as recorded in specimen data: "chiuta" (chiYao).

Var. **pedicellaris** (Pax) Pax in Engler, Pflanzenr. [IV, fam. 147, iii] **47**: 57 (1911). —J. Léonard in F.C.B. **8**, 1: 101 (1962). —Troupin, Fl. Pl. Lign. Rwanda: 254 (1982); Fl. Rwanda **2**: 212 (1983). —Radcliffe-Smith in F.T.E.A., Euphorb. 1: 336 (1987). Type from Tanzania (Northern Province).

Cluytia richardiana var. *pedicellaris* Pax in Bot. Jahrb. Syst. **23**: 531 (1897).

Cluytia pedicellaris (Pax) Hutch. in F.T.A. **6**, 1: 806 (1912). —Brenan, Check-list For. Trees Shrubs Tang. Terr.: 203 (1949). —Dale & Greenway, Kenya Trees & Shrubs: 188 (1961).

Stems and leaves evenly pubescent.

Zambia. E: Nyika Plateau, o. fr. 22.viii.1977, *Pawek* 12913 (K; MO). **Zimbabwe**. E: Mutare (Umtali), Hoonga R., 3 km west of Bomponi Hill, male fl. 19.iii.1982, *Pope & Müller* 2063 (MO; PRE; SRGH). **Malawi**. N: Rumphi Distr., Nyika Plateau, male fl. 14.ii.1968, *Simon, Williamson & Ball* 1805 (K; LISC; MAL; PRE; SRGH). C: Dedza Distr., Dedza–Mphunzi road, male fl. 18.iii.1968, *Jeke* 156 (K; LISC; MAL; SRGH). S: Zomba Plateau, male fl. 5.vi.1946, *Brass* 16288 (BM; K; MO; NY; PRE; SRGH). **Mozambique**. Z: Serra do Gurué, o. fr. 30.ix.1941, *Torre* 3530 (LISC). T: Zóbuè, fr. 3.x.1942, *Mendonça* 564 (LISC).

Also in Zaire (Kivu Province), Rwanda, Burundi, Ethiopia, Uganda, Kenya and Tanzania. On steep rocky slopes and ridgetops, in grassland and savanna, in *Brachystegia, Uapaca* woodland, on margin of evergreen forest, in rainforest regrowth and in montane forest; 720–2300 m.

Mendonça 564 (q.v.) and 2212, from Gurué, seem to show signs of introgression with *C. swynnertonii* S. Moore.

Rogasian L.A. Mahunnah & K. Mtotomwema (J. Econ. Tax. Bot. **7**, 1: 163–165 (1985)) reinstated *C. pedicellaris* at the species level on the basis of alleged pollen differences but in my view totally without foundation since it barely qualifies as a variety distinct from typical *C. abyssincia*.

3. **Clutia pulchella** L., Sp. Pl.: 1042 (1753). —Müller Argoviensis in De Candolle, Prodr. **15**, 2: 1045 (1866). —Pax in Engler, Pflanzenr. [IV, fam. 147, iii] **47**: 54 (1911) as "*Cluytia*". —Prain in Bull. Misc. Inform., Kew **1913**, 10: 404 (1913); in F.C. **5**, 2: 444 (1920). —Engler, Pflanzenw. Afrikas (Veg. Erde 9) **3**, 2: 124 (1921). —Burtt Davy, Fl. Pl. Ferns Transvaal **2**: 304 (1932). —Drummond in Kirkia **10**: 252 (1975). —K. Coates Palgrave, Trees Southern Africa, ed. 2, rev.: 431 (1983). Type from South Africa (Cape Province).

Var. **obtusata** Sond. in Linnaea **23**: 129 (1850). —Müller Argoviensis in De Candolle, Prodr. **15**, 2: 1046 (1866). —Prain in F.C. **5**, 2: 445 (1920). —Hutchinson, Botanist in Southern Africa: 667 (1946). Types from South Africa.

Very similar to *C. abyssinica* Jaub. & Spach, but differing chiefly in its leaves which are mostly less than twice as long as broad, broadly ovate to suborbicular, and rounded at the apex.

Botswana. SE: Ootse Mt., st. 1.i.1979, *Woollard* 489 (SRGH). **Zimbabwe**. C: Selukwe, Ferny Creek, male fl. 8.xii.1953, *Wild* 4294 (K; LISC; MO; SRGH). S: Mberengwa N Mt., y. fr. 13.iii.1982, *Burrows* 1849 (SRGH). **Mozambique**. M: Goba, near R. Maiuana, male fl. 3.xi.1960, *Balsinhas* 186 (K; LISC; LMA; PRE; SRGH).

Also in Lesotho and South Africa (Cape Province, KwaZulu-Natal, Free State, Transvaal). In hot dry localities, on rocky hillsides, along seasonal watercourses, also in riverine forest and mixed deciduous woodland; 500–1400 m.

In the var. *obtusata* Sond. the twigs are not warty as in the var. *pulchella*.

Clutia pulchella is a southern vicariad of *C. abyssinica*.

4. **Clutia galpinii** Pax in Bull. Herb. Boissier **6**: 736 (1898) as "*Cluytia*", pro parte excl. *Galpin* 961. —Prain in Bull. Misc. Inform., Kew **1913**, 10: 403 (1913); in F.C. **5**, 2: 443 (1920). —Burtt Davy, Fl. Pl. Ferns Transvaal **2**: 304 (1932). —Hutchinson, Botanist in Southern Africa: 667 (1946). Type from South Africa (Transvaal).

Clutia pulchella f. *genuina* (Müll. Arg.) Pax in Engler, Pflanzenr. [IV, fam. 147, iii] **47**: 54 (1911), pro parte, nequaquam *C. pulchella* L.

Similar to *C. pulchella* but plants are smaller (c. 15 cm high) and less branched, with petioles not more than 3 mm long, and leaf blades more markedly glaucous, turning pink; and the fruiting pedicels not more than 4 mm long.

Botswana. SE: Ootse Hill, 48 km SE of Gaborone, male fl. & fr. 6.iv.1974, *Mott* 220b (SRGH; UBLS).

Also in South Africa (Transvaal); 1280–1400 m.

5. **Clutia punctata** Wild in Kirkia **4**: 142 (1964). Type: Zimbabwe, Chimanimani, fr. 20.viii.1954, *Wild* 4582 (SRGH, holotype; K; LISC; PRE).

Similar to *C. abyssinica* Jaub. & Spach, but the plant completely glabrous, the stems tuberculate and the leaf blades not more than 4.5 × 3 cm and rounded to truncate at the base with reticulate venation.

Zimbabwe. E: Chimanimani, Uncontoured Peak, male fl. 23.ii.1957, *Goodier* 167 (K; LISC; SRGH).

Endemic to the Chimanimani Mountains. Montane grassland among quartzite crags on rocky summits and steep slopes; 2135–2285 m.

6. **Clutia brassii** Brenan in Mem. N.Y. Bot. Gard. **9**, 1: 71 (1954). Type: Malawi, Mulanje Mt., west slope, male fl. 18.vii.1946, *Brass* 16866 (K, holotype; MO; NY).

A glabrous shrub up to 2 m high, dioecious. Stems sparingly branched, closely beset with longitudinal usually centrally punctate tubercles, greenish. Stipules absent. Petioles up to 1 mm long, or leaves subsessile. Leaves numerous and closely spaced, often imbricate, evergreen. Leaf blades 0.5–3 × 0.5–2 cm, suborbicular to broadly ovate, obtuse or rounded at the apex, cordate, with entire cartilaginous slightly revolute margins, firmly chartaceous to subcoriaceous, pellucid-punctate; midrib prominent above and beneath, lateral nerves in 4–8 pairs, not prominent. Male fascicles few-flowered; bracts minute, numerous, imbricate, scaly, pallid. Male flowers: pedicels c. 2 mm long; sepals c. 2 × 1 mm, elliptic-obovate, rounded at the apex, convex, greenish, each with 3 unequal clavate glands near the base; petals c. 2 × 1.5 mm, rhombic, cream-coloured, each with a solitary rhomboid gland at the base; staminal column 1.5 mm high, free filaments 0.75 mm long, anthers 0.5 mm long, white; pistillode 0.75 mm high, cylindric, peltate-capitate above. Female flowers not known. Fruits solitary on pedicels 4–5 mm long; fruiting sepals 2.5 × 1.5 mm, each with a flattened obovate or bifid gland at the base, otherwise as in the male; petals 2 × 1 mm, spathulate, eglandular. Fruits 5 × 5 mm, rugulose, olivaceous. Seeds c. 3 × 2.5 × 2 mm, minutely puncate; caruncle 1x 1.5 mm, yellowish-brown.

Malawi. S: Mulanje Mt., west slope, fr. 18.vii.1946, *Brass* 16867 (BM; K; MO; NY; PRE; SRGH).

Endemic to Mt. Mulanje. In submontane shrubby grassland amongst rocks; 1830–2120 m.

7. **Clutia conferta** Hutch. in F.T.A. **6**, 1: 805 (1912) as "*Cluytia*". —Engler, Pflanzenw. Afrikas (Veg. Erde 9) **3**, 2: 126 (1921). Type: Malawi, Tuchila Plateau, male fl. & fr. ix.1901, *Purves* 100 (K, holotype).

A shrub or woody herb up to 2 m high, sparingly branched, monoecious. Branches usually simple, longitudinally ridged, sparingly pubescent in the furrows, otherwise glabrous; leaf scars prominent on stem and branches. Stipules absent. Petioles 0.5–1 mm long, or leaves subsessile. Leaves crowded towards the tips of the branches, imbricate; blades 0.8–2.2 × 0.4–1 cm, oblanceolate-oblong, rounded to emarginate at the apex, rounded to very shallowly cordate at the base, chartaceous to subcoriaceous, pellucid-punctate and pusticulate; margin entire and cartilaginous; lateral nerves in c. 3 pairs, ascending, indistinct. Male fascicles 2–3-flowered; bracts minute, chaffy, chestnut-brown. Male flowers: pedicels c. 1.5–2 mm long; sepals 2 × 1 mm, obovate-oblanceolate, rounded, greenish-yellow, each with a 2–3-lobed gland towards the base and 1 or 2 ± spherical glands at the base; petals 1.5 × 1 mm, broadly spathulate, cream-coloured, with a pair of spherical glands at the base; staminal column 1 mm high, free filaments 0.5 mm long, anthers 0.5 mm long, white; pistillode 0.5 mm high, cylindric, shallowly 3-lobed at the apex. Female flowers not known. Fruits solitary on pedicels 2–5 mm long in axils below the male flowers; fruiting sepals 2.5 × 1 mm, otherwise as in the male; petals 2 × 1 mm, oblanceolate, eglandular. Fruits 4 × 4.5 mm, pusticulate-verruculose, brownish. Seeds 3.5 × 2 × 1.5 mm, ellipsoid, smooth; caruncle 0.5 × 1 mm, yellowish-brown.

Malawi. S: Mulanje Mt., Sapitwa, male fl. & fr. 13.ii.1979, *Blackmore, Brummitt & Banda* 413 (BM; K; MAL).

Endemic to Mt. Mulanje. On submontane rocky slopes in low tussock grassland with low shrubs, and between boulders with *Philippia* scrub; 2133–2590 m.

8. **Clutia robusta** Pax in Engler, Pflanzenw. Ost-Afrikas **C**: 241 (1895); Pflanzenr. [IV, fam. 147, iii] **47**: 60 (1911). —Hutchinson in F.T.A. **6**, 1: 811 (1912) as "*Cluytia*". —Engler, Pflanzenw. Afrikas (Veg. Erde 9) **3**, 2: 124 (1921). —Brenan, Check-list For. Trees Shrubs Tang. Terr.: 203 (1949). —Dale & Greenway, Kenya Trees & Shrubs: 189 (1961). —Agnew, Upl. Kenya Wild Fls: 219 (1974). —Radcliffe-Smith in F.T.E.A., Euphorb. 1: 337 (1987). —Beentje, Kenya Trees, Shrubs Lianas: 190 (1994). Types from Tanzania (Northern Province).

Cluytia brachyadenia Volkens ex Pax in Engler, Pflanzenr. [IV, fam. 147, iii] **47**: 61 (1911); Pflanzenw. Afrikas (Veg. Erde 9) **3**, 2: 124 (1921) as "*Cluytia*". —Brenan, Check-list For. Trees Shrubs Tang. Terr.: 202 (1949). Types from Tanzania (Northern Province).

Cluytia stenophylla Pax & K. Hoffm. in Engler, Pflanzenr. [IV, fam. 147, iii] **47**: 63 (1911) as "*Cluytia*". —Brenan, Check-list For. Trees Shrubs Tang. Terr.: 203 (1949), pro parte, quoad spec. Keniensia.

Similar to *C. paxii*, but the plants somewhat taller (up to 5 m tall) and more or less glabrous; the petioles longer (up to 1.5 cm long); the leaves larger (up to 11 × 3.5 cm) and oblanceolate; and usually only 2 glands at the base of each petal in the male flowers.

Zimbabwe. E: Nyanga (Inyanga), female fl. & fr. 4.ix.1954, *Wild* 4606 (K; LISC; MO; PRE; SRGH).

Also in N Somalia, Ethiopia, Uganda, Kenya and Tanzania. Submontane evergreen forest margins, kloof forests and grassland; 1575–2300 m.

The Zimbabwean material is generally more glabrous and its leaves are broader in relation to their length than is the case in the East African material, but is otherwise indistinguishable from it. Although *C. robusta* is reported from SW Tanzania in east Africa, I have only seen material from E Uganda, S Kenya and N Tanzania, where it is particularly associated with the large volcanic mountains. I have not seen material from Zambia, Malawi or Mozambique. The disjunct distribution might best be marked by describing the Zimbabwean plant as a distinct subspecies, but I prefer not to introduce another name into such a plastic group.

Dame Alice and the Misses Godman 252 (BM) from Old Zimbabwe, 29.viii.1929, may belong here, but the male petals are entirely devoid of basal glands.

9. **Clutia paxii** Knauf ex Pax in Bot. Jahrb. Syst. **30**: 341 (1901) as "*Cluytia*". —Pax in Engler, Pflanzenr. [IV, fam. 147, iii] **47**: 60 (1911) as "*Cluytia*". —Hutchinson in F.T.A. **6**, 1: 809 (1912) as "*Cluytia*". —Eyles in Trans. Roy. Soc. South Africa **5**: 398 (1916). —Engler, Pflanzenw. Afrikas (Veg. Erde 9) **3**, 2: 124 (1921). —Brenan, Check-list For. Trees Shrubs Tang. Terr.: 203 (1949); in Mem. N.Y. Bot. Gard. **9**, 1: 73 (1954). —J. Léonard in F.C.B. **8**, 1: 98 (1962). —White, F.F.N.R.: 195 (1962). —Drummond in Kirkia **10**: 252 (1975). —Troupin, Fl. Pl. Lign. Rwanda: 254, fig. 87/3 (1982); Fl. Rwanda **2**: 212, fig. 64/3 (1983). —Radcliffe-Smith in F.T.E.A., Euphorb. 1: 338 (1987). Type from Tanzania (Southern Highlands Province).

Cluytia stenophylla Pax & K. Hoffm. in Engler, Pflanzenr. [IV, fam. 147, iii] **47**: 63 (1911). —Brenan, Check-list For. Trees Shrubs Tang. Terr.: 203 (1949), pro parte, quoad spec. Malawiensem.

Clutia phyllanthoides S. Moore in J. Linn. Soc., Bot. **40**: 198 (1911). Type: Zimbabwe, Chimanimani, male fl. 23.ix.1906, *Swynnerton* 1722 (BM, holotype; K; SRGH).

Cluytia gracilis Hutch. in F.T.A. **6**, 1: 809 (1912). Type: Malawi, Kondowe to Karonga, male fl. vii.1896, *Whyte* 362 (K, holotype).

Clutia robusta sensu Eyles in Trans. Roy. Soc. South Africa **5**: 398 (1916), non Pax.

A virgate much-branched subshrub or shrub up to 2.5 m tall, sparingly pubescent, dioecious. Stipules absent. Petioles 0.5–3(5) mm long. Leaf blades 0.5–3(6) × 0.3–1.5(2.5) cm, obovate to elliptic-lanceolate, rounded or obtuse, sometimes subacute at the apex, rounded or cuneate at the base, chartaceous, sparingly pubescent to subglabrous on both surfaces, bright green above, greyish-green beneath, later turning bright red; lateral nerves in 3–8 pairs, often indistinct. Male fascicles many-flowered; bracts 0.5 mm long, deltate, chaffy, chestnut-brown. Male flowers: pedicels c. 1 mm long; sepals 2–2.5 × 1.3 mm, oblanceolate, rounded, each with 3 clavate glands below the middle and 2–3 spherical glands at the base, greenish-white; petals 2 × 1 mm, spathulate, each with 2–3 spherical glands at the base, cream-coloured, later becoming finely brownish-mottled; staminal column 1.5 mm high, free filaments 0.5 mm long, anthers 0.5 mm long, pale primrose-yellow; pistillode 0.5 mm high, cylindric, papillose, dark brown. Female flowers solitary or in pairs; pedicels 1–2.5(5) mm long, not extending in fruit; sepals 2 × 1 mm, oblong, obtuse, each with 1–3 flattened rounded or truncate glands at the base, green; petals 2 × 1 mm, oblanceolate, eglandular, greenish-yellow; ovary 1.5 mm in diameter, glabrous; styles 1 mm long, divaricate. Fruit 4 × 4 mm, glabrous, green at first, later dull reddish-brown. Seeds 3 × 2 × 1.5 mm, compressed-ellipsoid, smooth; caruncle 0.5 × 1 mm, pale yellow.

Zambia. E: Nyika Plateau, male fl. & fr. 22.x.1958, *Robson & Angus* 258 (BM; K; LISC; SRGH). **Zimbabwe**. E: Chimanimani, male fl. 18.viii.1950, *Crook* M84a (K; MO; PRE; SRGH). **Malawi**. N: Nyika Plateau, male fl. 11.viii.1946, *Brass* 17168 (K; MO; NY; PRE; SRGH). **Mozambique**. MS: Tsetserra, male fl. 6.vi.1971, *Biegel* 3569 (K; LISC; MO; PRE; SRGH).

Also in Zaire (Kivu Province), Rwanda, Burundi and Tanzania. Submontane evergreen forest margins, kloof and gully forest and submontane grassland with ericoid scrub; 1525–2400 m.

Radcliffe-Smith, Pope & Goyder 5910, 20.v.1989 on the Zambian side of the Nyika Plateau, has leaves up to 7 cm long and only 2 glands at the base of the male petals. It therefore seems to be intermediate between *C. paxii* and *C. robusta* as here understood.

10. **Clutia hirsuta** (Sond.) Müll. Arg. in De Candolle, Prodr. **15**, 2: 1046 (1866). —Pax in Engler, Pflanzenr. [IV, fam. 147, iii] **47**: 73 (1911) as "*Cluytia*". —Prain in F.C. **5**, 2: 449 (1920). —Engler, Pflanzenw. Afrikas (Veg. Erde 9) **3**, 2: 128 (1921). —Burtt Davy, Fl. Pl. Ferns Transvaal **2**: 304 (1932). —Drummond in Kirkia **10**: 252 (1975). Type from South Africa (Cape Province).

Cluytia hirsuta E. Mey. in Drège, Zwei Pflanzengeogr. Dokum.: 174 (1843), nomen tantum.

Cluytia heterophylla var. *hirsuta* Sond. in Linnaea **23**: 129 (1850).

Cluytia inyangensis Hutch. in F.T.A. **6**, 1: 804 (1912). —Eyles in Trans. Roy. Soc. South Africa **5**: 397 (1916). —Engler, Pflanzenw. Afrikas (Veg. Erde 9) **3**, 2: 125 (1921). Type: Zimbabwe, Manica, Nyanga (Inyanga) Mts., male fl. xii.1899, *Cecil* 181 (K, holotype).

Cluytia volubilis Hutch. in F.T.A. **6**, 1: 809 (1912). —Eyles in Trans. Roy. Soc. South Africa **5**: 398 (1916). —Engler, Pflanzenw. Afrikas (Veg. Erde 9) **3**, 2: 124 (1921). Type: Zimbabwe, Chimanimani, male fl. & fr. 28.ii.1907, *Johnson* 188 (K, holotype).

A perennial herb 30–90(120) cm high, sparingly to densely pubescent and/or hirsute, dioecious; stems several, simple or sparingly branched, arising from a woody rootstock. Stipules absent, or rarely developed and then 1–2 mm long, oblong-lanceolate, subglabrous. Petioles 1–2 mm long, or leaves sessile. Leaf blades 0.5–4(5) × 0.2–2 cm, elliptic-lanceolate or elliptic-oblanceolate, acute, subacute or obtuse at the apex, attenuate, cuneate or rounded-cuneate at the base, entire, chartaceous, densely villous on both surfaces at first, later becoming sparingly to evenly pubescent or hirsute; lateral nerves in 3–5(6) pairs, looped. Male fascicles 1–several-flowered; bracts 1 mm long, triangular-lanceolate, chaffy, chestnut-brown. Male flowers: pedicels c. 1 mm long; sepals 2.5 × 1 mm, oblanceolate-oblong, rounded, each with a tripartite scale below the middle, and 1–2 subspherical glands at the base, yellowish-green; petals 2 × 1 mm, oblanceolate, rounded, each with 3–5 subspherical glands at and near the base, yellowish, often finely brownish-mottled; staminal column 1.5 mm high, free filaments 0.5 mm long, anthers 0.67 mm long, yellow; pistillode 0.5 mm high, apically expanded. Female flowers solitary; pedicels 2 mm long, not or only slightly extending in fruit, stouter than in the male; sepals c. 3.5–4 × 1 mm, elliptic-lanceolate, subacute, each with 2–3 clavate glands at the base; petals 3 × 1 mm, oblanceolate-spathulate, subeglandular at the base; ovary c. 2 mm in diameter, densely fulvous-sericeous; styles 1.25 mm long, suberect-divaricate. Fruit 5 × 5 mm, sparingly to evenly pubescent, pale green. Seeds 3.5 × 2.67 × 2 mm, compressed-ellipsoid, smooth; caruncle 0.5 × 1 mm, creamy-white.

Zimbabwe. N: Mazowe Distr., Umvukwes, Ruorka Ranch, male fl. 17.xii.1952, *Wild* 3495 (K; LISC; MO; PRE; SRGH). C: Wedza Mt., male fl. 27.ii.1964, *Wild* 633 (K; LISC; PRE; SRGH). E: Mutare Distr., Watsomba, Kukwanisa Farm, male fl. 9.i.1967, *Biegel* 1675 (K; LISC; MO; SRGH). **Mozambique**. MS: Báruè, Serra de Choa, 9 km – Catandica, fr. 24.v.1971, *Torre & Correia* 18607 (COI; LISC; LMA).

Also in Lesotho and South Africa (Cape Province, KwaZulu-Natal, Free State, Transvaal). Submontane grassland, also in kloof forest; 975–2400 m.

This is a very variable species in many respects. The indumentum can be very dense, as in the type of *C. inyangensis* Hutch., or quite sparse, as in *Fries, Norlindh & Weimarck* 2735, Nyanga, near the Pungwe R. 6.xi.1930 (K; SRGH; UPS). The colour on drying can vary from yellow-green, as in the latter, which thereby superficially resembles *C. monticola* var. *stelleroides* vegetatively, to a dull brownish-green or greyish-brown (many collections). Furthermore, branching specimens can begin to resemble small-leaved herbaceous versions of *C. swynnertonii.*

11. **Clutia swynnertonii** S. Moore in J. Linn. Soc., Bot. **40**: 197 (1911). —Hutchinson in F.T.A. **6**, 1: 811 (1912). —Eyles in Trans. Roy. Soc. South Africa **5**: 398 (1916). —Engler, Pflanzenw. Afrikas (Veg. Erde 9) **3**, 2: 128 (1921). —Drummond in Kirkia **10**: 253 (1975). Syntypes: Zimbabwe, Chipete, male fl. 29.x.1905 (BM); v.1906 (K), *Swynnerton* 197 (BM; K); near Chirinda, male fl. 27.v.1906, *Swynnerton* 530c (BM; K; SRGH); 14.vi.1906, *Swynnerton* 530a (BM; K; SRGH); fr. 19.vi.1906, *Swynnerton* 530b (BM).

Very like *C. hirsuta,* but a much-branched shrub up to 4 m tall without a woody rootstock, with the leaf blades up to 11 × 4(5) cm and with the female flowers 1–3 per axil on pedicels up to 6 mm long in fruit.

Zimbabwe. E: Stapleford For. Res., male fl. 11.vi.1934, *Gilliland* 263 (BM; FHO; K; PRE; SRGH). S: Masvingo Distr., Great Zimbabwe, male fl. 8.vii.1972, *Chiparawasha* 498 (K; LISC; PRE; SRGH). **Malawi**. S: Mulanje Mt., Malosa Valley, male fl. 16.ix.1983, *Johnston-Stewart* 99 (FHO). **Mozambique**. MS: Gorongosa Mt., male fl. 23.vii.1970, *Müller & Gordon* 1425 (K; LISC; PRE; SRGH).

Not known elsewhere. Submontane grassland and evergreen forest margins, often with *Philippia* scrub, also in *Uapaca kirkiana* woodland and open mixed savanna woodland, and riverine vegetation; 370–2130 m.

Vernacular name as recorded in specimen data: "n'Yagatsari" (eastern Zimbabwe).

Torre & Correia 13429 & 15415 (both LISC), from Serra de Choa, Báruè, Mozambique, have basal shoots reminiscent of *C. hirsuta.* Field study may reveal that *C. swynnertonii* and *C. hirsuta* are two states of the same taxon.

12. **Clutia angustifolia** Knauf ex Pax in Bot. Jahrb. Syst. **30**: 340 (1901) as "*Cluytia*". —Pax in Engler, Pflanzenr. [IV, fam. 147, iii] **47**: 64 (1911). —Hutchinson in F.T.A. **6**, 1: 810 (1912) as "*Cluytia*". —Engler, Pflanzenw. Afrikas (Veg. Erde 9) **3**, 2: 125 (1921) as "*Cluytia*". —Brenan, Check-list For. Trees Shrubs Tang. Terr.: 202 (1949). —J. Léonard in F.C.B. **8**, 1: 96 (1962). —Radcliffe-Smith in F.T.E.A., Euphorb. 1: 339 (1987). Type from Tanzania (Southern Highlands Province).

Cluytia lasiococca Pax & K. Hoffm. in Engler, Pflanzenr. [IV, fam. 147, vii, Addit. v] **63**: 405 (1914); Pflanzenw. Afrikas (Veg. Erde 9) **3**, 2: 124 (1921). —Brenan, Check-list For. Trees Shrubs Tang. Terr.: 202 (1949). Type from Tanzania (Southern Highlands Province).

Clutia sp. aff. *C. swynnertonii* S. Moore; Brenan in Mem. N.Y. Bot. Gard. **9**, 1: 73 (1954).

Very like *C. swynnertonii* but differing in being a monoecious (occasionally dioecious) perennial herb or subshrub not usually exceeding 1.5 m in height, and in having narrowly lanceolate leaves up to 10 × 2.5 cm.

Zambia. N: Kasulo, male, female fls. & fr. 2.iii.1955, *Richards* 4744 (K). **Malawi**. N: Mzimba Distr., Mzuzu, Marymount, male, female fls. & fr. 11.vii.1976, *Pawek* 11488 (K; MAL; MO; PRE; SRGH; UC). C: Ntchisi Mt., male fl. & fr. 31.vii.1946, *Brass* 17058 (K; MO; NY; PRE; SRGH). S: Kirk Range, Goche, male & female fls. 30.i.1959, *Robson* 1353 (BM; K; LISC; MAL; PRE; SRGH). **Mozambique**. T: Angónia, Dómuè, male fl. & fr. 9.iii.1964, *Torre & Paiva* 11110 (LISC).

Also in Zaire (Shaba Province), Burundi and SW Tanzania. Submontane *Brachystegia* woodland, also on evergreen forest margins and clearings, in submontane grassland, ravine forests and riverine vegetation; 610–2200 m.

13. **Clutia monticola** S. Moore in J. Linn. Soc., Bot. **40**: 197 (1911). —Hutchinson in F.T.A. **6**, 1: 803 (1912) as "*Cluytia*". —Eyles in Trans. Roy. Soc. South Africa **5**: 398 (1916). —Prain in F.C. **5**, 2: 451 (1920). —Engler, Pflanzenw. Afrikas (Veg. Erde 9) **3**, 2: 125 (1921). —Burtt Davy, Fl. Pl. Ferns Transvaal **2**: 304 (1932). Syntypes: Zimbabwe, Chimanimani Distr., Mount Pene, male fl. 28.ix.1906, *Swynnerton* 2012 (BM; K); 14.x.1908, *Swynnerton* 6159 (BM; K).

A glabrous perennial herb, dioecious. Stems several simple or subsimple, up to c. 60 cm high, arising from a woody stock. Stipules absent. Petioles 0.5–3 mm long, or leaves subsessile. Leaf blades 0.7–3 × 0.5–2 cm, suborbicular-ovate in lower leaves, elliptic-ovate in upper leaves, or all elliptic-oblong, rounded, obtuse or subacute at the apex, broadly cuneate or rounded at the base, membranous to firmly chartaceous, pellucid-punctate and pusticulate; margins entire, slightly revolute; lateral nerves in 3–5 pairs, looped. Male fascicles 1–5-flowered; bracts 0.5–1 mm, triangular, chaffy. Male flowers: pedicels up to 1.7 cm long, capillary; sepals 2.5 × 1.5 mm, elliptic-obovate, rounded, each with c. 5 clavate glands near the base, pale yellowish-green; petals 1.5 × 1 mm, triangular-rhombic to obovate-spathulate, each with c. 5 ± spherical glands at the base, whitish; staminal column c. 1 mm high, free filaments 0.67 mm long, anthers 0.67 mm long, white; pistillode c. 1 mm high, cylindric, slightly expanded apically. Female flowers solitary; pedicels 1–1.3 cm long, extending to c. 2.5 cm in fruit; sepals c. 2 × 1 mm, elliptic-oblong, each with a small ± flattened 2-lobed basal gland, otherwise as in the male; petals 2 × 1 mm, rhombic-elliptic, eglandular, otherwise as in the male; ovary c. 1.5 mm in diameter, smooth, glabrous; styles c. 1 mm long, spreading. Fruit 4–4.5 × 4.5–5 mm, punctate, glabrous, ochreous. Seeds 2.5 × 2 × 1.5 mm, obovoid, minutely punctate; caruncle c. 1 × 1 mm, thin, papyraceous, straw-coloured.

Leaf blades, at least the lower ones, suborbicular-ovate; petals of male flowers obovate-spathulate · var. *monticola*
Leaf blades all elliptic-ovate to elliptic-oblong; petals of male flowers triangular-rhombic · var. *stelleroides*

Var. **monticola**

Leaf blades, at least the lower ones, suborbicular-ovate; petals of male flowers obovate-spathulate.

Zimbabwe. E: Nyanga (Inyanga), above the R. Pungwe, male fl. & fr. 17.xii.1930, *Fries, Norlindh & Weimarck* 3860a (K; PRE; SRGH; UPS). **Mozambique**. MS: Rotanda R., Serra de Messambuzi, male fl. 26.xi.1965, *Torre & Correia* 13316 (LISC).

Also in South Africa (KwaZulu-Natal, Free State, Transvaal) and Swaziland. Submontane grassland, often as a pyrophytic suffrutex; 1375–2135 m.

Var. **stelleroides** (S. Moore) Radcl.-Sm. in Kew Bull. **47**: 114 (1992). Tab. **22**. Type: Zimbabwe, northern Chimanimani Distr., male fl. 8.x.1908, *Swynnerton* 6214 (BM, holotype).

Clutia stelleroides S. Moore in J. Linn. Soc., Bot. **40**: 198 (1911). —Hutchinson in F.T.A. **6**, 1: 804 (1912). —Eyles in Trans. Roy. Soc. South Africa **5**: 398 (1916). —Engler, Pflanzenw. Afrikas (Veg. Erde 9) **3**, 2: 125 (1921).

Leaf blades all elliptic-ovate or elliptic-oblong; petals of male flowers triangular-rhombic.

Zimbabwe. E: Mutare Distr., Engwa, fr. 3.ii.1955, *Exell, Mendonça & Wild* 155 (BM; LISC; SRGH).

Known only from this locality. In submontane grassland; 1525–1830 m.

14. **Clutia sessilifolia** Radcl.-Sm. in Kew Bull. **47**: 115 (1992). Type: Zimbabwe, Chimanimani Mts., "Uncontoured Peak", male fl. 23.ii.1957, *Goodier* 180 (K, holotype; LISC; SRGH).

A sparse, ± glabrous shrublet up to 1 m high, dioecious. Stems smooth. Stipules absent. Leaves sessile; leaf blades 1–4.5 × 0.7–2.5 cm, those of the lateral shoots much smaller than those of the main stem, elliptic-oblong, obtuse or rounded at the

Tab. 22. CLUTIA MONTICOLA var. STELLEROIDES. 1, portion of branch, male plant (× ⅔); 2, male flower (× 8); 3, male flower, showing glands and staminal column (× 8), 1–3 from *Swynnerton* 6214; 4, portion of branch, female plant (× ⅔); 5, female flower (× 8); 6, fruit (× 6), 4–6 from *Exell, Mendonça & Wild* 155. Drawn by Christine Grey-Wilson. From Kew Bull.

apex, cordate, slightly revolute on the margin, chartaceous, pellucid-punctate, pale yellow, drying brownish; lateral nerves in c. 6 pairs, not prominent. Male fascicles several-flowered; bracts imbricate, 0.5 × 0.75 mm, broadly triangular-ovate, acuminate, sparingly pubescent. Male flowers: pedicels 1.5 mm long, jointed; sepals 2 × 1.3 mm, elliptic-ovate, rounded, each with 3 pale clavate glands below the middle and c. 4 dark spherical glands at the base, yellowish-green; petals 1.5 × 1 mm, obtrullate-spathulate, obtuse, each with c. 5 dark spherical glands at the base; staminal column 1 mm high, free filaments 0.5 mm long, anthers c. 0.5 mm long; pistillode 0.75 mm high, cylindric, slightly expanded at the apex. Female flowers, fruit and seeds unknown.

Zimbabwe. E: Chimanimani, summit of Binga Mt., male fl. 25.iv.1964, *Whellan* 2102 (SRGH).
Endemic to the Chimanimani Mts. Occasional on rocky mountain summits and in boulder scree; 1980–2440 m.

19. CHAETOCARPUS Thwaites

Chaetocarpus Thwaites in Hooker's J. Bot. Kew Gard. Misc. **6**: 300, t. 10A (1854), nom. conserv.

Dioecious trees or shrubs, with a hirsute and sericeous indumentum. Leaves alternate, shortly petiolate, stipulate, simple, entire, penninerved, eglandular. Flowers in axillary fascicles, the male fascicles many-flowered, the female fascicles few-flowered; bracts minute. Male flowers small; pedicels short, jointed; sepals 4–6, free, imbricate; petals absent; disk glands extrastaminal, free; stamens 5–16, borne on a column, anthers extrorse, dorsifixed, connective broad, the cells longitudinally dehiscent; pistillode trifid, hairy. Female flowers small; pedicels short; sepals ± as in the male flowers; petals absent; disk hypogynous, cupular, crenellate; ovary 3-celled, with 1 ovule per cell; styles 3, ± free, bifid or bipartite, stigmas fimbriate or papillose. Fruit ellipsoid-subglobose, dehiscing into bivalved cocci; exocarp tuberculate, setiferous, or both; endocarp woody; columella persistent. Seeds compressed-ovoid; testa crustaceous; caruncle bipartite, fleshy; albumen fleshy; cotyledons broad, flat.

A pantropical genus of 10–12 species, only one of which is African.

Chaetocarpus africanus Pax in Bot. Jahrb. Syst. **19**: 113 (1894); in Engler, Pflanzenr. [IV, fam. 147, iv] **52**: 10 (1912). —Prain in F.T.A. **6**, 1: 947 (1912). —Engler, Pflanzenw. Afrikas (Veg. Erde 9) **3**, 2: 133 (1921). —De Wildeman, Pl. Bequaert. **3**, 4: 506 (1926). —J. Léonard in F.C.B. **8**, 1: 127 (1962). Tab. **23**. Types from Angola (Lunda) and Zaire (Kasai).

A straggling or scandent shrub, or weak-stemmed small tree up to c. 4.5 m high. Twigs hirsute. Stipules 2–7 mm long, elliptic-lanceolate, acute, sericeous. Petioles 3–5 mm long, sericeous-pubescent. Leaf blades 2–15 × 1.5–4 cm, elliptic to elliptic-lanceolate, acutely acuminate and mucronulate at the apex, rounded to cuneate at the base, thinly coriaceous, sparingly sericeous on both surfaces, later glabrescent above, dark green and shiny above, paler beneath; lateral nerves in 7–11 pairs, tertiary nerves laxly reticulate. Male flowers: pedicels 1 mm long; sepals 4(6), 1.5–2 × 1–1.5 mm, suborbicular, unequal, rounded, appressed-pubescent, creamy-white; disk glands 6, unequal, ellipsoid; staminal column 1.5 mm high, stamens 8–10, filaments 1.5–2 mm long, pubescent, anthers minute; pistillode 0.5 mm long. Female flowers: pedicels 1–2 mm long, lengthening slightly in fruit; sepals slightly larger than those of the male flowers, otherwise resembling them; disk c. 1 mm in diameter; ovary 2–3 mm in diameter, densely setiferous; styles c. 2 mm long, bifid, pubescent without, stigmas fimbriate, black. Fruit 7–10 × 6–7 mm, tuberculate, setiferous, reddish when fresh, brown when dry; endocarp c. 0.5 mm thick. Seeds 4 × 3 mm, smooth, shiny, black; caruncle bright red or orange.

Zambia. W: 6.5 km north of Kalene Hill, Mwinilunga, o. fr. 22.ix.1952, *White* 3322 (FHO; K); male fl. 12.xii.1963, *Robinson* 5951 (K; PRE; SRGH); Kalene Hill Mission, male fl. 18.v.1969, *Mutimushi* 3154 (FHO; K; NDO; SRGH).

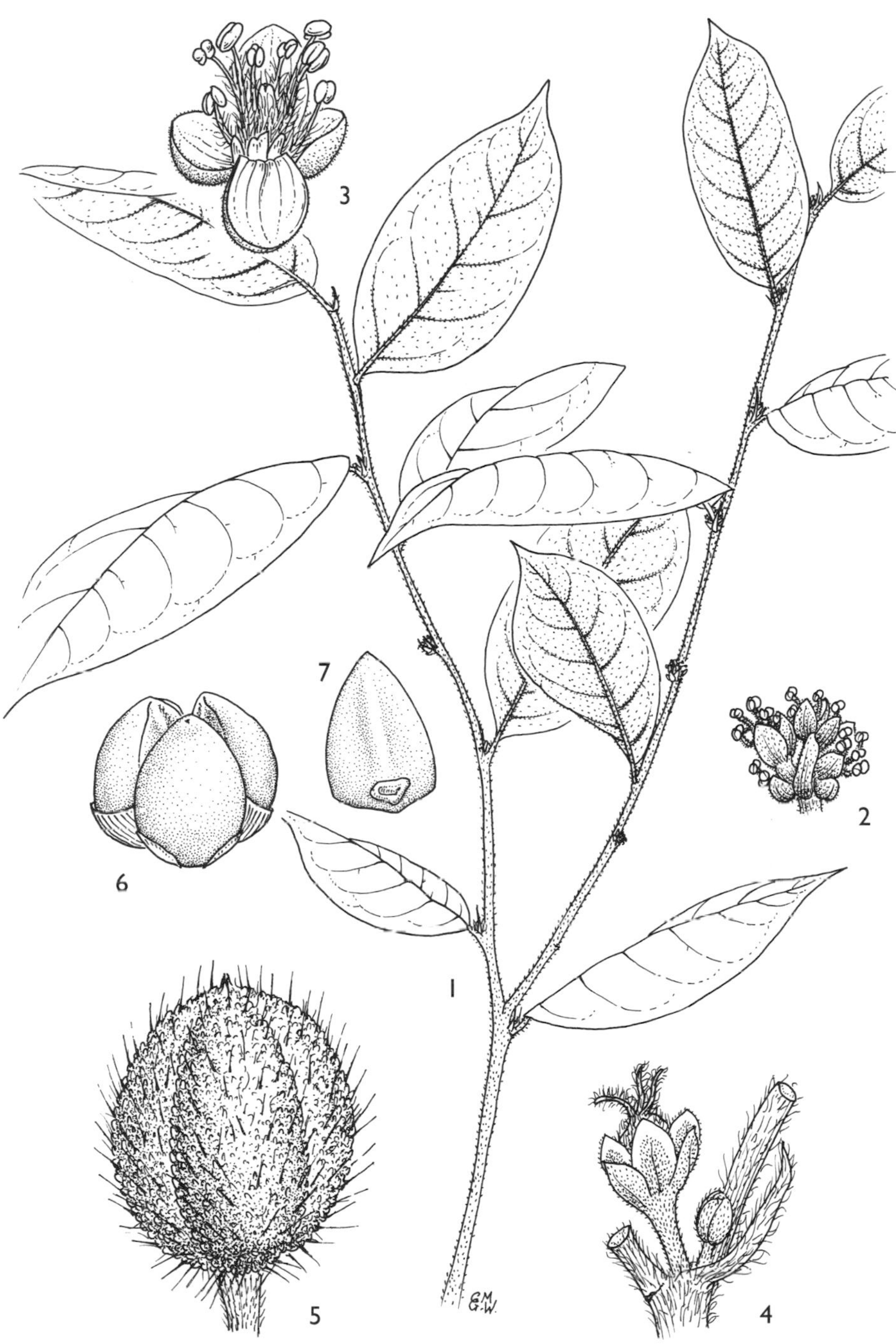

Tab. 23. CHAETOCARPUS AFRICANUS. 1, distal portion of flowering branch ($\times \frac{2}{3}$); 2, male inflorescence (× 4); 3, male flower (× 6), 1–3 from *Mutimushi* 3154; 4, female flower (× 6), from *Gossweiler* 13675 (Angola); 5, young fruit (× 4), from *Richards* 16961; 6, cluster of seeds with caruncles (× 4); 7, single seed, caruncle removed (× 4), 6 & 7 from *Mutimushi* 3156. Drawn by Christine Grey-Wilson.

Also in Gabon, Cabinda, Congo, Zaire and Angola. Riverine evergreen forest and thicket, and in understorey of adjacent woodland, mostly on sandy alluvium; 1280–1290 m.

The description of female flowers, fruits and seeds was based on Zairean and Angolan material.

20. PSEUDAGROSTISTACHYS Pax & K. Hoffm.

Pseudagrostistachys Pax & K. Hoffm. in Engler, Pflanzenr. [IV, fam. 147, vi] **57**: 96 (1912); [IV, fam. 147, xvii, Addit. vii] **85**: 180 (1924). —J. Léonard in F.C.B. **8**, 1: 183 (1962).

Agrostistachys sensu Müll. Arg. in Flora **47**: 534 (1864); in De Candolle, Prodr. **15**, 2: 725 (1866), pro parte. —Hutchinson in F.T.A. **6**, 1: 828 (1912), non Dalz.

Dioecious shrubs or small trees. Indumentum simple (more or less confined to the inflorescences). Leaves alternate, petiolate, stipulate, stipellate or not, simple, entire, subentire or remotely and shallowly serrulate-denticulate, penninerved, with 2–4 glands at the top of the petiole or on the upper surface of the lamina at the base. Stipules fused into an elongate sheath, which on falling leaves a distinct annular scar. Inflorescences axillary or borne on old wood, racemose, solitary or fasciculate, shortly pedunculate or not, strobiliform at first, later laxly flowered; bracts broad, concave, ciliate, persistent; male inflorescences 1–more-flowered; female inflorescences usually 1-flowered; bracteoles 2. Male flowers pedicellate, pedicels jointed near the base; calyx closed in bud, later splitting into 2–5 membranous valvate lobes; petals 5–8, free, imbricate; receptacle convex, covered in contiguous disk glands, pubescent; stamens 20–55, free, erect in bud, arising from pits in the glandular surface, anthers fixed at apex, extrorse, the thecae almost free, pendulous from a thickened glandular apiculate connective, longitudinally dehiscent; pistillode reduced or absent. Female flowers: pedicels stouter than in the male, jointed near the base; calyx and petals resembling those of the male; disk hypogynous, annular, thickened, alveolate; staminodes often present, filiform, arising from the alveoli; ovary 3-locular, with 1 ovule per locule; styles 3, almost free, bipartite, the stigmas papillose. Fruit 3-locular, dehiscing into 3 bivalved cocci; endocarp thinly woody; columella persistent, apically tricornute. Seeds subglobose, ecarunculate; testa crustaceous, shiny; albumen thick, fleshy; embryo straight or slightly curved; cotyledons broad, flat.

A genus of 2 species confined to tropical Africa.

Pseudagrostistachys ugandensis (Hutch.) Pax & K. Hoffm. in Engler, Pflanzenr. [IV, fam. 147, xvi, Addit. vii] **85**: 180 (1924). —Eggeling & Dale, Indig. Trees Uganda, ed. **2**: 138 (1952). —J. Léonard in F.C.B. **8**, 1: 185 (1962). —Radcliffe-Smith in F.T.E.A., Euphorb. 1: 167 (1987). Tab. **24**. Types from Uganda (U4).

A much-branched small tree up to 9 m tall, mostly glabrous; stem c. 20 cm d.b.h. Petioles 1–3 cm long, with a pair of small subulate semipersistent stipels and/or a pair of sessile discoid glands at their apices. Leaf blades 15–30 × 6–11 cm, elliptic-oblong, very shortly obtusely acuminate at the apex, rounded or sometimes very shortly decurrent at the base, shallowly spinulose-denticulate on the margin, firmly chartaceous to thinly coriaceous, with a pair of sessile discoid glands on the upper surface at the base, drying reddish-brown beneath; lateral nerves in 16–25 pairs, the lower open, the upper looped, slightly prominent on leaf upper surface, prominent on leaf lower surface, tertiary nerves numerous and closely parallel, ± perpendicular to the lateral nerves, quaternary nerves reticulate. Stipular sheath terminal, 2–3 cm long, subulate, sharply acute, readily caducous. Inflorescences 1–3(7) cm long, axillary or sometimes borne on old wood, 2–3 or more per fascicle, not pedunculate; axis densely pubescent; bracts 2–3 mm long, broadly ovate, rounded, pubescent at the base, otherwise glabrous, chestnut-brown; bracteoles similar, but smaller and dorsally keeled. Male flowers fragrant; pedicels 0.5–1 cm long, partly or entirely pubescent, reddish; calyx lobes 3–5 × 1 mm, lanceolate, acute, pubescent outside at the apex, otherwise glabrous, becoming reflexed; petals 4–5 × 3–4 mm, elliptic-oblong to elliptic-obovate, ± rounded, glabrous, creamy-white; disk gland-mass

Tab. 24. PSEUDAGROSTISTACHYS UGANDENSIS. 1, portion of flowering branchlet (× $^2/_3$), from *Sangster* 1042; 2, male flower (× 4), from *Fyffe* 19; 3, female flower (× 6), from *Fyffe* 64; 4, columella after dehiscence, with one sepal of female calyx (× 2); 5, seeds and mericarps (× 1), 4 & 5 from *Drummond & Hemsley* 4564. Drawn by Christine Grey-Wilson. From F.T.E.A.

glabrous or sparingly pubescent; stamens 42–55, filaments 3.5–5 mm long, anthers 0.5 mm long, white; pistillode (when present) 2 mm long, tripartite, pubescent. Female flowers: pedicels 0.8–1.2 cm long, evenly to densely pubescent; calyx ± as in the male; petals 4–5 × 4–5 mm, broadly ovate-suborbicular, somewhat erose; disk evenly pubescent; staminodes ± 20, 1.5 mm long; ovary 2 × 3 mm, depressed-globose, densely tomentose; styles 2 mm long. Fruit 1.5–1.8 × 2–2.3 cm, depressed 3-lobed to subglobose, densely tomentose and intermixed with longer hairs, dull reddish-brown; columella 7–9 mm long. Seeds 8–10 × 7–8 × 7–8 mm, transversely ovoid-subglobose, smooth, slightly shiny, brownish, streaked with darker brown.

Zambia. W: Mwinilunga Distr., source of R. Lunga at Kanema, fr. 16.xi.1961, *Holmes* 1354 (FHO).

Also in Zaire and Uganda. Swamp forest (mushitu) interior.

21. CAPERONIA A.St.-Hil.

Caperonia A.St.-Hil., Hist. Pl. Remarq. Brésil: 244 (1824).

Monoecious, rarely dioecious, annual or perennial herbs, usually of marshy places. Indumentum simple, often hispid or glanduliferous. Leaves alternate, shortly petiolate, stipulate, simple, serrate, penninerved or sometimes palminerved; nerves craspedodromous. Inflorescences axillary, spicate or racemose, pedunculate, distally male, proximally female; bracts small, 1-flowered. Male flowers shortly pedicellate; calyx closed in bud, later splitting, at least partially, into 5 valvate lobes; petals 5, imbricate, free, subequal or unequal, inserted on the staminal column; disk absent; stamens 10, rarely fewer, partially fused into a column, biseriate, anthers introrse, medifixed, 2-celled, longitudinally dehiscent; pistillode cylindric, entire or 3-lobed, situated at the top of the staminal column. Female flowers shortly pedicellate to subsessile; sepals 5–10, imbricate, equal or unequal, accrescent; petals 5–6, free, imbricate, sometimes much reduced; disk absent; ovary sessile, 3-locular, with 1 ovule per loculus; styles 3, ± free or slightly connate at the base, deeply laciniate. Fruits 3-lobed, hispid or echinate, dehiscing septicidally into 3 bivalved cocci; endocarp thinly crustaceous. Seeds ovoid or globose, smooth, punctate or lineate, ecarunculate, but with a very thin closely adherent aril which may leave a false raphe; albumen fleshy; cotyledons broad, flat.

A genus of some 40 species in tropical America, Africa and Madagascar. Six species occur in Africa.

Young stems, inflorescence axes and female calyx lobes commonly glandular-hispid; stipules lanceolate, 1–4 × 0.7–1.5 mm, entire; male flower buds pubescent; female flower sepals accrescent to 2–5.5 × 1–2 mm ··· 1. *stuhlmannii*

Young stems, inflorescence axes and female calyx lobes eglandular; stipules ovate to ovate-lanceolate, 2–8 × 1.5–5 mm, fimbriate; male flower buds glabrous or more or less so; female flower sepals accrescent to 3–8 × 1–4 mm ··· 2. *fistulosa*

1. **Caperonia stuhlmannii** Pax in Bot. Jahrb. Syst. **19**: 81 (1894). —Engler, Pflanzenw. Ost-Afrikas **C**: 237 (1895). —Pax in Engler, Pflanzenr. [IV, fam. 147, vi] **57**: 38 (1912). —Prain in F.T.A. **6**, 1: 831 (1912); in F.C. **5**, 2: 457 (1920). —Engler, Pflanzenw. Afrikas (Veg. Erde 9) **3**, 2: 52 (1921). —J. Léonard in Bull. Jard. Bot. État **26**: 320 (1956). —Radcliffe-Smith in F.T.E.A., Euphorb. 1: 166 (1987). Tab. **25**. Syntypes: Mozambique, Zambézia Province, Quelimane (Quilimane), 12.i.1889, *Stuhlmann* 601 (B†) and from Tanzania.

Caperonia castaneifolia sensu Klotzsch in Peters, Naturw. Reise Mossambique **6**, 1: 99 (1861), non (L.) St.-Hil.

Caperonia palustris sensu Müll. Arg. in De Candolle, Prodr. **15**, 2: 754 (1866), pro parte quoad spec. Peters, Kirk. —Engler, Pflanzenw. Ost-Afrikas **C**: 237 (1895). —Prain in F.T.A. **6**, 1: 832 (1912), pro parte, non (L.) St.-Hil.

An erect, ascending or scrambling, tufted annual herb, up to 1.8 m tall, although commonly much shorter, often branched, glandular-hispid when young; stems fibrous, up to 7 mm thick. Petioles 0.1–2 cm long. Leaf blades 2–14 × 0.5–4 cm,

Tab. 25. CAPERONIA STUHLMANNII. 1, habit (× ²⁄₃), from *Polhill & Paulo* 2116; 2, leaf (× ²⁄₃), from *Polhill & Paulo* 11642; 3, male flower (× 6); 4, male flower, petals removed (× 6); 5, male flower, sepals, some petals and anthers removed (× 6); 6, petal (× 6); 7, female flower (× 6); 8, persistent female calyx and columella (× 6); 9, two fruit valves (× 6); 10, seed (× 6), 3–10 from *Polhill & Paulo* 2116. Drawn by Mary Millar Watt. From F.T.E.A.

narrowly oblong-lanceolate to broadly elliptic-lanceolate, acute or subacute at the apex, cuneate or rounded at the base, serrate on the margins, thinly chartaceous, sparingly appressed-pubescent to subglabrous on both surfaces, 3–5-nerved from the base; lateral nerves in 6–12 pairs. Stipules 1–4 × 0.7–1.5 mm, lanceolate, entire. Inflorescences 2–11 cm long, with the peduncle up to 5 cm long; axes glandular-hispid; bracts smaller than the stipules, but otherwise resembling them. Male flowers soon falling; pedicels 1–2 mm long; buds 1 mm in diameter, pubescent; calyx lobes subequal, 1.5–2 × 1 mm, ovate-lanceolate, greenish; petals unequal, three of them 2 × 1 mm, spathulate, the other two 1.5 × 0.5 mm, narrowly elliptic, unguiculate, undulate, white; staminal column 2 mm high, free parts of filaments 0.5 mm long, anthers 0.5 mm long; pistillode c. 1 mm long, slightly 3-lobed. Female flowers 1–3 per inflorescence; pedicels shorter and stouter than in male flowers; sepals usually 6, very unequal, 1–2 × 0.5–1 mm, accrescent to 2–5.5 × 1–2 mm, ovate-lanceolate to oblong-lanceolate, more or less acute, glandular-hispid, yellowish-green; petals 1.5 × 0.7 mm, elliptic, subequal, white, soon falling; ovary c. 1 mm in diameter, subglobose, densely glandular-hispid; styles 0.5–1 mm long, white. Fruit 4–4.5 × 6–7 mm, roundly 3-lobed, echinate, evenly glandular-hispid, green. Seeds 3–3.5 mm in diameter, globose, grey or greyish-brown, sometimes mottled blackish, white-lineate.

Zambia. C: Mfuwe, fr. 9.iv.1969, *Astle* 5688 (K; SRGH). E: Chipata Distr., 13°06'S, 31°47'E, o. fr. 21.iv.1970, *Abel* 136 (SRGH). S: near Chirundu Bridge, fl. & fr. 1.ii.1958, *Drummond* 5425 (K; LISC; PRE; SRGH). **Zimbabwe**. N: near Chirundu Bridge, fr. 31.iii.1961, *Drummond & Rutherford-Smith* 7508 (K; LISC; SRGH). S: Manjinji Pan, SE of Malapati, fr. 3.vi.1971, *Grosvenor* 613 (K; LISC; PRE; SRGH). **Malawi**. N: Hara R., fr. 25.viii.1969, *Mrs. Fitzpatrick* 50 (BM). S: near Girumba, lower Shire, fl. & fr. ii.1888, *Scott* s.n. (K). **Mozambique**. Z: foot of Morrumbala (Moramballa), fr. xii.1858, *Kirk* s.n. (K). T: Sisitso, fl. & fr. 15.vii.1950, *Chase* 2757 in *GHS* 29860 (BM; K; LISC; SRGH). MS: 5 km Chemba–Tambara, fl. & fr. 23.iv.1960, *Lemos & Macuácua* 145 (BM; COI; K; LISC; LMA; PRE; SRGH). GI: Xai-Xai (João Belo), fl. & fr. 14.viii.1957, *Barbosa & Lemos* in *Barbosa* 7846 (K; LISC; LMA). M: Incanhini, fr. 15.i.1898, *Schlechter* 12039 (BM; COI; K).

Also in Tanzania and South Africa (KwaZulu-Natal). River banks, floodplains and pans, in heavy black clay and sandy soils, in seasonally waterlogged or permanently moist ground, sometimes in standing water; 6–1830 m.

2. **Caperonia fistulosa** Beille in Bull. Soc. Bot. Fr., **55**, Mém 8b: 73 (1908). —Pax in Engler, Pflanzenr. [IV, fam. 147, vi] **57**: 37 (1912). —Engler, Pflanzenw. Afrikas (Veg. Erde 9) **3**, 2: 52 (1921). —J. Léonard in Bull. Jard. Bot. État **26**: 314 (1956); in F.C.B. **8**, 1: 167 (1962). —Radcliffe-Smith in F.T.E.A., Euphorb. 1: 166 (1987). Type from Mali.

Caperonia buchananii Baker in Bull. Misc. Inform., Kew **1912**: 103 (1912). —Prain in F.T.A. **6**, 1: 831 (1912). Type: Malawi, fr. 1891, *Buchanan* s.n. (K, holotype).

Caperonia palustris sensu Prain in F.T.A. **6**, 1: 832 (1912), pro parte quoad spec. Chevalier. —F.W. Andrews, Fl. Pl. Anglo-Egypt. Sudan, **2**: 57 (1952). — Keay in F.W.T.A., ed. 2, **1**, 2: 398 (1958), pro max. parte, non (L.) St.-Hil.

Caperonia sp. —P.G. Meyer in Merxmüller, Prodr. Fl. SW. Afrika, fam. 67: 8 (1967).

Similar to *C. stuhlmannii*, but differing as follows: plant not glanduliferous; the hollow stem somewhat thicker (up to 1 cm thick) and somewhat inflated; stipules 2–8 × 1.5–5 mm, ovate to ovate-lanceolate and slightly fimbriate; leaf blades very slightly shorter and broader; inflorescences slightly shorter; male flower buds up to 1.5 mm in diameter and glabrous or subglabrous; female sepals larger, accrescent to 3–8 × 1–4 mm; styles 1–2 mm long; fruit somewhat larger (5 × 7–9 mm), more coarsely echinate and hispid.

Caprivi Strip. 27 km east of Nyangana Mission Station, fl. & fr. 20.ii.1956, *de Winter & Marais* 4770 (PRE). **Botswana**. N: Okavango River, fr. 27.iv.1975, *Gibbs-Russell* 2823 (K; MO; PRE; SRGH). **Zambia**. B: Sesheke, fl. & fr. iv.1910, *Gairdner* 506 (K). N: Chembe, y. fl. 12.xi.1964, *Mutimushi* 1080 (K; NDO; SRGH). S: Nega Nega, Mazabuka Distr., fl. & fr. 22.i.1964, *van Rensburg* 2810 (K; SRGH). **Malawi**. C: Bua R., fr. 10.ii.1959, *Robson* 1538 (BM; K; LISC; SRGH).

From Mali eastwards to Somalia and south to Namibia, Botswana, Zambia and Malawi. River banks, floodplains, dambos and drying waterholes, in heavy black clay and sandy soils, in seasonally waterlogged or permanently moist ground, with grasses and reeds; 900–1030 m.

22. CHROZOPHORA Neck. ex A. Juss.

Chrozophora Neck. ex A. Juss., Euphorb. Gen.: 27 (1824) (as "*Crozophora*"), nom. conserv.

Monoecious, annual or perennial herbs or shrubs. Indumentum stellate. Leaves alternate, petiolate, stipulate, simple, subentire, dentate or sublobate, penni- or palminerved, often plicate or bullate when young, 2-glandular at the base of the lamina. Inflorescences pseudo-axillary, lateral or leaf-opposed, racemose or subpaniculate, male above, female below; bracts 1-flowered; female flowers on 1 or more 1–4-flowered peduncles. Male flowers: pedicels short; calyx closed in bud, later splitting into 5 valvate lobes; petals 5, distally imbricate, lightly coherent; disk 5-lobed, the lobes alternating with the petals; stamens 3–15, filaments connate into a column, 1–3-seriate, anthers oblong, erect, longitudinally dehiscent; pistillode absent. Female flowers: pedicels long, often further elongating and reflexing in fruit; sepals 5, open in bud; petals 5, smaller than in the male or 0; disk more or less as in the male; ovary 3-locular, with 1 ovule per loculus; styles 3, connate at the base, bifid, more or less erect. Fruits 3-lobed, dehiscing septicidally into 3 bivalved cocci, leaving a columella; endocarp thinly woody. Seeds ovoid or subglobose, tuberculate or smooth, enveloped in a thin aril, ecarunculate; albumen thick, fleshy; cotyledons broad and flat.

An Old World genus of 6–12 species from Portugal and Senegal east to Kazakhstan and Thailand. 5–7 species in Africa.

Chrozophora plicata (Vahl) A. Juss. ex Spreng., Syst. Veg. **3**: 850 (1826), excl. syn. Lam. —Klotzsch in Peters, Naturw. Reise Mossambique **6**, 1: 576 (corr.) (1864). —Müller Argoviensis in De Candolle, Prodr. **15**, 2: 747 (1866). —Engler, Pflanzenw. Ost-Afrikas **C**: 237 (1895). —Pax in Engler, Pflanzenr. [IV, fam. 147, vi] **57**: 19 (1912). —Prain in F.T.A. **6**, 1: 834 (1912); in Bull. Misc. Inform., Kew **1918**: 92 (1918). —Engler, Pflanzenw. Afrikas (Veg. Erde 9) **3**, 2: 51 (1921). —F.W. Andrews, Fl. Pl. Anglo-Egypt. Sudan **2**: 58 (1952). —Keay in F.W.T.A., ed. 2, **1**, 2: 398 (1958). —Radcliffe-Smith in F.T.E.A., Euphorb. 1: 161 (1987). Tab. **26**. Type from Egypt.

Croton tinctorius sensu Burman f., Fl. Ind.: 304, t. 62, f. 2 (1768), non L.

Croton plicatus Vahl, Symb. Bot. **1**: 78 (1790).

Croton obliquifolius Vis., Pl. Aegypt., No. 171, p. 39, t. 7, f. 2 (1836). Type as above.

Chrozophora tinctoria sensu Klotzsch in Peters, Naturw. Reise Mossambique **6**, 1: 99 (1861), non (L.) Raf.

Chrozophora plicata var. *prostrata* Müll. Arg. in De Candolle, Prodr. **15**, 2: 747 (1866), quoad spec. Afr. cit. tantum.

Chrozophora plicata var. *obliquifolia* (Vis.) Prain in F.T.A. **6**, 1: 835 (1912); in Bull. Misc. Inform., Kew **1918**: 94 (1918).

Chrozophora plicata var. *erecta* Prain in Bull. Misc. Inform., Kew **1918**: 94 (1918); in F.C. **5**, 2: 457 (1920); Burtt Davy, Fl. Pl. Ferns Transvaal: 304 (1932). Type: Mozambique, Gaza Province, Guijá Distr., E bank of the R. Limpopo, *Beijer* s.n. in *Tvl. Mus. Herb.* 26114 (PRE, holotype).

A prostrate, decumbent or ± erect, branched annual or perennial herb up to c. 50 cm high. Stems yellowish or pinkish. Petioles 1–7 cm long. Leaf blades 1.5–7 × 1–5.5 cm, broadly ovate to rhombic-ovate, sometimes shallowly 3-lobed, obtuse or rounded at the apex, asymmetrically cuneate, truncate or shallowly cordate at the base, plicate-undulate at first, later ± flat and shallowly repand-dentate or ± entire on the margin, thinly chartaceous, with 2 discoid dark purple glands at the base, 3–5-nerved from the base; leaf upper surface at first incubously bullate, later ± flat; lateral nerves in 2–3 pairs, tertiary nerves arachnoid, scarcely prominent on both surfaces. Stipules 1–2 mm long, subulate. Inflorescences 1.5–3.5 cm long; bracts resembling the stipules. Male flowers: pedicels c. 1 mm long; calyx lobes 3 × 1 mm, lanceolate, acute, stellate-pubescent without, glabrous within; petals 3 × 1 mm, elliptic-oblong, ± obtuse, lepidote without, glabrous within, yellowish-orange or pinkish; disk c. 1 mm in diameter; staminal column 4 mm high, anthers 15 in (2)3 series, 1 mm long. Female flowers: pedicels 3–7 mm long, extending to 1.5–2 cm in fruit; sepals 1.5–2 × 0.5 mm, linear-lanceolate, acute, stellate-pubescent without, simply puberulous within; petals minute or absent; disk more or less as in the male flowers; ovary 2 × 2 mm, 3-lobed, densely stellate-pubescent; styles 1.5–2 mm long, abaxially stellate-

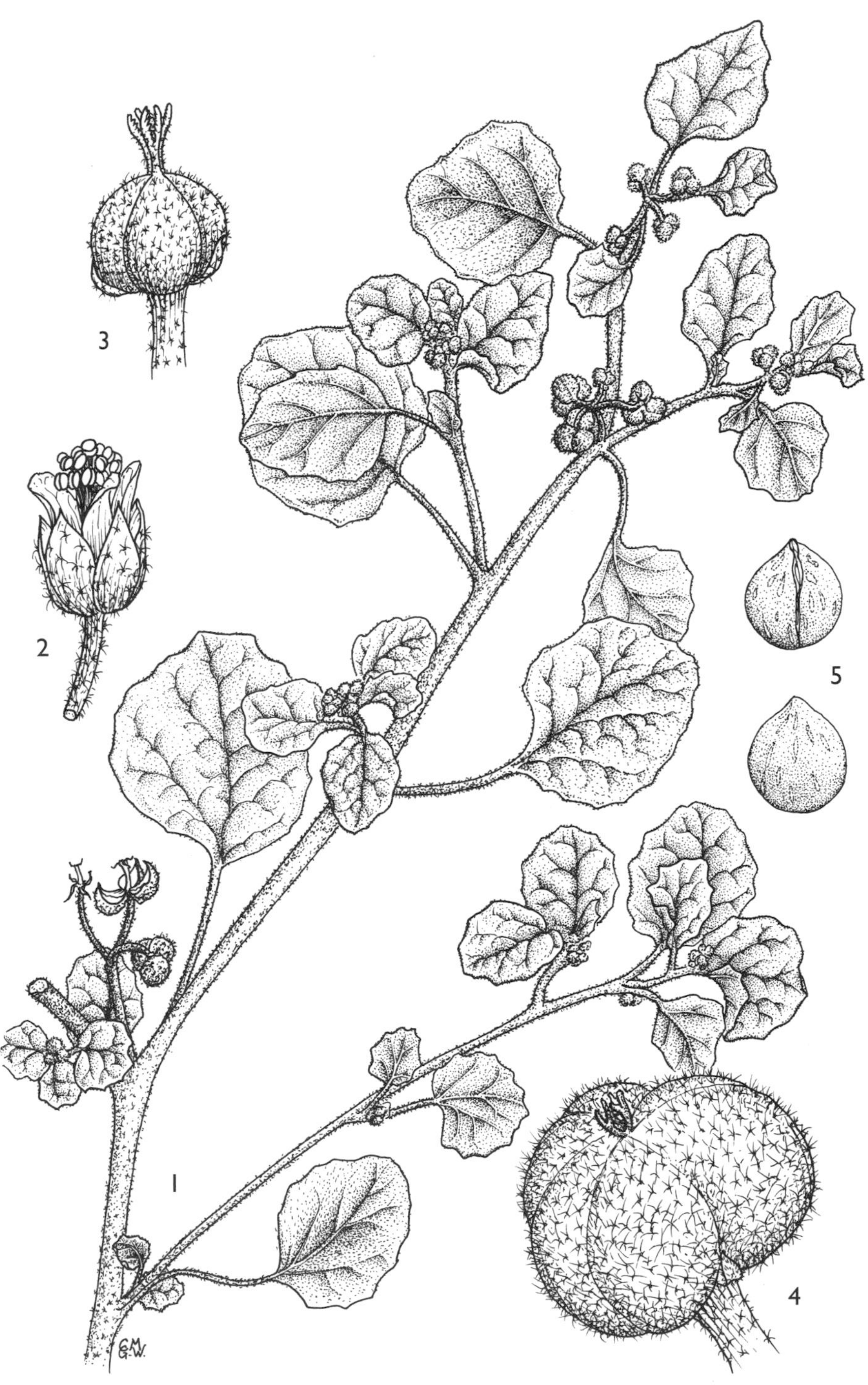

Tab. 26. CHROZOPHORA PLICATA. 1, habit (× $^2/_3$), from *Fanshawe* 8878; 2, male flower (× 6); 3, female flower (× 6); 4, fruit (× 6); 5, seeds (× 4), 2–5 from *van Rensburg* 2513. Drawn by Christine Grey-Wilson.

pubescent, adaxially coarsely-papillose, crimson. Fruit 4–5 × 7–9 mm, strongly 3-lobed, densely stellate-pubescent, green at first, later reddish- or bluish-purple. Seeds 3.5 × 3 mm, minutely tuberculate-rugulose or ± smooth; sarcotesta papery when dry, pale grey; sclerotesta dark brownish or blackish.

Zambia. B: "NW Rhodesia", fr. vii.1908, *Allen* 716 (K). N: Mfuwe, Luangwa Valley, fr. 16.vii.1968, *Astle* 5366 (K; SRGH). E: Chipata Distr., 13°06'S, 31°47'E, y. fr. 26.v.1970, *Abel* 118 (SRGH). S: Ngoma, Kafue Nat. Park, fr. 23.viii.1964, *Fanshawe* 8878 (K; LISC; NDO; SRGH). **Zimbabwe**. N: Hurungwe Distr., Mana Pools Game Reserve, Chewore Wilderness Area, fr. x.1971, *P.R. Guy* 1884 (K; MO; PRE; SRGH). S: Manjinji Pan, SW of Malapati, fl. 3.vi.1971, *Grosvenor* 614 (FHO; K; LISC; MO; PRE; SRGH). **Mozambique**. T: 11 km from Régulo Fortuna to Ancuaze (Dôa), fl. & y. fr. 20.vi.1949, *Barbosa & Carvalho* in *Barbosa* 3187 (K; LISC; LUAI). MS: Gorongosa Nat. Park, Urema floodplains, fl. vii.1970, *Tinley* 1934 (K; LISC; PRE; SRGH). GI: Guijá–Mapai, Limpopo R., fr. 8/9.v.1944, *Torre* 6602 (LISC; PRE).

Throughout tropical Africa from Senegal to Ethiopia and south to South Africa (Transvaal); also in Egypt, Syria, Palestine and W Arabia. Low altitude river floodplains, usually in damp or drying black clay and alluvium, on mudflats, sandbanks, and pan margins; 40–610 m.

23. **NECEPSIA** Prain

Necepsia Prain in Bull. Misc. Inform., Kew **1910**: 343 (1910).
Neopalissya Pax in Engler, Pflanzenr. [IV, fam. 147, vii] **63**: 16 (1914).

Dioecious or monoecious trees with a simple indumentum. Buds often perulate (furnished with protective scales). Leaves alternate, petiolate, stipulate, simple, crenate-serrate to subentire, chartaceous or subcoriaceous, penninerved, sparingly gland-dotted on the lower surface; petioles often pulvinate and geniculate. Inflorescences axillary, solitary or congested, pedunculate or sessile, spicate, racemose or subpaniculate, unisexual or bisexual; bracts rigid, more or less scarious. Male flowers shortly pedicellate; calyx closed in bud, ovoid, apiculate, later splitting into 4–5 valvate lobes; petals absent; stamens many, inserted on a globose receptacle, filaments free, anthers with slightly oblique pendulous thecae, apicifixed, introrse, longitudinally dehiscent, connective produced; disk glands free; pistillode absent. Female flowers subsessile or very shortly pedicellate; sepals (4)5(6), imbricate, persistent; petals absent; disk hypogynous, annular or cupuliform, entire or more or less crenulate, thick, pubescent or setose; ovary 3-celled, with 1 ovule per cell; styles 3, ± free or connate at the base, bifid or bipartite, papillose. Fruit tricoccous, or dicoccous by abortion, dehiscing septicidally into 3 bivalved cocci leaving a columella; endocarp thinly woody. Seeds subglobose, ecarunculate.

A genus of 3 species in tropical Africa and Madagascar.

Necepsia castaneifolia (Baill.) Bouchat & J. Léonard in Bull. Jard. Bot. Belg. **56**: 191 (1986). — Radcliffe-Smith in F.T.E.A., Euphorb. 1: 218 (1987). Types from Madagascar.
Palissya castaneifolia Baill., Étud. Gén. Euphorb.: 503 (1858).
Alchornea castaneifolia (Baill.) Müll. Arg. in Linnaea **34**: 167 (1865), non A. Juss.
Alchornea madagascariensis Müll. Arg. in De Candolle, Prodr. **15**, 2: 900 (1866). Types from Madagascar.
Neopalissya castaneifolia (Baill.) Pax in Engler, Pflanzenr. [IV, fam. 147, vii] **63**: 16 (1914). —Drummond in Kirkia **10**: 252 (1975). —K. Coates Palgrave, Trees Southern Africa, ed. 2, rev.: 426 (1983).

Subsp. **chirindica** (Radcl.-Sm.) Bouchat & J. Léonard. in Bull. Jard. Bot. Belg. **56**: 194 (1986). Tab. **27**. Type: Zimbabwe, Chirinda Forest, male, female & fr. 26.x.1947, *Wild* 2227 in *GHS* 18064 (K, holotype; BR; COI; SRGH).
Neopalissya castaneifolia subsp. *chirindica* Radcl.-Sm. in Kew Bull. **39**: 791 (1984).

A small tree to 9 m high. Bark smooth, dark. Twigs greyish-brown. Petioles 0.5–1(1.5) cm long, abaxially pulvinate at both apex and base, geniculate, pubescent. Leaf blade 5–17 × 2–6(7) cm, elliptic to elliptic-oblanceolate, acuminate apically, cuneate at the base, shallowly and somewhat remotely crenate-serrate to subentire on the margins, chartaceous, glabrous on upper surface, sparingly pubescent only along

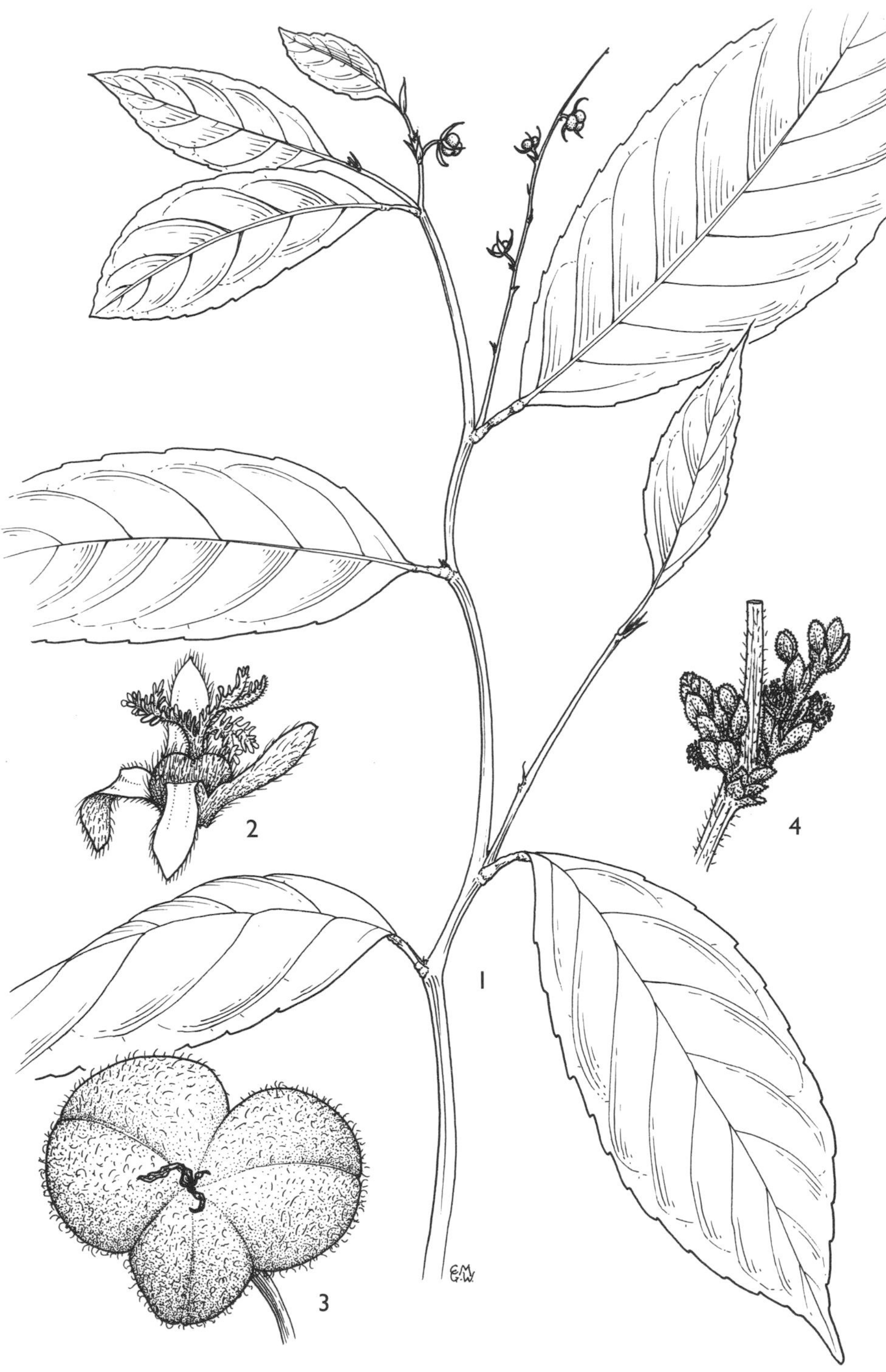

Tab. 27. NECEPSIA CASTANEIFOLIA subsp. CHIRINDICA. 1, distal portion of flowering branch (× $^{2}/_{3}$); 2, female flower (× 3), 1 & 2 from *Wild* 2227; 3, fruit (× 3), from *Sturgeon* in *GHS* 18199; 4, male inflorescence (× 3), from *Wild* 2227. Drawn by Christine Grey-Wilson.

the midrib and main nerves beneath, with a few scattered glands beneath, red, pink or mauve at first, later dark green; lateral nerves in (10)12–16(18) pairs, slightly prominent, tertiary nerves closely parallel. Stipules 6–8 mm long, linear-lanceolate, appressed-pubescent. Male inflorescences 0.6–3 cm long, racemose or subpaniculate; bracts 1 mm long, ovate, subacute; bracteoles minute; axis pubescent. Male flowers: pedicels 1.5 mm long; buds 1.5 mm long; calyx lobes ovate, yellow; stamens c. 1 mm long, anthers minute. Female inflorescences 5–12 cm long, peduncle up to 2 cm long. Female flowers: pedicels c. 1 mm long; sepals 5, 3 mm long, elliptic-ovate, obtuse, pubescent, white; disk c. 2 mm in diameter; ovary 3 mm in diameter, rounded-trigonous, somewhat verruculose, densely pubescent; styles 3 mm long, the segments linear, pubescent abaxially. Fruit 7–18 mm × c. 1.5 cm, 3-lobed, or 2-lobed by abortion, the lobes subglobose, slightly roughened, evenly pubescent, light green or pinkish. Young seeds round, green.

Zimbabwe. E: Chirinda Forest, fr. 22.x.1947, *Sturgeon* in *GHS* 18199 (K; SRGH); male, female, fr. x.1947, *Chase* 409 (BM; BR; COI); male fl. x.1947, *Chase* 428 in *GHS* 19257 (SRGH); female fl. & fr. vi.1954, *Pardy & Armitage* 2 in *GHS* 47677 (MO; SRGH); male fl. 12.xii.1955, *Armitage* 200/55 (K; SRGH); male, female fl. & fr. iii.1963, *Goldsmith* 1/63 (K; LISC; PRE; SRGH); female fl. & fr. 21.iii.1970, *Kelly* 200 (K; LISC; PRE; SRGH); male fl. 16.i.1974, *Bamps, Symoens & Vanden Berghen* 882 (BR; SRGH).

Subspecies endemic to Chirinda forest. Understorey tree of submontane evergreen forest; 1100–1160 m.

Subspecies *kimbozensis* (Radcl.-Sm.) Bouchat & J. Léonard is restricted to eastern Tanzania, while subsp. *capuronii* and *castaneifolia* are confined to Madagascar.

24. **PARANECEPSIA** Radcl.-Sm.

Paranecepsia Radcl.-Sm. in Kew Bull. **30**: 684 (1976).

Dioecious tree with a simple indumentum. Leaves alternate, crowded at the ends of the branchlets, shortly petiolate, stipulate, stipellate, simple, serrate, penninerved. Inflorescences racemose, axillary, solitary, lax; bracts 1–5-flowered. Male flowers: pedicels jointed; calyx closed in bud, later splitting into 3–5 valvate lobes which become reflexed; petals absent; disk glands free, interstaminal; stamens 25–40, filaments free, anthers dorsifixed, introrse, pendulous, longitudinally dehiscent; pistillode absent; receptacle subglobose. Female flowers: pedicels elongate, jointed; sepals 5–7, imbricate, unequal, accrescent; petals absent; staminodes 8–10; disk annular; ovary 3-celled, with 1 ovule per cell; styles 3, connate at the base, bifid, papillose. Fruit 3-lobed, septicidally dehiscent into 3 bivalved cocci; endocarp thinly woody. Seeds globose, ecarunculate.

A monotypic genus, endemic to E and S tropical Africa.

Paranecepsia alchorneifolia Radcl.-Sm. in Kew Bull. **30**: 684 (1976); in F.T.E.A., Euphorb. 1: 220 (1987). Tab. **28**. Type: Mozambique, Niassa Province, 30 km north of Chomba, male fl. 11.xi.1959, *Gomes e Sousa* 4514 (K, holotype; COI; EA; PRE; SRGH).

A much-branched shrub or small tree to 7 m high (with *Terminalia* branching). Bark ± smooth, light grey. Wood soft, white. Internodes variable in length. Leaves interspersed with bract-like scales in the terminal clusters. Petioles 2–5 mm long. Leaf blade 4–18(25) × 1.5–7(10) cm, oblanceolate, obtuse at the apex, narrowed to a rounded or cordulate base, shallowly glandular-serrate or crenate-serrate on the margins, glabrous on upper surface, pubescent at first on the midrib and with domatial hair tufts beneath; lateral nerves in 10–16 pairs. Scales 3 × 2 mm, triangular, acute, densely pubescent without, sparingly so within. Stipules 5–7 mm long, linear-lanceolate, pubescent. Stipels 3–4 mm long, setaceous, glabrous. Male inflorescences up to 7 cm long; axis ± glabrous; bracts 2–3 mm long, triangular-lanceolate, pubescent. Male flowers: pedicels 0.5–1 cm long; buds 1.5 mm in diameter, subglobose, whitish; calyx lobes 1.5 mm long, ovate-lanceolate, pale yellow; stamens 1.5–2 mm long, filaments purple, anthers 0.5 mm long, purplish-tinged; disk glands numerous, dark purple. Female inflorescences up to 11 cm long, 2–6-flowered, otherwise as in the male. Female flowers: pedicels 2–3 cm long, geniculate;

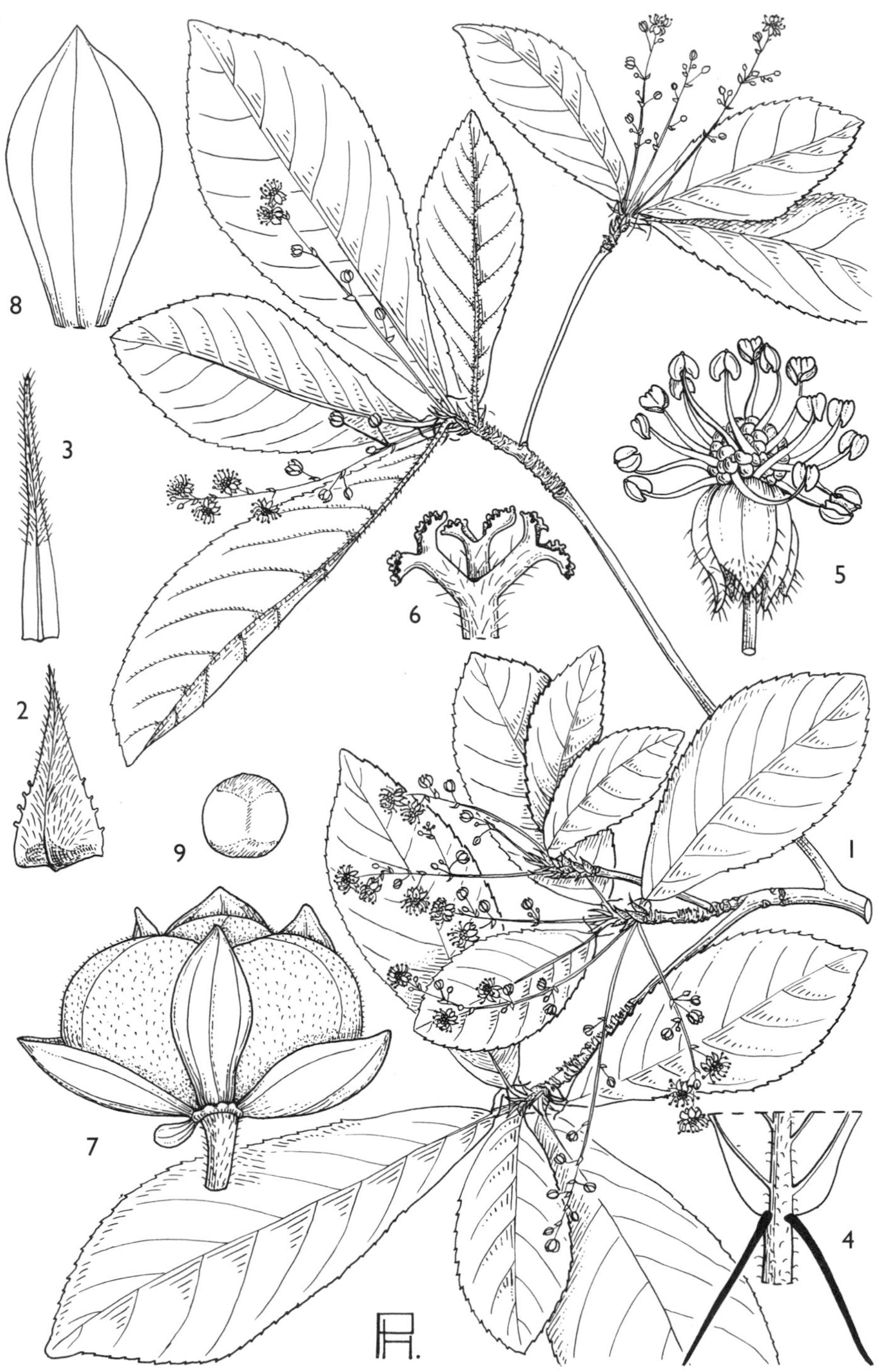

Tab. 28. PARANECEPSIA ALCHORNEIFOLIA. 1, portion of male flowering branch (× $^{2}/_{3}$), from *Gomes e Sousa* 4514 & 4514a; 2, bract (× 6); 3, stipule (× 6); 4, stipels (× 6), 2–4 from *Gomes e Sousa* 4514; 5, male flower, with some stamens removed (× 6), from *Semsei* 1046; 6, styles (× 12); 7, fruit (× 4); 8, fruiting sepal (× 6); 9, seed (× 2), 6–9 from *Rees* T56. Drawn by Pat Halliday. From F.T.E.A.

sepals in the fruiting stage 2–6 × 1–4 mm, broadly elliptic to elliptic-ovate, obtuse, strongly triplinerved, ± glabrous; staminodes minute; disk 1.5 mm in diameter, thin; ovary size not known; styles 1 mm long, pubescent at the base. Fruit 5–6 × 10 mm, strongly 3-lobed, minutely tuberculate, sparingly or evenly pubescent, grey-green. Seeds 4 mm in diameter, smooth, purplish-brown.

Mozambique. N: Negomano–Mueda road, c. 30 km north of Chomba, st. 4.iv.1960, *Gomes e Sousa* 4514/A (COI; K; SRGH); 30 km Chomba–Negomano, st. 13.iv.64, *Torre & Paiva* 11880 (LISC); Ridi R., west of Pemba (Porto Amélia), st. 1.x.1964, *Gomes e Sousa* 4831 (COI; K; PRE).

Also in Tanzania. Locally common in riverine forest fringing seasonal watercourses; 15–300 m.

Because of the absence of female and fruiting material in specimens from Mozambique, Tanzanian material was used in drawing up the above description. In Tanzania, this species occurs up to 450 m in altitude.

25. ARGOMUELLERA Pax

Argomuellera Pax in Bot. Jahrb. Syst. **19**: 90 (1894). —J. Léonard in Bull. Soc. Roy. Bot. Belgique. **91**, 2: 274 (1959).

Wetriaria sect. *Argomuellera* (Pax) Pax in Engler, Pflanzenr. [IV, fam. 147, vii] **63**: 50 (1914).

Monoecious or subdioecious shrubs or small trees with a simple indumentum. Leaves alternate, shortly petiolate or subsessile, stipulate, simple, dentate to subentire, eglandular, penninerved. Inflorescences unisexual or bisexual, racemose, rarely subpaniculate, axillary, usually solitary per axil. Male flowers: calyx closed in bud, later splitting into (2)3–4(5) valvate lobes which become reflexed; petals absent; disk glands free, interstaminal; stamens 15–120, filaments free, anthers basifixed, introrse, longitudinally dehiscent, connective broad; pistillode absent. Female flowers: sepals 5–6(9), ± biseriate, imbricate; petals absent; disk annular; ovary 3-celled, with 1 ovule per cell; styles 3, connate at the base, undivided, rather stout. Fruits 3-lobed, septicidally dehiscent into 3 bivalved cocci; endocarp thinly woody. Seeds globose, ecarunculate; albumen fleshy; cotyledons broad, flat.

A small genus of 10 species with 4 species in tropical Africa and 6 in Madagascar and the Comoros.

Argomuellera macrophylla Pax in Bot. Jahrb. Syst. **19**: 90 (1894); in Pflanzenw. Ost-Afrikas **C**: 238 (1895). —Prain in F.T.A. **6**, 1: 925 (1912). —Eyles in Trans. Roy. Soc. South Africa **5**: 395 (1916). —De Wildeman, Pl. Bequaert. **3**, 4: 473 (1926). —Robyns & Tournay, Fl. Sperm. Parc Nat. Alb. **1**: 451 (1948). —Brenan, Check-list For. Trees Shrubs Tang. Terr.: 200 (1949). —F.W. Andrews, Fl. Pl. Anglo-Egypt. Sudan **2**: 56 (1952). —Keay in F.W.T.A., ed. 2, **1**, 2: 405 (1958). —Topham, Check List For. Trees Shrubs Nyasaland Prot.: 49 (1958). —J. Léonard in Bull. Soc. Roy. Bot. Belgique **91**, 2: 275 (1959). —Dale & Greenway, Kenya Trees & Shrubs: 185 (1961). —White, F.F.N.R.: 194 (1962). —Drummond in Kirkia **10**: 252 (1975). —Troupin, Fl. Pl. Lign. Rwanda 249, fig. 85/1 (1982); Fl. Rwanda **2**: 208, fig. 61/1 (1983). —K. Coates Palgrave, Trees Southern Africa, ed. 2, rev.: 424 (1983). —Radcliffe-Smith in F.T.E.A., Euphorb. 1: 225 (1987). —Beentje, Kenya Trees, Shrubs Lianas: 186 (1994). Tab. **29**. Types from Zaire (Kasai/Shaba and Kivu) and Uganda.

Pycnocoma hirsuta Prain in Bull. Misc. Inform., Kew **1909**: 51 (1909). Types from Uganda.

Pycnocoma parviflora Pax in Bot. Jahrb. Syst. **43**: 81 (1909). Type from Tanzania.

Wetriaria macrophylla (Pax) Pax in Engler, Pflanzenr. [IV, fam. 147, vii] **63**: 50 (1914). Types as above.

An unbranched or sparingly-branched shrub or small tree 2–4(9) m high. Bark ± smooth. Young shoots and petioles pubescent, later glabrescent. Petioles 0.5–1.5 cm long. Leaf blade 10–40 × 5–12 cm, elliptic-oblanceolate to elliptic-oblong, acutely acuminate at the apex, attenuate to the acute base, sharply coarsely serrate on the margins, ± glabrous on upper surface, hirsute-pubescent along the midrib and main nerves beneath, bright red when young; lateral nerves in 20–30 pairs, looped, prominent beneath. Stipules up to 1.5 cm long, linear, pubescent. Inflorescences up to 20 cm long, with a peduncle up to 3.5 cm long; axis densely or evenly pubescent;

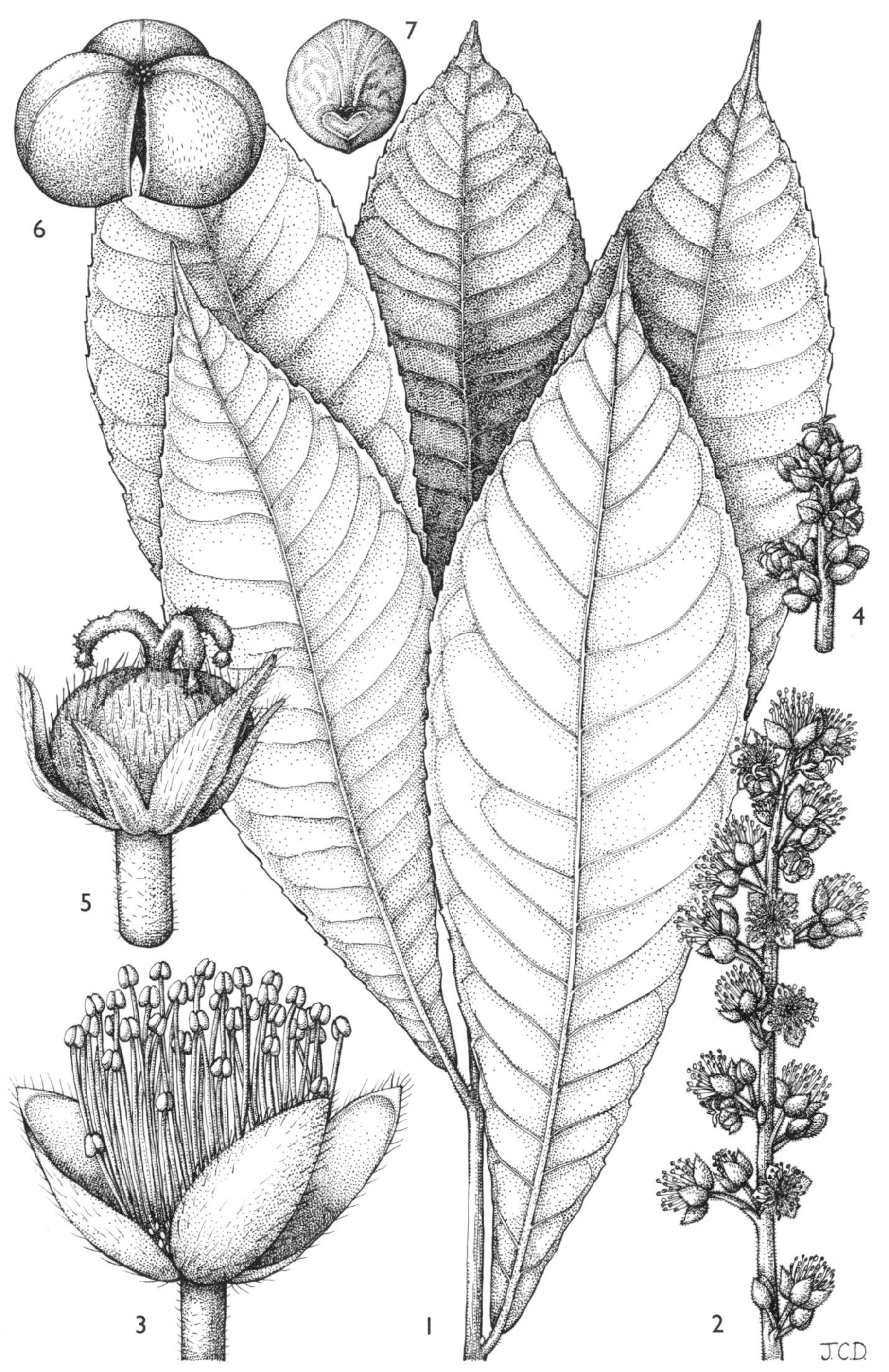

Tab. 29. ARGOMUELLERA MACROPHYLLA. 1, branchlet (× $^2/_3$); 2, bisexual inflorescence (× 1$^1/_3$); 3, male flower (× 6), 1–3 from *Lewis* 234; 4, portion of female inflorescence (× 1$^1/_3$), from *Drummond & Hemsley* 1304; 5, female flower (× 6); 6, fruit (× 3), 5 & 6 from *Ford* 321; 7, seed (× 4), from *Gillman* 444. Drawn by Judy Dunkley. From F.T.E.A.

bracts 2 mm long, spreading, triangular-ovate. Flowers sweetly scented. Male flowers: pedicels (2)3–5 mm long; calyx lobes 2–4, 4–5 mm long, becoming reflexed, ovate to lanceolate, white; disk glands numerous, bright orange; stamens 30–120, 5–7 mm long, anthers 0.5 mm long, pale yellow. Female flowers: pedicels (1)3–6 mm, jointed; sepals 5–9, 3–4 mm long, ± biseriate, triangular-ovate, sericeous-pubescent; disk crenellate; ovary 2 mm in diameter, somewhat 3-lobed, densely fulvous-hirsute; styles 3 mm long, spreading or recurved, abaxially pubescent, yellow. Fruit 6–7 × 10–13 mm, 3-lobed, smooth, hirsute, green. Seeds 5 × 4 mm, subglobose, smooth, brown-marbled.

Zambia. N: Mbala Distr., Lunzua Falls, fr. 12.i.1975, *Brummitt & Polhill* 13743 (K; SRGH). **Zimbabwe**. E: Chipinge Distr., Chirinda Forest, male fl. x.1965, *Goldsmith* 31/65 (K; LISC; PRE; SRGH). **Malawi**. S: Nsanje Distr., Matandwe For. Res., Malawi Hill, male fl. 11.viii.1960, *Willan* 46 (FHO; K; MAL; SRGH). **Mozambique**. Z: Serra Morrumbala, female fl. 13.xii.1971, *Müller & Pope* 2022 (K; LISC; SRGH). MS: Serra da Gorongosa, slopes of Mt. Nhandore, male fl. 22.x.1965, *Torre & Pereira* 12555 (COI; LISC; SRGH).

From Guinea eastwards to Ethiopia and south to Angola, Zimbabwe and Mozambique. Evergreen rain and mist forest, evergreen riverine forest and dry evergreen lakeshore forest, a common understorey tree of the Chirinda forest; 760–1300 m.

26. CEPHALOCROTON Hochst.

Cephalocroton Hochst. in Flora **24**, 1: 370 (1841).

Monoecious shrubs or subshrubs. Indumentum stellate. Leaves alternate, petiolate or subsessile, stipulate, simple, entire or toothed, palminerved. Inflorescences terminal, racemose, usually bisexual, with the male flowers aggregated into a dense globose terminal head, and with (0)1–7 female flowers at the base; bracts small or minute. Male flowers pedicellate; calyx closed in bud, globose, later splitting into 3–4 valvate lobes; petals absent; disk absent; stamens (4)6–8(10), biseriate (Africa) or uniseriate, filaments free, inflexed or not in bud, anthers oblong, dorsifixed, longitudinally dehiscent; pistillode entire or 2–3-fid. Female flowers pedicellate, the pedicels extending in fruit; sepals 4–6, free or united (not Africa) at the base, pinnatifid, bipinnatipartite or entire (not Africa), the segments gland-tipped or not, persistent, accrescent or not (not Africa); petals absent; disk annular, subentire, ovary 3-locular, with 1 ovule per loculus; styles 3, free or connate, multifid. Fruits 3-lobed, dehiscing septicidally into 3 bivalved cocci; endocarp thinly woody. Seeds ovoid-subglobose, smooth, ecarunculate; testa crustaceous; albumen fleshy; cotyledons broad, flat.

An Old World genus of 6 species, 3 African, 1 Madagascan, 1 Socotran and 1 Sri Lankan (see Radcliffe-Smith in Kew Bull. **28**: 123–132 (1973) and M.G. Gilbert in Kew Bull. **42**: 365 (1987). G.L. Webster in Ann. Missouri Bot. Gard. **81**, 1: 78–79 (1994), however, re-segregates *Adenochlaena* Baill. (2 spp.) and *Cephalocrotonopsis* Pax (1 sp.), and gives only 5 species for *Cephalocroton* s. str.

Cephalocroton mollis Klotzsch in Peters, Naturw. Reise Mossambique **6**, 1: 99 (1861). —Müller Argoviensis in De Candolle, Prodr. **15**, 2: 761 (1866). —Engler, Pflanzenw. Ost-Afrikas **C**: 240 (1895). —Pax in Engler, Pflanzenr. [IV, fam. 147, ii] **44**: 10 (1910). —Prain in F.T.A. **6**, 1: 846 (1912). —Engler, Pflanzenw. Afrikas (Veg. Erde 9) **3**, 2: 83 (1921). —O.B. Miller, Check-list For. Trees Shrubs Bech. Prot.: 31 (1948). —Topham, Check List For. Trees Shrubs Nyasaland Prot.: 50 (1958). —Radcliffe-Smith in F.T.E.A., Euphorb. 1: 283 (1987). Tab. **30**. Type: Mozambique, Vila de Sena, Peters s.n. (B†, holotype).

Cephalocroton pueschelii Pax in Bot. Jahrb. Syst. **43**: 84 (1909); in Engler, Pflanzenr. [IV, fam. 147, ii] **44**: 11 (1910). —Prain in F.T.A. **6**, 1: 845 (1912). —Engler, Pflanzenw. Afrikas (Veg. Erde 9) **3**, 2: 83 (1921). —O.B. Miller, Check-list For. Trees Shrubs Bech. Prot.: 31 (1948). —Brenan, Check-list For. Trees Shrubs Tang. Terr.: 201 (1949). —P.G. Meyer in Merxmüller, Prodr. Fl. SW. Afrika, fam. 67: 8 (1967). Type from Namibia (Ovamboland).

Cephalocroton depauperatus Pax & K. Hoffm. in Engler, Pflanzenr. [IV, fam. 147, ii] **44**: 12 (1910). —Prain in F.C. **5**, 2: 458 (1920). —Engler, Pflanzenw. Afrikas (Veg. Erde 9) **3**, 2: 83 (1921). —Burtt Davy, Fl. Pl. Ferns Transvaal: 304 (1932). Type from South Africa (Transvaal).

An erect, much-branched, woody perennial herb or shrub up to 3.6 m tall, with the stems arising from a woody rootstock. Stems evenly to densely stellate-pubescent at

Tab. 30. CEPHALOCROTON MOLLIS. 1, distal portion of flowering branch (× $^2/_3$), from *Richards* 19918; 2, male flower (× 6); 3, female flower (× 6); 4, distal portion of fruiting branch (× $^2/_3$); 5, fruit (× 1); 6, fruiting sepal, indumentum omitted (× 1); 7, seed (× 2), 2–7 from *Greenway & Polhill* 11528. Drawn by Pat Halliday. From F.T.E.A.

first, with or without glandular hairs, later glabrescent and greyish with white lenticels. Petioles 0.5–7 mm long. Leaf blades 1–10.5 × 0.5–5 cm, ovate-oblong to oblong-lanceolate, acute or obtuse at the apex, rounded, truncate or very shallowly cordate at the base, entire or subentire on the margins, chartaceous, 3–5-nerved from the base, evenly or sparingly stellate-pubescent to subglabrous on upper surface, more densely so beneath, sometimes slightly viscid, pale grey-green or almost glaucous, sometimes pinkish-tinged; lateral nerves in 4–6(10) pairs, not or scarcely prominent. Stipules 1–4 mm long, subulate-filiform. Plants rarely dioecious. Male flowering head 0.5–2 cm in diameter; peduncle 1–10 cm long, with a whorl of 1–4(7) female flowers at its base; bracts resembling the stipules. Male flowers: pedicels 3–4 mm long, ± glabrous; calyx lobes 4, 2 × 1–1.5 mm, ovate to elliptic-ovate, ± acute, usually glabrous, pale yellowish-green; stamens 5–6 mm long, filaments orange-yellow, anthers 1 mm long, pale yellow; pistillode 1 mm high, cylindric, scarcely lobulate. Female flowers: pedicels 3–5 mm long, extending to 7 cm in fruit, pubescent; sepals 6, 4 × 2 mm, extending to 2.5 × 1 cm in fruit, bipinnatipartite, the lateral lobes linear, the lobules subulate-filiform, pubescent, green; ovary 1.5 mm in diameter, 3-lobed, densely pubescent; styles 4–5 mm long, the segments filiform, glabrous, orange-yellow. Fruit 0.6–1.1 × 1.3–1.7 cm, evenly pubescent, green. Columella 6 mm high, broadly winged. Seeds 6–8 mm in diameter, fawn or light brown, dull.

Botswana. N: 15 km SW of Maun, fl. 24.i.1972, *Biegel & Gibbs-Russell* 3770 (K; LISC; MO; PRE; SRGH). SW: Kang, y. fl. 20.x.1975, *Mott* 773 (MO; PRE; SRGH; UBLS). SE: Kutse Game Res., fl. 28.iii.1978, *O.J. Hansen* 3391 (C; GAB; K; PRE; SRGH). **Zimbabwe**. E: Chipinge Distr., upper Rupembe, fl. & fr. 22.i.1957, *Phipps* 77 (COI; LISC; SRGH). S: Masvingo (Fort Victoria)–Beitbridge road, fl. 6.xii.1961, *Leach* 11317 (FHO; K; LISC; MO; PRE; SRGH). **Malawi**. S: Zomba Distr., Namadzi Stream, fl. 5.i.1979, *Banda, Blackmore, Patel & Hargreaves* 1354 (MAL; MO; SRGH). **Mozambique**. T: Moatize–Tete, fl. & fr. 14.i.1966, *Torre & Correia* 14043 (LISC). MS: between R. Urema and Inhaminga, o. fr. 6.v.1942, *Torre* 4066 (LISC). M: near Moamba, male fl. 18.xii.1944, *Torre* 6866 (LISC).

Also in Tanzania, Namibia and South Africa (Transvaal, KwaZulu-Natal). Hot dry lowveld, usually on sandy soils, short grassland, often with scattered shrubs and trees, dry water courses, grassy pans or depressions, also in mopane woodland and scrub, and in *Acacia/Colophospermum* woodland; 335–1066 m.

27. ALCHORNEA Swartz

Alchornea Swartz, Prodr.: 98 (1788).
Stipellaria Benth. in Hooker's J. Bot. Kew Gard. Misc. **6**: 2 (1854).
Lepidoturus Baill., Étud. Gén. Euphorb.: 448 (1858).

Dioecious or sometimes monoecious trees or shrubs, with a simple indumentum (or stellate, outside Flora Zambesiaca area). Leaves alternate, petiolate, stipulate, simple, crenate or serrate, sometimes remotely so, sometimes glandular beneath, stipellate or exstipellate, penni- or palminerved. Inflorescences usually unisexual, terminal or axillary, solitary or fascicled, spicate or paniculate, usually lax. Male flowers small, in bracteate clusters. Female flowers usually solitary per bract. Male flowers: calyx closed in bud, ± globose, later valvately 2–5-partite; petals and disk absent; stamens 8, rarely fewer, filaments free or ± connate at the base, anthers oblong, dorsifixed, introrse, the thecae ± parallel, longitudinally dehiscent; pistillode columnar, 2–3-lobed, or absent. Female flowers: sepals (3)4(6), imbricate; petals and disk absent; ovary (1)2–3(4)-celled, with 1 ovule per cell; styles free or ± free, usually simple, linear. Fruits 2–4-lobed or subglobose, smooth or warty, septicidally dehiscent into bivalved cocci; endocarp crustaceous. Seeds subglobose, ecarunculate; testa crustaceous; albumen fleshy; cotyledons broad, flat.

A pantropical genus of about 50 species in 3 sections. All 3 sections but only 6 species occur in tropical Africa.

1. Leaves exstipellate, shortly petiolate, penninerved ··········· 1. *hirtella*
– Leaves stipellate, usually long-petiolate, mostly palminerved ··········· 2

2. Inflorescences unisexual or bisexual, often paniculate; bracts of male flowers linear, arcuate; ovary and fruit warty · 4. *yambuyaënsis*
– Inflorescences unisexual, spicate; bracts of male flowers ovate, concave; ovary and fruit smooth · 3
3. Leaves acuminate or caudate, the acumen less than one quarter the length of the leaf, rounded to cuneate and rarely cordate at the base; male inflorescences up to 12 cm long; female inflorescences spicate · 2. *laxiflora*
– Leaves long-caudate-acuminate, the acumen one-third to half the length of the leaf, rounded, truncate or cordate at the base; male inflorescences not more than 3.5 cm long; female flowers solitary · 3. *occidentalis*

1. **Alchornea hirtella** Benth. in Hooker, Niger Fl.: 507 (1849). —Müller Argoviensis in De Candolle, Prodr. **15**, 2: 904 (1866). —Prain in F.T.A. **6**, 1: 917 (1912). —Pax in Engler, Pflanzenr. [IV, fam. 147, vii] **63**: 241 (1914). —Engler, Pflanzenw. Afrikas (Veg. Erde 9) **3**, 2: 78 (1921). —De Wildeman, Pl. Bequaert. **3**, 4: 475 (1926). —Robyns & Tournay, Fl. Sperm. Parc Nat. Alb. **1**: 460 (1948). —Brenan, Check-list For. Trees Shrubs Tang. Terr.: 199 (1949). —Eggeling & Dale, Indig. Trees Uganda, ed. 2: 115 (1952). —Keay in F.W.T.A., ed. 2, **1**, 2: 403 (1958). —Dale & Greenway, Kenya Trees & Shrubs: 184 (1961). —White, F.F.N.R.: 193 (1962). —Drummond in Kirkia **10**: 252 (1975). —K. Coates Palgrave, Trees Southern Africa, ed. 2, rev.: 425 (1983). —Radcliffe-Smith in F.T.E.A., Euphorb. 1: 255 (1987). —Beentje, Kenya Trees, Shrubs Lianas: 184 (1994). Types from Guinea and Liberia.

A spindly scandent, or straggly lax-branched shrub or small tree up to 12 m high, rarely taller, usually dioecious. Bark grey, ± smooth. Young shoots and petioles patent-hirsute (forma *hirtella*) or appressed-puberulous (forma *glabrata*). Stipules 3–5 mm long, setaceous. Petioles 0.3–2 cm long, often geniculate. Leaf blades 5–20 × 2–8 cm, elliptic-oblanceolate, shortly acuminate at the apex, attenuate to a cuneate or cordulate base, remotely glandular-crenate on the margins, penninerved, with a pair of glands on leaf upper surface towards the base, glabrous above, puberulous and later glabrescent beneath (forma *glabrata*) or with the midrib sparingly setose beneath (forma *hirtella*), dark green and shiny above, paler beneath; the lateral nerves in 6–14 pairs, bearing domatia on leaf lower surface. Male inflorescences up to 30 cm long, usually terminal, less often axillary or borne on old wood, broadly paniculate. Male flowers ± sessile; buds 0.5 mm long, subglobose, orange; sepals 2, cupular, reflexed, reddish; stamens 8, the united filaments forming a basal plate. Female inflorescences not usually more than 10 cm long, terminal, rarely cauliflorous, spicate, few-flowered, lax; bracts 2 mm long, biglandular at the base. Female flowers subsessile; sepals 5, 1.3 mm long, lanceolate, acute, minutely denticulate; ovary 1 × 1.5 mm, 3(4)-lobed, smooth or ± so, evenly appressed-pubescent; styles commonly up to 2 cm long, rarely longer, united at the base, linear-filiform, minutely papillose, reddish. Fruits 5 × 9 mm, 3-lobed, smooth or ± so, sparingly minutely puberulous, green. Seeds 4.5 × 4 mm, ovoid-subglobose, smooth, shiny, light brown, faintly-mottled.

Young shoots and petioles patent-hirsute; midrib sparingly setose beneath · · · · forma *hirtella*
Young shoots and petioles appressed-puberulous; midrib glabrescent beneath · · forma *glabrata*

forma **hirtella**

Young shoots and petioles patent-hirsute; midrib sparingly setose beneath.

Zambia. N: Chishinga Ranch, near Luwingu, male fl. 28.iv.1961, *Astle* 567 (K; SRGH); Kasama, male fl. 2.vii.1964, *Fanshawe* 8787 (K; NDO). W: Zambezi River, north of Kalene Hill, st. 20.ix.1952, *White* 3293 (FHO; K); Zambezi Rapids, Mwinilunga, female fl. 18.v.1969, *Mutimushi* 3286 (K; NDO).

From Guinea-Bissau east to Tanzania and south to Zambia. In swamp forest (mushitu) and evergreen riverine forest.

forma **glabrata** (Müll. Arg.) Pax & K. Hoffm. in Engler, Pflanzenr. [IV, fam. 147, vii] **63**: 242 (1914). —Robyns & Tournay, Fl. Sperm. Parc Nat. Alb. **1**: 460 (1948). —Brenan, Check-list For. Trees Shrubs Tang. Terr.: 199 (1949). —Eggeling & Dale, Indig. Trees Uganda, ed. 2

115 (1952). —Topham, Check List For. Trees Shrubs Nyasaland Prot.: 49 (1958). —Dale & Greenway, Kenya Trees & Shrubs: 184 (1961). —Troupin, Fl. Pl. Lign. Rwanda: 248, fig. 84/1 (1982); Fl. Rwanda **2**: 202, fig. 59/1 (1983). —Radcliffe-Smith in F.T.E.A., Euphorb. 1: 256 (1987). Type from Angola (Malanje).

Alchornea floribunda var. *glabrata* Müll. Arg. in J. Bot. **2**: 336 (1864); in De Candolle, Prodr. **15**, 2: 905 (1866). Type as above.

Alchornea glabrata (Müll. Arg.) Prain in Bull. Misc. Inform., Kew **1910**: 342 (1910); in F.T.A. **6**, 1: 916 (1912); in F.C. **5**, 2: 484 (1920). Type as above.

Young shoots and petioles appressed-puberulous; midrib glabrescent beneath.

Zambia. N: c. 49 km Mporokoso–Kasama, male fl. 17.x.1947, *Brenan & Greenway* 8134 (FHO; K; PRE). W: Mwinilunga, male fl. 19.v.1969, *Mutimushi* 3453 (K; NDO; SRGH). **Zimbabwe**. E: Chirinda Forest, male & female fl. x.1967, *Goldsmith* 89/67 (FHO; K; LISC; PRE; SRGH). **Malawi**. S: Mulanje Mt., foot of Great Ruo Gorge, between hydroelectric station and dam, male fl. 18.iii.1970, *Brummitt & Banda* 9209 (K; MAL; PRE; SRGH). **Mozambique**. Z: Serra Morrumbala, male fl. 9.xii.1971, *Müller & Pope* 1976 (LISC; SRGH). MS: Haroni/Makurupini Forest, male fl. & fr. 4.xii.1964, *Wild, Goldsmith & Müller* 6647 (BM; FHO; K; PRE; SRGH).

More or less throughout the more humid parts of tropical Africa. Often gregarious in understorey of low and medium altitude evergreen forest; also in swamp forest (mushitu) and spray zone of waterfalls; 400–2000 m.

2. **Alchornea laxiflora** (Benth.) Pax & K. Hoffm. in Engler, Pflanzenr. [IV, fam. 147, vii] **63**: 245, t. 37 (1914). —Engler, Pflanzenw. Afrikas (Veg. Erde 9) **3**, 2: 82, t. 36 (1921). —Brenan, Check-list For. Trees Shrubs Tang. Terr.: 199 (1949). —Eggeling & Dale, Indig. Trees Uganda, ed. 2: 115 (1952). —Keay in F.W.T.A., ed. 2, **1**, 2: 403 (1958). —Dale & Greenway, Kenya Trees & Shrubs: 184 (1961). —White, F.F.N.R.: 193 (1962). —Drummond in Kirkia **10**: 252 (1975). —Troupin, Fl. Pl. Lign. Rwanda: 249 (1982); Fl. Rwanda **2**: 202 (1983). —K. Coates Palgrave, Trees Southern Africa, ed. 2, rev.: 426 (1983). —Radcliffe-Smith in F.T.E.A., Euphorb. 1: 257 (1987). —Beentje, Kenya Trees, Shrubs Lianas: 185 (1994). Types from Sudan.

Lepidoturus laxiflorus Benth. in Hooker's Icon. Pl. **13**: 76, t. 1297 (1879). —Engler, Pflanzenw. Ost-Afrikas **C**: 238 (1895). —Prain in F.T.A. **6**, 1: 913 (1912). —De Wildeman, Pl. Bequaert. **3**, 4: 496 (1926). —Brenan in Mem. N.Y. Bot. Gard. **9**, 1: 74 (1954). Types as above.

Alchornea engleri Pax in Bot. Jahrb. Syst. **43**: 80 (1909). —Prain in F.T.A. **6**, 1: 918 (1912). —Pax & K. Hoffmann in Engler, Pflanzenr. [IV, fam. 147, vii] **63**: 247 (1914). —Brenan, Check-list For. Trees Shrubs Tang. Terr.: 199 (1949). Type from Tanzania (Uzaramo).

Alchornea schlechteri Pax in Bot. Jahrb. Syst. **43**: 221 (1909). —Prain in F.T.A. **6**, 1: 1057 (1913); in F.C. **5**, 2: 485 (1920). Syntypes: Mozambique, Maputo (Lourenço Marques), 30.xi.1897, *Schlechter* 11530 (male fl.), 11531 (fr. immat.) (B†, syntypes; K; MO; PRE, isosyntypes).

A deciduous, ± erect or straggling lax-branched subshrub, shrub or small tree 1–5(10) m high, monoecious, with male and female inflorescences on separate branches. Bark smooth, light grey, flaking. Twigs lenticellate, brownish. Buds perulate (furnished with protective scales), ovoid, chestnut-brown. Young shoots and petioles evenly or sparingly pubescent, puberulous or subglabrous. Stipules 2–8 mm long, linear to filiform. Stipels 1–3 mm long, subulate-filiform. Petioles 1–9 cm long, commonly bipulvinate. Leaf blades 5–17 × 3–8 cm, elliptic-lanceolate to oblong-oblanceolate, acuminate or caudate with the acumen less than one quarter the length of the blade, rounded or cuneate, rarely cordate at the base, shallowly crenate-serrate on the margins, thinly chartaceous, 3-nerved from the base, glandular at the base and sometimes also elsewhere beneath, sparingly pubescent on the midrib and main nerves at first, soon glabrescent except in the domatia, reddish when young. Male inflorescences up to 12 cm long, axillary, developing on the older twigs usually before the new leaf flush, spicate; bracts 1.5–5 × 1–2 mm, ovate, concave, scarious, chestnut-brown. Male flowers ± sessile; buds 0.7 mm long, subglobose; sepals 2–4(5), suborbicular and cupular or ovate, reflexed, yellowish-green or white; stamens 8–9, the united filaments forming a basal plate; pistillode absent or minute, columnar, 2–3-lobed. Female inflorescences usually not more than 10 cm long, terminal, spicate, few-flowered, lax; bracts 2–3 mm long, ovate-lanceolate, biglandular below the base. Female flowers sessile; sepals 5–6, minute, ovate, unequal, subacute, slightly denticulate; ovary 1 × 1.5 mm, subglobose, scarcely 3-lobed, smooth, evenly minutely appressed-puberulous; styles (2)3, up to 1.5 cm long, united at the base, linear-filiform, minutely papillose, red. Fruits 5–7 × 7–8 mm, 3-lobed, rarely 4-lobed, smooth, sparingly puberulous to subglabrous, dark green, brown or black. Seeds 4 × 3 mm, ovoid-subglobose, ± smooth or slightly rugulose, somewhat shiny, light brown or greyish.

Zambia. N: Mpika, female fl. 30.i.1955, *Fanshawe* 1900 (FHO; K; NDO; SRGH). C: Katondwe, female fl. & fr. 12.xi.1963, *Fanshawe* 8110 (FHO; K; NDO). E: Makutus, female fl. 28.x.1972, *Fanshawe* 11617 (K; NDO). **Zimbabwe**. N: Mutoko (Mtoko), Nyaderi Bridge, male fl. 5.xii.1968, *Müller & Burrows* 953 (K; PRE; SRGH). C: Sebakwe R., st. xii.1948, *Hodgson* 41/48 (SRGH). E: Chipinge, upper Tanganda R., st. 14.v.1962, *Chase* 7704 (BM; K; LISC; SRGH). S: Wanezi, Liebigs Ranch, West Nicholson, male & female fl. & fr. 27.x.1952, *Plowes* 1521 in *GHS* 40073 (K; LISC; PRE; SRGH). **Malawi**. N: Sanga, 16 km south of Nkhata Bay, male fl. 18.x.1976, *Pawek* 11913 (K; MAL; MO; SRGH; UC). C: Dedza Mt., fr. i.1961, *Chapman* 1138 (FHO; K; MAL; SRGH). S: Thyolo (Cholo) Mt., male fl. 24.ix.1946, *Brass* 17776 (BM; K; NY; PRE; SRGH). **Mozambique**. N: Monapo, female fl. & fr. 2.xii.1963, *Torre & Paiva* 9369 (LISC). Z: Namagoa Plantation, male & female fl. & fr. xi.1946, *Faulkner* 90 (K; PRE; SRGH). MS: 5 km Inhamitanga–Lacerdonia, fr. 4.xii.1971, *Müller & Pope* 1866 (K; LISC; PRE; SRGH). GI: Chibuto–Changane (Gomes da Costa), fr. 14.xi.1957, *Barbosa & Lemos* in *Barbosa* 8118 (COI; K; LUAI). M: Santaca, male fl. & fr. 16.x. & 19.xi.1948, *Gomes e Sousa* 3874 (COI; K; LISC; MO; PRE; SRGH).

From Nigeria eastwards to Ethiopia and south to South Africa (Transvaal). Locally common at low and medium altitudes in riverine vegetation and mixed deciduous woodland, often on rocky outcrops; also in understorey of submontane mixed evergreen forest, coastal forest and mixed deciduous coastal woodland, and in swamp forest (mushitu) and dambo margins; sea level–1000(1500) m.

Certain gatherings are either mixtures of this species with *A. yambuyaënsis*, or else present intermediate conditions between them. Thus in *Barbosa & Lemos* in *Barbosa* 8380 (K; LISC; LUAI) from Mozambique, the male shoot is *A. yambuyaënsis*, whilst the female shoot is *A. laxiflora*, whereas in *Torre* 3838 (LISC) also from Mozambique, the reverse situation is found.

3. **Alchornea occidentalis** (Müll. Arg.) Pax & K. Hoffm. in Engler, Pflanzenr. [IV, fam. 147, vii] **63**: 245 (1914). —Engler, Pflanzenw. Afrikas (Veg. Erde 9) **3**, 2: 82 (1921). —White, F.F.N.R.: 193 (1962). Tab. **31**, figs. A1–A4. Types from Angola (Malanje).

Lepidoturus occidentalis Müll. Arg. in J. Bot. **2**: 332 (1864); in De Candolle, Prodr. **15**, 2: 898 (1866). —Prain in F.T.A. **6**, 1: 914 (1912). Types as above.

A deciduous, slender ± erect lax-branched multistemmed shrub to 1.5 m high (when exposed to fire) or small tree up to 7 m high (when fire protected), monoecious, with male and female inflorescences on separate branches. Bark smooth, shiny, lenticellate, scaling, dark purplish-grey when old, pale grey or whitish when newly exposed. Twigs reddish-brown. Young shoots and petioles sparingly minutely puberulous or subglabrous. Stipules 2–4 mm long, filiform-setaceous. Stipels 1 mm long, setaceous. Petioles 0.5–2.5 cm. Leaf blades 3–9 × 1–4.5 cm, ovate to elliptic-ovate, long caudate-acuminate with the acumen one-third to half the length of the blade, rounded, truncate or cordate at the base, ± resembling the leaf of *Ficus religiosa* in outline, shallowly glandular-crenate or crenate-serrate on the margin, chartaceous, 3-nerved from the base, glandular towards the base or with few scattered glands on both surfaces, sparingly hirsute on upper surface, subglabrous except the domatia beneath. Male inflorescences up to 3.5 cm long, but usually not exceeding 2.5 cm in length, otherwise resembling those of *A. laxiflora*. Male flowers ± as in *A. laxiflora*. Female flowers solitary, terminal, pedicellate, the pedicels 3–4 mm long; sepals 5–6, 1–2 mm long, oblong, unequal, obtuse, ciliate; ovary 1 × 1.5 mm, subglobose, scarcely 3-lobed, smooth, glabrous; styles 3, up to 1.5 cm long, united for one fifth of their length, linear-filiform, ± smooth, red or purple. Fruits 4–5 × 5–6 mm, 3-lobed, smooth, glabrous, green or reddish. Seeds 4 × 3 mm, ovoid, shallowly malleate-rugulose or tuberculate, shiny, light brown.

Zambia. B: Kataba, fr. 12.xii.1960, *Fanshawe* 5970 (FHO; K; NDO). W: Mwinilunga Distr., Kalene Hill, male fl. 25.ix.1952, *Angus* 549 (BM; FHO; K). S: Kembe Forest, Kalomo, st. 19.xii.1963, *Bainbridge* 934 (FHO; K; NDO; SRGH).

Also in Zaire, Cabinda and Angola. In riverine evergreen forest on sandy alluvium, in understorey thicket of dry deciduous forest (mutemwa) and *Cryptosepalum* woodland on Kalahari Sand; also among rocky outcrops and in gully vegetation; 1000–1100 m.

Vernacular name as recorded in specimen data: "muzoboti" (Tok.)

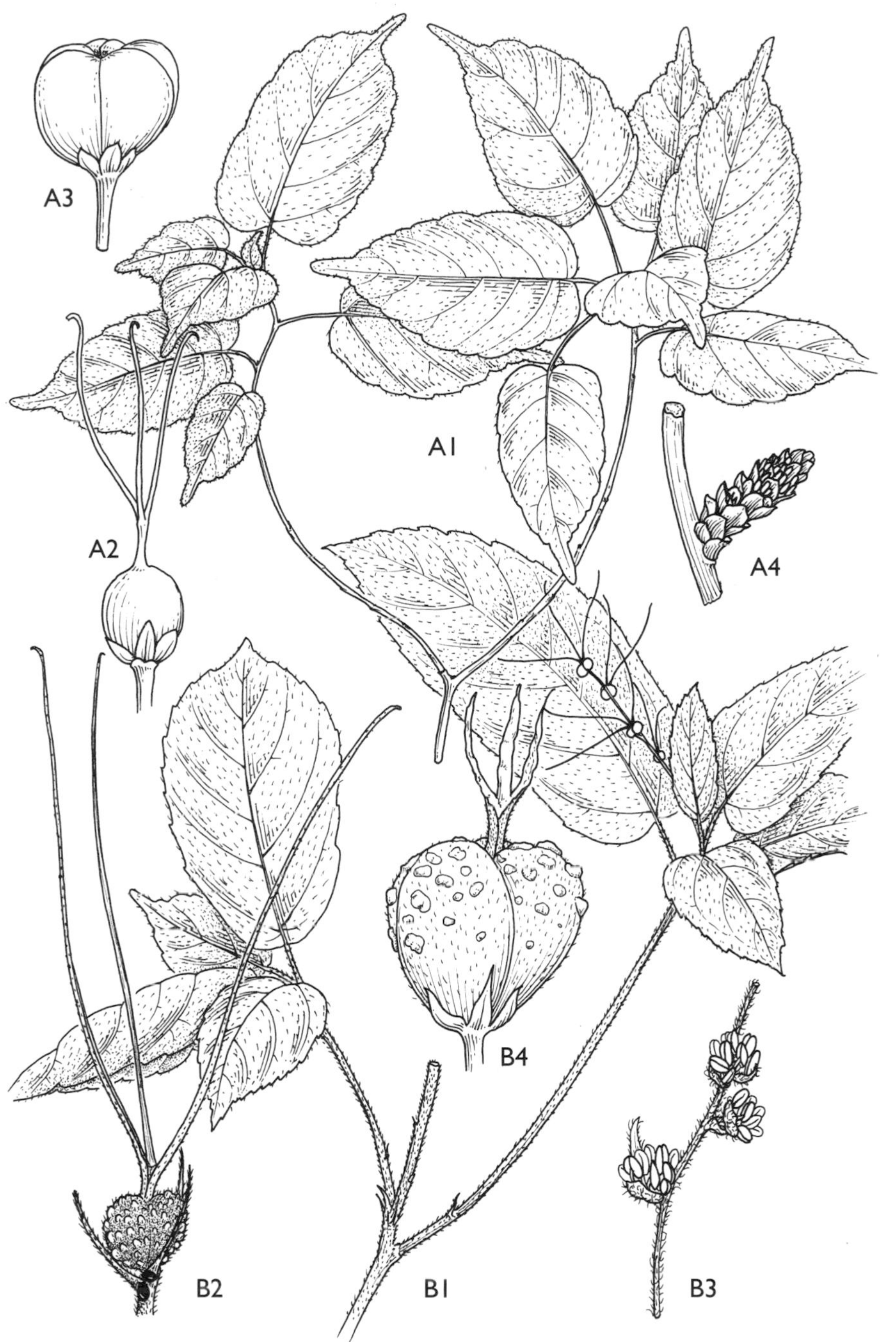

Tab. 31. A —ALCHORNEA OCCIDENTALIS. A1, branchlet (× ⅔); A2, female flower (× 4); A3, fruit (× 3), A1–A3 from *Fanshawe* 5970; A4, male flowers (× 4), from *White* 3341. B — ALCHORNEA YAMBUYAËNSIS. B1, portion of flowering branch (× ⅔); B2, female flower (× 4); B3, male flowers (× 4), B1–B3 from *Fanshawe* 324; B4, fruit (× 3), from *White* 3410. Drawn by Christine Grey-Wilson.

4. **Alchornea yambuyaënsis** De Wild., Ann. Mus. Congo Belge, Bot. Sér. 5, **2**: 280 (1908). —Prain in F.T.A. **6**, 1: 918 (1912). —Pax in Engler, Pflanzenr. [IV, fam. 147, vii] **63**: 250 (1914). —Engler, Pflanzenw. Afrikas (Veg. Erde 9) **3**, 2: 82 (1921). —Radcliffe-Smith in F.T.E.A., Euphorb. 1: 259 (1987). Tab. **31**, figs. B1–B4. Types from Zaire (Oriental).

Alchornea verrucosa Pax in Bot. Jahrb. Syst. **43**: 321 (1909). Types from Zaire.

Alchornea bangweolensis R.E. Fries, Wiss. Ergebn. Schwed. Rhod.-Kongo-Exped. 1911–1912, **1**, 1: 123, fig. 9 (1914). —Engler, Pflanzenw. Afrikas (Veg. Erde 9) **3**, 2: 82 (1921). —White, F.F.N.R.: 192 (1962). Syntypes: Zambia, Mansa Distr., Kawendimusi, ix.1911, *Fries* 769 (female fl.), 769a (male fl.) (UPS; K, photos of syntypes).

A ± erect or straggling, usually lax-branched shrub 2–3(6) m tall, monoecious or dioecious, with male and female flowers on the same or different inflorescences. Bark brown. Buds perulate (furnished with protective scales), ovoid, chestnut-brown. Young shoots and petioles evenly to densely puberulous and often also hirsute. Stipules 3–10 mm long, filiform-setaceous to linear. Stipels 1–4 mm long, filiform-setaceous. Petioles 0.2–6 cm long. Leaf blades 3–13 × 1.5–6 cm, elliptic-ovate to oblong-oblanceolate, rounded, obtuse, acute or acuminate at the apex, rounded or cordulate at the base, coarsely and shallowly crenate or crenate-serrate on the margins, soft to thinly chartaceous, obscurely 3-nerved from the base or almost penninerved, somewhat bullate, often with 2 or more small glands beneath at or towards the base, pubescent and/or hirsute along the midrib and main nerves on both surfaces, otherwise ± glabrous, dark green above, paler beneath. Male inflorescences up to 15 cm long, terminal or lateral, developing with the leaves, racemose or subpaniculate; bracts 2–5 mm long, linear-lanceolate to linear-setaceous, arcuate. Male flowers in dense bracteate clusters, pedicellate; buds 1 mm long, subglobose, pubescent; sepals 2–4, ovate-suborbicular, pale yellowish-green; stamens 7–8, the united filaments forming a basal plate; pistillode absent. Female inflorescences up to 8 cm long, but not usually more than 5 cm long, usually terminal, spicate, few-flowered, lax; bracts slightly larger than in the male inflorescences, biglandular at the base. Female flowers solitary, sessile; sepals 4–6, 2–3 mm long, lanceolate, acute, puberulous, greenish; ovary 1.5 × 1.5 mm, subglobose, verrucose, densely pubescent; styles 3–4, up to 2 cm long, united at the base, filiform, ± smooth or slightly papillose, crimson. Fruits 4–6 × 6–8 mm, 3-lobed, covered with conical warts in the upper half, otherwise smooth, pubescent, grey-green or yellow-green. Seeds 4 × 3 mm, ovoid-subglobose, shallowly tuberculate and lineate-rugulose, shiny, yellowish-brown.

Zambia. N: Kafulwe–Mporokoso, male & female fl. & fr. 3.xi.1952, *Angus* 701 (BM; FHO; K). W: Mwinilunga, R. Lunga, female fl. & y. fr. 2.xii.1937, *Milne-Redhead* 3489 (K; LISC; PRE). **Mozambique**. T: Boroma, male & female fl. 26.vi.1941, *Torre* 2923A (LISC).

Also in Zaire, Tanzania and Angola. Riverine and lakeshore vegetation, on sandy shores and fixed dunes, often gregarious in semi-evergreen thickets, and in lake side swamp forest (mushitu) understorey; also in high rainfall woodland; 1500 m.

Torre 3923 (LISC) from Mozambique (GI) is rather anomalous in that whilst it has the bisexual paniculate narrow bracteate terminal inflorescence characteristic of this species, the smooth fruits and leaf characters are those of *A. laxiflora*.

28. RICINUS L.

Ricinus L., Sp. Pl.: 1007 (1753); Gen. Pl. ed. 5: 437 (1754).

Monoecious, glabrous annual or perennial herb or shrub, often tree-like. Leaves alternate, petiolate, stipulate, peltate, palmately-lobed, the lobes glandular-serrate, penninerved. Petioles glanduliferous at apex and base. Stipules united to form a caducous sheath. Inflorescences paniculate, leaf-opposed or subterminal, male in the lower half, female in the upper or rarely all female; bracts soon caducous. Male flowers: pedicels jointed, bibracteolate; buds globose; calyx membranous, closed at first, later splitting into 3–5 valvate lobes; petals absent; disk absent; stamens up to c. 1000, the filaments variously united, anthers basifixed, the cells subglobose, longitudinally dehiscent; pistillode absent. Female flowers: pedicels considerably elongating in fruit; buds conical; sepals 5, valvate, soon caducous; petals absent; disk absent; ovary 3-celled, with 1 ovule per cell, echinate or smooth; styles 3, ± free or slightly connate at the base, bipartite, papillose-plumose, usually dark red. Fruit 3-

lobed, echinate or smooth, the spines accrescent, dehiscing into 3 bivalved cocci, leaving a prominent persistent columella. Seeds dorsiventrally compressed-ovoid, smooth, usually marmorate, carunculate; testa crustaceous; albumen fleshy; cotyledons broad, flat.

A monotypic genus originally native to NE tropical Africa, but now widely cultivated throughout the tropics, subtropics and warm temperate regions and often becoming naturalized.

Ricinus communis L., Sp. Pl.: 1007 (1753). —Peters, Naturw. Reise Mossambique **6**, 1: 98 (1861). —Müller Argoviensis in De Candolle, Prodr. **15**, 2: 1017 (1866). —Engler, Pflanzenw. Ost-Afrikas **C**: 240 (1895). —Prain in F.T.A. **6**, 1: 945 (1912). —Eyles in Trans. Roy. Soc. South Africa **5**: 397 (1916). —Pax in Engler, Pflanzenr. [IV, fam. 147, xi] **68**: 119 (1919). —Prain in F.C. **5**, 2: 487 (1920). —Engler, Pflanzenw. Afrikas (Veg. Erde 9) **3**, 2: 109 (1921). —De Wildeman, Pl. Bequaert. **3**, 4: 501 (1926). —Burtt Davy, Fl. Pl. Ferns Transvaal: 306 (1932). —O.B. Miller, Check-list For. Trees Shrubs Bech. Prot.: 33 (1948). —Brenan, Check-list For. Trees Shrubs Tang. Terr.: 225 (1949). —F.W. Andrews, Fl. Pl. Anglo-Egypt. Sudan **2**: 96 (1952). —Martineau, Rhod. Wild. Fl.: 45, t. 15 (1953). —Topham, Check List For. Trees Shrubs Nyasaland Prot.: 53 (1958). —Keay in F.W.T.A., ed. 2, **1**, 2: 410 (1958). —White, F.F.N.R.: 203 (1962).—P.G. Meyer in Merxmüller, Prodr. Fl. SW. Afrika, fam. 67: 42 (1967). —Mogg in Macnae & Kalk, Nat. Hist. Inhaca Isl., Moçamb.: 148 (1969). —Agnew, Upl. Kenya Wild Fls.: 218 (1974). —Drummond in Kirkia **10**: 252 (1975). — Biegel, Check List Ornam. Pl. Rhod. Parks & Gard.: 93 (1977). —Troupin, Fl. Rwanda **2**: 238, fig. 70/1 (1983). —Radcliffe-Smith in F.T.E.A., Euphorb. 1: 322 (1987). —Beentje, Kenya Trees, Shrubs Lianas: 221 (1994). Type a specimen either from the East or West Indies, Africa or Southern Europe.

An erect glabrous pruinose single-stemmed or bushy, tree-like herb up to 7 m high. Stems up to 10 cm thick at base, hollow, becoming ± woody, grey. Young shoots often reddish-tinged. Petioles 4–30 cm long, or longer; glands discoid or turbinate. Leaf blades (5)7–11-lobed, 7–35(100) cm long and wide, with the median lobe 2–8(20) cm wide, and the lateral lobes progressively smaller; lobes ovate-lanceolate to lanceolate, acuminate, coarsely glandular-serrate, dark green on upper surface, paler beneath, the nerves yellowish; lateral nerves in 15–25 pairs on the median lobe. Stipular sheath up to 2.7 cm long, ovate, reddish or purplish. Inflorescences 10–30 cm long; bracts c. 1 cm long, lanceolate; bracteoles similar, but smaller. Male flowers: pedicels up to 1.7 cm long; calyx lobes 5–8 × 2–5 mm, ovate, acute, pale green, often purplish-tinged; stamens 7–8 mm long, anthers 0.5 mm long, pale yellow. Female flowers: pedicels 0.5–1 cm long, extending to 4.5 cm in fruit; sepals c. 5 mm long, lanceolate, acuminate, often purplish-tinged; ovary 2 mm long and wide, 3-lobed to subglobose; styles up to 7 mm long. Fruit 1–2.3 cm long and wide, strongly 3-lobed, smooth, or sparingly to densely beset with narrowly-cylindric bristle-tipped processes 3–6 mm long, bluish-green. Seeds 7–21 × 5–15 × 4–8 mm, smooth, usually shiny, grey, silvery-white or beige, usually variously streaked, mottled, flecked or blotched with olive-brown, reddish-brown or brownish-black; caruncle 1–2 × 2–3 mm, depressed-conic.

1. Fruit 2–2.3 cm long and wide, echinate; seeds 1.5–2.1 cm long · · · · · · · var. *megalospermus*
– Fruit 1–1.5 cm long and wide, echinate or smooth; seeds 0.7–1.1 cm long · · · · · · · · · · 2
2. Fruit echinate · var. *communis*
– Fruit smooth or almost so · var. *africanus*

Var. **communis.** Tab. **32**.

Fruit 1–1.5 cm long and wide, echinate; seeds 0.7–1.1 cm long.

Caprivi Strip. Ngamiland, east of the Kwando (Cuando) R., y. fr. x.1945, *Curson* 1207 (PRE). **Botswana**. N: Nata, fl. & fr. 19.iv.1976, *Ngoni* 510 (K; MO; PRE; SRGH). SW: Mamuno, st. 13.ii.1970, *R.C. Brown* 28 (SRGH). SE: Tuli Block, Merryhill Farm, fl. 21.xii.1971, *Stephen & Wilson* 534 (PRE). **Zambia**. B: Mongu, fr. vi.1933, *Trapnell* 1258 (K). N: Isoka to Chinsali, fl. & fr. 21.v.1959, *Stewart* 175 (K). C: Mt. Makulu, fl. 9.vi.1956, *Angus* 1352 (BM; FHO; K). E: Chipata (Fort Jameson), fr. 24.iv.1952, *White* 2468 (FHO; K). S: Mazabuka, fl. & fr. 14.viii.1931, *Trapnell* in *CRS* 415 (K; PRE). **Zimbabwe**. N: Guruve (Sipolilo), fl. & fr. 18.xii.1980, *Nyariri* 826 (SRGH). W: 43 km west of Nyamandhlovu, fl. & fr. 20.iv.1972, *Grosvenor* 729 (SRGH). C: Marondera (Marandellas), fl. xi.1948, *Dehn* in *GHS* 23082 (SRGH). E: Mutare Distr., Odzani,

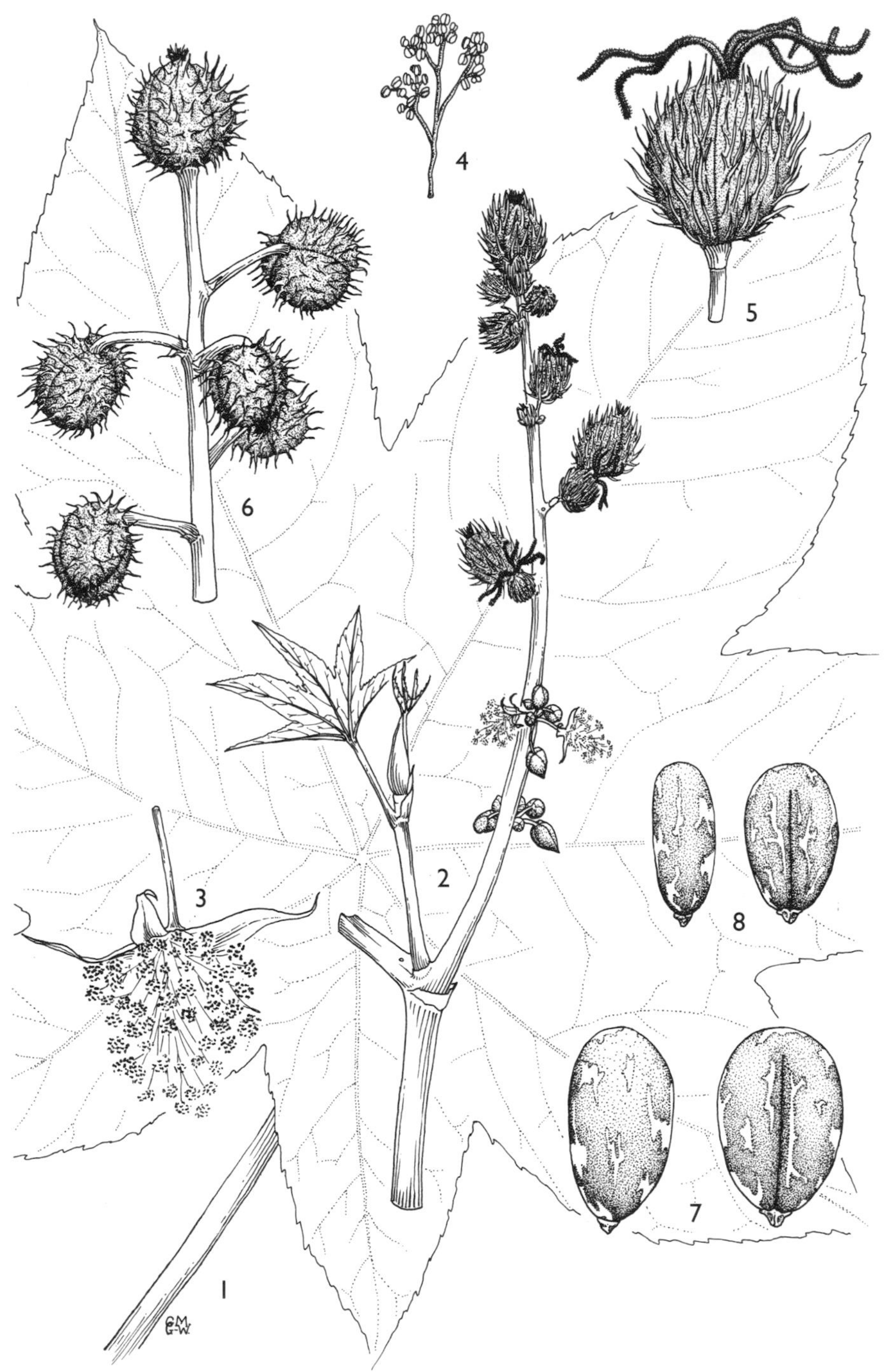

Tab. 32. RICINUS COMMUNIS var. COMMUNIS. 1, leaf (× $^{2}/_{3}$); 2, inflorescence (× $^{2}/_{3}$); 3, male flower (× 3); 4, stamen (× 4); 5, female flower (× 3); 6, infructescence (× $^{2}/_{3}$); 7, seeds (× 2), 1–7 from a cultivated plant; 8, seeds (× 2), from *Symon* 3185. Drawn by Christine Grey-Wilson. From Flore des Mascareignes.

y. fr. 28.x.1966, *Chase* 8443 (K; PRE; SRGH). S: Great Zimbabwe (Zimbabwe Ruins), y. fr. v.1925, *Earthy* '26361' (PRE). **Malawi**. N: Mzimba Distr., Mzuzu, fr. 7.xi.1975, *Pawek* 10341 (K; MAL; MO; PRE; SRGH; UC). C: Dedza Mountain, fl. 10.xi.1967, *Salubeni* 876 (K; LISC; MAL; SRGH). S: Mwanza, fl. & fr. 11.viii.1975, *Seyani* 638 (MAL; MO; SRGH). **Mozambique**. N: Angoche (António Enes), fl. & fr. 22.x.1965, *Mogg* 32349 (LISC; SRGH). Z/T: lower River Shire, fl. v.1861, *Meller* s.n. (K). T: Sisitso, fl. 11.vii.1950, *Chase* 2766 in *GHS* 29869 (BM; PRE; SRGH). MS: Marínguè, fr. 25.vi.1950, *Chase* 2476 in *GHS* 29427 (BM; SRGH). GI: Chibuto, fl. & fr. 6.viii.1958, *Barbosa & Lemos* 8302 (COI; K; LMU). M: Inhaca Island, fr. 18.vii.1958, *Mogg* 28068 (SRGH).

Throughout Africa, and cultivated and naturalized throughout the tropical, subtropical and warm temperate regions of the globe. Sandy river banks and bed of dry water courses, stony lakeshores; also in evergreen forest margins, and riverine vegetation, occasionally in mixed deciduous woodland; sea level to 1900 m.

The Castor Oil Plant, the oil from the seeds has many uses, especially as a purgative, lubricant, leather preservative and illuminant. The seeds are, however, poisonous in quantity.

Var. **africanus** (Willd.) Müll. Arg. in De Candolle, Prodr. **15**, 2: 1019 (1866). —Prain in F.T.A. **6**, 1: 946 (1912). —Pax in Engler, Pflanzenr. [IV, fam. 147, xi] **68**: 123 (1919). Type a cultivated specimen in Herb. Willd. No. 17930 (B-WILLD, holotype).

Ricinus africanus Willd., Sp. Pl. **4**: 565 (1805).

Fruit 1–1.5 cm long and wide, smooth or almost so; seeds 0.7–1.1 cm long.

Mozambique. GI: Inharrime, fl. & fr. 3.iv.1954, *Barbosa & Balsinhas* 5528 (BM; LISC).

Mediterranean and NE tropical Africa. Cultivated at Inharrime Experimental Station.

Var. **megalospermus** (Del.) Müll. Arg. in De Candolle, Prodr. **15**, 2: 1017 (1866). —Prain in F.T.A. **6**, 1: 946 (1912). —Pax, in Engler, Pflanzenr. [IV, fam. 147, xi] **68**: 121 (1919). Type a cultivated specimen in Geneva, 1862 (G–DC, neotype).

Ricinus megalospermus Del., Cent. Pl. Afr. Meroe Caill.: 89 (1826).

Fruit 2–2.3 cm long and wide, echinate; seeds 1.5–2.1 cm long.

Zambia. B: Mongu, fr. vi.1933, *Trapnell* 1259 (K). **Malawi**. N: Nkhata Bay Distr., 5.5 km SW of Chikangawa, fl. & fr. 4.vii.1978, *Phillips* 3439 (K; MO; SRGH).

South tropical Africa from Angola eastwards. Cultivated and becoming naturalized in grassland; 1060–1770 m.

29. LEIDESIA Müll. Arg.

Leidesia Müll. Arg. in De Candolle, Prodr. **15**, 2: 792 (1866).
Mercurialis sect. *Seidelia* Baill., Adansonia **3**: 158, 175 pro parte (1864).

Delicate monoecious annual herbs. Indumentum simple. Leaves alternate, occasionally subopposite, petiolate, stipulate, simple, crenate-dentate, palminerved. Inflorescences terminal, racemose, bisexual, mostly male with a basal female flower, and also axillary glomerulate unisexual (female), sometimes consisting of a solitary bracteate female flower; male bracts several-flowered; female bracts resembling reduced foliage leaves. Male flowers pedicellate; calyx closed in bud, later splitting in to 3 valvate lobes; petals absent; disk absent; stamens 4–7, central, filaments short, ± connate at the base, anther-cells subglobose, distinct from the base, bivalved, the valves spreading; pistillode absent. Female flowers shortly pedicellate; calyx absent, or with 1 bracteiform lobe; petals absent; disk absent; ovary 2-locular, with 1 ovule per loculus; styles 2, ± free, filiform, undivided. Fruits 2-lobed or monococcous by abortion, dehiscing septicidally into 2 bivalved cocci, or into 2 valves; endocarp thinly crustaceous. Seeds ovoid-subglobose, ecarunculate; testa crustaceous; albumen fleshy; cotyledons broad and flat.

A monotypic African genus.

Leidesia procumbens (L.) Prain in Ann. Bot. (London) **27**: 400 (1913). —Pax in Engler, Pflanzenr. [IV, fam. 147, vii] **63**: 284, t. 44 (1914). —Engler, Pflanzenw. Afrikas (Veg. Erde 9) **3**, 2: 86, t. 40 (1921). Tab. **33**. Type from South Africa (Cape Province).

Tab. 33. LEIDESIA PROCUMBENS. 1, habit (× $^2/_3$); 2, male and female flowers (× 6); 3, fruit (× 6); 4, seeds (× 6), 1–4 from *Wild* 6447. Drawn by Christine Grey-Wilson.

Mercurialis procumbens L., Sp. Pl.: 1036 (1753).
Mercurialis capensis Spreng. ex Eckl. & Zeyh. in Linnaea **20**: 213 (1847), nomen tantum.
Leidesia capensis Müll. Arg. in De Candolle, Prodr. **15**, 2: 793 (1866), excl. syn. *Urtica capensis* L.f. —Prain in F.C. **5**, 2: 463 (1920). —Burtt Davy, Fl. Pl. Ferns Transvaal: 304 (1932).

A weak, rather soft diffusely branched annual herb; stems up to 30 cm high or in extent, glabrous, subpellucid, green . Lower leaves opposite, upper leaves alternate. Petioles 1–3 cm long. Leaf blades 1.5–4.3 × 1–3 cm, ovate, obtuse or subacute at the apex, truncate or wide-cuneate at the base, crenate-dentate on the margins except at the base, very thinly membranaceous, 3–5-nerved from the base, glabrous or almost so beneath; lateral nerves in 3–4 pairs, sparingly pubescent above. Stipules 0.5–1 mm long, subulate to linear-lanceolate. Inflorescences 1–2.5 cm long; male bracts 1–1.5 mm long; female bracts 4–5 × 1–1.5 mm. Male flowers: pedicels c. 1 mm long; calyx lobes 0.5 mm long, ovate, acute, glabrous, greenish-white; stamens minute. Female flowers ± sessile; calyx lobe, when present, resembling a stipule; ovary c. 1 mm in diameter, 2-lobed, apically setose, the setae with bulbous based hairs, green, tipped purplish; styles c. 1 mm long. Fruit usually 2 × 2–3 mm, 2-lobed, smooth, apically setose, stramineous, purplish tipped. Seeds 1.5 × 1.2 mm, reticulate, shiny, dark brown to black.

Zimbabwe. E: Mutare Distr., Engwa, fl. & fr. 2.iii.1954, *Wild* 4438 (K; LISC; MO; PRE; SRGH). **Mozambique**. MS: Tsetserra, fl. 9.ii.1955, *Exell, Mendonça & Wild* 321 (BM; LISC; SRGH).
Also in Zaire (Kivu), South Africa (Transvaal, KwaZulu-Natal, Cape Province) and Swaziland. Submontane evergreen forest, on damp floor; 1830–1980 m.

30. MACARANGA Thouars

Macaranga Thouars, Gen. Nov. Madag.: 26 (1806).

Dioecious or rarely monoecious trees or shrubs, usually with a simple indumentum. Leaves alternate, petiolate, stipulate, simple or lobed, entire or toothed, peltate or not, palminerved and/or penninerved, gland-dotted on lower surface. Inflorescences axillary or subterminal, racemose or paniculate, shortly pedunculate; bracts entire or lobed, glandular or not; male bracts many-flowered, female bracts usually 1-flowered; flowers shortly pedicellate or subsessile. Male flowers: calyx closed in bud, globose or clavate, later valvately 2–4-partite; petals and disk absent; stamens 2–5 or 10–20, filaments short, free or ± connate at the base, anthers small, terminal, apicifixed, (2)3–4-celled, (2)3–4-valved, transversely or cruciately dehiscent; pistillode absent. Female flowers: calyx truncate or shallowly toothed, later becoming broadly cupular and splitting; petals and disk absent; ovary 1–3-celled, with 1 ovule per cell; styles 1–3, often rather stout, entire, free or connate at the base. Fruits not lobed, or less often 2–3-lobed, dehiscing into bivalved cocci or subindehiscent, smooth or verrucose, usually glandular. Seeds ± globose; exotesta fleshy; endotesta crustaceous; albumen fleshy; cotyledons broad, flat.

A palaeotropical genus of 280 species, especially well represented in New Guinea. 40 species occur in tropical Africa.

1. Leaves usually 3-lobed, sometimes 5-lobed; lateral nerves craspedodromous (running from midrib to margin without dividing) · 1. *schweinfurthii*
– Leaves not lobed; lateral nerves brochidodromous (looped) · 2
2. Leaves 3(5)-nerved from the base; bracts eglandular; stamens 13–17; fruits warty · 2. *mellifera*
– Leaves 5–11-nerved from the base; bracts glandular; stamens 2–4; fruits smooth · · · · · · 3
3. Leaves broadly ovate, often cordate, up to 40 cm long, not glaucous on lower surface; distal male bracts sinuate-lobulate · 3. *capensis*
– Leaves rhombic-ovate to triangular-ovate, cuneate or rounded, not more than 15 cm long, glaucous on lower surface; distal male bracts ± entire · · · · · · · · · · · · · 4. *kilimandscharica*

1. **Macaranga schweinfurthii** Pax in Bot. Jahrb. Syst. **19**: 92 (1894). —Prain in F.T.A. **6**, 1: 935 (1912). —Pax in Engler, Pflanzenr. [IV, fam. 147, vii] **63**: 313 (1914). —Engler, Pflanzenw. Afrikas (Veg. Erde 9) **3**, 2: 89 (1921). —De Wildeman, Pl. Bequaert. **3**, 4: 482 (1926). —

Eggeling & Dale, Indig. Trees Uganda, ed. 2: 133 (1952). —F.W. Andrews, Fl. Pl. Anglo-Egypt. Sudan **2**: 85 (1952). —Keay in F.W.T.A., ed. 2, **1**, 2: 407 (1958). —Dale & Greenway, Kenya Trees & Shrubs: 208 (1961). —White, F.F.N.R.: 200 (1962). —Troupin, Fl. Pl. Lign. Rwanda: 266 (1982); Fl. Rwanda **2**: 230 (1983). —Radcliffe-Smith in F.T.E.A., Euphorb. 1: 241 (1987). —Beentje, Kenya Trees, Shrubs Lianas: 212 (1994). Types from Zaire.

Macaranga rosea Pax in Bot. Jahrb. Syst. **26**: 328 (1899). —Prain in F.T.A. **6**, 1: 935 (1912). —De Wildeman, Pl. Bequaert. **3**, 4: 481 (1926). Type from Cameroon.

Macaranga lecomtei Beille in Bull. Soc. Bot. Fr. 55, Mém 8: 78 (1908). Syntypes from Congo and Central African Republic.

Macaranga calophylla Pax in Bot. Jahrb. Syst. **43**: 221 (1909). Type from Zaire.

A rapidly growing tree up to 18 m high. Stem with straight spines up to 6 cm long. Bark grey-green, smooth, flaking in minute papery scales. Young growth fulvous- to ferruginous-pubescent at first, later glabrescent. Stipules 2–3 × 1–2 cm, ovate, foliaceous, subacute or obtuse, deciduous. Petioles up to 22 cm long, bipulvinate. Leaf blades up to 40 × 30 cm, usually 3-lobed, rarely 5-lobed, sometimes not lobed, deeply cordate and auriculate at the base, repand-dentate on the margins, thinly coriaceous, sparingly puberulous to subglabrous on both surfaces, evenly gland-dotted beneath; lobes forwardly directed, triangular, often emarginate with the mid vein shortly produced; 7–9-nerved from the base, lateral nerves in 5–8 pairs in the median lobe, craspedodromous, tertiary nerves subparallel. Male inflorescences 8–23 cm long, paniculate; bracts 2–4 × 3–4 mm, broadly ovate-suborbicular, subacute or acute, subentire, erose, densely puberulous, many-flowered. Male flowers subsessile, strongly sweetly scented; calyx lobes 3–4, 1 mm long, elliptic-ovate, greenish-yellow; stamens (2)4(5), filaments 1 mm long, ± free, anthers 0.5 mm across. Female inflorescences 10 cm long, paniculate, little branched; bracts 1-flowered. Female flowers shortly pedicellate to subsessile; calyx c. 2.5 mm across, subtruncate to shallowly 4-lobed; ovary 2 × 2.5 mm, 2–3-lobed, tomentellous; styles 2–3, 4 mm long, slightly connate at the base, cylindric, spreading, papillose. Fruit 8–9 × 9–10(17) mm, 2-lobed or 1-chambered, shallowly verruculose-rugulose, puberulous, later glabrescent. Seeds 7–8 × 4.5 mm, ellipsoid, smooth, endotesta dark grey-brown, exotesta chestnut-brown.

Zambia. W: 6.5 km north of Kalene Hill, male fl. & fr. 26.ix.1952, *White* 3388 (FHO; K); Mwinilunga, st. 5.ix.1955, *Holmes* 1162 (K; NDO); st. 10.vi.1963, *Edwards* 722 (K; PRE; SRGH).

From Nigeria eastwards to the Sudan and south to Angola and Zambia. Swamp forest (mushitu).

2. **Macaranga mellifera** Prain in J. Linn. Soc., Bot. **40**: 201 (1911); in F.T.A. **6**, 1: 943 (1912). —Pax in Engler, Pflanzenr. [IV, fam. 147, vii] **63**: 387 (1914). —Eyles in Trans. Roy. Soc. South Africa **5**: 395 (1916). —Engler, Pflanzenw. Afrikas (Veg. Erde 9) **3**, 2: 93 (1921). —Topham, Check List For. Trees Shrubs Nyasaland Prot.: 51 (1958). —Drummond in Kirkia **10**: 252 (1975). —K. Coates Palgrave, Trees Southern Africa, ed. 2, rev.: 429 (1983). Tab. **34**. Syntypes: Zimbabwe, Chirinda outskirts, male fl. x.1905, *Swynnerton* 5 (BM; K); female fl. 29.x.1910, *Swynnerton* 2120 (BM; K); Mt. Pene, male fl. 12/4.x.1908, *Swynnerton* 6039 (BM).

A tree up to 18(30) m high, with ascending or spreading branches and a compact crown. Bark smooth, grey. Wood soft, white. Young shoots, petioles and inflorescence axes ferruginous-tomentellous and gland-dotted at first, soon glabrescent. Stipules 3 mm long, lanceolate, acute, deciduous. Petioles 3–8(14) cm long, pulvinate. Leaf blades (5)8–12(24) × (3)4–10(15) cm, ovate to elliptic-ovate, rounded to obtuse or subacute and shortly acuminate at the apex, cuneate to rounded or truncate and somewhat cordulate-auriculate at the base, 2-glandular on upper surface at the base, entire or subentire, sometimes remotely and shallowly glandular-denticulate, subcoriaceous, glabrous above, densely gland-dotted beneath and glabrous except for the densely hirsute lower third of the midrib and basal nerves; 3(5)-nerved from the base, lateral nerves in 6–7(8) pairs, looped, tertiary nerves parallel. Male inflorescences 4–7 cm long, densely paniculate; bracts small, 1–1.5 mm long, ovate, acute, entire, eglandular, many-flowered. Male flowers sessile, honey-scented; calyx lobes 3–4, 1.5 × 1 mm, ovate, acute or obtuse, glandular, yellow; stamens 13–17, the filaments of the inner 3–6 fused almost to the top, those of the outer ones united only at the base. Female inflorescences 2–5(7) cm long, racemose; bracts usually 1–2-flowered. Female flowers pedicellate, pedicels c. 2 mm long; calyx c. 1 mm across, cupular, ± truncate or scarcely

Tab. 34. MACARANGA MELLIFERA. 1, distal portion of female branch, and mature leaf (× $^{2}/_{3}$); 2a, upper portion of female inflorescence (× 3); 2b, female flower (× 4), 1 & 2 from *Fries, Norlindh & Weimarck* 2695; 3, upper portion of male inflorescence (× 3); 4, male flowers (× 6), 3 & 4 from *Chapman* 456; 5, fruits (× 3); 6, seeds (× 3), 5 & 6 from *Wild* 4408. Drawn by Christine Grey-Wilson.

3-lobed, green; ovary less than 1 mm in diameter, subglobose, unilocular, densely granulate-glandular and with or without a few fleshy warts; style lateral, 3–5 mm long, massive, subulate, thick, straight or recurved, ± smooth. Fruit 4 × 4 mm, subglobose, sparingly verrucose, densely granulate-glandular, pale yellowish-green. Seed 3.5 × 3.5 mm, subglobose, endotesta rugulose, dull, with a purplish-black fleshy exotesta.

Zimbabwe. E: Mutare (Umtali), fr. 28.i.1955, *Chase* 5455 (BM; COI; K; PRE; SRGH). S: Bikita, fr. 16.xii.1953, *Wild* 4408 (K; SRGH). **Malawi**. S: Mt. Mulanje, Lukulezi Valley, male 8.x.1957, *Chapman* 456 (BM; FHO; K; MAL; PRE). **Mozambique**. N: Ribáuè, Serra de Mepáluè, fr. 5.xii.1967, *Torre & Correia* 16375 (LISC). Z: Gurué, female 3.i.1968, *Torre & Correia* 16865 (LISC). MS: Chimanimani Mts., male 28.ix.1966, *Grosvenor* 243 (K; LISC; PRE; SRGH).

Not known from elsewhere. Submontane evergreen rainforest and forest margins, occasionally subdominant, also in kloof and gully forest, and on granite outcrops; 1100–1890 m.

The distribution of this species appears to parallel that of *Hibiscus burtt-davyi* Dunkley quite closely.

Vernacular names as recorded in specimen data include: “msoswe” (?); “mugarahanga” (Manyika); “nkwakwa” (chiYao).

3. **Macaranga capensis** (Baill.) Benth. ex Sim, For. Fl. Cape Col.: 314, t. 139 (1907). —Prain in F.C. **5**, 2: 486 (1920). —Pax in Engler, Pflanzenr. [IV, fam. 147, vii] **63**: 346 (1914). —Engler, Pflanzenw. Afrikas (Veg. Erde 9) **3**, 2: 93 (1921). —Brenan in Mem. N.Y. Bot. Gard. **9**, 1: 74 (1954). —Dale & Greenway, Kenya Trees & Shrubs: 206 (1961). —White, F.F.N.R.: 200 (1962). —Drummond in Kirkia **10**: 252 (1975). —K. Coates Palgrave, Trees Southern Africa, ed. 2, rev.: 428 (1983). —Radcliffe-Smith in F.T.E.A., Euphorb. 1: 243 (1987). —Beentje, Kenya Trees, Shrubs Lianas: 212 (1994). Type from South Africa.

Mappa capensis Baill., Étude Gén. Euphorb.: 430 (1858); in Adansonia **3**: 155 (1863). Type as above.

Mallotus capensis (Baill.) Müll. Arg. in Linnaea **34**: 189 (1865); in De Candolle, Prodr. **15**, 2: 966 (1866). —Schinz & Junod in Mém. Herb. Boissier, No. 10: 47 (1900). Type as above.

Macaranga ruwenzorica Pax in Bot. Jahrb. Syst. **43**: 322 (1909). —Prain in F.T.A. **6**, 1: 938 (1912). —Pax in Engler, Pflanzenr. [IV, fam. 147, vii] **63**: 345 (1914). —Engler, Pflanzenw. Afrikas (Veg. Erde 9) **3**, 2: 93 (1921). —Robyns & Tournay, Fl. Sperm. Parc Nat. Alb. **1**: 461 (1948). —Dale & Greenway, Kenya Trees & Shrubs: 208 (1961). Type from Zaire.

Macaranga inopinata Prain in F.T.A. **6**, 1: 944 (1912). —Pax in Engler, Pflanzenr. [IV, fam. 147, vii] **63**: 346 (1914). —Engler, Pflanzenw. Afrikas (Veg. Erde 9) **3**, 2: 93 (1921). —Brenan, Check-list For. Trees Shrubs Tang. Terr.: 217 (1949). Type from Tanzania.

Macaranga multiglandulosa Pax & K. Hoffm. in Engler, Pflanzenr. [IV, fam. 147, vii] **63**: 343 (1914). —Engler, Pflanzenw. Afrikas (Veg. Erde 9) **3**, 2: 93 (1921). —Brenan, Check-list For. Trees Shrubs Tang. Terr.: 218 (1949). Type from Tanzania.

Macaranga usambarica Pax & K. Hoffm., in Engler, Pflanzenr. [IV, fam. 147, vii] **63**: 344 (1914). —Engler, Pflanzenw. Afrikas (Veg. Erde 9) **3**, 2: 93 (1921). —Brenan, Check-list For. Trees Shrubs Tang. Terr.: 218 (1949). —Topham, Check List For. Trees Shrubs Nyasaland Prot.: 52 (1958). Types from Tanzania.

A tree up to 25 m high; crown rounded, spreading; stem ± buttressed. Trunk and branches with short spines. Bark light grey with white horizontal markings, smooth. Young growth densely ferruginous-tomentellous at first, later glabrescent. Stipules 1–2 cm × 4–8 mm, lanceolate, acutely acuminate, deciduous. Petioles up to 30 cm long, pulvinate at the base. Leaf blades up to 25 × 20 cm, broadly ovate to triangular-ovate, shortly acuminate at the apex, shallowly to deeply cordate or cordulate at the base, or if peltate then truncate or rounded at the base, entire or glandular-denticulate on the margins, firmly chartaceous, glabrescent on upper surface, hirsute beneath at least along the midrib and main nerves, densely gland-dotted beneath; 5–11-nerved from the base, lateral nerves in 6–14 pairs, looped, tertiary nerves parallel. Male inflorescences 8–12 cm long, paniculate; bracts 4–10 mm long, triangular ovate, adaxially glandular, strongly reflexed, many-flowered. Male flowers subsessile; calyx lobes 2–3, 0.75 mm long, ovate, yellowish-green; stamens 2–4, filaments ± free, 1 mm long, anthers minute, creamy-white. Female inflorescences 4–8 cm long, paniculate or subpaniculate; bracts 1–5-flowered. Female flowers shortly pedicellate, the pedicels c. 1–2 mm long; calyx urceolate, splitting into 2–4 lobes and becoming flattened as the fruit matures; ovary 1–1.5 × 0.75–1 mm, unlobed or rarely 2-lobed, densely yellowish-granulate-glandular; styles 1 or 2 and ± free, 2–3 mm long, recurved, fimbriate-papillose. Fruit 3–6 × 4–7(11) mm, subglobose, rarely 2-lobed, densely yellowish-green granulate-glandular at first, later becoming smooth, glabrous and green. Seeds 2.5–4 mm in diameter, subglobose, ± smooth, dull, red or black.

Zambia. N: Mpika, male fl. 29.i.1955, *Fanshawe* 1881 (FHO; K; NDO; SRGH). E: Kangampande Mt., Nyika, st. 6.v.1952, *White* 2731 (FHO; K). **Zimbabwe**. E: Chirinda Forest, female fl. i.1967, *Goldsmith* 8/67 (FHO; K; LISC; PRE; SRGH). S: La Rhone Farm, south of L. Mutirikwi (Kyle), Masvingo (Ft. Victoria), male fl. 17.xii.1970, *Müller & Pope* 1710 (K; PRE; SRGH). **Malawi**. N: Viphya Plateau, fr. 17.x.1977, *Pawek* 13135 (DAV; K; MAL; MO; SRGH; UC). C: Ntchisi Mt., male fl. 19.ii.1959, *Robson* 1669 (BM; K; LISC; MAL; PRE; SRGH). S: Thyolo (Cholo) Mt., y. fr. 22.ix.1946, *Brass* 17736 (BM; K; NY; PRE; SRGH). **Mozambique**. N: Malema, fr. 5.ii.1964, *Torre & Paiva* 10461 (LISC). Z: Serra Morrumbala, male fl. 9.xii.1971, *Müller & Pope* 1975 (K; LISC; SRGH). T: Ulóngué (Vila Coutinho), st. 19.xii.1980, *Macuácua* 1494 (K; LMA; PRE). MS: between Chimoio (Vila Pery) and Quedas do R. Revué, male fl. 24.v.1949, *Pedro & Pedrógão* 5886 (SRGH). M: Maputo, st. 4.iv.1947, *Hornby* 2617 (PRE; SRGH).

From S Ethiopia south to the Eastern Cape Province. Common in medium altitude submontane mixed evergreen forests, forest patches in grassland, and in regenerating forests, in the subcanopy and on forest margins, and at lower altitudes in riverine and gully forests, also in high rainfall woodland on mountain slopes and on swamp forest (mushitu) margins; 305–2133 m.

Vernacular names as recorded in specimen data include: "bwabwa" (Mulanje area); "chewaleeka", "chihaleeka" (chiYao); "mbalika" (chiWalika); "mepu" (chiYao); "mfukusa" (Nchese).

4. **Macaranga kilimandscharica** Pax in Engler, Pflanzenw. Ost-Afrikas **C**: 238 (1895); in Bot. Jahrb. Syst. **23**: 526 (1897). —Prain in F.T.A. **6**, 1: 938 (1912). —Pax & K Hoffm. in Engler, Pflanzenr. [IV, fam. 147, vii] **63**: 344 (1914). —Engler, Pflanzenw. Afrikas (Veg. Erde 9) **3**, 2: 93 (1921). —De Wildeman, Pl. Bequaert. **3**, 4: 478 (1926). —Robyns & Tournay, Fl. Sperm. Parc Nat. Alb. **1**: 461 (1948). —Brenan, Check-list For. Trees Shrubs Tang. Terr.: 217 (1949). —Eggeling & Dale, Indig. Trees Uganda, ed. 2: 132 (1952). —F.W. Andrews, Fl. Pl. Anglo-Egypt. Sudan **2**: 85 (1952). —Brenan in Mem. N.Y. Bot. Gard. **9**, 1: 74 (1954). —Topham, Check List For. Trees Shrubs Nyasaland Prot.: 52 (1958). —Dale & Greenway, Kenya Trees & Shrubs: 208 (1961). —Radcliffe-Smith in F.T.E.A., Euphorb. 1: 245 (1987). —Beentje, Kenya Trees, Shrubs Lianas: 212 (1994). Types from Tanzania.

Macaranga nyassae Pax & K. Hoffm. in Engler, Pflanzenr. [IV, fam. 147, vii] **63**: 343 (1914). —Engler, Pflanzenw. Afrikas (Veg. Erde 9) **3**, 2: 93 (1921). Type from Tanzania.

Macaranga mildbraediana Pax & K. Hoffm. in Engler, Pflanzenr. [IV, fam. 147, vii] **63**: 343 (1914). —Engler, Pflanzenw. Afrikas (Veg. Erde 9) **3**, 2: 93 (1921). Type from Zaire.

Macaranga neomildbraediana Lebrun in Ann. Soc. Sci. Bruxelles **54**, Sér. B: 160 (1934). —Robyns & Tournay, Fl. Sperm. Parc Nat. Alb. **1**: 461 (1948). —Troupin, Fl. Pl. Lign. Rwanda: 264, fig. 91/1 (1982); Fl. Rwanda **2**: 230, fig. 69/1 (1983). Type from Zaire.

Macaranga capensis var. *kilimandscharica* (Pax) Friis & M.G. Gilbert in Kew Bull. **41**: 68 (1986).

Very similar to *M. capensis*, but with leaf blades rhombic-ovate to triangular-ovate, more gradually acuminate at the apex, cuneate or rounded at the base, not more than 15 × 10 cm, and generally somewhat glaucous beneath; and with inflorescence axes more obviously zigzag towards the apex, and the distal bracts of the male flowers ± entire.

Zambia. E: Nyika Plateau, Chowo Forest, fr. 9.ix.1976, *Pawek* 11785 (K; MAL; MO). **Malawi**. N: Chitipa Distr., Misuku Hills, Mughesse Forest, y. male fl. 28.xii.1972, *Pawek* 6198 (K; MAL; MO; SRGH; UC). S: Zomba Mt., st. 22.vi.1961, *Chapman* 1389 (FHO; MAL).

Also in Sudan, Ethiopia, Zaire (Kivu Province), Uganda, Kenya, Rwanda, Burundi and Tanzania. Submontane mixed evergreen forest, in subcanopy and on forest margins, also in gullies by stream sides; 2000–2100 m.

Doubtfully distinct from *M. capensis*, with which it intergrades. *Armitage* 33/55 in *GHS* 53659 (PRE; SRGH) from Zimbabwe (E) and *Chase* 6837 (K; SRGH) from Mozambique (MS) are intermediate in most respects between the two. However, I have refrained from following Friis & M.G. Gilbert (1986) in uniting them since most collections are readily assignable to one or the other.

31. ERYTHROCOCCA Benth.

Erythrococca Benth. in Hooker, Niger Flora: 506 (1849).
Claoxylon sect. *Athroandra* Hook.f. in J. Linn. Soc., Bot. **6**: 21 (1862).
Claoxylon sect. *Adenoclaoxylon* Müll. Arg. in Flora **47**: 436 (1864).
Chloropatane Engl. in Bot. Jahrb. Syst. **26**: 383 (1899).
Athroandra (Hook.f.) Pax & K. Hoffm. in Engler, Pflanzenr. [IV, fam. 147, vii] **63**: 76 (1914).

Dioecious shrubs with a simple indumentum. Buds perulate (furnished with protective scales), the perulae crustaceous, persistent. Leaves alternate, petiolate, stipulate, simple, glandular-toothed, penninerved. Stipules often accrescent, persistent, hardened, sometimes spinescent, occasionally deciduous. Inflorescences axillary, solitary or fasciculate, sessile or pedunculate, glomerulate, racemose or subpaniculate, usually bracteate and bracteolate. Flowers small or minute; pedicels capillary, jointed. Male flowers: calyx closed in bud, later splitting into 3–4(5) valvate lobes; petals absent; disk glands extrastaminal, free or connate, interstaminal and free, or both, generally hirsute; stamens 2–60, filaments free, anthers erect, 2-celled, extrorse, basifixed, the cells obovoid-subglobose, ± free, apically dehiscent; pistillode absent. Female flowers: calyx 2–4-partite, the lobes imbricate; petals absent; disk glands 2–3, free or contiguous, with or without minute intercalated supernumerary glands, or disk shallowly urceolate; ovary 2–3-locular, with 1 ovule per loculus; styles 2–3, free or slightly connate at the base, spreading, stigmas smooth, papillose, lobulate, fimbriate, laciniate or plumose. Fruits 1–3-coccous, loculicidally dehiscent; cocci often subglobose; endocarp coriaceous. Seeds subglobose, thinly arillate; aril usually scarlet; testa crustaceous, foveolate-reticulate or ± smooth; albumen fleshy; embryo axile, radicle conical, cotyledons broad, flat.

A genus of some 50 species, mostly confined to tropical Africa.

1. Flowers in glomerules ··· 2
– Flowers in racemes ··· 3
2. Leaves often purplish-tinged at first; stamens 8–13 ··· 1. *kirkii*
– Leaves not purplish-tinged at all; stamens 2–5 ··· 4. *menyharthii*
3. Stipules thorn-like, 2–4 mm long ··· 4
– Stipules not thorn-like, 0.5–1 mm long ··· 5
4. Leaves elliptic-obovate to elliptic-ovate; male peduncles puberulous ··· 2. *zambesiaca*
– Leaves elliptic to elliptic-lanceolate; male peduncles ± glabrous ··· 3. *berberidea*
5. Leaves drying blackish-purple; ovary glabrous; stigmas papillose-lobulate ··· 7. *polyandra*
– Leaves drying yellowish, greenish or brownish; ovary glabrous or pubescent ··· 6
6. Stamens 2–35; stigmas subentire, lobulate, fimbriate or laciniate ··· 7
– Stamens 25–40; stigmas smooth or ± so ··· 9
7. Leaves elliptic-lanceolate, dark green; stamens 30–35 ··· 6. *ulugurensis*
– Leaves elliptic to ovate, dull or yellowish-green; stamens 2–25 ··· 8
8. Stamens 2–5; extrastaminal glands present; ovary 3-lobed, evenly puberulous; stigmas laciniate; fruit subglabrous ··· 4. *menyharthii*
– Stamens (9)10–25; extrastaminal glands absent; ovary 2-lobed, evenly to densely appressed sericeous-pubescent; stigmas fimbriate to fimbriate-lobulate; fruit evenly pubescent to subglabrous ··· 5. *trichogyne*
9. Leaves up to 18 × 10 cm; stigmas suborbicular, minute, much shorter than the cylindric stylar column ··· 10. *welwitschiana*
– Leaves up to 10 × 5 cm; stigmas divaricate-spreading, later recurved, longer than the fused portion of the style ··· 10
10. Leaves drying medium to dark green; male flowers 3–4 mm across; female racemes 10–12-flowered ··· 8. *atrovirens*
– Leaves drying brownish or brownish-green; male flowers 5–6 mm across; female racemes 5–6-flowered ··· 9. *angolensis*

1. **Erythrococca kirkii** (Müll. Arg.) Prain in Ann. Bot. (London) **25**: 609 (1911); in F.T.A. **6**, 1: 853 (1912). —Pax in Engler, Pflanzenr. [IV, fam. 147, vii] **63**: 96 (1914). —Engler, Pflanzenw. Afrikas (Veg. Erde 9) **3**, 2: 72, 74, t. 30 (1921). —Brenan, Check-list For. Trees Shrubs Tang. Terr.: 208 (1949). —Dale & Greenway, Kenya Trees & Shrubs: 195 (1961). —Radcliffe-Smith in F.T.E.A., Euphorb. 1: 266 (1987). —Beentje, Kenya Trees, Shrubs Lianas: 196 (1994). Type from Tanzania/Mozambique border, Rovuma Bay, lat. 10°S, Mar. 1861, *Kirk* s.n. (K, holotype).

Claoxylon kirkii Müll. Arg. in Flora **47**: 436 (1864); in De Candolle, Prodr. **15**, 2: 776 (1866). —Engler, Pflanzenw. Ost-Afrikas **C**: 238 (1895). —Sim, For. Fl. Port. E. Afr.: 105 (1909).

A shrub commonly to 1.5 m tall, sometimes taller. Bark rough. Twigs pale green or grey-green, evenly lenticellate. Young growth evenly to sparingly puberulous or glabrous. Petioles 0.5–2.5 cm long, purplish. Leaves 5–15 × 2–8 cm, ovate or elliptic to elliptic-lanceolate, shortly obtusely acuminate at the apex, coarsely crenate on the margin, cuneate or rounded-cuneate at the base, soft when fresh, chartaceous and brittle when dried, sparingly puberulous only along the midrib beneath at first, soon becoming completely glabrous, dull, pale or dark green, often purplish-tinged when young; lateral nerves in 6–8 pairs, looped well within the margin, scarcely prominent above, slightly so beneath. Stipules 1 mm wide, broadly triangular, slightly accrescent, becoming mammillate-umbonate but scarcely spinulose. Male inflorescences 1–2 cm long, densely glomerulate, sessile or shortly pedunculate, the peduncles glabrous; bracts minute, ciliate. Male flowers: pedicels extending to 1 cm at anthesis, slender, flexuous, glabrous; calyx lobes 3, 1 × 1.3 mm, broadly triangular, subacute, glabrous, pale greenish-white, sometimes mauve-tinged; extrastaminal disk of several irregularly connate, flat, pubescent glands; interstaminal glands erect, rhomboidal, truncate, pubescent, purple; stamens 8–13, 2 of which are central, the rest peripheral, 0.5 mm long, white or purplish-brown. Female inflorescences similar to those of the male, but fewer flowered. Female flowers: pedicels 3–4 mm long, stouter than in male; calyx lobes smaller than in male, ciliate, yellow; disk glands 3, scale-like; ovary 1 mm in diameter, 3-lobed, sparingly puberulous to subglabrous, purple; styles 1 mm long, free, stigmas fimbriate-laciniate, white. Fruit 3 × 6 mm, tricoccous, or by abortion di- or monococcous, glabrous or subglabrous, greenish, reddish or purplish. Seeds 2.5–3 mm in diameter, shallowly foveolate-reticulate, yellow, orange or red.

Mozambique. MS: Cheringoma Coastal Area, st. v.1973, *Tinley* 2937 (LISC; PRE; SRGH).
Also known from Kenya and Tanzania. In coastal forest.

2. **Erythrococca zambesiaca** Prain in Bull. Misc. Inform., Kew **1911**: 90 (1911); in Ann. Bot. (London) **25**: 613 (1911); in F.T.A. **6**, 1: 857 (1912). —Pax in Engler, Pflanzenr. [IV, fam. 147, vii] **63**: 90 (1914) as "*sambesiaca*". —Engler, Pflanzenw. Afrikas (Veg. Erde 9) **3**, 2: 71, 72 (1921). Type: Malawi S/Mozambique Z, lower Shire R., Chiromo, Jan. 1895, *Scott-Elliot* 2795 (K, holotype; BM).

A large shrub to small tree, commonly 1.5–3 m high. Young shoots, petioles and inflorescence axes evenly to sparingly puberulous. Petioles 2–8 mm long. Leaves 1.5–9.5 × 1–4 cm, elliptic-obovate to -ovate, obtuse or subacute at the apex, shallowly glandular crenate-serrate to subentire on the margin, cuneate or attenuate into the petiole, membranous, sparingly puberulous on the midrib and main nerves beneath, otherwise glabrous; 3–4-nerved from the base, lateral nerves in 3–4 pairs, ascending, scarcely looped, slightly prominent. Stipules 2–4 mm long, narrowly conical to subulate, spinescent, glabrous, thorn-like. Male inflorescences up to 4 cm long, with the peduncle to 1.5 cm long, interruptedly racemose; bracts 1 mm long, ovate, ciliate. Male flowers: pedicels slender, lengthening to 1 cm at anthesis, glabrous, jointed near the base; buds 1 mm in diameter, depressed-subglobose, apiculate, glabrous; calyx lobes 4–6, 1 × 0.6 mm, ovate; extrastaminal disk glands 6, flattened, truncate, puberulous at the apex, ± fused into a shallow cup; interstaminal glands liguliform, apically pubescent; stamens 15–30, c. 1 mm long, anthers minute. Female inflorescences up to 8 mm long, with the peduncle c. 1–3 mm long, racemose, few-flowered. Female flowers: pedicels 2 mm long, sparingly pubescent; calyx lobes 3, c. 1 × 1.25 mm, broadly ovate-suborbicular, subacute or obtuse, glabrous, ciliate, pale greenish-yellow; disk glands 3, petaloid, slightly smaller than the calyx lobes, triangular-ovate, pubescent, thick, creamy-yellow; ovary less than 1 mm in diameter, 3-lobed, evenly pubescent; styles 3, c. 1 mm long, free, spreading, laciniate, ± glabrous, persistent. Fruit 7 × 4 mm, tricoccous, sparingly puberulous, very pale green. Seeds 3.5 mm in diameter, globose, ± smooth.

Malawi. S: Chikwawa Distr., Lengwe Game Reserve, male fl. 7.i.1970, *Hall-Martin* 449 (K; PRE). **Mozambique**. MS: Gorongosa Nat. Park, Sangarassa For., male & female fl. i.1972, *Tinley* 2331 (K (male & female); LISC (female); PRE (female); SRGH (male)).

Not known elsewhere. In dry *Newtonia*/ *Xylia* forest understorey on sandy soil, in bushy clump savanna, in termitaria thickets and on alluvium; 70–110 m.

Very close to *E. natalensis* Prain from KwaZulu-Natal, from which it differs in having puberulous male peduncles, pedicels jointed near the base, and only 6 extrastaminal disk glands.

3. **Erythrococca berberidea** Prain in Bull. Misc. Inform., Kew **1911**: 92 (1911); in Ann. Bot. (London) **25**: 613 (1911); in F.C. **5**, 2: 459 (1920). —Pax in Engler, Pflanzenr. [IV, fam. 147, vii] **63**: 90 (1914). —Engler, Pflanzenw. Afrikas (Veg. Erde 9) **3**, 2: 71, 72 (1921). —K. Coates Palgrave, Trees Southern Africa, ed. 2, rev.: 422 (1983). —Radcliffe-Smith in F.T.E.A., Euphorb. 1: 271 (1987). Type from South Africa (KwaZulu-Natal).

A shrub to c. 3 m tall. Twigs pale grey. Young growth sparingly pubescent to subglabrous. Petioles 4–7 mm long. Leaf blades 2–7.5 × 1–2.5 cm, elliptic to elliptic-lanceolate, acute or subacute at the apex, shallowly and indistinctly glandular-serrate on the margin, cuneate and eglandular at the base, chartaceous, sparingly pubescent along the midrib beneath, otherwise glabrous, drying greenish or brownish; lateral nerves in 3–4 pairs, looped within the margin, scarcely prominent above or beneath. Stipules 1–3 mm long, thorn-like, conical, straight or slightly curved, accrescent, stramineous. Male inflorescences 2–4 cm long with peduncles 1–1.5 cm long, interruptedly racemose, glabrous or subglabrous; bracts c. 1 mm long, triangular, persistent. Male flowers: pedicels up to c. 1 cm long at anthesis, capillary, subglabrous; buds 1–1.5 mm in diameter, oblate, apiculate, basally truncate, glabrous; calyx lobes 3–4, 2 × 1.5 mm, ovate, obtuse or shortly acuminate, pale green; disk glands numerous, 5 of which are extrastaminal, the remainder interstaminal, flattened, rounded, ciliate; stamens 15–18, comprising 8–10 outer, and the rest central, filaments very short, anthers minute, white. Female inflorescences 1.5–3 cm long, the peduncles c. 1.5 cm long, few-flowered, glabrous; bracts ± as in male. Female flowers: pedicels 2–3 mm long, jointed near the base, glabrous; calyx lobes 3, c. 1 × 1 mm, ovate, subacute, ciliate, green; disk glands 3, petaloid, c. 1 × 0.5 mm, oblong-lanceolate, rounded; ovary 1 mm in diameter, 3-lobed, sparingly pubescent to subglabrous, green; styles 3, united at the base, 1–1.5 mm long, reflexed-spreading, stigmas laciniate. Fruit 5 × 9 mm, tricoccous, or di- or monococcous by abortion, glabrescent, green at first, drying purplish. Seeds 4 mm in diameter, shallowly foveolate-reticulate, aril orange.

Mozambique. GI: 10 km Chipenhe–Mainguelane, female fl. & fr. immat. 19.ix.1980, *Nuvunga, Boane & Conjo* 324 (K; LMU).

Also in Tanzania and South Africa (KwaZulu-Natal). Coastal forest margins, woodland and scrub.

On account of the scrappy nature of the Mozambican material, South African material was used to draw up the description of male and female flowers, fruits and seeds.

4. **Erythrococca menyharthii** (Pax) Prain in Ann. Bot. (London) **25**: 616 (1911); in F.T.A. **6**, 1: 860 (1912). —Pax in Engler, Pflanzenr. [IV, fam. 147, vii] **63**: 93 (1914). —Engler, Pflanzenw. Afrikas (Veg. Erde 9) **3**, 2: 71, 73 (1921). —O.B. Miller, Check-list For. Trees Shrubs Bech. Prot.: 32 (1948). —Brenan, Check-list For. Trees Shrubs Tang. Terr.: 208 (1949). —White, F.F.N.R.: 197 (1962). —P.G. Meyer in Merxmüller, Prodr. Fl. SW. Afrika, fam. 67: 12 (1967). —Drummond in Kirkia **10**: 252 (1975). —K. Coates Palgrave, Trees Southern Africa, ed. 2, rev.: 423 (1983). —Radcliffe-Smith in F.T.E.A., Euphorb. 1: 273 (1987). —Beentje, Kenya Trees, Shrubs Lianas: 197 (1994). Tab. **35**. Type: Mozambique, Tete Province, Boroma (Boruma), 1891, *Menyharth* 889b (Z, holotype).

Claoxylon menyharthii Pax in Bull. Herb. Boissier sér. 2, **9**: 877 (1901).

Claoxylon virens N.E. Br. in Bull. Misc. Inform., Kew **1909**: 140 (1909). Syntypes: Botswana, Ngamiland, Khwebe (Kwebe) Hills, *Lugard* 53, 94 (GRA; K); Mrs. *Lugard* 51 (GRA; K).

A straggly subshrub 0.5–3 m tall with a loose, open crown. Branches not usually exceeding 4 cm in diameter at the base. Bark pale grey, smooth, flaking. Wood soft. Twigs pale grey, white-lenticellate. Young growth softly puberulous or pilose. Petioles 1–6 mm long. Leaf blades 2–10 × 1–6 cm, ovate, ovate-lanceolate or elliptic-ovate, acutely to obtusely acuminate at the apex, subentire to coarsely and usually remotely serrate-dentate on the margin, cuneate, rounded or truncate at the base usually with 2–4 subulate stipelliform glands at or near the base, membranous at first, later thinly chartaceous, sparingly softly pubescent on upper surface, especially on the midrib and main nerves, more evenly so beneath, yellowish-green; lateral nerves in 4–5 pairs, ascending, looped or not, not prominent. Stipules 1–2.5 mm long, subulate-filiform, not or scarcely aculeate and hardened, yellowish-brown. Male inflorescences 1–5 cm long, with the peduncle up to 2 cm long, interruptedly

Tab. 35. ERYTHROCOCCA MENYHARTHII. 1, flowering branch (× $^{2}/_{3}$); 2, male flower (× 12), 1 & 2 from *Mutimushi* 3938; 3, fruiting branch (× $^{2}/_{3}$); 4, fruits (× 2); 5, seeds (× 2), 3–5 from *Martin* 491/32. Drawn by Christine Grey-Wilson.

racemose to subglomerulate; bracts 0.5 mm long, triangular-ovate, ciliate. Male flowers: pedicels slender, flexuous, lengthening to c. 5 mm at anthesis, sparingly pubescent; buds 1 mm in diameter, ovoid-conic, apiculate, evenly pubescent; calyx lobes 4–5, 1–1.2 × 0.5–0.8 mm, ovate-lanceolate, acute, pale yellowish-green, translucent; extrastaminal disk of 8–9 unequal free or irregularly-fused glands, glabrous or with 1–2 hairs each; interstaminal glands (0) 1–2, rarely more, subglobose to subclavate; stamens 2–3(5), rarely more, subsessile, anthers less than 0.5 mm across. Female inflorescences 1–3 cm long, with the peduncle up to 2.5 cm long, simply racemose, few-flowered. Female flowers: pedicels 1–2 mm long, pubescent; calyx lobes 3(4), 1–1.5 × 1–1.2 mm, suborbicular-ovate, subacute or obtuse, sparingly pubescent without, glabrous within, pale green; disk glands 3, 0.5 mm long, triangular-lanceolate, obtuse, glabrous; ovary 1 mm in diameter, 3-lobed, evenly puberulous; styles 3, c. 1 mm long, united only at the base, spreading or recurved, stigmas laciniate. Fruit 4 × 9 mm, generally tricoccous, rarely di- or monococcous by abortion, sparingly pubescent, rarely subglabrous, pale green at first, darker later. Seeds 3–4 mm in diameter, reticulate, aril reddish-orange.

Caprivi Strip. 67 km west of Katima Mulilo, fr. 17.ii.1969, *de Winter* 9211 (K; PRE; SRGH). **Botswana**. N: Linyanti R., fr. 14–15.i.1979, *P.A. Smith* 2611 (K; MN; PRE; SRGH). **Zambia**. B: Lonze Forest, fr. 19.xii.1952, *Angus* 955 (FHO; K). C: Katondwe, male fl. 6.xii.1969, *Mutimushi* 3938 (K; NDO). E: Nyika, y. fr. 30.xii.1962, *Fanshawe* 7340 (K; NDO). S: north of Siburu Forest, Kalomo, fr. 14.ii.1963, *Bainbridge* 733 (FHO; K; NDO; SRGH). **Zimbabwe**. N: Hurungwe Distr., Musukwi (Msukwe) R., male fl. 20.xi.1953, *Wild* 4221 (K; LISC; MO; PRE; SRGH). W: Hwange (Wankie) Nat. Park, fr. 25.ii.1967, *Rushworth* 234 (K; LISC; PRE; SRGH). **Mozambique**. N: Nampula, Monapo, female fl. & fr. 11.ii.1984, *de Koning et al.* 9547 (K; LMU). T: Chicoa, Serra de Songa, Cahora Bassa–R. Zambezi, male fl. 30.xii.1965, *Torre & Correia* 13887 (LISC). MS: Búzi, Mucheve For. Res., male fl. 4.xi.1967, *Carvalho* 958 (LISC). GI: between Funhalouro and Inhambane, st. 22.v.1941, *Torre* 2717 (LISC). M: between Umbelúzi and Matutuíne (Bela Vista), male fl. 20.xi.1940, *Torre* 2089 (LISC).

Also in Kenya, Tanzania, Angola, Namibia and South Africa (Transvaal). Often as a constituent of thickets on sandy soil; understorey thicket of *Baikiaea* woodland on Kalahari Sand (mutemwa), dense riverine thicket and thicket on floodplain termitaria; also on rocky outcrops and hillsides, dry mopane woodland on sandy soil, and submontane forest margins; 300–1050 m.

Vernacular name as recorded in specimen data: "dipokio".

5. **Erythrococca trichogyne** (Müll. Arg.) Prain in Ann. Bot. (London) **25**: 617 (1911); in F.T.A. **6**, 1: 863 (1912). —Pax in Engler, Pflanzenr. [IV, fam. 147, vii] **63**: 95 (1914). —Engler, Pflanzenw. Afrikas (Veg. Erde 9) **3**, 2: 71, 73 (1921). —Drummond in Kirkia **10**: 252 (1975). —Troupin, Fl. Pl. Lign. Rwanda: 260 (1982); Fl. Rwanda **2**: 220 (1983). —K. Coates Palgrave, Trees Southern Africa, ed. 2, rev.: 423 (1983). —Radcliffe-Smith in F.T.E.A., Euphorb. 1: 274 (1987). —Beentje, Kenya Trees, Shrubs Lianas: 197 (1994). Type from Angola (Cuanza Norte).

Claoxylon trichogyne Müll. Arg. in J. Bot. **2**: 334 (1864); in De Candolle, Prodr. **15**, 2: 778 (1866).

Erythrococca sp. 1 [*White* 3579, 3819 (both FHO; K)] in White, F.F.N.R.: 198 (1962).

An erect much-branched shrub or small spreading tree up to 6 m high. Bole light brown, rough. Wood hard, white. Branches whitish or pale to dark grey. Young growth pubescent. Perulae (protective scales of buds) creamy-white. Petioles 3–10 mm long. Leaf blades 1–11 × 0.5–5.5 cm, elliptic to elliptic-ovate, acuminate, coarsely and often somewhat irregularly glandular crenate-serrate or dentate to subentire, cuneate at the base, often with 1–2 small subglobose glands at or near the base, membranous, sparingly pubescent to subglabrous on upper surface, often more evenly pubescent beneath, especially along the midrib and main nerves, dull green; lateral nerves in 4–6 pairs, ascending, looped or not, scarcely prominent above, slightly so beneath. Stipules c. 0.5 mm long, conical, not or scarcely aculeate and hardened, yellowish-brown. Male inflorescences 1–2 cm long, racemose; bracts minute, linear, fugacious. Male flowers: pedicels 1–4 mm long, subglabrous, jointed above the base; buds 1–2 mm in diameter, subglobose, shortly apiculate, subglabrous; calyx lobes 3–4, c. 1.5 × 1.5 mm, ovate, subacute or obtuse, pale green or yellowish-green; extrastaminal disk absent; stamens (9) 10–25, filaments very short, anthers 0.67 mm across, yellow; interstaminal glands minute, clavate, each tipped with a single hair. Female inflorescences few-flowered, and peduncles extending to up to 4 cm long in fruit, otherwise as in male. Female flowers: pedicels 1–2(5) mm long, thicker than in male, pubescent, jointed below the flowers; calyx lobes 2–4, c. 1.5 × 0.5 mm, lanceolate, acute, pale green; disk glands 2, 0.5 mm long, compressed-ovoid, obtuse to truncate, glabrous; ovary 1.5 mm across, 2-lobed, evenly to densely appressed sericeous-pubescent; styles 2, 1 mm long, united at the base, spreading, stigmas fimbriate to fimbriate-lobulate. Fruit 5–10 mm across, pendulous, dicoccous, rarely monococcous by abortion, subglabrous or sparingly to evenly appressed-pubescent, pale green. Seeds 2.5–4 mm in diameter, foveolate-reticulate; aril sticky, bright orange-red; testa black.

Stamens 10–25; female pedicels 1–2 mm long in fruit; ovary densely appressed sericeous-pubescent; stigmas fimbriate-lobulate; fruit evenly pubescent · · · · · · · · · · · var. *trichogyne*
Stamens 9; female pedicels 3–5 mm long in fruit; ovary evenly appressed-pubescent; stigmas fimbriate; fruit subglabrous · var. *psilogyne*

Var. **trichogyne**

Stamens 10–25; female pedicels 1–2 mm long in fruit; ovary densely appressed sericeous-pubescent; stigmas fimbriate-lobulate; fruit evenly pubescent.

Botswana. N: 6 km south of Shakawe, fr. (SRGH), st. (MO) 24.iv.1975, *Müller & Biegel* 2258 (MO; SRGH). SE: Kwaneng Distr., Matlolakgang Ranch, male fl. 27.ix.1978, *O.J. Hansen* 3467 (C; GAB; K; PRE; SRGH; WAG). **Zambia**. N: Kafulwe Mission, female fl. 2.ix.1952, *White* 3579 (BM; FHO; K; MO). W: Ndola, fr. 24.xii.1954, *Fanshawe* 1744 (K; NDO; SRGH). C: Lusaka, male fl. 6.xii.1957, *Fanshawe* 4113 (K; NDO). S: Namwala Distr., Ngoma, Kafue Nat. Park, male fl. 16.xi.1961, *B.L. Mitchell* 11/27 (SRGH). **Zimbabwe**. N: Mazowe Distr., Citrus Estate Road, Iron Duke Mine, male fl. 9.xii.1960, *Rutherford-Smith* 425 (K; LISC; MO; PRE; SRGH). W: southwest Matopos, Maleme Valley, Mt. Mumongwe, fr. 5.i.1963, *Wild* 5973 (K; LISC; MO; PRE; SRGH). C: Shurugwi (Selukwe), y. fr. 8.xii.1953, *Wild* 4305 (K; LISC; MO; PRE; SRGH). E: Murahwa's Hill, female fl. 6.ii.1963, *Chase* 7953 (K; LISC; PRE; SRGH). S: Masvingo (Ft. Victoria), Muzero Farm, fr. 30.xii.1970, *Chiparawasha* 276 (SRGH). **Malawi**. C: Lilongwe Distr., Nature Sanctuary, fr. 28.i.1985, *Patel & Banda* 1989 (K; MAL). **Mozambique**. N: Monapo, fr. 10.ii.1984, *Groenendijk et al.* 989 (K; LMU).

Also in Zaire (Kivu Province), Rwanda, Burundi, Ethiopia, Uganda, Kenya, Tanzania, Angola and South Africa (Transvaal). Often as a constituent of thickets; riverine thickets, kopje and rocky outcrop thickets, termite mound thicket, and dry evergreen thicket (mateshi) of high rainfall *Brachystegia* woodland, also in rain forest and submontane *Brachystegia* woodland; 900–1500 m.

Although in common with all *Erythrococca*, this species is almost 100% dioecious, one specimen has been seen (*Mutimushi* 3813, from Kitwe, W Zambia, 25.x.1969 (K)) in which not only are some of the flowers female on this predominantly male gathering, but there are even some hermaphrodite flowers as well.

Var. **psilogyne** Radcl.-Sm. in Kew Bull. **47**: 678 (1992). Type: Malawi, N Prov., Chitipa Distr., Misuku Hills, Mughesse Forest, fr. 2.i.1978, *Pawek* 13504 (K, holotype; MAL; MO; PRE; SRGH; UC).

Stamens 9; female pedicels 3–5 mm long in fruit; ovary evenly appressed-pubescent; stigmas fimbriate; fruit subglabrous.

Zambia. E: Nyika Plateau, Chowo Forest, fr. 8.i.1982, *Dowsett-Lemaire* 265 (K). **Malawi**. N: Rumphi Distr., Nyika Plateau, male fl. 8.i.1974. *Pawek* 7878 (K; MAL; MO; SRGH; UC).

Not known elsewhere. Montane evergreen rainforest, in understorey and on margins; 1770–2200 m.

6. **Erythrococca ulugurensis** Radcl.-Sm. in Kew Bull. **33**: 239 (1978); in F.T.E.A., Euphorb. 1: 276 (1987). Tab. **36**. Type from Tanzania (Morogoro District).

A slender shrub up to c. 2.5 m tall, with arching spreading branches. Twigs dark greyish-brown. Young growth evenly pubescent. Petioles 1–5 mm long, sometimes with 1–2 conic-hemispheric yellowish stipelliform glands at the apex. Leaf blades 1–7.5 × 0.5–2.5 cm, elliptic-lanceolate, acutely long-acuminate apically, coarsely glandular serrate-dentate to subentire on the margins, cuneate at the base, membranous, sparingly pubescent on both surfaces, more evenly so along the midrib

Tab. 36. ERYTHROCOCCA ULUGURENSIS. 1, flowering branch (× 1); 2, male flower and bud (× 9), 1 & 2 from *Harris et al.* 5157; 3, fruiting branch (× 1); 4, seeds (× 3), 3 & 4 from *Robson* 1671. Drawn by Christine Grey-Wilson.

and main nerves, drying dark green; lateral nerves in 3–4 pairs, ascending, not or weakly looped, scarcely prominent above or beneath. Stipules 0.5 mm long, conic-cylindric, not aculeate, scarcely becoming hardened, orange-brown. Male inflorescences 1–1.5 cm long, racemose, few-flowered, pedunculate, appressed-pubescent; bracts 0.7 mm long, linear-setaceous. Male flowers: pedicels 2–4 mm long, slender, glabrous, jointed at the base; buds less than 1 mm across, conic-quadrangular, apiculate, truncate at the base, glabrous; calyx lobes 4, 2 × 1.5 mm broadly ovate, subacute, pale greenish-cream; extrastaminal disk glands c. 10, minute, rounded, tipped with a few hairs; interstaminal glands much shorter than the stamens, truncate, each tipped with a hair tuft; stamens 30–35, 10–12 outer, the rest central, 1 mm long, anthers minute, yellow; sometimes the whole centre of the male flower replaced by the shiny, black, quandrangular sclerotium of a fungus, ? *Sclerotinia sp.*, in which case the buds are swollen to 3 × 3 mm. Female inflorescences 2–3.5 cm long, racemose, 1–4-flowered, pedunculate, evenly to sparingly appressed-pubescent; bracts 1 mm long, narrowly lanceolate, pubescent. Female flowers: pedicels 3–6 mm long, extending a little in fruit, sparingly appressed-pubescent, jointed below the middle; calyx lobes 4, 2 × 1.5 mm, broadly ovate, subacute or obtuse, subglabrous, pale green; disk 1.5–3 mm in diameter, urceolate, thin, shallowly 3-lobed, sparingly ciliate; ovary 1–1.5 mm in diameter, 3-lobed, glabrous, subglabrous or evenly pubescent; styles 3, 1.5 mm long, shortly united at the base, spreading, stigmas proximally subentire, distally laciniate. Fruit 3–4 × 7–8 mm, 3-lobed, glabrous or subglabrous, pale green. Seeds c. 4 mm in diameter, foveolate-reticulate-muricate, aril orange.

Malawi. C: Ntchisi Mt., fr. 19.ii.1959, *Robson* 1671 (BM; K; LISC; SRGH); Ntchisi For. Res., diseased fls. 11.v.1984, *Banda & Kaunda* 2183 (K; MAL).

Also in Tanzania. Evergreen forest understorey; 1450 m.

Owing to the imperfection of the Malawi material, Tanzanian material was used in the drawing up of this description.

7. **Erythrococca polyandra** (Pax & K. Hoffm.) Prain in Ann. Bot. (London) **25**: 618 (1911); in F.T.A. **6**, 1: 864 (1912). —Pax in Engler, Pflanzenr. [IV, fam. 147, vii] **63**: 96 (1914). —Engler, Pflanzenw. Afrikas (Veg. Erde 9) **3**, 2: 71, 73 (1921). —Brenan, Check-list For. Trees Shrubs Tang. Terr.: 208 (1949). —Drummond in Kirkia **10**: 252 (1975). —K. Coates Palgrave, Trees Southern Africa, ed. 2, rev.: 423 (1983). —Radcliffe-Smith in F.T.E.A., Euphorb. 1: 277 (1987). Type from Tanzania (Lushoto District).
Claoxylon polyandrum Pax & K. Hoffm. in Bot. Jahrb. Syst. **45**: 237 (1910).

A shrub or small tree up to 10 m high. Bark grey, thin, slightly fissured at the base. Twigs pale greyish-brown or whitish. Young growth sparingly pubescent to subglabrous. Petioles 0.5–1.5 cm long. Leaf blades 2–13 × 1–7 cm, ovate to elliptic-ovate, acutely acuminate apically, shallowly and somewhat remotely glandular-crenate or crenate-serrate to subentire on the margins, wide-cuneate or rounded with 1–6 small hemispherical orange glands at the base, membranous, sparingly pubescent along the midrib and main nerves at first, but soon glabrescent, drying purplish-black when young, or at least the midrib and nerves drying purplish when more mature; lateral nerves in 4–7 pairs, the lower ones ascending, weakly-looped, not prominent. Stipules 0.5–0.7 mm long, subulate, not or scarcely aculeate and hardened, yellowish-brown. Male inflorescences 1–4 cm long, the peduncles up to 3 cm long, racemose; bracts c. 1 mm long, narrowly triangular-lanceolate to linear, fugacious. Male flowers: pedicels 2–4 mm long, slender, glabrous, jointed near the base; buds c. 2 mm in diameter, oblate, glabrous; calyx lobes 3, c. 2 × 2 mm, broadly ovate, subacute or obtuse, green or purplish-tinged; disk glands mostly 12, clavate, ± truncate, 6 alternating with stamens of the outer whorl, but scarcely extrastaminal, the rest interstaminal, glabrous or crowned with papillae, purplish; stamens 14–17, c. 1 mm long, filaments stout, pinkish-purple, anthers 0.5 mm across, pink. Female inflorescences up to 5.5 cm long, racemose, few-flowered, subglabrous; bracts minute, broadly triangular. Female flowers: pedicels 2–6 mm long, thick, jointed near the middle; calyx lobes 2–3, c. 1 × 1 mm, triangular, acute, glabrous, pale green; disk glands 2–3, 0.5 × 0.7 mm, scale-like, rounded, glabrous; ovary usually c. 1 mm in diameter, 2-lobed, glabrous, drying blackish; styles 2–3, c. 1 mm long, united at the base, divaricate, stigmas papillose-lobulate. Fruit 4–5 × 8–9 mm, usually dicoccous, glabrous, green at first, drying purplish or reddish. Seeds 4 mm in diameter, foveolate-reticulate, aril bright orange or scarlet, drying reddish-purple.

Zimbabwe. E: Chirinda Forest, fr. xi.1965, *Goldsmith* 16/65 (FHO; K; LISC; MO; PRE; SRGH). **Mozambique**. Z: Serra Morrumbala, y. fr. 13.xii.1971, *Müller & Pope* 2034 (K; LISC; SRGH). MS: 25 km from Lacerdonia, male fl. 6.xii.1971, *Müller & Pope* 1930 (K; LISC; SRGH).

Also in Equatorial Guinea and Tanzania. Low altitude and submontane mixed evergreen forest, understorey; 200–1980 m.

Vernacular name as recorded in specimen data: "tabetabe" (Gorongosa area, Mozambique).

Zimbabwean and Mozambican material differs from Tanzanian material in the male flowers having more stamens, but the Mozambican material is more like Tanzanian material in the nature of the male disk and the shape and colour of the female calyx lobes.

8. **Erythrococca atrovirens** (Pax) Prain in Ann. Bot. (London) **25**: 623 (1911); in F.T.A. **6**, 1: 869 (1912). —Brenan, Check-list For. Trees Shrubs Tang. Terr.: 208 (1949). —F.W. Andrews, Fl. Pl. Anglo-Egypt. Sudan **2**: 63 (1952). —Dale & Greenway, Kenya Trees & Shrubs: 194 (1961). —Radcliffe-Smith in F.T.E.A., Euphorb. 1: 277 (1987). —Beentje, Kenya Trees, Shrubs Lianas: 196 (1994). Type from Zaire (Orientale Province).

Claoxylon atrovirens Pax in Bot. Jahrb. Syst. **19**: 85 (1894).

Athroandra atrovirens var. *schweinfurthii* (Pax) Pax & K. Hoffm. in Mildbr., Wiss. Ergebn. Deutsch. Zentr.-Afr.-Exped. 1907–1908, **II**: 452 (1912); in Engler, Pflanzenr. [IV, fam. 147, vii] **63**: 83 (1914). —Engler, Pflanzenw. Afrikas (Veg. Erde 9) **3**, 2: 69, 71 (1921). —Robyns & Tournay, Fl. Sperm. Parc Nat. Alb. **1**: 455 (1948). Type from Sudan (Equatoria Province).

A shrub up to 2.5 m high. Wood soft. Twigs pale greyish-brown. Young growth evenly to densely yellowish-pubescent. Petioles 5–10 mm long. Leaf blades 5–10 × 3–5 cm, ovate, acuminate, subentire, rounded or truncate at the base, membranous, sparingly pubescent on upper surface but more evenly so along the midrib and main nerves, evenly so beneath and densely so along the midrib and main nerves, medium green above, paler beneath; lateral nerves in 4–6 pairs, ascending, not or weakly-looped, not prominent above, slightly so beneath. Stipules minute, conic-cylindric, not aculeate nor becoming hardened, yellowish-brown. Male inflorescences up to 5 cm long, racemose, long-pedunculate, pendulous; bracts minute, glabrous. Male flowers: pedicels up to 1.3 cm long, capillary, glabrous or sparingly pubescent, jointed near the base; buds 1.5–2 mm in diameter, oblate, slightly apiculate, truncate at the base, glabrous or subglabrous; calyx lobes 4(5), 1.5 × 1.5 mm, broadly triangular-ovate, acute or subacute, greenish-white, translucent; extrastaminal glands absent or rarely developed, interstaminal glands numerous, minute, rounded, hair-tipped; stamens 25–40, 7–10 outer, the rest central, 0.7 mm long, anthers yellowish-white. Female inflorescences 2–3 cm long, the peduncle up to 2 cm long, racemose, 10–12-flowered, sparingly pubescent; bracts 1 mm long, narrowly triangular-lanceolate, acute, glabrous. Female flowers: pedicels 1–3 mm long, glabrous, jointed near the base; calyx lobes 2(3), 1.5 × 1 mm, broadly ovate, enclosing the ovary, acute, glabrous, pale green; disk glands 2, 0.5 × 0.5–1 mm, suborbicular; ovary 1 mm in diameter, 2-lobed, glabrous; styles 2(3), 1 mm long, united at the base, divaricate-spreading, later recurved, stigmas smooth. Fruit 4–6 × 5 or 10 mm, didymous, or by abortion subglobose, glabrous, pale green. Columella laterally compressed. Seeds 3–4 mm in diameter, coarsely shallowly foveolate-reticulate, aril orange-red.

Zambia. N: Kawambwa, female fl. 16.xi.1957, *Fanshawe* 4077 (FHO; K; NDO).

Also in Cameroon, Zaire, S Sudan, Uganda, Kenya and Tanzania. In dry evergreen thicket (mateshi) of *Marquesia* woodland.

The description of the male flowers, fruit and seed is based on East African material, owing to the deficiency of the Zambian material.

9. **Erythrococca angolensis** (Müll. Arg.) Prain in Ann. Bot. (London) **25**: 625 (1911); in F.T.A. **6**, 1: 871 (1912). Type from Angola (Malanje Province).

Claoxylon angolense Müll. Arg. in J. Bot. **2**: 333 (1864); in De Candolle, Prodr. **15**, 2: 777 (1866).

Athroandra angolensis (Müll. Arg.) Pax & K. Hoffm. in Engler, Pflanzenr. [IV, fam. 147, vii] **63**: 83 (1914). —Engler, Pflanzenw. Afrikas (Veg. Erde 9) **3**, 2: 69, 71 (1921).

A deciduous shrub to c. 60 cm tall. Twigs brownish, pubescent. Young growth densely pubescent. Petioles 3–5 mm long. Leaf blades 4–6 × 1.5–3 cm, elliptic-lanceolate, cuspidate-acuminate apically, minutely and somewhat irregularly denticulate on the margins, cuneate to rounded at the base, membranaceous, sparingly pubescent on upper surface, more evenly so beneath, especially along the midrib and main nerves, dull green, drying brownish-green; lateral nerves in c. 5 pairs, ascending, somewhat irregularly looped, not prominent above, scarcely so beneath. Stipules 0.5 mm long, conical, obtuse, shiny, yellowish-brown. Male inflorescences c. 2 cm long, with peduncles c. 1 cm long, racemose, few-flowered, pubescent; bracts minute, subulate, glabrous. Male flowers: pedicels up to 1 cm long, slender, glabrous, jointed at the base; buds 1.5–2.5 mm in diameter, subglobose, scarcely apiculate, subtruncate at the base, glabrous; calyx lobes 4–5, 2–2.5 × 1–2 mm, triangular-ovate, acute or subacute, pale green; extrastaminal glands 10–15, minute, rounded, hair-tipped; interstaminal glands numerous, smaller than the extrastaminal ones; stamens 25–40, 12–15 outer, the rest central, less than 1 mm long, anthers yellowish-white. Female inflorescences 2–3 cm long, the peduncle up to 2 cm long, racemose, 4–6-flowered, sparingly pubescent; bracts 0.5 mm long, triangular-ovate, acute, glabrous. Female flowers: pedicels 1–2 mm long, glabrous, jointed near the base; calyx lobes 2, 1–1.5 × 1.5–2 mm, transversely ovate, obtuse, glabrous, yellow-green; disk urceolate, 2-lobed; ovary 1–1.5 mm in diameter, 2-lobed, glabrous; styles 2, 1 mm long, united at the base, divaricate-spreading, later recurved, stigmas ± smooth. Fruit 5–6 × 9–10 mm, didymous, glabrous, pale green. Columella laterally compressed. Seeds 5–6 mm in diameter, shallowly foveolate-reticulate; aril bright orange.

Zambia. B: Kataba (Katuba) Valley, male fl. 24.x.1953, *Gilges* 204 (PRE; SRGH). W: Mwinilunga Distr., Sinkabolo Dambo, male fl. 12.xi.1937, *Milne-Redhead* 3205 (K; LISC; PRE; SRGH); female fl. 12.xi.1937, *Milne-Redhead* 3206.

Also in Angola. In *Cryptosepalum* woodland on Kalahari Sand; 1070 m.

The description of the fruit and seed is based on Angolan material, owing to the deficiency of the Zambian material.

10. **Erythrococca welwitschiana** (Müll. Arg.) Prain in Ann. Bot. (London) **25**: 622 (1911); in F.T.A. **6**, 1: 868 (1912). —White, F.F.N.R.: 198 (1962). Type from Angola (Cuanza Norte).

Claoxylon welwitschianum Müll. Arg. in J. Bot. **2**: 333 (1864); in De Candolle, Prodr. **15**, 2: 776 (1866).

Athroandra welwitschiana (Müll. Arg.) Pax & K. Hoffm. in Engler, Pflanzenr. [IV, fam. 147, vii] **63**: 81 (1914). —Engler, Pflanzenw. Afrikas (Veg. Erde 9) **3**, 2: 68, 70 (1921).

A shrub or slender small tree up to 4.5 m tall, with short spreading branches. Bark pale olive-grey with straw-coloured lenticels. Young growth pubescent, deep green and glossy. Petioles 1–1.5 cm long, with 2 minute hemispherical orange-brown glands at the junction with the blade. Leaf blades 1.5–18 × 1–10 cm, elliptic-ovate to oblong-elliptic, acuminate apically, shallowly and remotely crenate-serrate to subentire on the margins, cuneate to obliquely subcordate at the base, membranaceous to thinly chartaceous, sparingly pubescent along the midrib and main nerves at first, later glabrescent, deep green and glossy on upper surface, paler beneath; lateral nerves in 5–8 pairs, weakly-looped, impressed above, prominent beneath. Stipules 0.5 mm long, subulate, not aculeate, scarcely indurate, yellow-brown. Male inflorescences up to 5 cm long, the peduncle up to 2 cm long, racemose, few-flowered, slender, pubescent; bracts 0.5–1 mm long, filiform, glabrous. Male flowers: pedicels up to 2 cm long, capillary, glabrous, jointed at the base; buds c. 2 mm in diameter, oblate, scarcely apiculate, truncate-concave at the base, glabrous; calyx lobes 4, 2 × 1–2 mm, elliptic-ovate to broadly-ovate, subacute, pale yellowish-green; extrastaminal disk glands c. 20, minute, rounded, hair-tipped; interstaminal glands numerous, similar to the extrastaminal ones; stamens c. 30–40, c. 12 outer, the rest central, c. 0.5 mm long, anthers minute, yellow. Female inflorescences extending to 11 cm in fruit, otherwise resembling the male. Female flowers: pedicels up to 4 mm long; calyx lobes 2(3), c. 1 × 1.5 mm, broadly ovate, rounded, glabrous; disk shallowly 2(3)-lobed; ovary 2-lobed, glabrous; styles 2, fused into a cylindrical column, c. 1 mm high, barely perceptibly 2-lobed at the top, stigmas minute, smooth. Fruit 3–4 × 4 or 8 mm, didymous, or by abortion subglobose, glabrous. Seeds c. 4 mm in diameter, muricate-reticulate-rugulose.

Zambia. N: Kawambwa Boma, male fl. 30.x.1952, *White* 3553 (FHO; K). W: Solwezi, male fl. x.1960, *Holmes* 1338 (K; LISC; NDO).

Also in Angola. Swamp forest (mushitu) understorey, and in dense *Parinari excelsa* forest.

The description of the female flowers, fruit and seed is based on Angolan material, owing to the deficiency of the Zambian material.

32. MICROCOCCA Benth.

Micrococca Benth. in Hooker, Niger Flora: 503 (1849).
Claoxylon sect. *Micrococca* (Benth.) Müll. Arg. in Linnaea **34**: 166 (1865).

Dioecious or monoecious annual or perennial herbs or shrubs (or small trees outside the Flora Zambesiaca area). Indumentum simple. Leaves alternate (sometimes opposite in lowermost), petiolate, stipulate, stipellate, simple, crenate or serrate, membranous, penninerved, often purplish-tinged. Inflorescences axillary, spicate or racemose with the flowers in male, female or bisexual bracteate clusters along the axis, usually the terminal flower female. Male flowers pedicellate; calyx closed in bud, later splitting into 3(4) valvate lobes; petals absent; disk of 3–30 free interstaminal glands or absent; stamens 3–30, usually biseriate, filaments free, short, anthers extrorse, basifixed, the cells obovoid, free, longitudinally dehiscent; pistillode absent. Female flowers pedicellate; sepals 3–5, imbricate; petals absent; disk glands 2–6, often resembling staminodes and alternating with the cocci; ovary 2–4-locular, with 1 ovule per loculus; styles 2–4, free, spreading, linear, usually fimbriate or plumose-laciniate. Fruits 2–4-coccous, or monococcous by abortion, dehiscing septicidally and loculicidally; pericarp thin; endocarp crustaceous; columella persistent, crustaceous or woody. Seeds subglobose, ecarunculate, thinly arillate; testa crustaceous, often foveolate-reticulate; albumen fleshy; cotyledons suborbicular, sometimes scarcely larger than the radicle.

A genus of 12 species in the Old World tropics, 5 of which occur in Africa.

1. Annual herb · 1. *mercurialis*
– Shrub or small tree · 2
2. Petiole 2–13 mm long; leaf blade shallowly and irregularly crenate to subentire; disk of male flower present; female flower calyx lobes 5 · 2. *scariosa*
– Petiole up to 5 cm long; leaf blade sharply serrate; disk of male flower absent; female flower calyx lobes 3 · 3. *capensis*

1. **Micrococca mercurialis** (L.) Benth. in Hooker, Niger Flora: 503 (1849). —Engler, Pflanzenw. Ost-Afrikas **C**: 238 (1895). —Prain in F.T.A. **6**, 1: 878 (1912). —Pax in Engler, Pflanzenr. [IV, fam. 147, vii] **63**: 133, t. 18 D–F (1914). —Engler, Pflanzenw. Afrikas (Veg. Erde 9) **3**, 2: 76, t. 31 (1921). —De Wildeman, Pl. Bequaert. **3**, 4: 470 (1926). —Robyns & Tournay, Fl. Sperm. Parc Nat. Alb. **1**: 458 (1948). —F.W. Andrews, Fl. Pl. Anglo-Egypt. Sudan **2**: 86 (1952). —F.W.T.A., ed. 2, **1**, 2: 402 (1958). —White, F.F.N.R.: 201 (1962). —Agnew, Upl. Kenya Wild Fls.: 213 (1974). —Radcliffe-Smith, F.T.E.A., Euphorb. 1: 261 (1987). Type from India.

Tragia mercurialis L., Sp. Pl.: 980 (1753).

Claoxylon mercuriale (L.) Thwaites, Enum. Pl. Zeyl.: 271 (1861). —Müller Argoviensis in De Candolle, Prodr. **15**, 2: 790 (1866).

Mercurialis alternifolia Lam., Encycl. Méth. Bot. **4**: 120 (1797). Type from Senegal.

An erect or procumbent annual herb up to 50 cm tall, rarely taller. Stems crisped-pubescent in rows. Petioles 0.5–2.5 cm long. Leaf blades 2–7 × 1–3.5 cm, elliptic-ovate, subacute or obtuse apically, crenate, ± rounded at the base, membranaceous, sparingly pubescent to subglabrous, punctate, dull, pale green, sometimes purplish-tinged; lateral nerves in 4–7 pairs, ascending, camptodromous. Stipules 0.5 mm long, glandular. Stipels minute, glandular. Inflorescences 1–7.5 cm long, usually bisexual; bracts c. 1 mm long, ovate-lanceolate, green with hyaline margins. Male flowers: pedicels 2 mm long, capillary, glabrous; calyx lobes 0.75 mm long, obovate-suborbicular, ± glabrous, pale yellowish-green or whitish; disk glands spathulate, pubescent, purple; stamens 3–20, 1–multiseriate, filaments 0.5 mm long, purple, anthers minute, yellow. Female flowers: pedicels 1–5 mm long, extending to 3 cm in fruit, pubescent; sepals 1–2 mm long, ovate-lanceolate, acute, sparingly pubescent, green with a narrow hyaline margin; disk glands 1 mm long, linear-filiform; ovary 1 mm in diameter, 2-lobed to 4-lobed, minutely papillose, strigose, purple; styles 1 mm long, glabrous, ochreous-tawny. Fruit 2–3 × 3.5–5.5 mm, roundly 3-lobed, rarely 2- or 4-lobed, sparingly strigose to subglabrous, dark green or bluish-green becoming dull purple on drying; columella 1.5 mm long. Seeds 2 × 1.5 mm, ovoid-subglobose, reticulate, reddish-brown, greyish or blackish, paler in the pits.

Botswana. N: Aha Hills, fr. 13.iii.1965, *Wild & Drummond* 6955 (K; LISC; PRE; SRGH). **Zambia**. B: Masese, fr. 10.iii.1960, *Fanshawe* 5432 (PRE; SRGH). W: Mindolo, Kitwe, fr. 28.iv.1963, *Mutimushi* 301 (K; NDO; SRGH). C: Luangwa Valley Game Res. South, fl. 18.ii.1967, *Prince* 260 (K; LISC; PRE; SRGH). **Zimbabwe**. N: Hurungwe Distr., Nyanyana R., fl. & fr. 2.ii.1958, *Drummond* 5437 (K; LISC; PRE; SRGH). W: Hwange Distr., Gwayi (Gwai)–Lutope R. junction, fr. 27.ii.1963, *Wild* 6033 (K; LISC; MO; PRE; SRGH). E: Chipinge Distr., Lower Save (Sabi), Hippo Mine, fr. 12.iii.1957, *Phipps* 589 (K; PRE; SRGH). S: Beitbridge Distr., Bubye R., fr. 25.ii.1961, *Wild* 5385 (K; MO; PRE; SRGH). **Malawi**. S: Elephant Marsh, Shire Valley, fr. 16.ii.1888, *G.F. Scott* s.n. (K). **Mozambique**. N: 22 km Montepuez–Nantulo, fr. 8.iv.1964, *Torre & Paiva* 11759 (LISC).

More or less throughout tropical Africa; also in Yemen, Madagascar, India, Sri Lanka, W Malaysia and N Australia.

Low stony hills and kopjes, also in seasonally waterlogged floodplain clay and alluvium, often in mopane woodland and scrub; frequently as a weed of disturbed ground and cultivation; 365–1300 m.

2. **Micrococca scariosa** Prain in Bull. Misc. Inform., Kew **1912**: 192 (1912); in F.T.A. **6**, 1: 878 (1912). —Pax in Engler, Pflanzenr. [IV, fam. 147, vii] **63**: 132 (1914). —Engler, Pflanzenw. Afrikas (Veg. Erde 9) **3**, 2: 76 (1921). —Brenan, Check-list For. Trees Shrubs Tang. Terr.: 220 (1949). —Radcliffe-Smith, F.T.E.A., Euphorb. 1: 263 (1987). —Beentje, Kenya Trees, Shrubs Lianas: 215 (1994). Tab. **37**. Types from Tanzania.

A small tree up to 4 m high. Twigs sparingly lenticellate, puberulous and greenish at first, later glabrescent and dark grey. Petioles 2–13 mm long, puberulous. Leaf blade 5.5–21 × 2–8 cm, elliptic-ovate, shortly acuminate apically, remotely shallowly crenate to subentire, rounded or rounded-cuneate at the base, chartaceous, glabrous on both surfaces except for the pubescent domatia; basal glands minute, subulate; lateral nerves in 5–10 pairs, looped. Stipules 3–6 mm long, subulate, puberulous, soon falling. Male inflorescences up to 11 cm long; axis angular; bracts 2.5 × 1.5 mm, ovate-lanceolate, acute, entire, chaffy, many-flowered; bracteoles similar, but smaller. Male flowers shortly pedicellate; calyx lobes 2–3(4), 1 × 0.7 mm, suborbicular-ovate, subacute, glabrous; disk glands minute, numerous, angular, pilose; stamens 17, 1 mm long, filaments twisted, anthers 0.3 mm wide. Female inflorescences up to 8 cm long; bracts somewhat longer and narrower than in the male; bracteoles each subtending a solitary flower. Female flowers: pedicels 3 mm long, extending to 5–7 mm in fruit; calyx lobes 5, 1 mm long, triangular, acute; disk glands 6, 0.5 mm long, transversely ovate, scale-like, free, contiguous; ovary 1 mm in diameter, 3-lobed, ± smooth, venose, glabrous; styles 1.5–2 mm long, tawny. Fruit 6–7 × 11–12 mm, roundly 3-lobed, ± smooth, venose, glabrous, sometimes tinged pinkish. Seeds 6 × 5 mm, subglobose, smooth, dark reddish-brown, mottled pale greyish or whitish.

Mozambique. MS: Cheringoma Distr., Durunde Forest, fr. 29.v.1948, *Mendonça* 4428 (LISC); Inhaminga, male fl. viii.1973, *Earle* P66E (SRGH).

Also in Kenya and Tanzania. In primeval forest.

The description of the female flowers above was based on East African material.

3. **Micrococca capensis** (Baill.) Prain in Ann. Bot. (London) **25**: 630 (1911). —Pax in Engler, Pflanzenr. [IV, fam. 147, vii] **63**: 135 (1914). —Prain in F.C. **5**, 2: 460 (1920). —Engler, Pflanzenw. Afrikas (Veg. Erde 9) **3**, 2: 76 (1921). —Hutchinson, Botanist in Southern Africa: 667 (1946). —K. Coates Palgrave, Trees Southern Africa, ed. 2, rev.: 421 (1983). Type from South Africa.

Claoxylon capense Baill., Étud. Gén. Euphorb.: 493 (1858); in Adansonia **3**: 161 (1862). —Müller Argoviensis in De Candolle, Prodr. **15**, 2: 786 (1866). —Sim, For. Fl. Port. E. Afr.: 105 (1909).

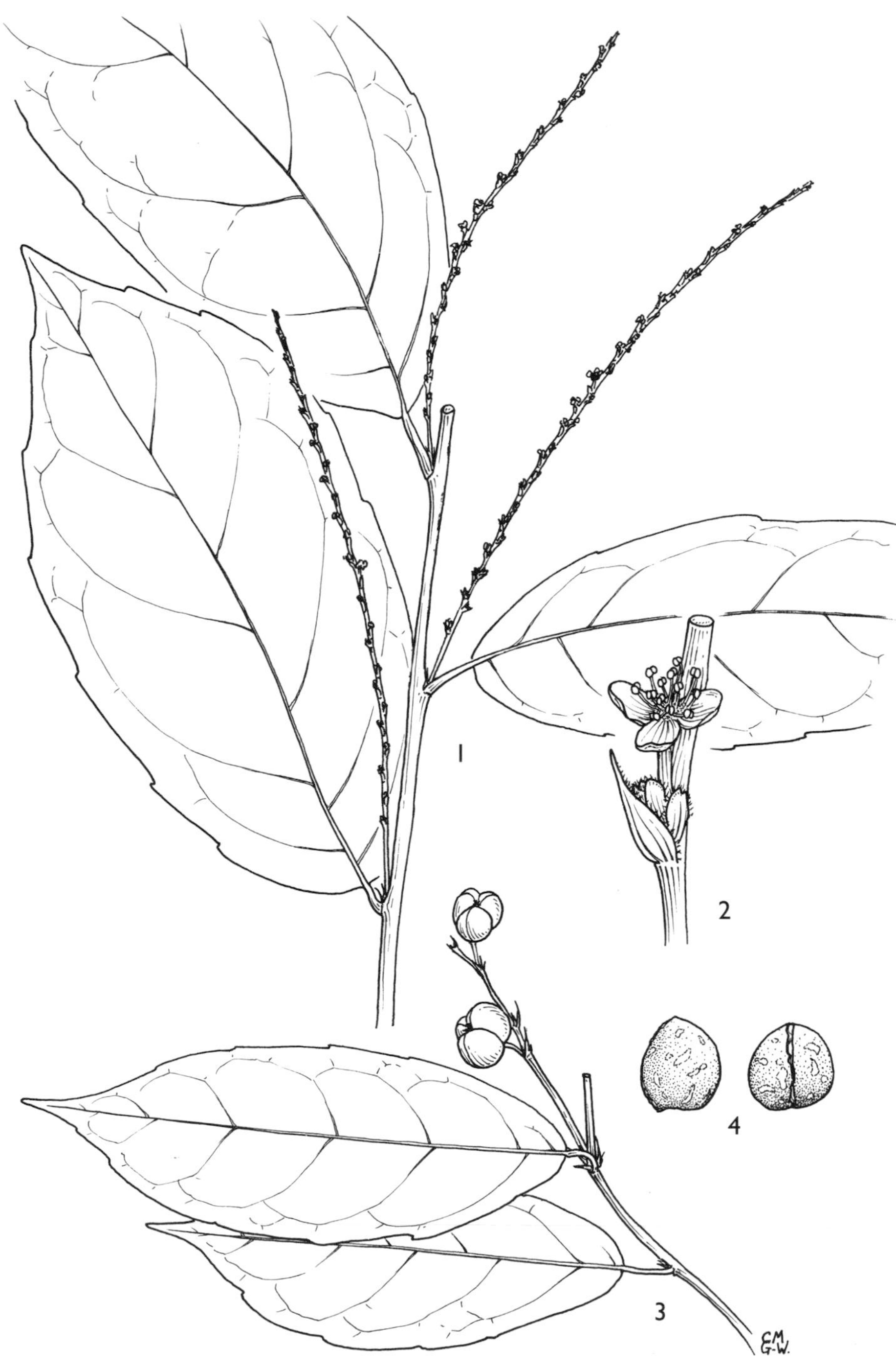

Tab. 37. MICROCOCCA SCARIOSA. 1, portion of flowering branch (× $^2/_3$); 2, male flower (× 6), 1 & 2 from *Drummond & Hemsley* 3595/2; 3, distal portion of fruiting branch (× $^2/_3$); 4, seeds (× 2), 3 & 4 from *Drummond & Hemsley* 3595/5. Drawn by Christine Grey-Wilson.

A shrub or small tree up to 5 m high. Young shoots, petioles and axes sparingly appressed-pubescent, soon glabrescent. Petioles up to 5 cm long. Leaf blades 6–15 × 2–6 cm, elliptic-lanceolate, acutely acuminate, sharply glandular-serrate on the margin, attenuate or cuneate at the base, firmly membranous, 3-nerved from the base, sparingly pubescent at first, soon glabrescent, dark green and shiny above, paler and duller beneath, sometimes purplish-tinged; basal glands usually absent; lateral nerves in 4–6 pairs, ascending, looped. Stipules minute, subulate, glabrous, soon falling. Racemes unisexual or bisexual, with the flowers dioecious or monoecious, pedunculate. Male racemes up to 17 cm long; bracts 1 mm long, triangular-ovate, pubescent, 2–20-flowered, the flower clusters remote, occasionally the lowest pedunculate. Male flowers: pedicels up to 2 mm long; buds 1.5 mm in diameter, subglobose, apiculate, glabrous; calyx lobes 3, 1.5 × 1.5 mm, suborbicular-ovate, greenish-white; disk glands absent; stamens 20–30, filaments very short, anther cells 0.3 mm long. Bisexual racemes resembling the male ones, but with the addition of 1–2 female flowers at the base, in the middle or at the top of the axis. Female racemes up to 20 cm long; bracts 2 mm long, narrowly lanceolate, sparingly pubescent, 1-flowered, the flowers remote. Female flowers: pedicels up to 7 mm long; calyx lobes 3, 1 × 0.75 mm, triangular-lanceolate, slightly accrescent in fruit, acute, very sparingly pubescent; disk glands 3, 1 mm long, ligulate, pinkish-purple; ovary 1.5 mm in diameter, 3-lobed, the lobes opposite the calyx lobes, smooth, glabrous, purplish; styles 3, less than 1 mm long, ± free; stigmas papillose, pinkish. Fruit 4 × 7–8 mm, roundly 3-lobed, + smooth, glabrous, purplish; columella 3 mm long. Seeds 3.5 × 2.5 mm, ovoid, shallowly and finely reticulate-rugulose, pinkish-brown, pale grey in the depressions.

Mozambique. M: Maputo (Lourenço Marques), *Sim*; Maputo, *Sim* (literature records only).
Also in South Africa (KwaZulu-Natal and Eastern Cape Province). No habitat data from the Flora Zambesiaca area.
The above description was based on South African material.

33. MALLOTUS Lour.

Mallotus Lour., Fl. Cochinch.: 635 (1790).

Dioecious, or rarely monoecious trees or shrubs often with a stellate, or sometimes simple or mixed, indumentum. Leaves opposite or alternate, sometimes anisophyllous, petiolate, stipulate, sometimes lobed, entire or toothed, with 2 or more basal glands on the upper surface, and often pellucid-gland-dotted on upper and lower surfaces, penni- or palminerved. Inflorescences unisexual, spicate, racemose or paniculate, terminal, subterminal or axillary, few- to many-flowered; male flowers commonly in fascicles along the axis, female flowers 1–2 per bract. Male flowers: calyx usually globose in bud, later valvately (2)3–4(5)-partite; petals absent; disk glands absent or numerous, free, dispersed amongst the stamens; stamens numerous, on a slightly elevated receptacle, filaments free, anthers subdorsifixed with a variable connective, longitudinally dehiscent; pistillode absent or rarely minute. Female flowers: calyx shallowly to deeply imbricately or valvately 3–5(10)-lobed, persistent, or subspathaceous and caducous; petals absent; disk absent; ovary (2)3(4)-locular, with 1 ovule per loculus; styles free or connate at the base, simple, recurved, papillose or plumose. Fruits globose or (2)3(4)-lobed, smooth or echinate, dehiscing septicidally into bivalved cocci leaving a more or less 3-winged persistent columella; endocarp crustaceous. Seeds globose or ovoid, ecarunculate, the outer testa slightly fleshy or soft, the inner testa crustaceous, smooth or rugulose, albumen fleshy, cotyledons broad, flat.

An Old World genus of some 140 species in the Indo-Pacific area, with 2 species in tropical Africa.

Mallotus oppositifolius (Geisel.) Müll. Arg. in Linnaea **34**: 194 (1865); in De Candolle, Prodr. **15**, 2: 976 (1866). —Engler, Pflanzenw. Ost-Afrikas **C**: 238 (1895). —Prain in F.T.A. **6**, 1: 928 (1912). —Pax in Engler, Pflanzenr. [IV, fam. 147, vii] **63**: 158, t. 23 (1914). —Engler, Pflanzenw. Afrikas (Veg. Erde 9) **3**, 2: 77, t. 33 (1921). —De Wildeman, Pl. Bequaert. **3**, 4:

471 (1926). —Robyns & Tournay, Fl. Sperm. Parc Nat. Alb. **1**: 459 (1948). —Brenan, Check-list For. Trees Shrubs Tang. Terr.: 218 (1949). —F.W. Andrews, Fl. Pl. Anglo-Egypt. Sudan **2**: 85 (1952). —Keay in F.W.T.A., ed. 2, **1**, 2: 402 (1958). —Dale & Greenway, Kenya Trees & Shrubs: 210 (1961). —White, F.F.N.R.: 201 (1962). —K. Coates Palgrave, Trees Southern Africa, ed. 2, rev.: 424 (1983). —Radcliffe-Smith in F.T.E.A., Euphorb. 1: 236 (1987). —Beentje, Kenya Trees, Shrubs Lianas: 213 (1994). Type from Dahomey.
Croton oppositifolius Geisel., Monogr. Croton.: 23 (1807).

An open shrub or small tree up to 10 m tall, often sarmentose, dioecious. Bark ± smooth to slightly roughened and flaky, whitish- or greenish-grey or brown. Young shoots fulvous-stellate-pubescent. Older twigs glabrescent, often purplish-brown. Leaves opposite, one of each pair long-petiolate; petiole up to 8 cm long in the longer of the pair, up to 2 cm long in the shorter, slightly pulvinate, pubescent; leaf blade not lobed or 3-lobed, up to 17 × 13 cm, broadly ovate to ovate-oblong or ovate-lanceolate, the two leaves of each opposite pair somewhat unequal, obtusely or subacutely acuminate at the apex, shallowly cordate to truncate or rounded-cuneate at the base, subentire and shallowly glandular-denticulate to repand-dentate or sinuate on the margin, membranous to chartaceous, usually 3-nerved from the base, sometimes 5-nerved, with up to 4 discoid glands on the upper surface near the base, sparingly yellow-pellucid-gland-dotted above, evenly so beneath, sparingly stellate-pubescent to almost glabrous on both surfaces, sometimes with an admixture of simple hairs along the midrib and main nerves beneath, and in the domatia, or with multicellular simple hairs, dark green above, paler beneath; lateral nerves in 4–7 pairs. Stipules 1 mm long, subulate, soon deciduous. Male racemes up to 11 cm long; axis sparingly to evenly pubescent, sparingly gland-dotted; bracts 0.5 mm long, triangular, 3–5-flowered. Flowers fragrant, with an odour of *Convallaria*. Male flowers: pedicels 3–7 mm long, jointed, pubescent; buds subglobose, apiculate; sepals 2 mm long, elliptic, subacute, sparingly pubescent without, glabrous and gland-dotted within, strongly reflexed, pale yellow-green; disk absent; stamens 2 mm long, filaments greenish-white, anthers 0.3 mm, whitish; pistillode absent. Female racemes resembling the males, but the bracts 1.5 mm long, lanceolate, 1–2-flowered. Female flowers: pedicels shorter than in the male flowers at first, but elongating in fruit; buds ovoid-ellipsoid; calyx lobes 3–5(6), 2 mm long, united at the base or for half their length, ovate or lanceolate, acute, pubescent and gland-dotted at the base without, glabrous within, recurved, green; ovary 1 mm in diameter, shallowly 3(4)-lobed to subglobose, densely pubescent and glandular; styles 1.5 mm long, ± free, plumose. Fruits 5–7 × 7–9 mm, deeply 3(4)-lobed, sparingly to evenly pubescent or puberulous and gland-dotted. Seeds 3.5–4 × 3 mm, subglobose, smooth, shiny, greyish-olive-brown.

1. Leaves 3-lobed, the median lobe usually larger than the 2 lateral lobes · · · · · var. *lindicus*
– Leaves not lobed; margins subentire, denticulate, repand-dentate or sinuate · · · · · · · · · 2
2. Leaves 3-nerved at the base; stellate hairs present · · · · · · · · · · var. *oppositifolius* f. *glabratus*
– Leaves 5-nerved at the base; multicellular hairs present · · · var. *oppositifolius* f. *polycytotrichus*

Var. **oppositifolius**

Leaves not lobed; margins subentire, denticulate, repand-dentate or sinuate.

forma **glabratus** (Müll. Arg.) Pax in Engler, Pflanzenr. [IV, fam. 147, vii] **63**: 160 (1914). — Brenan, Check-list For. Trees Shrubs Tang. Terr.: 218 (1949). —Radcliffe-Smith in F.T.E.A., Euphorb. 1: 237 (1987). Types from Nigeria and Madagascar.
Mallotus oppositifolius var. *glabratus* Müll. Arg. in Linnaea **34**: 194 (1865); in De Candolle, Prodr. **15**, 2: 977 (1866). —Prain in F.T.A. **6**, 1: 929 (1912).

Leaves 3-nerved at the base, sparingly scurfily stellate-pubescent along the midrib and main nerves beneath and otherwise glabrous, or else almost completely glabrous. Fruits subglabrous.

Zambia. B: Kabompo, fl. 10.xi.1952, *Gilges* 265 (K; PRE; SRGH); Kabompo R., near Kabompo Boma, fl. 6.x.1952, *Angus* 616 (BM; FHO; K; PRE). S: Kafue Hook, fl. 15.ii.1960, *Fanshawe* 5384 (FHO; K; NDO). **Malawi**. S: Mulanje, Litchenya For. Res., st. 7.ix.1970, *Müller* 1564 (K; SRGH).

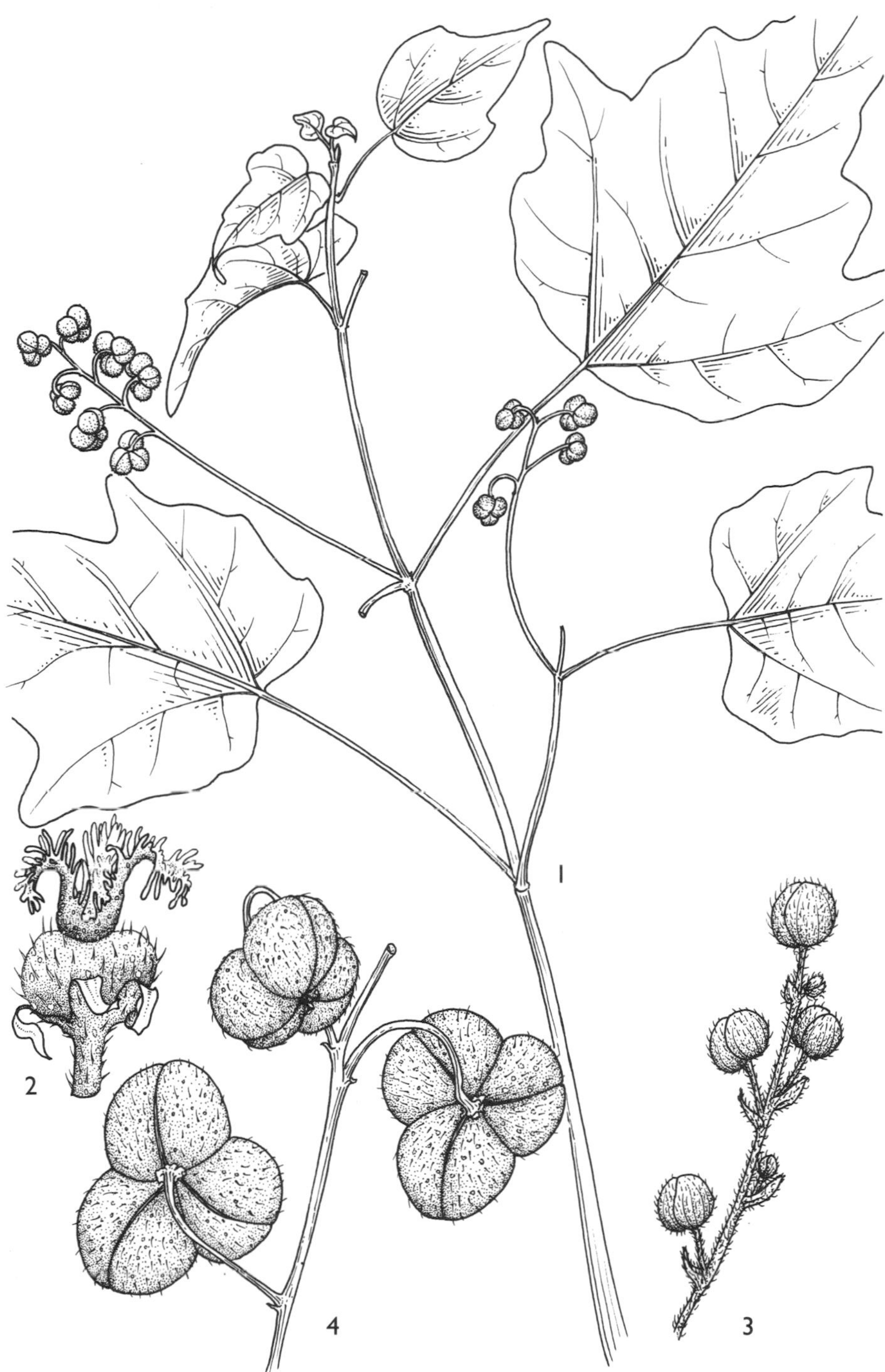

Tab. 38. MALLOTUS OPPOSITIFOLIUS var LINDICUS. 1, distal portion of fruiting branch (× $^{2}/_{3}$); 2, female flower (× 8), 1 & 2 from *Torre & Paiva* 9350A; 3, portion of male inflorescence (× 4), from *Torre & Paiva* 9350; 4, fruits (× 4), from *Torre & Paiva* 9350A. Drawn by Christine Grey-Wilson.

Mozambique. N: Nampula, 33 km on road to Murrupula, fl. 3.x.1942, *Mendonça* 1222 (LISC). Z: Quelimane Distr., road to Pebane (Pibane), fl. & fr. 29.xii.1947, *Faulkner* Kew 154 (K).

From the Gambia to the Sudan and southwards to Angola, Zambia, Malawi and Mozambique. Evergreen riverine forest and submontane mixed evergreen forest; 30–1100 m.

The typical form is more restricted in distribution, ranging from the Ivory Coast to Ethiopia and southwards to Tanzania. It has the leaves hirsute to pubescent beneath.

forma **polycytotrichus** Radcl.-Sm. in Kew Bull. **40**: 658 (1985). Type: Zimbabwe, Chipinge Distr., 3 km west of Mt. Selinda, 28.iv.1976, *Müller* 2467 (SRGH, holotype, FHO; PRE).

Leaves 5-nerved at the base; multicellular simple hairs present on the lower surface.

Zimbabwe. E: Chipinge Distr., 3 km west of Mt. Selinda, st. 28.iv.1976, *Müller* 2467 (SRGH; FHO; PRE).

Endemic to Mt. Selinda. In mixed evergreen forest; c. 1000 m.

Var. **lindicus** (Radcl.-Sm.) Radcl.-Sm. in Kew Bull. **39**: 790 (1984); in F.T.E.A., Euphorb. 1: 239 (1987). Tab. **38**. Type from Tanzania (Southern Province).

Mallotus oppositifolius f. *lindicus* Radcl.-Sm. in Kew Bull. **29**: 437 (1974). Type as above.

Leaves 3-lobed, the median lobe usually larger than the two lateral lobes; the lobes subentire, dentate or sinuate, subglabrous to evenly hirsute or pubescent.

Mozambique. N: Mogincual, fl. 27.xi.1963, *Torre & Paiva* 9350, fr. 9350-A (LISC); 4 km from Cabo Delgado lighthouse going towards Palma, fl. 17.iv.1964, *Torre & Paiva* 12099 (LISC).

Also in southern Tanzania. In the shrub layer of coastal mixed forest; 10–100 m.

Vollesen in MRC3171 from Tanzania is intermediate in leaf shape between the two varieties.

34. ACALYPHA L.

Acalypha L., Sp. Pl.: 1003 (1753); Gen. Pl., ed. 5: 436 (1754).
Calyptrospatha Klotzsch ex Baill., Étud. Gén. Euphorb.: 440 (1858); in Peters, Naturw. Reise Mossambique **6**, 1: 96, t. 18 (1861).

Monoecious, or occasionally dioecious, procumbent or erect, annual or perennial herbs or shrubs, often superficially resembling certain *Urticaceae*. Indumentum simple (Old World) or stellate (New World), often glandular. Leaves alternate, petiolate or subsessile, stipulate, sometimes stipellate, simple, crenate or serrate, palmi- or penninerved. Inflorescences terminal, axillary or both, unisexual or bisexual, spicate, racemose or paniculate. Male flowers small, shortly pedicellate, glomerulate, usually subtended by inconspicuous bracts. Female flowers sessile or shortly pedicellate, solitary or 3–5 together, subtended by conspicuous accrescent foliaceous or inconspicuous bracts. Allomorphic female flowers terminal, median or basal in the inflorescences of some species, ebracteate [as a rule, allomorphic female flowers consist of long-pedicellate or subsessile, ebracteate, 5-sepalous structures, each having a unilocular ovary, the axis of which is turned through almost 180°, so that the single style is sub-basal, sometimes protruding between 2 of the sepals. Each flower thus produces only a single-seeded mericarp, which may or may not be provided with a dehiscence suture – see Radcliffe-Smith in Kew Bull. **28**, 3: 525–529 (1974)]. Male flowers: calyx closed in bud, later splitting into 4 valvate lobes; petals absent; disk absent; stamens usually 8 on a slightly raised receptacle, filaments free, broad, anther cells distinct, spreading or pendulous, oblong or linear, later becoming flexuous-vermiform; pistillode absent. Female flowers: calyx lobes 3–4(5), imbricate, small; petals absent; disk absent; ovary 2–3-locular, with 1 ovule per loculus; styles free or connate, usually laciniate, reddish and rather showy, rarely entire or bilobed. Allomorphic female flowers: calyx ± as in the normal female flowers; ovary 1(2)-locular; styles sub-basal. Fruits 3-lobed, small, soon dehiscing septicidally into 3 bivalved cocci. Fruits of the allomorphic flowers 1(2)-lobed. Seeds ellipsoid or subglobose, small, carunculate or not; testa crustaceous; albumen fleshy; cotyledons broad, flat.

A pantropical genus of 450 species, some 50 of which are African.

1. Male and female flowers usually in the same inflorescence · 2
– Male and female flowers usually in separate inflorescences · 15
2. Inflorescences female above, male below · 3
– Inflorescences male above, female below · 4
3. Unarmed shrub with brownish bark; leaves usually obtusely acuminate at the apex · 15. *acrogyna*
– Thorny shrub with pale grey bark; leaves usually obtuse to rounded at the apex · 16. *sonderiana*
4. Leaves gland-dotted beneath · 5
– Leaves not gland-dotted beneath · 6
5. Leaf blades ovate, up to 7 cm long; female bracts usually several per inflorescence, c. 1 × 1.5 cm · 17. *fruticosa*
– Leaf blades lanceolate, up to 15 cm long; female bracts 1–2 per inflorescence, up to c. 2.5 × 3 cm · 18. *pubiflora*
6. Leaves generally obovate-oblanceolate, broadest above the middle · · · · · · · 20. *neptunica*
– Leaves generally ± ovate, broadest at or below the middle · 7
7. Female bracts deeply 3–7-lobed or 3–7-partite · 8
– Female bracts shallowly lobed or toothed or fimbriate · 9
8. Woody perennial herb or subshrub up to 3 m high · · · · · · · · · · · · · · · · 21. *psilostachya*
– Weak annual herb usually to not more than 30 cm high · · · · · · · · · · · · 22. *brachystachya*
9. Female bracts fimbriate · 10
– Female bracts shallowly lobed, dentate or crenate · 11
10. Teeth of the female bracts linear-filiform, not contiguous, erect · · · · · · · · · · · · 23. *ciliata*
– Teeth of the female bracts falcate-lanceolate, often ± contiguous, apically-directed · 24. *fimbriata*
11. Shrubs or small trees · 19. *glabrata*
– Herbs, or very rarely subshrubs · 12
12. Female bracts repand-dentate; allomorphic female flowers fimbriate, long-pedicellate · 25. *indica*
– Female bracts crenate, or if dentate, then not repand; allomorphic female flowers muricate, or if fimbriate, then ± sessile or only shortly pedicellate · 13
13. Female bracts accrescent to 1.5 × 4 mm, glandular-hairy; allomorphic female flowers fimbriate · 28. *lanceolata*
– Female bracts accrescent to 13 × 15 mm, not or scarcely glandular; allomorphic female flowers muricate · 14
14. Female bracts crenate, usually 1–3 per inflorescence; allomorphic female flower puberulous · 26. *segetalis*
– Female bracts dentate, usually more than 3 per inflorescence; allomorphic female flowers glabrous, but with each tubercle tipped with a single hair · · · · · · · · · · · · · · · · 27. *crenata*
15. Plants dioecious · 16
– Plants monoecious · 24
16. At least some leaves verticillate; male inflorescences long-pedunculate · · · · · · · · 31. *sp. A*
– All leaves alternate · 17
17. Female flowers solitary, axillary · 18
– Female flowers in dense axillary or terminal spikes · 20
18. Leaves toothed, up to 10 × 3 cm · 4. *fuscescens*
– Leaves not exceeding 3 × 1.2 cm · 19
19. Stems pilose or pubescent; leaves entire, pubescent beneath · · · · · · · · · · · · · · 5. *clutioides*
– Stems densely pilose; leaves sometimes denticulate, pilose beneath · · · · · · 6. *dikuluwensis*
20. Leaves sessile or shortly petiolate · 21
– Leaves long petiolate · 22
21. Leaves coarsely toothed all round; female bracts ovate-lanceolate · · · · · · · 1. *caperonioides*
– Leaves serrate or crenate-serrate towards the apex; female bracts orbicular or transversely ovate · 2. *polymorpha*
22. Stipelliform glands present; female inflorescences ovoid or subcylindric, up to 4 cm long · 8. *welwitschiana*
– Stipelliform glands absent; female inflorescences cylindric, up to 30 cm long · · · · · · · 23
23. Female bracts accrescent to 1.5 × 2.5 cm; styles up to 3 mm long · · · · · · · · · · · 10. *ornata*
– Female bracts minute, not accrescent; styles up to 7 mm long · · · · · · · · · · · · 30. *hispida*
24. Female inflorescences paniculate; bracts not accrescent · · · · · · · · · · · · · · · 14. *racemosa*
– Female inflorescences spicate or racemose; bracts usually accrescent · · · · · · · · · · · · · 25

25. Vegetative stems viscid, with numerous gland-tipped hairs · · · · · · · · · · · · · · · · 12. *allenii*
 – Vegetative stems not viscid, or rarely with the occasional glandular hair below the inflorescence · 26
26. Female bracts with gland-tipped hairs, especially on the margin · · · · · · · · · · · · · · · · · 27
 – Female bracts eglandular · 29
27. Leaf blades broadly ovate to elliptic-ovate, serrate · 10. *ornata*
 – Leaf blades lanceolate to linear-lanceolate, serrate or dentate · · · · · · · · · · · · · · · · · · 28
28. Stipelliform glands present at base of leaf blade; plant pubescent or villous; leaves serrate · 11. *villicaulis*
 – Stipelliform glands absent; plant glabrous or almost so; leaves remotely dentate · 13. *paucifolia*
29. Female inflorescences usually 1–few-flowered, axillary · · · · · · · · · · · · · · · · · · 3. *ambigua*
 – Female inflorescences many-flowered, spicate, axillary or terminal · · · · · · · · · · · · · · · 30
30. Female inflorescences lax flowered · 29. *wilkesiana*
 – Female inflorescences dense flowered · 31
31. Male inflorescences shortly pedunculate to subsessile · 32
 – Male inflorescences long-pedunculate, the peduncles up to 9 cm long · · · · · · · · · · · · · 33
32. Leaves usually narrowed to the base; infructescences cylindric · · · · · · · · · · · · 7. *chirindica*
 – Leaves more or less rounded at the base; infructescences ovoid or subcylindric · 8. *welwitschiana*
33. Annual herb · 9. *nyasica*
 – Perennial herbs with several stems arising from a woody stock · · · · · · · · · · · · · · · · · · 34
34. Leaves coarsely toothed all round; female bracts ovate-lanceolate · · · · · · · · 1. *caperonioides*
 – Leaves serrate or crenate-serrate towards the apex; female bracts orbicular or transversely ovate · 2. *polymorpha*

1. **Acalypha caperonioides** Baill., Adansonia **3**: 157 (1863). —Prain & Hutchinson in Bull. Misc. Inform., Kew **1913**: 23 (1913). —Eyles in Trans. Roy. Soc. South Africa **5**: 396 (1916). —Prain in F.C. **5**, 2: 481 (1920). —Engler, Pflanzenw. Afrikas (Veg. Erde 9) **3**, 2: 97 (1921). —Pax in Engler, Pflanzenr. [IV, fam. 147, xvi] **85**: 76 (1924). —Burtt Davy, Fl. Pl. Ferns Transvaal: 306 (1932). —Suessenguth & Merxmüller, Contrib. Fl. Marandellas Distr. **43**: 83 (1951). Type from South Africa (Transvaal).

Acalypha peduncularis E. Mey. in Drège ex Meisn. apud Krauss var. *glabrata* Sond. in Linnaea 23: 115 (1850). Type from South Africa (Transvaal).

Acalypha peduncularis var. *caperonioides* (Baill.) Müll. Arg. in De Candolle, Prodr. **15**, 2: 846 (1866).

Acalypha peduncularis sensu Hutch. in F.T.A. **6**, 1: 884 (1912), pro parte, non *E. Mey.*
Acalypha punctata sensu Eyles in Trans. Roy. Soc. South Africa **5**: 396 (1916) non Meisn.

A perennial herb up to 40 cm high, from a branched rhizome system associated with a fist-sized woody rootstock; stems few–many, usually simple, sparingly pubescent to villous. Petioles 1–2 mm long, or leaves subsessile. Leaf blades 1.5–6 × 1–2 cm, ovate to elliptic-ovate or ovate-lanceolate, acute or obtuse at the apex, sharply and coarsely serrate on the margin, rounded to shallowly cordate at the base, firmly chartaceous to thinly coriaceous, 3–7-nerved from the base, sparingly pubescent and hirsute with bulbous-based hairs on both surfaces; lateral nerves in 1–3 pairs, ascending, slightly prominent beneath, ± so above. Stipules 2–4 mm long, linear to narrowly-oblong. Plants dioecious, rarely monoecious. Male racemes up to 13 cm long, axillary, solitary, densely flowered on a peduncle up to 7 cm long; bracts resembling the stipules. Male flowers: pedicels c. 1 mm long; buds depressed, 4-lobed, sparingly puberulous, reddish-brown; stamens 5–7, anthers white. Female spikes terminal, up to 4 × 3 cm in fruit; bracts few, up to 2 × 1.2 cm in fruit, ovate-lanceolate, acuminate, dentate with up to 5 narrowly lanceolate teeth on each side, foliaceous, up to 11-nerved from the base, sparingly hirsute with bulbous-based hairs and sometimes with a few stipitate or sessile glands along the nerves abaxially, 1-flowered. Female flowers sessile; sepals ovate-lanceolate, acute, ciliate; ovary c. 1 mm in diameter, 3-lobed, with stipitate glands and a few bulbous-based hairs towards the apex; styles 2–3 cm long, united in the lower sixth to one-quarter, the style arms very sparingly fimbriate, crimson. Fruits 3 × 4 mm, 3-lobed, sparingly stipitate-glandular, reddish. Seeds c. 3 × 2 mm, ovoid-subglobose, brownish-grey, with a small yellowish caruncle.

Zimbabwe. N: 5 km west of Mazowe (Mazoe), female fl. 9.xi.1962, *Drummond & Angus* 3391 (FHO; K). W: Besna Kobila Farm, male fl. i.1957, *Miller* 4071 (K; SRGH). C: Harare, fr. 4.xi.1968, *Biegel* 2675 (K; LISC; PRE; SRGH). E: Tarka For. Res., male fl. 27.xi.1967, *Simon & Ngoni* 1353 (K; LISC; PRE; SRGH). **Malawi**. N: Viphya Plateau, fr. 12.xi.1972, *Pawek* 5966 (K; MO). C: Ciwao Hill, Chongoni, male & female fl. 19.i.1959, *Robson* 1263 (BM; K; LISC; SRGH). S: Kirk Range, female fl. & fr. 30.i.1959, *Robson* 1346 (BM; K; LISC; SRGH). **Mozambique**. MS: Tandara, male fl. 18.xi.1965, *Torre & Correia* 13101 (LISC). M: Namaacha, male fl. 20.xi.1966, *Moura* 136 (COI).

Also from South Africa (Transvaal, KwaZulu-Natal, Free State, Cape). Plateau and montane grassland and *Brachystegia* woodland, sometimes on dambo margins; particularly noticeable after grass fires; 915–2230 m.

This species is somewhat variable in leaf characters. The leaves of *Simon & Ngoni* 1353 (K; SRGH) are narrower than usual, and the *Robson* gatherings cited above have leaves resembling those of *A. polymorpha*. In South Africa, the species also shows intermediates with *A. peduncularis*, *A. punctata* and *A. wilmsii*.

2. **Acalypha polymorpha** Müll. Arg. in J. Bot. **2**: 335 (1864); in De Candolle, Prodr. **15**, 2: 835 (1866), pro parte, quoad var. *elliptica*, var. *sericea* & var. *oblongifolia*. —Hutchinson in F.T.A. **6**, 1: 894 (1912). —Engler, Pflanzenw. Afrikas (Veg. Erde 9) **3**, 2: 96 (1921). —Pax in Engler, Pflanzenr. [IV, fam. 147, xvi] **85**: 71 (1924). —De Wildeman, Pl. Bequaert. **3**, 4: 493 (1926). Types from Angola (Huíla Province).

Acalypha crotonoides Pax in Bot. Jahrb. Syst. **19**: 97 (1894). —Hutchinson in F.T.A. **6**, 1: 895 (1912). —R.E. Fries, Wiss. Ergebn. Schwed. Rhod.-Kongo-Exped. **1**, 1: 122 (1914). —Engler, Pflanzenw. Afrikas (Veg. Erde 9) **3**, 2: 97 (1921). Type from Angola (Lunda Province).

Acalypha stuhlmannii Pax in Bot. Jahrb. Syst. **19**: 99 (1894). —Engler, Pflanzenw. Ost-Afrikas **C**: 239 (1895). —Hutchinson in F.T.A. **6**, 1: 905 (1912). —R.E. Fries, Wiss. Ergebn. Schwed. Rhod.-Kongo-Exped. **1**, 1: 123 (1914). —Engler, Pflanzenw. Afrikas (Veg. Erde 9) **3**, 2: 98 (1921). —Pax in Engler, Pflanzenr. [IV, fam. 147, xvi] **85**: 74 (1924). —Brenan, Check-list For. Trees Shrubs Tang. Terr.: 198 (1949). —Agnew, Upl. Kenya Wild Fls.: 215 (1974). —Troupin, Fl. Rwanda **2**: 198, fig. 56/4 (1983). —Radcliffe-Smith, F.T.E.A., Euphorb. 1: 194 (1987). Type from Tanzania (Lake Province).

Acalypha crotonoides var. *caudata* Hutch. in R.E. Fries, Wiss. Ergebn. Schwed. Rhod.-Kongo-Exped. **1**, 1: 123 (1914). Type: Zambia, Western Province, Ndola, fl. 1.ix.1911, *R.E. Fries* 507 (UPS, holotype).

Acalypha shirensis Hutch. ex Pax in Engler, Pflanzenr. [IV, fam. 147, xvi] **85**: 32 (1924). Type: Malawi, without precise locality or date, *Buchanan* 55 (BM, holotype).

Acalypha goetzei Pax & Hoffm. in Engler, Pflanzenr. [IV, fam. 147, xvi] **85**: 78 (1924). Type from Tanzania (Southern Highlands Province).

An erect densely caespitose perennial herb up to 45 cm high, from a woody rootstock; stems simple, numerous, crisped-puberulous, tomentose, villous or hirsute. Petioles 0.1–1 cm long. Leaf blades 1–11 × 1–3.5 cm, very variable, ranging from suborbicular (at the base of the stem) to ovate, obovate, elliptic, oblong or narrowly oblanceolate or lanceolate, obtuse to subacute, acute or shortly caudate-acuminate at the apex, serrate or crenate-serrate on the margin, especially in the upper half, attenuate, cuneate to rounded or shallowly cordate at the base, chartaceous, sparingly pubescent above and beneath, later glabrescent, 3–5(7)-nerved from the base, tertiary nerves subparallel; lateral nerves in 3–6 pairs, ascending, fairly prominent beneath. Stipules 1–3 mm long, linear-subulate. Male racemes up to 12 cm long with a peduncle up to 5 cm long, solitary, axillary, erect, occasionally with 1–2 female flowers at the base or scattered along the axis; bracts resembling the stipules. Male flowers shortly pedicellate; buds 4-lobed to subglobose, somewhat apiculate, sparingly to evenly pubescent; calyx lobes red, or white and edged with red; anthers white. Female inflorescences 1–6 × 1.5–2.5 cm, spicate, usually solitary and terminal, occasionally axillary, capitate or cylindric, subsessile or shortly pedunculate, sometimes with a terminal male appendage; bracts accrescent to 1 × 1.5 cm, transversely ovate, regularly 8–11-fid almost to the base, the segments 3–5 mm long, triangular-lanceolate, pubescent, 1-flowered, greenish-brown. Female flowers sessile; sepals 3, 1 mm long, linear-lanceolate, acute, ciliate; ovary c. 1 mm in diameter, subglobose, densely whitish-pubescent; styles up to 1.5 cm long, united at the base, pectinate, red or sometimes whitish. Fruits 3–4 × 3.5–4.5 mm, 3-lobed, pubescent, greenish. Seeds 2.5 × 2 mm, ovoid-subglobose, smooth, dark grey, with a small whitish elliptical caruncle.

Zambia. N: Mbala (Abercorn), L. Chila, female fl. 9.xii.1954, *Siame* 529 (K; MO). W: 17.5 km Mwinilunga–Kalene Hill, female fl. & fr. 21.xi.1972, *Strid* 2545 (C; K; MO). C: 40 km

Mumbwa–Kafue Hook, fr. 7.xi.1959, *Drummond & Cookson* 6186 (K; LISC; PRE; SRGH). S: Kalomo, fl. 8.x.1955, *Gilges* 459 (K; PRE; SRGH). **Zimbabwe**. N: Dawson's, fr. 18.xii.1952, *Wild* 3955 in *GHS* 40771 (K; LISC; MO; PRE; SRGH). C: Makoni, fl. 30.xi.1930, *Fries, Norlindh & Weimarck* 3347 (K; LD; PRE). E: Tarka For. Res., fr. xi.1968, *Goldsmith* 165/68 (K; LISC; PRE pro parte; SRGH pro parte). **Malawi**. N: Katoto, 5 km west of Mzuzu, fr. 10.xii.1977, *Pawek* 13278 (K; MAL; MO; PRE; SRGH; UC). C: Chongoni For. Res., fl. 17.i.1959, *Robson & Jackson* 1238 (BM; K; LISC; SRGH). S: Zomba Plateau, male fl. 20.xi.1979, *Banda & Salubeni* 1602 (MAL; MO; SRGH). **Mozambique**. N: near Litunde, fr. 15.xii.1934, *Torre* 455 (COI; LISC).

Also from Uganda, Kenya, Tanzania, Rwanda and Burundi. High rainfall plateau woodland, submontane grassland and *Brachystegia* woodland, usually in open woodland with grass but also in woodland with dense ground cover; also on dambo margins and tall grasslands; particularly noticeable after bush fires; 350–2470 m.

Goldsmith 165/68, cited above, is a mixed gathering; the PRE and SRGH sheets include material *of A. caperonioides*. *Richards* 3599 (K) and *Sanane* 383 (K) from N Zambia, *Richards* 17182 (K) from W Zambia and *Corby* 522 in *GHS* 30013 (K; SRGH) from C Zimbabwe are intermediate between this species and *A. ambigua*.

3. **Acalypha ambigua** Pax in Bot. Jahrb. Syst. **19**: 96 (1894). —Hutchinson in F.T.A. **6**, 1: 906 (1912). —Engler, Pflanzenw. Afrikas (Veg. Erde 9) **3**, 2: 96 (1921). —Pax in Engler, Pflanzenr. [IV, fam. 147, xvi] **85**: 73 (1924). —P.G. Meyer in Merxmüller, Prodr. Fl. SW. Afrika, fam. 67: 5 (1967). —Radcliffe-Smith in F.T.E.A., Euphorb. 1: 194 (1987). Type from Angola (Malanje Province).

Acalypha polymorpha var. *angustifolia* Müll. Arg. in J. Bot. **2**: 335 (1864); in De Candolle, Prodr. **15**, 2: 836 (1866). Types from Angola.

Acalypha polymorpha var. *depauperata* Müll. Arg. in J. Bot. **2**: 335 (1864); in De Candolle, Prodr. **15**, 2: 836 (1866). Types from Angola.

Acalypha angustissima sensu R.E. Fries, Wiss. Ergebn. Schwed. Rhod.-Kongo-Exped. **1**, 1: 122 (1914), non Pax (1894).

An erect caespitose perennial herb up to 50 cm high, arising from a stout woody rootstock; stems several, simple or sparingly branched, pubescent. Petioles 1 mm long, or leaves subsessile. Leaf blades 3–9 × 0.3–1.5 cm, linear to linear-lanceolate or linear-oblanceolate, acute or subacute at the apex, with margins toothed in the upper half, otherwise ± entire, attenuate at the base, firmly chartaceous, sparingly pubescent above and beneath, bright green; lateral nerves in 5–6 pairs, ascending, fairly prominent beneath. Stipules 2–3 mm long, linear-subulate. Inflorescences up to 9 cm long, axillary, unisexual or bisexual, if the latter then consisting of a male spike with 1–2 female flowers at the base, otherwise spikes all male and many-flowered or all female and 1–few-flowered; male bracts resembling the stipules; female bracts accrescent to up to 5–6 × 8 mm, suborbicular, almost cupular, irregularly 13–19-fid almost to halfway, the segments linear to lanceolate, pubescent, 1-flowered. Male flowers very shortly pedicellate to subsessile; buds 4-lobed to subglobose, sparingly puberulous, creamy-yellow, sometimes pinkish- or reddish-tinged; anthers red. Female flowers sessile; sepals 3–4, 1 mm long, ovate to ovate-lanceolate, acute, ciliate; ovary 1 mm in diameter, 3-lobed, sericeous-pubescent, green; styles up to c. 1 cm long, free, pectinately laciniate, crimson. Fruits 3–4 × 4–5 mm, 3-lobed, sericeous-pubescent. Seeds 2–2.5 × 1.7–2 mm, ovoid-subglobose, smooth, grey, with a flattened elliptical caruncle.

Zambia. B: Kabompo R., fl. 25.x.1966, *Leach & Williamson* 13453 (K; SRGH). N: Kawambwa, fr. 8.xi.1952, *White* 3636 (FHO; MO). W: 16 km Ndola–Luanshya, fl. 20.xi.1959, *Wild* 4861 (K; LISC; MO; PRE; SRGH). C: Kabwe (Broken Hill)–Kapiri Mposhi, fl. 1.xii.1952, *Angus* 900 (FHO; K). **Zimbabwe**. C: Driffield, Marondera (Marandellas), fl. 23.xi.1945, *Rattray* 1374 in *GHS* 20746 (K; SRGH).

Also in Zaire (Shaba), Burundi, western Tanzania, Angola and Namibia. Plateau miombo, high rainfall wooded grassland and dambo margins; particularly noticeable after bush fires; 1200–1400 m.

In western Zambia, intermediates occur between this species and *A. fuscescens* (e.g. *Milne-Redhead* 2576 (K), 2598 (K) and 2643 (K; PRE) and *Fanshawe* 305 (K; NDO), and *A. polymorpha* (e.g. *Milne-Redhead* 779 (K), *Cruse* 267 (K) and *Linley* 203 (K; SRGH)).

4. **Acalypha fuscescens** Müll. Arg. in De Candolle, Prodr. **15**, 2: 821 (1866). —Hutchinson in F.T.A. **6**, 1: 884 (1912). —Engler, Pflanzenw. Afrikas (Veg. Erde 9) **3**, 2: 96 (1921). —Pax in Engler, Pflanzenr. [IV, fam. 147, xvi] **85**: 95 (1924). Type from Angola (Malanje).

A perennial herb up to c. 30 cm tall, arising from a woody rootstock; stems several, simple, puberulous to pubescent. Petiole not more than 2 mm long, or leaves sessile. Leaf blades 3–10 × 1.5–3 cm, narrowly elliptic-lanceolate or lanceolate, the lowermost sometimes obovate, acute at the apex (the lowermost usually obtuse), serrate on the margins, tapered or rounded at the base, chartaceous, sparingly setose with bulbous-based hairs on the upper surface, more evenly pubescent beneath, drying dull brown, ± penninerved with 9–10 pairs of lateral nerves, fairly prominent beneath, tertiary nerves closely parallel. Stipules 3 mm long, linear. Plants dioecious or sometimes monoecious. Male racemes up to 2 cm long, axillary, solitary, sometimes with a female flower at the base; bracts resembling the stipules. Male flowers: pedicels less than 0.5 mm long; buds subglobose, puberulous, purplish-tinged; anthers minute. Female flowers solitary or in pairs in the axils of the upper leaves, sessile; bracts c. 4–5 × 6 mm when mature, palmatifid, c. 20-fid to halfway, the segments 1.5–3 mm long, linear-lanceolate to subulate, alternately long and short, pubescent, eglandular, 1-flowered. Female flowers: sepals oblong, minute, acute, pubescent; ovary c. 1 mm in diameter, globose, densely fulvous-tomentose; styles 3–8 mm long, free, pectinately lacinulate, minutely puberulous, red. Fruits 4 × 4 mm, subglobose, pubescent. Seeds 2 × 1.5 mm, ovoid, smooth, dark brown.

Zambia. N: Kawambwa, fr. 8.xi.1952, *White* 3636 (FHO; K; PRE). W: Ndola, fl. & fr. 22.iii.1954, *Fanshawe* 1014 (K; NDO; SRGH). C: 40 km Mumbwa–Kafue Hook, fl. 7.xi.1959, *Drummond & Cookson* 6191 (K; SRGH). S: Tara Protected Forest Area, female fl. 26.i.1960, *White* 6392 (FHO; K; SRGH).

Also in Angola. Plateau miombo, chipya woodland and open *Brachystegia–Protea* woodland on dambo margins; particularly noticeable after bush fires; 1220 m.

White 3636 (K) and *Drummond & Cookson* 6191 (K) vary in the direction of *A. ambigua*, whereas *Shepherd* 85 (K) approaches *A. polymorpha*.

5. **Acalypha clutioides** Radcl.-Sm. in Kew Bull. **28**: 287 (1973). Tab. **39**. Type: Zambia, Mwinilunga Distr., 8 km from Kalene Hill Mission on Zaire border road, fr. 11.xi.1962, *Richards* 17180 (K, holotype).

A densely caespitose perennial herb or suffrutex up to 50 cm tall, from a stout woody rootstock; stems many, simple, densely leafy, pilose or pubescent. Petioles 0.5–2 mm long, or leaves subsessile. Leaf blades 0.5–3 × 0.4–1.2 cm, elliptic-ovate to ovate-lanceolate, acuminate, entire, rounded to shallowly cordate at the base, sparingly pilose above, pubescent along the midrib and main nerves beneath, chartaceous, 5-nerved from the base, with tertiary nerves parallel and fairly prominent beneath; lateral nerves in 1–3 pairs, slightly prominent above, fairly prominent beneath. Stipules 1–1.5 mm long, linear, reddish-brown. Plants dioecious. Male racemes up to 5 cm long, axillary, solitary, densely flowered, on peduncles up to 2 cm long; bracts 1.5 mm long, linear-spathulate. Male flowers: pedicels 1 mm long; buds subglobose, sparingly pubescent, greenish-brown; anthers minute, cream-coloured. Female flowers axillary, solitary; peduncles 2 mm long; bracts 5 × 8 mm, palmatipartite with lobes linear-lanceolate and acuminate, pubescent without, glabrous within; flower subsessile within the bract; sepals 3, 2 mm long, ovate-lanceolate, acute, ciliate; ovary 2 mm in diameter, 3-lobed, pubescent, hirsute and tuberculate at the apex; styles 4 mm long, free, fimbriate, red. Fruits 3 × 3 mm, 3-lobed, pubescent-pilose, yellowish. Seeds 2 × 1.5 mm, ovoid, dark brown.

Zambia. W: north of Kalene Hill, near Zairean border, female fl. 24.ix.1952, *Angus* 535B (FHO; K); top of Kalene Hill, female fl. 15.xii.1963, *Robinson* 6050 (K; LISC; PRE; SRGH); Mwinilunga, male fl. 16.v.1969, *Mutimushi* 3163 (K; NDO; SRGH).

Also in Zaire (Shaba) and possibly Angola (Moxico). Watershed grassland plains, on Kalahari Sand, in sandy dambos and in moist deep black soil, or in *Uapaca* woodland; 1200–1400 m.

6. **Acalypha dikuluwensis** P.A. Duvign. & Dewit in Bull. Soc. Roy. Bot. Belgique **96**, 2: 1 (1963). Type from Zaire (Shaba Province).

Very like *A. clutioides*, but distinguished from it in the following features; stems not more than 25 cm high, more densely pilose but less densely leafy than in *A. clutioides*;

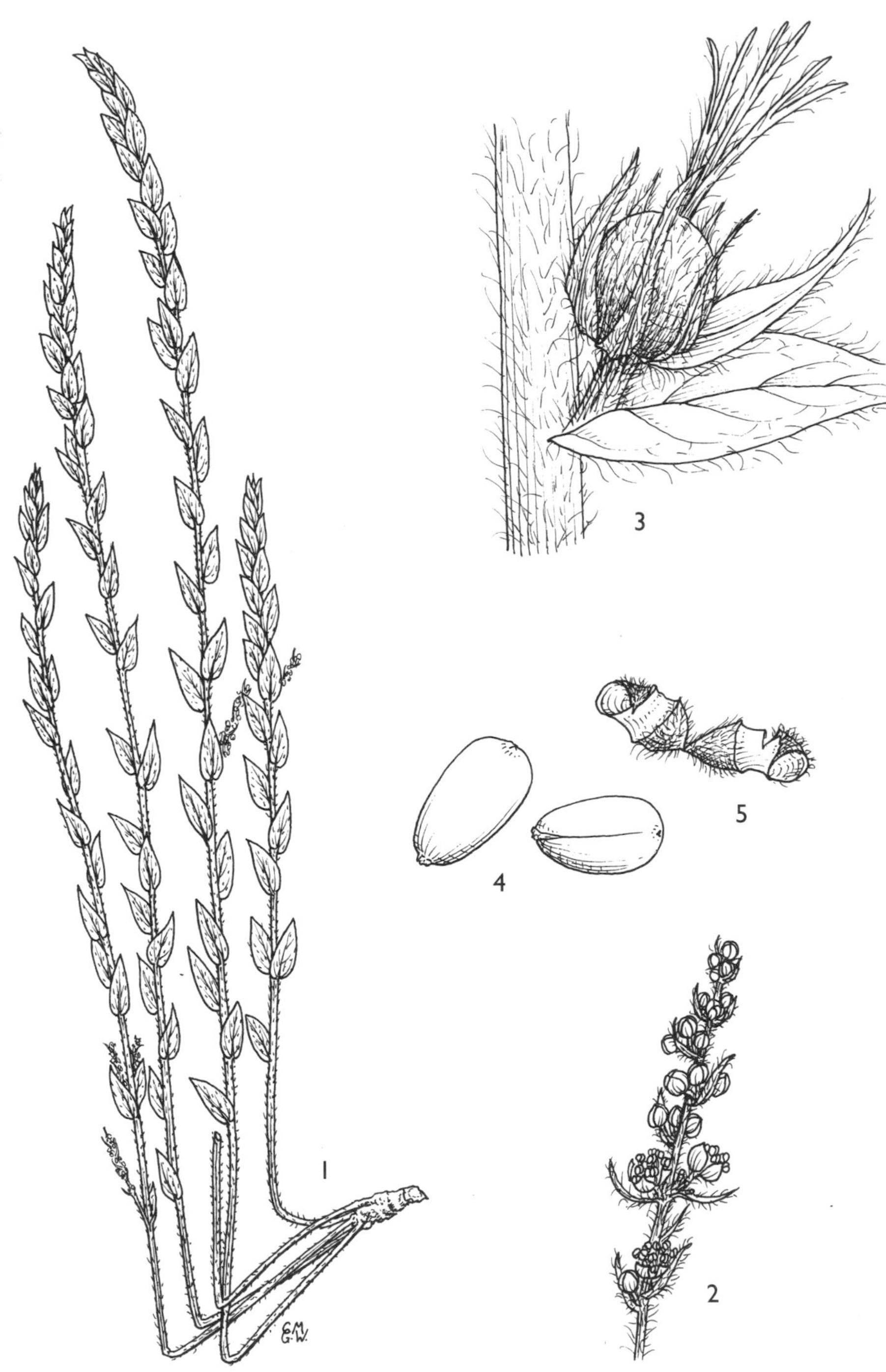

Tab. 39. ACALYPHA CLUTIOIDES. 1, habit (× $^{2}/_{3}$); 2, male raceme (× 6), 1 & 2 from *Mutimushi* 3163; 3, female flower (× 6); 4, seeds (× 6); 5, fruit valves (× 6), 3–5 from *Richards* 16942. Drawn by Christine Grey-Wilson.

lower leaves more broadly ovate to almost suborbicular; leaves sometimes irregularly, shallowly and remotely denticulate on the margins, and pilose beneath; female flowers sometimes geminate in the leaf axils; female bracts palmatifid; styles connate at the base and the ovary globose.

Zambia. W: southwest of Dobeka (Jobeka) Bridge, fl. 13.x.1937, *Milne-Redhead* 2741 (K).

Also in Zaire. Watershed grassland, on termite mounds in open at edge of boggy dambos.

This species may even be considered to be conspecific with *A. clutioides*, in which case its name has priority by 10 years. In Zaire it is a polycuprophyte, only being found on a substrate where the copper content of the soil is above 2,500 ppm., and it can tolerate concentrations of up to 50,000 ppm.

7. **Acalypha chirindica** S. Moore in J. Linn. Soc., Bot. **40**: 199 (1911). —Hutchinson in F.T.A. **6**, 1: 885 (1912). —R.E. Fries, Wiss. Ergebn. Schwed. Rhod.-Kongo-Exped. **1**, 1: 122 (1914). —Eyles in Trans. Roy. Soc. South Africa **5**: 396 (1916). —Engler, Pflanzenw. Afrikas (Veg. Erde 9) **3**, 2: 98 (1921). —Pax in Engler, Pflanzenr. [IV, fam. 147, xvi] **85**: 81 (1924). —Brenan, Check-list For. Trees Shrubs Tang. Terr.: 196 (1949); in Mem. N.Y. Bot. Gard. **9**, 1: 74 (1954). —Topham, Check List For. Trees Shrubs Nyasaland Prot.: 48 (1958). —White, F.F.N.R.: 192 (1962). —Drummond in Kirkia **10**: 252 (1975). —Radcliffe-Smith in F.T.E.A., Euphorb. 1: 189 (1987). Tab. **40**. Type: Zimbabwe, near Chirinda, 1160 m., fl. xii.1905, *Swynnerton* 381 (BM, holotype).

Acalypha psilostachyoides Pax in Annuario Reale Ist. Bot. Roma **6**: 183 (1896), pro minore parte, quoad spec. Buchanan.

A deciduous, much-branched, straggling or scrambling, slender-stemmed shrub or small tree up to 5 m tall, pubescent or pilose; branches over-arching and pendent. Petioles 2–10(25) mm long. Leaf blades 3–13(15) × 1–5(8) cm, ovate-lanceolate to asymmetrically rhombic-lanceolate, acuminate at the apex, tapered to a narrowly cordate base, crenate-serrate on the margin, sometimes subentire towards the base, firmly membranous, 3–5-nerved from the base, sparingly pubescent or pilose on both surfaces and more evenly so on the midrib and main nerves beneath, pale to mid green; lateral nerves in 5–7 pairs. Stipules 2–9 mm long, linear-lanceolate, long-acuminate, chestnut-brown. Plants monoecious. Male racemes 5–6 cm long, axillary, solitary, scarcely pedunculate, sometimes galled. Male flowers: pedicels 0.5 mm long; buds tetragonal, pubescent, greenish- or reddish-brown; anthers cream-coloured. Female spikes 2–7 × 1 cm, terminal, cylindric, on a peduncle up to 7 mm long; bracts accrescent to 6–12 × 7–13 mm, transversely-ovate, pilose, eglandular, green, 1-flowered, 7–17-toothed, the teeth narrowly lanceolate, c. one-third to half the length of the bract. Female flowers ± sessile; sepals 3, c. 1 mm long, ovate, pubescent; ovary c. 1 mm in diameter, 3-lobed, densely pubescent; styles free, 5 mm long, laciniate, crimson to maroon. Fruits 1.5–2 × 2.5 mm, 3-lobed, pubescent. Seeds 1.8 × 1.4 mm, ovoid, ± smooth, purplish-grey.

Zambia. B: Zambezi (Balovale), fl. 10.xi.1952, *Gilges* 271 (K; PRE; SRGH). N: Kasama, fl. 26.xi.1952, *Angus* 848 (FHO; K; MO). W: Ndola, fr. 4.i.1955, *Fanshawe* 1774 (K; NDO; SRGH). C: Rufunsa, fl. 12.ix.1953, *Fanshawe* 294 (K; NDO). E: Chadiza, fl. 25.xi.1958, *Robson* 694 (BM; FHO; K; LISC; PRE; SRGH). S: Siamambo For. Res., fr. 17.i.1960, *White* 6294 (FHO; K). **Zimbabwe**. N: Mutoko Distr., Nyatsene, fl. 31.xii.1961, *Wild* 5577 (K; MO; PRE; SRGH). E: Mutare (Umtali), Murahawa's Hill, fl. & fr. 28.xii.1964, *Chase* 8219 (K; LISC; MO; SRGH). **Malawi**. N: Mzuzu, fl. 7.xii.1972, *Pawek* 6074 (K; MO; SRGH; UC). C: Nkhota Kota Distr., Chankoma R., fl. 19.xi.1963, *Salubeni* 141 (SRGH). S: Nswadzi R., fl. 29.ix.1946, *Brass* 17869 (K; MO; NY; SRGH). **Mozambique**. N: R. Mucuburi, fl. & fr. 25.xii.1936, *Torre* 1097 (COI; LISC). Z: Namagoa, fl. 20.ii.1949, *Faulkner* 384 (K). T: Ulóngüè, fl. 8.xii.1980, *Macuácua* 1402 (K; LUAI; PRE). MS: Barúè, fl. 8.xii.1965, *Torre & Correia* 13416 (LISC).

Also in Zaire (Shaba) and Tanzania. Frequent in understorey of mixed evergreen rainforest, dense evergreen riverine forest and in gully and escarpment ravine forests; in closed canopy *Brachystegia* woodlands and dense kopje vegetation; also in understorey thicket of dry deciduous forest on Kalahari Sand (mutemwa), termite mound thickets and pemba thicket; 500–1675 m.

Vernacular name as recorded in specimen data: "mpachulu" (chiNyanja).

In Malawi, this species hybridizes with *A. ornata* S. Moore, to produce *A.* × *malawiensis* Radcl.-Sm. Type: Liwonde, Shire River, 26.ii.1961, *Richards* 14463 (K) [in Kew Bull. **44**: 444 (1989)]. The following collections have been seen:

Malawi. N: Kaseye Mission, 16 km east of Chitipa, fl. 26.xii.1977, *Pawek* 13384 (MO). S: Shire R., Liwonde, fl. & failed fr. 26.ii.1961, *Richards* 14463 (K); Ndirande Mt., fl. 23.xi.1977, *Brummitt, Seyani & Banda* 15174 (K).

Tab. 40. ACALYPHA CHIRINDICA. 1, portion of flowering branch (× ⅔); 2, female flower (× 6); 3, portion of male inflorescence (× 6), 1–3 from *Chase* 8219; 4, abnormal male inflorescence (× 2), from *Chase* 1227. Drawn by Christine Grey-Wilson.

8. **Acalypha welwitschiana** Müll. Arg. in J. Bot. **2**: 334 (1864); in De Candolle, Prodr. **15**, 2: 834 (1866). —Hutchinson in F.T.A. **6**, 1: 892 (1912). —Engler, Pflanzenw. Afrikas (Veg. Erde 9) **3**, 2: 96 (1921). —Pax in Engler, Pflanzenr. [IV, fam. 147, xvi] **85**: 69 (1924). —Radcliffe-Smith in F.T.E.A., Euphorb. 1: 188 (1987). Type from Angola (Malanje).
Acalypha angolensis Müll. Arg. in J. Bot. **2**: 335 (1864); in De Candolle, Prodr. **15**, 2: 835 (1866). —Hutchinson in F.T.A. **6**, 1: 886 (1912). —Engler, Pflanzenw. Afrikas (Veg. Erde 9) **3**, 2: 96 (1921). —Pax in Engler, Pflanzenr., [IV, fam. 147, xvi] **85**: 71 (1924). Type from Angola (Malanje).

A many-stemmed lax pubescent or puberulous shrub or subshrub up to 2 m high, or an erect or procumbent suffrutex with annual stems arising from a woody stock. Petioles 0.2–5 cm long, often with a number of stipelliform glands at the apex. Leaf blades 1–13 × 0.5–7.5 cm, elliptic-ovate to lanceolate, acute or obtuse at the apex, crenate to crenate-serrate on the margin, rounded to cordulate at the base, chartaceous, 5-nerved from the base, sparingly to evenly pubescent on both surfaces; lateral nerves in 3–6 pairs. Stipules 1.5–3 mm long, filiform. Plants monoecious or sometimes apparently dioecious. Male racemes up to 9 cm long, solitary, axillary, yellow-green, pink-tinged, shortly pedunculate or subsessile, occasionally terminating in a mass of female flowers, rarely with one or two female flowers at the base or halfway up. Male flowers very shortly pedicellate; buds quadrangular-subglobose, minute, sparingly pubescent, pinkish; anthers white. Female spikes 1.5–4 × 1.5–2 cm, terminal or axillary or both, capitate, ovoid or subcylindric, shortly pedunculate or subsessile; bracts c. 0.6–1 × 1–1.3 cm in the fruiting stage, flabelliform, c. 15–25-fid to halfway, the segments 3–5 mm long, linear-lanceolate to subulate, pilose, eglandular, pinkish-green, 1-flowered. Female flowers sessile; sepals 3, c. 0.5 mm long, ovate-lanceolate, subglabrous; ovary c. 1 mm in diameter, somewhat 3-lobed, glabrous or sericeous-hispid in the upper half; styles 3–5 mm long, free, slender, pectinately lacinulate, maroon or crimson. Fruits c. 2–3 × 2–3 mm, somewhat 3-lobed, glabrous or pubescent. Seeds 2 × 1 mm, ovoid, smooth, shiny, dark purplish-brown.

Zambia. N: Kawambwa, fl. 29.i.1957, *Fanshawe* 2958 (K; NDO; SRGH). W: 1.6 km east of Mwinilunga, female fl. 27.xi.1937, *Milne-Redhead* 3418 (K; LISC; PRE). C: Kapiri, fl. 10.ii.1964, *Fanshawe* 8312 (K; NDO). **Malawi**. N: Kawalazi Tea Estate, fr. 3.vi.1973, *Pawek* 6798 (K; MAL; MO; SRGH; UC). S: Mulanje Mt., Chambe Plateau, fr. 20.v.1971, *Salubeni* 1612 (K; MO; SRGH). **Mozambique**. N: Ribáuè, Serra de Mepáluè, female & male fl. 23.i.1964, *Torre & Paiva* 10182 (LISC). Z: Gurué, female fl. & fr. 6.xi.1967, *Torre & Correia* 15970 (LISC).

Also in Burundi, Tanzania and Angola. Closed canopy plateau woodlands, escarpment woodlands and Kalahari Sand woodlands; in evergreen riverine forest and thicket, swamp forest (mushitu) margins and dambo margins often on termite mounds; also in submontane evergreen forest and gully forest; 1000–2000 m.

Richards 22785 from Chenga–Nakonde (Nacondy) R., Zambia, male fl. 29.xi.1967 (K) is intermediate between this species and *A. fruticosa* var. *eglandulosa.*

Whyte s.n. from Zomba, Malawi, ix.1891, and *Whyte* 65 from Mt. Mulanje (Milanje), x.1891, both at the BM, were to have been the types of Hutchinson's unpublished *A. whytei.*

9. **Acalypha nyasica** Hutch. in F.T.A. **6**, 1: 894 (1912). —Pax in Engler, Pflanzenr. [IV, fam. 147, xvi] **85**: 70 (1924). —Radcliffe-Smith in F.T.E.A., Euphorb. 1: 190 (1987). Tab. **41**. Type: Malawi, Likoma Island, Lake Malawi (L. Nyasa), *Johnson* s.n. (K, lectotype chosen by Radcliffe-Smith (1987)).

An annual herb with an erect lead shoot to c. 15 cm high and numerous prostrate, decumbent or ascending lateral shoots up to 30 cm long. Stems puberulous or pubescent. Petioles 0.3–1(2) cm long. Leaf blades 1–5 × 0.4–2.5 cm, ovate-lanceolate or ovate-trullate, subacute or obtuse at the apex, finely crenulate to subentire on the margin, cuneate or rounded at the base, membranous, sparingly puberulous or pubescent on both surfaces, especially on the midrib and main nerves, or else ± glabrous, 3–5-nerved from the base; lateral nerves in 4–6 pairs. Stipules c. 1 mm long, subulate. Male spikes axillary, up to 13 cm long, with a peduncle up to 9 cm long; bracts minute. Female spikes terminal, or apparently lateral when a lateral shoot arises from beneath one, 1–6 × 0.5–1 cm, cylindric or sometimes ovoid, subsessile or shortly pedunculate; bracts 7 × 8–9 mm, ± reniform, pubescent, eglandular, often reddish-tinged, 1-flowered, 16–20-toothed, the teeth c. 3 mm long, linear. Allomorphic female flowers solitary, axillary. Male flowers sessile; buds

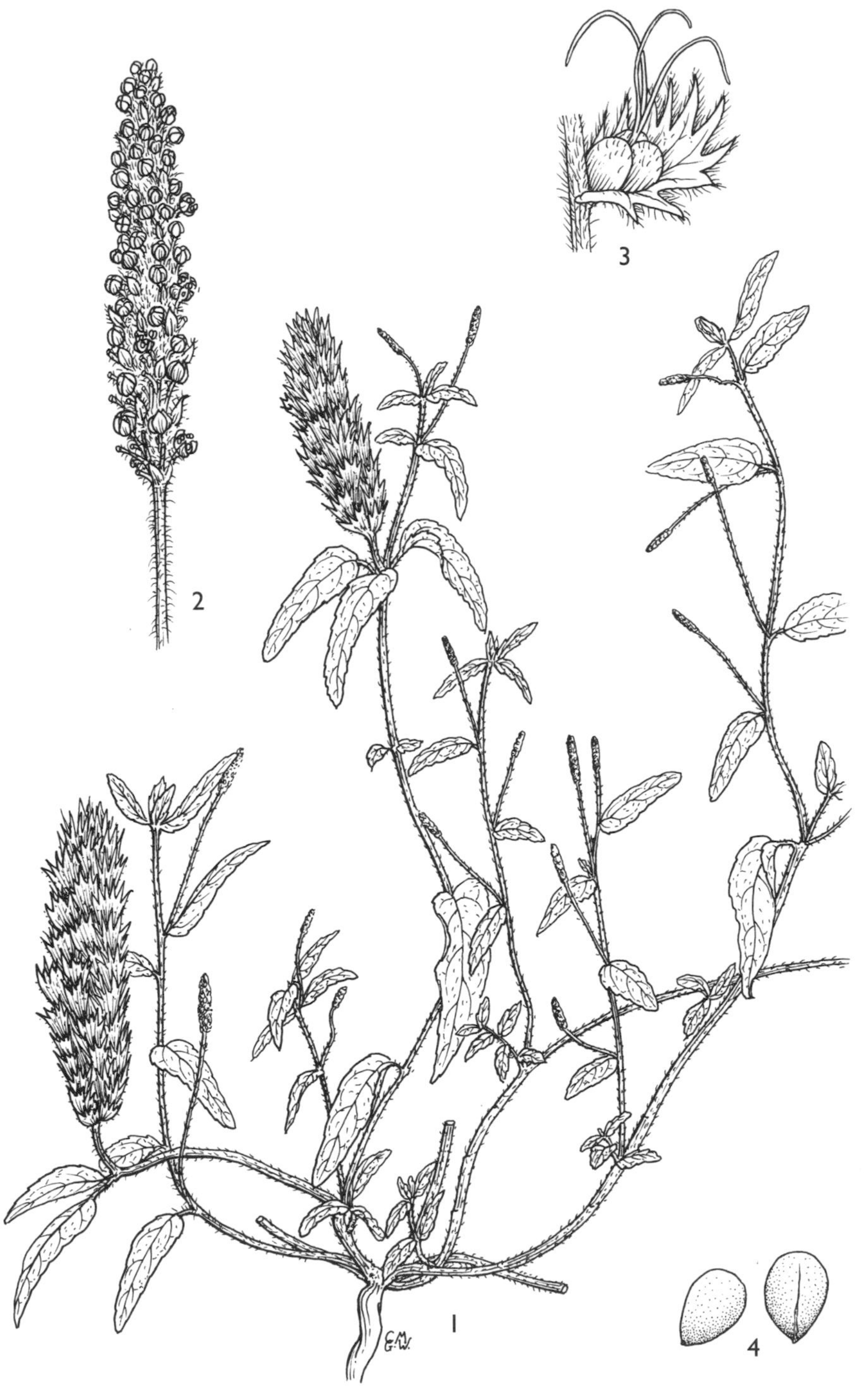

Tab. 41. ACALYPHA NYASICA. 1, habit (× ⅔); 2, male inflorescence (× 8); 3, female flower (× 4); 4, seeds (× 6), 1–4 from *Grosvenor & Renz* 1290. Drawn by Christine Grey-Wilson.

subglobose, minute, sparingly pubescent to subglabrous, reddish; anthers creamy-white. Female flowers sessile; sepals 3, 1 mm long, triangular, acute, ciliate; ovary 0.5 mm in diameter, deeply 3-lobed, sparingly pubescent at the apex, otherwise ± glabrous; styles free, 7–8 mm long, slender, filiform, simple, glabrous, dark red. Allomorphic female flowers long-pedicellate, the pedicels up to 3 cm long, capillary. Fruits 1.5 × 2.5 mm, deeply 3-lobed, sparingly pubescent at the apex, otherwise glabrous. Fruits of the allomorphic flowers 2 × 1.5 mm, obovoid, tuberculate, minutely puberulous. Seeds 1.3 × 1 mm, ovoid-subglobose, smooth, pinkish-brown, with a small white velar caruncle.

Malawi. N: Kaporo, fl. 2.i.1974, *Pawek* 7727 (K; MAL; MO; UC). C: Salima Distr., Grand Beach Hotel, fl. & fr. 20.iii.1977, *Grosvenor & Renz* 1290 (K; PRE; SRGH). S: Nsessi R., fr. xii.1887, *Scott* s.n. (K). **Mozambique**. N: Eráti, fr. 30.iii.1961, *Balsinhas & Marrime* in *Balsinhas* 337 (BM; COI; K; LISC; LMA; PRE; SRGH). Z: c. 7 km Mocuba–Mugeba, fr. 16.v.1949, *Barbosa & Carvalho* 2649 (K; LMA).

Also in Tanzania. On lakeshore sand dunes, also in *Brachystegia* woodland on sandy soils; 350–850 m.

Barbosa & Carvalho 2649 (K; LMA) is intermediate between this species and *A. welwitschiana.*

10. **Acalypha ornata** Hochst. ex A. Rich., Tent. Fl. Abyss. **2**: 247 (1851). —Müller Argoviensis in De Candolle, Prodr. **15**, 2: 833 (1866). —Engler, Pflanzenw. Ost-Afrikas **C**: 239 (1895). —Hutchinson in F.T.A. **6**, 1: 890 (1912). —Prain & Hutchinson in Bull. Misc. Inform., Kew **1913**: 19 (1913). —R.E. Fries, Wiss. Ergebn. Schwed. Rhod.-Kongo-Exped. **1**, 1: 122 (1914). —Eyles in Trans. Roy. Soc. South Africa **5**: 396 (1916). —Engler, Pflanzenw. Afrikas (Veg. Erde 9) **3**, 2: 96 (1921). —Pax in Engler, Pflanzenr. [IV, fam. 147, xvi] **85**: 58 (1924). —De Wildeman, Pl. Bequaert. **3**, 4: 492 (1926). —Robyns & Tournay, Fl. Sperm. Parc Nat. Alb. **1**: 464 (1948). —Brenan, Check-list For. Trees Shrubs Tang. Terr.: 196 (1949). —F.W. Andrews, Fl. Pl. Anglo-Egypt. Sudan **2**: 50 (1952). —Keay in F.W.T.A., ed. 2, **1**, 2: 409 (1958). —Topham, Check List For. Trees Shrubs Nyasaland Prot.: 49 (1958). —White, F.F.N.R.: 192 (1962). —P.G. Meyer in Merxmüller, Prodr. Fl. SW. Afrika, fam. 67: 5 (1967). —Agnew, Upl. Kenya Wild Fls.: 216 (1974). —Drummond in Kirkia **10**: 252 (1975). —Troupin, Fl. Pl. Lign. Rwanda: 246, fig. 83/2 (1982); Fl. Rwanda **2**: 198, fig. 58/2 (1983). —Radcliffe-Smith in F.T.E.A., Euphorb. 1: 190 (1987). —Beentje, Kenya Trees, Shrubs Lianas: 184 (1994). Type from Ethiopia (Tigray).

Acalypha nigritiana Müll. Arg. in Flora **47**: 440 (1864); in Linnaea **34**: 19 (1865); in De Candolle, Prodr. **15**, 2: 834 (1866). —Hutchinson in F.T.A. **6**, 1: 890 (1912). Type from Nigeria.

Acalypha ornata var. *bracteosa* Müll. Arg. in Flora **47**: 441 (1864); in Linnaea **34**: 19 (1865); in De Candolle, Prodr. **15**, 2: 833 (1866). Type: Mozambique, Zambézia, foot of Morrumbala Mountain (Moramballa), fr. xii.1858, *Kirk* Acal. (2) (K, holotype).

Acalypha ornata var. *pilosa* Müll. Arg. in Flora **47**: 441 (1864); in Linnaea **34**: 19 (1865); in De Candolle, Prodr. **15**, 2: 833 (1866). Type: Mozambique, Manica e Sofala, Shiramba, fl. i.1860, *Kirk* Acal. (2) [this is a different collection from *Kirk* Acal. (2) above] (K, holotype).

Acalypha ornata var. *glandulosa* Müll. Arg. in Linnaea **34**: 19 (1865); in De Candolle, Prodr. **15**, 2: 833 (1866). Type from Angola (Malanje Province).

Acalypha grantii Baker & Hutch. in Bull. Misc. Inform., Kew **1911**, 5: 230 (1911). —Hutchinson in F.T.A. **6**, 1: 885 (1912). Types from Sudan and Uganda.

Acalypha swynnertonii S. Moore in J. Linn. Soc., Bot. **40**: 200 (1911). Type: Mozambique, Manica e Sofala, Chibabava, lower Búzi R., 120 m, fl. 1.xii.1906, *Swynnerton* 799 (BM, holotype; K).

Acalypha moggii Compton in J. S. African Bot. **41**, 1: 48 (1975). Type from South Africa (Transvaal).

A large much-branched woody herb, lax shrub, prolific vine or small tree up to 5 m tall; stems glabrous or pubescent, pilose or villous, often purplish. Petioles 2–15 cm long. Leaf blades (3)5–16 × (2)3–10 cm, ovate or elliptic-ovate, caudate-acuminate at the apex, serrate on the margins, cuneate to rounded, truncate or shallowly cordate at the base, membranous, sparingly pilose to subglabrous on both surfaces, usually more evenly to densely pilose along the midrib and main nerves, 5–7-nerved from the base; lateral nerves in 4–7 pairs. Stipules 4–10 mm long, linear-lanceolate to subulate-filiform, soon falling. Plants monoecious, or rarely dioecious. Male racemes solitary, axillary, up to 12 cm long, limp, scarcely or shortly pedunculate, densely flowered; bracts 1–4 mm long, linear-oblong. Male flowers shortly pedicellate; buds subglobose, minute, almost glabrous, green, pink- or reddish-flushed; anthers yellowish-white. Female spikes terminal, very rarely also axillary, up to 19 cm long and 2 cm wide in fruit, narrowly cylindric, rigid, subsessile; bracts accrescent to 1–1.5 × 2–2.5 cm, transversely ovate-rhombic, chartaceous, sparingly pilose, stipitate-glandular on and near the margin, greenish, 1-flowered, c. 20-toothed; terminal tooth linear-lanceolate, the rest triangular. Female flowers sessile; sepals 3, 1 mm long, triangular, ciliate; ovary less than 1 mm in diameter, 3-lobed, pubescent at the apex; styles free, 3 mm long, laciniate, greenish-white or crimson. Fruits 3 × 4 mm, 3-lobed, sparingly pubescent at the apex. Seeds 1.8 × 1.5 mm, ovoid-subglobose, smooth, purplish-grey; caruncle elliptic, appressed, ochreous.

Caprivi Strip. Mpilila Island, y. fr. 15.i.1959, *Killick & Leistner* 3392 (PRE; SRGH). **Botswana**. N: Okavango R., Old Mohembo, fl. 16.iii.1965, *Wild & Drummond* 7072 (K; LISC; PRE; SRGH). **Zambia**. B: Sesheke Boma, fl. & fr. 3.i.1952, *White* 1987 (BM; FHO; K; MO). N: Lake Mweru, near Kafulwe Mission, fl. 2.xi.1952, *White* 3575 (BM; FHO; K; MO). C: Old Feira Road, fl. 30.xii.1972, *Strid* 2722 (C; MO). E: Mkhania, fl. 28.ii.1969, *Astle* 5556 (K; MO; SRGH). S: Sinazongwe, Zeze, fl. 29.xii.1958, *Robson & Angus* 993 (BM; K; LISC; PRE; SRGH). **Zimbabwe**. N: 5 km from Chirundu, fl. 8.i.1958, *Goodier* 542 (COI; K; PRE; SRGH). W: Victoria Falls Rainforest, fr. 26.iii.1974, *Gonde* 80/74 (K; MO; PRE; SRGH). E: Mutare Distr., Murahwa's Hill, fl. 18.iii.1964, *Chase* 8139 (K; LISC; MO; SRGH). S: Mtilikwe R., fl. 26.i.1949, *Wild* 2765 in *GHS* 22614 (K; LISC; SRGH). **Malawi**. N: 21 km west of Karonga, male fl. 31.xii.1976, *Pawek* 12125 (K; MAL; MO; SRGH; UC). C: Mua, lower Livulezi R., fl. 20.i.1959, *Robson & Jackson* 1279, 1279a (BM; K; LISC; SRGH). S: Namwera Road, fl. 23.xi.1978, *Banda et al.* 1319 (MAL; MO; SRGH). **Mozambique**. N: Namacuto, fr. 30.i.1984, *Groenendijk et al.* 855 (K; LMA). Z: Namagoa, fl. & fr. 18.i.1949, *Faulkner* 365 (K; SRGH). T: Estima, fr. 30.i.1972, *Macêdo* 4750 (LMA; LISC; MO; SRGH). MS: Lucite R., between Gogoi and Dombe, fl. 21.iv.1974, *Pope & Müller* 1229 (K; LISC; PRE; SRGH).

Also from Nigeria east to Eritrea and south to Angola and Namibia. Riverine and lakeshore forest and thicket, usually on sandy soil, in *Acacia* and mopane woodland beside rivers and as a constituent of termitaria thicket, also in submontane evergreen riverine and gully forest, occasionally on swamp forest margins; 20–1700 m.

Vernacular names as recorded in specimen data include: "mpalachulu" (chiNyanja); "chipasulu" (chiYao).

In Malawi, this species hybridizes with *A. chirindica* S. Moore, to produce *A.* × *malawiensis* Radcl.-Sm. (in Kew Bull. **44**: 444 (1989)). The following collections have been seen:

Malawi. N: Kaseye Mission, 16 km east of Chitipa, fl. 26.xii.1977, *Pawek* 13384 (MO). S: Shire R., Liwonde, fl. & failed fr. 26.ii.1961, *Richards* 14463 (K); Ndirande Mt., fl. 23.xi.1977, *Brummitt, Seyani & Banda* 15174 (K).

11. **Acalypha villicaulis** Hochst. in Sched. Schimperi Iter. Abyss., sectio secunda, coll. no.: 737 (1842). —A. Rich., Tent. Fl. Abyss. **2**: 248 (1851). —Müller Argoviensis in De Candolle, Prodr. **15**, 2: 845 (1866). —Engler, Pflanzenw. Ost-Afrikas **C**: 239 (1895). —Hutchinson in F.T.A. **6**, 1: 893 (1912). —Eyles in Trans. Roy. Soc. South Africa **5**: 397 (1916). —Engler, Pflanzenw. Afrikas (Veg. Erde 9) **3**, 2: 97 (1921). —Pax in Engler, Pflanzenr. [IV, fam. 147, xvi] **85**: 80 (1924). —De Wildeman, Pl. Bequaert. **3**, 4: 494 (1926). —Brenan, Check-list For. Trees Shrubs Tang. Terr.: 197 (1949). —F.W. Andrews, Fl. Pl. Anglo-Egypt. Sudan **2**: 50 (1952). —Agnew, Upl. Kenya Wild Fls.: 216 (1974). —Troupin, Fl. Rwanda **2**: 200, fig. 56/1 (1983). —Radcliffe-Smith, F.T.E.A., Euphorb. 1: 192 (1987). Types from Ethiopia (Tigray Province).

Acalypha hirsuta Hochst. in Sched. Schimperi Iter. Abyss., sectio secunda, coll. no. 781 (1842). Type from Ethiopia.

Acalypha petiolaris Hochst. ex Krauss in Flora **28**: 83 (1845). —Müller Argoviensis in De Candolle, Prodr. **15**, 2: 847 (1866). —Schinz & Junod in Mém. Herb. Boissier, No. 10: 47 (1900). —Prain & Hutchinson in Bull. Misc. Inform., Kew **1913**: 18 (1913). —Eyles in Trans. Roy. Soc. South Africa **5**: 396 (1916). —Prain in F.C. **5**, 2: 474 (1920). —Engler, Pflanzenw. Afrikas (Veg. Erde 9) **3**, 2: 97 (1921). —Pax in Engler, Pflanzenr. [IV, fam. 147, xvi] **85**: 81 (1924). —Burtt Davy, Fl. Pl. Ferns Transvaal: 305 (1932). —Martineau, Rhod. Wild Fl.: 45 (1953). Type from South Africa (KwaZulu-Natal).

Acalypha sidifolia A. Rich., Tent. Fl. Abyss. **2**: 249 (1851), as "*sidaefolia*". —Müller Argoviensis in De Candolle, Prodr. **15**, 2: 844 (1866). —Hutchinson in F.T.A. **6**, 1: 909 (1912). —Pax in Engler, Pflanzenr. [IV, fam. 147, xvi] **85**: 81 (1924). Type from Ethiopia (Tigray Province).

Acalypha senensis Klotzsch in Peters, Naturw. Reise Mossambique **6**, 1: 96 (1861). —Müller Argoviensis in De Candolle, Prodr. **15**, 2: 845 (1866). —Engler, Pflanzenw. Ost-Afrikas **C**: 239 (1895). —Hutchinson in F.T.A. **6**, 1: 888 (1912). —Prain & Hutchinson in Bull. Misc. Inform., Kew **1913**: 18 (1913). —R.E. Fries, Wiss. Ergebn. Schwed. Rhod.-Kongo-Exped. **1**, 1: 122 (1914). —Eyles in Trans. Roy. Soc. South Africa **5**: 396 (1916). —Prain in F.C. 5, 2: 473 (1920). —Engler, Pflanzenw. Afrikas (Veg. Erde 9) **3**, 2: 97 (1921).

—Pax in Engler, Pflanzenr. [IV, fam. 147, xvi] **85**: 79 (1924). —De Wildeman, Pl. Bequaert. **3**, 4: 493 (1926). —Burtt Davy, Fl. Pl. Ferns Transvaal: 305 (1932). —Bremekamp & Obermeyer in Ann. Transvaal Mus. **16**, 3: 421 (1935). —Robyns & Tournay, Fl. Sperm. Parc Nat. Alb. **1**: 464 (1948). —Brenan, Check-list For. Trees Shrubs Tang. Terr.: 197 (1949). —Suessenguth & Merxmüller, Contrib. Fl. Marandellas Distr. **43**: 84 (1951). —F.W. Andrews, Fl. Pl. Anglo-Egypt. Sudan **2**: 50 (1952). —Martineau, Rhod. Wild Fl.: 45, t. 15 (1953). —Brenan in Mem. N.Y. Bot. Gard. **9**, 1: 74 (1954). —Keay in F.W.T.A., ed. 2, **1**, 2: 409 (1958). —Topham, Check List For. Trees Shrubs Nyasaland Prot.: 49 (1958). —P.G. Meyer in Merxmüller, Prodr. Fl. SW. Afrika, fam. 67: 6 (1967). —Mogg in Macnae & Kalk, Nat. Hist. Inhaca Isl., Moçamb., rev. ed.: 148 (1969). Type: Mozambique, Manica e Sofala, Vila de Sena, *Peters* s.n. (B†, holotype).

Acalypha zambesica Müll. Arg. in Flora **47**: 440 (1864); in De Candolle, Prodr. **15**, 2: 845 (1866). —Engler, Pflanzenw. Ost-Afrikas **C**: 239 (1895). Type: Mozambique, Zambézia Province, Morrumbala (Moramballa), xii.1858, *Kirk* Acalypha (3) (K, holotype).

Acalypha villicaulis var. *minor* Müll. Arg. in Abh. Naturwiss. Vereine Bremen **7**: 26 (1882). Type from Kenya.

Acalypha haplostyla Pax in Bot. Jahrb. Syst. **19**: 98 (1894). Syntypes from Zaire.

Acalypha chariensis Beille in Bull. Soc. Bot. Fr., Mém **8**: 80 (1908). Syntypes from Central African Republic.

A perennial herb or subshrub up to 2 m tall, from a woody rootstock; stems several, erect, ascending or decumbent, often many-branched, villous, hirsute or pubescent. Petioles 0.5–6 cm long. Leaf blades 2–17 × 0.7–5 cm, ovate, lanceolate, oblong to linear-lanceolate, acuminate at the apex, serrate on the margin, rarely irregularly lobed, cordate or sometimes rounded or truncate at the base, often with a pair of stipelliform glands at the base, firmly membranous, 7-nerved from the base, sparingly to evenly hirsute-pubescent on both surfaces, pale green to sage-green; lateral nerves in 5–12(15) pairs. Stipules 2–7 mm long, linear, subulate or filiform. Male racemes axillary, solitary, up to 12 cm long on a peduncle up to 5 cm long, dense flowered; axis crisped-puberulous; bracts linear, 1.5 mm long. Male flowers: pedicels 1 mm long; buds tetragonal-subglobose, sparingly pubescent; calyx lobes yellowish- or creamy-green; anthers yellow. Female spike up to 5 cm long with a peduncle up to 1.5 cm, terminal, sometimes also with 1 or more axillary in the upper axils, or terminating short axillary shoots; bracts accrescent to up to 8 × 15 mm, transversely ovate, coarsely dentate, sparingly to densely stipitate-glandular, ciliate, greenish, the glands yellow, 1-flowered. Female flowers sessile; sepals 3–4, 1 mm long, ovate-lanceolate, ciliate; ovary 0.8 mm in diameter, 3-lobed to subglobose, pubescent in the upper half, with or without gland-tipped hairs; styles up to 1.6 cm long, free, pectinate or with few or no lateral segments, crimson. Fruits 4 × 5 mm, 3-lobed, sparingly pubescent and sometimes also with a few gland-tipped pairs at the apex, otherwise glabrous. Seeds 2.5 × 2 mm, ovoid-subglobose, smooth, dark grey, with a depressed-hemispherical caruncle c. 1 mm across.

Caprivi Strip. Linyanti area, fl. 28.xii.1958, *Killick & Leistner* 3166 (PRE; SRGH). **Botswana**. N: Gwetshaa Island, Okavango Swamp, fl. 25.ii.1973, *P.A. Smith* 444 (K; MO; PRE; SRGH). SE: Gaborone, University Campus, fl. 12.ii.1974, *Mott* 157 (K; UBLS). **Zambia**. B: Kaoma (Mankoya), fl. 20.xi.1959, *Drummond & Cookson* 6660 (SRGH). N: South Downs, fl. 12.ii.1961, *Linley* 73 (K; MO; SRGH). W: Chondwe For. Res., east of Ndola, fl. 2.ii.1952, *Angus* 911 (BM; FHO; K; MO). C: Lusaka, Stewart Park, fl. 7.v.1961, *Lusaka Natural History Club* 39 (K; MO; SRGH). E: near Chadiza, fl. & y. fr. 25.xi.1958, *Robson* 707 (BM; K; LISC; PRE; SRGH). S: Monze, fl. 3.i.1964, *van Rensburg* 2701 (K; SRGH). **Zimbabwe**. N: Umvukwe Mts., fl. 24/27.iv.1948, *Rodin* 4454 (K; MO; PRE; SRGH; UC). W: Gwampa For. Res., fl. i.1956, *Goldsmith* 77/56 (K; LISC; PRE; SRGH). C: Rusape, The Springs, y. fl. 30.xi.1930, *Fries, Norlindh & Weimarck* 3326a (K; LD; PRE). E: Stapleford, Nyamakwarara Valley, fl. 1.xi.1967, *Mavi* 380 (K; LISC; PRE; SRGH). S: Runde (Lundi) Hotel, fl. xii.1955, *Davies* 1707 (K; LISC; MO; PRE; SRGH). **Malawi**. N: Livingstonia Escarpment, fl. 31.xii.1973, *Pawek* 7686 (K; MAL; MO; SRGH; UC). C: Chimaliro For. Res., Phaso Road, fr. 10.i.1975, *Pawek* 8888 (K; MAL; MO; SRGH). S: Namwera, Sister Martha Hospital, st. 23.viii.1976, *Pawek* 11661 (K; MAL; MO; SRGH; UC). **Mozambique**. N: Nampula, fl. 14.iii.1937, *Torre* 1283 (COI; LISC). Z: Namagoa Plantation, fl. 19.ix.1948, *Faulkner* K 296 (K). T: 3.6 km Casula–Chiuta, fl. 8.vii.1949, *Barbosa & Carvalho* in *Barbosa* 3499 (K; LISC; LMA). MS: Gondola–Nhamatanda (Vila Machado), fl. 22.iii.1960, *Wild & Leach* 5215 (K; PRE; MO; SRGH). GI: Bilene, fl. & fr. 5.ii.1948, *Torre* 7254 (LISC). M: Ressano Garcia, y. fl. 22.xii.1897, *Schlechter* 11882 (BM; MO).

Throughout tropical Africa from Senegal east to Ethiopia and southwards to Namibia, South Africa (Transvaal) and Swaziland. Widespread in open woodland, miombo woodland, mopane woodland, *Acacia* woodland and woodland on Kalahari Sand, and in grassland, usually with tall grasses, in submontane and wooded grasslands and dambos, also in coastal and lakeshore grassland, sometimes in riverine and submontane forest; 30–2135 m.

A very variable species as regards habit, indumentum, petiole length and leaf blade shape, size and colour.

Vernacular names as recorded in specimen data include: "cigaga" (Malawi-Mua area); "munkulukuncha" (Tok.); "nyalisa" (chiNyanja); "nyalisi" (chiYao).

12. **Acalypha allenii** Hutch. in Bull. Misc. Inform., Kew **1911**: 229 (1911); in F.T.A. **6**, 1: 889 (1912). —Eyles in Trans. Roy. Soc. South Africa **5**: 395 (1916). —Pax in Engler, Pflanzenr., [IV, fam. 147, xvi] **85**: 54 (1924). —Suessenguth & Merxmüller, Contrib. Fl. Marandellas Distr. **43**: 83 (1951). Tab. **42**. Syntypes: Zimbabwe, Gwaai Forest, fl. i.1906, *C.E.F. Allen* 238 (K; SRGH) [the SRGH sheet of *Allen* 238 gives data as "Malindi, ix.1906"]; Hwange Distr., Wankie Line, Umgusa, y. fr. xii.1902, *Eyles* 1130 (PRE; SRGH).

A perennial caespitose herb or suffrutex 15–50(90) cm tall; stems annual from a rhizomatous woody rootstock, several, erect, sometimes decumbent or semiprostrate, simple or occasionally few-branched, channelled, viscid glandular-pubescent, burned back each year. Petioles 1–3 cm long. Leaf blades 2–13.5 × 1–5(7) cm, ovate-rhombic, ovate-lanceolate or oblong-lanceolate, subacute or obtuse and sometimes mucronulate at the apex, serrate to subentire on the margins, rounded to truncate or cordulate at the base, chartaceous, sparingly to evenly hirsute or pubescent on both surfaces, especially along the midrib and main veins, sometimes also sparingly glandular-pubescent on upper surface, 5–9-nerved from the base; tertiary nerves parallel; lateral nerves in 6–9(13) pairs. Stipules 5–8 mm long, linear-subulate, glandular-pubescent. Male racemes axillary, solitary, up to 18 cm long on a peduncle up to 5 cm long, dense flowered; axis glandular-pubescent; bracts 3 mm long, linear, glandular-pubescent. Male flowers: pedicels 1 mm long, slender; buds tetragonal-subglobose, subglabrous, greenish, sometimes reddish-tinged; anthers yellowish-cream. Female spike terminal, rarely also 1–2 spikes axillary in the upper axils, up to 10 cm long with a peduncle 3–5 cm long; bracts accrescent to 1.3 × 2 cm, transversely ovate, acuminate, repand-dentate, densely stipitate-glandular on the margin, evenly so without, glabrous within, viscid, yellowish-green, 1-flowered. Female flowers sessile; sepals 3, 1 mm long, ovate, glandular-ciliate; ovary 1 mm in diameter, 3-lobed, pubescent towards the apex and also with 1–2 stipitate glands; styles 2–4 mm long, ± free or united at the base, laciniate, glabrous or sparingly pubescent, green, yellow or red. Fruits 3 × 4 mm, deeply 3-lobed, sparingly pubescent, dark green, sordid. Seeds 2.5 × 2 mm, subglobose, smooth, dark grey or pale straw-coloured, ecarunculate.

Botswana. N: Leshomo (Leshumo) Valley, fr. i.1876, *Holub* s.n. (K). **Zambia**. B: Mongu–Lealui, fr. 10.i.1960, *Gilges* 940 (K; SRGH). N: Kapamba R., fl. 6.i.1966, *Astle* 4252 (K; SRGH). W: 16 km Ndola–Luanshya, fl. 19.xi.1959, *Wild* 4860 (K; MO; PRE; SRGH). C: Kabwe Distr., 32 km west of Kapiri Mposhi on Mpika road, fl. 1.xii.1952, *Angus* 897 (BM; FHO; K; MO; PRE). E: Mwangazi Valley, fl. 26.xi.1958, *Robson* 725 (BM; K; LISC; PRE; SRGH). S: Muckle Neuk, c. 17 km north of Choma, fl. 28.xi.1954, *Robinson* 1002 (K; SRGH). **Zimbabwe**. N: Kandeya N.R., Mavuradonha Mt., fl. 17.i.1960, *Phipps* 2296 (K; MO; PRE; SRGH). W: Bulawayo, fl. & y. fr. xi.1957, *Miller* 4761 (COI; K; LISC; PRE; SRGH). C: Dombi Estate, 20 km north of Marondera (Marandellas), fl. 13.xii.1970, *Biegel* 3438 (K; LISC; PRE; SRGH). E: Mutare (Umtali), fr. 3.i.1963, *Chase* 7924 (K; LISC; MO; PRE; SRGH). **Malawi**. C: Lilongwe Distr., Bunda Agricultural College, 23 km south of Lilongwe, fr. 19.ii.1970, *Brummitt* 8624 (K). **Mozambique**. T: Ulóngué, y. fr. 3.xii.1980, *Macuácua* 1383 (K; LMA; PRE). MS: 2 km Manica–Chimoio (Vila Pery), fl. 25.xi.1965, *Torre & Correia* 13246 (LISC).

Not known from elsewhere. Miombo, plateau woodland and mixed deciduous woodlands, also in wooded grassland, usually on granite sand, stony soil and Kalahari Sand; 750–1525 m.

13. **Acalypha paucifolia** Baker & Hutch. in Bull. Misc. Inform., Kew. **1911**: 230 (1911). —Hutchinson in F.T.A. **6**, 1: 891 (1912). —Pax in Engler, Pflanzenr. [IV, fam. 147, xvi] **85**: 63 (1924). —Radcliffe-Smith in F.T.E.A., Euphorb. 1: 194 (1987). Type: Mozambique, Niassa Province, near Lake Nyasa, 1902, *Johnson* 457 (K, holotype).

Very similar to *A. allenii* but differs in having the glandular hairs confined to the inflorescence, the stems terete, and the leaves linear-lanceolate, more remotely toothed and usually ± glabrous. Also very like *A. villicaulis* but differs in having stems

Tab. 42. ACALYPHA ALLENII. 1, flowering stem (× $^{2}/_{3}$); 2, portion of male inflorescence (× 6); 3, portion of female inflorescence (× 6), 1–3 from *Corby* 573; 4, fruit (× 6); 5, seeds (× 4), 4 & 5 from *Chase* 7924. Drawn by Christine Grey-Wilson.

glabrous or ± so; stipelliform glands absent from the base of the leaf blade; leaf margin shallowly repand-dentate, and the female inflorescence laxer.

Zambia. N: Chilongowelo, fl. 11.xii.1954, *Richards* 3578 (K). E: 8 km west of Chikomeni, Lukusuzi Nat. Park, fl. 3.xii.1970, *Sayer* 840 (SRGH). **Malawi**. N: 11 km south of Euthini (Eutini), fr. 29.xii.1975, *Pawek* 10627 (K; MAL; MO; PRE; SRGH; UC). C: Kasungu–Bua, y. fr. 13.i.1959, *Robson* 1127 (BM; K; LISC; SRGH). S: Namwera Road, fl. 23.xi.1978, *Banda, Salubeni & Masiye* 1304 (MAL; MO; SRGH). **Mozambique**. N: Lichinga (Vila Cabral), fr. 12.xii.1934, *Torre* 457 (COI; LISC).

Also from W and S Tanzania. In *Brachystegia* woodland, and in damp soil in riverine woodland shade; 1000–1465 m.

The *Richards* collection cited above approaches *A. villicaulis*, whereas *Wright* 279 (K), collected in eastern Zambia near the border with Malawi on 12.xii.1960, approaches *A. allenii.*

14. **Acalypha racemosa** Wall. ex Baill., Étud. Gén. Euphorb.: 443 (1858). —F.W. Andrews, Fl. Pl. Anglo-Egypt. Sudan **2**: 49 (1952). —Keay in F.W.T.A., ed. 2, **1**: 409 (1958). —Agnew, Upl. Kenya Wild Fls.: 216 (1974). —Troupin, Fl. Rwanda **2**: 198, fig. 56/5 (1983). —Radcliffe-Smith in F.T.E.A., Euphorb. 1: 187 (1987). —Beentje, Kenya Trees, Shrubs Lianas: 184 (1994). Type from S India.

Acalypha paniculata Miq., Fl. Ind. Bat. **1**, 2: 406 (1859). —Müller Argoviensis in De Candolle, Prodr. **15**, 2: 802 (1866). —Engler, Pflanzenw. Ost-Afrikas **C**: 239 (1895). —Hutchinson in F.T.A. **6**, 1: 886 (1912). —R.E. Fries, Wiss. Ergebn. Schwed. Rhod.-Kongo-Exped. **1**, 1: 122 (1914). —Eyles in Trans. Roy. Soc. South Africa **5**: 396 (1916). —Engler, Pflanzenw. Afrikas (Veg. Erde 9) **3**, 2: 95, t. 44 (1921). —Pax in Engler, Pflanzenr. [IV, fam. 147, xvi] **85**: 14, t. 3 (1924). —De Wildeman, Pl. Bequaert. **3**, 4: 493 (1926). —Robyns & Tournay, Fl. Sperm. Parc Nat. Alb. **1**: 463 (1948). —Brenan, Check-list For. Trees Shrubs Tang. Terr.: 196 (1949). Type from Java.

A laxly branched perennial herb or subshrub up to 1.5 m tall. Stems puberulous or pubescent, sometimes also pilose or hirsute. Petioles up to 10 cm long. Leaf blades 3–13 × 1.5–9 cm, ovate, acuminate at the apex, crenate-serrate on the margins, rounded to shallowly cordate at the base, membranous, sparingly pubescent or hirsute or subglabrous on both surfaces, also somewhat puberulous along the midrib and main nerves, 5–9-nerved from the base; lateral nerves in 3–5 pairs. Stipules 3–4 mm long, filiform-setaceous, pubescent. Male inflorescences up to 12 cm long, spicate, axillary, solitary; bracts 0.5 mm long. Male flowers subsessile or shortly pedicellate; buds rounded tetragonal, glabrous, reddish. Female inflorescence up to 25 cm long and 6 cm wide, paniculate, terminal, rarely sometimes spicate and axillary; panicle branches subspicate or racemose; bracts minute, not accrescent, 1–3-flowered. Female flowers sessile or shortly pedicellate; sepals 5, 0.5 mm long, lanceolate, acute, sparingly pubescent; ovary 0.5 mm in diameter, somewhat 3-lobed, papillose and stipitate-glandular; styles 1–2 mm long, ± free, deeply 4–6-partite, the segments filiform, red. Fruits 1.5 × 2 mm, 3-lobed, granulate near the apex and sparingly stipitate-glandular. Seeds 1.2 × 0.8 mm, ovoid, minutely alveolate-reticulate, purplish-grey, ecarunculate.

Zimbabwe. E: between Rusitu (Lusitu) and Haroni Rivers, fr. 24.iv.1962, *Wild* 5728 (K; MO; PRE; SRGH). **Malawi**. S: Nsanje (Port Herald) Distr., Nantibule R. Drift, fr. 19.vii.1975, *Seyani & Patel* 380 (MO; SRGH). **Mozambique**. N: south slopes of Serra Merripa, fr. 5.ii.1964, *Torre & Paiva* 10466 (LISC). Z: 5.3 km from the 1st Crossroads from Mocuba on the Milange–Quelimane Road, o. fl. 20.v.1949, *Barbosa & Carvalho* in *Barbosa* 2763 (K; LMA). MS: c. 25 km southwest of Dombe, female fl. & y. fr. 22.iv.1974, *Pope & Müller* 1243 (K; LISC; PRE; SRGH).

From the Ivory Coast to Eritrea and south to Angola; also in southern India, Sri Lanka and Indonesia. Understorey of low altitude evergreen forest; 320–1220 m.

15. **Acalypha acrogyna** Pax in Bot. Jahrb. Syst. **43**: 323 (1909). —Hutchinson in F.T.A. **6**, 1: 897 (1912). —Engler, Pflanzenw. Afrikas (Veg. Erde 9) **3**, 2: 95 (1921). —Pax in Engler, Pflanzenr. [IV, fam. 147, xvi] **85**: 93 (1924). —Robyns & Tournay, Fl. Sperm. Parc Nat. Alb. **1**: 466 (1948). —Brenan, Check-list For. Trees Shrubs Tang. Terr.: 197 (1949). —White, F.F.N.R.: 192 (1962). —Drummond in Kirkia **10**: 252 (1975). —Radcliffe-Smith in F.T.E.A., Euphorb. 1: 195 (1987). Types from Zaire (Kivu Province).

A much-branched scrambling shrub up to 3 m tall, unarmed, slender-stemmed, with long shoots and short lateral shoots (brachyblasts). Bark of twigs brownish. Shoots puberulous or pubescent at first, later glabrescent. Petioles 2–5(10) mm long, densely pubescent. Leaf blades 1.5–6(9) × 1–3(4) cm, elliptic-ovate to elliptic-lanceolate, usually obtusely acuminate at the apex, crenate, crenate-serrate or sometimes subentire on the margins, rounded or cordate at the base, membranous, sparingly pubescent or subglabrous, sometimes evenly pubescent along the midrib and main nerves and with tufts of hairs in the axils beneath, 3(5)-nerved from the base or ± penninerved; lateral nerves in 5–7 pairs. Stipules 2–3 mm long, narrowly triangular-lanceolate, keeled, sparingly pubescent. Inflorescences up to 7 cm long, spicate or racemose, usually on the brachyblasts, usually mostly male with 1–3 female flowers at the top; bracts 1 mm long, ovate-oblong, entire, ciliate, brownish. Male flowers shortly pedicellate; buds depressed-globose, sparingly pubescent, greenish-yellow. Female flowers shortly pedicellate; sepals 5, 1 mm long, ovate-lanceolate, acute, ciliate; ovary 1 mm in diameter, somewhat 3-lobed, echinate, densely pubescent; styles 3–4 mm long, ± free, pectinate, reddish. Fruits 2.5 × 4 mm, 3-lobed, echinate, sparingly pubescent. Seeds 2 × 1.75 mm, subglobose, smooth, brown, ecarunculate.

Zambia. N: L. Mweru, Kafulwe Mission, fl. 6.xi.1952, *Angus* 731 (BM; FHO; K; MO). C: Kasanka National Park, 12°35'S, 30°15'E, fl. 19.xi.1993, *Harder et al.* 1959 (K; MO). **Zimbabwe**. C: Shurugwi (Selukwe), fr. 8.xii.1953, *Wild* 4308 (K; LISC; MO; PRE; SRGH). E: Chipinge, west of the Musirizwe/Bwazi R. confluence, st. 29.i.1975, *Pope, Biegel & Gibbs-Russell* 1413 (MO; PRE; SRGH). S: Masvingo (Fort Victoria), y. fr. 25.x.1965, *West* 6936 (K; SRGH). **Mozambique**. MS: Búzi, fl. 12.xi.1967, *Carvalho* 965 (LISC; LMA).

Also in Zaire (Kivu), Sudan, Ethiopia, Uganda and Tanzania. Understorey of riverine and lakeshore evergreen forest and in gully forest; 500–1200 m.

The *Carvalho* specimen cited above with its greyish bark and stiff almost spiny lateral shoots, thus approaches *A. sonderiana* Müll. Arg.

16. **Acalypha sonderiana** Müll. Arg. in Linnaea **34**: 9 (1865); in De Candolle, Prodr. **15**, 2: 804 (1866). —Prain & Hutchinson in Bull. Misc. Inform., Kew **1913**: 12 (1913). —Prain in F.C. **5**, 2: 467 (1920). —Engler, Pflanzenw. Afrikas (Veg. Erde 9) **3**, 2: 95 (1921). —Pax in Engler, Pflanzenr. [IV, fam. 147, xvi] **85**: 93 (1924). —K. Coates Palgrave, Trees Southern Africa, ed. 2, rev.: 430 (1983). —Radcliffe-Smith in F.T.E.A., Euphorb. 1: 196, in adnot. (1987). Types from South Africa (KwaZulu-Natal).

A. ? petiolaris Sond. in Linnaea **23**: 117 (1850), non Hochst. ex Krauss (1845).

Very like *A. acrogyna*, but a thorny shrub with stout branches and grey bark, with the leaves usually obtuse or rounded at the apex.

Mozambique. M: Goba, fl. 22.xi.1942, *Mendonça* 3040 (LISC); 3.xi.1960, *Balsinhas* 182 (K; LISC; LMA; PRE; SRGH); 8.xi.1961, *Lemos & Balsinhas* 219 (K; LISC; LMA; PRE; SRGH).

Also in South Africa (KwaZulu-Natal). In low altitude hot dry localities in *Androstachys* woodland, beside seasonal watercourses.

17. **Acalypha fruticosa** Forssk., Fl. Aegypt.-Arab.: 161 (1775). —Müller Argoviensis in De Candolle, Prodr. **15**, 2: 822 (1866). —Engler, Pflanzenw. Ost-Afrikas **C**: 239 (1895). —Hutchinson in F.T.A. **6**, 1: 895 (1912). —Engler, Pflanzenw. Afrikas (Veg. Erde 9) **3**, 2: 96 (1921). —Pax in Engler, Pflanzenr. [IV, fam. 147, xvi] **85**: 169 (1924). —Gardner, Trees & Shrubs of Kenya Col.: 53 (1936). —Robyns & Tournay, Fl. Sperm. Parc Nat. Alb. **1**: 468 (1948). —Brenan, Check-list For. Trees Shrubs Tang. Terr.: 197 (1949). —F.W. Andrews, Fl. Pl. Anglo-Egypt. Sudan **2**: 50 (1952). —Brenan in Mem. N.Y. Bot. Gard. **9**, 1: 74 (1954). —P.G. Meyer in Merxmüller, Prodr. Fl. SW. Afrika, fam. 67: 5 (1967). —Agnew, Upl. Kenya Wild Fls.: 215 (1974). —Radcliffe-Smith in F.T.E.A., Euphorb. 1: 206 (1987). —Beentje, Kenya Trees, Shrubs Lianas: 183 (1994). Type from Yemen Arab Republic.

Var. **fruticosa**

Acalypha betulina Retz., Obs. Bot. **5**: 30 (1789). —A. Rich., Tent. Fl. Abyss. **2**: 249 (1851). Type from India.

Acalypha chrysadenia Suess. & Friedrich in Mitt. Bot. Staatss. München **8**: 333 (1953). Type from Namibia.

An aromatic shrub up to 4 m tall. Stems pubescent and greenish at first, later glabrescent and reddish-brown. Petioles 0.5–3 cm long. Leaf blades 2–7 × 1–4.5 cm, ovate to rhombic-ovate, shortly caudate-acuminate at the apex, crenate-serrate to dentate on the margin, rounded to wide-cuneate or subtruncate at the base, membranous to thinly chartaceous, sparingly or evenly yellowish-pellucid gland-dotted beneath, sparingly to evenly pubescent on both surfaces, and usually more densely so along the midrib and main nerves beneath, 5(7)-nerved from the base; lateral nerves in 2–4 pairs. Stipules 3–4 mm, narrowly lanceolate, puberulous, chestnut-brown. Plants usually monoecious. Inflorescences rarely exceeding 2 cm in length, spicate, axillary, usually androgynous with a densely congested terminal male portion and with 1–4 bracteate female flowers at or near the base; male bracts 1 mm long, ovate, densely white-pubescent; female bracts foliaceous, accrescent to c. 8–10 × 10–15 mm, broadly ovate to reniform, crenate or repand-dentate, sparingly yellow gland-dotted and often fairly prominently ribbed on the lower surface, sparingly pubescent, 1-flowered. Male flowers subsessile; buds tetragonous-subglobose, densely pubescent or white-tomentose. Female flowers sessile; sepals 3, 1 mm long, ovate-lanceolate, ciliate; ovary 0.7 mm in diameter, 3-lobed to subglobose, ± smooth, yellow-glandular in the grooves, densely pubescent; styles 4 mm long, ± free, laciniate, pink or red. Fruits 2 × 3 mm, 3-lobed, yellow gland-dotted, evenly pubescent-pilose. Seeds 1.5–2 × 1–1.3 mm, ellipsoid-ovoid, smooth, brown, with an elliptic vulviform caruncle.

Malawi. S: Chikwawa, fl. 2.x.1946, *Brass* 17896 (K; MO; NY; SRGH); Lengwe Nat. Park, fl. 13.xii.1970, *Hall-Martin* 1091 (K; SRGH). **Mozambique**. MS: Gorongosa Nat. Park, fr. iii.1972, *Tinley* 2466 (K; LISC; PRE; SRGH).

From Sudan, Ethiopia, Somalia, Uganda, Kenya, Tanzania, Burundi and Namibia; also in S Arabia, S India, Sri Lanka and Myanmar (Burma). Riverine thicket and in dry forest, on sand and alluvium; 100–200 m.

Var. *villosa* Pax ex Hutch. is found in E and NE Africa and the Yemen, whilst var. *eglandulosa* Radcl.-Sm. occurs in E and NE Africa and Burundi. *Richards* 22785 from Zambia is intermediate between var. *eglandulosa* Radcl.-Sm. and *A. welwitschiana.*

18. **Acalypha pubiflora** Baill., Adansonia **1**: 268 (1861). —Müller Argoviensis in De Candolle, Prodr. **15**, 2: 866 (1866). —Engler, Pflanzenw. Ost-Afrikas **C**: 239 (1895). —Hutchinson in F.T.A. **6**, 1: 896 (1912). —Engler, Pflanzenw. Afrikas (Veg. Erde 9) **3**, 2: 98 (1921). —Pax in Engler, Pflanzenr. [IV, fam. 147, xvi] **85**: 136 (1924). —Drummond in Kirkia **10**: 252 (1975). Tab. **43**. Type: Mozambique, Inhambane, fr. 1842–8, Peters s.n. (B†, holotype; K).
Calyptrospatha pubiflora (Baill.) Klotzsch in Peters, Naturw. Reise Mossambique **6**, 1: 97, t. 18 (1861).

A much-branched shrub up to 6 m tall, similar to *A. fruticosa* but with lanceolate, long-acuminate, shallowly crenate-serrate leaves up to 15 × 5 cm; leaves ± penninerved, with 5–7 pairs of lateral nerves; male flower buds sometimes glabrous; female bracts much larger, spathaceous, accrescent to 1.5–2.5 × 1.5–3 cm in fruit.

Botswana. SE: Old Palapye, fl. 26.x.1980, *Woollard* 824 (GAB; SRGH; PRE). **Zimbabwe**. W: Mt. Silozwane Mt. (Silorzwe), southwest Matopos, fr. 8.i.1963, *Wild* 5968 (BM; K; LISC; PRE; SRGH). S: Chikadziwa, fl. xi.1956, *Davies* 2205 (K; MO; PRE; SRGH). **Malawi**. S: Lengwe Nat. Park, y. male fl. 24.v.1981, *Sherry* 237 (MAL). **Mozambique**. ? Z: fr. *Kirk* s.n. (K). MS: Boka, lower Búzi, fr. 20.xii.1906, *Swynnerton* 736 (BM; K; SRGH). GI: Inhambane, fr. 1842–8, *Peters* s.n. (B†; K). M: Magude, Mapulanguene, fr. 30.xi.1944, *Mendonça* 3176 (LISC).

Also from South Africa (northern Transvaal). Lowveld deciduous woodland, often at the base of rocky outcrops or in riverine vegetation, in granite sand; 30–1220 m.

Subsp. *australica* Radcl.-Sm. was recently discovered in NW Australia [Kew Bull. **45**, 4: 678 (1990)].

19. **Acalypha glabrata** Thunb., Prodr. Fl. Cap., part 2: 117 (1800). —Müller Argoviensis in De Candolle, Prodr. **15**, 2: 857 (1866). —Prain & Hutchinson in Bull. Misc. Inform., Kew **1913**: 12 (1913). —Prain in F.C. **5**, 2: 468 (1920). —Engler, Pflanzenw. Afrikas (Veg. Erde 9) **3**, 2: 97 (1921). —Pax in Engler, Pflanzenr. [IV, fam. 147, xvi] **85**: 128 (1924). —Burtt Davy, Fl. Pl. Ferns Transvaal: 305 (1932). —Hutchinson, Botanist in Southern Africa: 667 (1946). —Mogg in Macnae & Kalk, Nat. Hist. Inhaca Isl., Moçamb., rev. ed.: 148 (1969). —Drummond in Kirkia **10**: 252 (1975). —K. Coates Palgrave, Trees Southern Africa, ed. 2, rev.: 429 (1983). Type from South Africa (Cape Province).

A many-stemmed, much-branched spreading shrub or small tree up to 5 m tall. Bark silver-grey. Stems minutely puberulous, soon glabrescent, or (var. *pilosa*)

Tab. 43. ACALYPHA PUBIFLORA. 1, distal portion of fruiting branch (× $^{2}/_{3}$); 2, inflorescence (× 4), 1 & 2 from *Goodier* 1710; 3, mature female bract with fruit removed (× 4), from *Wild* 5968. Drawn by Christine Grey-Wilson.

densely pubescent. Petioles 0.5–5 cm long, slender. Leaf blades 1–8(9) × 0.5–5(5.5) cm, elliptic-ovate to rhombic-ovate, obtuse or acuminate at the apex, crenate or crenate-serrate except at the base, wide-cuneate to subtruncate at the base, membranous, glabrous or sparingly to evenly (var. *pilosa*) pubescent or villous on both surfaces, and sometimes also puberulous along the midrib above, 3(5)-nerved from the base; lateral nerves in 2–4 pairs. Stipules 3–4 mm long, linear-lanceolate, ± glabrous, persistent, light brown. Plants usually monoecious. Inflorescences up to 3.5 cm long, spicate, axillary, androgynous mostly with clusters of male flowers and a solitary basal female flower, or the lower inflorescences all male and with solitary axillary female flowers in the upper axils; male bracts minute; female bracts accrescent to 5 × 10 mm, spathaceous, transversely ovate, shallowly 5–7-lobed, nervose, 1-flowered. Male flowers: pedicels 1 mm long, capillary; buds somewhat knobbly, glabrous; anthers white. Female flowers sessile; sepals 3, minute, ovate, acute, pilose; ovary 0.5 mm in diameter, 3-lobed to subglobose, densely setose; styles 3–5 mm long, shortly united at the base, laciniate, whitish. Fruits 2 × 3 mm, 3-lobed, echinulate, pubescent. Seeds 1.5 × 1.5 mm, globose, smooth.

Young shoots puberulous, later glabrescent; leaves glabrous or sparingly pubescent on lower surface · var. *glabrata*
Young shoots densely pubescent; leaves evenly pubescent or villous · · · · · · · · · · · · var. *pilosa*

Var. **glabrata**

Acalypha glabrata var. *genuina* Müll. Arg. in Linnaea **34**: 33 (1865); in De Candolle, Prodr. **15**, 2: 857 (1866). Type from South Africa (Cape Province).

Acalypha glabrata var. *latifolia* (Sond.) Müll. Arg. in Linnaea **34**: 33 (1865). —Prain & Hutchinson in Bull. Misc. Inform., Kew **1913**: 13 (1913). Type from South Africa (KwaZulu-Natal).

Acalypha betulina E. Mey. in Drège, Zwei Pflanzengeogr. Dokum.: 161 (1843), pro parte. —Ecklon & Zeyher in Linnaea **20**: 213 (1847). —Sonder in Linnaea **23**: 116 (1850), non Retz. Types from South Africa (Cape Province and KwaZulu-Natal).

Acalypha betulina var. *latifolia* Sond. in Linnaea **23**: 117 (1850).

Young shoots puberulous, later glabrescent; leaves glabrous or sparingly pubescent on lower surface.

Zimbabwe. E: Mutare (Umtali), Murahwa's Hill, fl. 15.xii.1962, *Chase* 7914 (K; LISC; MO; PRE; SRGH). S: Bikita, Benga, fl. 10.ii.1965, *West* 6321 (SRGH). **Malawi.** S: Blantyre Limbe, Malabvi For., fl. 6.x.1960, *Chapman* 978 (FHO). **Mozambique**. M: Goba, fl. 7.xi.1960, *Balsinhas* 214 (COI; K; LISC; LMA; PRE; SRGH).

Also in South Africa. Low altitude evergreen forest margins, usually on rocky hillsides, also in coastal woodlands with *Androstachys, Spirostachys, Hymenocardia* and *Afzelia* species; (100)1160–1585 m.

Var. **pilosa** Pax in Bull. Herb. Boissier **6**: 733 (1898). —Prain & Hutchinson in Bull. Misc. Inform., Kew **1913**: 14 (1913). —Pax in Engler, Pflanzenr. [IV, fam. 147, xvi] **85**: 129 (1924). Type from South Africa (KwaZulu-Natal).

Ricinocarpus glabratus var. *genuinus* f. *pilosior* Kuntze, Revis. Gen. Pl. **3**, 2: 291 (1898). Types from South Africa (Cape Province).

Acalypha glabrata var. *pilosior* (Kuntze) Prain & Hutch. in Bull. Misc. Inform., Kew **1913**: 15 (1913); in F.C. **5**, 2: 468 (1920). —Pax in Engler, Pflanzenr. [IV, fam. 147, xvi] **85**: 128 (1924). —O.B. Miller, Check-list For. Trees Shrubs Bech. Prot.: 31 (1948).

Young shoots densely pubescent; leaves evenly pubescent or villous.

Botswana. SE: 3 km north of Kanye, fl. 27.x.1978, *O.J.Hansen* 3507 (C; GAB; K; PRE; SRGH; UPS; WAG).

Also in South Africa. Dry wooded grassland with shrubs and scattered small trees, on stony hillsides and beside seasonal watercourses; 1100–1350 m.

20. **Acalypha neptunica** Müll. Arg. in Abh. Naturwiss. Vereine Bremen **7**: 26 (1882). —Engler, Pflanzenw. Ost-Afrikas **C**: 239 (1895). —Hutchinson in F.T.A. **6**, 1: 907 (1912). —Engler, Pflanzenw. Afrikas (Veg. Erde 9) **3**, 2: 98 (1921). —Pax in Engler, Pflanzenr. [IV, fam. 147, xvi] **85**: 109 (1924). —Brenan, Check-list For. Trees Shrubs Tang. Terr.: 198 (1949). —F.W.

Andrews, Fl. Pl. Anglo-Egypt. Sudan **2**: 52 (1952). —Keay in F.W.T.A., ed. 2, **1**, 2: 410 (1958). —Agnew, Upl. Kenya Wild Fls.: 216 (1974), excl. fig. —Troupin, Fl. Pl. Lign. Rwanda: 244, fig. 83/3 (1982); Fl. Rwanda **2**: 198, fig. 58/3 (1983). —Radcliffe-Smith in F.T.E.A., Euphorb. 1: 210 (1987). —Beentje, Kenya Trees, Shrubs Lianas: 184 (1994). Type from Zanzibar.

Var. **pubescens** (Pax) Hutch. in F.T.A. **6**, 1: 908 (1912). —Pax in Engler, Pflanzenr. [IV, fam. 147, xvi] **85**: 109 (1924). —Radcliffe-Smith in F.T.E.A., Euphorb. 1: 211 (1987). Types from Zaire (Kivu Province).

Acalypha mildbraediana var. *pubescens* Pax in Bot. Jahrb. Syst. **43**: 324 (1909).

Acalypha subsessilis var. *mollis* Baker & Hutch. in Bull. Misc. Inform., Kew **1911**: 231 (1911); in F.T.A. **6**, 1: 907 (1912). —Pax in Engler, Pflanzenr. [IV, fam. 147, xvi] **85**: 110 (1924). —Brenan, Check-list For. Trees Shrubs Tang. Terr.: 198 (1949). Types from Kenya (Coast Province) and Zanzibar.

Acalypha neptunica var. *vestita* Pax & K. Hoffm. in Engler, Pflanzenr. [IV, fam. 147, xvi] **85**: 109 (1924). —Brenan, Check-list For. Trees Shrubs Tang. Terr.: 198 (1949). Types from Tanzania (Tanga Province).

Similar to *A. glabrata*, but with larger leaf blades (up to 15 × 7 cm in Malawi material) which are generally obovate-oblanceolate and broadest above the middle.

Malawi. S: Chikwawa Distr., Lengwe National Park, fl. 8.ii.1970, *Hall-Martin* 510 (SRGH); Lengwe National Park, 13.xii.1970, *Hall-Martin* 1092 (SRGH).

Also from Ghana eastwards to Ethiopia and south to Zaire (Shaba). In thickets and dry forest understorey, on sandy soil or associated with termitaria; 914–1066 m.

Var. *neptunica* ranges from Ghana to Sudan and Tanzania.

21. **Acalypha psilostachya** Hochst. ex A. Rich., Tent. Fl. Abyss. **2**: 246 (1851). —Müller Argoviensis in De Candolle, Prodr. **15**, 2: 865 (1866). —Engler, Pflanzenw. Ost-Afrikas **C**: 239 (1895). —Hutchinson in F.T.A. **6**, 1: 899 (1912). —Engler, Pflanzenw. Afrikas (Veg. Erde 9) **3**, 2: 98 (1921). —Pax in Engler, Pflanzenr. [IV, fam. 147, xvi] **85**: 134 (1924). —Robyns & Tournay, Fl. Sperm. Parc Nat. Alb. **1**: 467 (1948). —Brenan, Check-list For. Trees Shrubs Tang. Terr.: 198 (1949) —F.W. Andrews, Fl. Pl. Anglo-Egypt. Sudan **2**: 51 (1952). —Brenan in Mem. N.Y. Bot. Gard. **9**, 1: 74 (1954). —White, F.F.N.R.: 192 (1962). —Agnew, Upl. Kenya Wild Fls.: 215 (1974). —Troupin, Fl. Pl. Lign. Rwanda: 246 (in obs.) (1982); Fl. Rwanda, **2**: 198, fig. 56/2 (1983). —Radcliffe-Smith in F.T.E.A., Euphorb. 1: 203 (1987). Type from Ethiopia (Gonder Province).

An erect or scandent woody perennial herb or subshrub up to 3 m tall. Stems tomentose, puberulous, hirsute or glandular (var. *glandulosa*), greenish at first, often becoming reddish-brown. Petioles up to 11 cm long, fairly slender. Leaf blades 2–15 × 1–8 cm, ovate to ovate-lanceolate, acuminate or often caudate at the apex, crenate or serrate, rounded or cordate at the base, membranous to thinly chartaceous, sparingly to evenly pubescent or hirsute, especially on the midrib and main nerves on both surfaces, or else subglabrous, 5–9-nerved from the base; lateral nerves in 4–6 pairs. Stipules 2–4 mm long, linear-lanceolate or setaceous. Inflorescences up to 9 cm long, spicate, axillary, solitary or paired, comprised mostly of male flowers, or at least half are male flowers, with 1–several female flowers at the base or in the lower half, rarely terminated by a solitary allomorphic female flower; male bracts minute; female bracts deeply digitately 5–7-partite, the lobes accrescent to up to 8 × 12 mm, linear-lanceolate to ovate-lanceolate, glabrous, pubescent or hirsute, sometimes glandular (var. *glandulosa*), 1-flowered. Male flowers subsessile; buds rounded to 4-angled, pubescent, reddish; anthers yellowish. Female flowers sessile; sepals 3, 0.5 mm long, ovate, ciliate; ovary 0.3 mm in diameter, somewhat 3-lobed, pubescent and tomentose; styles 2–3 mm long, free, laciniate, white. Allomorphic female flowers (when present) obovoid, with an apical fimbriate whorl, sparingly puberulous. Fruits 1.5 × 2.5 mm, 3-lobed, ± smooth or slightly tubercled, pubescent above, tomentose below. Seeds 1 × 1 mm, ovoid-subglobose, smooth, brown, with a small velar caruncle.

Stems, inflorescence axes and female bracts tomentose, pubescent or hirsute, sometimes subglabrous; glands absent · var. *psilostachya*

Stems, inflorescence axes and female bracts variously indumented, but also with an admixture of subsessile or stipitate glands · var. *glandulosa*

Var. **psilostachya**

Stems, inflorescence axes and female bracts tomentose, pubescent or hirsute, sometimes subglabrous; glands absent.

Zambia. N: Mpika, fl. 26.i.1955, *Fanshawe* 1860 (K; NDO). E: Nyika Plateau, c. 3 km southwest of Rest House, y. fl. 22.x.1958, *Robson & Angus* 272 (BM; FHO; K; LISC; PRE; SRGH). **Malawi**. N: Viphya Plateau, 50 km southwest of Mzuzu, fl. 9.x.1976, *Pawek* 11896 (K; MAL; MO). C: Dedza, Chongoni Forestry School, fr. 18.i.1967, *Salubeni* 510 (K; LISC; PRE; SRGH). S: Thyolo (Cholo) Mt., fl. 19.ix.1946, *Brass* 17652 (K; MO; NY; SRGH). **Mozambique**. N: Lichinga (Vila Cabral), fl. 8.i.1935, *Torre* 460 (COI; LISC). Z: Gurué Mts., fl. 30.vii.1979, *de Koning* 7474 (K; LMA).

Also from Sudan and Ethiopia southwards to Angola. Understorey and margins of submontane evergreen forest and in medium to high altitude riverine forest, also in swamp forest (mushitu) and submontane grassland; 900–3050 m.

Var. **glandulosa** Hutch. in F.T.A. **6**, 1: 900 (1912). —Pax in Engler, Pflanzenr. [IV, fam. 147, xvi] **85**: 134 (1924) (erroneously cited as forma *glandulosa*). —Brenan, Check-list For. Trees Shrubs Tang. Terr.: 198 (1949). —Troupin, Fl. Rwanda **2**: 198 (1983). —Radcliffe-Smith in F.T.E.A., Euphorb. 1: 204 (1987). Type: Malawi, Nyika Plateau, 1200–1850 m, 1896, *Whyte* s.n. (K, lectotype; chosen by Radcliffe-Smith (1987)).

Acalypha bequaertii Staner in Bull. Jard. Bot. État **15**, 2: 141 (1938). Type from Zaire (Shaba).

Stems, inflorescence axes and female bracts variously indumented, but also with an admixture of subsessile or stipitate glands.

Zambia. N: near Mbala (Abercorn), fr. 27.iii.1960, *Angus* 2180 (FHO; K; SRGH). **Malawi**. N: 30.5 km southwest of Mzuzu, fl. 27.iii.1976, *Pawek* 10921 (K; MAL; MO; PRE; SRGH; UC). S: Zomba Plateau, fl. 18.xii.1978, *Salubeni & Masiye* 2390 (MAL; MO; SRGH). **Mozambique**. Z: Milange, fl. & fr. 12.xi.1942, *Mendonça* 1378 (LISC).

Also from Uganda, Kenya, Tanzania, Burundi, and Zaire (Shaba). Understorey of evergreen submontane forest and forest margins, also in ravine and gully forests, submontane grassland and swamp forest (mushitu); 520–2320 m.

Torre 462 from N Mozambique (Maniamba), with only one or two glands here and there on the plant body, represents an intermediate condition between var. *psilostachya* and var. *glandulosa*.

22. **Acalypha brachystachya** Hornem., Hort. Bot. Hafn. **2**: 909 (1815). —Müller Argoviensis in De Candolle, Prodr. **15**, 2: 870 (1866). —Engler, Pflanzenw. Ost-Afrikas **C**: 239 (1895). —Hutchinson in F.T.A. **6**, 1: 899 (1912). —Engler, Pflanzenw. Afrikas (Veg. Erde 9) **3**, 2: 99 (1921). —Pax in Engler, Pflanzenr. [IV, fam. 147, xvi] **85**: 101 (1924). —De Wildeman, Pl. Bequaert. **3**, 4: 388 (1926). —Robyns & Tournay, Fl. Sperm. Parc Nat. Alb. **1**: 466 (1948). —F.W. Andrews, Fl. Pl. Anglo-Egypt. Sudan **2**: 51 (1952). —Keay in F.W.T.A., ed. 2, **1**, 2: 409 (1958). —Troupin, Fl. Rwanda **2**: 196 (1983). —Radcliffe-Smith in F.T.E.A., Euphorb. 1: 203 (1987). Type from China (specimen cult. in Copenhagen Bot. Gard. in 1806).

Acalypha elegantula Hochst. ex A. Rich., Tent. Fl. Abyss. **2**: 246 (1851). Type from Ethiopia (Tigray Province).

Very like *A. psilostachya*, but a weak annual herb not usually exceeding 30 cm in height, with shorter petioles (up to c. 5 cm long), smaller leaves (up to c. 6 × 4 cm) and much shorter inflorescences (0.5–2 cm long) which are often almost entirely female; the female bracts are deeply 3–5-partite, with the lobes accrescent to c. 5 mm in length, ± linear, and stipitate-glandular.

Malawi. N: "Tanganyika Plateau", fr. 1896, *Whyte* s.n. (K). C: near Grand Beach Hotel, Salima, fl. 16.ii.1959, *Robson & Steele* 1632 (BM; K; LISC; SRGH). S: Zomba, Govt. Hostel grounds, fr. 31.iii.1977, *Brummitt* 15023 (K; SRGH). **Mozambique**. N: Serra Merripa, fr. 4.ii.1964, *Torre & Paiva* 10439 (LISC).

Widespread in the Old World tropics. Deciduous woodland in shade, also as a weed of waste places; 500–1000 m.

23. **Acalypha ciliata** Forssk., Fl. Aegypt.-Arab.: 162 (1775). —Müller Argoviensis in De Candolle, Prodr. **15**, 2: 873 (1866), pro parte excl. syn. —Engler, Pflanzenw. Ost-Afrikas **C**: 239 (1895), pro parte. —Hutchinson in F.T.A. **6**, 1: 901 (1912), pro parte excl. syn. —Engler, Pflanzenw. Afrikas (Veg. Erde 9) **3**, 2: 99 (1921), pro parte. —Pax in Engler, Pflanzenr. [IV,

fam. 147, xvi] **85**: 98 (1924), pro parte excl. syn. —De Wildeman, Pl. Bequaert. **3**, 4: 288 (1926). —F.W. Andrews, Fl. Pl. Anglo-Egypt. Sudan **2**: 51 (1952), pro parte. —Keay in F.W.T.A., ed. 2, **1**, 2: 410 (1958), pro parte excl. syn. —P.G. Meyer in Merxmüller, Prodr. Fl. SW. Afrika, fam. 67: 5 (1967). —Agnew, Upl. Kenya Wild Fls.: 215 (1974). —Radcliffe-Smith in F.T.E.A., Euphorb. 1: 197 (1987). Type from Arabia (Yemen Arab Republic).

An erect, slender, unbranched annual herb up to 1.3 m tall. Stems puberulous. Petioles up to 7 cm long. Leaf blades 2–10 × 1–5.5 cm, ovate, elliptic-ovate or ovate-lanceolate, caudate-acuminate at the apex, crenate-serrate, cuneate or rounded at the base, membranous, subglabrous or sparingly pubescent on both surfaces, and more evenly so along the midrib and main nerves beneath, 3–5-nerved from the base; lateral nerves in 4–6 pairs. Stipules 1 mm long, narrowly lanceolate, ciliate. Inflorescences up to 3 cm long, spicate, axillary, often paired, male in the upper half, female in the lower, sometimes terminated by a solitary allomorphic female flower; male bracts minute; female bracts accrescent to up to 6 × 10 mm, transversely ovate, multifid-laciniate, the teeth up to 4 mm long, linear to filiform, separate, erect, ciliate, the bracts as a whole sparingly pubescent or ± glabrous, ribbed, 1-flowered. Male flowers subsessile; buds minute, tetragonal, granular-tuberculate, greenish; anthers yellow. Female flowers sessile; sepals 3, 1 mm long, ovate-lanceolate, ciliate; ovary 0.3 mm in diameter, somewhat 3-lobed, sparingly pubescent above; styles 2 mm long, ± free, laciniate, white. Allomorphic female flowers obovoid, rugulose, with or without a pair of fimbriate whorls near the top, pubescent. Fruits 1.5 × 2.5 mm, 3-lobed, smooth, subglabrous. Seeds 1.3 × 1 mm, ovoid, ± smooth, brown, shiny, with a flattened, elliptic caruncle.

Caprivi Strip. Andara Mission Station, fr. 24.ii.1956, *de Winter & Marais* 4834 (PRE). **Botswana**. N: Mutsoi, fl. & fr. 6.iv.1967, *Lambrecht* 116 (K; LISC; PRE; SRGH). SE: Mahalapye R., fl. & fr. 3.iii.1978, *O.J. Hansen* 3362 (C; GAB; K; PRE). **Zambia**. B: Sesheke, fr. iv.1910, *Gairdner* 578 (K). C: 21 km SSW of Lusaka, Chipongwe Cave, fl. & fr. 26.ii.1995, *Bingham* 10430 (K). S: Nega Nega, fr. 27.ii.1963, *van Rensburg* 1478 (K; SRGH). **Zimbabwe**. N: Mensa Pan, fl. 29.i.1958, *Drummond* 5332 (BR; COI; K; LISC; PRE; SRGH). E: Hippo Mine, lower Save (Sabi), fl. & fr. 12.iii.1957, *Phipps* 576 (BR; K; LISC; PRE; SRGH). S: Beitbridge, fr. 25.ii.1961, *Wild* 5382 (K; MO; PRE; SRGH). **Malawi**. N: Ngala, north of Chilumba, fr. 14.iv.1976, *Pawek* 11007 (K; MAL; MO). S: Koko Bay, fl. & fr. 11.i.1980, *Masiye, Tawakali & Salubeni* 222 (BR; MAL; MO; SRGH). **Mozambique**. T: 30 km Chicoa–Magoé, fr. 17.ii.1970, *Torre & Correia* 18012 (LISC (partly); SRGH). MS: near south bank of Save (Sabi) R., fr. 30.vi.1950, *Chase* 2465 & A (BM; BR; SRGH). GI: Guijá, between Caniçado and Mabalane, fr. 12.v.1948, *Torre* 7802 (LISC).

In tropical Africa, also from Burkina Faso (Upper Volta) eastwards to Ethiopia and Somalia and south to Namibia, but absent from the Congo Basin; Yemen; Pakistan; India eastwards to Orissa; Sri Lanka. Riverine vegetation, thicket and *Acacia* woodland, also on pan margins in floodplain mopane, and in lakeshore thicket vegetation, growing in alluvial soils and sand at waters edge; 300–1000 m.

24. **Acalypha fimbriata** Schumach. & Thonn., Beskr. Guin. Pl.: 409 (1827). —Hochstetter ex A. Richard, Tent. Fl. Abyss. **2**: 245 (1851). Types from Ghana.

Acalypha vahliana Müll. Arg. in Linnaea **34**: 43 (1865); in De Candolle, Prodr. **15**, 2: 873 (1866), excl. syn. —Engler, Pflanzenw. Ost-Afrikas **C**: 239 (1895). Type from Ghana.

Acalypha ciliata sensu Prain & Hutch. in Bull. Misc. Inform., Kew **1913**: 15 (1913). —Eyles in Trans. Roy. Soc. South Africa **5**: 396 (1916). —Prain in F.C. **5**, 2: 470 (1920). —Burtt Davy, Fl. Pl. Ferns Transvaal: 305 (1932), non Forssk.

Very like *A. ciliata* in all respects except for the female bracts, which show a constant and consistent difference in form throughout the range of both species; in *A. fimbriata*, the teeth of the bracts are up to 1.5 mm long, falcate-lanceolate and curved towards the apex, often lying alongside each other and almost contiguous; furthermore the teeth are minutely puberulous, and the bracts as a whole are frequently sparingly long-hispid. The allomorphic female flowers always have the fimbriate whorls, instead of just sometimes as in *A. ciliata*.

Caprivi Strip. Shamapi Island, west of Kake Camp, fl. & fr. 21.i.1956, *de Winter* 4362 (K; PRE). **Zambia**. N: Mwawe Dambo, fr. 17.iv.1961, *Phipps & Vesey-FitzGerald* 3256 (BR; K; LISC; MO; PRE; SRGH). W: Chingola, fr. 1.iii.1956, *Fanshawe* 2810 (K; NDO; SRGH). C: Kamaila For. Rest House, fr. 9.ii.1975, *Brummitt, Hooper & Townsend* 14282 (BR; K; SRGH). E: Sasare, fr. 9.xii.1958, *Robson* 878 (BM; BR; K; LISC; PRE; SRGH). S: Siantambo, fr. 7.ii.1963, *Mataundi* 17/68 (BR; LISC; PRE; SRGH). **Zimbabwe**. N: Andrew's Kraal, fr. 21.ii.1958, *Phipps* 860 (BR; K; LISC; MO;

PRE; SRGH). W: Kazungula, fr. iv.1955, *Davies* 1125 (K; SRGH). C: Chegutu (Hartley), fr. 29.i.1952, *R.M. Hornby* 3252 (K; PRE; SRGH). E: Mutare (Umtali), fr. 10.iii.1957, *Chase* 6359 (BR; COI; K; PRE; SRGH). S: Masvingo (Fort Victoria)–Beitbridge Road, fl. & fr. 15.iii.1967, *Mavi* 197 (K; LISC; SRGH). **Malawi**. N: Nkhata Bay, fr. 31.iii.1974, *Pawek* 8278 (K; MAL; MO; PRE; SRGH; UC). C: Lilongwe, fr. 23.iii.1970, *Brummitt & Little* 9326 (K; PRE; SRGH). S: Zomba, fr. 1901, *Purves* 112 (K). **Mozambique**. N: Nampula, fr. 7.iv.1961, *Balsinhas & Marrime* 354 (BM; COI; K; LISC; LUAI; PRE; SRGH). T: Angónia Distr., E Serra Dómuè, fr. 3.iii.1980, *Macuácua & Matéus* 1168 (LMA; PRE). MS: Chimoio, Serra de Bandula, fr. 13.iii.1948, *Garcia* 600 (LISC).

Also in tropical Africa from Senegal to Sudan (Equatoria) and south to Angola and South Africa (Transvaal). In ground layer of high rainfall miombo, riverine and lakeshore woodland, thicket, dense vegetation at base of rocky outcrops and in mopane woodland on alluvium, also in dambos and moist grassland, frequently seen as a weed of waste ground; 200–1370 m.

Although for over a century this species has been confused with the preceding, and has been regarded as synonymous with it, following Mueller's muddled treatment in De Candolle, Prodr. **15**, 2, (1866) there is no doubt about the constancy of their distinctness [see Radcliffe-Smith in Kew Bull. **44**, 3: 439–440 (1989)].

25. **Acalypha indica** L., Sp. Pl.: 1003 (1753). —Müller Argoviensis in De Candolle, Prodr. **15**, 2: 868 (1866). —Engler, Pflanzenw. Ost-Afrikas **C**: 239 (1895). —Hutchinson in F.T.A. **6**, 1: 903 (1912). —Prain & Hutchinson in Bull. Misc. Inform., Kew **1913**: 15 (1913). —Eyles in Trans. Roy. Soc. South Africa **5**: 396 (1916). —Prain in F.C. **5**, 2: 470 (1920). —Engler, Pflanzenw. Afrikas (Veg. Erde 9) **3**, 2: 98 (1921). —Pax in Engler, Pflanzenr. [IV, fam. 147, xvi] **85**: 33 (1924). —Burtt Davy, Fl. Pl. Ferns Transvaal: 305 (1932). —Bremekamp & Obermeyer in Ann. Transvaal Mus. **16**, 3: 421 (1935). —Hutchinson, Botanist in Southern Africa: 667 (1946). —F.W. Andrews, Fl. Pl. Anglo-Egypt. Sudan **2**: 52, t. 19 (1952). —Martineau, Rhod. Wild. Fl.: 45 (1953). —P.G. Meyer in Merxmüller, Prodr. Fl. SW. Afrika, fam. 67: 5 (1967). —Agnew, Upl. Kenya Wild Fls.: 215 (1974). —Radcliffe-Smith in F.T.E.A., Euphorb. 1: 199 (1987). Type from S India.

Acalypha bailloniana Müll. Arg. in Linnaea **34**: 44 (1865); in De Candolle, Prodr. **15**, 2: 869 (1866). —Engler, Pflanzenw. Ost-Afrikas **C**: 239 (1895). Type from Zanzibar.

Acalypha somalium Müll. Arg. in Abh. Naturwiss. Vereine Bremen **7**: 27 (1880). Type from Somalia.

Acalypha indica var. *bailloniana* (Müll. Arg.) Hutch. in F.T.A. **6**, 1: 904 (1912). —Pax in Engler, Pflanzenr. [IV, fam. 147, xvi] **85**: 35 (1924).

An erect usually simple-stemmed annual, or sometimes a woody subperennial herb or subshrub up to 1.2 m tall. Stems ribbed, and pubescent along the ribs. Petioles up to 8.5 cm long. Leaf blades 2–7 × 1–5 cm, ovate to ovate-rhombic, acute or subacute at the apex, serrate except towards the base, cuneate at the base, membranous, sparingly puberulous or pubescent to subglabrous or glabrous on both surfaces, sometimes more evenly puberulous along the midrib and main nerves beneath, 5-nerved from the base; lateral nerves in 4–5 pairs. Stipules 2 mm long, linear-filiform, ciliate. Inflorescences up to 9 cm long, spicate, axillary, solitary or paired, mostly female with a few male flowers above, or else all female, often terminated by a long-pedicellate allomorphic female flower; male bracts minute; female bracts accrescent to up to 1.2 × 1.4 cm, transversely ovate to suborbicular, repand-dentate, many ribbed, sparingly pubescent along the ribs, otherwise usually ± glabrous, up to 5-flowered. Male flowers subsessile; buds tetragonal, slightly granulate, yellowish-green; anthers white. Female flowers sessile; sepals 3, 1 mm long, triangular-ovate, ciliate; ovary 0.5 mm in diameter, somewhat 3-lobed, sparingly pubescent above; styles 2 mm long, united at the base, laciniate, white. Allomorphic female flowers obovoid, with a pair of fimbriate whorls near the top, pubescent. Fruits 1.5 × 2 mm, 3-lobed, tuberculate, pubescent. Seeds 1.3 × 1 mm, ovoid, ± smooth, grey, with a flattened linear caruncle.

Botswana. N: Duba (Xesabe) Island, fl. 28.vi.1974, *P.A. Smith* 1060 (K; MO; PRE; SRGH). SW: Ghanzi Pan Farm, fl. 11.iii.1970, *Brown* 8800 (K; PRE). SE: Tlalamabele–Mosu area, near Soa Pan, fr. 9.i.1974, *Ngoni* 290 (MO; PRE; SRGH). **Zambia**. B: Zambezi–Sinkapu R. junction, fl. 14.viii.1957, *Angus* 1664 (K; MO; PRE; SRGH). S: Lusitu, fl. & fr. 19.v.1960, *Fanshawe* 5690 (K; NDO; SRGH). **Zimbabwe**. N: 25 km west of Kanyemba, fr. 2.ii.1966, *Müller* 351 (K; MO; SRGH). W: Hwange (Wankie), Deka R., fr. 21.vi.1934, *Eyles* 8080 (BM; K; SRGH). E: Hot Springs, fr. iv.1947, *Chase* 339 in *GHS* 16608 (BM; K; SRGH). S: Chipinda Pools, Runde (Lundi) R., fr. 22.i.1961, *Goodier* 1025 (K; LISC; MO; PRE; SRGH). **Malawi**. S: Nsanje, railway station, fr. 29.v.1970, *Brummitt* 11153 (K; SRGH). **Mozambique**. T: Chicoa, fr. 2.iii.1972, *Macêdo* 4982 (LISC; LMA; SRGH). MS: 19 km Tambara–Lupata, fr. 16.v.1971, *Torre & Correia* 18505 (COI;

LISC; LMU; LUA; PRE). GI: Inhambane, fr. iv.1936, *Gomes e Sousa* 1723 (K). M: Polana, fl. & fr. 8.ii.1920, *Borle* 310 (K; PRE).

More or less throughout the Old World tropics, and also introduced into the warmer parts of the New World. Hot, low to medium altitude, sandy margins of rivers, seasonal water courses and pans, usually in shade of thickets, also on rocky hillsides and rocky outcrops, often in disturbed ground and as a weed of cultivation; 125–1000 m.

26. **Acalypha segetalis** Müll. Arg. in J. Bot. **2**: 336 (1864); in De Candolle, Prodr. **15**, 2: 877 (1866). —Hutchinson in F.T.A. **6**, 1: 904 (1912). —Prain & Hutchinson in Bull. Misc. Inform., Kew **1913**: 15 (1913). —Eyles in Trans. Roy. Soc. South Africa **5**: 396 (1916). —Prain in F.C. **5**, 2: 471 (1920). —Engler, Pflanzenw. Afrikas (Veg. Erde 9) **3**, 2: 99 (1921). —Pax in Engler, Pflanzenr. [IV, fam. 147, xvi] **85**: 36 (1924). —De Wildeman, Pl. Bequaert. **3**, 4: 494 (1926). —Burtt Davy, Fl. Pl. Ferns Transvaal: 305 (1932). —Keay in F.W.T.A., ed. 2, **1**, 2: 410 (1958). —P.G. Meyer in Merxmüller, Prodr. Fl. SW. Afrika, fam. 67: 6 (1967). —Radcliffe-Smith in F.T.E.A., Euphorb. 1: 200 (1987). Types from Angola (Cuanza Norte Province).

Acalypha sessilis var. *brevibracteata* Müll. Arg. in Flora **47**: 465 (1864). Type from Nigeria.

Acalypha sessilis var. *exserta* Müll. Arg. in Flora **47**: 465 (1864). Type from Nigeria.

Acalypha gemina var. *brevibracteata* (Müll. Arg.) Müll. Arg. in De Candolle, Prodr. **15**, 2: 866 (1866).

Acalypha gemina var. *exserta* (Müll. Arg.) Müll. Arg. in De Candolle, Prodr. **15**, 2: 866 (1866).

Acalypha sessilis sensu De Wild. & Durand in Mém. Herb. Boissier, 2e Sér. No. 20: 47 (1900), non Poir.

Very like *A. indica*, but usually much more branched, with a number of decumbent-ascending stems arising from the base, and seldom exceeding 50 cm in height; inflorescences usually shorter (less than 2.5 cm long) with only 1–3 female bracts per inflorescence; bracts crenate; the few-flowered male portion of the inflorescence pedunculate; male flowers often reddish; allomorphic female flowers usually long-pedicellate but most commonly arising from the base of the inflorescence, muricate and evenly whitish-puberulous.

Botswana. N: Xaudum (Khardoum) Valley, fl. & fr. 14.iii.1965, *Wild & Drummond* 7012 (BM; K; LISC; PRE; SRGH). SW: west Kalahari, north of Nata on Kasane road, fr. 15.iii.1976, *Vahrmeijer* 3116 (PRE). **Zambia**. N: Kasama Distr., Chambeshi, fr. 4.i.1963, *Astle* 1921 (MO; SRGH). W: Mpongwe, fl. 2.ix.1963, *Fanshawe* 7960 (K; NDO; SRGH). E: Petauke, fl. 17.xii.1958, *Robson* 973 (BM; K; PRE; SRGH). S: Kafue Flats, fr. 17.vi.1960, *Angus* 2422 (FHO; K; SRGH). **Zimbabwe**. N: Oswa Road, fr. 17.i.1969, *Grosvenor* 469A (K; PRE; SRGH). W: Matetsi Safari Area, fl. & fr. 4.ii.1980, *Gonde* 294 (K; MO; PRE; SRGH). C: Chegutu (Hartley), fr. 24.ii.1969, *Mavi* 969 (K; PRE; SRGH). E: Nyanga (Inyanga), fr. 15.i.1931, *Norlindh & Weimarck* 4350 (K; LD; PRE). S: Makaholi Experiment Station, fr. 13.iii.1978, *Senderayi* 207 (K; PRE; SRGH). **Malawi**. C: Kasungu Distr., fr. 28.ii.1979, *Salubeni, Banda & Tawakali* 2513 (MAL; MO; SRGH). S: Magomero, fr. 14.i.1981, *Salubeni* 2881 (PRE; SRGH). **Mozambique**. GI: Chibuto, fr. 12.ii.1959, *Barbosa & Lemos* in *Barbosa* 8373 (COI; K; LMA). M: Incanhini, fl. 15.i.1898, *Schlechter* 12043 (COI).

Also from Sierra Leone eastwards to Ethiopia and south to Namibia and South Africa (Free State, KwaZulu-Natal). On floodplain alluvium and black clay soils, in pans and moist grassy dambos, also on rocky outcrops, in miombo and mopane woodlands, often as a weed of cultivation and disturbed ground; ± sea level to 1500 m.

27. **Acalypha crenata** Hochst. ex A. Rich., Tent. Fl. Abyss. **2**: 246 (1851). —Müller Argoviensis in De Candolle, Prodr. **15**, 2: 871 (1866), pro parte excl. var. *glandulosa* Müll. Arg. —Engler, Pflanzenw. Ost-Afrikas **C**: 239 (1895), pro parte. —Hutchinson in F.T.A. **6**, 1: 903 (1912). —Engler, Pflanzenw. Afrikas (Veg. Erde 9) **3**, 2: 99 (1921). —Pax in Engler, Pflanzenr. [IV, fam. 147, xvi] **85**: 97 (1924). —F.W. Andrews, Fl. Pl. Anglo-Egypt. Sudan **2**: 52 (1952). —Keay in F.W.T.A., ed. 2, **1**, 2: 410 (1958). —Agnew, Upl. Kenya Wild Fls.: 215 (1974). —Radcliffe-Smith in F.T.E.A., Euphorb. 1: 200 (1987). Type from Ethiopia (Tigray Province).

Acalypha abortiva Hochst. ex Baill., Étud. Gén. Euphorb.: 443 (1858). Type from Sudan.

Acalypha indica var. *abortiva* (Hochst. ex Baill.) Müll. Arg. in Linnaea **34**: 42 (1865); in De Candolle, Prodr. **15**, 2: 868 (1866).

Acalypha vahliana sensu Oliv. in Trans. Linn. Soc. London, **29**: 147 (1875), pro parte quoad spec. *Speke & Grant*, non Müll. Arg. (1865, 1866).

Very like *A. indica*, especially with regard to the number and shape of the female bracts, although similar to *A. segetalis* in habit, but the leaves are more rounded at the base, and the muricate allomorphic female flowers are usually subsessile and may be terminal or lateral on the inflorescence, and are glabrous except for a single white hair at the apex of each tubercle.

Botswana. N: Chobe Distr., Mpandamatenga (Pandamatenga), fl. & fr. 9.iv.1986, *Riches* 6 (K). **Zambia**. N: Luangwa Valley Game Res. South, fr. 15.ii.1967, *Prince* 216 (K; LISC; SRGH). W: Ndola, fl. & fr. 21.ii.1954, *Fanshawe* 861 (K; NDO; SRGH). C: Mt. Makulu Res. Station, fl. & fr. 26.ii.1960, *Angus* 2151 (FHO; K; SRGH). S: Mazabuka Distr., Kafue Flats, fr. 23.i.1963, *Angus* 3544 (FHO). **Zimbabwe**. N: Mukumbura R. (Mkumburu R.), fr. 23.i.1960, *Phipps* 2400 (K; PRE; SRGH). W: Hwange Distr., Kazuma Range, o. fr. 11.v.1972, *Gibbs-Russell* 1975 (SRGH). C: Gwebi R. bank, fr. 22.ii.1961, *Rutherford-Smith* 558 (MO; SRGH). E: Chisumbanje, fl. & fr. i.1969, *Wall* in *GHS* 191974 (SRGH). S: Gona-re-Zhou, fr. 30.v.1971, *Grosvenor* 648 (SRGH). **Malawi**. N: Karonga, fr. 7.iii.1953, *J.Williamson* 189 (BM). C: north of Chitala, fr. 12.ii.1959, *Robson* 1571 (BM; K; LISC; SRGH). **Mozambique**. MS: Nhamatanda (Vila Machado), fr. 26.ii.1968, *Mendonça* 3824 (LISC).

Also from Cape Verde eastwards to Somalia and south to East Africa. In mopane woodland on damp black clay soils, in moist grassland and on sandy river banks, often in waste ground and is a weed of cultivation; 510–1340 m.

28. **Acalypha lanceolata** Willd., Sp. Pl. **4**: 524 (1805). —Agnew, Upl. Kenya Wild Fls.: 215 (1974). —Troupin, Fl. Rwanda **2**: 196 (1983). —Radcliffe-Smith in F.T.E.A., Euphorb. 1: 202 (1987). Type from Sri Lanka.

Var. **glandulosa** (Müll. Arg.) Radcl.-Sm. in Kew Bull. **44**: 444 (1989). Type from Zanzibar.

Acalypha crenata var. *glandulosa* Müll. Arg. in Linnaea **34**: 43 (1865); in De Candolle, Prodr. **15**, 2: 871 (1866).

Acalypha boehmerioïdes var. *glandulosa* (Müll. Arg.) Pax & K. Hoffm. in Engler, Pflanzenr. [IV, fam. 147, xvi] **85**: 97 (1924).

Acalypha glomerata Hutch. in Bull. Misc. Inform., Kew **1911**: 229 (1911); in Bull. Misc. Inform., Kew **1913**: 15 (1913); in F.T.A. **6**, 1: 902 (1912). Types from Bongoland, Uganda, Zanzibar, Tanzania and Mozambique: Mozambique Province, fr. 5.x.1894, *O. Kuntze* 5480 (NY, syntype; K).

Very like *A. crenata* in general habit, but the leaf bases cuneate, the inflorescences crowded towards the shoot apices, the female bracts smaller (accrescent to 1.5 × 4 mm), closely enfolding the ovary and fruit, and are evenly beset with gland-tipped hairs, and the allomorphic female flowers have 2 conspicuous lateral fimbriate whorls and are rugulose and puberulous, but not muricate.

Mozambique. N: Moçambique, fr. 5.x.1894, *Kuntze* 5480 (K; NY). GI: Nhachengue (Inhachengo), fr. 26.ii.1955, *Exell, Mendonça & Wild* 631 (BM; LISC; SRGH). M: Maputo (Lourenço Marques), fr. 26.iii.1910, *Howard* 5676 (K).

The species is widespread in the Old World tropics; the var. *glandulosa* only in tropical Africa; the var. *lanceolata* only in tropical Asia. In shade of semi-evergreen coastal forest; at or near sea level.

29. **Acalypha wilkesiana** Müll. Arg. in De Candolle, Prodr. **15**, 2: 817 (1866). —Pax in Engler, Pflanzenr. [IV, fam. 147, xvi] **85**: 153 (1924). —De Wildeman, Pl. Bequaert. **3**, 4: 495 (1926). —Brenan, Check-list For. Trees Shrubs Tang. Terr.: 197 (1949). —Biegel, Check List Ornam. Pl. Rhod. Parks & Gard.: 18 (1977). —Troupin, Fl. Rwanda **2**: 200, fig. 57/2, 3 (1983). —Radcliffe-Smith in F.T.E.A., Euphorb. 1: 212 (1987). Type from Polynesia (Fiji).

Acalypha godseffiana Mast. in Gard. Chron., Ser. 3, **23**: 241, f. 87 (1898). Type from New Guinea.

Acalypha marginata J.J. Sm. in Med. Dept. Landb. **10**: 19 (1910), non Spreng. (1826). Type ? from Java.

Acalypha godseffiana var. *marginata* (J.J. Sm.) Hort. sec. Pax & K. Hoffm. in Engler, Pflanzenr. [IV, fam. 147, xvi] **85**: 155 (1924). —De Wildeman, Pl. Bequaert. **3**, 4: 489 (1926).

Acalypha amentacea subsp. *wilkesiana* (Müll. Arg.) F.R. Fosberg. Sm. Con. Bot. **45**: 10 (1980).

A densely branched rounded shrub mostly 1–2 m tall, monoecious, tomentellous to pubescent. Petioles 1–5 cm long. Leaf blades usually up to 15 × 10 cm, sometimes larger, elliptic-ovate to broadly ovate, obtusely acuminate at the apex, crenate-dentate on the margins, cuneate or rounded at the base, 5–7-nerved from the base, membranous, sparingly pubescent along the midrib and main veins on both surfaces at first, later glabrescent, green, copper or bronze and variously variegated with purple, red, pink, cream-coloured or white, or else pink-, cream- or white-margined; lateral nerves in 6–10 pairs. Stipules 7 mm long, narrowly lanceolate, acutely acuminate. Inflorescences axillary, usually solitary, spicate, unisexual, on the same or different shoots. Male spikes up to 12 cm long, densely but interruptedly flowered; bracts minute, many-flowered. Female spikes up to 7 cm long, lax flowered; bracts c. 5 × 4 mm, ± ovate, later accrescent, dentate, with 3–6 teeth on each side, sparingly pubescent to subglabrous, 1-flowered. Male flowers sessile or ± so; buds tetragonal, subglabrous, reddish; anthers yellowish. Female flowers sessile; sepals 3–4, 1 mm long, ovate, subacute; ovary 1.5 mm in diameter, subglobose, tomentose; styles c. 6 mm long, united at the base, deeply laciniate, red. Fruits 1.5–2 × 4 mm, 3-lobed, pubescent. Seeds not set in Africa.

Zambia. C: Lusaka, Chilanga, Mr. Sanders' Garden (Munda Wanga), fl. 28.xii.1960, *White* 6052 (FHO). **Zimbabwe**. C: Harare, female fl. 17.xii.1974, *Biegel* 4718 (K; SRGH). **Mozambique**. N: Nampula, male fl. v.1944, *Gomes e Sousa* 3 (K; PRE). M: Maputo (Lourenço Marques), male & female fl. 7.vi.1972, *Balsinhas* 2447 (K).

Native of Polynesia, cultivated in the Flora Zambesiaca area, occasionally occurs as a garden escape.

Many cultivars exist of this species, of which the following are of note: cultivars *macrophylla, illustris, triumphans, circinata, hoffmannii, monstrosa, heterophylla* and *obovata*.

30. **Acalypha hispida** Burm. f., Fl. Ind.: 203, t. 61, f. 1 (1768). —Müller Argoviensis in De Candolle, Prodr. **15**, 2: 815 (1866). —Pax in Engler, Pflanzenr. [IV, fam. 147, xvi] **85**: 140 (1924). —Brenan, Check-list For. Trees Shrubs Tang. Terr.: 196 (1949). —Troupin, Fl. Rwanda **2**: 196, fig. 57/1 (1983). —Radcliffe-Smith in F.T.E.A., Euphorb. 1: 212 (1987). Type is the illustration in Burman (1768) which was based on cultivated material.

A much-branched shrub up to 2 m tall, dioecious. Young shoots and petioles tomentose, later sparingly puberulous or glabrescent. Petioles up to 15 cm long; leaf blades up to 20 × 15 cm, broadly ovate or rhombic-ovate, shortly acuminate at the apex, serrate on the margins, rounded or cuneate at the base, 5–7-nerved from the base, thinly chartaceous, sparingly pubescent to subglabrous on both surfaces, more evenly pubescent along the midrib and main veins; lateral nerves in 6–9 pairs. Stipules 6–7 mm long, lanceolate, sparingly pubescent, brown. Male inflorescences unknown. Female inflorescences up to 30 cm long, spicate, axillary, solitary, dense-flowered, bright red on account of the masses of styles; axis sparingly pubescent; female bracts minute, ovate, acute, entire, not accrescent. Female flowers sessile; sepals 3–4, 0.7 mm long, triangular-ovate, acute, ciliate; ovary 1 mm in diameter, 3-lobed to subglobose, densely pubescent; styles 5–7 mm long, ± free to the base, laciniate, bright red. Mature fruit and seeds not known.

Mozambique. M: Maputo (Lourenço Marques), fl. 11.vi.1971, *Balsinhas* 1906 (LISC; LMA). Cultivated garden ornamental. Native of New Guinea and the Bismarck Archipelago.

31. **Acalypha sp. A**

An erect sparingly-branched perennial herb or subshrub up to 60 cm tall, presumably dioecious; stems few, from a woody rootstock, pubescent. Leaves alternate or verticillate. Petioles 2–6 mm long. Leaf blades 4–8.5 × 1–2 cm, lanceolate, subacute at the apex, finely serrulate on the margins, rounded at the base, chartaceous, 3-nerved from the base, sparingly to evenly pilose on both surfaces, drying purplish-brown; lateral nerves in 9–12 pairs, slightly prominent on both surfaces. Stipules 2–3 mm long, subulate. Male inflorescences up to 14 cm long, on a peduncle up to 6 cm long, axillary or subterminal; bracts minute. Male flowers in spirally arranged clusters; pedicels 0.5 mm long; buds depressed-subglobose, apiculate, glabrous, yellowish. Female inflorescences, flowers, fruit and seeds unknown.

Mozambique. Z: c. 51 km Namacurra–Milange, o. male fl. 3.ii.1966, *Torre & Correia* 14410 (LISC).

Known only from this collection. In secondary forest with *Brachystegia boehmii, Julbernardia globiflora, Albizia adianthifolia, Millettia stuhlmannii* etc. on sandy clay soil; 40 m.

Species of uncertain position

Acalypha helenae Buscalioni & Muschler, Bot. Jahrb. Syst. **49**: 477 (1913), from Lake Bangweulu in northern Zambia, based on *Aosta* 908, cannot be accurately identified from Muschler's description. According to Pax (in Engl., Pflanzenreich, Heft. **85**: 175 (1924)) "e descriptione valde manca vix ad genus *Acalypham* pertinet".

35. PTEROCOCCUS Hassk.

Pterococcus Hassk. in Flora **25**, 2: Beibl. **3**: 41 (1842), nom. conserv., non Pall. (1773).
Sajorium Endl. Gen. Suppl. **3**: 98 (1843).
Plukenetia sect. *Hedraiostylus* (Hassk.) Müll. Arg. in De Candolle, Prodr. **15**, 2: 772 (1866).
Plukenetia sect. *Pterococcus* (Hassk.) Benth. & Hook., Gen. Pl. **3**, 1: 327 (1880), pro parte.
Pseudotragia Pax in Bull. Herb. Boissier Sér. 2, **8**: 635 (1908).

Monoecious usually scandent herbs or subshrubs, with a simple indumentum. Leaves alternate, petiolate, stipulate, simple, toothed or lobed, palminerved. Inflorescences terminal, lateral, or leaf-opposed, racemose, male inflorescence with a solitary female flower at the base; bracts small. Male flowers pedicellate; calyx globose in bud, later splitting into 4 valvate lobes; petals absent; disk absent; stamens 8–18, borne on a prominent receptacle, filaments free, short, anthers basifixed, cruciately 4-locular, longitudinally dehiscent; interstaminal glands minute, often indistinct; pistillode absent. Female flowers long-pedicellate, the pedicel extending in fruit; sepals 4, imbricate; petals absent; disk absent; ovary 4-celled, with 1 ovule per cell, and 4-winged; styles 4, connate into a short, stout column, stigmas 4, obcordate-obovate, borne in a cross-shaped arrangement. Fruit strongly 4-lobed, the lobes winged to cornute, dehiscing into 4 bivalved cocci; exocarp coriaceous; endocarp woody; columella short. Seeds laterally compressed, winged, rugulose, ecarunculate; testa crustaceous; albumen fleshy; cotyledons suborbicular, flat, cordate and 3-nerved at the base.

A tropical genus of 3 species, 1 Asiatic and 2 from Angola and south tropical Africa.

Pterococcus africanus (Sond.) Pax & K. Hoffm. in Engler, Pflanzenr. [IV, fam. 147, ix] **68**: 22 (1919). —Engler, Pflanzenw. Afrikas (Veg. Erde 9) **3**, 2: 101 (1921). —P.G. Meyer in Merxmüller, Prodr. Fl. SW. Afrika, fam. 67: 41 (1967). Tab. **44**. Type from South Africa (Transvaal).

Plukenetia africana Sond. in Linnaea **23**: 110 (1850). —Müller Argoviensis in De Candolle, Prodr. **15**, 2: 773 (1866). —Prain in F.T.A. **6**, 1: 951 (1912); in F.C. **5**, 2: 496 (1920). —Burtt Davy, Fl. Pl. Ferns Transvaal: 307 (1932). —Hutchinson, Botanist in Southern Africa: 667 (1946). Type as above.

Sajorium africanum (Sond.) Baill., Étud. Gén. Euphorb.: 483 (1858).

Plukenetia hastata Müll. Arg. in Flora **47**: 469 (1864); in De Candolle, Prodr. **15**, 2: 772 (1866). —Pax in Engler, Pflanzenw. Ost-Afrikas **C**: 240 (1895). —Prain in F.T.A. **6**, 1: 950 (1912); in F.C. **5**, 2: 497 (1920). —Burtt Davy, Fl. Pl. Ferns Transvaal: 307 (1932). Type: Mozambique, Manica e Sofala Province, between Chupanga (Shupanga) and Sena, y. fr. i.1859, *Kirk* s.n. (K, holotype).

Pseudotragia schinzii Pax in Bull. Herb. Boissier, Sér. 2, **8**: 635 (1908). Type from Namibia.

Pseudotragia scandens Pax in Bull. Herb. Boissier, Sér. 2, **8**: 636 (1908). Type from Namibia.

A trailing or climbing puberulous perennial herb or suffrutex, from a woody rootstock; stems several, up to 80 cm long, prostrate, decumbent or ascending, pale green. Stipules 1 mm long, lanceolate. Petioles 1–25 mm long. Leaf blades 1.5–12 × 0.1–5 cm, lanceolate or linear, acute or rarely obtuse at the apex, cuneate to truncate and hastate or sagittate at the base, serrulate to denticulate or subentire on the margins, membranous, 5–7-nerved from the base, glabrescent on the upper surface, sometimes purplish-tinged; lateral nerves in 2–9 pairs. Inflorescences 1–10 cm long, usually leaf-opposed; bracts 1.5–2.5 mm long, narrowly elliptic-lanceolate, sublobate. Male flowers: pedicels 2 mm long, jointed; calyx lobes 1–1.5 mm long, ovate, acute, glabrous, greenish-cream or yellowish; stamens 10–18, minute, filaments broadened at the base, anthers 0.3 mm across. Female flower: pedicel 1–2 mm long, extending to 1–2 cm in fruit; sepals 2 mm long, elliptic-ovate, acute, sparingly pubescent; ovary 1 mm in diameter, 4-winged, densely strigose; stylar column 1 mm tall, stigmas 1.5–2

Tab. 44. PTEROCOCCUS AFRICANUS. 1, portion of stem (× $^2/_3$); 2, inflorescence (× 2); 3, male flower (× 6); 4, fruit (× 3), 1–4 from *Fanshawe* 5957; 5, seeds (× 6), from *Wild & Drummond* 7049. Drawn by Christine Grey-Wilson.

mm across, minutely papillose. Fruit 0.75 × 1.5–2 cm, cross-shaped when viewed from above, the cocci horned, rugulose, sparingly strigose-pubescent, green. Seeds 6 × 3 mm, lenticular, rounded-triangular in outline, irregularly winged, pale green.

Caprivi Strip. 16 km Katima–Singalamwe (Finaughty's Road), y. fr. 30.xii.1958, *Killick & Leistner* 3188 (K; SRGH). **Botswana**. N: Xaudum (Khardoum) Valley, fr. 14.iii.1965, *Wild & Drummond* 7049 (K; LISC; PRE; SRGH). SW: 80 km north of Kang, fl. 18.ii.1960, *Wild* 5062 (BM; K; PRE; SRGH). SE: Kweneng Distr., 15 miles from turn off towards Ngware, fl. & y. fr. 20.x.1977, *O.J. Hansen* 3251 (C; GAB; K; PRE; SRGH; WAG). **Zambia**. S: Machili, fl. & fr. 9.xii.1960, *Fanshawe* 5957 (K; NDO; SRGH). **Zimbabwe**. N: Gokwe Distr., Sengwa Research Station, fr. 19.ix.1975, *P.R. Guy* 2367 (K; PRE; SRGH). W: Gwampa Forest Reserve, fr. i.1956 *Goldsmith* 65/56 (K; LISC; SRGH). E: Dott's Drift, fl. 16.xi.1959, *Goodier* 661 (K; LISC; PRE; SRGH). S: near the Mozambique border opposite Sango (Vila de Salazar), o. fr. 26.iv.1961, *Drummond & Rutherford-Smith* 7535 (K; LISC; SRGH). **Mozambique**. Z: Mocuba–Olinga (Maganja da Costa), y. fr. 20.xi.1967, *Torre & Correia* 16131 (LISC, & photo). MS: c. 10 km north of Mwanza, between Dondo and Inhaminga, fl. 4.xii.1971, *Pope & Müller* 517 (K; LISC; SRGH). GI: Vila Eduardo Mondlane (Malvérnia), subst. 26.iv.1961, *Thompson* 2 (K; LISC; PRE; SRGH). M: Magude, Chobela, fl. & y. fr. 30.xii.1947, *Torre* 7014 (LISC).

Also in ? Angola, Namibia and South Africa (Transvaal). In deciduous woodlands on sand, including miombo and Kalahari Sand woodlands, also in wooded grassland on dry sandy soils; 130–1100 m.

36. TRAGIELLA Pax & K. Hoffm.

Tragiella Pax & K. Hoffm. in Engler, Pflanzenr. [IV, fam. 147, ix] **68**: 104 (1919).
Tragia sensu Sond. in Linnaea **23**: 107 (1850), pro parte non L.
Sphaerostylis sensu Croizat in J. Arnold Arbor. **22**, 3: 430 (1941), non Baill.

Monoecious erect or twining perennial herbs with a simple indumentum, mixed with urticating (stinging) bristles. Leaves alternate, petiolate or subsessile, stipulate, simple or palmately-lobed, serrate, palminerved. Inflorescences leaf-opposed or terminal on short axillary shoots, racemose, usually mostly male with 1–2 female flowers at the base. Bracts conspicuous, persistent, 1-flowered. Flowers shortly pedicellate. Male flowers: calyx closed in bud, later splitting into 3 valvate lobes, with or without a basal tube constricted below the limb; petals absent; disk absent; stamens 3–4, alternating with the calyx lobes, filaments short, thickened and sometimes connate at the base, connective thickened; anthers dorsifixed, introrse, 2-celled, the cells parallel, longitudinally dehiscent; pistillode present, small, circular or 3-lobed. Female flowers: calyx lobes 6, uniseriate, imbricate, pinnatifid, usually with a narrow central rhachis and one terminal and several dark green lateral lobules which are often longer than the width of the calyx-lobe rhachis, accrescent, the calyx-lobe rhachis later becoming hardened and sometimes almost woody; petals absent; disk absent; ovary 3-celled, with 1 ovule per cell; styles 3, almost completely united to form a hollow conical or infundibuliform column, or a subglobose mass. Fruit 3-lobed, dehiscing into 3 bivalved cocci; pericarp thin, crustaceous; endocarp thick, woody; columella trifid, not persistent. Seeds globose, mottled, ecarunculate. Cotyledons broad, flat.

An eastern African genus of 5 species ranging from Sudan and Somalia south to Eastern Cape Province.

1. Plants erect; leaves narrowly lanceolate to linear-lanceolate; petiole not more than 3 mm long; stylar column conical ··· 1. *friesiana*
– Plants climbing; leaves elliptic- to oblong-lanceolate; petiole 0.5–8 cm long; stylar column infundibuliform, or styles a subglobose mass ··· 2
2. Twining perennial herb; stipules membranous, green; leaves sharply toothed with rounded sinuses or -sharply biserrate; stylar column infundibuliform ··· 2. *natalensis*
– Liane; stipules coriaceous, buff; leaves crenate-serrate with the teeth forwardly directed; stylar mass subglobose to turbinate-obpyriform ··· 3. *anomala*

1. **Tragiella friesiana** (Prain) Pax & K. Hoffm. in Engler, Pflanzenr. [IV, fam. 147, ix] **68**: 106 (1919). —Engler, Pflanzenw. Afrikas (Veg. Erde 9) **3**, 2: 101 (1921). Tab. **45**. Type: Zambia, Kunkute, near Mporokoso, fl. x.1911, *R.E. Fries* 1182 (UPS, holotype; K, fragment & drawing of holotype).

Tab. 45. TRAGIELLA FRIESIANA. 1, upper portion of stem (× $^{2}/_{3}$); 2, part of inflorescence (× 4); 3, male flower (× 8); 4, fruit (× 4); 5, calyx after fruit dehiscence (× 4), 1–5 from *Sanane* 1327. Drawn by Christine Grey-Wilson.

Tragia friesiana Prain in R.E. Fries, Wiss. Egebn. Schwed. Rhod.-Kongo-Exped. **1**, 1: 125 (1914).
Sphaerostylis friesiana (Prain) Croizat in J. Arnold Arbor. **22**, 3: 430 (1941).

An erect, weakly urticating (stinging), branched perennial herb up to 50 cm tall, monoecious; stems one or more arising from a woody rootstock, the indumentum puberulous and with a mixture of long hairs and stinging bristles. Stipules 3–4 mm long, oblong-lanceolate, puberulous. Petiole 1–3 mm long, or leaves subsessile. Leaf blade 2–5 × 0.3–1.5 cm, narrowly lanceolate to linear-lanceolate, acute at the apex, shallowly and somewhat remotely serrate-denticulate on the margins, rounded-cuneate at the base, thinly chartaceous, sparingly pubescent and setose on the upper surface, more evenly so beneath, 3–5-nerved from the base, with the inner pair of nerves running ± one-third the length of the blade; tertiary nerves reticulate; lateral nerves in 3–5 pairs, not or scarcely prominent above, fairly prominent beneath. Inflorescences up to 10 cm long, the peduncle up to 3 cm long; male bracts 2–2.5 mm long, elliptic, subentire; male bracteoles 1 mm long, linear-lanceolate; female bracts 2–2.5 mm long, ovate-lanceolate, sometimes shallowly toothed; female bracteoles resembling the male bracts. Male flowers: pedicels less than 0.5 mm long; buds turbinate-subglobose; calyx lobes 1 × 1 mm, suborbicular, cucullate, sparingly pubescent to glabrous, yellow; stamens 3, 0.5 mm long; pistillode 3-lobed, the lobes rounded. Female flowers: pedicels 0.5 mm long, extending to 3–5 mm long in fruit; calyx lobes 6, 1.5 × 1.5 mm, accrescent to 4 × 2 mm, pinnatifid with 6–8 lateral lobules, the lobules narrowly oblong-lanceolate and shorter than the width of the calyx-lobe rhachis, sparingly puberulous and setulose, green, the calyx-lobe rhachis loriform or oblong, sparingly puberulous without, glabrous within, becoming hardened and stramineous within; ovary 1.5 mm in diameter, strongly 3-lobed, puberulous, setose on the keels; stylar column 1 mm high, 1 mm wide at the base, conical, sparingly puberulous, persistent, stigmas minute, smooth, recurved. Fruit 5 × 9–10 mm, strongly 3-lobed, smooth, evenly pubescent, sparingly setose on the keels. Seeds 4 mm in diameter, globose, buff, spotted and mottled with dark purplish-brown, with scattered pale circular patches fringed with minute whitish papillae.

Zambia. N: Mbala (Abercorn)–Kambole road, fl. 10.ix.1960, *Richards* 13208 (K); Chilwa Stream, Chitembwa (Chitimbwa) Road, fr. 20.viii.1970, *Sanane* 1327 (K; SRGH).

Not known from elsewhere. In high rainfall miombo woodland ground cover, appearing after early season fires; 1500–1525 m.

This species presents a remarkable parallel to *Tragia lasiophylla* Pax & K. Hoffm. resembling it in many vegetative respects. The form of the styles especially serve to distinguish the two, however.

2. **Tragiella natalensis** (Sond.) Pax & K. Hoffm. in Engler, Pflanzenr. [IV, fam. 147, ix] **68**: 105, t. 24 A–E (1919). —Engler, Pflanzenw. Afrikas (Veg. Erde 9) **3**, 2: 101, t. 46 A–E (1921). —F.W. Andrews, Fl. Pl. Anglo-Egypt. Sudan **2**: 100 (1952). —Brenan in Mem. N.Y. Bot. Gard. **9**, 1: 75 (1954). —Agnew, Upl. Kenya Wild Fls.: 218 (1974). —Radcliffe-Smith in F.T.E.A., Euphorb. 1: 318 (1987). Type from South Africa (KwaZulu-Natal).

Tragia natalensis Sond. in Linnaea **23**: 107 (1850). —Müller Argoviensis in De Candolle, Prodr. **15**, 2: 942 (1866). —Eyles in Trans. Roy. Soc. South Africa **5**: 397 (1916). —Prain in F.T.A. **6**, 1: 974 (1913); in F.C. **5**, 2: 506 (1920). —Burtt Davy, Fl. Pl. Ferns Transvaal: 309 (1932). Type as above.

Tragia involucrata sensu Jacq. ex E. Mey. in Drège, Zwei Pflanzengeogr. Dokum.: 226 (1843), non L. (1753).

Tragia mitis var. *oblongifolia* Müll. Arg. in Flora **47**: 435 (1864); in De Candolle, Prodr. **15**, 2: 942 (1866). Syntypes: Mozambique, Morrumbala (Moramballa), 1067 m, fr. 3.xii.1858, *Kirk* s.n., *Tragia* (2), (K); the other syntype from Eastern Cape Province.

Tragia ambigua S. Moore in J. Linn. Soc., Bot. **40**: 202 (1911). Type: Zimbabwe, Chirinda Forest, 1128–1219 m, 14.vi.1906, *Swynnerton* 795 (BM, holotype; K; SRGH).

Tragia ambigua var. *urticans* S. Moore in J. Linn. Soc., Bot. **40**: 203 (1911). Type: Zimbabwe, Chirinda Forest, 1158 m, 26.v.1906, *Swynnerton* 446 (BM, holotype; K; SRGH).

Sphaerostylis natalensis (Sond.) Croizat in J. Arnold Arbor. **22**, 3: 430 (1941).

A twining or scrambling, urticating perennial herb, monoecious; stems up to 3 m long; indumentum a mixture of short soft hairs, longer stiffer hairs and stinging bristles. Stipules 4–7.5 mm long, linear-lanceolate, membranous, sparingly pubescent without, glabrous within, green. Petioles 0.5–8 cm long, indumentum similar to that of the stems. Leaf blades 3–14 × 1.5–8 cm, elliptic-ovate to oblong-lanceolate or lanceolate, shortly acutely acuminate at the apex, closely and sharply patent serrate with rounded sinuses or sharply patent biserrate with rounded sinuses, truncate to cordate at the base, membranaceous, 5–7(9)-nerved from the base, sparingly to evenly puberulous and weakly hispid on the upper surface, evenly to densely pubescent beneath, and setose along the midrib and main nerves on both surfaces; lateral nerves in 3–5 pairs, not or scarcely prominent above or beneath. Inflorescences 2–5.5 cm long, the peduncle 1–3 cm long, ± laxly few-flowered; male bracts 2–3 mm long, patent-reflexed, oblanceolate-oblong to subspathulate, entire or subentire, pubescent; male bracteoles 1.5 mm long, linear-oblanceolate, very fugacious; female bracts 4–6 × 6–8 mm, rhombic-ovate to transversely ovate, serrate or dentate, pubescent; female bracteoles 2, 5 mm long, linear to linear-oblanceolate, pubescent. Male flowers: pedicels 2 mm long; buds subglobose; calyx tube constricted; calyx lobes 1 × 1–1.5 mm, broadly ovate, pubescent without, glabrous within, pale greenish-yellow; stamens less than 1 mm long; pistillode minute, depressed-hemispherical. Female flowers: pedicels 1 mm long, 2 mm long in fruit, densely pubescent; calyx lobes 6, 2 × 1 mm, accrescent to 8 × 3 mm, pinnatifid, with 11–21 lobules, the terminal lobule lanceolate, the lateral lobules subulate, ± equalling the width of the calyx-lobe rhachis, hispid and setose, greenish, the calyx-lobe rhachis narrowly loriform, sparingly pubescent without, glabrous within, yellowish green, becoming hardened and stramineous within; ovary 1.5–2 mm in diameter, 3-lobed, densely appressed-pubescent, setose on the keels; stylar column 3 mm high, infundibuliform, appressed-pubescent, stigmas short, granulate, recurved. Fruit 4.5 × 7 mm, deeply 3-lobed, smooth, evenly appressed-pubescent and hispid, setose on the keels. Seeds 3 mm in diameter, globose, dark grey with whitish wavy lines, and spotted, flecked and mottled with dark purplish-brown.

Zimbabwe. E: Umvumvumu R. Valley near Cashel, fl. & fr. 24.xii.1947, *Chase* 481 in *GHS* 19205 (BM; K; PRE; SRGH). **Malawi**. C: Dedza–Kasumbu's Court, fl. 28.i.1961, *Chapman* 1139 (SRGH). S: Thyolo (Cholo) Mt., fl. 22.ix.1946, *Brass* 17732 (BM; K; PRE; SRGH). **Mozambique**. N: Malema (Entre Rios), Serra Merripa, o. fr. 5.ii.1964, *Torre & Paiva* 10476 (LISC). Z: Massingire, Serra da Morrumbala, fl. & fr. 15.v.1943, *Torre* 5320 (LISC). MS: 22 km west of Dombe, fl. 23.iv.1974, *Pope & Müller* 1254 (K; LISC; SRGH). M: Namaacha, fl. 10.v.1969, *Balsinhas* 1479 (LISC).

Also in Sudan, Uganda, Kenya, Tanzania, South Africa (Transvaal, KwaZulu-Natal and Eastern Cape Province), and Swaziland. In undergrowth and margins of evergreen rain forest and riverine forest; 760–1220 m.

3. **Tragiella anomala** (Prain) Pax & K. Hoffm. in Engler, Pflanzenr. [IV, fam. 147, ix] **68**: 106, t. 24F (1919). —Engler, Pflanzenw. Afrikas (Veg. Erde 9) **3**, 2: 101, t. 46F (1921). —Radcliffe-Smith in F.T.E.A., Euphorb. 1: 321 (1987). Syntypes: Malawi, Misuku Hills (Masuku Plateau), 1980–2135 m, fl. vii.1896, *Whyte* 269 (K); the other syntype from Tanzania (Southern Highlands).

Tragia anomala Prain in Bull. Misc. Inform., Kew **1912**, 4: 194 (1912); in F.T.A. **6**, 1: 975 (1913).

Sphaerostylis anomala (Prain) Croizat in J. Arnold Arbor. **22**, 3: 430 (1941).

Tragia sp. 1, White in F.F.N.R.: 205 (1962).

A liane extending up to 6 m in length, monoecious. Young stems minutely puberulous and/or retrorsely hispid intermixed with stinging bristles. Stipules 3–5 mm long, linear-lanceolate, sparingly pubescent without, glabrous within, thinly coriaceous, buff, persistent. Petioles 0.5–7 cm long, indumentum similar to that of the stems. Leaf blades 3–11 × 1–5 cm, ovate-oblong to oblong or oblong-lanceolate, shortly ± acute acuminate at the apex, crenate-serrate with the teeth forwardly directed on the margins, shallowly wide-cordate to ± deep and narrowly cordate at the base, membranaceous, 3–5-nerved from the base, sparingly pubescent or hispid on both surfaces, very sparingly setose along the midrib and main nerves on both surfaces; lateral nerves in 3–6 pairs, scarcely to slightly prominent above and beneath. Inflorescences 3–5 cm long, the peduncle 1–2 cm long, ± laxly flowered; male bracts 2–2.5 mm long, obovate-subspathulate, entire, adaxially concave, puberulous without, glabrous within; male bracteoles 1–1.5 mm long, subulate or linear, fugacious; female bracts 3.5 × 2.5 mm, ovate, entire or with a few small sharp lateral teeth; female bracteoles 2.5–3 mm long, linear-lanceolate, entire or subentire. Male flowers: pedicels 3 mm long, minutely puberulous; buds turbinate, truncate at the apex; calyx tube very short, scarcely constricted; calyx lobes c. 1 × 1 mm, broadly ovate, sparingly puberulous without, glabrous within; stamens minute; pistillode minute, circular. Female flowers: pedicels 0.5 mm long, densely pubescent, extending to 3 mm in fruit; calyx lobes 6, 2.5 × 0.5 mm, accrescent to 9 × 4 mm, pinnatifid, with 15–19 lobules, the terminal lobule lanceolate, the lateral lobules narrowly triangular-lanceolate and ± equalling the width of the calyx-lobe rhachis, densely setose, green, the calyx-lobe rhachis narrowly lanceolate, sparingly puberulous without, glabrous within, green without, becoming somewhat hardened and stramineous within; ovary 1.5 mm in diameter, 3-lobed, densely appressed-setose; stylar mass 1.5–2 × 2–2.3 mm, subglobose to turbinate-obpyriform, stipitate, appressed-puberulous, stigmas short, separate, smooth, recurved and closely applied to the rim of the stylar mass and later retracted within it. Fruit 5 × 9 mm, 3-lobed, smooth, sparingly appressed-pubescent, and setose on the keels. Seeds 4.5 mm in diameter, globose, buff, closely spotted, flecked and mottled with dark brown.

Zambia. E: Nyika Plateau, Mt. Kangampande, fl., y. & o. fr. 7.v.1952, *White* 2761 (FHO; K). **Malawi**. N: Misuku Hills, Mugesse (Mughesse), fr. 12.ix.1977, *Pawek* 12994 (K; MAL; MO).

Also in Tanzania. In montane evergreen forest understorey; 1580–2150 m.

The genus *Sphaerostylis* was established by Baillon for a plant from Madagascar with no stinging hairs, suborbicular stipules, entire leaves and entire female calyx lobes. The only feature it has in common with *Tragiella* is in the nature of the style, but this would not seem to be sufficient to justify Croizat's reduction of the latter thereunder.

37. TRAGIA L.

Tragia L., Sp. Pl.: 980 (1753); Gen. Pl., ed. 5: 421 (1754).

Monoecious or dioecious erect, suberect or scandent perennial herbs or shrubs; indumentum simple, usually intermixed with stinging (urticating) bristles. Leaves alternate, petiolate or sessile, stipulate, usually simple, rarely lobed, dentate, serrate or entire, often cordate, palminerved. Inflorescences terminal, leaf-opposed or lateral, rarely axillary, racemose, male or female or usually mostly male with 1(3) female flowers at the base, pedunculate. Bracts conspicuous, persistent, the male bracts 1–3-flowered, the female bracts 1-flowered. Pedicels jointed, 2-bracteolate. Male flowers: calyx closed in bud, later splitting into 3(6) valvate lobes; petals absent; disk absent or obscure; stamens (1)3(5), (rarely more outside tropical Africa), alternating with the calyx lobes, filaments short, usually ± free, anthers 2-celled, dorsifixed or basifixed, usually introrse, the cells parallel, longitudinally dehiscent; pistillode minute, often 3-radiate. Female flowers: calyx lobes 3–6, 1- or 2-seriate, imbricate, pinnatifid or ± palmatifid, usually with a narrow central rhachis and one terminal and several dark green lateral lobules which are often longer than the width of the calyx-lobe rhachis (entire in the Neotropics), accrescent (tropical Africa), the calyx-lobe rhachis later becoming hardened and sometimes almost woody; petals absent; disk absent; ovary 3-celled, with 1 ovule per cell; styles 3, connate below, free above, entire. Fruit 3-lobed, dehiscing into 3 bivalved cocci; endocarp crustaceous; columella 3-fid. Seeds globose, marbled, ecarunculate; testa crustaceous; albumen fleshy, cotyledons broad and flat.

A genus of 125 species in the tropics and subtropics of both hemispheres, but especially well represented in the Neotropics and in Africa.

1. Male bracts 3 or more-flowered · 2
– Male bracts 1(2)-flowered · 8
2. Plants prostrate; leaves ovate; inflorescences unisexual · · · · · · · · · · · · · · · · · · 1. *prostrata*
– Plants erect · 3
3. Inflorescences bisexual · 6. *rhodesiae*
– Inflorescences unisexual or plants dioecious · 4

4. Leaves ovate · · · 5. *descampsii*
– Leaves linear-lanceolate, lanceolate or triangular-lanceolate · · · 5
5. Leaf margins finely and closely serrate throughout · · · 4. *lukafuensis*
– Leaf margins coarsely and/or irregularly toothed to subentire or entire, never finely serrate · · · 6
6. Leaves not exceeding 4 cm in length, entire or at most with a few apical teeth · · 3. *mazoensis*
– Leaves up to 8.5 cm in length, coarsely toothed at least at the base · · · 7
7. Stems densely hirsute, at least when young · · · 2. *shirensis* var. *shirensis*
– Stems sparingly puberulous to subglabrous · · · 2. *shirensis* var. *glabriuscula*
8. Sepals of female flowers 6, ± equal · · · 9
– Sepals of female flowers 3 · · · 25
9. Female sepals terminated by a broad laminula · · · 7. *tenuifolia*
– Female sepals terminated by a narrow laminula · · · 10
10. Plant erect · · · 11
– Plant scandent, or twining at least above · · · 16
11. Leaves deeply incised-dentate or laciniate · · · 8. *incisifolia*
– Leaf margins coarsely to finely crenate-serrate or subentire · · · 12
12. Leaves truncate or cordate at the base · · · 13
– Leaves rounded or cuneate at the base · · · 14
13. Leaf margins coarsely sharply serrate; female calyx lobes accrescent to 10 mm long · · · 9. *minor*
– Leaf margins shallowly or remotely serrate to subentire; female calyx lobes accrescent to 5 mm long · · · 10. *angolensis*
14. Leaves serrate to denticulate on the margin, pubescent · · · 11. *lasiophylla*
– Leaves coarsely dentate to subentire on the margin, usually ± glabrous, apart from the stinging bristles · · · 15
15. Leaves not lobed · · · 12. *hildebrandtii*
– Leaves 3-lobed at the base, at least in the lower part of the plant · · · 13. *plukenetii*
16. Leaves lobed · · · 17
– Leaves not lobed · · · 19
17. Petioles of the lower leaves up to 11 cm long; female calyx lobes accrescent to 3 mm long · · · 14. *petiolaris*
– Petioles rarely more than 8 cm long; female calyx lobes accrescent to 6–8 mm long · · 18
18. Leaves shallowly roundly lobed at the base · · · 15. *okanyua*
– Leaves markedly lobed · · · 13. *plukenetii*
19. Leaves coarsely and sharply serrate or biserrate on the margin, tomentose on the lower surface · · · 19. *prionoides*
– Leaves crenate-serrate to subentire on the margin, or if coarsely serrate, then not tomentose on the lower surface · · · 20
20. Whole plant twining · · · 21
– Plant erect at first, stems twining at the tips · · · 24
21. Calyx lobes of female flowers accrescent to 2–3 mm long · · · 22
– Calyx lobes of female flowers accrescent to 5–14 mm long · · · 23
22. Petioles of the lower leaves up to 11 cm long · · · 14. *petiolaris*
– Petioles rarely exceeding 5 cm in length · · · 17. *micromeres*
23. Leaf base narrowly cordate; leaves pubescent and setose · · · 16. *brevipes*
– Leaf base widely cordate; leaves usually setose only · · · 18. *benthamii*
24. Leaves truncate or cordate at the base · · · 10. *angolensis*
– Leaves rounded or cuneate at the base · · · 12. *hildebrandtii*
25. Plant erect · · · 26
– Plant with stems twining at least at the top · · · 27
26. Leaves not lobed · · · 20. *gardneri*
– Leaves 3-lobed at the base · · · 21. *dioica*
27. Stipules falcate, 5–10 mm long · · · 28. *stipularis*
– Stipules lanceolate, 2–7 mm long · · · 28
28. Leaves ovate-lanceolate, oblong-lanceolate or elliptic-lanceolate, never 3-lobed ·27. *furialis*
– Leaves narrowly lanceolate to triangular-ovate, generally 3-lobed at the base · · · 29
29. Bracts of male flowers ovate, glandular · · · 26. *adenanthera*
– Bracts of male flowers lanceolate, eglandular · · · 30
30. Lobules of the female calyx lobes shorter than the width of the rhachis · · · 31
– Lobules of the female calyx lobes as long as or longer than the width of the rhachis · · 32

31. Leaves triangular-ovate to triangular; female calyx lobes 18–22-lobulate · · · · · 22. *glabrata*
– Leaves triangular-lanceolate to linear-lanceolate; female calyx lobes 11–13-lobulate · 25. *kirkiana*
32. Leaves ovate-lanceolate, usually 3-lobed, pubescent and setose on upper surface · 23. *rupestris*
– Leaves triangular-ovate, not 3-lobed, ± glabrous on upper surface · · · · · · 24. *wahlbergiana*

1. **Tragia prostrata** Radcl.-Sm. in Kew Bull. **42**: 398 (1987). Tab. **46**. Type: Zambia, Mporokoso Distr., Lumangwe, male fl. 14.xi.1957, *Fanshawe* 3994 (K, holotype; NDO).

Prostrate, sparingly-branched non-urticating perennial herb, dioecious; stems up to 50 cm long arising from a woody rootstock, pubescent and hispid. Stipules 5 × 3 mm, ovate-lanceolate, occasionally somewhat falcate, entire. Petiole short, 3–9 mm long. Leaf blade (2)3–5 × 1.5–3 cm, ovate, subacute at the apex, shallowly denticulate on the margins, cordate at the base, chartaceous, 7(9)-nerved from the base; lateral nerves in 3–4 pairs, scarcely prominent above, prominent beneath, camptodromous; indumentum sparingly puberulous on both surfaces, and also hirsute along the midrib and main veins beneath. Inflorescences erect, up to 17 cm high with peduncles up to 3 cm long, leaf-opposed or terminal; bracts 5 × 1 mm, linear-lanceolate, several flowered; bracteoles 1–3 mm long, linear. Male flowers: pedicels 1 mm long; buds ovoid-subglobose; calyx lobes 1.5 × 1 mm, ovate, subacute, puberulous without, glabrous within, pale yellow; stamens 1 mm long; pistillode flat, 3-lobed, the lobes truncate at the apex. Female inflorescence and flowers, fruit and seeds unknown.

Zambia. N: Mporokoso Distr., Lumangwe Forest Reserve, male fl. 14.iv.1989, *Radcliffe-Smith, Pope & Goyder* 5673 (K).

Not known from elsewhere. High rainfall miombo and chipya woodland, and in tall grassland with scattered trees; c. 1000 m.

A very distinctive species on account of its prostrate habit, lack of stinging hairs and its large bracts. Its closest relative would appear to be *T. bongolana* Prain from the Sudan.

Richards 12092 from Lumangwe Falls may be conspecific, but differs in having smaller stipules, some leaves obtuse at the apex, the nerves less prominent beneath, shorter male bracts and only 1 male flower per bract.

2. **Tragia shirensis** Prain in Bull. Misc. Inform., Kew **1912**, 5: 239 (1912); in F.T.A. **6**, 1: 991 (1913). —Pax in Engler, Pflanzenr. [IV, fam. 147, ix] **68**: 73 (1919). —Engler, Pflanzenw. Afrikas (Veg. Erde 9) **3**, 2: 105 (1921). Type: Malawi, Shire Highlands, near Blantyre, male fl. & fr. communicated 1887, *Last* s.n. (K, holotype).

An erect, branched, urticating perennial herb to 60 cm tall, dioecious or sometimes monoecious; stems several from a woody rootstock, densely ± stinging-setose and hirsute or sparingly puberulous to subglabrous (var. *glabriuscula*). Stipules 3–6 mm long, linear-lanceolate. Petiole short, 1–10(20) mm long, or leaves ± sessile. Leaf blade (2)5–8.5 × 0.3–2 cm, lanceolate to linear-lanceolate or ± linear, acute or subacute at the apex, coarsely toothed at least towards the base, otherwise subentire, cuneate or rounded to truncate or shallowly cordate and somewhat broadened at the base, membranaceous, 5(7)-nerved from the base, hirsute and with urticating bristles along the midrib and nerves above and more densely so beneath, except in var. *glabriuscula*; lateral nerves in 4–6 pairs, apically directed. Inflorescences terminal, unisexual; males up to 35 cm long, females not more than 15 cm long usually much shorter; male bracts 2–5 mm long, lanceolate, ciliate, 3–more-flowered; bracteoles 1–2 mm long, linear-lanceolate; female bracts resembling the male bracts, but 1-flowered. Male flowers: pedicels 2–4 mm long; calyx lobes 3(5), 2–4 × 1–1.5 mm, elliptic, acute, cucullate, glabrous, green; stamens 3(4), 1.5 mm long; pistillode 3(4)-radiate. Female flowers: pedicels 2 mm long, extending slightly in fruit; calyx lobes 3 or 6, when 6 equal or unequal, 3–4 × 2 or 3–4 mm, elliptic or suborbicular in outline, slightly accrescent, each lobe 7–9 lobulate, the lobules linear or lanceolate and ± equalling the width of the calyx-lobe rhachis, dark green, the calyx-lobe rhachis stramineous, glabrous within and without but setose on the margins; ovary 1.5 × 2–2.5 mm, 3-lobed, densely setose on the keels; styles 3, up to 5 mm long, column 2

Tab. 46. TRAGIA PROSTRATA. 1, portion of prostrate stem (× $^2/_3$); 2, portion of male inflorescence (× 4); 3, male flower (× 8), 1–3 from *Fanshawe* 3994. Drawn by Christine Grey-Wilson.

mm long, glabrous, stigmas ± smooth, tightly coiled. Fruit 5 × 9–10 mm, strongly 3-lobed, smooth, setose on the keels, otherwise ± glabrous. Seeds 4 mm in diameter, globose, whitish, greyish or pinkish, flecked and mottled with chestnut-brown, and with scattered white rings of papillae.

Stems densely hirsute and stinging-setose; leaves hirsute and uniformly setose on under surface . . . . var. *shirensis*
Stems sparingly puberulous to subglabrous, scarcely stinging-setose; leaves hirsute and sparingly setose on under surface . . . . var. *glabriuscula*

Var. **shirensis**

Stems densely hirsute and stinging-setose; leaves hirsute and uniformly setose beneath.

Malawi. S: Shire Highlands, near Blantyre, male fl. & fr. commissioned 1887, *Last* s.n. (K). **Mozambique**. Z: Mocuba–Nicuadala, male & female fl. 6.x.1941, *Torre* 3610 (LISC); Olinga (Maganja da Costa)–Namacurra, fr. 26.i.1966, *Torre & Correia* 14152 (LISC).

Not known from elsewhere. In open *Brachystegia* woodland, and in wooded grassland, on sandy soil; 40–200 m.

The nearest relative is *T. rogersii* Prain from the southeast Transvaal, in which the leaves are more usually cordate and uniformly sharply serrate, and the female calyx lobes larger and densely hispid and setose.

Var. **glabriuscula** Radcl.-Sm. in Kew Bull. **42**: 399 (1987). Type: Mozambique, Nampula Distr., Mutivaze–Namina, female fl. 27.iv.1937, *Torre* 1455 (COI, holotype; LISC).

Stems sparingly puberulous to subglabrous, scarcely stinging-setose; leaves hirsute and sparingly setose beneath.

Mozambique. N: Nampula Distr., Mutivaze–Namina, female fl. 27.iv.1937, *Torre* 1455 (COI; LISC).

Not known from elsewhere. In dry bushland.

This variety occurs on the eastern limit of the range of the species. It is almost *Caperonia*-like in overall appearance. It is only known from the type collection.

3. **Tragia mazoensis** Radcl.-Sm. in Kew Bull. **42**: 396 (1987). Tab. **47**. Type: Zimbabwe, Makonde Distr., west side of Umvukwes Range, c. 24 km south of Kildonan, female fl. 25.ii.1959, *Drummond* 5848 (K, holotype; LISC; PRE; SRGH).

An erect, leafy, urticating perennial herb up to 40 cm tall, dioecious; stems several from a woody rootstock, branched, puberulous. Stipules 1.5–2 mm long, linear-lanceolate. Petioles short, 1–2 mm long, or leaves subsessile. Leaf blade 1–4 × 0.2–1 cm, linear-lanceolate, acute or obtuse at the apex, entire or with only a few apical teeth on the margins, truncate or shallowly cordate at the base, chartaceous, 3–5-nerved from the base; lateral nerves in 4–7 pairs, not prominent above, prominent beneath, brochidodromous; indumentum sparingly puberulous on both surfaces, and with urticating bristles along the midrib and main nerves beneath. Inflorescences terminal (male) or leaf-opposed (female), the male inflorescences up to 9 cm long, the females not exceeding 4 cm; male bracts resembling the stipules, 2–3-flowered; male bracteoles 1 mm long, filiform; female bracts 2 mm long, ovate, 1-flowered; female bracteoles resembling the male bracts. Male flowers: pedicels 1–1.5 mm long; buds globose; calyx lobes 2 × 1 mm, elliptic-ovate, subacute, subglabrous without, glabrous within; stamens 1 mm long; pistillode 3-radiate, the segments apically dilated. Female flowers; pedicels less than 1 mm long, extending slightly in fruit; calyx lobes 6, 2 mm long and wide, accrescent to 6 × 3 mm, each lobe 7–11 lobulate, the lobules linear and ± equalling the width of the calyx lobe rhachis, setose, green, the calyx-lobe rhachis loriform, subglabrous without, glabrous within; ovary 1 mm in diameter, 3-lobed, densely setose; styles 3, 1–2 mm long, connate to halfway, subglabrous or sparingly pubescent, stigmas smooth, recurved. Fruit 6 × 8 mm, 3-lobed, smooth, densely stinging-setose. Mature seeds as yet unknown.

Tab. 47. TRAGIA MAZOENSIS. 1, habit (× ²⁄₃); 2, portion of male inflorescence (× 4); 3, male flower (× 8), 1–3 from *Wild* 7973; 4, female inflorescence (× 4), from *Drummond* 5848; 5, fruit (× 4), from *Wild* 7788. Drawn by Christine Grey-Wilson.

Zimbabwe. N: Mazowe Distr., Vanad Pass, Great Dyke, male fl. 13.xii.1974, *Wild* 7973 (K; PRE; SRGH); Makonde Distr., Mutorashanga, fr. 18.i.1970, *Wild* 7788 (SRGH).

Endemic to this serpentine dyke. On open treeless grassy hillsides of serpentine dyke, often on termite mounds.

Closely related to *T. lukafuensis*, from which it differs chiefly in the ± entire leaf margins.

4. **Tragia lukafuensis** De Wild. in Ann. Mus. Congo, Ser. IV, i: 206 (1903). —Prain in F.T.A. **6**, 1: 987 (1913). —Pax in Engler, Pflanzenr. [IV, fam. 147, ix] **68**: 73 (1919). —Engler, Pflanzenw. Afrikas (Veg. Erde 9) **3**, 2: 105 (1921). Type from Zaire (Shaba (Katanga)).

An erect sparingly-branched urticating perennial herb up to 45 cm tall, dioecious or monoecious; stems few from a small woody rootstock, puberulous and/or hirsute. Stipules 3–5 mm long, linear-lanceolate, pubescent without, glabrous within. Petioles short, 2–12 mm long. Leaf blade 2–7 × 1–4 cm, lanceolate to triangular-lanceolate, acute at the apex, finely and closely serrate on the margins, truncate to shallowly or more deeply cordate at the base, thinly chartaceous, 5–7-nerved from the base, evenly to sparingly pubescent and with urticating bristles along the midrib and main nerves on both surfaces; lateral nerves in 4–5 pairs, slightly prominent. Inflorescences terminal, axillary or leaf-opposed, unisexual, the male inflorescences up to 12 cm long, the females not exceeding 4 cm; male bracts 1–2 mm long, linear, 2–3-flowered; male bracteoles minute, subulate; female bracts 3 mm long, lanceolate; female bracteoles resembling the male bracts. Male flowers: pedicels 1 mm long; buds obovoid-ellipsoid; calyx lobes 2 × 1 mm, oblong, obtuse, pubescent without, glabrous within, greenish-yellow; stamens 1.5 mm long; pistillode 3-lobed. Female flowers: pedicels 1 mm long, extending to c. 5 mm in fruit; calyx lobes 6, 2 × 1.5 mm, accrescent to 5 × 3 mm, each lobe 9–13 lobulate, the lobules linear and ± equalling the width of the calyx-lobe rhachis, setose, greenish-purple, the calyx-lobe rhachis loriform, pubescent without, subglabrous within, becoming hardened, stramineous; ovary 2 mm in diameter, 3-lobed, pubescent, setose on the keels, yellow; styles 3, 2 mm long, connate to halfway, sparingly puberulous, stigmas ± smooth, recurved. Fruit c. 3.5 × 5 mm, 3-lobed, smooth, sparingly pubescent, and stinging-setose on the keels. Seeds 3 mm in diameter, globose, light brown.

Zambia. N: Mporokoso Distr., 21 km northeast of Chiengi, female fl. 13.x.1949, *Bullock* 1267 (K); 8 km south of Chiengi, male fl. fr. 5.xi.1952, *Angus* 718 (BM; FHO; K). W: Ndola, female fl. 24.ix.1953, *Fanshawe* 306 (K; NDO; SRGH).

Also in Zaire. In mixed deciduous woodland on stony hill slopes, and in waste places.

The Mporokoso material accords better with the type than does the Ndola material as regards indumentum (puberulous rather than hirsute), leaf margin (serrate rather than crenate-serrate) and leaf base (shallowly rather than deeply cordate), but these differences may not be significant.

5. **Tragia descampsii** De Wild. in Ann. Mus. Congo, Ser. IV, i: 207 (1903). —Prain in F.T.A. **6**, 1: 979 (1913). —Pax in Engler, Pflanzenr. [IV, fam. 147, ix] **68**: 73 (1919). —Engler, Pflanzenw. Afrikas (Veg. Erde 9) **3**, 2: 105 (1921). —Radcliffe-Smith in F.T.E.A., Euphorb. 1: 297 (1987). Type from Zaire (Shaba).

Similar to *T. lukafuensis*, but differing chiefly in having the leaf blades ovate and generally subacute or obtuse at the apex; the male calyx lobes ovate to ovate-lanceolate; the female calyx lobes accrescent to 2 cm long, with 21–25 lobules, and much more densely setose; and fruits somewhat larger (4.5 × 8 mm).

Zambia. S: Mumbwa Distr., Kafue National Park, Chunga, male fl. 10.xi.1961, *B.L. Mitchell* 10/84 (SRGH).

Also in Zaire and Tanzania. Amongst grasses of open deciduous woodland, with *Julbernardia*, *Ochna* and *Ricinodendron species*.

T. descampsii, like *T. lukafuensis*, appears to be rather variable in indumentum, particularly as regards the male flower buds, which in the Tanzanian material are densely to evenly hirsute, in the Zairean sparingly pubescent and in the Zambian minutely puberulous.

6. **Tragia rhodesiae** Pax in Bot. Jahrb. Syst. **39**: 665 (1907). —Prain in F.T.A. **6**, 1: 979 (1913). —Eyles in Trans. Roy. Soc. South Africa **5**: 397 (1916). —Pax in Engler, Pflanzenr. [IV, fam.

147, ix] **68**: 73 (1919). —Engler, Pflanzenw. Afrikas (Veg. Erde 9) **3**, 2: 105 (1921). Type: Zimbabwe, Mashonaland, near Harare (Salisbury), fl. 16.ix.1905, *Engler* 3073 (B†, holotype; K, drawing of holotype).

An erect urticating perennial herb up to 75 cm tall, monoecious; stems few, branched, arising from a woody rootstock, indumentum a mixture of short soft hairs and longer stiffer hairs and sometimes with some stinging bristles. Stipules 3–5 mm long, lanceolate, acute, pubescent without, glabrous within. Petioles short, 2–8(20) mm long. Leaf blade 2–10 × 1–5 cm, ovate-lanceolate to lanceolate, acute or acutely to obtusely acuminate at the apex, serrate to crenate-serrate on the margins, usually narrowly and fairly deeply cordate at the base, membranous, 5–7-nerved from the base, evenly to sparingly pubescent, at least with urticating bristles along the midrib and main nerves beneath, sometimes also above; lateral nerves in 3–4 pairs, somewhat prominent above, prominent beneath. Inflorescences up to 13 cm long, with the peduncle up to 8 cm long, terminal, axillary or leaf-opposed, bisexual; male bracts 3–4 mm long, linear-lanceolate, 2–3-flowered; male bracteoles 1–2 mm long, linear; female bracts 5 × 1 mm, lanceolate; female bracteoles resembling the male bracts. Male flowers: pedicels 1–1.5 mm long; buds obovoid-subglobose, slightly apiculate; calyx lobes 2 × 1 mm, ovate, cucullate, subglabrous; stamens c. 1 mm long; pistillode 3-lobed, the lobes truncate. Female flowers: pedicels c. 1 mm long, jointed above the bracteoles, extending to 5–6 mm long in fruit and becoming geniculate; calyx lobes 6, 2 × 1 mm, accrescent to 8 × 4 mm, each lobe 9–11(13) lobulate, the lobules lanceolate and slightly shorter than the width of the calyx lobe rhachis, setose, green, white at the base, the calyx-lobe rhachis loriform, sparingly pubescent without, glabrous within, becoming hardened in the lower half, green without, stramineous within; ovary 1.5 mm in diameter, 3-lobed, subglabrous except for the setose keels; styles 3, 3 mm long, connate to halfway, subglabrous, stigmas ± smooth, recurved. Fruit 6 × 9 mm, 3-lobed, the keels shouldered, smooth, subglabrous, sparingly setose on the keels. Seeds 4 mm in diameter, globose, greyish- or pinkish-brown, mottled or flecked with dark brown or chestnut-brown, and with small circular grey patches ringed with white papillae.

Zambia. W: Ndola, fr. 29.ix.1954, *Fanshawe* 1582 (K; NDO). C: Chilanga, fl. 7.x.1963, *Angus* 3784 (FHO; K). **Zimbabwe**. N: Mazowe (Mazoe) Valley, fr.i.1957, *McLaren* s.n. in *GHS* 72553 (K; SRGH). C: Harare, fr. 23.xi.1968, *Biegel* 2690 (K; PRE; SRGH). **Malawi**. C: Dedza, near Bembeke Mission, fr. 15.xi.1967, *Salubeni* 898 (K; LISC; SRGH).

Known only from these localities. In high rainfall and chipya woodlands and in plateau mixed deciduous woodlands, and wooded grassland; 1160–1500 m.

Very variable as to the colour of the leaves on drying (grey- or yellow-green, grey- or yellow-brown or dark brown), possibly depending on edaphic factors.

7. **Tragia tenuifolia** Benth. in Niger Fl.: 502 (1849). —Müller Argoviensis in De Candolle, Prodr. **15**, 2: 945 (1866). —Prain in F.T.A. **6**, 1: 973 (1913). —Pax in Engler, Pflanzenr. [IV, fam. 147, ix] **68**: 96 (1919). —Engler, Pflanzenw. Afrikas (Veg. Erde 9) **3**, 2: 107 (1921). —De Wildeman, Pl. Bequaert. **3**, 4: 499 (1926). —Robyns & Tournay, Fl. Sperm. Parc Nat. Alb. **1**: 470 (1948). —Keay in F.W.T.A., ed.2, **1**, 2: 412 (1958). —Radcliffe-Smith in F.T.E.A., Euphorb. 1: 295 (1987). Type from São Tomé (St. Thomas' Island).

Tragia manniana Müll. Arg. in Flora **47**: 436 (1864); in De Candolle, Prodr. **15**, 2: 941 (1866). Types from Sierra Leone and Nigeria.

Tragia klingii Pax in Bot. Jahrb. Syst. **19**: 105 (1894). Types from Sierra Leone and Togo.

Tragia zenkeri Pax in Bot. Jahrb. Syst. **23**: 528 (1897). —Prain in F.T.A. **6**, 1: 973 (1913). Types from Cameroon.

Tragia calvescens Pax in Bot. Jahrb. Syst. **43**: 324 (1909). Types from Zaire.

A weakly urticating climbing subshrub with sparingly puberulous stems, monoecious. Stipules 3–4 × 1–1.5 mm, ovate-lanceolate, acute, subglabrous, ciliate. Petioles 1–7.5 cm long, sparingly puberulous and stinging-setose. Leaf blade 4–14 × 1–5 cm, elliptic-oblong to elliptic-oblanceolate, shortly acutely acuminate at the apex, subentire to shallowly and remotely denticulate on the margins, subtruncate to cordate at the base, membranaceous, 5–9-nerved from the base, sparingly pubescent and setose on the upper surface, and sparingly hirsute and setose along the midrib and main nerves beneath; lateral nerves in 4–6 pairs, slightly prominent. Inflorescences 2–3 cm long, with the peduncle 1.5–2.5 cm long, leaf-opposed, bisexual, densely flowered; male bracts 2 mm long, oblong, subglabrous; male bracteoles 1 mm long, subulate; female bracts 3 mm long, ovate; female bracteoles resembling the male bracts. Male flowers: pedicels 0.5–1 mm long; buds obovoid-turbinate; calyx lobes 1 × 1 mm, broadly ovate-suborbicular, apiculate, subglabrous; stamens 0.5 mm long; pistillode trigonous, depressed-conic. Female flowers: pedicels 0.5–1 mm long; calyx lobes 6, 2 mm long, accrescent to 5 mm long in fruit, spathulate, with 3 small linear subglabrous green lobules on each side, a terminal ovate foliaceous laminula, and a strap-shaped rhachis pubescent and greenish without, glabrous and stramineous within, becoming slightly hardened; ovary 2 mm in diameter, 3-lobed, uniformly densely pubescent, setose on the keels; styles 3, 2 mm long, connate to halfway, ± glabrous, stigmas shallowly rugulose, tightly coiled. Fruit 3–4 × 7–8 mm, 3-lobed, evenly pubescent, sparingly setose on the keels. Seeds c. 3 mm in diameter, globose, silvery-grey, streaked with beige and mottled chestnut-brown.

Zimbabwe. E: Chipinge Distr., Chiredza Gorge, fl. & fr. iii.1962, *Goldsmith* 70/62 (K; LISC; SRGH); Mt. Selinda, Chirinda Forest, fl. & fr. 21.iv.1976, *Müller* 2450 (SRGH).

From Guinée eastwards to Sudan and south to Cabinda and Zaire. In mixed evergreen rainforest understorey.

The Zimbabwe plants differ from those of other populations in having slightly smaller female flowers, fruits and seeds, and there are also differences in indumentum (e.g. female calyx not setose). It may represent a distinct subspecies.

8. **Tragia incisifolia** Prain in Bull. Misc. Inform., Kew **1912**, 5: 237 (1912). —Pax in Engler, Pflanzenr. [IV, fam. 147, ix] **68**: 87 (1919). —Prain in F.C. **5**, 2: 504 (1920). —Engler, Pflanzenw. Afrikas (Veg. Erde 9) **3**, 2: 106 (1921). —Burtt Davy, Fl. Pl. Ferns Transvaal: 307 (1932). Types from South Africa (Transvaal).

An erect sparingly branched urticating perennial herb to c. 30(40) cm tall, monoecious; stems few from a slender woody rootstock, pubescent. Stipules 2–3 mm long, linear-lanceolate, ciliate, otherwise glabrous, becoming reflexed. Petiole up to 12 mm long, or leaves subsessile. Leaf blade 1–5 × 0.5–2 cm, ovate to lanceolate in outline, acute at the apex, deeply incised-dentate or laciniate, cuneate to rounded or truncate at the base, thinly chartaceous, 3–5(7)-nerved from the base, sparingly setose along the midrib and main nerves on both surfaces, otherwise usually subglabrous; lateral nerves in 2–4 pairs, slightly prominent. Inflorescences up to 9 cm long, the peduncle up to 5 cm long, terminal or leaf-opposed, bisexual with 1–4 female flowers at the base; male bracts and female bracteoles resembling the stipules; male bracteoles 1 mm long, linear; female bracts 3 mm long, lanceolate. Male flowers: pedicels 1 mm long; buds obovoid; calyx lobes 1.5 × 0.5–1 mm, lanceolate to broadly ovate, glabrous, yellowish-green; stamens c. 1 mm long; pistillode 3-lobed. Female flowers: pedicels c. 1.5 mm long, scarcely extending in fruit, very stout, densely pubescent; calyx lobes 6, biseriate, 3 × 2 mm, accrescent to 4–5 × 3–4 mm, each lobe (7)9–13 lobulate, the lobules linear and equalling or exceeding the width of the calyx-lobe rhachis, setose, green, the calyx-lobe rhachis ± triangular-lanceolate, ± glabrous, becoming hardened, stramineous; ovary 1.5–2 mm in diameter, 3-lobed, densely setose on the keels, otherwise ± glabrous; styles 3, c. 2 mm long, connate at the base, glabrous, stigmas shallowly papillose, recurved at the tips. Fruit 5 × 8 mm, 3-lobed, the keels slightly shouldered, smooth, setose on the keels, otherwise ± glabrous. Seeds 3.5–4 mm in diameter, subglobose, pale grey, flecked with light brown and with scattered circular patches of white papillae.

Zimbabwe. E: east bank of lower Save (Sabi), fr. 28.i.1948, *Wild* 2343 (K; SRGH). **Mozambique**. M: Boane–Impamputo, fl. & fr. 8.xi.1961, *Lemos & Balsinhas* in *Lemos* 212 (COI; K; LISC; LUAI; PRE; SRGH).

Also in South Africa (Transvaal, KwaZulu-Natal). In low altitude grassland and open *Acacia* woodland, often on river banks and amongst basalt rocks; 350–560 m.

In *Davies* 2360 (K), from Gwanda Distr., Zimbabwe, the inner female calycine whorl is very much reduced or absent, whilst the leaves are less deeply incised than in other material.

9. **Tragia minor** Sond. in Linnaea **23**: 108 (1850). —Pax in Engler, Pflanzenr. [IV, fam. 147, ix] **68**: 87 (1919). —Prain in F.C. **5**, 2: 504 (1920). —Engler, Pflanzenw. Afrikas (Veg. Erde 9) **3**, 2: 106 (1921). —Burtt Davy, Fl. Pl. Ferns Transvaal: 307 (1932). Type from South Africa (Transvaal).

Tragia rupestris var. *minor* (Sond.) Müll. Arg. in De Candolle, Prodr. **15**, 2: 940 (1866) pro parte quoad typ.

An erect urticating perennial herb up to 50 cm tall, monoecious; stems 1–2, arising from a woody rootstock, subsimple or basally sparingly branched, hispid. Stipules 3–4 mm long, lanceolate, ciliate, otherwise glabrous, spreading. Petiole 1–6 mm long, densely hispid. Leaf blade 2.5–6.5 × 1–3 cm, lanceolate, acute or subacute at the apex, coarsely sharply serrate or crenate-serrate on the margins, shallowly cordate at the base, thinly chartaceous, 5–7-nerved from the base, sparingly hispid and setose along the midrib and main nerves on both surfaces, otherwise ± glabrous; lateral nerves in 4–5 pairs, fairly prominent beneath. Inflorescences up to 7.5 cm long, the peduncle up to 4.5 cm long, terminal or leaf-opposed, bisexual, with 1–3 female flowers at the base; male bracts and female bracteoles 2–4 mm long, elliptic-lanceolate; male bracteoles 2 mm long, linear-oblanceolate; female bracts 5 × 2 mm, lanceolate. Male flowers: pedicels 1.5 mm long; buds obovoid; calyx lobes 2 × 1.5 mm, elliptic-ovate, subglabrous; stamens 1 mm long; pistillode 3-lobed. Female flowers: pedicels c. 1 mm long, extending slightly in fruit; calyx lobes 6, 2 × 1 mm, accrescent to 8–10 × 4–5 mm, each lobe 11–13-lobulate, the lobules ascending, linear-lanceolate and equalling or exceeding the width of the rhachis, densely setose, green, the rhachis lanceolate, hispid and setose without, glabrous within, becoming hardened, stramineous; ovary 2 mm in diameter, 3-lobed, hispid and setose on the keels; styles 3, 3 mm long, erect, connate at the base, glabrous, stigmas smooth, ± straight. Fruit 5 × 8 mm, 3-lobed, smooth, hispid and setose on the keels, otherwise glabrous. Seeds 3.5 mm in diameter, subglobose, grey, occasionally flecked with brown, and with irregularly aggregated circular densely papillose whitish patches.

Mozambique. M: Namaacha, fr. 9.i.1948, *Torre* 7078 (LISC); Namaacha, o. fl. 13.i.1948, *Torre* 7136 (LISC); Namaacha, fr. 7.iii.1967, *Moura* 218 (COI).

Also in South Africa (Transvaal, KwaZulu-Natal) and Swaziland. On stony (rhyolitic) soils in grassland and grassed bushland.

10. **Tragia angolensis** Müll. Arg. in J. Bot. **2**: 333 (1864); in De Candolle, Prodr. **15**, 2: 940 (1866). —Prain in F.T.A. **6**, 1: 990 (1913). —R.E. Fries, Wiss. Ergebn. Schwed. Rhod.-Kongo-Exped. **1**, 1: 125 (1914). —Pax in Engler, Pflanzenr. [IV, fam. 147, ix] **68**: 82 (1919). —Engler, Pflanzenw. Afrikas (Veg. Erde 9) **3**, 2: 106 (1921). Types from Angola (Cuanza Norte/Malanje, Huíla).

An erect, not or slightly urticating perennial herb up to 50 cm tall, monoecious; stems several from a branching woody rootstock, simple or sparingly branched, erect at first (in new growth following a burn) becoming drooping, trailing or twining at the top, puberulous and/or pubescent. Stipules 3–4 mm long, linear-lanceolate to lanceolate, ciliate, otherwise minutely papillose-puberulous, erect or spreading. Petiole 1–5(10) mm long, or leaves subsessile. Leaf blade 2–8.5 × 0.2–2 cm, linear to lanceolate or narrowly triangular-lanceolate, acute or subacute at the apex, shallowly or remotely serrate to subentire on the margins, truncate or shallowly cordate at the base, thinly chartaceous, 5-nerved from the base, subglabrous or sparingly pubescent along the midrib and main nerves on both surfaces and otherwise ± glabrous, dark green on upper surface, somewhat glaucous beneath, sometimes reddish-tinged; lateral nerves in 4–5(7) pairs, not prominent above, scarcely to slightly so beneath. Inflorescences up to 5 cm long, with a peduncle up to 2.5 cm long, terminal or leaf-opposed, bisexual with 1 female flower at the base, or some inflorescences entirely male; male bracts and female bracteoles 1–2 mm long, elliptic-lanceolate; male bracteoles 0.5 mm long, linear-subulate; female bracts 2–2.5 mm long, ovate, sometimes 3-dentate. Male flowers: pedicels 1–1.5 mm long; buds 3-lobed to subglobose; calyx lobes 1.5 × 1 mm, broadly ovate, apiculate, glabrous, creamy-green or reddish tinged; stamens less than 1 mm long, yellow; pistillode triangular, the angles rounded. Female flowers: pedicels 1 mm long, extending to 2 mm and becoming somewhat geniculate in fruit; calyx lobes 6, 1.5–2.5 × 0.5–1 mm, accrescent to 5 × 2 mm, the 3 inner calyx lobes becoming hardened and stramineous, the 3 outer herbaceous and green, each lobe 9–13(15)-lobulate, the lobules oblong-linear, ± perpendicular to, and shorter than the width of the calyx-lobe rhachis, shortly setose, green, the calyx-lobe rhachis lanceolate to oblong-loriform, minutely papillose-puberulous without, glabrous within; ovary 1.5 mm in diameter, 3-lobed, puberulous, shortly setose on the keels; styles 3, 3 mm long, erect, connate to over halfway, minutely puberulous, stigmas shallowly verruculose, straight, glabrous. Fruit 4.5–5 × 7.5–9 mm, 3-lobed, smooth, uniformly puberulous, and shortly setose especially on the keels, grey-green. Seeds 3.5 mm in diameter, globose, brownish-grey with scattered greyish-white circular blotches.

Zambia. W: Kitwe, fl. & fr. 7.xii.1955, *Fanshawe* 2644 (K; NDO; SRGH). C: 10 km east of Lusaka, fl. & fr. 1.xi.1955, *King* 198 (K). S: Muckle Neuk, 19 km north of Choma, fl. 28.xi.1954, *Robinson* 1001 in *GHS* 49420 (K; SRGH).

Also in Angola. In *Brachystegia* woodland on Kalahari Sand, in plateau miombo and dry evergreen thicket, also in dambos, grassland beside streams and on laterite outcrops; 1200–1280 m.

In habit this species links other *Tragia* species with *Tragiella* via *T. friesiana* (Prain) Pax & K. Hoffm.

11. **Tragia lasiophylla** Pax & K. Hoffm. in Engler, Pflanzenr. [IV, fam. 147, ix] **68**: 86 (1919). — Engler, Pflanzenw. Afrikas (Veg. Erde 9) **3**, 2: 106 (1921). —Radcliffe-Smith in F.T.E.A., Euphorb. 1: 298 (1987). Type from Tanzania (Western Province).

Erect sparingly branched urticating perennial herb 40–50(80) cm tall, monoecious; stems 1 or more from a woody rootstock, pubescent. Stipules 4 mm long, lanceolate, ciliate, otherwise subglabrous, erect or spreading. Petiole 2–5(10) mm long, densely pubescent. Leaf blade 4–12 × 1–2.5 cm, narrowly triangular-lanceolate to elliptic-lanceolate, acute or subacute at the apex, serrate to denticulate on the margins, rounded or cuneate at the base, thinly chartaceous, 3(5)-nerved from the base, with the inner pair of nerves often running more than halfway up the blade, sparingly pubescent and also hirsute and setose along the midrib and main nerves on both surfaces; tertiary nerves reticulate; lateral nerves in 3–7 pairs, scarcely prominent above, fairly prominent beneath. Inflorescences up to 11 cm long, with a peduncle up to 6 cm long, either terminal on the main stem and/or on the axillary leafy shoots or else leaf-opposed, bisexual, mostly male with 1–3 female flowers at the base; male bracts 2 mm long, lanceolate, ciliate; male bracteoles 1–1.5 mm long, linear; female bracts c. 3 mm long, lanceolate to ovate-lanceolate, sometimes 3-dentate to 3-fid; female bracteoles resembling the male bracts, but slightly larger. Male flowers: pedicels 1.5–2 mm long; buds ellipsoid-subglobose; calyx lobes 1.5 × 1 mm, elliptic-ovate, acute, cucullate, glabrous, yellow-green; stamens c. 1 mm long; pistillode 3-lobed, the lobes truncate. Female flowers: pedicels less than 1 mm long, extending to 2–3 mm long in fruit; calyx lobes 6, unequal, 1.5 × 1–1.5 mm, with 2 being broader than the other 4, accrescent to 4.5–6 × 2–3 mm, each lobe (7)9–13-lobulate, the lobules linear to linear-lanceolate, longer than or equalling or shorter than the width of the calyx-lobe rhachis, sparingly pubescent and/or shortly setose, green, the calyx-lobe rhachis loriform or elliptic-lanceolate, sparingly puberulous to subglabrous without, glabrous within, becoming hardened and stramineous proximally, remaining ± herbaceous distally; ovary 1–1.5 mm in diameter, 3-lobed, densely setose; styles 3, 2–3 mm long, extending to 4–5 mm long, ± erect, slightly divergent, connate for one-third of their length, later for almost three-quarters of their length by the extension of the column, subglabrous; stigmas minutely verruculose, recurved at the tips. Fruit 5 × 7–8 mm, 3-lobed, smooth, ± evenly setose. Seeds 4 mm in diameter, globose, greyish-buff, closely flecked and blotched with dark purplish-brown, and with scattered circular patches fringed with whitish papillae.

Zambia. B: Kaoma (Mankoya), fl. 11.x.1963, *Fanshawe* 8047 (K; NDO). N: Mbala, Lake Chila, fl. & y. fr. 19.xi.1958, *Napper* 953 (EA; K). **Malawi**. N: Mzimba Distr., South Rukuru River, fl. & fr. 20.i.1978, *Pawek* 13652 (K; MO). S: Blantyre, y. fl. 1895, *Buchanan* s.n. in *Herb. J.M. Wood* 7099 (B†; K; PRE).

Also in Tanzania. In Kalahari Sand woodland and in grassland in open woodland, usually in sandy soil; (800)1170–1250 m.

In habit this species also links other *Tragia* species with *Tragiella* via *T. friesiana* (Prain) Pax & K. Hoffm.

12. **Tragia hildebrandtii** Müll. Arg. in Abh. Naturwiss. Vereine Bremen, **7**: 26 (1880). —Pax in Engler, Pflanzenw. Ost-Afrikas **C**: 239 (1895). —Prain in F.T.A. **6**, 1: 977 (1913). — Radcliffe-Smith in F.T.E.A., Euphorb. 1: 297 (1987). Type from Kenya (Coast Province).

Tragia cannabina var. *hildebrandtii* (Müll. Arg.) Pax & K. Hoffm. in Engler, Pflanzenr. [IV, fam. 147, ix] **68**: 86 (1919).

Similar to *T. lasiophylla*, but stems sometimes twining; leaves coarsely and remotely dentate to subentire and glabrous except for the midrib, nerves and margins; the female calyx lobes equal or subequal, with the lateral lobules longer than the width of the narrowly strap-shaped calyx-lobe rhachis and more densely setose.

Malawi. S: Gombwa, above Elephant Marsh, R. Shire, fr. ii.1888, *Scott* s.n. (K).
Also in Ethiopia, Kenya and Tanzania. Floodplain grassland on alluvial soils.
Scott s.n. (K) is cited by Prain as *T. cannabina* var. *intermedia* Prain, but in that taxon at least the lower leaves are deeply 3-partite.

13. **Tragia plukenetii** Radcl.-Sm. in Kew Bull. **37**: 688 (1983); in F.T.E.A., Euphorb. 1: 296 (1987). Type from India.

Croton hastatus L., Sp. Pl.: 1005 (1753)non *Tragia hastata* (Klotzsch) Müll. Arg. in Fl. Bras. **11**, 2: 407 (1874). Type from India.

Croton urens L., Sp. Pl.: 1005 (1753) non *Tragia urens* L., Sp. Pl. ed. 2: 1391 (1763). Type from India.

Tragia cannabina L.f., Suppl. Pl.: 415 (1781). —Prain in F.T.A. **6**, 1: 976 (1913). —Pax in Engler, Pflanzenr. [IV, fam. 147, ix] **68**: 84 (1919). —Engler, Pflanzenw. Afrikas (Veg. Erde 9) **3**, 2: 106 (1921). —Chiov., Fl. Somala **2**: 400 (1932). —F.W. Andrews, Fl. Pl. Anglo-Egypt. Sudan **2**: 98 (1952), nom. illegit. superfl. Type from India.

Tragia involucrata var. *intermedia* Müll. Arg. in De Candolle, Prodr. **15**, 2: 944 (1866). Type from Sri Lanka.

Tragia involucrata var. *cannabina* (L.f.) Müll. Arg. in De Candolle, Prodr. **15**, 2: 944 (1866).

Tragia tripartita Beille in Mém. Soc. Bot. Fr. **55**, Sér. 8: 83 (1908), non Schweinf. (1868). Types from Chad.

Tragia cannabina var. *intermedia* (Müll. Arg.) Prain in F.T.A. **6**, 1: 976 (1913). —Pax in Engler, Pflanzenr. [IV, fam. 147, ix] **68**: 86 (1919). —F.W. Andrews, Fl. Pl. Anglo-Egypt. Sudan **2**: 98 (1952).

Tragia cannabina var. *hastata* (L.) Pax & K. Hoffm. in Engler, Pflanzenr. [IV, fam. 147, ix] **68**: 85 (1919).

Very like *T. hildebrandtii*, but differing chiefly in having all the leaves deeply 3-lobed or 3-partite, or occasionally at least some of the leaves with 1–2 lateral lobes.

Zimbabwe. E: Chipinge Distr., near Hippo Mine, east bank of lower Save (Sabi) R., fr. 12.iii.1957, *Phipps* 581 (SRGH).
From Nigeria eastwards to Somalia, India and Sri Lanka. Floodplain grassland, on black clay soils, often a weed of irrigated fields; 365 m.

14. **Tragia petiolaris** Radcl.-Sm. in Kew Bull. **37**: 687 (1983); in F.T.E.A., Euphorb. 1: 300 (1987). Type from Tanzania (Central Province).

A climbing urticating perennial herb, monoecious; stems pubescent and puberulous. Stipules 2 mm long, linear-lanceolate, acute. Petiole 0.5–11 cm long, slender. Leaf blade 2–10 × 0.7–5 cm, ovate to ovate-lanceolate, sometimes indistinctly lobed, acutely acuminate at the apex, irregularly dentate to shallowly sinuate on the margins or subentire, cuneate to rounded or truncate to shallowly wide-cordate at the base, membranaceous, 3–5-nerved from the base, sparingly hirsute-pubescent on the upper surface, evenly pubescent and shortly setose especially along the midrib and main nerves beneath; lateral nerves in 3–6 pairs, not or scarcely prominent above or beneath. Inflorescences 1–3 cm long, with the peduncle up to 1.5 cm long, leaf-opposed on short lateral shoots, delicate, bisexual, with a single female flower at the base; male bracts 1.5 mm long, linear, ciliate; male bracteoles 0.5 mm long, subulate; female bracts lanceolate, and bracteoles linear, otherwise ± as in the male. Male flowers: pedicels 1 mm long, jointed; buds subglobose, apically depressed; calyx lobes 0.8 × 0.5 mm, elliptic-ovate, cucullate, subglabrous, green; stamens less than 0.5 mm long; pistillode triangular. Female flowers: pedicels 0.5 mm long, extending to 1.5 mm in fruit; calyx lobes 6, 1 × 0.5 mm, accrescent to 2–3 × 1 mm, each lobe (5)7–9-lobulate, the lobules oblong-linear, shorter than the width of the calyx-lobe rhachis, sparingly shortly setose, green; the calyx-lobe rhachis linear-lanceolate, sparingly pubescent without, glabrous within, becoming hardened and whitish within; ovary 1 mm in diameter, 3-lobed, densely setose; styles 3, 1 mm long, connate for two-thirds of their length, glabrous, stigmas ± smooth, recurved. Fruit 3.5 × 5.5 mm, 3-lobed, smooth, evenly minutely puberulous, sparingly setose. Seeds 2.5 mm in diameter, subglobose, greyish-buff, mottled reddish-brown, with scattered whitish papillose patches.

Zambia. B: Masese, o. fl. & fr. 10.iii.1960, *Fanshawe* 5566 (K; NDO). N: Mbala Distr., Chimka Bay, L. Tanganyika, fl. 28.xii.1963, *Richards* 18704 (K). W: Chingola, fr. 1.iii.1956, *Fanshawe* 2812 (K; NDO).

Also in Tanzania. In shade of dense thicket on Kalahari Sand (mutemwa) and dry evergreen thicket (mateshi) in plateau woodland; 780 m.

The closest relative of this species is not African, but is *T. montana* (Thwaites) Müll. Arg. from Sri Lanka.

15. **Tragia okanyua** Pax in Bull. Herb. Boissier **6**: 735 (1898). —Prain in F.T.A. **6**, 1: 986 (1913). —Eyles in Trans. Roy. Soc. South Africa **5**: 397 (1916). —Pax in Engler, Pflanzenr. [IV, fam. 147, ix] **68**: 78 (1919). —Prain in F.C. **5**, 2: 505 (1920). —Engler, Pflanzenw. Afrikas (Veg. Erde 9) **3**, 2: 105 (1921). —Burtt Davy, Fl. Pl. Ferns Transvaal **2**: 308 (1932). —Hutchinson, Botanist in Southern Africa: 667 (1946). —Brenan in Mem. N.Y. Bot. Gard. **9**, 1: 75 (1954). —P.G. Meyer in Merxmüller, Prodr. Fl. SW. Afrika, fam. 67: 46 (1967). —Mogg in Macnae & Kalk, Nat. Hist. Inhaca Isl., Moçamb., rev. ed.: 148 (1969) sphalm. "*okangua*". —Radcliffe-Smith in F.T.E.A., Euphorb. 1: 304 (1987). Type from Angola (Huíla Province).

Tragia angustifolia sensu Engl., Pflanzenw. Ost-Afrikas **C**: 239 (1895). —Schinz & Junod in Mém. Herb. Boissier, No. 10: 47 (1900), non Nutt. (1837), nec Benth. (1849).

Tragia cordifolia sensu N.E. Br. in Bull. Misc. Inform., Kew **1909**: 141 (1909), non Vahl (1790), nec Benth. (1849).

Tragia madandensis S. Moore in J. Linn. Soc., Bot. **40**: 203 (1911). Type: Mozambique, Gazaland, Madanda Forests, 120 m, 5.xii.1906, *Swynnerton* 794 (BM, holotype).

A climbing urticating perennial herb, monoecious; stems up to 3 m long from a woody rootstock, twining (sometimes erect at first), sparingly to densely pubescent, hirsute and/or setose. Stipules (3)4–6 mm long, lanceolate, ciliate, otherwise ± glabrous, erect at first, soon spreading then becoming reflexed. Petiole 0.3–8(10) cm long, pubescent, hirsute and/or setose. Leaf blade 1.5–11.5(14) × 0.5–6(10) cm, ovate-lanceolate to oblong-lanceolate, usually ± distinctly 3-lobed, shortly acutely or subacutely acuminate at the apex, serrate on the margins and often more coarsely so on the lateral lobes, widely cordate at the base, thinly chartaceous, 5(7)-nerved from the base, sparingly hirsute to subglabrous on the upper surface, evenly to sparingly pubescent beneath, and also setose along the midrib and main nerves on both surfaces; lateral nerves in 3–5 pairs, scarcely prominent above, somewhat prominent beneath. Inflorescences up to 14 cm long, the peduncle up to 10 cm long, terminal or leaf-opposed, bisexual, with 1–2 female flowers at the base; male bracts 1.5–2 mm long, lanceolate or elliptic, ciliate, the lowermost sometimes 2-flowered; male bracteoles 1–1.5 mm long, linear-lanceolate; female bracts 3 mm long, ovate, dentate at the apex; female bracteoles resembling the male bracts. Male flowers: pedicels c. 1 mm long, puberulous; buds obovoid-subglobose; calyx lobes 1 × 1 mm, ovate, apiculate, sparingly pubescent without, glabrous within, greenish; stamens 3(4), c. 1 mm long, anthers yellow; pistillode trifid, minutely puberulous. Female flowers: pedicels 1 mm long, extending to 3 mm long in fruit; calyx lobes 6, rarely fewer, ± equal, 2 × 1 mm, accrescent to 6–8 × 5–7 mm, each lobe 9–13-lobulate on the margins, the lobules linear-lanceolate, usually longer than the width of the calyx-lobe rhachis, densely white-setose, green, the calyx-lobe rhachis linear to lanceolate, sparingly pubescent and setose without, ± glabrous within, becoming indurated, yellow, becoming stramineous; ovary 1 mm in diameter, 3-lobed, sparingly setose; styles 3, 1.5–2 mm long, ± erect, connate at the base, subglabrous, stigmas ± smooth, straight at first, later recurved, green. Fruit 5 × 8–9 mm, 3-lobed, smooth, sparingly setose on the keels, yellow-green. Seeds c. 4 mm in diameter, ovoid-subglobose, greyish, lightly flecked and mottled with reddish-brown, and with scattered circular patches fringed with papillae.

Caprivi Strip. Linyanti, fl. & fr. 27.xii.1958, *Killick & Leistner* 3148 (K; PRE; SRGH). **Botswana**. N: south of Toromoja School on bank of Botletle R., fr. 29.iv.1971, *Pope* 417 in *Peterhouse School* 280 (K; PRE; SRGH). SE: Dikgomodikae (Dikomo di Ki), o. fr. 26.ii.1960, *Wild* 5182 (K; PRE; SRGH). **Zambia**. B: Lusu, fl. 8.vii.1962, *Fanshawe* 6920 (FHO; K; NDO). W: Ndola, fl. 12.iv.1973, *Fanshawe* 11833 (K; NDO). C: 57 km west of Lusaka, fr. 5.v.1957, *Noaks* 227 (SRGH). E: Nyimba–Petauke,

y. fr. 14.xii.1958, *Robson* 950 (BM; K; LISC; PRE; SRGH). S: Magoye For. Res., Mazabuka, fr. 16.i.1952, *White* 1930 (FHO; K). **Zimbabwe**. N: Kariba, Gache Gache R. (Gashe Gashe R.), fr. 14.iii.1959, *Chase* 7060 (K; SRGH). W: Gwampa For. Res., fr. i.1956, *Goldsmith* 56/56 (K; LISC; PRE; SRGH). C: 29 km SSE of Kwekwe (Que Que), 19.ii.1965, *Biegel* 921 (K; LISC; PRE; SRGH). E: Chipinge Distr., E Save (Sabi), fr. 22.i.1957, *Phipps* 112 (K; PRE; SRGH). S: Ndanga Distr., Chipinda Pools, fl. & fr. 24.i.1961, *Goodier* 77 (K; LISC; SRGH). **Malawi**. N: Ngala, 27 km north of Chilumba, fl. & fr. 14.iv.1976, *Pawek* 10999 (K; MAL; MO; UC). S: lower Mwanza R., fl. 4.x.1946, *Brass* 17949 (K; NY; SRGH). **Mozambique**. N: Pemba (Porto Amélia), fr. 26.viii.1948, *Barbosa* 1899 (LISC). Z: Morrumbala (Morambala) Marsh, fl. & fr. ii.1888, *Scott* s.n. (K). T: Tete, fr. 19.v.1948, *Mendonça* 4325 (LISC). GI: Chibuto–Caniçado, fl. 11.x.1957, *Barbosa & Lemos* in *Barbosa* 7996 (COI; K; LISC; LUAI). M: Sábiè Distr., Bundoio, fl. & fr. 8.ii.1945, *Sousa* 7 (LISC).

Also in Tanzania, Angola, Namibia and South Africa (Transvaal). Widespread but not common at low to medium altitudes, often in riverine vegetation and rocky outcrops, in mopane woodland, deciduous woodland on Kalahari Sand, plateau woodland and lakeshore thicket; sea level to 1350 m.

Tragia okanyua grades almost imperceptibly into a number of other species with respect to leaf shape and indumentum. Thus *Jacobsen* 387 (SRGH) from Zimbabwe (N), *Gilges* 713 (K; SRGH) from Zambia (S) and *Torre* 7791 (LISC) from Mozambique (GI) are intermediate between it and *Tragia brevipes* Pax, *Robson* 950 (K) from Zambia (E) converges on *T. benthamii*, *Prince* 502 (K) from Zambia (C), *Bainbridge* 674 (K) from Zambia (S), and *Wild* 4028 (K; SRGH) are intermediate between *T. okanyua*, *T. brevipes* and *T. benthamii*, *Barbosa* 1899 (LISC) from Mozambique (N) is intermediate between it and *T. kirkiana*, and *Wild* 5182 (K; SRGH) from Botswana is intermediate between *T. okanyua*, *T. kirkiana* and *T. dioica*.

Vernacular names as recorded in specimen data include: "lubabanzovu", "luemya" (Tok.).

16. **Tragia brevipes** Pax in Bot. Jahrb. Syst. **19**: 103 (1894); in Engler, Pflanzenw. Ost-Afrikas **C**: 240 (1895). —Prain in F.T.A. **6**, 1: 983 (1913). —Pax in Engler, Pflanzenr. [IV, fam. 147, ix] **68**: 77 (1919). —Engler, Pflanzenw. Afrikas (Veg. Erde 9) **3**, 2: 105 (1921). —Robyns & Tournay, Fl. Sperm. Parc Nat. Alb. **1**: 470 (1948). —Agnew, Upl. Kenya Wild Fls.: 216 (1974). —Troupin, Fl. Rwanda **2**: 242, fig. 66/2 (1983). —Radcliffe-Smith in F.T.E.A., Euphorb. 1: 302 (1987). Syntypes from Tanzania (Lake Province) and Zaire (Kivu Province).

Tragia mitis var. *genuina* Müll. Arg. in De Candolle, Prodr. **15**, 2: 942 (1866), pro parte.

Tragia mitis sensu Oliv. in Trans. Linn. Soc. London, **29**: 147 (1875), non Hochst. ex A. Rich.

Tragia velutina Pax in Bot. Jahrb. Syst. **19**: 104 (1894); in Engler, Pflanzenw. Ost-Afrikas **C**: 240 (1895). Type from Tanzania (Lake Province).

Tragia volkensii Pax in Engler, Pflanzenw. Ost-Afrikas **C**: 240 (1895). Type from Tanzania (Northern Province).

Very similar to *T. okanyua*, from which it differs in having ovate leaves which are not lobed at the base, but which are narrowly and deeply cordate, and in having the female calyx lobes accrescent to up to 1.5 cm long and generally much more densely setose.

Zambia. B: Masese, fr. 11.iii.1973, *Chisumpa* 11 (K; NDO). N: Mwambeshi R., fl. & fr. 3.v.1957, *Richards* 9528 (K; SRGH). W: Mindolo Dam, fr. 26.ii.1966, *Mutimushi* 1276 (SRGH). E: Mkania (Mkhania), 32 km SE of Mfuwe, fr. 28.ii.1969, *Astle* 5558 (K; SRGH). S: Kazungula, fr. 5.i.1957, *Gilges* 713 (K sheet only; SRGH sheet = *Tragia okanyua*). **Zimbabwe**. N: Chirundu, fl. & fr. 15.iii.1966, *Simon* 703 (K; LISC; SRGH). W: near Dopi Pan, Hwange (Wankie) Nat. Park, fl. 12.iii.1969, *Rushworth* 1676 (SRGH). **Malawi**. N: Viphya Plateau, fl. 15.v.1976, *Pawek* 11265 (K; MAL; MO; UC). C: Ntchisi Mt., fl. 21.ii.1959, *Robson & Steele* 1705 (BM; K; LISC; SRGH). S: Chikwawa Bridge, fl. & fr. 25.ii.1970, *Brummitt & Banda* 8754 (K; PRE; SRGH).

From Cameroon eastwards to Somalia and south to Zimbabwe. In understorey of riverine forest and thicket, lakeshore thickets and thickets on Kalahari Sand (mutemwa), also in submontane evergreen forest margins and high rainfall miombo; 90–1800 m.

The following specimens are intermediate between this and *Tragia micromeres* Radcl.-Sm.; *Fanshawe* 3276 (K) and *Richards* 15146 (K) from Zambia, and *Pawek* 2292 (K) and *Brummitt* 8925 (K) from Malawi.

Mitchell 2812 (K) from the South Luangwa National Park in Zambia approaches *Tragia okanyua*.

17. **Tragia micromeres** Radcl.-Sm. in Kew Bull. **47**: 681 (1992). Type: Zambia, Samfya, near shore of L. Bangweulu, 11°21'S, 29°33'E, 21.iv.1989, *Radcliffe-Smith, Pope & Goyder* 5752 (K, holotype; LISC; MAL; NDO; PRE; SRGH).

A scrambling weakly urticating perennial herb, monoecious; stems, petioles and inflorescence axes evenly retrorsely puberulous and sparingly retrorsely- or patent-hirsute. Stipules 1.5–2 × 0.5–1 mm, lanceolate, sparingly pubescent. Petiole up to

5 cm long. Leaf blade 2–8 × 1–5 cm, ovate to ovate-lanceolate, scarcely or not at all 3-lobed, acutely acuminate at the apex, shallowly cordate at the base, sharply to indistinctly serrate on the margins, membranous, 5(7)-nerved from the base, sparingly hirsute on upper surface, ± evenly setose along the midrib and main nerves beneath; lateral nerves in 3–4 pairs, scarcely prominent above, somewhat prominent beneath. Inflorescences up to 5 cm long, the peduncle up to 2 cm long, leaf-opposed or supra-axillary, bisexual with only 1 female flower at the base; male bracts 1 mm long, linear-lanceolate; female bracts 1.5 mm long, 3-lobed, all ciliate and 1-flowered; male bracteoles 0.5 mm long, subulate; female bracteoles resembling the male bracts. Male flowers: pedicels 1–2 mm long, jointed at the base; buds subglobose; calyx lobes 1 × 0.75 mm, elliptic-obovate, apiculate, sparingly pubescent without, glabrous within; stamens 3, 0.5 mm long, anthers with a broad connective; pistillode 3-lobed, the lobes rounded. Female flowers: pedicels 0.5 mm long, scarcely extending in fruit; calyx lobes 6, 1.5 × 1 mm, accrescent to 3 × 1.5 mm, each lobe 7-lobulate on the margins, the lobules 0.5–1 mm long, linear-lanceolate, equalling or longer than the width of the calyx-lobe rhachis, puberulous and setose, dark green, the calyx-lobe rhachis linear-lanceolate, sparingly pubescent without, glabrous and stramineous within, becoming thinly woody in part; ovary 1.25 mm in diameter, 3-lobed, densely setose; styles 3, 2.25 mm long, erect, connate for two-thirds of their length, glabrous, stigmas smooth, strongly recurved. Fruit 4 × 7 mm, 3-lobed, smooth, sparingly setose, urticating, pale yellow-green. Seeds 2.75 mm in diameter, globose, pale brownish-grey, shining and irregularly dark purplish-brown mottled and flecked, and sinuously whitish-lined.

Zambia. N: Samfya, near shore of Lake Bangweulu, 11°21'S, 29°33'E, fr. 21.iv.1989, *Radcliffe-Smith, Pope & Goyder* 5752 (K; LISC; MAL; NDO; PRE; SRGH); *Radcliffe-Smith, Pope & Goyder* 5754 (K; SRGH).

Known only from this locality. On fixed dunes with *Alchornea yambuyaënsis, Phyllanthus polyanthus, Croton polytrichus* subsp. *brachystachyus, Hymenocardia ulmoides, Thecacoris trichogyne* and *Bridelia duvigneaudii*; 1067 m.

18. **Tragia benthamii** Baker in Bull. Misc. Inform., Kew **1910**: 128 (1910). —Prain in F.T.A. **6**, 1: 984 (1913). —R.E. Fries, Wiss. Ergebn. Schwed. Rhod.-Kongo-Exped. **1**, 1: 124 (1914). —F.W. Andrews, Fl. Pl. Anglo-Egypt. Sudan **2**: 99 (1952). —Keay in F.W.T.A., ed. 2, **1**, 2: 412 (1958). —Radcliffe-Smith in F.T.E.A., Euphorb. 1: 303 (1987). Types from Ghana and Bioko.

Tragia cordifolia sensu auct. incl. Benth. in Niger Fl.: 501 (1849). —Müller Argoviensis in Flora **47**: 436 (1864); in De Candolle, Prodr. **15**, 2: 944 (1866). —Pax in Engler, Pflanzenr. [IV, fam. 147, ix] **68**: 76 (1919). —Engler, Pflanzenw. Afrikas (Veg. Erde 9) **3**, 2: 105 (1921). —De Wildeman, Pl. Bequaert. **3**, 4: 498 (1926), non Vahl (1790).

Tragia mitis var. *kirkii* Müll. Arg. in Flora **47**: 435 (1864); in De Candolle, Prodr. **15**, 2: 942 (1866). Type: Mozambique, between Lupata and Tete (Tette), ii.1859, *Kirk* s.n. (K, holotype).

Very like *T. brevipes*, from which it differs in having the leaf blades more shallowly and widely cordate at the base, and ± glabrous apart from the stinging hairs.

Botswana. SW: Kuke, y. fl. 22.ii.1970, *Brown* 8682 (K). **Zambia**. B: Masese, fr. 14.v.1962, *Fanshawe* 6816 (FHO; K). N: Mbala, fl. 30.iii.1960, *Fanshawe* 5595 (FHO; K; NDO). S: Mazabuka, fr. 6.iii.1963, *van Rensburg* 1613 (K; SRGH). **Zimbabwe**. N: near Nyarandi R., fr. 31.i.1970, *Pope* 227 (K; LISC; PRE; SRGH). S: Triangle Sugar Estate, fl. 4.iii.1970, *Mavi* 1081 (K; PRE; SRGH). **Malawi**. S: Lengwe Game Reserve, fr. 10.ii.1970, *Hall-Martin* 505 (K; PRE). **Mozambique**. N: Ilha do Ibo, fl. & fr. 1884/5, *Carvalho* s.n. (COI). MS/T: Lupata–Tete, fr. ii.1859, *Kirk* s.n. (K).

From the Ivory Coast eastwards to Ethiopia and south to Angola, Botswana, Zimbabwe and Mozambique. Floodplain thickets and termitaria in floodplain grassland, also on rocky outcrops and pan margins in mopane woodland; near sea level to c. 1000 m.

Fanshawe 5595 (cited above) and *Cameron* 1 (K) from Namasi, Malawi, are somewhat intermediate between this and *T. micromeres*, whilst the one Botswana specimen seen (*Brown* 8682) seems to fall between the 3 species *T. okanyua, T. brevipes* and *T. benthamii*, but is perhaps closer to the latter than the others.

19. **Tragia prionoides** Radcl.-Sm. in Kew Bull. **42**: 397 (1987). Tab. **48**. Type: Zimbabwe, Mutare Distr., Zimunya's Reserve, Rowa Township, fl. & fr. 2.v.1954, *Chase* 5232 in *GHS* 46927 (SRGH, holotype; BM; K; LISC).

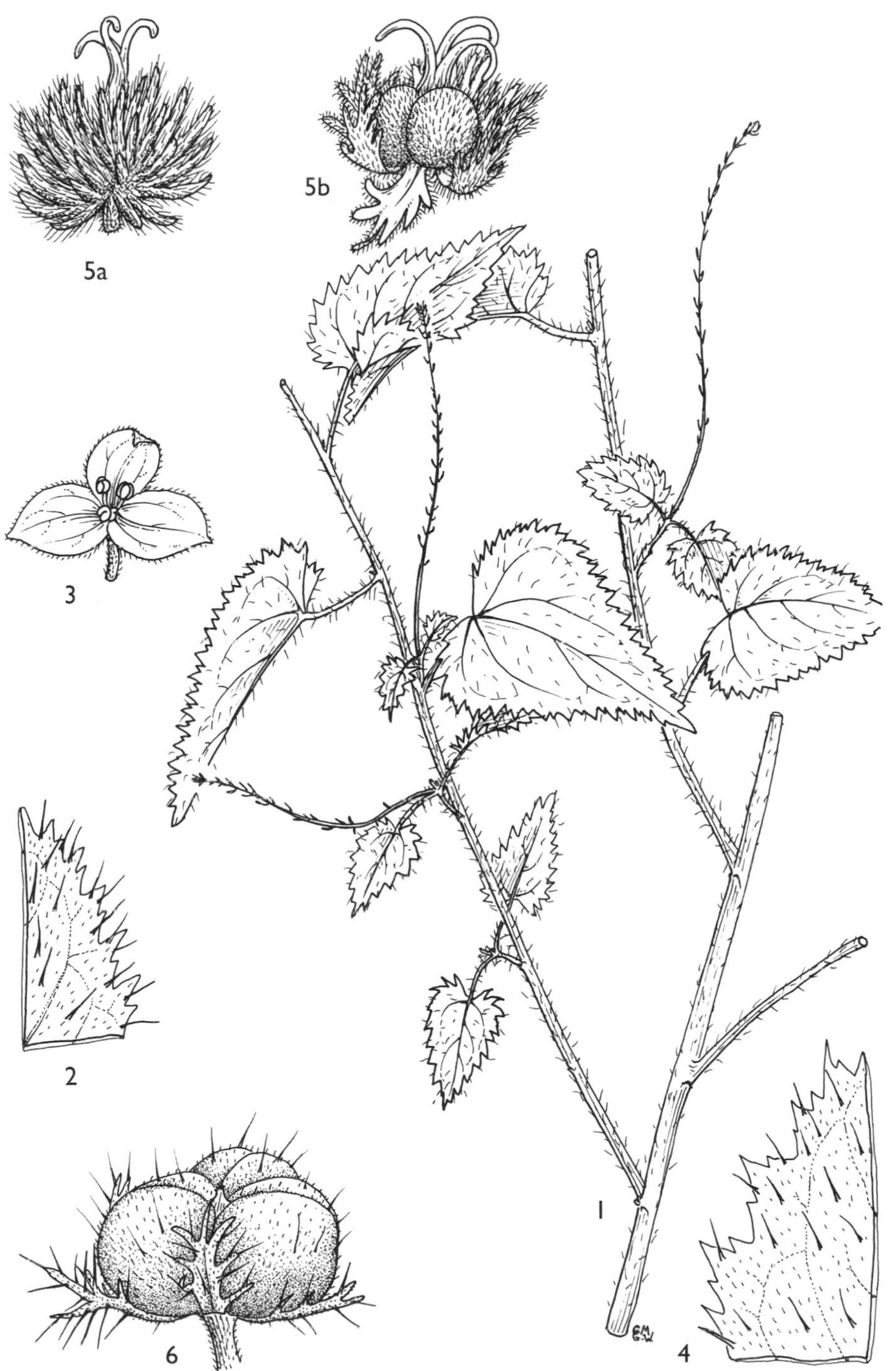

Tab. 48. TRAGIA PRIONOIDES. 1, portion of stem (× $^{2}/_{3}$); 2, leaf margin, bidentate (× 2); 3, male flower (× 8), 1–3 from *Eyles* 816; 4, leaf margin, tridentate (× 2); 5a, female flower (× 8); 5b, female flower with pistil displayed (× 8); 6, fruit (× 4), 4–6 from *Chase* 5232. Drawn by Christine Grey-Wilson.

A climbing branched severely urticating perennial herb or shrub, up to 9 m tall when supported, monoecious. Stems and petioles tomentellous, or retrorsely puberulous and hispid and sometimes stinging-setose. Stipules 5–6 mm long, linear-lanceolate. Petiole up to 13.5 cm long. Leaf blade 2–18.5 × 1–14 cm, triangular-ovate, acute at the apex, coarsely serrate or biserrate on the margins, deeply or widely cordate at the base, chartaceous, 5–7-nerved from the base, sparingly pubescent and setose on upper surface, whitish-tomentose and setose along the midrib and main nerves beneath; lateral nerves in 3–5 pairs, scarcely prominent above, slightly so beneath, camptodromous or craspedodromous. Inflorescences up to 11 cm long, on peduncles up to 3 cm long, leaf-opposed, mostly male with 1(2) female flowers at the base; axis tomentellous; male bracts 2–3 mm long, linear, 1-flowered; male bracteoles 1–2 mm long, subulate; female bracts and bracteoles resembling those of the male, but slightly larger. Male flowers: pedicels 2 mm long; buds turbinate; calyx lobes 1.5 × 1 mm, elliptic-ovate, obtuse, sparingly puberulous without, glabrous within; stamens 1 mm long; pistillode flat, 3-lobed, the lobes truncate at the apex. Female flowers: pedicels very short, extending to 3 mm long in fruit; calyx lobes 6, 2 × 1.5 mm, accrescent to 7 × 7 mm, each lobe 7-lobulate, the lobules linear, longer than the width of the calyx-lobe rhachis, puberulous and setose, green, the calyx-lobe rhachis lanceolate, sparingly puberulous without, glabrous within, stramineous, at length woody proximally; ovary 1.5 mm in diameter, 3-lobed, densely setose; styles 3, 1.5 mm long, connate at the base, glabrous, stigmas smooth, recurved. Fruit 4–5 × 7–8 mm, 3-lobed, smooth, puberulous and setose. Seeds 3 mm in diameter, globose, greyish-brown, unevenly finely dark brown-granulate and with scattered patches of minute whitish papillae.

Zimbabwe. W: Matobo Distr., Lucydale, fl. & fr. 1.i.1948, *West* 2533 in *GHS* 25562 (K; SRGH). C: Makoni Distr., Forest Hill, fl. & o. fr. vi.1917, *Eyles* 816 (BM; K; SRGH). E: Mutare Distr., Zimunya's Reserve, Rowa Township, fl. & fr. 2.v.1954, *Chase* 5232 (BM; K; LISC; SRGH). S: Masvingo (Ft. Victoria), fl. iv.1921, *Eyles* 2990 (SRGH).

Also in South Africa (Transvaal). On granite outcrops amongst boulders and in wooded kloofs; c. 1000–1500 m.

Although in the past confused with members of the *T. okanyua-brevipes-benthamii* complex, this plant is very distinct on account of its ragged saw-like leaf margin, the greyish tomentum on the leaf lower surface and its granulate seeds. The stinging hairs are recorded as being particularly virulent.

20. **Tragia gardneri** Prain in Bull. Misc. Inform., Kew **1909**, 2: 52 (1909); in F.T.A. **6**, 1: 994 (1913). —Eyles in Trans. Roy. Soc. South Africa **5**: 397 (1916). —Pax in Engler, Pflanzenr. [IV, fam. 147, ix] **68**: 91 (1919). —Engler, Pflanzenw. Afrikas (Veg. Erde 9) **3**, 2: 107 (1921). Type: Zimbabwe, Gweru (Gwelo), fr. i.1905, *Gardner* 34 (K, holotype).

An erect, branched, urticating perennial herb up to 50 cm tall, monoecious; stems several from a woody rootstock, sparingly puberulous and stinging-setose. Stipules 2–4 mm long, lanceolate. Petiole 1–4 mm long, or leaves sessile. Leaf blade 1–6 × 0.5–2.5 cm, ovate to ovate-lanceolate or narrowly lanceolate, acute or subacute at the apex, crenate-serrate to sharply serrate on the margins except at the base, truncate or subtruncate at the base, chartaceous, 3(5)-nerved from the base, sparingly finely puberulous on upper surface, setose along the midrib and main nerves beneath; lateral nerves in 5–7 pairs, scarcely prominent on leaf upper and lower surfaces. Inflorescences up to 11 cm long, on peduncles up to 5 cm long, terminal or leaf-opposed, comprised mostly of male flowers with 2 female flowers at the base; axis pubescent and setose; male bracts 2 mm long, narrowly lanceolate, entire; male bracteoles 1 mm long, linear-lanceolate; female bracts and bracteoles 2–3 mm long, elliptic-lanceolate, sometimes denticulate. Male flowers: pedicels 1 mm long; buds ellipsoid-subglobose; calyx lobes 1.5 × 1 mm, ovate, subacute, glabrous; stamens 3(4), 1 mm long; pistillode slightly raised, usually 3-radiate. Female flowers: pedicels 2 mm long, not or scarcely extending in fruit in length; calyx lobes 3, c. 1.5 × 1.5 mm, accrescent to 4.5 × 6 mm, each lobe c. 20-lobulate, the lobules shortly lanceolate, much shorter than the width of the calyx-lobe rhachis, puberulous and setose, green, the calyx-lobe rhachis transversely-ovate to suborbicular, minutely puberulous and setose without, glabrous within, stramineous, at length somewhat woody proximally; ovary 1.5 mm in diameter, 3-lobed, densely setose; styles 2–3 mm long, erect, connate at the base, glabrous, stigmas smooth, straight. Fruit 5 × 8 mm, 3-lobed, smooth, puberulous, densely setose on the keels. Seeds 3.5 mm in diameter, globose, dark grey, mottled chestnut, ± smooth but white-papillose alongside the hilum.

Botswana. N: Pandamatenga, male fl. & fr. ix.1989, *Phillips* s.n. (GAB). **Zimbabwe**. W: Gwampa Forest Reserve, fr. i.1956, *Goldsmith* 64/56 (K; LISC; PRE; SRGH). C: Gwebi, fl. 11.x.1960, *Rutherford-Smith* 228 (K; LISC; SRGH).

Not known from elsewhere. In dambo and vlei grassland, in sand and black clay soils, sometimes on termitaria; c. 1500 m.

A very distinct species, only distantly related to *Tragia dioica*.

21. **Tragia dioica** Sond. in Linnaea **23**: 109 (1850). —Prain in F.T.A. **6**, 1: 993 (1913). —Eyles in Trans. Roy. Soc. South Africa **5**: 397 (1916). —Pax in Engler, Pflanzenr. [IV, fam. 147, ix] **68**: 88 (1919). —Prain in F.C. **5**, 2: 507 (1920). —Engler, Pflanzenw. Afrikas (Veg. Erde 9) **3**, 2: 106 (1921). —Burtt Davy, Fl. Pl. Ferns Transvaal: 309 (1932). —P.G. Meyer in Merxmüller, Prodr. Fl. SW. Afrika, fam. 67: 45 (1967). Type from South Africa (Transvaal).

Tragia rupestris var. *lobata* Müll. Arg. in De Candolle, Prodr. **15**, 2: 941 (1866). Type as for *T. dioica* Sond.

Tragia schinzii Pax in Bull. Herb. Boissier **6**: 734 (1898). Types from Namibia and South Africa (Transvaal).

Tragia dioica var. *lobata* (Müll. Arg.) Pax and K. Hoffm. in Engler, Pflanzenr. [IV, fam. 147, ix] **68**: 88 (1919).

Tragia dioica var. *schinzii* (Pax) Pax in Engler, Pflanzenr. [IV, fam. 147, ix] **68**: 88 (1919); Engler, Pflanzenw. Afrikas (Veg. Erde 9) **3**, 2: 106 (1921).

An erect or suberect urticating perennial herb up to 60 cm tall, usually much less, monoecious or polygamo-dioecious; stems several from a woody rootstock, puberulous to tomentose and stinging-setose. Stipules 3–4 mm long, lanceolate, puberulous. Petioles 0.2–2 cm long. Leaf blades 3–7 × 0.3–1.5 cm, 1–3.5 cm wide at the base, usually narrowly lanceolate and often strongly 3-lobed at the base, acute to subacute or obtuse at the apex, median lobe entire or serrate or crenate-serrate on the margins towards the apex, lateral lobes coarsely serrate or dentate, truncate or shallowly wide cordate at the base, thinly chartaceous, 5–7-nerved from the base, sparingly pubescent on both surfaces and sparingly to densely setose along the midrib and main nerves on both surfaces; lateral nerves in 4–8 pairs, scarcely prominent above, slightly so beneath. Inflorescences up to 8.5 cm long, on peduncles up to 5.5 cm long, terminal and/or leaf-opposed, usually comprised of male flowers with 2–3 female flowers below, rarely all male or all female; male bracts 2 mm long, elliptic-lanceolate; male bracteoles 1 mm long, linear; female bracts 3 mm long, ovate; female bracteoles resembling the male bracts. Male flowers: pedicels 1 mm long; buds 3-lobed to subglobose; calyx lobes 1.5 × 1 mm, ovate, subacute, subglabrous, yellow; stamens 3, less than 1 mm long; pistillode 3-lobed, the lobes truncate. Female flowers: pedicels 1 mm long, extending to up to 4 mm in fruit; calyx lobes 3, sometimes with 1–4 small additional lobules, 2 × 2 mm, accrescent to 7 × 8 mm, each lobe deeply pectinately 9–15-lobulate, the lobules linear-lanceolate, ± equalling the width of the calyx-lobe rhachis, sparingly puberulous and densely setose, green, the calyx-lobe rhachis ± ovate, sparingly puberulous without, glabrous within, green at the base and otherwise yellowish without, whitish and somewhat woody at length within; ovary 1.5 mm in diameter, 3-lobed, setose; styles 3, 2–3 mm long, suberect, connate to about halfway, sparingly pubescent at the base, otherwise glabrous, stigmas smooth, ± recurved at the tips. Fruit 0.6 × 1 cm, 3-lobed, smooth, sparingly puberulous, setose on the keels. Seeds 3.5 mm in diameter, subglobose, pale grey, sparingly flecked with chestnut-brown, and with scattered circular yellow patches fringed with white papillae.

Botswana. N: Nyei Pan, fr. 25.xii.1967, *Lambrecht* 472 (SRGH). SW: 8 km southwest of Ghanzi Camp, fr. 11.i.1970, *Brown* 7587 (GAB; K; SRGH). SE: Mochudi, fl. i.1915, *Harbor* in *PRE* 17020 (PRE). **Zimbabwe**. S: Nottingham Ranch, c. 24 km WNW of Beitbridge, fl. & fr. 25.iii.1959, *Drummond* 5998 (K; LISC; PRE; SRGH).

Also in Namibia and South Africa (Transvaal). In hot dry country in scrub mopane and calcrete pans; c. 950 m.

This species is sometimes confused with *T. okanyua* when only suberect, when the basal leaf lobes are not as strongly developed as usual and when the female calyx has supernumerary lobes.

22. **Tragia glabrata** (Müll. Arg.) Pax & K. Hoffm. in Engler, Pflanzenr. [IV, fam. 147, ix] **68**: 94 (1919). —Engler, Pflanzenw. Afrikas (Veg. Erde 9) **3**, 2: 107 (1921). Types from South Africa (KwaZulu-Natal and Eastern Cape Province).

Tragia capensis E. Mey. in Drège, Zwei Pflanzengeogr. Dokum.: 226 (1843), ex Sond. in Linnaea **23**: 110 (1850) non Thunb. (1794).

Tragia meyeriana var. *glabrata* Müll. Arg. in De Candolle, Prodr. **15**, 2: 938 (1866).

Tragia durbanensis Kuntze, Revis. Gen. Pl. **3**, 3: 293 (1898). —Prain in F.C. **5**, 2: 510 (1920). Types as for *T. glabrata.*

A much-branched climbing perennial herb, weakly urticating, monoecious; stems from a woody rootstock, up to 2.5 m long, twining, subglabrous or sparingly puberulous, rarely evenly hispid. Stipules 3–4 mm long, ovate-lanceolate, glabrous. Petiole 0.5–3 cm long. Leaf blade 2–6 × 1.5–4.5 cm, triangular-ovate to triangular, sometimes with 2 weakly developed lateral lobes at the base, acute to subacute or obtuse at the apex, crenate to crenate-serrate or crenate-dentate on the margins, wide-cordate at the base, membranaceous, 5–7-nerved from the base, very sparingly puberulous and with a few scattered setae to almost glabrous on both surfaces; lateral nerves in 3–4 pairs, scarcely prominent above or beneath. Inflorescences up to 5.5 cm long with peduncles up to 2 cm long, terminal on lateral shoots and leaf-opposed on the main stems, composed mostly of male flowers with 1–2 female flowers below or else all male; male bracts 1 mm long, lanceolate; male bracteoles 0.7 mm long, linear-lanceolate; female bracts 2 × 1.5 mm, ovate; female bracteoles resembling the male bracts. Male flowers: pedicels 1 mm long; buds 3-lobed to subglobose; calyx lobes 1 × 1 mm, suborbicular, obtuse, subglabrous, green; stamens 3, 0.5 mm long; pistillode 3-lobed, minute. Female flowers: pedicels 0.5 mm long, extending to 2–3 mm in fruit; calyx lobes 3, 1.5 × 2 mm, accrescent to 4 × 6 mm, each lobe palmately 18–22-lobulate, the lobules linear-lanceolate, shorter than the width of the calyx-lobe rhachis, sparingly puberulous and setose, green, the calyx-lobe rhachis transversely ovate-suborbicular, sparingly puberulous and setose without, glabrous within, yellowish, only becoming slightly hardened; ovary 1.5 mm in diameter, 3-lobed, setose; styles 3, 2–3 mm long, connate to halfway, glabrous, stigmas smooth, recurved. Fruit 5 × 8 mm, 3-lobed, smooth, setose on the keels, otherwise ± glabrous. Seeds 4 mm in diameter, globose, pale greyish-fulvous, with chestnut-brown flecks, and with scattered circular patches fringed with papillae.

Stems subglabrous or sparingly puberulous · var. *glabrata*
Stems evenly hispid · var. *hispida*

Var. **glabrata**

Stems subglabrous or sparingly puberulous.

Mozambique. GI: Inhambane, Ponta da Barra, fl. 21.vii.1963, *Mogg* 32615 (LISC). M: 5 km north of Maputo (Lorenzo Marques), fl. & fr. 29.iii.1948, *Rodin* 4165 (K; MO; UC); Inhaca Island, fl. 14.vii.1957, *Barbosa* 7703 (K; LUAI; PRE); Marracuene, fl. 16.xi.1964, *Marques & Balsinhas* 72 (COI; SRGH).

Also in South Africa (KwaZulu-Natal and Cape Province). In dune forest on sand; sea level to 10 m.

Var. **hispida** Radcl.-Sm. in Kew Bull. **42**: 396 (1987). Type: Mozambique, Maputo Province, between Umbelúzi and Namaacha, fr. 16.x.1940, *Torre* 1777 (LISC, holotype).

Stems evenly hispid.

Mozambique. M: between Umbelúzi and Namaacha, fr. 16.x.1940, *Torre* 1777 (LISC).

Not known from elsewhere. In dry open bushland.

23. **Tragia rupestris** Sond. in Linnaea **23**: 108 (1850). —Pax in Engler, Pflanzenr. [IV, fam. 147, ix] **68**: 89 (1919). —Prain in F.C. **5**, 2: 509 (1920). —Engler, Pflanzenw. Afrikas (Veg. Erde 9) **3**, 2: 106 (1921). —Burtt Davy, Fl. Pl. Ferns Transvaal: 309 (1932). Type from South Africa (Transvaal).

Similar to *T. glabrata*, but the plants sometimes dioecious, and the leaves generally narrower and more evenly pubescent and setose on the upper surface, and the lobules of the female calyx lobes as long as or longer than the width of the calyx-lobe rhachis.

Mozambique. GI: Guijá–Mobaze, male fl. 15.iii.1948, *Torre* 7490 (LISC); Bazaruto Island, male, female fl. & fr. 21.x.1958, *Mogg* 28507 (BM; K; SRGH; LISC). M: Maputo (Lourenço Marques), fr. 6.xii.1947, *Barbosa* 666 (LISC); Sábiè, fl. 1.xii.1944, *Mendonça* 3191 (LISC).
Also in South Africa (KwaZulu-Natal, Transvaal) and Swaziland. Coastal dry open woodland and thickets on sand flats; near sea level to 50 m.

24. **Tragia wahlbergiana** Prain in J. Bot. **51**: 169 (1913); in F.C. **5**, 2: 509 (1920). —Burtt Davy, Fl. Pl. Ferns Transvaal: 309 (1932). Types from South Africa (Transvaal).
Tragia rupestris var. *glabrata* Sond. in Linnaea **23**: 108 (1850). —Müll. Arg. in De Candolle, Prodr. **15**, 2: 940 (1866). —Pax in Engler, Pflanzenr. [IV, fam. 147, ix] **68**: 89 (1919). Type from South Africa (Transvaal).
Tragia rupestris var. *minor* sensu Müll. Arg. in De Candolle, Prodr. **15**, 2: 940 (1866) pro parte quoad spec. Wahlberg, excl. syn. (Sond., 1850).
Tragia affinis Müll. Arg. ex Prain in Bull. Misc. Inform., Kew **1912**: 334 (1912), non Robinson & Greenman (1894). Type from South Africa (Transvaal).

Very close to *T. rupestris*, from which it differs in having the leaves triangular-ovate, but not 3-lobed, and almost completely glabrous on the upper surface; and to *T. glabrata*, from which it differs in having the female calyx lobes more deeply lobulate.

Mozambique. M: Marracuene, Rikatla, fl. & o. fr. xi.1917–18, *Junod* 174 (LISC).
Also in South Africa (Transvaal). No habitat details available for the Flora Zambesiaca area.

25. **Tragia kirkiana** Müll. Arg. in Flora **47**: 538 (1864); in De Candolle, Prodr. **15**, 2: 939 (1866). —Pax in Engler, Pflanzenw. Ost-Afrikas **C**: 239 (1895). —Prain in F.T.A. **6**, 1: 998 (1913). —Eyles in Trans. Roy. Soc. South Africa **5**: 397 (1916). —Pax in Engler, Pflanzenr. [IV, fam. 147, ix] **68**: 93 (1919). —Engler, Pflanzenw. Afrikas (Veg. Erde 9) **3**, 2: 107 (1921). —Radcliffe-Smith in F.T.E.A., Euphorb. 1: 315 (1987). Type: Mozambique, Zambézia Province, banks of the R. Shire at Morrumbala (Moramballa), fl. 15.i.1863, *Kirk* s.n. (K, holotype).
Tragia angustifolia sensu Benth. (1849), non Nutt. (1837), var. *hastata* Müll. Arg in Flora **47**: 435 (1864); in De Candolle, Prodr. **15**, 2: 939 (1866). Syntypes: Mozambique, Zambézia Province, foot of Morrumbala (Moramballa), fr. xii.1858, *Kirk* s.n. "*Tragia* (3)" (K); Manica e Sofala Province, Chupanga (Shupanga), o. fr. i.1859, *Kirk* s.n. (K).
Tragia stolziana Pax & K. Hoffm. in Engler, Pflanzenr. [IV, fam. 147, ix] **68**: 78 (1919). —Engler, Pflanzenw. Afrikas (Veg. Erde 9) **3**, 2: 106 (1921). Type from Tanzania (Southern Highlands Province).

A climbing or trailing urticating perennial herb, monoecious; stems up to 3 m long from a woody rootstock, sparingly puberulous and/or setose to subglabrous. Stipules 4–6 mm long, lanceolate, ciliate, becoming strongly reflexed. Petiole 0.5–4(7) cm long. Leaf blade 3–12 × 0.6–5 cm, triangular-lanceolate to linear-lanceolate, often hastate with two diverging lateral lobes at the base, acutely caudate-acuminate at the apex, sharply serrate to crenate-serrate on the margins, more coarsely so towards the base, deeply cordate, thinly chartaceous, 5–7-nerved from the base, subglabrous to sparingly pubescent or hispid above and beneath, and also setose along the midrib and main nerves beneath; lateral nerves in 4–6 pairs, scarcely prominent above, slightly so beneath. Inflorescences up to 19 cm long with a peduncle up to 15 cm long, although commonly much less, leaf-opposed, sometimes terminal on lateral shoots, comprised mostly of male flowers usually with 2 female flowers below; male bracts 1.5–2 mm long, lanceolate to ovate-lanceolate, eglandular; male bracteoles 1 mm long, lanceolate to linear; female bracts 4–5 × 2 mm, ovate to ovate-lanceolate; female bracteoles 2 mm long, elliptic. Male flowers: pedicels 1–1.5 mm long; buds subglobose; calyx lobes c. 1 × 1 mm, ovate-suborbicular, apiculate, subglabrous, greenish-yellow; stamens 3, 1 mm long, strongly incurved; pistillode trifid, the segments bifid. Female flowers: pedicels 0.5 mm long, extending to 3 mm in fruit; calyx lobes 3, 1.5 × 1.5 mm, accrescent to 7 × 7–8 mm, each lobe palmately 11–13-lobulate, the lobules linear-lanceolate, shorter than the width of the calyx-lobe rhachis, sparingly minutely puberulous, densely setose, greenish, the calyx-lobe rhachis obovate, subglabrous without, glabrous within, yellow-green, becoming slightly hardened within; ovary 1 mm in diameter, 3-lobed, densely setose on the keels, otherwise sparingly minutely puberulous to subglabrous, styles 3, 2–3 mm long, to 4 mm long in fruit, connate for up to half their length, subglabrous, stigmas smooth, recurved at the tips. Fruit 4 × 8 mm, strongly 3-lobed, the keels carinate, smooth, setose and hispid on the keels, otherwise sparingly minutely puberulous to subglabrous. Seeds 3–3.5 mm in diameter, subglobose, grey, flecked and mottled with dark brown, and with scattered fulvous circular patches densely fringed with white papillae.

Zambia. W: Luanshya, fr. 11.xii.1963, *Fanshawe* 8168 (K; NDO); Miengwe For. Reserve, y. fl. 3.xi.1955, *Fanshawe* 2571 (K; NDO). **Zimbabwe**. N: Kandeya Nat. Res., fr. 17.i.1960, *Phipps* 2269 (K; SRGH). C: Poole Farm, fr. 4.i.1954, *Hornby* 3330 in *GHS* 46153 (SRGH). S: Masvingo Distr., Mutirikwi (Kyle) Nat. Park, fr. 26.xi.1970, *Basera* 190 (SRGH). **Malawi**. N: Chelinda R., fr. 5.ii.1978, *Pawek* 13788 (K; MAL; MO; SRGH; UC). C: Dedza Distr., Sosola Rest House, Mua, y. fl. 15.xii.1969, *Salubeni* 1438 (SRGH). S: Zomba Distr., Chikomwe Hill, fr. 5.i.1979, *Patel* 380 (MAL; SRGH). **Mozambique**. N: Monapo, fr. 24.xii.1963, *Torre & Paiva* 9266 (LISC). Z: Mopeia Velha–Quelimane, fr. 7.xii.1971, *Pope & Müller* 539 (LISC; SRGH). MS: Gorongosa, Chitengo, y. fl. 8.xi.1965, *Balsinhas* 1044 (COI).

Also in Kenya, Tanzania and South Africa (Transvaal). Riverine forest and thicket, gully forest, plateau and high rainfall miombo, also on rocky outcrops; 180–1280 m.

Hornby 3330 (SRGH) has some female flowers with more than 3 perianth lobes, and with the lobules of different lengths. It is somewhat intermediate between this species and *T. rupestris*. Some specimens with a more or less erect habit (e.g. *Pope & Müller* 539), show a tendency in the direction of *T. dioica*.

26. **Tragia adenanthera** Baill., Adansonia **1**: 275 (1861). —Müller Argoviensis in De Candolle, Prodr. **15**, 2: 938 (1866). —Pax in Engler, Pflanzenw. Ost-Afrikas **C**: 239 (1895). —Prain in F.T.A. **6**, 1: 997 (1913). —Pax in Engler, Pflanzenr. [IV, fam. 147, ix] **68**: 95 (1919). —Engler, Pflanzenw. Afrikas (Veg. Erde 9) **3**, 2: 107 (1921). —Radcliffe-Smith in F.T.E.A., Euphorb. 1: 316 (1987). Type from Zanzibar.

Very similar to *T. kirkiana*, but with the leaf blades generally more triangular-ovate; the male bracts ovate and evenly glandular, the filaments shorter, erect, and with the connective apically produced; the female bracts and bracteoles often crenate-dentate; and the female calyx lobes with the lobules longer than or equalling the width of the calyx-lobe rhachis. The flowers are often reddish-tinged, and the fruit orange-red.

Malawi. S: Zomba Plain, fl. & fr. xii.1896, *Whyte* s.n. (K).

Also in Tanzania. Wooded grassland; 760 m.

27. **Tragia furialis** Bojer [Hort. Maurit.: 286 (1837), nom. nud.] ex Prain in F.T.A. **6**, 1: 995 (1913). —Pax in Engler, Pflanzenr. [IV, fam. 147, ix] **68**: 94 (1919). —Engler, Pflanzenw. Afrikas (Veg. Erde 9) **3**, 2: 107 (1921). —Radcliffe-Smith in F.T.E.A., Euphorb. 1: 316 (1987). Type is a plant that was cultivated in Mauritius from material from the Comoro Islands.

Tragia angustifolia var. *furialis* (Bojer ex Prain) Müll. Arg. in De Candolle, Prodr. **15**, 2: 939 (1866).

Tragia scheffleri Baker in Bull. Misc. Inform., Kew **1908**: 439 (1908). —Prain in F.T.A. **6**, 1: 996 (1913). —Agnew, Upl. Kenya Wild Fls.: 216 (1974). Type from Tanzania (Usambara Mts.).

Tragia furialis var. *scheffleri* (Baker) Pax & K. Hoffm. in Engler, Pflanzenr. [IV, fam. 147, ix] **68**: 94 (1919). —Engler, Pflanzenw. Afrikas (Veg. Erde 9) **3**, 2: 107 (1921).

A climbing urticating perennial herb or shrub, monoecious; stems up to 3 m long, sparingly pubescent and/or setose. Stipules 3–5 mm long, lanceolate, ciliate, otherwise glabrous, becoming strongly reflexed. Petiole 1–7 cm long. Leaf blades 5–13 × 2–6 cm, ovate-lanceolate to elliptic-lanceolate or oblong-lanceolate, acutely acuminate at the apex, sharply serrate on the margins, subtruncate to shallowly and widely or deeply and narrowly cordate at the base, membranaceous, 5–7-nerved from the base, sparingly puberulous along the midrib and main nerves above and beneath and also sparingly setose mostly between the midrib and main nerves above, and along them at the base beneath; lateral nerves in 3–5 pairs, scarcely prominent above, slightly so beneath. Inflorescences up to 8.5 cm long on a peduncle up to 2.5 cm long, leaf-opposed or axillary, or terminal on short axillary leafy shoots, consisting mostly of male flowers commonly with only 1 female flower at the base; male bracts 1–1.5 mm long, narrowly lanceolate, sparingly glandular; male bracteoles usually less than 1 mm long, subulate; female bracts up to 3 mm long, lanceolate, ciliate and glandular; female bracteoles resembling the male bracts. Male flowers: pedicels 1 mm long; buds 3-lobed, sparingly puberulous; calyx lobes less than 1 × 1 mm, broadly ovate-suborbicular, apiculate, sparingly puberulous without, glabrous within; stamens 3, less than 1 mm long; pistillode minute or absent. Female flowers: pedicels less than 1 mm long, extending to 2 mm long in fruit; calyx lobes 3, 1 × 2 mm, accrescent to 6 × 6–7 mm, each lobe palmately 11–13-lobulate, the lobules linear-lanceolate, equalling or longer than the width of the calyx-lobe rhachis, sparingly pubescent, densely setose, greenish, the calyx-lobe rhachis ovate, sparingly puberulous without, subglabrous within, yellowish, becoming slightly hardened within; ovary 1.5 mm in diameter, strongly 3-lobed, setose on the keels, otherwise sparingly appressed-puberulous; styles 3, 2–3 mm long, connate for c. half their length, glabrous, stigmas smooth, recurved at the tips. Fruit 5 × 9 mm, strongly 3-lobed, smooth, sparingly setose on the keels, otherwise subglabrous, green. Seeds 3 × 2.5 mm, ovoid-subglobose, pale grey, intricately flecked and mottled with reddish-brown, and with scattered pale grey circular patches densely fringed with yellowish papillae.

Zimbabwe. E: Chimanimani Distr., lower Haroni, fl. & fr. 25.x.1959, *Ball* 828 (K; SRGH). **Mozambique**. MS: Chimoio, Gondola, fr. 19.ii.1948, *Garcia* 283 (LISC).

Also in Kenya, Tanzania, Comoro Island and Madagascar. In low altitude evergreen forest (*Newtonia*), and riparian forest and thicket; 330–400 m.

Tragia scheffleri is generally distinguished from *T. furialis* in having fewer stinging hairs. However, as all states exist with respect to this character, sometimes on the same individual, it is not deemed to be worthy even of varietal status.

28. **Tragia stipularis** Radcl.-Sm. in Kew Bull. **37**: 688 (1983); in Kew Bull. **42**, 2: 399 (1987); in F.T.E.A., Euphorb. 1: 311 (1987). Type from Tanzania (Lake Province).

A weakly urticating perennial herb up to 90 cm tall, suberect at first, later twining at the top, monoecious. Stems puberulous and weakly hirsute. Stipules erect, 5–10 × 4–7 mm, falcate-ovate to falcate-lanceolate, acute, entire, rounded on one side at the base, many-nerved from the base, glabrous but for the ciliate margins. Petiole 1–3 cm long. Leaf blade 3–6 × 2–3.5 cm, triangular ovate to triangular-lanceolate, acute at the apex, crenate-serrate on the margins, shallowly wide-cordate at the base, chartaceous, 5-nerved from the base, sparingly pubescent on upper surface, sparingly weakly hirsute along the midrib and main nerves beneath; lateral nerves in 2–4 pairs, not prominent above, prominent beneath. Inflorescences up to 4 cm long on a peduncle up to 2 cm long, leaf-opposed, or terminal on short axillary shoots, consisting mostly of male flowers with 1–2 female flowers at the base; male bracts 1.5–2 mm long, lanceolate or elliptic-lanceolate; male bracteoles 1 mm long, linear-lanceolate; female bracts c. 3 × 2 mm, ovate-lanceolate to elliptic-lanceolate, sometimes bifid at the apex; female bracteoles resembling the male bracts. Male flowers: pedicels 1 mm long; buds turbinate; calyx lobes 1.3 mm long, elliptic, apiculate, cucullate, subglabrous; stamens 3, 1 mm long; pistillode trifid, the segments bifid. Female flowers subsessile, but developing a pedicel c. 2 mm long in fruit; calyx lobes 3, 1.5 × 1.5 mm, accrescent to 7 × 7 mm, each lobe palmately 13–15-lobulate, the lobules linear, shorter than the width of the calyx-lobe rhachis, hirsute and setose, green, the calyx-lobe rhachis broadly ovate-suborbicular, sparingly puberulous without, glabrous within, yellowish, becoming hardened and whitish within; ovary 1 mm in diameter, 3-lobed, densely setose on the keels, otherwise puberulous; styles 3, 2–3 mm long, connate for half their length, puberulous, stigmas ± smooth, spreading. Fruit 4 × 7 mm, 3-lobed, smooth, setose on the keels, otherwise sparingly puberulous. Seeds 3.5 mm in diameter, subglobose, grey, streaked and flecked with dark purplish-brown, with scattered tawny spots and with a mass of white papillae near the hilum.

Zambia. S: 16 km Livingstone–Kazungula, fr. 11.iii.1960, *White* 7725 (FHO; K); Lochinvar Nat. Park, fr. 8.v.1981, *Ellenbrach* 080581–6 (SRGH).

Also in Uganda and Tanzania. Floodplain grassland and *Colophospermum mopane* woodland, on black clay basalt derived soil, often on termitaria; 1000 m.

A very distinct species, isolated from the other *Tragia* species in the Flora Zambesiaca area on account of the remarkable stipules.

Ctenomeria capensis (Thunb.) Harv. (*Tragia capensis* Thunb.) is reported by Mogg, in Macnae & Kalk, Nat. Hist. Inhaca Isl., Moçamb., rev. ed.: 148 (1969) from Inhaca Island, but I have seen no specimen. It occurs ± throughout South Africa and is distinguished from *Tragia* sens. str. by its large number of stamens (c. 40–50, instead of 1–5).

38. DALECHAMPIA Plum. ex L.

Dalechampia Plum. ex L., Sp. Pl.: 1054 (1753); Gen. Pl. ed. 5: 473 (1754).

Monoecious, usually scandent herbs or subshrubs with a simple, sometimes urticating indumentum. Leaves alternate, petiolate, stipulate, simple, 3–5-lobed or 3–5-foliolate, entire or toothed, glandular-stipellate, palminerved. Inflorescences terminal or axillary, usually solitary and long-pedunculate, enveloped by 2 large foliaceous bracts; bracts subequal, sessile, stipulate, entire, toothed or 3-lobed, heterochromous; male flowers 7–20 in a pedunculate, involucrate pleiochasium (aggregation of dichasial cymes in umbel-like arrangement) composed of 5(7) sessile (1)3-flowered bracteate cymes accompanied by a mass of fused bracts and/or aborted flowers encrusted with a waxy secretion; female flowers 1–3 in a subsessile 2–3-bracteate cyme abaxial to the male peduncle; flowers copiously nectariferous. Male flowers: pedicels jointed; calyx globose in bud, later splitting into 4–6 valvate lobes; petals absent; disk absent; stamens usually 10–30, rarely more, filaments partially united to form a column, anthers longitudinally dehiscent; pistillode absent. Female flowers: pedicels short, elongating in fruit; sepals 5–12, imbricate, narrow, usually pinnatifid, later accrescent and becoming hardened, enclosing the fruit; petals absent; disk absent; ovary 3-celled, with 1 ovule per cell; styles completely connate into a long column obtuse, dilated or obliquely excavated at the apex. Fruit 3-lobed, dehiscing into 3 bivalved cocci; endocarp crustaceous or woody; columella persistent. Seeds globose, ecarunculate; albumen fleshy; cotyledons broad and flat.

A pantropical genus of 110 species, of which only 6 or 7 are African.

1. Leaves 3-foliolate · 1. *galpinii*
– Leaves 3–5(7)-lobed, -fid, or -partite · 2
2. Leaves commonly 5-partite · 2. *capensis*
– Leaves commonly 3-lobed or 3-fid · 3. *scandens*

1. **Dalechampia galpinii** Pax in Bull. Herb. Boissier **6**: 736 (1898); in Engler, Pflanzenr. [IV, fam. 147, xii] **68**: 22 (1919). —Prain in F.C. **5**, 2: 498 (1920). —Engler, Pflanzenw. Afrikas (Veg. Erde 9) **3**, 2: 110 (1921). —Burtt Davy, Fl. Pl. Ferns Transvaal: 307 (1932). Tab. **49**. Type from South Africa (Transvaal).

A rhizomatous trailing or climbing perennial herb to over 2 m in length, sparingly pubescent. Stipules 2–4 mm long, lanceolate. Petioles 0.7–3 cm long. Leaves 3-foliolate; median leaflet simple, 2–6 × 0.5–2 cm, elliptic-ovate to elliptic-oblong or lanceolate, obtuse or acute at the apex, serrate to subentire, lateral nerves in 9–14 pairs; lateral leaflets asymmetrically 2-lobed, the inner lobe 1–5 cm long, the outer 0.5–2.5 cm long, or else the outer lobe scarcely developed, or leaves ± 5-partite; base of leaf deeply cordate, with a pair of minute subulate glandular stipels near the apex of the petiole; leaf blades sparingly pubescent chiefly along the midribs and main veins beneath and on the margins. Inflorescences axillary; peduncles 2–11 cm long; bract stipules resembling the foliar stipules; bracts 0.7–2 × 0.7–2 cm, ovate in outline, the abaxial slightly larger than the adaxial, 3-lobed or sometimes 5-lobed, the lobes triangular or lanceolate, acute, slightly serrate or subentire, truncate to shallowly cordate at the base, 7-nerved from the base, pubescent along the nerves without, ± glabrous within, yellowish-green to creamy-white. Male peduncles 1.5 mm long; involucre 6 mm across, 7-flowered; mass of fused bracts and aborted flowers adaxial, flattened. Male flowers: pedicels 2 mm long; calyx lobes 5, 2 mm long, ovate-lanceolate, subglabrous; stamens c. 20–25. Female bracts 4, very unequal, the adaxial 2.5 × 2.5 mm, suborbicular, the abaxial 3 × 4 mm, 2-lobed, the lateral bracts 2 × 1 mm, oblong, all ciliate. Female flowers subsessile or pedicels very short, but extending to 3–5 mm in fruit; sepals 9–11, sepal rhachis 1.5 mm long,

Tab. 49. DALECHAMPIA GALPINII. 1, portion of climbing stem (× $^{2}/_{3}$); 2, persistent female calyx (× 4), 1 & 2 from *Hilary & Robertson* 500; 3, inflorescence (× 4); 4, male flowers (× 8), 3 & 4 from *Acocks* 23349. Drawn by Christine Grey-Wilson.

extending to 6–7(10) mm in fruit, linear, lateral lobules 2–5 pairs, eglandular, pubescent and with urticating hairs; ovary 1 mm in diameter, pubescent; stylar column 4–5 mm long, dilated and excavated at the apex. Fruit 4 × 8 mm, smooth, sparingly pubescent, reddish-brown or blackish. Seeds 3.5 × 3 mm, ± smooth, brownish, streaked and mottled with pale grey.

Botswana. SE: 8 km south of Gaborone (Gaberones), fl. & fr. 18.i.1974, *Mott* 126C (SRGH; UBLS). **Zimbabwe**. W: Filabusi, fl. i.1933, *Eyles* 8013 (K; SRGH). **Mozambique**. M: Goba–Changalane, fl. & fr. 1.iv.1945, *A.E. de Sousa* 142 (LISC; PRE).

Also in Swaziland and South Africa (Transvaal, KwaZulu-Natal). Hot dry low altitude wooded grassland with scattered deciduous trees and shrubs or thicket, often on stony or rocky hillsides. Climbing or trailing over herbs and low shrubs; near sea level to 1220 m.

2. **Dalechampia capensis** A. Spreng., Tent. Suppl. [Syst. Veg.]: 18 (1828). —Müller Argoviensis in De Candolle, Prodr. **15**, 2: 1243 (1866). —Pax in Engler, Pflanzenr. [IV, fam. 147, xii] **68**: 36 (1919). —Prain in F.C. **5**, 2: 499 (1920). —Engler, Pflanzenw. Afrikas (Veg. Erde 9) **3**, 2: 110 (1921). —Burtt Davy, Fl. Pl. Ferns Transvaal: 307 (1932). —Mogg in Macnae & Kalk, Nat. Hist. Inhaca Isl., Moçamb., rev. ed.: 148 (1969). —Radcliffe-Smith in F.T.E.A., Euphorb. 1: 287 (1987). Type from South Africa (Cape).

Dalechampia kirkii Prain in Bull. Misc. Inform., Kew **1912**, 8: 363 (1912). —Pax in Engler, Pflanzenr. [IV, fam. 147, xii] **68**: 36 (1919), in adnot. —Prain in F.C. **5**, 2: 498 (1920). —Burtt Davy, Fl. Pl. Ferns Transvaal: 307 (1932). Type from South Africa (Transvaal).

A prostrate creeping or climbing perennial herb; stems up to 3.5 m long from a woody rootstock, hirsute and pubescent. Stipules 5–6 mm long, lanceolate. Petioles 3–6.5 cm long. Leaves 5-partite; the median lobe 5.5–11 × 1–2.5 cm, lanceolate to elliptic-lanceolate, acute at the apex, serrate to subentire, constricted at the base, lateral nerves in 10 pairs; the lateral lobes slightly successively smaller; base of the leaf deeply cordate; stipels 1.5 mm long; pubescence chiefly confined to the midrib, main veins and margins; dark green on leaf uper surface, lighter beneath. Inflorescences axillary; peduncles 4–16 cm long, leafless or with a small tripartite leaf near the base; bract stipules the same as the foliar stipules; bracts 3.5–6 × 1.5–4 cm, ovate in outline, 3-lobed, the abaxial somewhat more deeply lobed than the adaxial, the lobes lanceolate, acute or subacute, glandular-serrate, rounded-cuneate, truncate or shallowly cordate at the base, 7–9-nerved from the base, pubescent without, glabrous within, pale lemon-yellow to greenish-yellow. Male peduncles 5 mm long, involucre 8 mm across, (7)9–11-flowered; mass of fused bracts and aborted flowers adaxial, flattened. Male flowers: pedicels 3 mm long; calyx lobes 5, 2 mm long, ovate or lanceolate, subglabrous; stamens c. 25. Female bracts 4, very unequal; the adaxial bracts 1.5 × 2 mm and bifid, the abaxial bracts 2 × 3 mm and bifid to 5 × 3 mm and ovate, the lateral bracts 1–2 × 0.5–1 mm, elliptic; bracts all ciliate. Female flowers subsessile, but developing pedicels up to 1 cm long in fruit; sepals 10–12, sepal rhachis 1 mm long and extending to 1–1.5 cm in fruit, linear, lateral lobules of the sepals 10–11-paired, gland-tipped, pubescent and with urticating hairs; ovary 1 mm in diameter, pubescent; stylar column 0.8–1 cm long, dilated and excavated at the apex. Fruit 5 × 9 mm, ± smooth, hispidulous, reddish-brown or blackish. Seeds 4 × 4 mm, ± smooth, brownish, streaked or mottled with pale grey.

Zambia. N: Mpui–Mbala, fl. 7.xii.1954, *Richards* 3553 (K; SRGH); Mbala–Mpulungu, fl. 5.iv.1955, *Richards* 5316 (K; SRGH).

Also in Tanzania (Ufipa District), Swaziland, South Africa (Cape, KwaZulu-Natal and Transvaal). In dense high rainfall plateau woodland, often in stony or rocky places, creeping amongst grass or climbing over bushes, also by roadsides; 1200–1830 m.

Menyharth 1072 (K), pro max. parte, from Boruma, Tete Province, Mozambique, st. 1891 and *Magadza* 23 (SRGH), from Mwenda Research Station, Northern Province, Zimbabwe, fr. 30.xii.1965, are probably referable here, but the stems are stouter, the leaves are 5-fid to 5-lobed or up to 7-lobed, the petioles 7–8 cm long, and the fruits more evenly pubescent.

Miller B982 (PRE) from Kanye, Botswana, fr. i.1950, is intermediate between this species and *Dalechampia scandens*. More aberrant, however, is *Barbosa* 2197 (LISC) from Palma–Nangade in Mozambique's Niassa Province, fr. 17.ix.1948. The leaves are 5-partite, but the margins of the segments and of the bracts are sharply denticulate rather than serrate, and are undulate in the sinuses as in some *Manihot* species. The bracts are very deeply cordate. There are only 6 calyx lobes in the female flowers, and the lobules lack the terminal clavate glands of *D. capensis*, but have a covering of smaller glands along their margins instead. This may be deserving of varietal status, but more material needs to be seen before confirming this opinion.

The description of the fruits and seeds above was based on South African material. The bimodal distribution of this species is of interest to note.

3. **Dalechampia scandens** L., Sp. Pl.: 1054 (1753). —Müller Argoviensis in De Candolle, Prodr. **15**, 2: 1244 (1866). —Prain in F.T.A. **6**, 1: 954 (1912). —Pax in Engler, Pflanzenr. [IV, fam. 147, xii] **68**: 32 (1919). —Engler, Pflanzenw. Afrikas (Veg. Erde 9) **3**, 2: 110 (1921). —Keay in F.W.T.A., ed. 2, **1**, 2: 413 (1958). —P.G. Meyer in Merxmüller, Prodr. Fl. SW. Afrika, fam. 67: 11 (1967). —Radcliffe-Smith in F.T.E.A., Euphorb. 1: 287 (1987). Type based on material of tropical American origin cultivated in Uppsala, Sweden.

A slender pubescent and often urticating trailing or climbing perennial herb; stems up to 3 m from a woody stock. Stipules 2–7 mm long, linear-lanceolate. Petioles 1–9 cm long. Leaves up to 13 × 14 cm, 3-fid or 3-lobed, rarely 5-lobed, shallowly to deeply cordate at the base with 5–9 basal nerves; lobes 3–10 × 1.5–4 cm, subequal, obovate to elliptic-oblanceolate or lanceolate, acutely acuminate to obtuse at the apex, serrate to subentire, slightly or not constricted at the base, puberulous or pubescent along the nerves, sometimes with scattered urticating hairs, or else subglabrous; lateral nerves in 5–9 pairs; stipels 1–3 mm long. Inflorescences axillary; peduncles 1–10 cm long, leafless or with a small leaf near the base; bract stipules 1.5–5 mm long, lanceolate; bracts (1)2–4 × (1)2.5–4 cm, ovate in outline, 3-lobed, often shallowly so, the lobes triangular-lanceolate, acute or subacute, glandular-serrate to subentire, cordate at the base, 7–9-nerved from the base, glabrous or pubescent without and within, cream-coloured to pale green. Male peduncles 3–5 mm long, involucre 5–8 mm across, 13-flowered; mass of fused bracts and aborted flowers adaxial, flattened. Male flowers: pedicels 3 mm long; calyx lobes 5, 2 mm long, elliptic-ovate, subglabrous; stamens 20–30, rarely fewer. Female bracts 2–3(4), ovate or oblong; the adaxial bract 2 mm long, repand-denticulate; the abaxial bracts 3 mm long, ± entire; the lateral bracts smaller, undulate, all ciliate. Female flowers: pedicels 2 mm long, extending to 1.5 cm in fruit; sepals 6–9, sepal rhachis 2.5 mm long, extending to 0.8–1.5 cm in fruit, linear, lateral lobules in 6–10 pairs, gland-tipped, pubescent and with urticating hairs; ovary c. 1 mm in diameter, densely pubescent; stylar column 5–7 mm long, abruptly dilated and excavated at the apex. Fruit 5 × 8–10 mm, ± smooth, pubescent, brownish. Seeds 3–3.5 × 3–3.5 mm, ± smooth, pale grey or dark reddish-brown and streaked and mottled with silvery-grey.

Leaves serrate, commonly evenly to densely pubescent on lower surface ······ var. *cordofana*
Leaves subentire, commonly glabrous or subglabrous, or if pubescent, then only along the midrib and main nerves ······································· var. *natalensis*

Var. **cordofana** (Hochst. ex Webb) Müll. Arg. in De Candolle, Prodr. **15**, 2: 1245 (1866). —Prain in F.T.A. **6**, 1: 954 (1912). —Pax in Engler, Pflanzenr. [IV, fam. 147, xii] **68**: 34 (1919). —Engler, Pflanzenw. Afrikas (Veg. Erde 9) **3**, 2: 110 (1921). —Keay in F.W.T.A., ed. 2, **1**, 2: 413 (1958). —Radcliffe-Smith in F.T.E.A., Euphorb. 1: 289 (1987). Type from Sudan.

Dalechampia parvifolia Lam., Encycl. Méth. Bot. **2**: 258 (1786). Type in P-JU, erroneously signified as from China.

Dalechampia scandens var. *parvifolia* (Lam.) Müll. Arg. in De Candolle, Prodr. **15**, 2: 1245 (1866). —Prain in F.T.A. **6**, 1: 954 (1912). Type as above:

Dalechampia cordofana Hochst. ex Webb in Hooker, Niger Fl.: 178 (1849) sphalm. *cordafana*. Type from Sudan.

Leaves serrate, commonly evenly to densely pubescent on lower surface.

Mozambique. Z: Praia de Pebane, fr. 24.x.1942, *Torre* 4673 (LISC). T: 33 km Chicoa–Magoé, fl. & fr. 17.ii.1970, *Torre & Correia* (LISC; LMA; LMU). MS: Cheringoma, fl. 28.v.1948, *Barbosa* 1693 (LISC). M: Goba, fr. 1.iv.1945, *A.E. de Sousa* 149 (LISC; PRE).

From the Cape Verde Islands to Pakistan, and south to Angola and Swaziland. In evergreen coastal scrub and sand dune thickets, in dry sandy ground, also in humus rich soil on riverine forest margins and in clearings in old cultivations; sea level to 300 m.

Some material presents an intermediate indumentum state between this and var. *natalensis* (Müll. Arg.) Pax & K. Hoffm., thus in *R.M. Hornby* 2587 (K; PRE; SRGH) from Maputo, fl. 22.i.1947, *Wild & Leach* 5195 (K; SRGH) st. 23.iii.1960 and *Noel* 2472 (K; LISC; SRGH) post fr. 8.ix.1962, both from Macuti (Mozambique (MS)), and *de Koning* 7727 (K) post fr. 29.xi.1979 from Matutuíne (Mozambique (M)), the indumentum is confined to the midrib and nerve network, but the leaf lobe margins are more or less serrate.

Var. **natalensis** (Müll. Arg.) Pax & K. Hoffm. in Engler, Pflanzenr. [IV, fam. 147, xii] **68**: 35 (1919). —Engler, Pflanzenw. Afrikas (Veg. Erde 9) **3**, 2: 110 (1921). Type from South Africa (KwaZulu-Natal).

Dalechampia volubilis E. Mey. in Drège, Zwei Pflanzengeogr. Dokum.: 158, 177 (1843), nom. nud.; in Baill., Étud. Gén. Euphorb.: 487 (1858), nomen tantum. —Prain in F.C. **5**, 2: 499 (1920). Type from South Africa (KwaZulu-Natal).

Dalechampia capensis Sond. in Linnaea **23**: 106 (1850), pro parte, non A. Spreng.

Dalechampia natalensis Müll. Arg. in De Candolle, Prodr. **15**, 2: 1243 (1866).

Leaves subentire, commonly glabrous or subglabrous, or if pubescent, then only along the midrib and main nerves.

Mozambique. GI: Inharrime, Ponta Zavora, fr. 4.iv.1959, *Barbosa & Lemos* in *Barbosa* 8506 (COI; K; LISC). M: Maputo (Lourenço Marques), Catembe, fr. 18.iv.1920, *Borle* 954 (COI; K; PRE; SRGH).

Also in South Africa (KwaZulu-Natal). In coastal forest margins and dune thickets, also in riverine forest, on dry sandy soil; sea level to 200 m.

Some material presents an intermediate indumentum state between this and the preceding variety, thus *Torre* 1602 (COI; LISC) and 7511 (LISC), both from Manjacaze, Mozambique (GI), and *Balsinhas* 1282 (COI) from Matutuíne, Salamanga, Mozambique (M), have the leaves more or less pubescent beneath, but the leaf margins are subentire. Furthermore, in *Groenendijk & Dungo* 1572 (LMU), fr. 12.xii.1984, from Inhaca Island, the leaf margins are shallowly serrate, whereas the leaves are almost glabrous beneath.

Var. *scandens* occurs in West Indies, Central America and northern South America, and differs in lacking glandular hairs on the bracts of the female flowers.

39. HEVEA Aubl.

Hevea Aubl., Hist. Pl. Guiane **2**: 871, t. 335 (1775). —Müller Argoviensis in De Candolle, Prodr. **15**, 2: 716 (1866). —Pax in Engler, Pflanzenr. [IV, fam. 147, i] **42**: 117 (1910).

Siphonia L.C. Rich. in Schreb., Gen. Pl. **2**: 656 (1791).

Monoecious laticiferous trees. Leaves alternate, stipulate, long-petiolate, trifoliolate; petiole with an apical gland; leaflets petiolulate, entire, penninerved; stipules small, fugacious. Inflorescences axillary, subtended by leaves or bracts, paniculate, androgynous, the central flower of each cymule female, the rest male, protogynous; bracts small, entire. Male flowers: calyx ovoid or globose in bud, 5-lobed, the lobes valvate; petals absent; disk glands 5, free or united; stamens 5–10, the filaments united into a column, anthers sessile in 1–2 series, those of the outer series alternisepalous; pistillode at the apex of the staminal column, entire. Female flowers: calyx as in the male; petals absent; disk minute or absent; ovary ovoid, 3(4), short, stigmas thick, spreading, entire or slightly 2-lobed. Fruits large, trigonous, septicidally dehiscent into 3 bivalved cocci; endocarp woody. Seeds large, oblong-ovoid, smooth, blotched, ecarunculate; testa crustaceous; endosperm scanty or absent; cotyledons thick, fleshy, ± equal in size.

A genus of 10 species with several varieties from Amazonian South America. One species is widely cultivated.

Hevea brasiliensis (A. Juss.) Müll. Arg. in Linnaea **34**: 204 (1865); in De Candolle, Prodr. **15**, 2: 718 (1866). —Pax in Engler, Pflanzenr. [IV, fam. 147, i] **42**: 121 (1910). —Hutchinson in F.T.A. **6**, 1: 743 (1912). —De Wildeman, Pl. Bequaert. **3**, 4: 505 (1926). —Brenan, Check-list For. Trees Shrubs Tang. Terr.: 215 (1949). —Keay in F.W.T.A., ed. 2, **1**, 2: 392 (1958). —Radcliffe-Smith in F.T.E.A., Euphorb. 1: 183 (1987). Tab. **50**. Type from Brazil.

Siphonia brasiliensis A. Juss., Euph. Gen. Tent.: 113, t. 12/38B (1824).

Hevea sp., Eyles in Trans. Roy. Soc. South Africa **5**: 397 (1916).

Tree up to 20 m tall. Bark pale grey. Branches ± erect. Twigs glabrous. Stipules 1 mm long, lanceolate. Petioles 6–20(30) cm long, glabrous; petiolules 1–1.5 cm long; petiole gland reniform; leaflets 7–20(25) × 3–8(10) cm, the median leaflet larger than the others, obovate to elliptic, acuminate at the apex, attenuate or cuneate at the base, chartaceous, glabrous above and beneath, somewhat glaucous beneath; midrib impressed above, prominent beneath; lateral nerves in 15–25 pairs,

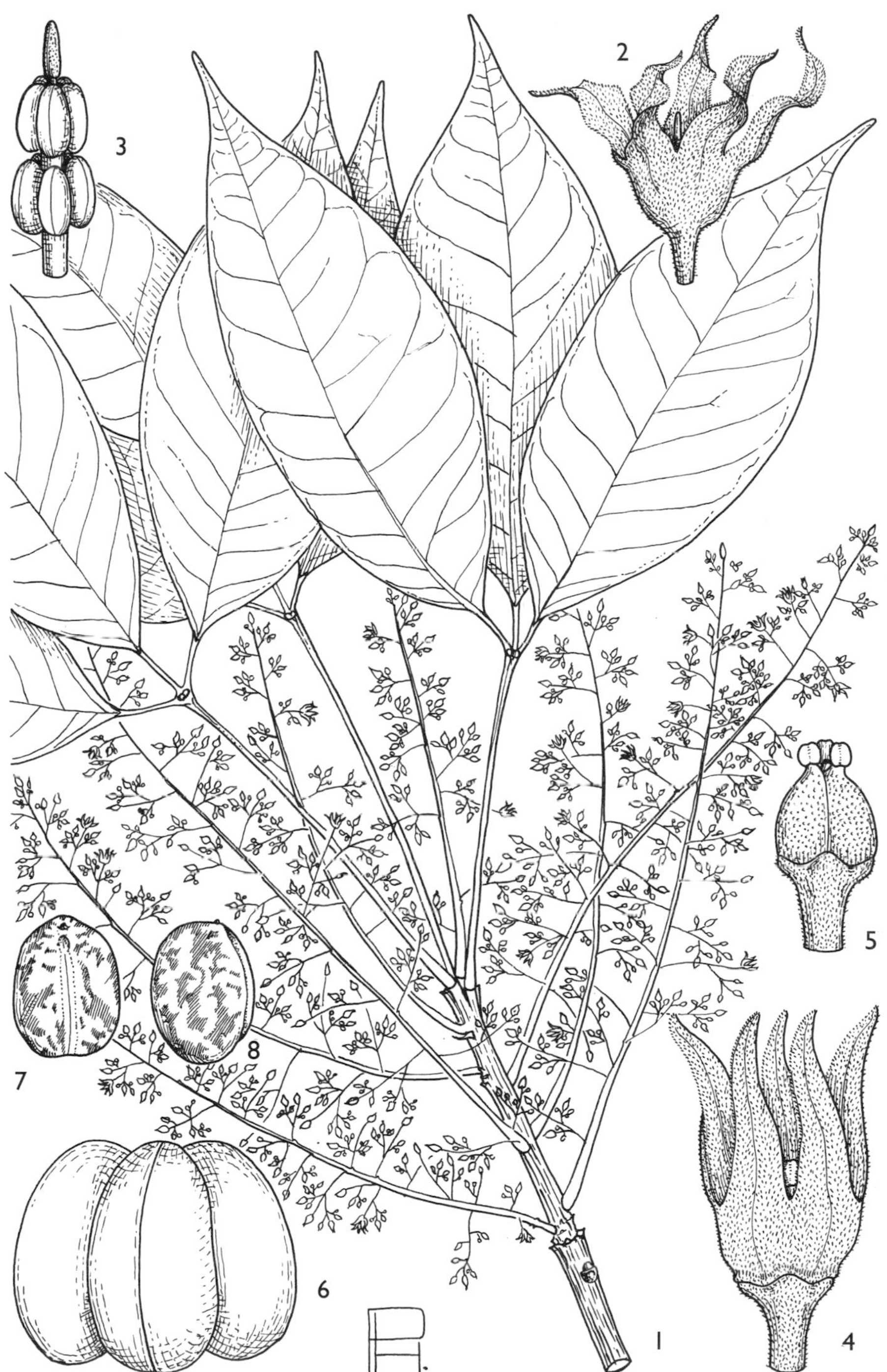

Tab. 50. HEVEA BRASILIENSIS. 1, flowering branch (× $^{2}/_{3}$); 2, male flower (× 6); 3, staminal column (× 12); 4, female flower (× 6); 5, pistil (× 6), 1–5 from *Verdcourt & Greenway* 150; 6, fruit (× $^{2}/_{3}$); 7, seed, adaxial view (× $^{2}/_{3}$); 8, seed, lateral view (× $^{2}/_{3}$), 6–8 from *Melville* s.n. Drawn by Pat Halliday. From F.T.E.A.

looped just within the margin. Inflorescences up to 20 cm long, subterminal below the apical tuft of leaves; axis sparingly pubescent; bracts 0.5 mm long, deltate, fugacious. Male flowers: buds ovoid, acuminate; calyx lobes 2 × 0.5 mm, narrowly lanceolate, somewhat contorted, acuminate, subacute, tomentellous; tube 1 mm long; disk ± annular; staminal column 1.5 mm high, anthers 10, in 2 whorls of 5; pistillode narrowly conical, puberulous. Female flowers: buds as in the male; calyx lobes larger than in the male, otherwise similar; tube 2 mm long; disk scarcely visible; ovary 2 × 2 mm, subglobose, minutely papillose; stigmas 0.3 mm long, ± sessile, grooved. Fruits c. 4 × 4.5 cm, shallowly trigonous, glabrescent. Seeds 2.3 × 1.5 cm, ovoid-cylindric, pale grey, mottled and streaked with darker grey.

Malawi. S: Zomba Bot. Gdn., leaf & seed 25.ix.1962, *Chapman* 1676 (FHO; SRGH).

The author recalls passing through a plantation of *H. brasiliensis* between Nkhota-Kota and Chintheche near the Northern/Central Provincial border in May 1989, but no specimen was collected. The trees were in fruit at the time.

Widely cultivated in the wetter parts of Africa and the humid tropics generally, the Pará Rubber is the source of the best natural rubber; native of Brazil and Colombia.

40. MANIHOT Mill.

Manihot Mill., Gard. Dict. abr. ed. 4, **2** (1754). —Müller Argoviensis in De Candolle, Prodr. **15**, 2: 1057 (1866).
Janipha Kunth, Nov. Gen. et Spec. **2**: 84, t. 109 (1817).

Monoecious often glaucous and pruinose laticiferous trees or shrubs, rarely herbs. Indumentum, when present, simple. Leaves alternate, petiolate, stipulate, simple, sometimes peltate, entire or more usually deeply palmately 3–11-lobed or -partite; the lobes entire or lobulate, penninerved. Inflorescences terminal, rarely axillary, racemose or paniculate, bisexual with 1–few female flowers and several male flowers; bracts sometimes foliaceous. Male flowers pedicellate; buds ovoid, ellipsoid or fusiform; calyx campanulate or tubular, 5-lobed, the lobes imbricate or contorted, often pigmented; petals absent; disk deeply lobed or consisting of separate glands; stamens 10, biverticillate, free, inserted between the lobes or glands of the disk; anthers dorsifixed, longitudinally dehiscent; pistillode small, trifid, or absent. Female flowers long-pedicellate; tepals 5, free, imbricate, caducous; petals absent; disk entire or lobulate, pigmented; ovary 3-locular with 1 ovule per loculus; styles 3, short, connate at the base, often dilated or lobed. Fruits subglobose, ± smooth or longitudinally 6-winged, dehiscing septicidally into 3 bivalved cocci; endocarp woody; columella persistent. Seeds carunculate; testa crustaceous; albumen fleshy; cotyledons broad, flat.

A neotropical genus of c. 100 species, some of which are cultivated and have become more or less naturalized in the Old World tropics. See Rogers & Appan, Flora Neotropica Monograph No. **13**, Manihot, Manihotoides (1973).

1. Leaves distinctly peltate ········ 3. *glaziovii*
– Leaves not or scarcely peltate ········ 2
2. Fruits not winged ········ 2. *grahamii*
– Fruits distinctly 6-winged ········ 3
3. Leaf lobes oblanceolate, gradually acuminate; inflorescence paniculate, 2–11 cm long; food plants ········ 1. *esculenta*
– Leaf lobes elliptic-obovate, abruptly acuminate; inflorescence racemose, 3–5 cm long; rubber plants ········ 4. *dichotoma*

1. **Manihot esculenta** Crantz, Inst. Rei Herb. **1**: 167 (1766). —Brenan, Check-list For. Trees Shrubs Tang. Terr.: 219 (1949). —F.W. Andrews, Fl. Pl. Anglo-Egypt. Sudan **2**: 86 (1952). —Keay in F.W.T.A., ed 2, **1**, 2: 413 (1958). —J. Léonard in F.C.B. **8**, 1: 121 (1962). —White, F.F.N.R.: 201 (1962).—Mogg in Macnae & Kalk, Nat. Hist. Inhaca Isl., Moçamb., rev. ed.: 148 (1969). —Rogers & Appan, Fl. Neotrop. Monogr. No. **13**: 25 (1973). —Biegel, Check List Ornam. Pl. Rhod. Parks & Gard.: 74 (1977). —Troupin, Fl. Rwanda **2**: 232, fig. 70/2 (1983). —Radcliffe-Smith in F.T.E.A., Euphorb. 1: 367 (1987). Tab. **51**. Type from S America.
Jatropha manihot L., Sp. Pl.: 1007 (1753). Type from S America.
Manihot utilissima Pohl, Pl. Bras. **1**: 32, t. 24 (1827). —Müller Argoviensis in De Candolle, Prodr. **15**, 2: 1064 (1866). —Engler, Pflanzenw. Ost-Afrikas **C**: 240 (1895). —Pax in Engler,

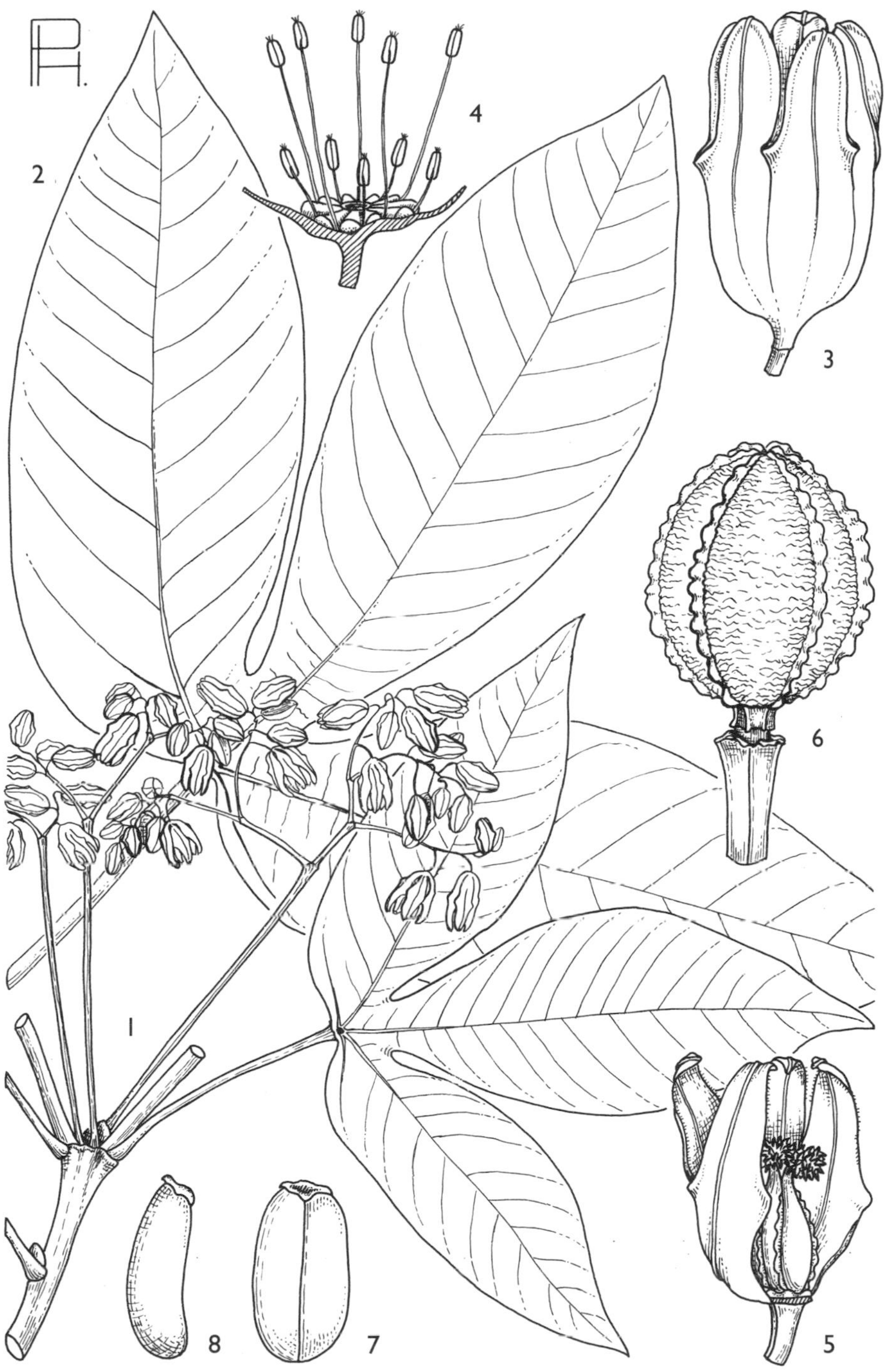

Tab. 51. MANIHOT ESCULENTA. 1, distal portion of flowering branch (× ²⁄₃); 2, leaf (× ²⁄₃); 3, male flower (× 3); 4, stamens (× 3), 1–4 from *Pirozynski* 257; 5, female flower, with one sepal removed (× 3), from *Last* s.n.; 6, fruit (× 2); 7, seed, adaxial view (× 2); 8, seed, lateral view (× 2), 6–8 from *Pirozynski* 257. Drawn by Pat Halliday. From F.T.E.A.

Pflanzenr. [IV, fam. 147, ii] **44**: 67 (1910). —Prain in F.T.A. **6**, 1: 842 (1912). —Engler, Pflanzenw. Afrikas (Veg. Erde 9) **3**, 2: 133 (1921). Type from S America.

Manihot dulcis (J.F. Gmel.) Pax in Engler, Pflanzenr. [IV, fam. 147, ii] **44**: 71 (1910). Type from S America.

A brittle-stemmed dichotomously branched shrub or small tree up to 5 m tall. Root tubers up to 50 cm long, farinaceous. Bark smooth. Latex whitish, watery. Young shoots glaucous. Petiole 4–25 cm long, often reddish. Leaf blades 6–25 cm across, deeply palmatipartite, 3–7-lobed, rarely simple, shallowly cordate, sometimes very slightly peltate with 1–2 mm width of lamina below the petiole-insertion; the lobes usually oblanceolate, the median 6.5–15 × 2–6 cm, gradually acutely acuminate at the apex, narrowed at the base, entire, sparingly pubescent near the midrib or subglabrous, dark green above, glaucous beneath; lateral nerves up to 20 pairs. Stipules 4–5 × 2 mm, triangular-lanceolate, soon falling. Inflorescence 2–11 cm long, paniculate; bracts resembling the stipules. Male flowers: pedicels 5 mm long, slender; buds 2 × 1 mm, cylindric-conic; calyx lobes 6 × 4 mm, triangular, subacute, glabrous without, pubescent within near the apex, greenish, tinged orange or crimson and sometimes purple-veined; stamen filaments slender, the longer 7 mm long, the shorter 2.5 mm long, glabrous, white, anthers 1.5 mm long, with an apical hair tuft, pale yellow; disk 10-lobed, concave, the lobes acute. Female flowers; pedicels 7 mm long, extending to 2.5 cm and thickening to 3 mm in fruit, decurved; sepals 1 × 0.5 cm, triangular-ovate, subacute; disk shallowly 5-lobed; ovary 2 × 2 mm, hexagonal, pink; styles 2 mm long, botryoidal. Fruit 1.3–1.7 × 1.3–1.5 cm, ellipsoid-subglobose, rugulose, 6-winged, the wings undulate-subcrenate, greenish. Seeds 1.1 cm × 5.5 mm × 3.5 mm, ellipsoid, depressed-pentagonal, somewhat shiny, pale grey, sometimes mottled, caruncle 3 mm wide.

Zambia. W: Kitwe, y. fr. 20.iv.1969, *Fanshawe* U9 (NDO; SRGH). C: Mt. Makulu Res. Station, near Chilanga, fl. 13.vi.1961, *Angus* 2900 (FHO). **Zimbabwe**. C: Gweru Distr., Senka, female fl. 6.v.1967, *Biegel* 2138 (MO; SRGH). **Mozambique**. MS: Chimoio (Vila Pery), no date, st. *Esselen* s.n. (PRE). GI: Inharrime, Nhacongo, fl. 1960, *Claudio* s.n. (K; LUAI; PRE; SRGH). M: Maputo, Matutuíne (Bela Vista), st. 7.ii.1948, *Torre* 7294 (LISC).

Widely cultivated throughout tropical Africa for the tuberous roots which form a staple food in many tropical regions. The tubers contain prussic acid, a potent poison which is destroyed by cooking. Native of E tropical S America, as far as can be ascertained.

Vernacular names include: “manioc”, “mandioca”, “cassava” etc.

2. **Manihot grahamii** Hook., Icon. Pl. **6**, t. 530 (1843). —Rogers & Appan, Fl. Neotrop. Monogr. No. **13**: 88 (1973). —Radcliffe-Smith in F.T.E.A., Euphorb. 1: 367 (1987). Type from Brazil (Paraná).

Janipha loeflingii var. *multifida* Grah., Edinb. Philos. J. **29**: 172 (1840). Type as above.

Manihot palmata var. *multifida* (Grah.) Müll. Arg. in De Candolle, Prodr. **15**, 2: 1062 (1866).

Very similar to *M. esculenta*, but differing in having the leaf blades 12–30 cm wide, 6–13-lobed, with the lobes narrowly oblanceolate-oblong and often subpanduriform, the median lobe 6.5–16 × 1–4 cm; the disk in both sexes bright orange; the fruits slightly larger (1.8 × 1.9 cm) and without wings; the seeds biconvex with a sharp lateral ridge.

Zimbabwe. W: Bulawayo, fr. 23.i.1979, *Parsons* in *GHS* 263941 (SRGH); 25.i.1979, *Judge* in *GHS* 263942 (SRGH).

Occasionally cultivated in the Old World tropics, chiefly as an ornamental, but also in Cassava breeding research. Native of E tropical S America from Surinam to N Argentina.

3. **Manihot glaziovii** Müll. Arg. in Martius, Fl. Bras. **11**, 2: 446 (1874). —Pax in Engler, Pflanzenr. [IV, fam. 147, ii] **44**: 89 (1910). —Prain in F.T.A. **6**, 1: 839 (1912). —Eyles in Trans. Roy. Soc. South Africa **5**: 397 (1916). —Engler, Pflanzenw. Afrikas (Veg. Erde 9) **3**, 2: 133 (1921). —Brenan, Check-list For. Trees Shrubs Tang. Terr.: 219 (1949). —Keay in F.W.T.A., ed. 2, **1**, 2: 413 (1958). —J. Léonard in F.C.B. **8**, 1: 121 (1962), in adnot. —White, F.F.N.R.: 201 (1962). —Rogers & Appan, Fl. Neotrop. Monogr. No. **13**: 177 (1973). —Biegel, Check List Ornam. Pl. Rhod. Parks & Gard.: 74 (1977). —Troupin, Fl. Rwanda **2**: 232 (1983). —Radcliffe-Smith in F.T.E.A., Euphorb. 1: 370 (1987). Type from Brazil (Rio de Janeiro).

Manihot glaziovii var. *alienigena* Prain in F.T.A. **6**, 1: 840 (1912). Type: Mozambique, Manica e Sofala Province, Búzi (Nova Lusitania), Búzi R., fl. & fr. 18.i.1907, *W.H. Johnson* 67 (K, holotype).

A glabrous tree up to 10 m tall, often with a wide-spreading rounded crown. Bark smooth and pale grey, or papery, peeling and purplish-brown. Young shoots glaucous. Petiole 7–13.5 cm long. Leaf blades 10–28 cm across, deeply palmatipartite, 3–5-lobed, rarely simple, shallowly to moderately cordate, peltate with 0.7–2 cm width of lamina below the petiole insertion; the lobes usually obovate, the median 3.5–20 × 2–11.5 cm, usually subacute and mucronulate at the apex, narrowed at the base, entire, dull dark bluish-green on upper surface, pale bluish-green beneath; lateral nerves up to 25 pairs. Stipules 5–7 mm long, linear-lanceolate, soon falling. Inflorescence 2–15 cm long, paniculate; bracts resembling the stipules. Male flowers: pedicels 7–9 mm long, slender; buds 2–3 × 1.5 mm, ellipsoid-ovoid; calyx lobes 8–9 × 6 mm, triangular, obtuse, greenish-white and mauve-tinged or pale yellow with reddish markings; stamen filaments slender, the longer 1 cm long, the shorter 5 mm long, anthers 2 mm long; disk 10-lobed, concave, the lobes rounded. Female flowers: pedicels 0.8–1.4 cm long, extending to 2–3 cm and thickening to 3 mm in fruit, decurved; sepals 1.2 cm × 3.5–4 mm, lanceolate, ± acute; disk shallowly 5-lobed; ovary 4 × 4 mm, subglobose, smooth; styles 2 mm long, flabelliform, strongly decurved, stigma tuberculate, shiny. Fruit 1.5–2 × 1.4–2.2 cm, globose, smooth, becoming muricate-tuberculate on drying. Seeds 1.2–1.5 × 0.9–1 × 0.6–0.7 cm, biconvex, laterally ridged, slightly shiny, whitish, mottled ochreous; caruncle 2 × 1.5 mm, rectangular.

Zambia. N: Kankomba School, Luapula Valley, y. fr. 13.iii.1958, *Mundenda* 13 (FHO). C: Chilanga, fl. 15.ii.1958, *Angus* 1849 (FHO; K; PRE; SRGH). E: Chadiza, fl. 11.ii.1957, *Angus* 1512 (BM; FHO; K). S: Mazabuka, fr. 1931, *Stevenson* 203/31 (K). **Zimbabwe**. C: Harare (Salisbury), fr. immat. iv.1910, *Mundy* 765 (SRGH). **Malawi**. S: Zomba, male fl. 5.i.1979, *Salubeni* 2401 (MAL; MO; SRGH). **Mozambique**. MS: Manica, fl. 5.iv.1958, *Chase* 6871 (K; LISC; PRE; SRGH). M: Maputo, Matutuíne (Bela Vista), fr. 27.iv.1948, *Torre* s.n. (LISC).

Widely cultivated in the tropics generally both as a rubber-plant and as an ornamental.

Native of E Brazil (the states of Ceará, Paraiba, Pernambuco and Bahia).

Vernacular names include: "Ceará rubber".

4. **Manihot dichotoma** Ule in Notizbl. Königl. Bot. Gart. Berlin **5**, 41: 2 (1907). —Hooker, Icon. Pl. **29**: t. 2876, 2877 (1909). —Pax in Engler, Pflanzenr. [IV, fam. 147, ii] **44**: 83 (1910). —Prain in F.T.A. **6**, 1: 841 (1912). —Engler, Pflanzenw. Afrikas (Veg. Erde 9) **3**, 2: 133 (1921). —Brenan, Check-list For. Trees Shrubs Tang. Terr.: 219 (1949). —Keay in F.W.T.A., ed. 2, **1**, 2: 413 (1958). —Rogers & Appan, Fl. Neotrop. Monogr. No. **13**: 187 (1973). —Radcliffe-Smith in F.T.E.A., Euphorb. 1: 369 (1987). Types from Brazil (Bahia).

Similar to *M. glaziovii*, but differing in having the petiole 6–9 cm long; the 3–7-lobed leaf blades deeply cordate, not peltate, with the lobes 5–8 × 3.5–4.5 cm and abruptly acutely acuminate; the inflorescence 3–5 cm long and racemose; the fruits 2.5–4 × 2–3.5 cm and 6-winged, the undulate wings 3–5 mm wide; the seeds 1.5–2.2 × 1–1.2 × 0.6–0.7 cm, plano-convex, not mottled, with a caruncle c. 6 mm wide.

Zimbabwe. E: Mutare (Umtali), st. 5.i.1953, *Chase* 4750 in *GHS* 41858 (BM; K; SRGH).

Once sporadically cultivated in tropical Africa for its rubber; grown in Zimbabwe as a hedge plant.

Native of E Brazil (Bahia state).

Vernacular names include: "Jequié manicoba".

41. ADENOCLINE Turcz.

Adenocline Turcz. in Bull. Soc. Imp. Naturalistes Moscou **16**: 59 (1843).
Mercurialis sect. *Adenocline* (Turcz.) Baill., Adansonia **3**: 159 (1864).

Dioecious or monoecious, glabrous annual or perennial herbs, often slender and weak. Leaves alternate or opposite, petiolate or sessile, stipulate, sometimes stipellate, simple, entire or toothed, palminerved. Inflorescences axillary or terminal, cymose, racemose or paniculate, unisexual or rarely bisexual; male inflorescences many-flowered, female few-flowered. Male flowers shortly pedicellate, open in bud; sepals 5, scarcely imbricate at the base; petals absent; disk glands interstaminal, turbinate, sometimes confluent; stamens (4)10(14), 2-seriate, the outer ones alternate with the

Tab. 52. ADENOCLINE ACUTA. 1, distal portion of stem (× $^{2}/_{3}$); 2, male flowers (× 6), 1 & 2 from *Brummitt* 9963; 3, female flower (× 6); 4, fruit (× 6); 5, seeds (× 6), 3–5 from *Brummitt* 9097. Drawn by Christine Grey-Wilson.

sepals, filaments free, short, anthers basifixed, the cells distinct, globose, divaricate, dehiscing from the apex; pistillode absent. Female flowers: pedicels often reflexed; sepals more or less as in the male, but longer; petals absent; disk glands 3, alternating with the carpels; ovary 3-locular, with 1 ovule per loculus; styles 3, more or less connate at the base, patent-recurved, bifid or bipartite. Fruits 3-lobed, small, dehiscing septicidally into 3 bivalved cocci; endocarp thin. Seeds ovoid-subglobose, ecarunculate; testa thinly crustaceous; albumen fleshy; cotyledons scarcely broader than the radicle.

An African genus of 8 species, 7 of which are endemic to South Africa.

Adenocline acuta (Thunb.) Baill., Étud. Gén. Euphorb.: 457 (1858). —Müller Argoviensis in De Candolle, Prodr. **15**, 2: 1141 (1866). —Pax in Engler, Pflanzenr. [IV, fam. 147, vii, Addit. v] **63**: 410, t. 67 (1914). —Engler, Pflanzenw. Afrikas (Veg. Erde 9) **3**, 2: 135, t. 69 (1921). —Milne-Redhead in Kew Bull. **5**: 349 (1951). —Brenan in Mem. N.Y. Bot. Gard. **9**, 1: 73 (1954). Tab. **52**. Type from South Africa (Cape Province).

Acalypha acuta Thunb., Fl. Cap., ed 2, **2**: 546 (1823).

Adenocline mercurialis Turcz. in Bull. Soc. Imp. Naturalistes Moscou **16**, 2: 60 (1843); **25**, 2: 179 (1857). —Prain in F.C. **5**, 2: 492 (1920). —Burtt Davy, Fl. Pl. Ferns Transvaal: 306 (1932). Type from South Africa (Cape Province).

A much-branched annual or perennial herb, or suffrutex, dioecious; stems weak, trailing or scrambling, up to 3 m in extent, hollow, cylindric, ribbed, green. Leaves opposite. Petioles 1–5 cm long. Leaf blades 2–8 × 1–4 cm, triangular-ovate to ovate-lanceolate, acutely acuminate at the apex, serrate, cuneate to rounded, truncate or cordate, membranous, 5–7-nerved from the base, light green; lateral nerves in 1–4 pairs. Stipules dissected, the segments up to 4 mm long, subulate-filiform, persistent. Stipels resembling the stipules, but not more than 1 mm long. Male inflorescences up to 15 cm long and 20 cm across, laxly paniculate; bracts 1–3 cm long, narrowly-lanceolate, ± entire, cuneate-attenuate, grading below into the leaves. Female inflorescences resembling the male, but generally somewhat smaller. Male flowers: pedicels up to 3 mm long; sepals 1–2 mm long, linear, greenish; disk glands minute; stamens 6–12, c. 1 mm long, anthers yellowish. Female flowers: pedicels up to 5 mm long; sepals 2–2.5 mm long, linear-lanceolate, green; disk glands c. 0.5 mm long, compressed-cylindric-infundibuliform; ovary c. 1 mm in diameter, 3-lobed; styles 1.2 mm long, bifid, inflexed at the tip. Fruit 2 × 3 mm, 3-lobed, apically tricuspidate, smooth, green. Seeds 1.5 mm in diameter, foveolate, dark purplish brown with whitish pits.

Zimbabwe. E: Nyanga Distr., Cheshire, male fl. 4.ii.1931, *Norlindh & Weimarck* 4844 (K; LD; PRE); Mutare Distr., Inyamatshira Mt., male fl. 29.iii.1953, *Chase* 4891 (BM; K; LISC; MO; PRE; SRGH); Chimanimani Distr., Bridal Veil Falls, male fl. 30.v.1963, *Chase* 8021 (BM; COI; LISC; SRGH). **Malawi**. C: Dedza Mt., male fl. 20.iii.1955, *Exell, Mendonça & Wild* 1085 (BM; LISC; SRGH). S: Zomba Plateau, female fl. & fr. 4.vi.1946, *Brass* 16206 (K; MO; NY; SRGH).

Also in South Africa (Transvaal, KwaZulu-Natal, Cape Province) and Swaziland. In montane grassland, rainforest margins and on shady riverbanks in forest, also in thickets and on wooded granite kopjes; 1340–2300 m.

42. SUREGADA Roxb. ex Rottler

Suregada Roxb. ex Rottler in Ges. Naturf. Freunde Berlin Neue Schriften **4**: 206 (1803).
Gelonium Roxb. ex Willd., Sp. Pl. **4**, 2: 831 (1806).

Dioecious, sometimes monoecious, usually glabrous trees or shrubs. Leaves alternate, shortly petiolate, stipulate, simple, entire, denticulate or crenate-serrate, vesicular-punctate, penninerved. Stipules soon falling, leaving prominent scars. Flowers fasciculate or glomerulate in leaf-opposed cymules, or subsolitary, somewhat gummy at first; bracts minute. Male flowers usually pedicellate; sepals (4)5–6(7), free, imbricate, unequal, glandular or not; petals absent; disk glands usually interstaminal, confluent; stamens 6–30 (African species), free, anthers dorsifixed, extrorse, longitudinally dehiscent at the back; pistillode absent. Female flowers pedicellate; sepals (4)5–6(8), free, imbricate, subequal, glandular or not; petals absent; disk annular, sometimes lobed; staminodes 5–10; ovary 2–4-celled, with 1 ovule per cell; styles 2–4, united at the base, bifid or multifid, short. Fruit 2–4-lobed or subglobose, dehiscing into bivalved cocci, loculicidal or indehiscent; exocarp smooth or reticulate; endocarp woody, coriaceous or crustaceous; columella persistent. Seeds ecarunculate; exotesta pulpy; endotesta crustaceous; albumen fleshy; cotyledons broad, flat.

A palaeotropical genus of c. 40 species, 8 of which are African.

1. Leaves abruptly acutely acuminate with the vesicles central in each reticulation, often drying bright green or yellowish-green; sepals ciliate, usually with a dorsal gland, not hooded; stamens 6–14; endotesta foveolate ····················· 1. *zanzibariensis*
– Leaves rounded, obtuse or obtusely acuminate at the apex, with the vesicles covering each reticulation, usually drying a dull grey-green; sepals glabrous, eglandular, hooded or horned; stamens 12–30; endotesta smooth ································ 2
2. Leaves 3–14 × 1–7 cm, obtuse or obtusely acuminate; lateral nerves in 7–15 pairs; sepals hooded ··· 2. *procera*
– Leaves 1.5–6.5 × 1–2.5 cm, obtuse or rounded at the apex; lateral nerves in 5–6 pairs; sepals horned ··· 3. *africana*

1. **Suregada zanzibariensis** Baill., Adansonia **1**: 254 (1861). —J. Léonard in Bull. Jard. Bot. État **28**, 4: 446 (1958). —Dale & Greenway, Kenya Trees & Shrubs: 221 (1961). —Drummond in Kirkia **10**: 253 (1975). —K. Coates Palgrave, Trees Southern Africa, ed. 2, rev.: 434 (1983). —Radcliffe-Smith in F.T.E.A., Euphorb. 1: 377 (1987). —Beentje, Kenya Trees, Shrubs Lianas: 223 (1994). Type from Zanzibar.

Gelonium zanzibariense (Baill.) Müll. Arg. in De Candolle, Prodr. **15**, 2: 1130 (1866). —Pax in Engler, Pflanzenr. [IV, fam. 147, iv] **52**: 21, t. 5 (1912). —Prain in F.T.A. **6**, 1: 948 (1912). —Engler, Pflanzenw. Afrikas (Veg. Erde 9) **3**, 2: 135, t. 70 (1921). —Brenan, Check-list For. Trees Shrubs Tang. Terr.: 215 (1949). —Topham, Check List For. Trees Shrubs Nyasaland Prot.: 51 (1958) sphalm. *"zanzibarense"*. —Mogg in Macnae & Kalk, Nat. Hist. Inhaca Isl., Moçamb., rev. ed.: 148 (1969). Type as above.

Gelonium adenophorum sensu Sim, For. Fl. Port. E. Afr.: 105 (1909), non Müll. Arg. (1866).

Gelonium serratum Pax & K. Hoffm. in Engler, Planzenr. [IV, fam. 147, iv] **52**: 23 (1912). —Prain in F.C. **5**, 2: 495 (1920). Type from South Africa (KwaZulu-Natal).

A glabrous shrub or small tree up to 10 m tall, rarely taller, generally dioecious; branches horizontal. Bark smooth, grey, often with fine longitudinal markings. Twigs greenish. Stipules 1 × 1 mm, triangular-ovate, subacute. Petioles 3–7 mm long. Leaf blades 2–13 × 1.5–7 cm, obovate to elliptic-oblanceolate, abruptly acutely acuminate at the apex, cuneate or ± rounded at the base, entire or sharply denticulate, coriaceous, glossy, paler beneath than above, often drying bright green or yellowish-green; lateral nerves in 5–9 pairs, looped; tertiary and quaternary nerves reticulate with raised pustules in the centre of each reticulation. Male flowers: pedicels 1–2 mm long; sepals 5, 2.5–3 × 2–2.5 mm, suborbicular, each with a yellow gland on the abaxial face, ciliate, the outer greenish, the inner creamy-white; glands confluent; stamens (6)14, usually in an outer whorl of 9 and an inner of 5, filaments 1–2 mm long, anthers 0.5–0.7 mm long. Female flowers: pedicels somewhat stouter than in the male flowers; sepals ± as in the male flowers; disk 5-angular or shallowly 10-lobed; staminodes small, subulate or occasionally staminiform; ovary c. 1.5 mm in diameter, 3-celled, subglobose, smooth; styles 3, 1 mm long, bifid, stigmas subulate. Fruit 7 × 8 mm, (1)3-celled, 3-lobed to subglobose, smooth, green. Seeds c. 4 × 4 mm, broadly compressed-ovoid, foveolate; exotesta red, grey when dry, endotesta red or black.

Zimbabwe. S: Gonarezhou (Gona-Re-Zhou), fl. 2.vi.1971, *Grosvenor* 605 (K; LISC; PRE; SRGH). **Mozambique**. N: Cabo Delgado, Palma, female fl. & y. fr. 18.iv.1964, *Torre & Paiva* 12142 (LISC). Z: Pebane, fr. 12.i.1968, *Torre & Correia* 17103 (LISC). MS: Chinizíua, fl. 19.x.1957, *Gomes e Sousa* 4415 (COI; K; LISC; PRE; SRGH). GI: Bazaruto Island, male fl. 20.x.1958, *Mogg* 28776 (BM; K; LISC; PRE; SRGH). M: Matutuíne, fr. 13.i.1983, *Groenendijk & de Koning* 185 (LMU; MO).

Also in Somalia, Kenya, Tanzania (incl. Zanzibar and Pemba), Madagascar and South Africa (KwaZulu-Natal). Low altitude *Brachystegia* and closed *Androstachys* woodlands on sandy soil, coastal plain mixed evergreen forest understorey and riverine forest, and coastal mixed deciduous woodland with *Spirostachys, Hymenocardia* and *Uapaca,* and in dense *Millettia* thickets, also in saltmarshes; sea level to 220 m.

2. **Suregada procera** (Prain) Croizat in Bull. Jard. Bot. Buitenzorg, Ser. 3, **17**, 2: 216 (1942). —J. Léonard in Bull. Jard. Bot. État **28**, 4: 447 (1958). —Dale & Greenway, Kenya Trees & Shrubs: 220 (1961). —J. Léonard in F.C.B. **8**, 1: 125 (1962). —White, F.F.N.R.: 204 (1962). —Drummond in Kirkia **10**: 253 (1975). —K. Coates Palgrave, Trees Southern Africa, ed. 2, rev.: 433 (1983). —Radcliffe-Smith in F.T.E.A., Euphorb. 1: 378, t. 71 (1987). —Beentje, Kenya Trees, Shrubs Lianas: 223 (1994). Type from Kenya.

Gelonium procerum Prain in Bull. Misc. Inform., Kew **1911**, 5: 233 (1911); in F.T.A. **6**, 1: 948 (1912). —Eyles in Trans. Roy. Soc. South Africa **5**: 398 (1916). —Eggeling & Dale, Indig. Trees Uganda, ed. 2: 129 (1952). Type as above.

A glabrous tree up to 24 m tall, generally dioecious; branches lax, horizontal. Wood white, close-grained. Bark grey-brown, smooth or minutely fissured. Twigs greenish. Stipules 2 × 1 mm, triangular, acute. Petioles 3–10 mm long. Leaf blades 3–14 × 1–7 cm, elliptic to elliptic-ovate, obtuse or obtusely acuminate at the apex, asymmetrically cuneate at the base and decurrent onto the petiole, entire or shallowly crenate-serrate, coriaceous, glossy, dark green above, paler beneath, the pustules paler still, drying a dull grey-green; lateral nerves in 7–11(15) pairs, weakly or not looped; tertiary and quaternary nerves reticulate with raised pustules covering each reticulation. Male flowers: pedicels 1–2 mm long; sepals 5–7, 2.5–3 × 2.5–3 mm, suborbicular-ovate, not ciliate, eglandular, the outer slightly hooded and greenish, the inner flat and ± cream coloured; glands confluent; stamens (16)20–30, usually with an outer whorl of 15–20 and an inner of 5–15, filaments 3–4 mm long, white, anthers 1 mm long, yellow. Female flowers: pedicels 2–3 mm long, extending to 5–6 mm in fruit, stouter than in the male flowers; sepals 6–8, ± as in the male flowers; disk crenate; staminodes usually subulate; ovary 2 mm in diameter, usually 3-celled, 3-lobed to subglobose, smooth; styles usually 3, 1–2 mm long, bifid, stigmas oblong, slightly lobulate. Fruit 1 × 1.5 cm, usually 3-celled, obovoid-trigonous, ± smooth, dark green later turning brown. Seeds 6 × 5 mm, ± ovoid, smooth; exotesta pearly-grey on drying, endotesta dark grey-brown.

Zambia. W: Mufulira, fr. 8.ix.1954, *Fanshawe* 1545 (K; NDO; SRGH). **Zimbabwe**. E: Chirinda For. Res., female fl. ix.1966, *Goldsmith* 71/66 (K; LISC; PRE; SRGH). S: 6 km north of Runde R. (Lundi R.) on Masvingo (Ft. Victoria) road, Chirongwe Hills, male fl. 10.v.1970, *Pope* 286 (K; SRGH). **Malawi**. N: Wenya Hills, st. 17.iv.1963, *Chapman* 1884 (SRGH). C: Ntchisi Mt., male fl. 21.ii.1959, *Robson* 1704 (BM; K; LISC; SRGH). S: Mulanje Distr., Maudzi Hill, female fl. 6.ix.1983, *Seyani & Balaka* 1374 (K; MAL). **Mozambique**. MS: eastern foothills of Chimanimani, 15 km south of Mussapa R., fl. 26.iii.1973, *Müller* 2087 (K; MO; PRE; SRGH).

Also in Zaire, Sudan, Ethiopia, Uganda, Kenya, Tanzania and South Africa (Transvaal, KwaZulu-Natal). Subcanopy or understorey tree of low or medium altitude mixed evergreen forest, also in riverine and gully forest and swamp forest (mushitu); 300–2133 m.

S. procera and *S. zanzibariensis* are almost completely mutually exclusive geographically and altitudinally in the Flora Zambesiaca area.

3. **Suregada africana** (Sond.) Kuntze, Revis. Gen. Pl. **2**: 619 (1891). —J. Léonard in Bull. Jard. Bot. État **28**, 4: 446 (1958). —K. Coates Palgrave, Trees Southern Africa, ed. 2, rev.: 433 (1983). Tab. **53**. Syntypes from South Africa.

Ceratophorus africanus Sond. in Linnaea **23**: 121 (1850).

Suregada ceratophora Baill., Adansonia **3**: 154 (1862). Types as above.

Gelonium africanum (Sond.) Müll. Arg. in De Candolle, Prodr. **15**, 2: 1129 (1866). —Sim, For. Fl. Port. E. Afr.: 105 (1909). —Pax in Engler, Pflanzenr. [IV, fam. 147, iv] **52**: 21 (1912). —Prain in F.C. **5**, 2: 495 (1920). —Engler, Pflanzenw. Afrikas (Veg. Erde 9) **3**, 2: 136 (1921).

A glabrous shrub, or small slender lax tree up to 3 m tall, dioecious. Twigs grey. Stipules 1.5 × 1.5 mm, triangular, acute. Petioles 1–2 mm long. Leaf blades 1.5–6.5 × 1–2.5 cm, obovate, obtuse or rounded at the apex, rounded-cuneate at the base, subentire or crenate-serrate at the apex, firmly chartaceous, drying a dull greyish- or brownish-green; lateral nerves in 5–6 pairs, not or weakly looped; tertiary and quaternary nerves reticulate with shallow pustules covering each reticulation. Male flowers: pedicels 2–5 mm long, turbinate; sepals 3(4), 2 × 2 mm, suborbicular, not ciliate, eglandular, the outer hooded, cornute, greenish, the inner flat and cream coloured, petaloid; glands confluent, pubescent; stamens 12–14, usually in an outer whorl of 9–10 and an inner of 4–5, filaments 2 mm long, anthers 0.7 mm long. Female flowers: pedicels 3–4 mm long, not or scarcely extending in fruit, stouter

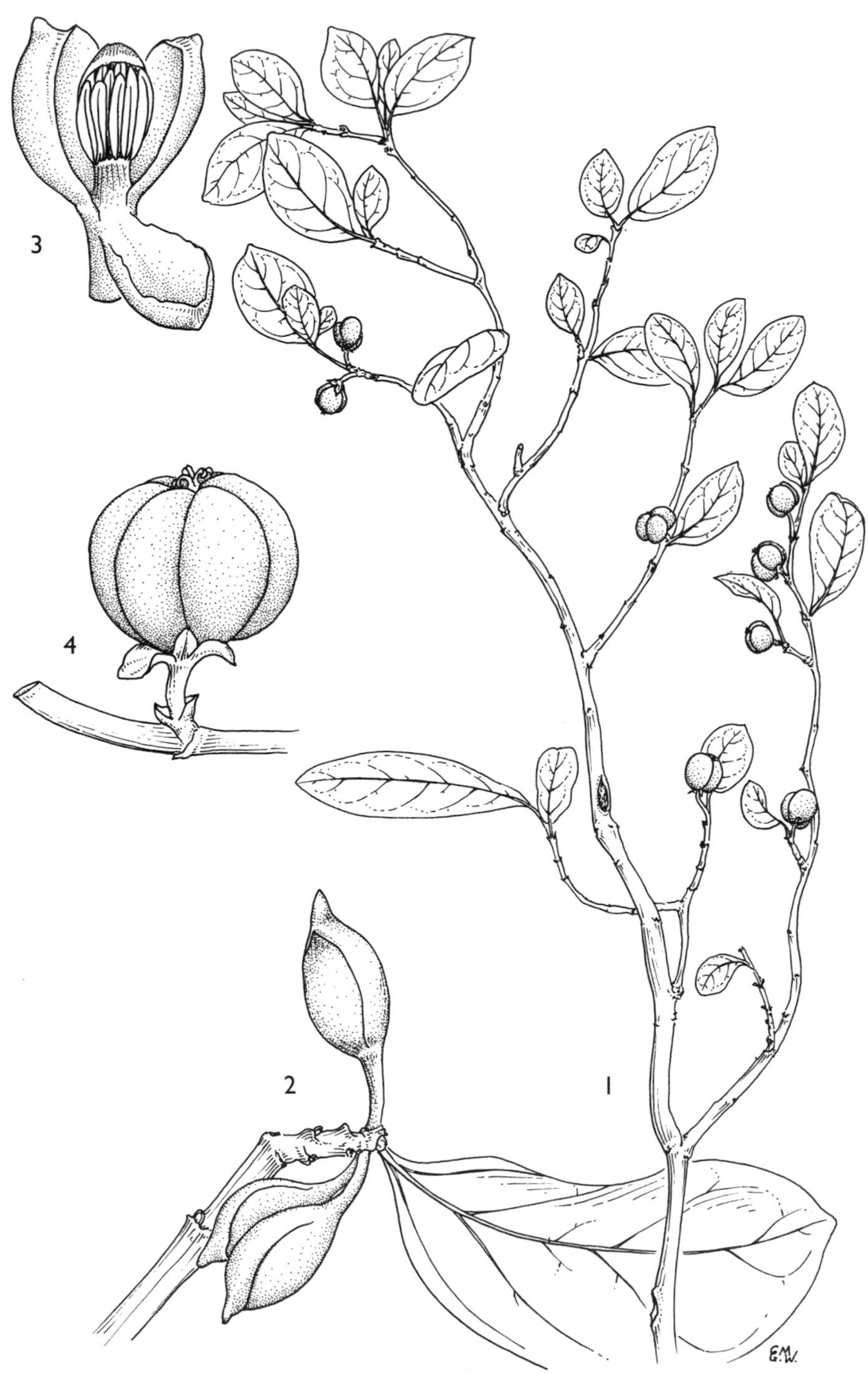

Tab. 53. SUREGADA AFRICANA. 1, distal portion of branch (× $^{2}/_{3}$), from *Balsinhas* 206; 2, male inflorescence (× 6); 3, male flower with stamens displayed (× 6), 2 & 3 from *Hornby* 2636; 4, fruit (× 4), from *Balsinhas* 206. Drawn by Christine Grey-Wilson.

than in the male flowers; sepals ± as in the male flowers, but with rather more prominent horns; disk subcrenate; staminodes minute, subulate; ovary 3-celled, subglobose, 1.5–2 mm in diameter, smooth; styles 3, 0.7 mm long, bifid, recurved. Fruit 7–8 × 8–10 mm, 3-celled, subturbinate-globose, 3-lobed, ± smooth. Seeds 5 × 5 mm, subglobose, smooth; exotesta whitish.

Mozambique. M: Ressano Garcia, female fl. 25.xii.1897, *Schlechter* 11932 (BM; K); Goba, male fl. 15.xi.1940, *Torre* 2034 (LISC); Maputo, y. male fl. 28.iii.1947, *R.M. Hornby* 2636 (K; SRGH); Goba, near R. Maivavo, y. fr. 5.xi.1960, *Balsinhas* 206 (BM; COI; K; LISC; LMA; PRE; SRGH); Goba, near Railway Station, male fl./subst. 3.x.1961, *Balsinhas* 518 (BM; COI; K; LMA; PRE).

Also in South Africa (Transvaal, KwaZulu-Natal, Cape Province) and Swaziland. Subcanopy and margins of coastal evergreen forest, and in riverine forest and mixed deciduous woodlands; near sea level to 300 m.

The description of the mature fruits and seeds was made from South African material.

43. JATROPHA L.

Jatropha L., Sp. Pl.: 1006 (1753), pro parte; Gen. Pl., ed. 5: 437 (1754).
Curcas Adans., Fam. Pl. **2**: 356 (1763).
Adenoropium Pohl, Pl. Bras. **1**: 12 (1827).

Monoecious, rarely dioecious, trees shrubs subshrubs or herbs with the stems arising from a thick perennial rootstock. Indumentum simple and/or glandular. Leaves alternate, stipulate, petiolate or sessile, simple, entire, lobed or partitely divided, penninerved or palminerved. Stipules subulate, bifid or multifid, usually glandular. Petioles usually eglandular. Inflorescences terminal or subterminal, often corymbiform, cymose, androgynous, protogynous, a solitary female flower terminating each major axis, lateral cymules male; bracts entire or glandular-stipitate. Male flowers: calyx usually 5-lobed, lobes imbricate; petals 5, free or sometimes laterally coherent, imbricate or contorted; disk usually of 5 free glands; stamens 8 (Flora Zambesiaca area), in 2 fused whorls (5 + 3), outer whorl opposite the petals, anthers dorsifixed, longitudinally dehiscent; pistillode absent. Female flowers: calyx and petals more or less as in the male flowers; disk annular, 5-lobed, or sometimes of 5 free glands; staminodes (when present) filiform; ovary usually 3-locular, with 1 ovule per loculus; styles usually united at the base, erect or spreading, stigmas 3, usually bifid and somewhat tumid. Fruit shallowly 3-lobed, dehiscing septicidally into 3 bivalved cocci, less often loculicidally into 3 valves, sometimes subdrupaceous, indehiscent; endocarp crustaceous or slightly woody; columella persistent. Seeds usually shiny, carunculate, the caruncle usually bifid, fimbriate; testa crustaceous; albumen fleshy; cotyledons broad, flat.

A genus of 156 species ranging throughout the tropics and subtropics, especially in America and Africa; also in extratropical N America and South Africa.

Jatropha integerrima Jacq. is reported as cultivated in Biegel, Check List Ornam. Pl. Rhod. Parks & Gard.: 66 (1977).

1. Leaves entire, penninerved ··· 2
– Leaves lobed or mostly lobed, often deeply so, and usually palminerved ··· 6
2. Leaf blades not more than 5 mm wide, and usually much less ··· 1. *baumii*
– Leaf blades 10–60 mm wide ··· 3
3. Plant prostrate or procumbent ··· 2. *seineri*
– Plant erect ··· 4
4. Plant pilose or villous ··· 5. *hirsuta*
– Plant glabrous ··· 5
5. Stipules multifid; petioles 5–20 mm long; leaf blades glandular-serrulate ··· 3. *prunifolia*
– Stipules usually simple; petioles 1–4 mm long or leaves subsessile; leaf blades usually subentire ··· 4. *latifolia*
6. Trees or shrubs ··· 7
– Perennial herbs with annual stems arising from an often reddish woody rootstock ··· 13
7. Petioles glanduliferous ··· 6. *gossypiifolia*
– Petioles eglandular ··· 8
8. Leaves peltate ··· 9. *podagrica*
– Leaves not peltate ··· 9

9. Leaf blades pentagonal or shallowly 5-lobed; male petals laterally coherent; fruits loculicidal · · · · · · 7. *curcas*
– Leaf blades deeply palmatilobed, palmatifid, palmatipartite or palmatisect; male petals free; fruits septicidal or subdrupaceous · · · · · · 10
10. Lobes of leaves 10–12, sometimes irregularly pinnatipartite · · · · · · 8. *multifida*
– Lobes of leaves 5–7, entire or serrate · · · · · · 11
11. Lobes of leaves serrate · · · · · · 12. *spicata*
– Lobes of leaves entire · · · · · · 12
12. Lobes of leaves unequal; leaf base truncate to shallowly cordate; stipules persistent; bracts stipitate-glandular at the base; fruits 12 × 10–12 mm; seeds 9.5 × 5 × 4 mm · · 10. *variifolia*
– Lobes of leaves subequal; leaf base deeply cordate; stipules fugacious; bracts entire, eglandular; fruits 17 × 15 mm; seeds 12–13 × 8–9 × 6–7 mm · · · · · · 11. *subaequiloba*
13. Plant usually scapose, or if not scapose, then petioles up to 28 cm long · · · · · · 14
– Plant not scapose; petioles not more than 2 cm long or leaves sessile or subsessile · · · · 15
14. Leaf blades glabrous on upper surface · · · · · · 13. *macrophylla*
– Leaf blades tomentose on upper surface · · · · · · 14. *scaposa*
15. Plant glabrous · · · · · · 16
– Plant, or at least the young stems, sparingly to densely pubescent or hirsute-setose · · · 19
16. Stipule-segments glandular at the apex; leaf lobes closely stipitate-glandular · · · · · · 17. *pachyrrhiza*
– Stipules and leaves eglandular · · · · · · 17
17. Leaf lobes entire, rarely irregularly lobulate or repand-dentate; outer whorl of stamens fused to column · · · · · · 19. *campestris*
– Leaf lobes closely and sharply denticulate or bidenticulate; outer whorl of stamens free · · 18
18. Stipules 1–2 mm long, simple, subulate, falling; anthers 1.25 mm long · · · · · · 20. *monroi*
– Stipules 4–7 mm long, sometimes laciniate, strap-shaped, persistent; anthers 3 mm long · · · · · · 21. *loristipula*
19. Leaf blades irregularly pinnatipartite · · · · · · 20
– Leaf blades palmatilobed, palmatifid, palmatipartite, subpalmatipartite or rarely elobate · · · · · · 21
20. Plant hirsute-setose; stems arising from the tuber; outer whorl of stamens fused to column · · · · · · 22. *schlechteri*
– Plant pubescent when young; stems arising from caudiculi (short underground stems); outer whorl of stamens ± free · · · · · · 23. *erythropoda*
21. Petals pubescent without, pilose within · · · · · · 16. *zeyheri*
– Petals completely glabrous · · · · · · 22
22. Plant pilose; leaf blades all 5-partite · · · · · · 15. *botswanica*
– Plant sparingly pubescent; leaf blades 3-lobed, 3-fid or 3-partite, rarely some elobate or somewhat 5-partite · · · · · · 18. *schweinfurthii*

1. **Jatropha baumii** Pax in Warb., Kunene-Sambesi Exped. Baum: 283 (1903); in Engler, Pflanzenr. [IV, fam. 147, i] **42**: 64 (1910). —Hutchinson in F.T.A. **6**, 1: 782 (1912). —Engler, Pflanzenw. Afrikas (Veg. Erde 9) **3**, 2: 115, 117 (1921). —Radcliffe-Smith in Kew Bull. **46**: 141 (1991). Type from Angola.

An erect, single-stemmed pubescent herb up to c. 20 cm high, though commonly much less, arising from a globose tuber; tuber verrucose, up to 6 cm in diameter (shrinking to c. 4 cm when dried) lying just below, or down to 8 cm below, ground level, with a reddish sap. Stems whitish-purple below, greenish-purple above, with a milky latex. Stipules c. 0.5 mm long, subulate-filiform, or absent. Petioles up to 1 mm long, or leaves subsessile. Leaf blades 0.5–3.5 × 0.1–0.5 cm, linear-lanceolate or linear-setaceous, the lower ones scale like, acute, entire, undulate, membranaceous, penninerved, sparingly pubescent to subglabrous on both surfaces, somewhat glaucous; the lateral nerves in up to 12 pairs. Inflorescences terminal or terminating short axillary shoots; bracts resembling the smaller foliage leaves, reddish. Male flowers: pedicels up to 2 mm long; calyx 3 mm long, green, lobes c. 1 × 1 mm, triangular-ovate, obtuse, entire, glabrous, reddish; petals 5 × 1.5 mm, spathulate-oblanceolate, rounded, glabrous without, appressed-pubescent within below, pale cream-coloured or whitish; disk glands 5, free, 0.5 mm long, obcordate, compressed, emarginate; stamens 8, ± free, 3–4 mm long, filaments glabrous, anthers 1.5 mm long, yellow. Female flowers: pedicels 2 mm long, stouter than in the male; calyx 3.5 mm long, lobes 2–2.5 × 1 mm, triangular-lanceolate, acute, entire, glabrous; petals 6 × 1 mm, linear-oblanceolate, glabrous, otherwise as in male; disk 1.5 mm in diameter, deeply 5-lobed, the lobes flattened, rounded; ovary 2 × 2 mm, 3-lobed to subglobose, apiculate, sparingly minutely puberulous; stylar column 1 mm tall, style arms 1.5 mm long, spreading, bipartite, stigmatic surface papillose. Fruit 8–10 × 9–10 mm, 3-lobed to subglobose, smooth, subglabrous or pubescent, green. Seeds 6.5–7 × 4.5 × 3.5 mm, ellipsoid-subcylindric, smooth, grey, mottled brown; caruncle 1 × 2.5 mm, depressed-hemispherical, 2-lobed, brownish, edged with cream-buff.

Zambia. B: Sesheke Distr., c. 3 km NW of Masese Valley, fl. & y. fr. 11.viii.1947, *Brenan & Keay* 7674 (FHO; K). S: Machili, fl. 14.xi.1960, *Fanshawe* 5884 (NDO; PRE; SRGH, pro parte).
Also in Angola. In open sandy plains near river; 800–1100 m.

2. **Jatropha seineri** Pax in Bot. Jahrb. Syst. **43**: 84 (Feb. 1909). —Pax in Engler, Pflanzenr. [IV, fam. 147, i] **42**: 65 (1910). —Hutchinson in F.T.A. **6**, 1: 781 (1912). —Engler, Pflanzenw. Afrikas (Veg. Erde 9) **3**, 2: 116, 117 (1921). —P.G. Meyer in Merxmüller, Prodr. Fl. SW. Afrika, fam. 67: 37 (1967). Tab. **54**. Type: Caprivi Strip, fl. 21.x.1906, *Seiner* 109 (B†, holotype).
Jatropha humilis N.E. Br. in Bull. Misc. Inform., Kew **1909**: 139 (Apr. 1909). Syntypes: Botswana, Ngamiland Distr., Khwebe (Kwebe), xii.1896, *Lugard* 56, 159 (K).

A multistemmed, pubescent, glaucous suffrutex, with a copious white, milky, latex; stems prostrate or straggling, up to 42 cm long, but commonly much less, branching from a scaly underground stem which arises from a tuber; tuber up to 10 cm in diameter and resembling a coconut in shape and size, lying 15–70 cm below ground level, with red flesh. Stipules c. 1 mm long, subulate-filiform, or absent. Petioles 2–12 mm long. Leaf blades 3–7.5(10) × 1–3 cm, lanceolate, very rarely with 2 small basal lobes, acute at the apex, entire or crisped-undulate, cuneate to rounded or occasionally shallowly cordate at the base, membranaceous, penninerved, sparingly pubescent or densely tomentose (var. *tomentella*) on both surfaces, sometimes purplish beneath; the lateral nerves in 9–15(18) pairs, brochidodromous. Inflorescences terminating the short main orthotropic axis and sometimes also the plagiotropic axes (the axes of leafy or floriferous lateral shoots of limited growth); peduncle 1.5–3 cm long; bracts ovate, lanceolate, linear or subulate, the lowest up to 1.2 cm long, the remainder progressively smaller upwards, glandular or not. Male flowers: pedicels short, extending to 3 mm at maturity; calyx 2–2.5 mm long, lobes 1–1.25 mm wide, triangular-ovate, subacute or obtuse, subentire, glabrous, green; petals 4–5 × 1.5–1.75 mm, spathulate-oblanceolate, rounded, glabrous without, pilose within below, cream-coloured, pale yellow, pale pink, whitish or greenish, carmine at the base; disk glands minute, 5, free, compressed, truncate; stamens 8, c. 4 mm long, 5 outer free, 3 inner connate in the lower half, filaments pilose, anthers 2 mm long, yellow. Female flowers: pedicels 2–3 mm long, stouter than in the male; calyx 4 mm long, lobes 2.5–3 × 1 mm, triangular-lanceolate, acute, entire, ciliate, pubescent; petals 6 × 2 mm, oblong-oblanceolate, glabrous, otherwise as in male; disk deeply 5-lobed, the lobes unequal, flattened, truncate; ovary 2 × 2 mm, 3-lobed to subglobose, pubescent; styles 3 mm long, united at the base, divaricate, stigmas bifid, rugulose. Fruit 9–11 × 9–12 mm, oblately spheroidal, shallowly 3-lobed, minutely rugulose, subglabrous or sparingly to evenly pubescent, pale green. Seeds 6.8–7.2 × 4.4–4.9 × 3.5–4 mm, ovoid-ellipsoid, smooth, pale yellowish-grey, sometimes brown-flecked, with a large 2-lobed flabelliform caruncle 2 × 3.5–4.5 mm, chestnut at the base shading to buff.

Leaf blades sparingly pubescent above and beneath; inflorescence axes slender, less than 2 mm thick; fruit subglabrous to sparingly pubescent ······················· var. *seineri*
Leaf blades densely tomentose above and beneath; inflorescence axes stout, up to c. 4 mm thick; fruit evenly pubescent ··································· var. *tomentella*

Var. **seineri**

Leaf blades sparingly pubescent above and beneath; inflorescence axes slender, less than 2 mm thick; fruit subglabrous to sparingly pubescent.

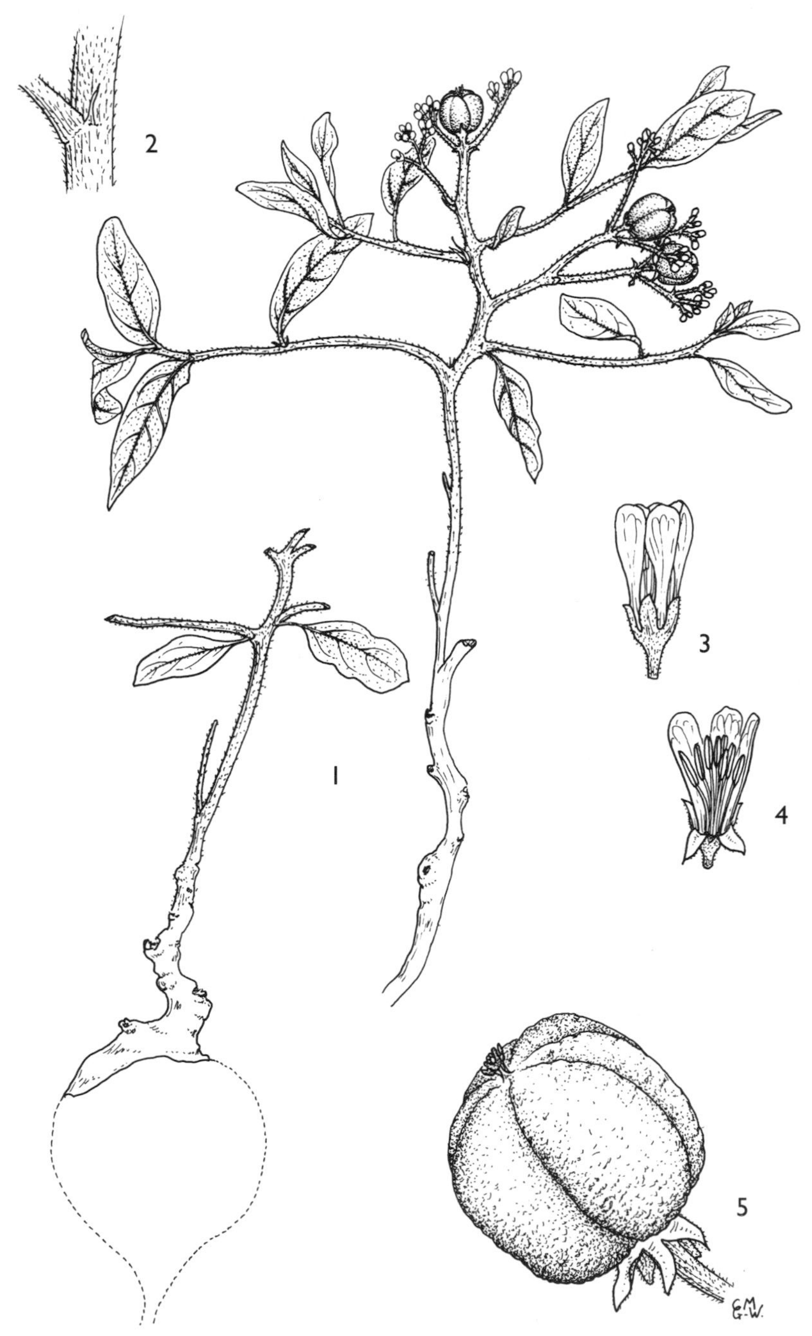

Tab. 54. JATROPHA SEINERI. 1, habit, underground stem and tuber outline displayed (× $^{2}/_{3}$); 2, stipule (× 4); 3, male flower (× 4); 4, male flower, with 2 petals removed (× 4); 5, fruit (× 4), 1–5 from *Chisumpa* 116. Drawn by Christine Grey-Wilson.

Caprivi Strip. fl. 21.x.1906, *Seiner* 109 (B†). **Botswana**. N: Maun, fr. 3.xii.1967, *Lambrecht* 431 (K; PRE; SRGH). **Zambia**. B: Kande Lake, 13 km NE of Mongu, fl. & o. fr. 11.xi.1959, *Drummond & Cookson* 6344 (K; LISC; MO; PRE; SRGH). S: Livingstone, fl. & fr. 7.xi.1973, *Chisumpa* 116 (K; NDO). **Zimbabwe**. N: Gokwe, fl. & fr. 10.i.1964, *Bingham* 1049 (SRGH). W: Hwange (Wankie) Nat. Park, y. fr. 27.xi.1968, *Rushworth* 1305 (SRGH).

Also in Angola and Namibia. Kalahari Sand woodland, and grassland with *Lonchocarpus* and *Grewia* shrubs and scattered trees on Kalahari Sand, also in basaltic soil and clayey loam beside the Zambezi R.; 915 m.

Var. **tomentella** Radcl.-Sm. in Kew Bull. **46**: 154 (1991). Type: Zambia, Mazabuka, Central Research Station, fl. & y. fr. 23.x.1931, *Trapnell* in *C.R.S.* 454 (K, holotype; PRE).

?*Jatropha decumbens* Pax & K. Hoffm. in Engler, Pflanzenr. [IV, fam. 147, vii, Addit. v] **63**: 398 (1914). Type from Namibia, n.v.

Leaf blades densely tomentose above and beneath; inflorescence axes stout, c. 4 mm thick; fruit evenly pubescent.

Zambia. S: Mazabuka, Central Res. Station, fl. & o. fr. 4.xi.1930, Vet. Officer in *C.R.S.* 94 (PRE).

Not known elsewhere. On heavy red soil in wooded grassland; c. 1005 m.

Vernacular names as recorded in specimen data include: “kepea”, “musanungu”.

3. **Jatropha prunifolia** Pax in Engler, Pflanzenw. Ost-Afrikas **C**: 240 (1895); in Engler, Pflanzenr. [IV, fam. 147, i] **42**: 54 (1910). —Hutchinson in F.T.A. **6**, 1: 782 (1912). —Engler, Pflanzenw. Afrikas (Veg. Erde 9) **3**, 2: 114, (1921). —Brenan, Check-list For. Trees Shrubs Tang. Terr.: 217 (1949). —Radcliffe-Smith in F.T.E.A., Euphorb. 1: 351 (1987). Type from Tanzania (Tanga Province).

A glabrous woody herb or subshrub up to 75 cm high; stems simple, erect, somewhat fleshy, arising from a soft-woody rootstock. Inner bark red. Stipules 5–7 mm long, multifid; the segments filiform, gland-tipped. Petioles 0.5–2 cm long. Leaf blades 2–8.5 × 1–4.5 cm, elliptic-ovate to elliptic-lanceolate, rarely 1(2)-lobed at the base, obtuse or subacute at the apex, glandular-serrulate, rounded or cuneate at the base, membranaceous, 5–7-nerved from the base, the inner pair of nerves often running two-thirds the length of the blade, the main lateral nerves in c. 5 pairs. Inflorescences terminal, subcapitate; peduncle c. 3 cm long; bracts up to 1 mm long, narrowly triangular-lanceolate, densely stipitate-glandular. Male flowers: pedicels up to 3 mm long; calyx 2 mm long, the lobes c. 1 × 1 mm, ovate, rounded, subentire, reddish; petals 3.5 × 1.5 mm, obovate-oblanceolate, rounded, glabrous without, pilose within below, greenish-white; disk glands 5, free, turbinate, truncate; stamens 8, the 5 outer 3 mm long and two-thirds united, the 3 inner 3.5 mm long and three-quarters united, filaments glabrous, anthers 0.6 mm long. Female flowers: pedicels 1 mm long, stout; calyx 3 mm long, lobes 2.5 × 1 mm, triangular-lanceolate, acute, subentire, with 1–2 lateral glands, glabrous; petals 5–6 × 1.5 mm, narrowly oblanceolate, otherwise as in the male; disk deeply 5-lobed, the lobes flattened, rounded; ovary 2 × 2 mm, ellipsoid-subglobose, scarcely 3-lobed, glabrous; styles 1 mm long, united at the base, divaricate, stigmas thick, capitate, bifid, minutely papillose. Fruit 8–11 × 9–10 mm, ellipsoid, shallowly 3-lobed, ± smooth, glabrous, green. Seeds 7 × 3 × 3 mm, ellipsoid-subcylindric, smooth, brownish-buff, dark brown-mottled, with a fimbriate caruncle 2.5 mm wide.

Zambia. N: near Kapamba R., fl. 10.i.1966, *Astle* 4322 (K; NDO; SRGH). W: Mwinilunga, y. fl. & fr. 7.xi.1955, *Holmes* 1319 (K; NDO). E: west of Sasare, fl. & fr. 10.xii.1958, *Robson* 886 (BM; K; LISC; SRGH).

Also in Kenya and Tanzania. Occasional among limestone rubble, and on stony hills in miombo woodland; 610–700 m.

The variation outlined in F.T.E.A., Euphorb. 1: 351 (1987) can all be accommodated within the range of the species, without the need for formal recognition of any infraspecific taxa.

4. **Jatropha latifolia** Pax in Bot. Jahrb. Syst. **23**: 531 (1897); in Engler, Pflanzenr. [IV, fam. 147, i] **42**: 61 (1910). —Prain in F.C. **5**, 2: 423 (1920). —Engler, Pflanzenw. Afrikas (Veg. Erde 9) **3**, 2: 115, 117 (1921). —Burtt Davy, Fl. Pl. Ferns Transvaal: 302 (1932). Type from South Africa (Transvaal).

Var. **subeglandulosa** Radcl.-Sm. in Kew Bull. **46**: 146 (1991). Type: Mozambique, Maputo, Namaacha, o. fr. 12.i.1948, *Torre* 7118 (LISC, holotype).

A completely glabrous perennial herb up to 1 m high; stems usually simple, arising from a stout woody rootstock. Stipules c. 1 mm long, usually simple, setaceous, caducous. Petioles 1–4 mm long, or leaves subsessile. Leaf blades 4–12 × 2–6 cm, simple, elliptic, acute or subacute at the apex, rounded at the base, subentire or irregularly minutely glandular-denticulate on the margin, coriaceous; lateral nerves in 9–12 pairs. Inflorescences 10–13 cm long, terminal, subcorymbiform; peduncle up to 8 cm long; bracts up to 8 mm long, linear-lanceolate, entire. Male flowers: pedicels c. 2 mm long; calyx tube 1 mm long, the lobes 3 × 1.5 mm, ovate-lanceolate, obtuse or subacute, entire or almost so; petals 7–9 × 3–3.5 mm, oblong-spathulate, sparingly ferrugineous-pubescent within at the base, otherwise glabrous, creamy-white; disk glands 5, free, 0.5 mm long, obliquely truncate; stamens 8, the 5 outer 6 mm long and fused for half their length, the 3 inner 8 mm long and fused for three-eigths their length, anthers 1.2 mm long. Female flowers: pedicels stout, extending to 6 mm and becoming decurved in fruit; calyx tube very short, the lobes 4.5–5 × 1.2 mm, narrowly lanceolate, acute, entire; petals ± as in the male flowers; disk deeply 5-lobed, the lobes c. 1 mm across, transversely oblong, truncate; ovary c. 2 × 2 mm, 3-lobed to subglobose, glabrous; styles 4 mm long, united at the swollen base, ± erect, stigmas thick, capitate, bifid. Fruit c. 13 × 12.5 mm, 3-lobed to subglobose, shallowly tuberculate. Seeds 10 × 5.5 × 4.5 mm, ellipsoid-subcylindric, smooth, shiny, yellowish-brown, with a large fluted caruncle 3 × 5 mm, concolorous with the seed except for a pale buff fringe.

Mozambique. M: Namaacha, fr. 12.i.1948, *Torre* 7118 (LISC).

Known only from this locality. In wooded grassland.

The typical variety and the var. *angustata* Prain occur in the eastern Transvaal, while var. *swazica* Prain is recorded from Swaziland and KwaZulu-Natal. Var. *subeglandulosa* differs from the typical var. and var. *angustata* in not having the leaves, bracts and calyx lobes closely and minutely glandular-denticulate, and from the var. *swazica* in having the leaf blades elliptic and coriaceous, and in having the bracts quite entire.

Since the Mozambique specimen bore old fruit only, the descriptions of the male and female flowers in the above description are based on South African material of this species.

5. **Jatropha hirsuta** Hochst. ex Krauss in Flora **1845**, 6: 82 (1845). —Müller Argoviensis in De Candolle, Prodr. **15**, 2: 1088 (1866). —Pax in Engler, Pflanzenr. [IV, fam. 147, i] **42**: 62, t. 24 (1910). —Prain in F.C. **5**, 2: 424 (1920). —Engler, Pflanzenw. Afrikas (Veg. Erde 9) **3**, 2: 116, 117, t. 55 (1921). —Burtt Davy, Fl. Pl. Ferns Transvaal: 303 (1932). —Radcliffe-Smith in Kew Bull. **46**: 142 (1991). Type from South Africa (KwaZulu-Natal).

A pilose to villous perennial herb up to 30 cm high; stems usually simple, arising from a stout woody rootstock. Stipules 3–4 mm long, 4–6-fid; the segments filiform, gland-tipped. Leaves sessile or ± sessile; leaf blades up to 10 × 5 cm, simple, elliptic, acute or subacute at the apex, rounded or cuneate at the base, glandular-denticulate, evenly pilose to hirsute on both surfaces; lateral nerves in 8–9 pairs. Inflorescences 4–5 cm long, terminal, subcorymbiform; peduncle c. 2 cm long; bracts up to 1 cm long, linear, sparingly glandular-stipitate, sparingly pubescent. Male flowers: pedicels 4–5 mm long; calyx tube c. 1 mm long, the lobes 5–6 × 1–1.5 mm, linear-lanceolate, acute, glandular-stipitate, pubescent without, glabrous within; petals 8 × 5 mm, obovate, obtuse, sparingly pubescent, cream-yellow; disk glands 5, free, 0.5 mm long, rounded; stamens 8, the 5 outer 5 mm long and fused for half their length, the 3 inner 6.5 mm long and fused for two-thirds their length, anthers 1 mm long. Female flowers: pedicels more robust than in male flowers; calyx lobes and petals longer than in male flowers, but otherwise similar to them; disk glands 5, free, 0.5 mm long, turbinate, truncate; ovary 2 × 1.5 mm, 3-lobed-ellipsoid, smooth, hirsute; styles c. 2 mm long, ± free, suberect, stigmas bifid, capitate, crenate-lobulate. Fruit 1.6 × 1.3 cm, ellipsoid to 3-lobed, shallowly rugulose, hirsute. Seeds 9 × 4 × 3 mm, ellipsoid, brown, slightly shiny, with a knob-like fluted caruncle 2 × 3 mm.

Mozambique. M: Maputo, Estatuene, posto de Fronteira, male fl. 21.iv.1955, *Mendonça* 4514 (LISC).

Also in South Africa (KwaZulu-Natal). In fallow land.

Var. *glabrescens* Prain also occurs in KwaZulu-Natal and var. *oblongifolia* Prain in the Transvaal and Swaziland.

Since the Mozambique specimen bore male flowers only, the rest of the above description has been based on material from KwaZulu-Natal.

6. **Jatropha gossypiifolia** L., Sp. Pl.: 1006 (1753). —Müller Argoviensis in De Candolle, Prodr. **15**, 2: 1086 (1866). —Pax in Engler, Pflanzenw. Ost-Afrikas **C**: 240 (1895); Pflanzenr. [IV, fam. 147, i] **42**: 26 (1910). —Hutchinson in F.T.A. **6**, 1: 783 (1912). —Engler, Pflanzenw. Afrikas (Veg. Erde 9) **3**, 2: 112 (1921). —Brenan, Check-list For. Trees Shrubs Tang. Terr.: 217 (1949). —Keay in F.W.T.A., ed. 2, **1**, 2: 397 (1958). —J. Léonard in F.C.B. **8**, 1: 91 (1962). —Biegel, Check List Ornam. Pl. Rhod. Parks & Gard.: 66 (1977). —Radcliffe-Smith in F.T.E.A., Euphorb. 1: 354 (1987). Types from Central America.

Var. **elegans** (Pohl) Müll. Arg. in De Candolle, Prodr. **15**, 2: 1087 (1866). —Pax in Engler, Pflanzenr. [IV, fam. 147, i] **42**: 26 (1910). Type from Central America.
Adenoropium elegans Pohl, Pl. Bras. **I**: 15 (1827).
Jatropha elegans (Pohl) Klotzsch in Seem., Bot. Voy. Herald: 102 (1853).

A much-branched shrub usually up to c. 2 m tall. Young shoots thick, glabrous, exuding brownish latex. Stipules 5 mm long, multifid; the segments filiform, gland-tipped. Petioles up to 8.5 cm long, adaxially beset with trifid, bifid or simple stipitate glands, sparingly pubescent or subglabrous. Leaf blades 6–10 × 8–14 cm, 3(5)-palmatifid or palmatipartite, shallowly cordate and 3–5-nerved from the base, membranous, glabrous or subglabrous on both surfaces, reddish-brown or purplish-green; lobes obliquely or symmetrically obovate-oblanceolate, the median lobe c. 5–7.5 × 3–5 cm, acute or subacute at the apex and glandular-mucronulate, glandular-denticulate and ciliate on the margins; lateral nerves in 9–12 pairs per lobe. Inflorescences up to 11.5 cm long, with a peduncle up to 6 cm long; axis pubescent; bracts up to 1.5 × 0.3 cm, the largest below, linear-lanceolate, acutely acuminate, stipitate-glandular and ciliate on the margin, otherwise glabrous. Male flowers: pedicels 1 mm long; calyx c. 4 mm long, lobes 3 × 1.25 mm, elliptic-lanceolate, acute, stipitate-glandular and ciliate on the margin, otherwise subglabrous; petals 4 × 2 mm, obovate, unguiculate, rounded, glabrous, dark wine-red; disk glands 5, free, turbinate, truncate; stamens 8, 2–3 mm long, the 5 outer united to halfway, the 3 inner united for three quarters their length, filaments glabrous, anthers ovoid, 0.5 mm long. Female flowers: pedicels 1.5 mm long; calyx and petals as in the male flowers; disk shallowly 5-lobed, the lobes c. 1 mm across, truncate or retuse; ovary 1.75 × 1.5 mm, shallowly 3-lobed to subglobose, 6-ribbed, glabrous or sparingly pubescent; styles 1.75 mm long, united for c. one-quarter their length, somewhat divaricate, stigmas capitate, shortly bifid, smooth. Fruit 1 × 1 cm, shortly subcylindric to 3-lobed, shallowly verruculose, glabrous to subglabrous or sparingly pubescent, septicidally and loculicidally dehiscent. Seeds 7 × 5 × 3.5 mm, somewhat compressed ellipsoid-ovoid, smooth, light brown, with a flabellately multifid caruncle c. 1–1.5 × 2–3 mm, slightly darker brown.

Zambia. N: Filisi (Felisi), fl. 3.vi.1950, *Bullock* 2916 (K). **Zimbabwe**. N: Makonde Distr., Mhangura (Mangula), fr. xi.1964, *Jacobsen* 2939 (PRE). W: Bulawayo, o. fr. i.1955, *Orpen* in *GHS* 49910 (K; SRGH). E: Old Umtali Mission, fl. 23.vii.1955, *Chase* 5680 (BM; SRGH). **Malawi**. S: Chikwawa Distr., Ngabu, 0.4 km from Nsanje Road, fl. & fr. 21.iv.1980, *Blackmore, Dudley & Brummitt* 1309 (BM; K; MAL). **Mozambique**. N: Messinge, Mecaloja, fl. & fr. 12.ix.1934, *Torre* 579 (COI; LISC). T: Tete, fr. 23.xii.1980, *Macuácua* 1499 (K; LMA; PRE). MS: 5 km west of Dombe, fr. 21.iv.1974, *Pope & Müller* 1234 (K; LISC; MO; PRE; SRGH). GI: Chibuto, Maniquenique, fl. & fr. 15.vi.1960, *Lemos & Balsinhas* 103 (K; LMA).

Cultivated, escaping and becoming naturalized throughout tropical Africa, as well as elsewhere in the Old World tropics. On sandy soil in *Hyparrhenia* grassland, on heavily grazed land, near rivers and drainage ditches, on arkose outcrops, in waste land and maize fields; cultivated as a quick growing hedge and boundary plant or ornamental; 215–1220 m.

Native of tropical America from Mexico to Paraguay.

7. **Jatropha curcas** L., Sp. Pl.: 1006 (1753). —Müller Argoviensis in De Candolle, Prodr. **15**, 2: 1080 (1866). —Pax in Engler, Pflanzenw. Ost-Afrikas **C**: 240 (1895). —Sim, For. Fl. Port. E. Afr.: 107 (1909). —Pax in Engler, Pflanzenr. [IV, fam. 147, i] **42**: 77, t. 30 (1910). —

Hutchinson in F.T.A. **6**, 1: 791 (1912). —Prain in F.C. **5**, 2: 420 (1920). —Engler, Pflanzenw. Afrikas (Veg. Erde 9) **3**, 2: 118 (1921). —Burtt Davy, Fl. Pl. Ferns Transvaal: 302 (1932). —Chiov., Fl. Somala **2**: 394 (1932). —Brenan, Check-list For. Trees Shrubs Tang. Terr.: 216 (1949). —Keay in F.W.T.A., ed. 2, **1**, 2: 397 (1958). —Topham, Check List For. Trees Shrubs Nyasaland Prot.: 51 (1958). —Dale & Greenway, Kenya Trees Shrubs: 206 (1961). —J. Léonard in F.C.B. **8**, 1: 90 (1962). —White, F.F.N.R.: 200 (1962). —Verdcourt & Trump, Common Pois. Pl. E. Afr.: 56 (1969). —Agnew, Upl. Kenya Wild Fls.: 218 (1974). —Drummond in Kirkia **10**: 252 (1975). —Biegel, Check List Ornam. Pl. Rhod. Parks & Gard.: 66 (1977). —Troupin, Fl. Rwanda **2**: 228 (1983). —K. Coates Palgrave, Trees Southern Africa, ed. 2, rev.: 430 (1983). —Radcliffe-Smith in F.T.E.A., Euphorb. 1: 356 (1987). —Beentje, Kenya Trees, Shrubs Lianas: 210 (1994). Type from tropical America.

Curcas purgans Medik., Ind. Pl. Hort. Manhem. **1**: 90 (1771). —Peters, Naturw. Reise Mossambique **6**, 1: 98 (1861). Type is presumed to be a plant cultivated in Mannheim, Germany, but nothing is known as to the whereabouts of Medikus's types.

Jatropha afrocurcas Pax in Bot. Jahrb. Syst. **43**: 83 (1909); Pax in Engler, Pflanzenr. [IV, fam. 147, i] **42**: 79 (1910). Type from Tanzania (Lake Province).

An eglandular shrub or small tree up to 7 m tall. Bark light brown to olive grey-green, smooth, papery, peeling. Wood soft. Branchlets thick, semisucculent. Latex copious, watery, sticky. Stipules 0.5 mm long, subulate, fugacious. Petioles up to 20 cm long. Leaf blades 5–15 × 5–15 cm, ovate, pentagonal or shallowly 5-lobed, shallowly to deeply cordate at the base, sparingly puberulous along the main nerves beneath and otherwise subglabrous when young, later glabrescent, 7–9-nerved from the base; the median lobe usually acutely acuminate, the lateral lobes subacute to obtuse or rounded. Inflorescences up to 10 cm long, subterminal or supra-axillary, often paired, subcorymbiform, sparingly puberulous to subglabrous; peduncle up to 6 cm long; bracts up to 8 mm long, linear-lanceolate, acute, entire, sparingly puberulous to subglabrous. Male flowers: pedicels up to 4 mm long; calyx 5 mm long, lobes 3.5 × 2.5 mm, elliptic-ovate, rounded, entire, eglandular, sparingly puberulous, green; petals 7 × 3 mm, laterally coherent for about half their length, oblong, rounded, glabrous without, pilose within, greenish-yellow; disk glands 5, free, c. 1 mm tall, turbinate-cylindric, truncate; stamens 8, the 5 outer 5 mm long and fused for two-fifths their length, the 3 inner 6 mm long and fused for half their length, filaments glabrous, anthers c. 2 mm long, linear. Female flowers: pedicels 3 mm long, extending to 1 cm in fruit, stouter than in the male, puberulous; calyx lobes ± free to the base, unequal, 5 × 2 cm or 7 × 3 cm, elliptic-lanceolate, obtuse, entire, eglandular, puberulous; petals 6.5 × 3 mm, free, elliptic-oblong, rounded, glabrous without, pilose within, greenish-yellow; disk glands 5, free, 1 × 1 mm, somewhat flattened, suborbicular; staminodes 10, up to 1 mm long, often minute, orange-tipped; ovary 2.5 × 2 mm, ovoid-ellipsoid, somewhat 3-lobed, tapered upwards into the stylar column, smooth, glabrous; styles 2.5 mm long, united for c. two-thirds their length, divaricate, stigmas capitate, unequally 4-fid, smooth. Fruits 2.5–3 × 2–2.5 cm, ellipsoid, slightly 3-lobed, ± smooth, loculicidally dehiscent, green. Seeds 1.5–1.7 × 0.9–1 × 0.7–0.8 cm, ellipsoid to subcylindric, dull, blackish, minutely buff-punctate, caruncle c. 2 × 4 mm, depressed-conic, mauve-grey when dry.

Zambia. B: Zambezi Distr., near Chavuma, fl. 13.x.1952, *White* 3478 (FHO; K; MO). N: Kawambwa, fl. 16.xi.1957, *Fanshawe* 4063 (K; NDO). E: Kazandwe, Nsefu, fr. 5.iv.1968, *R. Phiri* 124 (K). S: 37 km west of Namwala, fr. 24.vi.1952, *White* 2981 (FHO; K). **Zimbabwe**. N: Mutoko Distr., Kaunye–Mushimbo, fl. 15.ii.1978, *Pope* 1559 (MO; SRGH). C: Diana's Vow, 30 km Rusape–Nyanga (Inyanga), fl. 14.viii.1972, *Mavi* 1423 (K; PRE; SRGH). E: Mutare Distr., Umtasa South Reserve, Manika Bridge, fl. 3.xii.1950, *Chase* 3328 in *GHS* 34017 (BM; K; MO; PRE; SRGH). **Malawi**. C: Dedza Distr., Bembeke Mission, fr. 21.ii.1961, *Chapman* 1153 (MAL). S: Zomba Distr., near Makwapala, Naisi Stream, fl. & y. fr. 30.xi.1936, *Lawrence* 205 (K; MAL). **Mozambique**. T: Boruma, fl. ii.1892, *Menyharth* 925 (K; Z). MS: Mavita–Chimoio (Vila Pery), R. Munhinga, fr. 25.iv.1948, *Barbosa* 1569 (LISC). GI: Xai-Xai (João Belo), fr. 18.iii.1948, *Torre* 7523 (LISC). M: outskirts of Maputo, fl. i.1946, *Pimenta* 6006A (LISC).

Widely cultivated and naturalized in the Old World tropics generally. As a street tree, and in villages for oil and as living hedges and stockades, often becoming naturalized. Often on rocky outcrops, and in sandy soils; sea level to 1220 m.

Planted for the oil-rich seeds. The seeds are poisonous when chewed, the oil is used medicinally and for anointing the body.

Native of tropical America from Mexico and the West Indies to Chile.

The "Physic Nut"; the vernacular name in chiShona is "munjirimono", and in the Maputo area is recorded in specimen data as "galamaluco".

8. **Jatropha multifida** L., Sp. Pl.: 1006 (1753). —Peters, Naturw. Reise Mossambique **6**, 1: 98 (1861). —Müller Argoviensis in De Candolle, Prodr. **15**, 2: 1089 (1866). —Pax in Engler, Pflanzenw. Ost-Afrikas **C**: 240 (1895); in Engler, Pflanzenr. [IV, fam. 147, i] **42**: 40 (1910). —Hutchinson in F.T.A. **6**, 1: 784 (1912). —Engler, Pflanzenw. Afrikas (Veg. Erde 9) **3**, 2: 112 (1921). —Brenan, Check-list For. Trees Shrubs Tang. Terr.: 217 (1949). —Keay in F.W.T.A., ed. 2, **1**, 2: 397 (1958). —Dale & Greenway, Kenya Trees & Shrubs: 206 (1961). —J. Léonard in F.C.B. **8**, 1: 89 (1962). —White, F.F.N.R.: 200 (1962). —Biegel, Check List Ornam. Pl. Rhod. Parks & Gard.: 66 (1977). —Troupin, Fl. Rwanda **2**: 228, fig. 70/3 (1983). —Radcliffe-Smith in F.T.E.A., Euphorb. 1: 354 (1987). —Beentje, Kenya Trees, Shrubs Lianas: 210 (1994). Type: The illustration of this species in Dillenius' "Hortus Elthamensis" (1732), (chosen as lectotype by Radcliffe-Smith in F.T.E.A.).

A glabrous eglandular shrub usually up to 2 m high, sometimes taller. Branchlets thick. Latex white. Stipules 1.5–2 cm long, multifid, the segments setaceous. Petioles commonly up to 20 cm long. Leaf blades up to 20 × 30 cm, (9)10–12-palmatipartite, bright green above, glaucous beneath; the lobes up to 15–17 × 7 cm, narrowly lanceolate or irregularly pinnatipartite, acutely long-acuminate, sometimes aristate, entire, confluent into a narrowly cordate basal disk; lateral nerves numerous. Inflorescences commonly up to 25 cm long, with a peduncle up to 22 cm long, densely corymbiform to subcapitate; bracts up to 5 mm long, lanceolate, acute, entire. Male flowers: pedicels up to 1 cm long, coral-red; calyx c. 2 mm long; calyx lobes c. 1 × 1 mm, broadly ovate, rounded and erose at the apex, otherwise entire, red; petals c. 5 × 2 mm, oblong-oblanceolate, rounded, scarlet; disk glands 5, free, turbinate, rounded; stamens 8, 4 mm long, filaments free, anthers 2 mm long, oblong-linear, sagittate. Female flowers: pedicels 2 mm long; calyx and petals resembling those of the male, but c. 1.5 times larger; disk 2.5 mm in diameter, flattened, 5-lobed, with lobes rounded-truncate; ovary 3 × 2 mm, trigonous-ellipsoid; styles 2.5 mm long, united at the base, stigmas capitate, 2-lobed. Fruit 2.5 × 2.8 cm when dried, somewhat larger when fresh, 3-lobed to pyriform, depressed at the apex, abruptly tapered to the base, the lobes slightly keeled, tardily septicidally dehiscent to subdrupaceous. Seeds 1.7 × 1.5 × 1.3 cm, broadly ovoid-subglobose, buff, mottled brownish, ecarunculate.

Zambia. C: Lusaka Distr., Chilanga, Mr. Sanders' Garden (Munda Wanga), fl. 28.xii.1959, *White* 6060 (FHO). E: Chipata (Fort Jameson) township, cult. y. fl. 25.v.1961, *Angus* 2881 (FHO). **Zimbabwe**. C: Gweru (Gwelo), fl. 4.viii.1942, *Govt. Analyst* in *GHS* 9201 (K; SRGH). **Malawi**. S: Blantyre, fl. 11.xii.1965, *Berrie* 171 (K; MAL). **Mozambique**. N: Namatil (Nametil), o. fl. & fr. 17.v.1937, *Torre* 1556 (LISC). M: Maputo, Jardim Tunduru (Vasco da Gama), fl. & y. fr. 26.ii.1973, *Balsinhas* 2470 (K).

Widely cultivated and often naturalized in the Old World tropics generally. Cultivated as a hedge plant and garden ornamental; 1370 m.

Native of tropical America from Mexico to Paraguay.

The "Coral Tree".

9. **Jatropha podagrica** Hook. in Bot. Mag. **74**, t. 4376 (1848). —Müller Argoviensis in De Candolle, Prodr. **15**, 2: 1093 (1866). —Pax in Engler, Pflanzenr. [IV, fam. 147, i] **42**: 44 (1910). —Keay in F.W.T.A., ed. 2, **1**, 2: 397 (1958). —Biegel, Check List Ornam. Pl. Rhod. Parks & Gard.: 66 (1977). —Radcliffe-Smith in F.T.E.A., Euphorb. 1: 354 (1987). Type from Panama.

A completely glabrous shrub up to 2 m tall; stem swollen at the base. Latex watery. Stipules up to 5 mm long, branched, glandular, becoming indurated. Petioles 10–20 cm long. Leaf blades 10–20 × 10–20 cm, peltate, 3–5-lobed, bright green above, paler beneath; the lobes broadly ovate to obovate, subacute, entire; the median lobe 8–12 × 6–11 cm, the laterals the same or slightly smaller. Inflorescences 20–25 cm long, with a peduncle 17–22 cm long, densely corymbiform to subcapitate; bracts up to 2 mm long, triangular, subacute, subentire. Male flowers: pedicels 1–2 mm long; calyx 1.5 mm long; calyx lobes 0.5 × 1 mm, transversely ovate, emarginate, entire; petals 5–6 × 2 mm, obovate-oblong, obtuse, scarlet; disk urceolate; stamens 6–8, 5 mm long, filaments connate at the base, anthers 2 mm long, orange. Female flowers: pedicels 1–2 mm long, stout; calyx lobes 1.5–2 × 1–1.5 mm, ovate to triangular-ovate, obtuse, entire; petals 6–7 mm long; disk glands free, flattened, ± truncate; ovary 2–3 × 2 mm, ellipsoid; styles 1.5 mm long, united at the base, stigmas capitate, 2-lobed. Fruit 1.4 × 1.3 cm when dried, ellipsoid-ovoid, 3-lobed, apically and basally truncate, septicidally and loculicidally dehiscent. Seeds 1.2 × 0.6 × 0.4 cm, ellipsoid, triangular-convex in section, brown, with a small fluted caruncle.

Malawi. S: Cape Maclear, fl. & y. fr. 24.ix.1966, *Binns* 373 (MAL).
Widely cultivated in the tropics generally and often becoming naturalized. Garden escape naturalized in *Brachystegia* woodland.
Native of C America.
"White rhubarb".

10. **Jatropha variifolia** Pax in Engler, Pflanzenr. [IV, fam. 147, i] **42**: 54 (1910). —Prain in F.C. **5**, 2: 420 (1920). —Engler, Pflanzenw. Afrikas (Veg. Erde 9) **3**, 2: 113 (1921). —Burtt Davy, Fl. Pl. Ferns Transvaal: 302 (1932). Tab. **55**. Type from South Africa (Transvaal).
Jatropha heterophylla Pax in Bot. Jahrb. Syst. **28**: 25 (1899), non Steud. (1840), nec. Heyne ex Hook.f. (1887).

A 1–2-stemmed, sparingly branched, somewhat succulent, glabrous shrub or subshrub up to 2 m tall, with a thickened rootstock. Stipules 4 × 6 mm, flabellately multifid; the segments up to c. 3 mm long, filiform-setaceous, gland-tipped. Petioles up to 9 cm long. Leaf blades up to 18 × 20 cm (in Flora Zambesiaca area), palmately 5-fid to 5-partite, truncate to shallowly cordate at the base, acutely or obtusely shortly acuminate at lobe apices, entire, membranaceous, dark green and shiny on upper surface, paler and duller beneath; the median lobe up to 13 × 6 cm, oblong-lanceolate to elliptic-lanceolate, the lateral lobes progressively smaller; lobe midribs all arising at the lamina base, lateral nerves in up to c. 15 pairs in the median lobe, fewer in the lateral lobes. Inflorescences up to 30 cm long, terminal or subterminal, subcorymbiform; peduncle up to 13 cm long; bracts up to 1.5 cm long, linear-lanceolate, acute, glandular to long-stipitate. Male flowers: pedicels slender, 1–3 mm long; calyx tube 1.5 mm long; calyx lobes 2 × 1 mm, lanceolate, subacute, entire, terracotta; petals 5 × 2 mm, oblanceolate, rounded, becoming reflexed, glabrous, yellow; disk glands 5, free, minute, erect, turbinate, truncate; stamens 8, the 5 outer 5 mm long and fused for two fifths their length, the 3 inner 6 mm long and fused for two-thirds their length, filaments glabrous, anthers 1 mm long. Female flowers: pedicels c. 5 mm long, extending a little in fruit; calyx tube c. 2 mm long, lobes 3 × 1.25 mm, and extending a little in fruit, otherwise resembling the male; petals as in the male, but yellow-green; disk 5-lobed, the lobes short, broad and somewhat irregularly lobulate; ovary 2 × 2 mm, 3-lobed to subglobose, smooth, glabrous; styles 1.25 mm long, ± free, erect or spreading, stigmas small, capitate, bifid, smooth. Fruits c. 12 × 10–12 mm, 3-lobed to subcylindric, shallowly nervose and verruculose, glabrous; columella 7 mm long. Seeds 9.5 × 5 × 4 mm, somewhat compressed ellipsoid-subcylindric, slightly shiny, olive-brown; caruncle 2 × 3–3.5 mm, hemispherical, 2-lobed, the lobes multifid, greyish-brown.

Mozambique. M: Goba, near R. Maiuána, y. fr. 7.xi.1960, *Balsinhas* 215 (K; LMA; PRE).
Also in South Africa (Transvaal, KwaZulu-Natal and Eastern Cape Province). In dense mixed dry deciduous woodland and thicket, and in wooded grassland on dry stony soils.

11. **Jatropha subaequiloba** Radcl.-Sm. in Kew Bull. **46**: 154 (1991). Type: Mozambique, Cabo de São Sebastião, fl. 10.xi.1958, *Mogg* 29126 (SRGH, holotype; LISC).

A glabrous shrub up to 4 m tall. Stipules very fugaceous. Petioles 4–6.5 cm long. Leaf blades 10–12 × 11–13.5 cm, 5-lobed, deeply cordate at the base, 5-nerved from the base; lobes subequal, 4–7 × 2–4 cm, oblong-oblanceolate, acutely acuminate, entire; lateral nerves of each lobe in 12–14 pairs, brochidodromous. Inflorescences up to 10 cm long, terminal, subcorymbiform; bracts 1–4 mm long, triangular-lanceolate, eglandular, entire. Male flowers: pedicels 0.5 mm long; calyx 3 × 4 mm, campanulate; calyx lobes 1.5 × 1.5 mm, triangular-ovate, obtuse, entire; petals 5 × 1.5 mm, oblanceolate-spathulate, obtuse, greenish; disk glands 5, free, minute; staminal column 4 mm high; stamens 8, the 5 outer shorter than the 3 inner, anthers 0.75 mm long. Female flowers not known. Fruits 1.7 × 1.5 cm, 3-lobed to subglobose, shallowly tuberculate-rugulose. Seeds 12–13 × 8–9 × 6–7 mm, ovate-quadrate, dark brown, obscurely mottled with darker brown and buff, slightly shiny, with a knob-like, sulcate, shortly fimbriate brown caruncle 4 × 5 mm.

Mozambique. GI: Bazaruto I., Ponta Estone, fr. 28.x.1958, *Mogg* 28677 (LISC).
Not known elsewhere. Coastal dune forest, in sandy soil and clay loam, also in freshwater marshland; sea level to 5 m.

Tab. 55. JATROPHA VARIIFOLIA. 1, distal portion of branch (× $^{2}/_{3}$); 2, male flower (× 4), 1 & 2 from *Balsinhas* 215; 3, young fruit (× 4), from *Leach & Bayliss* 11934; 4, seeds (× 2); 5, fruit valves (× 2), 4 & 5 from *de Koning & Nuvunga* 8587. Drawn by Christine Grey-Wilson.

12. **Jatropha spicata** Pax in Bot. Jahrb. Syst. **19**: 109 (1894); in Engler, Pflanzenw. Ost-Afrikas **C**: 240 (1895); in Engler, Pflanzenr. [IV, fam. 147, i] **42**: 36 (1910). —Hutchinson in F.T.A. **6**, 1: 790 (1912). —Engler, Pflanzenw. Afrikas (Veg. Erde 9) **3**, 2: 112, 113 (1921). —Dale & Greenway, Kenya Trees & Shrubs: 206 (1961). —Agnew, Upl. Kenya Wild Fls.: 218 (1974). —Radcliffe-Smith in F.T.E.A., Euphorb. 1: 359 (1987). —Beentje, Kenya Trees, Shrubs Lianas: 211 (1994). Type from Kenya (Coast Province).

Jatropha pseudoglandulifera Pax in Engler, Pflanzenr. [IV, fam. 147, i] **42**: 34 (1910). —Hutchinson in F.T.A. **6**, 1: 790 (1912). —Engler, Pflanzenw. Afrikas (Veg. Erde 9) **3**, 2: 112, 113 (1921). —P.G. Meyer in Merxmüller, Prodr. Fl. SW. Afrika, fam. 67: 37 (1967). Types from Angola (Luanda Province).

Jatropha kilimandscharica Pax & K. Hoffm. in Engler, Pflanzenr. [IV, fam. 147, i] **42**: 40 (1910). —Hutchinson in F.T.A. **6**, 1: 785 (1912). —Engler, Pflanzenw. Afrikas (Veg. Erde 9) **3**, 2: 112, 113 (1921). Type from Kenya.

Jatropha messinica E.A. Bruce in Bothalia **6**, 1: 226 (1951). —Drummond on Kirkia **10**: 252 (1975). Type from South Africa (Transvaal).

A brittle-stemmed, semi-succulent, glabrous shrub up to 1.25 m tall. Stipules up to 1 cm long, multifid; the segments filiform-setaceous, gland-tipped. Petioles up to 7.5 cm long. Leaf blades up to 10 × 14 cm (in Flora Zambesiaca area), 5–7-palmatipartite to palmatisect, truncate to wide-cordate at the base, acute to subacute or obtuse at lobe apices, glandular-serrate on margins, deep green; the median lobe up to 9.5 × 2.5 cm, elliptic-lanceolate, constricted at the base, the lateral lobes progressively smaller, the outermost linear-lanceolate and not basally constricted; lateral nerves in up to c. 25 pairs in the median lobe, fewer in the lateral lobes. Inflorescences up to 11 cm long (in Flora Zambesiaca area), terminal, laxly subcorymbiform, cymose but with the ultimate branches pseudospicate at maturity; peduncle up to 4.5 cm long; bracts up to 7 mm long, linear-lanceolate, acute, sparingly glandular-denticulate. Male flowers shortly pedicellate to subsessile; calyx tube 1.5 mm long; calyx lobes 1.5–2 × 1 mm, lanceolate, obtuse, entire or subentire; petals 5 × 2.5 mm, obovate, rounded, with a small patch of pubescence within at the base, otherwise glabrous, yellow; disk glands 5, free, c. 0.5 mm high, erect, turbinate, truncate; stamens 8, the 5 outer 4 mm long and fused for two-thirds their length, the 3 inner 5 mm long and fused for c. four-fifths their length, filaments glabrous, anthers 1 mm long, red. Female flowers: pedicels 2 mm long; calyx tube c. 0.5 mm, lobes 2.5–3 × 1 mm, extending a little in fruit, linear-lanceolate, obtuse, entire or almost so; petals 6 × 2.25 mm, elliptic, obtuse, sometimes mucronulate, completely glabrous, greenish-yellow; disk 5-lobed, the lobes semicircular, thin, flat, pink-tinged; ovary 2 × 1.5 mm, 3-lobed-ellipsoid, smooth, glabrous; styles 2.5 mm long, united for one-quarter their length, erect, stigmas small, capitate, bifid, verruculose. Fruits 10–12 × 8–9 mm, 3-lobed-subcylindric, smooth, glabrous, pale green; columella 7 mm long. Seeds 9–10 × 4–5 × 3–3.5 mm, ellipsoid-cylindric, slightly shiny, greyish-brown; caruncle 2–2.5 × 3.5–4.5 mm, hemispherical, 2–many-lobed, the lobes deeply fimbriate, dark brown with a creamy-buff fringe when dried.

Zimbabwe. E: Chipinge Distr., Save (Sabi) Valley, Experimental Station, o. fr. iv.1960, *Soane* 246 (K; LISC; PRE; SRGH). S: Beitbridge Distr., near R. Limpopo, o. fl. & y. fr. 15.ii.1955, *Exell, Mendonça & Wild* 428 (BM; LISC; SRGH).

Also in Kenya, Tanzania, Angola and South Africa (northern Transvaal). Hot, dry, low altitude river valleys, on rocky basaltic hills, shallow gravelly soils in sparse mopane/*Terminalia*/*Commiphora*/*Kirkia* woodland or scrub, also in riverine mopane and *Acacia* woodlands on sand; 365–580 m.

13. **Jatropha macrophylla** Pax & K. Hoffm. in Engler, Pflanzenr. [IV, fam. 147, i] **42**: 80 (1910). —Hutchinson in F.T.A. **6**, 1: 792 (1912). —Engler, Pflanzenw. Afrikas (Veg. Erde 9) **3**, 2: 118 (1921). Type: Malawi, *Buchanan* 670 (BM, holotype).

An erect perennial herb up to 1 m high, arising from a massive reddish rootstock c. 30 cm long which emits a red juice from the outer layers when cut or damaged. Stems, petioles and inflorescences glabrous or rarely sparingly puberulous, emitting a milky-watery latex when cut. Stipules absent or very fugacious. Petioles up to 28 cm long. Leaf blades 7–25(28) × 7–28(30) cm, 5(7)-lobed, truncate or wide-cordate at the base, membranaceous, glabrous on upper surface, sparingly pilose to subglabrous along the main nerves beneath, yellow-green; leaf lobes up to 13 × 8.5

cm, usually broadly oblong-obovate, rounded to obtuse or sometimes shortly acuminate at the apex, entire; nerves (5)7(9) from the base, lateral nerves in up to 15 pairs per lobe. Inflorescences up to 25 cm long, terminal or subterminal, subcorymbiform, peduncle up to 12 cm long; bracts up to 6 × 3 mm, triangular, acute, entire. Male flowers: pedicels c. 1 mm long, bibracteolate; calyx tube very short; calyx lobes c. 1.5 × 0.9 mm, ovate-oblong, obtuse, entire, eglandular, greenish-yellow; petals 5 × 2.25 mm, oblong-elliptic, obtuse, glabrous, yellow; disk glands 5, free, 0.5 mm in diameter, subglobose, punctulate, deep yellow; stamens 8, the 5 outer c. 4 mm long and fused for c. half their length, the 3 inner c. 4.5 mm long and fused for more than half their length, filaments glabrous, green, anthers c. 1 mm long, ellipsoid, pale yellow. Female flowers: pedicels c. 2 mm long, stout; calyx tube very short, calyx lobes unequal, 1.5 × 0.75 mm or c. 2 × 2 mm, broadly ovate or ovate-oblong, obtuse, entire, eglandular; petals 7–8 × 2.5–3 mm, free, elliptic-oblong, obtuse, glabrous, yellowish-green; disk c. 2.5 mm in diameter, shallowly 5-lobed; staminodes not seen; ovary c. 2 × 2 mm, 3-lobed to subglobose, smooth, glabrous; styles not seen. Fruits c. 1.2–1.3 × 1.1–1.2 cm, obovoid-subglobose, slightly 3-lobed, shallowly tuberculate-rugulose, pale green. Seeds 11.5 × 6 × 5 mm, including the caruncle, ellipsoid-subcylindric, shiny, evenly or mottled purplish-brown; caruncle 2.5 × 3.5 mm, ± hemispherical, 2-lobed, the lobes shortly fimbriate, golden-brown when dry, sometimes with a white fringe.

Zambia. E: Luangwa R., fr. 25.iii.1955, *Exell, Mendonça & Wild* 1187 (BM; LISC; SRGH). **Malawi**. N: Chitipa Distr., c. 20 km south of Karonga on Chilumba Road, leaf, fl. & fr. 26.v.1989, *Radcliffe-Smith, Pope & Goyder* 5953 (K; MAL; PRE; SRGH). C: Salima Distr., Lifidzi, fl. & y. fr. 15.iv.1985, *Kwatha & Tawakali* 40 (MAL). S: Mulanje Distr., Phalombe Plain, Khongoloni Dambo, fl. & y. fr. 31.xii.1980, *Patel* 791 (MAL; PRE; SRGH).

Also in S Tanzania. In dry river valleys in mopane woodland, and in dry lake shore vegetation on hillside mixed deciduous woodland; 110–1067 m.

Vernacular name recorded in specimen data: “namicura”.

14. **Jatropha scaposa** Radcl.-Sm. in Kew Bull. **46**: 151 (1991). Type: Mozambique, Cheringoma Distr., between Durundi and Inhaminga, fl. & y. fr. 11.ix.1942, *Mendonça* 192 (LISC, holotype).

An erect scapose perennial herb up to 36 cm tall; leafy stems and flowering scapes usually arising separately from a stout woody rootstock. Stems and scapes pubescent or glabrous. Stipules 1–2 mm long, setaceous. Petioles up to 7.5 cm long. Leaf blades up to 15 × 17 cm, 5-lobed, wide-cordate at the base, membranous, densely tomentose on both surfaces at first, later evenly pubescent; the median lobe up to 8 × 3.5 cm, the laterals slightly smaller, all lobes elliptic-lanceolate, acutely acuminate, entire, with rounded sinuses between the lobes; nerves 5(7) from the base, lateral nerves in up to 18 pairs per lobe. Inflorescences up to 22 × 15 cm, paniculate; scape up to 15 cm long; bracts up to 7 mm long, triangular-lanceolate, acute, entire, with or without a pair of lateral linear lobes at the base. Male flowers: pedicels 1.5–2 mm long, strongly jointed at the apex; calyx tube very short; calyx lobes 2 × 1.5 mm, ovate, obtuse, entire, eglandular; petals 5 × 3 mm, ovate-oblong, obtuse, sparingly pubescent within at the base, otherwise glabrous, lemon-yellow; disk glands 5, free, 0.6 × 0.75 mm, turbinate, apically truncate and tuberculate-rugulose; stamens 8, fused together in a staminal column 6 mm tall with an outer whorl of 5 anthers and an inner of 3, anthers c. 1 mm long. Female flowers: pedicels up to 6 mm long, stout; calyx tube c. 1 mm long, calyx lobes 2.5 × 2 mm, otherwise as in the male flowers; petals 6–7 × 2.5–3 mm, oblong-oblanceolate, obtuse, glabrous, yellow-green; disk glands 5, free, 0.5 × 1 mm, flat, quadrate, truncate; staminodes absent; ovary c. 2 × 2 mm, 3-lobed to ellipsoid, smooth, glabrous; styles 1.5–2 mm long, free, suberect, stigmas bifid, capitate, smooth. Fruits c. 1.3–1.5 × 1.5 cm, 3-lobed to subglobose, shallowly tuberculate-rugulose. Mature seeds not seen.

Mozambique. N: Mogincual Distr., between Liúpo and Corrane, fl. & fr. leafless, 18.x.1948, *Barbosa* 2492 (LISC). MS: Cheringoma Distr., R. Urema, male fl. 10.ix.1944, *Mendonça* 2015 (LISC). M: Mucomaze, fl. ix.1908, *Cartagena* 784 (BM).

Not known elsewhere. Coastal plain in *Brachystegia/Isoberlinia/Combretum/Pterocarpus* woodland, in sandy soil.

The Maputo gathering may represent a distinct variety, with leaves on the flowering shoots.

15. **Jatropha botswanica** Radcl.-Sm. in Kew Bull. **46**: 142 (1991). Type: Botswana, between Bisoli (Bosoli) and Francistown, y. fr. 8.iii.1965, *Wild & Drummond* 6818 (K, holotype; LISC; PRE; SRGH).

A few-branched pilose perennial herb up to 30 cm tall, arising from a large reddish swollen rootstock. Stipules 1–2 mm long, divided into a few setaceous gland-tipped segments, pink, soon falling. Petioles 1–2 cm long. Leaf blades (2.5)7–9 × (2)5–7 cm, palmately 5-partite or spuriously palmatipartite, cuneate at the base, sparingly pubescent to subglabrous on upper surface, evenly or sparingly pubescent beneath especially along the midribs; the median lobe (1)2–6 × (0.5)1–2 cm, the next 1–5 × 0.5–1 cm and the outer 0.5–3.5 × 0.2–1 cm; all lobes narrowly oblong-oblanceolate, obtuse at the apex, irregularly dentate and sometimes shallowly pinnately lobulate, sparingly glandular-stipitate on the margin; midribs of the lateral lobes diverging above the base, lateral nerves in 10–16 pairs in the median lobe, 9–13 in the next and 6–9 in the outer lobes. Inflorescences up to 8.5 cm long, terminal, pseudopaniculate; peduncle up to 4 cm long; bracts c. 5 mm long, linear, glandular-stipitate. Male flowers: pedicels 1.5 mm long; calyx tube 0.5 mm long; calyx lobes 2–2.5 × 1 mm, ovate-lanceolate, obtuse, glandular-stipitate, pubescent without, glabrous within; petals 5.5 × 1.5 mm, narrowly oblong-oblanceolate, truncate, glabrous, yellow; disk glands 5, free, conoidal, acute; stamens 8, the 5 outer 3 mm long and fused for half their length, the 3 inner 4 mm long and fused for three-quarters their length, filaments glabrous, anthers c. 1 mm long. Female flowers: pedicels 2 mm long; calyx tube 1 mm long; calyx lobes 3 × 2 mm, otherwise as in the male; petals ± as in the male, but slightly larger; disk 5-lobed, the lobes transversely oblong, truncate, somewhat thickened; ovary 2.5 × 2.5 mm, 3-lobed to subglobose, smooth, pubescent; styles 2 mm long, free, erect; stigmas capitate, 2-lobed, subcylindric, smooth. Fruits 9 × 8 mm, obovoid-3-lobed, smooth, evenly pubescent; columella 7 mm long. Mature seeds not known.

Botswana. N: between Bisoli (Bosoli) and Francistown, y. fr. 8.iii.1965, *Wild & Drummond* 6818 (K; LISC; PRE; SRGH). SE: 60 km NW of Serowe, o. fr. & y. fl. 24.iii.1965, *Wild & Drummond* 7284 (K; LISC; SRGH).

Not known elsewhere. In dry mopane scrub on bare black soil and in *Acacia* woodland in depressions.

16. **Jatropha zeyheri** Sond. in Linnaea **23**: 117 (1850). —Müller Argoviensis in De Candolle, Prodr. **15**, 2: 1088 (1866). —Pax in Engler, Pflanzenr. [IV, fam. 147, i] **42**: 68 (1910). —Prain in F.C. **5**, 2: 426 (1920). —Engler, Pflanzenw. Afrikas (Veg. Erde 9) **3**, 2: 116, 117 (1921). —Burtt Davy, Fl. Pl. Ferns Transvaal: 303 (1932). —Bremekamp & Obermeyer in Ann. Transvaal Mus. **16**, 3: 421 (1935). —Hutchinson, Botanist in Southern Africa: 667 (1946). Type from South Africa (Transvaal).

Jatropha brachyadenia Pax & K. Hoffm. in Engler, Pflanzenr. [IV, fam. 147, i] **42**: 66 (1910). Type from South Africa (Transvaal).

An evenly to densely pubescent, sometimes creeping, perennial herb up to 30 cm tall; stems simple or sparingly branched, somewhat fleshy, arising from a stout woody rootstock. Sap watery, greenish. Stipules 2 mm long, multifid with filiform gland-tipped segments, stramineous, subpersistent. Petioles 1–5 mm long. Leaf blades (3)5–12 × (2)4–10 cm (in Flora Zambesiaca area), all or most leaves deeply 3–5-palmatipartite or subpalmatipartite, cuneate to truncate or rarely shallowly cordate at the base, sparingly to evenly pubescent on both surfaces, but evenly to densely so on the midribs; the median lobe 3.5–9 × 0.3–2.5 cm, narrowly elliptic-oblanceolate, the lateral lobes markedly progressively shorter and usually more narrowly lanceolate; all lobes acute or subacute, rarely obtuse, subentire or rarely dentate, often closely and densely glandular-stipitate, sometimes more sparsely so; lateral nerves in 15–20 pairs in the median lobe, fewer in the lateral lobes. Inflorescences up to 10 cm long, terminal, subcorymbiform; peduncle up to 2.5 cm long; bracts 0.5–1 cm, linear-lanceolate, glandular-stipitate, the lowest sometimes larger and foliaceous. Male flowers: pedicels 2 mm long; calyx tube 0.5 mm long; calyx lobes 2–3 × 1 mm, lanceolate, acute, sparingly to uniformly glandular-stipitate, pubescent without, glabrous within; petals 4.5 × 1.5–2 mm, oblanceolate, rounded, pubescent without, pilose within, cream-coloured; disk glands 5, free, c. 1 × 0.5 mm, pulviniform, rounded-truncate; stamens 8, the 5 outer 3 mm long and fused for one-third their length, the 3 inner 3.5–4 mm long and fused for half their length, filaments glabrous, anthers 1 mm long, thecae free but not divergent at the base. Female flowers: pedicels 2 mm long, stout; calyx tube 1.5–2 mm; calyx lobes 4–7 × 2 mm, triangular-lanceolate, evenly glandular-stipitate, pubescent without, glabrous within; petals 7–8 × 3 mm, oblong-lanceolate, otherwise as in the male; disk glands 5, free, c. 1 × 1.25 mm, suborbicular, somewhat thickened; ovary 2.5 × 2.5 mm, 3-lobed to subglobose, densely hirsute; styles 3 mm long, united at the base, suberect, stigmas capitate, bipartite, verruculose. Fruits 11–13 × 9–12 mm, 3-lobed to subcylindric, smooth, sparingly to evenly pubescent to densely pilose; columella 9–10 mm long. Seeds 9 × 5 × 3.5 mm, compressed-ellipsoid, slightly shiny, buff, mottled brown; caruncle 2 × 4 mm, depressed-hemispherical, 2-lobed, the lobes fimbriate-multifid, golden-brown with a paler fringe when dried.

Botswana. SE: 5 km north of Gaborone, fl. 3.ix.1977, *O.J. Hansen* 3172 (C; GAB; K; PRE; SRGH; WAG).

Also in South Africa (Transvaal, KwaZulu-Natal, Northern Cape Province) and Swaziland. Wooded grassland with scattered shrubs (*Dichrostachys* and *Terminalia*), mopane woodland on sand and *Acacia* sand veld, also in disturbed and cleared areas; 900–1220 m.

17. **Jatropha pachyrrhiza** Radcl. Sm. in Kew Bull. **46**: 150 (1991). Type: Zambia, Nchelenge Distr., 5.6 km NE of Chiengi, near L. Mweru, 08°41'S, 29°06'E, 13.x.1949, *Bullock* 1255 (K, holotype).

An erect perennial herb up to 20 cm tall; stems arising from a tuber up to 10 × 3 cm, glabrous. Stipules 4–6 mm long, deeply dissected, multifid with setaceous-filiform segments, glandular at the apex, reddish, persistent. Petioles 1–5 mm long. Leaf blades 2.5–3.5 × 2 cm, 3-partite or sometimes elobate, rounded at the base; lobes lanceolate, acute, closely stipitate-glandular, median lobe 2 × 0.6 cm, lateral lobes somewhat smaller; lamina 5-nerved from the base, lateral nerves of median lobe in c. 10 pairs. Inflorescences 3 cm long, terminal, subcorymbiform; lowest bracts 3–4 mm long, linear-lanceolate, upper bracts c. 1 mm long, triangular, the rest intermediate in shape and size, all closely glandular-stipitate. Male flowers: pedicels 1.5 mm long; calyx 2 × 2.5 mm, campanulate, the lobes 1 × 1 mm, triangular-ovate, subacute, entire; petals 3 × 1.5 mm, elliptic, obtuse, minutely ferrugineous-pubescent within at the base, otherwise glabrous, straw-yellow to buff in colour; disk glands 5, 0.3 mm long, turbinate, oblique and concave at the apex; staminal column 3 mm high, glabrous, stamens 8, the 5 outer c. 1 mm shorter than the 3 inner and fused for half their length, the inner fused for two-thirds of their length, anthers 0.8–1 mm long, yellow. Female flowers: pedicels 2.5 mm long, robust; sepals 2–2.5 × 1 mm, ovate-lanceolate, acute, sparingly shortly glandular-stipitate to subentire; petals 4 × 1 mm, narrowly elliptic, acute, subentire; disk glands 5, 0.6 mm wide, transversely ovate, truncate; ovary 2 × 1 mm, 3-lobed to subcylindric, glabrous; styles c. 1 mm long. Fruit 1.2 × 1 cm, 3-lobed to subcylindric, somewhat reticulate-nerved and shallowly irregularly granular-verruculose. Seeds 9 × 5 × 4 mm, compressed-cylindric, scarcely shiny, ochreous-brownish, with a 4-fid, fringed caruncle 3 mm wide, orange-chestnut in colour at the base shading to buff.

Zambia. N: 5.6 km NE of Chiengi, fl. 13.x.1949, *Bullock* 1255 (K). S: Choma Distr., Muyuni, south of Mapanza fr. 22.ix.1957, *Robinson* 2445 (PRE; SRGH).

Not known elsewhere. In dry sandy *Afzelia, Brachystegia* bush; 1067 m.

18. **Jatropha schweinfurthii** Pax in Bot. Jahrb. Syst. **19**: 110 (1894); in Engler, Pflanzenr. [IV, fam. 147, i] **42**: 70 (1910). —Hutchinson in F.T.A. **6**, 1: 794 (1912). —Engler, Pflanzenw. Afrikas (Veg. Erde 9) **3**, 2: 116 (1921). —F.W. Andrews, Fl. Pl. Anglo-Egypt. Sudan **2**: 85 (1952). —Radcliffe-Smith in F.T.E.A., Euphorb. 1: 358 (1987). Types from the Sudan (Bahr el Ghazal Province).

A semi-succulent perennial herb up to 60 cm tall; stems arising from a large woody tuberous rootstock, simple or sparingly branched, decumbent to erect, sparingly pubescent. Stipules 2 mm long, usually simple, filiform-setaceous, eglandular, pinkish, soon falling, or absent. Petioles up to 1 cm long, or leaves subsessile. Leaf blades 2–11(14) × 1.5–9(10) cm (in Flora Zambesiaca area), 3-partite, 3-fid or 3-

lobed, rarely elobate or somewhat 5-partite, or with only 1 lateral lobe; lamina cuneate to rounded or truncate or shallowly cordate at the base, sparingly pubescent to subglabrous on upper surface, more evenly pubescent along the midribs and main nerves beneath; the lateral lobes forwardly directed, the median lobe broadly ovate to narrowly oblanceolate, obtuse to subacute or acute at the apex, sharply somewhat irregularly denticulate, the teeth often aristate, slightly glandular or not; the lobe midribs arising above the lamina base, the median lobe with c. 9 pairs lateral nerves, the lateral lobes with fewer lateral nerves. Inflorescences 4–8 cm long, terminal or subterminal, thyrsoid-subcorymbiform, peduncle up to 5 cm long; bracts variable, the lowest up to 1.5 cm long, linear, foliaceous, denticulate, the upper shorter, linear-subulate, chaffy, entire. Male flowers: pedicels c. 1 mm long, jointed, bracteolate, glabrous; calyx tube 1.5 mm long, with lobes 1.5 or 2 × 1 mm, triangular-lanceolate, acute, entire, glabrous; petals 3.5 or 7 × 1.5 or 2 mm, narrowly oblong-oblanceolate, rounded, glabrous, pale yellow to white; disk glands 5, free, 0.75 mm long, compressed-subcylindric, rounded-truncate; stamens 8, the 5 outer 6 mm long and fused for about one-quarter of their length, the 3 inner 7 mm long and fused for c. one-third of their length, filaments glabrous, anthers 1.75 mm long, oblong, thecae free at the base. Female flowers: pedicels 2.5–3 mm long, stout, dilated at the apex, sparingly pubescent; calyx tube very short, calyx lobes unequal, 3 or 4–6 × 0.8–1 or 1.5 mm, oblong-linear, acute, subentire, eglandular, pubescent without, glabrous within; petals 5 or 6–7 × 1 or 1.7 mm, free, narrowly oblong-oblanceolate, subacute, glabrous, creamy-yellow or whitish; disk 5-lobed, the lobes transversely ovate, rounded, flattened; ovary c. 2.5 × 2 mm, ellipsoid to 3-lobed, smooth, sparingly pubescent to subglabrous; styles c. 2 mm long, united for one-quarter of their length, erect, stigmas subcapitate, cylindric, verruculose. Fruits 1.1 × 1 cm, ellipsoid to 3-lobed, smooth, glabrescent or glabrous (Flora Zambesiaca area). Seeds (7.5)9–10 × (4.5)5–5.5 × 3.5–4 mm, including the caruncle, compressed ellipsoid-subcylindric, shiny, chestnut-brown or black; caruncle 2 × 4 mm, hemispherical, 2-lobed, the lobes strongly fimbiate, orange or brown when dry, with a buff or whitish fringe.

Leaf blades elobate, 3-lobed or 3-fid, the median lobe usually broadly ovate to oblanceolate, the teeth slightly glandular; peduncle up to 5 cm long; male petals 3.5 × 1.5 mm; ovary subglabrous; fruit glabrous; seeds 10 × 5.5 × 4 mm, black · · · · · · · · · · subsp. *atrichocarpa*

Leaf blades 3-fid to 3-partite, the median lobe usually narrowly oblanceolate, the teeth eglandular; peduncle c. 2 cm long; male petals 7 × 2 mm; ovary sparingly pubescent; fruit glabrescent; seeds 7.5–9 × 4.5–5 × 3.5 mm, chestnut-brown · · · · · · · · · · · · subsp. *zambica*

Subsp. **atrichocarpa** Radcl.-Sm. in Kew Bull. **28**: 285 (1973); in F.T.E.A., Euphorb. 1: 359 (1987). Type from Tanzania (Southern Highlands Province).

Leaf blades elobate, 3-lobed or 3-fid, the median lobe usually broadly ovate to oblanceolate, the teeth slightly glandular; peduncle up to 5 cm long; male petals 3.5 × 1.5 mm; ovary subglabrous; fruit glabrous; seeds 10 × 5.5 × 4 mm, black.

Zimbabwe. N: Darwin Distr., Kandeya Native Reserve, fr. 17.i.1960, *Phipps* 2321 (K; SRGH).

Also in Tanzania. *Brachystegia boehmii* escarpment woodland, not common, also in soils with high copper oxide values; 975 m.

Subsp. **zambica** Radcl.-Sm. in Kew Bull. **46**: 154 (1991). Type: Zambia, Lusaka, fl. & y. fr. 8.xii.1968, *Fanshawe* 10488 (K, holotype; NDO; SRGH).

Leaf blades 3-fid to 3-partite, the median lobe usually narrowly oblanceolate, the teeth eglandular; peduncle c. 2 cm long; male petals 7 × 2 mm; ovary sparingly pubescent; fruit glabrescent; seeds 7.5–9 × 4.5–5 × 3.5 mm, chestnut-brown.

Zambia. C: Chilanga, Tom Puffett's Farm, fr. 6.i.1966, *Lawton* 1349 (K; NDO; SRGH).

Not known elsewhere. Locally frequent in shallow damp soil over limestone, in *Acacia* and *Combretum ghazalense* woodlands.

19. **Jatropha campestris** S. Moore in J. Linn. Soc., Bot. **40**: 196 (1911). —Hutchinson in F.T.A. **6**, 1: 785 (1912). —Engler, Pflanzenw. Afrikas (Veg. Erde 9) **3**, 2: 116, 117 (1921). Type: Mozambique, Mossurize (Umswirizwi) Flats, 1.xi.1905, *Swynnerton* 311 (BM, holotype).

A slightly succulent, glabrous, glaucous, shrubby forb-like perennial herb up to 50 cm high; stems arising from a rootstock, densely leafy. Young stems reddish. Stipules 0.5–1 mm long, simple, filiform-subulate, or bifid, eglandular, hyaline, soon falling, or absent. Petioles up to 1.2 cm long, or leaves sessile. Leaf blades (2)4–21 × (1)2–12 cm, 3–5(7)-palmatipartite to subpinnatipartite, rarely some also simple and ovate or elliptic, cuneate or attenuate at the base, membranaceous; the lobes up to 12 × 3 cm, usually narrowly oblong-oblanceolate, obtuse at the apex, entire or rarely irregularly pinnately lobulate or repand-dentate and hyaline on the margins; lateral nerves of the median lobe often in c. 20 pairs. Inflorescences 7–10(24) cm long, terminal, with a peduncle up to 7 cm long; bracts up to c. 5 mm long, narrowly lanceolate, acute, entire or subentire. Male flowers: pedicels c. 1 mm long, later up to 3 mm long; calyx 3 mm long, calyx lobes 1.5–2 × 1.5 mm, triangular-ovate, obtuse or rounded, entire, eglandular, pale green; petals 5 × 2 mm, oblanceolate-oblong, rounded, glabrous without, pilose within, yellow, pink-tinged; disk glands 5, free, contiguous, thick, angular, truncate; stamens 8, the 5 outer 4 mm long and fused for c. one-third of their length, the 3 inner 5 mm long and fused for c. two-fifths of their length, anthers 1.5 mm long, bright yellow. Female flowers: pedicels up to 4 mm long; calyx 3–4.5 mm long, calyx lobes 2.5–4 × 1 mm, linear or lanceolate, subacute or obtuse, entire, eglandular; petals 5–6 × 1.5 mm, linear, subacute or obtuse, yellowish; disk 2.5 mm in diameter, deeply 5-lobed, the lobes transversely oblong, 1 mm wide, flattened, retuse; ovary c. 2.5 × 2 mm, 3-lobed to ellipsoid; styles 2 mm long, united at the base, erect, stigmas bifid. Fruit c. 10–12 × 8–9(10) mm, 3-lobed to ellipsoid, shallowly tuberculate-rugulose. Seeds 9 × 5.5 × 4 mm, compressed oblong-ellipsoid, smooth, shiny, dark brown, with a multifid fimbriate caruncle 2 × 4.5 mm, blackish at the base shading through brown and buff to whitish.

Zimbabwe. C: Charter, fr. 27.xii.1926, *Eyles* 4591 (K; SRGH). S: Mwenezi Distr., 6.5 km south of Chipinda Pools, fl. & fr. 30.xii.1970, *Kelly* 331 (K; MO; PRE; SRGH). **Mozambique**. MS: Mossurize (Umswirizwi) Flats, 1.xi.1905, *Swynnerton* 311 (BM).

Not known elsewhere. Mopane woodland on sand, black soil depressions in mopane woodland, also in basaltic black clay and in sandveld; 335–1220 m.

20. **Jatropha monroi** S. Moore in J. Bot. **63**: 147 (1925). —Radcliffe-Smith in Kew Bull. **46**: 150 (1991). Type: Zimbabwe, Masvingo (Fort Victoria), fl. & y. fr. 1909/12, *Monro* 2187 (BM, holotype).

Jatropha cervicornis Suess. in Trans. Rhod. Sci. Ass. **43**: 84 (1951). Type: Zimbabwe, Marondera (Marandellas), 6.xi.1942, *Dehn* 704 (M, holotype).

A glabrous, somewhat glaucous, sparingly branched erect perennial herb up to 10 cm tall, probably arising from a tuberous rootstock. Stipules 1–2 mm long, simple, subulate-setaceous, eglandular, purplish, subpersistent. Petioles 3–5 mm long. Leaf blades up to 4 × 3 cm, 3–5-palmatipartite to subpinnatipartite, cuneate or attenuate at the base, chartaceous; the lobes up to 2.5 × 1 cm, narrowly oblanceolate-oblong, subacute and mucronulate at the apex, closely and sharply denticulate on the margins, sometimes also irregularly lobulate and hyaline on the margins; lateral nerves of the median lobe in 10–12 pairs. Inflorescences 2–3 cm long, terminal, with a peduncle up to 1 cm long; bracts 2–4 mm long, narrowly lanceolate, laterally lobulate at the base or subentire. Male flowers: pedicels 2–3 mm long; calyx 2–3 mm long, calyx lobes 1–1.5 × 0.75–1 mm, triangular-ovate, obtuse or rounded, subentire, reddish-purple; petals 4 × 1 mm, oblanceolate-oblong, obtuse-rounded, completely glabrous, yellow, pink-tinged; disk glands 5, 0.5 × 0.5 mm, free, erect, squarish, flat; stamens 8, the 5 outer free and 2.5 mm long, the 3 inner 3 mm long, united at the base, anthers 1.25 mm long. Female flowers: pedicels 2 mm long, stouter than in the male flowers; calyx 3.5–4 mm long, calyx lobes 3 × 1.25 mm, lanceolate, acute, subentire to irregularly denticulate or fimbriate, subeglandular; petals 7 × 1.5 mm, narrowly oblanceolate-spathulate, obtuse or rounded, completely glabrous; disk c. 2 mm in diameter, composed of 7 separate squarish flattened rounded or truncate glands; ovary c. 2.5 × 2.5 mm, subglobose, styles 2.5 mm long, united at the base, erect, stigmas bifid, rugulose. Fruit (immature) 10 × 8 mm, ellipsoid, minutely tuberculate-rugulose. Mature seeds unknown.

Zimbabwe. C: Marondera (Marandellas), 6.xi.1942, *Dehn* 704 (M). S: Masvingo (Fort Victoria), fl. & y. fr. 1909/12, *Monro* 2187 (BM).

Not known elsewhere. No ecological data available.

21. **Jatropha loristipula** Radcl.-Sm. in Kew Bull. **46**: 146 (1991). Type: Zimbabwe, c. 13 km north of Beitbridge, fl. & y. fr. 12.xi.1966, *Leach & Bullock* 13578 (SRGH, holotype).

An erect glabrous branched perennial herb up to 30 cm tall, perhaps arising from a tuberous rootstock. Stipules 4–7 mm long, strap-shaped, green, irregularly whitish-lacerate, persistent. Petioles 1–5 mm long. Leaf blades up to 13 × 12 cm, palmately 3- or 5-partite, attenuate or cuneate at the base; lobes unequal, the median up to 9 × 1.7 cm, the lateral lobes smaller, all lobes linear-oblanceolate, subacute, irregularly but closely and sharply bidenticulate, the teeth eglandular; the midribs of the lateral lobes diverging from the main midrib above the base of the lamina, the lateral nerves of the median lobe in c. 20 pairs, fewer in other lobes. Inflorescences terminal, up to 4 cm long, the peduncle c. 1 cm long; bracts 2–3 mm long, ovate, subentire or denticulate. Male flowers: pedicels 1.5 mm long; calyx tube 2 mm long, calyx lobes 1.3 × 1 mm, ovate, obtuse, entire; petals 7 × 2 mm, oblong-oblanceolate, obtuse or rounded, completely glabrous; disk glands 5, 0.75 × 0.75 mm, free, erect, squarish, flat; stamens 8, the 5 outer 5.5 mm long, free, the 3 inner 6 mm long, united at the base, anthers 3 mm long, sagittate. Female flowers: pedicels 3 mm long, stouter than in the male; calyx tube 0.5 mm long, calyx lobes 3 × 1.3 mm, ovate-lanceolate, obtuse, entire; petals 8 × 1 mm, linear, obtuse, completely glabrous; disk 5-lobed, the lobes c. 1 mm across, transversely ovate, flat, rounded; ovary 2 × 1.67 mm, 3-lobed to subcylindric, smooth; styles not known. Fruit 1.4 × 1.2 cm, 3-lobed to subcylindric, smooth. Mature seeds not seen.

Zimbabwe. S: c. 13 km north of Beitbridge, fl. & y. fr. 12.xi.1966, *Leach & Bullock* 13578 (SRGH).

Not known from elsewhere. In *Colophospermum mopane-Combretum* woodland; 305 m.

22. **Jatropha schlechteri** Pax in Bot. Jahrb. Syst. **28**: 24 (1899); in Engler, Pflanzenr. [IV, fam. 147, i] **42**: 67 (1910). —Prain in F.C. **5**, 2: 427 (1920). —Engler, Pflanzenw. Afrikas (Veg. Erde 9) **3**, 2: 116, 117 (1921). —Burtt Davy, Fl. Pl. Ferns Transvaal: 303 (1932). Type from South Africa (Transvaal).

A semi-succulent, sparingly hirsute-setose, perennial herb up to 40 cm tall; stems simple or sparingly branched, arising from a stout rootstock with red flesh. Stipules 1–5 mm long, simple, multifid or compound, eglandular, yellowish, subpersistent; the segments setaceous-filiform. Petioles 0.3–1 cm long (in Flora Zambesiaca area). Leaf blades (6)9–15 × (4)6–10 cm, irregularly pinnatipartite, attenuate-cuneate at the base, glabrous or subglabrous on upper surface, sparingly to evenly hirsute-setose beneath especially along the midrib and main lateral nerves; all lobes narrowly oblong-oblanceolate, subacute or obtuse at the apex, usually dentate to irregularly pinnately lobulate on the margins with the lobules often denticulate, often with subeglandular-aristate teeth on the margin; main lateral nerves of the midrib (very variable in thickness and length) in 15–30 pairs, lateral nerves of the segments (also very variable) (7)12–15 pairs. Inflorescences terminal, up to 15 cm long, laxly pseudopaniculate; peduncle up to 8 cm long; bracts 4–8 mm long, narrowly triangular-lanceolate, acute, weakly glandular-stipitate to subeglandular-fimbriate. Male flowers: pedicels 1 mm long, bracteolate; calyx tube c. 1 mm long, calyx lobes 2 × 1 mm, triangular-ovate, subacute or obtuse, entire or glandular-ciliate, glabrous; petals 6.5–8 × 2 mm, oblong-oblanceolate, emarginate, glabrous or pubescent without, pilose within, reddish; disk glands 5, free, 0.75 × 0.5 mm, rectangular, flattened, truncate; stamens 8, the 5 outer 5.5 mm long and fused for one-third of their length, the 3 inner 6 mm long and fused for half their length, filaments glabrous, anthers 1.3 mm long, thecae free and slightly divergent at the base. Female flowers: pedicels 2–3 mm long; calyx tube 1 mm long, calyx lobes 4 × 1.5 mm, oblong-lanceolate, obtuse, subentire to sparingly subeglandular-denticulate and/or fimbriate; petals similar to the male, but slightly longer and glabrous within; disk deeply 5-lobed, the lobes transversely oblong, truncate, slightly thickened; ovary 2.5

× 2 mm, 3-lobed to ellipsoid, smooth, ± glabrous or densely hirsute; styles c. 2 mm long, free or almost so, suberect, stigmas capitate, subcylindric, verruculose. Fruits 11–12 × 9–10 mm, 3-lobed to subcylindric, smooth, glabrous or pubescent; columella 9 mm long. Seeds 9 × 5 × 4 mm, slightly compressed ellipsoid-subcylindric, somewhat shiny, brown, mottled buff, or buff, mottled brown; caruncle 2–3 × 4.5 mm, ± hemispherical, 2-lobed, the lobes fimbriate-multifid, orange or brown with a creamy fringe when dried.

Male calyx lobes glandular-ciliate; ovary densely hirsute; fruit pubescent · · · · subsp. *schlechteri*
Male calyx lobes ± entire; ovary ± glabrous; fruit glabrous · · · · · · · · · · · · · · · · subsp. *setifera*

Subsp. **schlechteri**

Male calyx lobes glandular-ciliate; ovary densely hirsute; fruit pubescent.

Mozambique. M: Matutuíne (Bela Vista), Catuane, near R. Mazeminhama, fl. 5.x.1968, *Balsinhas* 1348 (LISC; PRE).
Also in South Africa (eastern Transvaal). In hard packed dark brown soil in heavily grazed grassland, after burning in *Themeda* grassland, in dry scrub and in *Acacia, Combretum* woodland.

Subsp. **setifera** (Hutch.) Radcl.-Sm. in Kew Bull. **46**: 152 (1991). Type from South Africa (Transvaal).
Jatropha setifera Hutch., Botanist in Southern Africa: 397 (1946).

Male calyx lobes ± entire; ovary ± glabrous; fruit glabrous.

Zimbabwe. S: Gwanda Distr., Quale R., fr. 17.xii.1956, *Davies* 2325 (K; LISC; MO; SRGH).
Also in South Africa (northern Transvaal). Dry sandveld with *Colophospermum mopane, Grewia, Terminalia, Commiphora,* also in seasonal pans in deep black basalt soils; 565 m.
Bruce Hargreaves (pers. comm.) reports both subspecies from the extreme east of Botswana (Masama and Semolale respectively), but I have seen no specimens.

23. **Jatropha erythropoda** Pax & K. Hoffm. in Engler, Pflanzenr. [IV, fam. 147, i] **42**: 66 (1910). —Hutchinson in F.T.A. **6**, 1: 783 (1912). —Prain in F.C. **5**, 2: 422 (1920). —Engler, Pflanzenw. Afrikas (Veg. Erde 9) **3**, 2: 116, 117 (1921). —Burtt Davy, Fl. Pl. Ferns Transvaal. 302 (1932). —Hutchinson, Botanist in Southern Africa: 667 (1946). —P.G. Meyer in Merxmüller, Prodr. Fl. SW. Afrika, fam. 67: 36 (1967). Type from Namibia.

An erect several-stemmed somewhat fleshy perennial herb up to 30 cm high; tubers up to 16 × 3.5 cm, elongate or ellipsoid, smooth, red, lying some 15 cm below ground level; stems glabrous below, pubescent above, rarely completely glabrous, reddish-tinged, arising at ground level from one or more small underground stems (caudiculi) produced from the tuber. Stipules c. 2 mm long, simple, setaceous-filiform, or bifid or trifid with filiform segments, eglandular, ochreous, subpersistent. Petioles 1–8 mm long. Leaf blades 2–8(12) × 1–4 cm, irregularly pinnatipartite, cuneate or rounded at the base, membranaceous, usually glabrous, rarely sparingly puberulous towards the base, glaucous; the lobes acute, lobulate, serrate or entire, with hyaline margins; lateral nerves often indistinct. Inflorescences up to 3 cm long, terminal, shortly pedunculate; bracts 2–5 mm long, lanceolate, acute, denticulate to subentire. Male flowers: pedicels up to 2 mm long; calyx c. 3 mm long; calyx lobes 1.25 × 1 mm, ovate to ovate-lanceolate, acute, subacute or obtuse, entire, subentire or minutely glandular-denticulate, glabrous or sparingly pubescent; petals 5–5.5 × 1.75 mm, oblanceolate, rounded, glabrous, creamy-white; disk glands 5, free, c. 0.75 mm, turbinate, truncate; stamens 8, 4.5–5 mm long, the 5 outer free, the 3 inner connate below, filaments glabrous, anthers 1.75–2 mm long. Female flowers: pedicels 2.5 mm long, stout; calyx c. 4 mm long, calyx lobes 3 × 0.8 mm, narrowly triangular-lanceolate, acute, irregularly glandular-denticulate or stipitate-glandular to subentire, glabrous or almost so; petals 6–8 × 1–1.5 mm, linear-oblanceolate, subacute, glabrous, white or pinkish; disk deeply 5-lobed, the lobes scale-like; ovary c. 2 × 2 mm, subglobose, glabrous; styles 2 mm long, united at the base, erect, stigmas bifid, papillose. Fruit 7–8(10) × 10–12 mm, depressed-globose, shallowly 3-lobed, very shallowly tuberculate-verruculose, glabrous. Seeds 7–8 × 4.5–5 × 3.5 mm, compressed-ellipsoid, smooth, pale yellowish-grey, with a 2-lobed flabelliform caruncle c. 2 × 4.5 mm, dark chestnut-brown at the base shading to buff.

Botswana. SW: 85 km north of Kang (Kan) on road to Ghanzi, y. fr. 19.ii.1960, *de Winter* 7386 (K; PRE). SE: 6.5 km NW of Derdepoort, fl. 30.xi.1954, *Codd* 8945 (PRE). **Zimbabwe**. W: Matobo Distr., Besna Kobila Farm, fl. i.1962, *Miller* 8156 (SRGH).

Also in Namibia and South Africa (Cape Province, Transvaal). Sandveld grassland, shallow pan margins and wooded grassland with *Acacia* and *Combretum*; 900–1400 m.

44. CODIAEUM A. Juss.

Codiaeum A. Juss., Euph. Gen. Tent.: 33, t. 9, f. 30 (1824). —Müller Argoviensis in De Candolle, Prodr. **15**, 2: 1116 (1866). —Pax in Engler, Pflanzenr. [IV, fam. 147, iii] **47**: 23 (1911), nom. conserv.

Monoecious or occasionally dioecious shrubs or small trees with a simple indumentum, or glabrous. Leaves alternate, petiolate, minutely stipulate or stipules absent, simple, entire or 3-lobed, occasionally the lamina contorted or interrupted along the midrib, penninerved, eglandular or ± so. Inflorescences axillary or subterminal, racemose, solitary or paired, usually unisexual, pedunculate; male bracts 1–6-flowered; female bracts l-flowered. Male flowers: pedicels slender, jointed; sepals (3)5(6), closely imbricate; petals (3)5(6), small, minute or absent; disk glands 5–15, free, alternipetalous; stamens 15–100, filaments free, anthers erect, extrorse, dorsifixed, connective broad, thecae apically confluent, longitudinally dehiscent; pistillode absent. Female flowers: pedicels shorter and stouter than in the male, not jointed; sepals smaller than in the male, otherwise ± similar; petals absent; disk shallowly cupular, subentire or shallowly 5-lobed; ovary 3-locular, 1 ovule per locule; styles 3, shortly connate at the base, spreading or recurved, usually simple. Fruit globose or 3-lobed, dehiscing septicidally into 3 bivalved cocci; pericarp thin; endocarp thinly woody; columella persistent. Seeds ovoid to subglobose, carunculate; testa smooth, shiny, marmorate, crustaceous; albumen fleshy; cotyledons broad, flat.

A small genus of 16 species ranging from Malaysia to N Australia and Polynesia. One species is widely cultivated in the tropics, often becoming naturalized.

Codiaeum variegatum (L). A. Juss., Euph. Gen. Tent.: 111, t. 9, f. 30 (1824). —Müller Argoviensis in De Candolle, Prodr. **15**, 2: 1119 (1866). —Pax in Engler, Pflanzenr. [IV, fam. 147, iii] **47**: 23 (1911). —Brenan, Check-list For. Trees Shrubs Tang. Terr.: 204 (1949). —Troupin, Fl. Rwanda **2**: 214, fig. 62/1 (1983). —Radcliffe-Smith in F.T.E.A., Euphorb. 1: 324 (1987). Type is a plate based on a specimen from Moluccas.

Var. **variegatum**

Croton variegatus L., Sp. Pl., ed. 3: 1424 (1763).

Croton pictus Lodd., Bot. Cab. **9**: t. 870 (1824). Type is an illustration by Loddige based on material from the "Indian Islands".

Codiaeum pictum (Lodd.) Hook. in Curtis's Bot. Mag., t. 3051 (1831).

Codiaeum variegatum var. *pictum* (Lodd.) Müll. Arg. in De Candolle, Prodr. **15**, 2: 1119 (1866). —Pax in Engler, Pflanzenr. [IV, fam. 147, iii] **47**: 24 (1911). —White, F.F.N.R: 195 (1962).

A much-branched shrub or small tree up to 6 m tall, but often much less. Twigs pale brownish-grey. Leaf scars ± circular. Young growth sparingly pubescent to subglabrous. Petioles 1–1.5 cm long. Leaf blades c. 10–45 × 1.5–10 cm, very variable in shape, simple, entire or 3-lobed, occasionally the lamina contorted or interrupted along the midrib, obtuse to acute or sometimes aristate at the apex, rounded-cuneate to attenuate at the base, coriaceous, glabrous on both surfaces, shiny and variously marked and pigmented but occasionally concolorous above, duller and paler beneath; lateral nerves in c. 20–40 pairs, often indistinct except when of a different colour from the rest of the blade, brochidodromous. Stipules minute or absent. Male inflorescences 10–25 cm long; axis glabrous; bracts 1–2 mm long. Male flowers: pedicels 0.5–1.5 cm long; sepals 3 × 2.5 mm, ovate-suborbicular, strongly concave, glabrous; petals 1 × 2 mm, transversely elliptic, shortly 2-lobed, white; disk glands 5, 1 mm long, turbinate, truncate; stamens 20–25, filaments 3–4 mm long, white, anthers 0.5 mm long, yellow. Female inflorescences 5–20 cm long, otherwise as in the male. Female flowers: pedicels 1–2 mm long, not extending in fruit; sepals 1 × 1 mm, ovate-suborbicular, often ciliolate; disk 1.25 mm in diameter; ovary 2 × 1.5 mm, ovoid-conical, glabrous; styles 2–3.5 mm long, simple, recurved, spreading or contorted. Fruit 5–7 × 6–9 mm, rounded to 3-lobed, ± smooth, glabrous, grey-green. Seeds 6 × 4 × 3.5 mm, ovoid-ellipsoid, ± smooth, slightly shiny, greyish, dark brown-mottled.

Leaf blade oblanceolate, continuous along the midrib · · · · · · · · · · · · · · · · · · f. *platyphyllum*
Leaf blade linear, often interrupted along the midrib · · · · · · · · · · · · · · · · · f. *appendiculatum*

forma **platyphyllum** Pax in Engler, Pflanzenr. [IV, fam. 147, iii] **47**: 24 (1911). Type from Indonesia.

Lamina oblanceolate, continuous along the midrib.

Mozambique. M: Maputo (Lourenço Marques), Parque José Cabral, male fl. 6.x.1972, *Balsinhas* 2438 (SRGH).
Cultivated as a garden ornamental.

forma **appendiculatum** Čelak. in Abh. Böhm. Ges. Wiss. Prag. **6**. F. XII: 21, t. 2 (1884). — Brenan, Check-list For. Trees Shrubs Tang. Terr.: 204 (1949).

Lamina linear, often interrupted along the midrib.

Zambia. C: Mt. Makulu Res. Station, near Chilanga, y. male and female fls. 18.xi.1963, *Angus* 3809 (FHO; K).
Cultivated as a garden ornamental.

Codiaeum variegatum var. *moluccanum* (Decne.) Müll. Arg., with uniformly green obovate-oblong leaves, is native from Java to N Australia and Fiji. The var. *variegatum* on the other hand (the so-called "croton" of horticulturalists), is unknown in a wild state, but is cultivated in one or other of its many forms ± throughout the tropics in both hemispheres, where it may become naturalized, and has become a popular greenhouse and house plant in temperate regions. Other formae of it which may be encountered are: forma *ambiguum* Pax (leaves lanceolate to narrowly lanceolate, flat, entire), Mozambique, Unango, 27.v.1948, *Pedro & Pedrógão* 3962 (LMA); f. *taeniosum* Müll. Arg. (leaves linear, c. 1 cm wide, flat, entire); f. *crispum* Müll. Arg. (leaves narrowly lanceolate to linear, undulate or contorted, entire); f. *lobatum* Pax (leaves lobed, most often 3-lobed, flat); and f. *cornutum* André (leaves variously shaped, with the midrib emergent before but prolonged well beyond the apex into an arista).

45. MILDBRAEDIA Pax

Mildbraedia Pax in Bot. Jahrb. Syst. **43**: 319 (1909); in Engler, Pflanzenr. [IV, fam. 147, iii] **47**: 11 (1911).
Jatropha Pax in Engler, Pflanzenw. Ost-Afrikas **C**: 240 (1895); in Bot. Jahrb. Syst. **23**: 529 (1827); in Bot. Jahrb. Syst. **33**: 284 (1903), pro parte non L.
Neojatropha Pax in Engler, Pflanzenr. [IV, fam. 147, i] **42**: 114 (1910).

Dioecious shrubs or small trees. Indumentum stellate, sometimes also simple. Leaves alternate, stipulate, long-petiolate, simple, entire or sometimes (in west and central Africa) deeply 2–3-lobed, subentire to repand-dentate, palminerved. Stipules subulate, deciduous. Inflorescences axillary or supra-axillary, cymose, few-flowered (in Flora Zambesiaca area), long-pedunculate; bracts resembling the stipules. Male flowers: calyx deeply 5(6)-lobed, or sepals free, imbricate; petals 5(6), free, imbricate; disk glands 5, free, fleshy, opposite the sepals; stamens 10–25, the outer more or less free, the inner united into a column, anthers small, introrse, dorsifixed, longitudinally dehiscent; pistillode absent. Female flowers: sepals and petals resembling those of the male flowers, but slightly larger; disk hypogynous, annular, shallowly 5-lobed; ovary 3-locular, with 1 ovule per loculus; styles 3, slightly connate at the base, deeply bifid to bipartite, spreading. Fruit 3-lobed, dehiscing septicidally into 3 bivalved cocci; endocarp thinly woody; columella persistent. Seeds ovoid-subglobose; testa crustaceous, smooth, somewhat shiny, mottled; caruncle foliaceous, appressed.

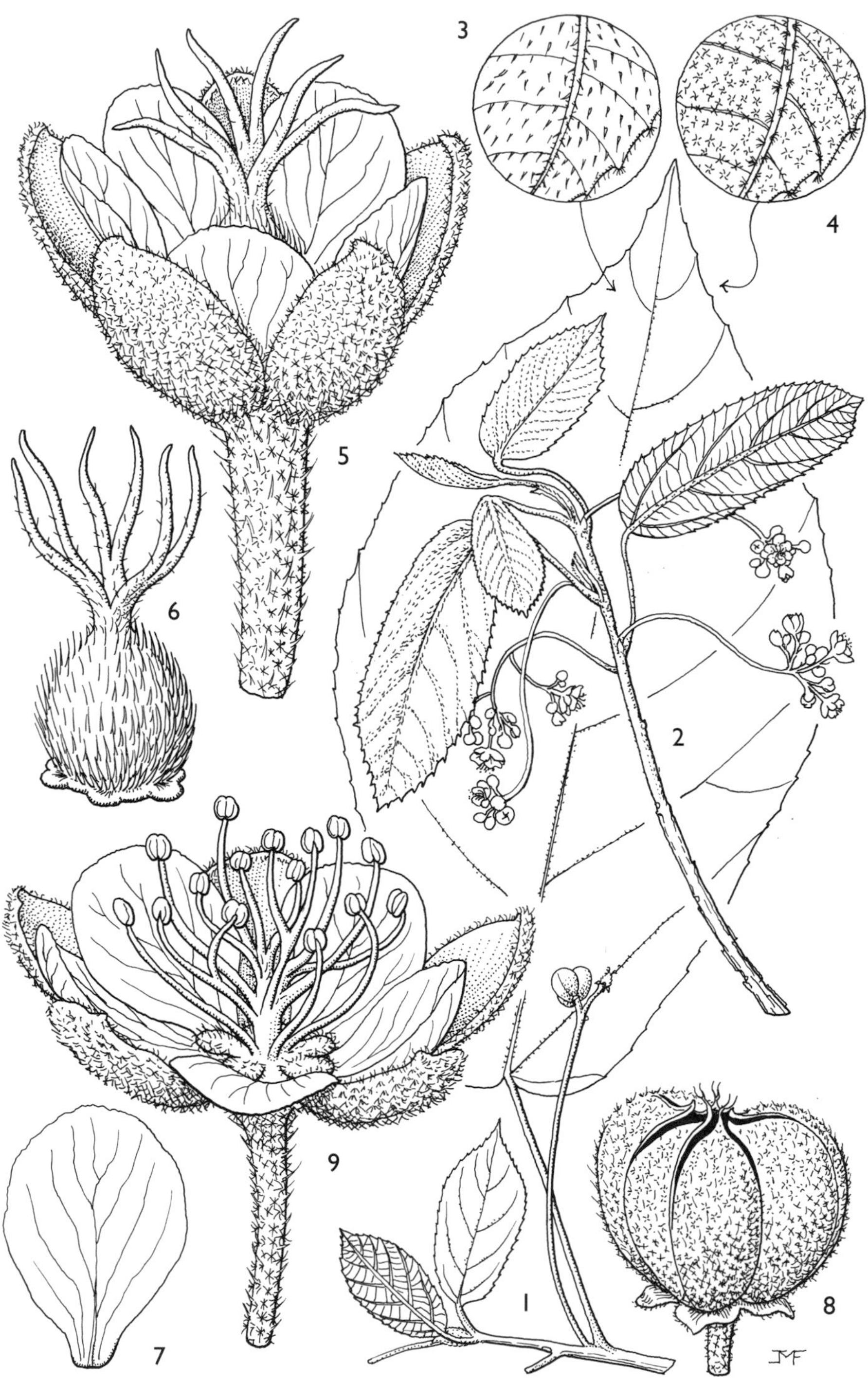

Tab. 56. MILDBRAEDIA CARPINIFOLIA var. STRIGOSA. 1, distal portion of fruiting branch (× ²⁄₃), from *Torre* 4016; 2, distal portion of branch with male flowers (× ²⁄₃); 3, upper surface of leaf (× 4); 4, lower surface of leaf (× 4); 5, female flower (× 10); 6, pistil (× 10); 7, petal (× 10); 8, fruit (× 4); 9, male flower (× 10), 2–9 from *Mendonça* 999. Drawn by J.M. Fothergill.

A small genus of 3–4 species confined to tropical Africa, one species occurring in the Flora Zambesiaca area.

Mildbraedia carpinifolia (Pax) Hutch. in F.T.A. **6**, 1: 801 (1912). —Engler, Pflanzenw. Afrikas (Veg. Erde 9) **3**, 2: 120 (1921). —Brenan, Check-list For. Trees Shrubs Tang. Terr.: 220 (1949). —Dale & Greenway, Kenya Trees & Shrubs: 210 (1961). —Radcliffe-Smith in F.T.E.A., Euphorb. 1: 340 (1987). —Beentje, Kenya Trees, Shrubs Lianas: 215 (1994). Type from Tanzania (Eastern Province).

Jatropha carpinifolia Pax in Engler, Pflanzenw. Ost-Afrikas **C**: 240 (1895); in Bot. Jahrb. Syst. **23**: 529 (1897).

Jatropha fallax Pax in Bot. Jahrb. Syst. **33**: 284 (1903). Type from Tanzania (Eastern Province).

Neojatropha carpinifolia (Pax) Pax in Engler, Pflanzenr. [IV, fam. 147, i] **42**: 114 (1910).

Neojatropha fallax (Pax) Pax in Engler, Pflanzenr. [IV, fam. 147, i] **42**: 115, t. 41 (1910).

Mildbraedia fallax (Pax) Hutch. in F.T.A. **6**, 1: 800 (1912). —Engler, Pflanzenw. Afrikas (Veg. Erde 9) **3**, 2: 120 (1921). —Brenan, Check-list For. Trees Shrubs Tang. Terr.: 220 (1949).

Var. **strigosa** Radcl.-Sm. in Kew Bull. **27**: 507 (1972); in F.T.E.A., Euphorb. 1: 341 (1987). Tab. **56**. Type from Tanzania (Eastern Province).

A rather weak, soft-wooded shrub, usually (0.5)1.5–4 m high. Bark of branches strongly longitudinally corky-ridged. Twigs densely scurfily stellate-pubescent and strigose. Stipules 5–12 mm long, filiform-subulate, sparingly to densely stellate-pubescent. Petioles (0.5)1–6.5 cm long, evenly stellate-pubescent. Leaf blades 2.5–20 × 1.5–12 cm, elliptic-obovate to oblong-oblanceolate, acutely acuminate at the apex, sharply somewhat irregularly denticulate on the margins, rounded to cordate at the base, chartaceous, scurfily stellate-pubescent and strigose on the midrib and main nerves on upper surface and sparingly strigose between them, evenly scurfily stellate-pubescent beneath; 3–5-nerved from the base, the lateral nerves in 6–7 pairs and the tertiary nerves subparallel. Male inflorescences up to 6 cm long with peduncles 4–5 cm long, 12–15-flowered; bracts 2 mm long, subulate. Male flowers: pedicels up to 5 mm long; sepals 3–4 × 1.5–2.5 mm, ovate, obtuse, greyish-pubescent on the outside except where overlapped, glabrous on the inside, pale green; petals 3–4 × 4 mm, broadly ovate-suborbicular, glabrous, white or greenish; disk glands 1 mm across, rounded, densely sericeous-pubescent; stamens 2–3 mm long, filaments glabrous, anthers 0.5 mm long. Female inflorescences 5–10 cm long with peduncles 4–9 cm long, 2–4-flowered; bracts 4–5 mm long; bracteoles 2 mm long, resembling the male bracts. Female flowers: pedicels 4–5 mm long, stout, thickened upwards, reddish-pubescent; sepals and petals as in the male flowers; disk c. 2 mm across, glabrous; ovary 2 mm in diameter, ovoid-subglobose, densely sericeous-strigose; styles 3 mm long, sparingly pubescent below, otherwise glabrous. Fruit 7 × 9–10 cm, evenly scurfily stellate-strigose. Seeds 5–6 × 4–5 mm, grey, mottled purplish-brown; caruncle applanate, 2 mm across.

Mozambique. N: Cabo Delgado, Tungué, between Nangade and Palma, male & female fl. 20.x.1942, *Mendonça* 999 (LISC). Z: Namacurra, 45 km from Nicuadala, y. fr. 2.ii.1966, *Torre & Correia* 14369 (LISC). MS: 8 km west of Machire, y. fr. 22.iv.1974, *Müller* 2102 (K; MO; PRE; SRGH).

Also in southern Tanzania. In mixed evergreen forest, forest clearings and in forest/woodland mixture, often on sandy soil, also in wooded grassland and by streamsides; 40–300 m.

The typical variety, in which the indumentum is stellate-pubescent and not strigose, is found in Kenya and NE Tanzania (including Zanzibar Island).

Vernacular name as recorded in specimen data: "que que" (Manica e Sofala).

46. CROTON L.

Croton L., Sp. Pl.: 1004 (1753), pro parte; Gen. Pl., ed. 5: 436 (1754).
Argyrodendron Klotzsch in Peters, Naturw. Reise Mossambique **6**, 1: 100 (1861).

Monoecious or sometimes dioecious trees or shrubs (or herbs or lianes outside the Flora Zambesiaca area) with a stellate and/or lepidote indumentum. Leaves alternate, sometimes subopposite or ± whorled, stipulate, petiolate, with 2 or more sessile or stipitate discoid glands at the petiole apex or the lamina base, simple, entire, toothed or lobed, penninerved or palminerved. Inflorescences usually terminal, racemose, androgynous or unisexual; bracts small. Male flowers shortly pedicellate; buds subglobose; sepals (4)5(6), valvate or imbricate, ± equal; petals (4)5(6), free; disk glands opposite the sepals, free or fused, fleshy; stamens 5–many, free, filaments inflexed in bud, later erect, anthers pendulous in bud, later erect, longitudinally dehiscent; receptacle usually pilose; pistillode absent. Female flowers pedicellate; sepals generally persistent, slightly accrescent, equal or unequal; petals often smaller than in male flowers, sometimes replaced by tufts of hairs or absent; staminodes sometimes present; disk annular, or of separate glands, or vestigial; ovary 2(3)4-locular, with 1 ovule per loculus; styles 1–several times bifid or bipartite, or multifid or multipartite. Fruits septicidally dehiscent into 3 bivalved cocci, loculicidally dehiscent into 3 valves, irregularly frangent (breaking up) or drupaceous and indehiscent; endocarp woody or crustaceous. Seeds carunculate; testa woody or crustaceous; albumen fleshy, copious; embryo straight; cotyledons broad and flat.

A pantropical genus of c. 750 species, especially well represented in east tropical S America.

1. Leaves entire or subentire, silvery-lepidote on under surface · · · · · · · · · · 2
– Leaves toothed, or if entire then stellate-pubescent on under surface · · · · · · · · 8
2. Leaves whorled or subopposite, rarely some alternate; racemes 0.5–2 cm long · · · · · · · · · · 1. *pseudopulchellus*
– Leaves generally all alternate; racemes 1–16 cm long · · · · · · · · · · 3
3. Petioles not more than 1 cm long; racemes up to 6 cm long · · · · · · · · · · 4
– Petioles up to 11 cm long (occasionally some less than 1 cm long); racemes up to 16 cm long · · · · · · · · · · 5
4. Leaves obtuse or emarginate at the apex; styles 2-partite · · · · · · · · · · 2. *menyharthii*
– Leaves acutely acuminate at the apex; styles 4-partite · · · · · · · · · · 3. *kilwae*
5. Leaves usually broadly ovate · · · · · · · · · · 6
– Leaves elliptic-ovate to oblong-lanceolate · · · · · · · · · · 7
6. Petals of male flowers pubescent; fruits 5 × 6–7 mm, septicidal · · · · · · · 5. *steenkampianus*
– Petals of male flowers lepidote; fruits 17–20 × 17–20 mm, subindehiscent to tardily loculicidal · · · · · · · · · · 7. *megalocarpoides*
7 Pedicels of male flowers 1–3(5) mm long; buds of male flowers 1.5–3 mm in diameter; fruits 8–10 × 8–11 mm, septicidal · · · · · · · · · · 4. *gratissimus*
– Pedicels of male flowers 3–10 mm long; buds of male flowers 4–5 mm in diameter; fruits (15)30–45 × (15)25–40 mm, loculicidal · · · · · · · · · · 6. *megalocarpus*
8. Leaves 3-lobed or 5–7-lobed · · · · · · · · · · 18. *aceroides*
– Leaves entire or toothed · · · · · · · · · · 9
9. Fruits echinate (beset with prickles) · · · · · · · · · · 15. *inhambanensis*
– Fruits more or less smooth · · · · · · · · · · 10
10. Female sepals accrescent to 10 × 5 mm in fruit, foliaceous · · · · · · · · · · 9. *scheffleri*
– Female sepals not or slightly accrescent (to up to 7 × 3 mm) · · · · · · · · · · 11
11. Leaves all or mostly penninerved · · · · · · · · · · 12
– Leaves all palminerved · · · · · · · · · · 13
12. Leaves gradually acuminate at the apex, stellate-pilose to glabrescent on the under surface · · · · · · · · · · 16. *leuconeurus* subsp. *leuconeurus*
– Leaves abruptly caudate-acuminate at the apex, glabrous on the under surface · · · · · · · · · · 17. *gossweileri*
13. Fruits 20–30 mm in diameter when dry, indehiscent · · · · · · · · · · 14. *megalobotrys*
– Fruits 4–15 mm in diameter when dry, dehiscent or not · · · · · · · · · · 14
14. Pedicels of female flowers up to 10 mm long, extending to up to 25 mm in fruit · · · · · · · · · · 8. *longipedicellatus* var. *longipedicellatus*
– Pedicels of female flowers up to 5 mm long, not or scarcely extending in fruit · · · · · · 15
15. Leaves glabrescent when mature; fruits subindehiscent · · · · · · · · · · 12. *sylvaticus*
– Leaves sparingly to densely stellate-pubescent or fimbriate-lepidote when mature; fruits dehiscent · · · · · · · · · · 16
16. Fruits irregularly frangent (breaking up) · · · · · · · · · 16. *leuconeurus* subsp. *mossambicensis*
– Fruits loculicidally or septicidally dehiscent · · · · · · · · · · 17

17. Racemes 7–35 cm long; fruits 8–15 mm in diameter, loculicidal or septicidal · · · · · · · · · · 13. *macrostachyus*
– Racemes not more than 17 cm long; fruits up to 7 mm in diameter, septicidal · · · · · · 18
18. Leaf blades sharply serrulate on the margin; styles deeply 2-partite; fruits evenly stellate-pubescent · 10. *polytrichus* subsp. *brachystachyus*
– Leaf blades entire, subentire or shallowly serrulate · 19
19. Styles deeply 2-partite; fruits evenly to sparingly fimbriate-lepidote, greenish becoming blackish · 8. *longipedicellatus* var. *brevipedicellatus*
– Styles twice 2-partite or 4-partite or occasionally twice 3-partite; fruits sparingly stellate-pubescent, pale green · 11. *madandensis*

1. **Croton pseudopulchellus** Pax in Bot. Jahrb. Syst. **34**: 371 (1904). —Hutchinson in F.T.A. **6**, 1: 757 (1912). —Prain in F.C. **5**, 2: 417 (1920). —Engler, Pflanzenw. Afrikas (Veg. Erde 9) **3**, 2: 47 (1921). —Burtt Davy, Fl. Pl. Ferns Transvaal: 302 (1932). —Hutchinson, Botanist in Southern Africa: 667 (1946). —Brenan, Check-list For. Trees Shrubs Tang. Terr.: 204 (1949). —Topham, Check List For. Trees Shrubs Nyasaland Prot.: 50 (1958). —Keay in F.W.T.A., ed. 2, **1**, 2: 394 (1958). —Dale & Greenway, Kenya Trees & Shrubs: 192 (1961). —White, F.F.N.R.: 196 (1962). —Mogg in Macnae & Kalk, Nat. Hist. Inhaca Isl., Moçamb., rev. ed.: 147 (1969). —Drummond in Kirkia **10**: 252 (1975). —K. Coates Palgrave, Trees Southern Africa, ed. 2, rev.: 418 (1983). —Radcliffe-Smith in F.T.E.A., Euphorb. 1: 137 (1987). —Beentje, Kenya Trees, Shrubs Lianas: 193 (1994). Type from Kenya (Coast Province).

Croton pulchellus sensu Müll. Arg. in De Candolle, Prodr. **15**, 2: 572 (1866), pro parte, quoad spec. Kirk (1859). —Schinz & Junod in Mém. Herb. Boissier, No. 10: 47 (1900). —Topham, Check List For. Trees Shrubs Nyasaland Prot.: 50 (1958), non Baill.

A much-branched shrub, or small tree up to 5 m tall, monoecious or dioecious. Bark smooth, brownish-grey, later becoming roughened. Wood hard. Young twigs densely silvery-lepidote and rusty-flecked. Stipules minute, obscured by the scales. Petioles 0.1–2(4) cm long. Leaves aromatic when crushed, whorled or subopposite; blades 1–8 × 0.5–4 cm, suborbicular-elliptic to elliptic-lanceolate, obtusely acuminate to emarginate at the apex, entire, cuneate to rounded at the base, with a pair of minute sessile discoid basal glands on the under surface often obscured by scales, firmly chartaceous, glabrous or minutely stellate-pubescent and bright or dark green on upper surface, densely silvery-lepidote and rusty-flecked beneath; midrib often impressed above, prominent beneath, lateral nerves in 4–10 pairs, not or scarcely prominent above, invisible beneath. Racemes 0.5–2 cm long, abbreviated and often apparently umbellate, terminal, androgynous, all male or all female; bracts minute, scale-like. Male flowers: pedicels 1–2 mm long in bud, 3–4 mm long in flower; sepals 5, 2.5 × 1.5 mm, triangular-ovate, densely silvery- or rusty-lepidote without, sparingly pubescent to subglabrous within; petals 5, 2 × 0.5 mm, elliptic-oblong to linear, glabrous without, pubescent within, ciliate, pale yellowish-cream or whitish; disk glands 5, truncate, thin, glabrous; stamens 16, filaments 3–4 mm long, glabrous above, pubescent below, white, anthers 1 mm long with a broad connective, yellow; receptacle pubescent. Female flowers: pedicels 2–2.5 mm long, stouter than in the male, densely lepidote; sepals 5, 3 × 2 mm, ovate, otherwise as in the male; petals as in the male; disk 5-lobed, the lobes truncate, glabrous; ovary 2.5 mm in diameter, globose, densely brown-lepidote; styles 3, 2 mm long, spreading, bicrurate (very deeply bipartite, almost bisected), the segments erect, linear-filiform, glabrous. Fruits 6–8 × 6–8 mm, shallowly rounded-trilobate, septicidal, densely silvery-green lepidote, brown-flecked. Seeds 4.5–5 × 2.5–3 × 1.5–2 mm, narrowly compressed-ellipsoid, smooth, shiny, dark brown or blackish; caruncle 1 mm wide, convex, creamy-white.

Zambia. B: Sisisi Forest, st. 23.xii.1952, *Angus* 1011 (BM; FHO; K; MO). **Zimbabwe**. N: Binga Distr., tributary of Busi R. 12 km NE of Lusulu, st. 16.xi.1982, *Craig, Mahlangu & Burrows* 16 (SRGH). W: Hwange (Wankie) Nat. Park, male fl. 16.xi.1968, *Rushworth* 1269 (K; LISC; PRE; SRGH). E: lower Save (Sabi) Valley, Dinde Hills, y. fl. 18.iii.1958, *Phelps* 228 (K; PRE; SRGH). S: Chiribira Falls, y. fl. 15.vi.1950, *Chase* 2360 (BM; K; LISC; MO; SRGH). **Malawi**. S: Nsanje Distr., Mwabvi Game Res. C., y. fl. 6.viii.1975, *Salubeni* 1972 (MO; SRGH). **Mozambique**. N: Angoche (António Enes), fr. & sd. 16.x.1965, *Mogg* 32224 (J; K; LISC; PRE; SRGH). Z: Pebane, y. fl. 4.x.1949, *Barbosa & Carvalho* in *Barbosa* 4297 (K; LMA). T: near Zimbabwe border on

Tete–Harare road, y. fl. 30.vi.1947, *R.M. Hornby* 2784 (K; SRGH). MS: lower Chiniziua R., male fl. vii.1972, *Tinley* 2660 (K; LISC; MO; PRE; SRGH). GI: Inharrime, Praia de Zavora, male fl. 22.vi.1960, *Lemos & Balsinhas* 170 (K; LISC; LMA; PRE; SRGH). M: Maputo, Costa do Sol, y. fl. 8.v.1946, *Gomes e Sousa* 3443 (COI; K; LISC; MO; PRE; SRGH).

Also in Mali, Nigeria, Somalia, Kenya, Tanzania, Angola, Namibia and South Africa (Transvaal, KwaZulu-Natal). In understorey of coastal dune forest and scrub, in *Androstachys johnsonii* thicket and dry forest, in *Baikiaea* mutemwa, mopane and *Burkea* woodlands on Kalahari Sand, in understorey of open *Brachystegia, Julbernardia, Kirkia, Lannea, Isoberlinia* woodland, usually in sandy soils, also on rocky outcrops and in riverine vegetation; sea level to 1220 m.

Although usually glabrous on the upper surface, the leaf blades can sometimes be minutely – almost microscopically – stellate-pubescent above. This has led to some confusion with the next species and also with *C. gratissimus* var. *subgratissimus*, but in these taxa the stellate hairs are coarser, the leaves are generally all alternate and the inflorescences elongate.

2. **Croton menyharthii** Pax in Bull. Herb. Boissier **6**: 733 (1898). —Hutchinson in F.T.A. **6**, 1: 753 (1912). —Prain in F.C. **5**, 2: 414 (1920). —Engler, Pflanzenw. Afrikas (Veg. Erde 9) **3**, 2: 47 (1921). —Burtt Davy, Fl. Pl. Ferns Transvaal: 301 (1932). —O.B. Miller, Check-list For. Trees Shrubs Bech. Prot.: 32 (1948). —Brenan, Check-list For. Trees Shrubs Tang. Terr.: 204 (1949). —Dale & Greenway, Kenya Trees & Shrubs: 192 (1961). —White, F.F.N.R.: 196 (1962). —P.G. Meyer in Merxmüller, Prodr. Fl. SW. Afrika, fam. 67: 10 (1967). —Drummond in Kirkia **10**: 252 (1975). —K. Coates Palgrave, Trees Southern Africa, ed. 2, rev.: 418 (1983). —Radcliffe-Smith in F.T.E.A., Euphorb. 1: 141 (1987). —Beentje, Kenya Trees, Shrubs Lianas: 192 (1994). Types: Mozambique, Maringa (Marenga), *Menyharth* 796, 797 (Z, syntypes).

Argyrodendron bicolor Klotzsch in Peters, Naturw. Reise Mossambique **6**, 1: 102 (1861), non *Croton bicolor* Roxb. (1832). Type: Mozambique, near Tete, 1842–8, *Peters* s.n. (B†, holotype).

Croton pulchellus sensu Müll. Arg. in De Candolle, Prodr. **15**, 2: 572 (1866), pro parte, quoad spec. Peters, Kirk (1860) atque syn. Klotzsch, non Baill. —sensu Topham, Check List For. Trees Shrubs Nyasaland Prot.: 50 (1958).

Croton kwebensis N.E. Br. in Bull. Misc. Inform., Kew **1909**: 140 (1909). Syntypes: Botswana, Ngamiland, Khwebe (Kwebe) Hills, *Lugard* 34 and *Mrs Lugard* 41 (K, syntypes).

A much-branched shrub or small tree up to 4 m tall, monoecious or sometimes dioecious. Bark smooth, brownish-grey. Young twigs densely stellate-lepidote. Stipules 1 mm long, subulate, lepidote, soon falling. Petioles 2–10 mm long. Leaf blades 1–9.2 × 0.5–4 cm, elliptic to ovate-lanceolate, obtuse or emarginate at the apex, entire, rounded to shallowly cordate at the base, with a pair of minute stipitate basal glands beneath, thinly chartaceous, sparingly stellate-pubescent and dark green on upper surface, silvery or yellowish-lepidote beneath, bright orange on dying; lateral nerves in 7–12 pairs, slightly prominent beneath. Racemes 1–6 cm long, terminal on short lateral shoots, androgynous or all male; bracts c. 1 mm long. Male flowers: pedicels 4–5 mm long; sepals 5, 2 × 1.5 mm, ovate, lepidote without, glabrous within, pale yellowish-green; petals 5, 2.5 × 0.75 mm, linear-oblanceolate to oblong, ciliate but otherwise glabrous, pale yellowish-cream; disk glands 5, truncate, fleshy, glabrous; stamens 15, filaments 2 mm long, glabrous except at the base, anthers 0.5 mm long with a broad connective, pale yellow; receptacle pilose. Female flowers: pedicels 2–4 mm long, stouter than in the male, densely lepidote; sepals 5, c. 2 × 1 mm, triangular-ovate, otherwise as in the male; petals absent; disk 5-lobed, the lobes truncate, glabrous; ovary 1.5 mm in diameter, subglobose, densely yellowish-lepidote; styles 3, 2–3 mm long, erect or spreading, deeply 2-partite, the segments linear-filiform, glabrous, pale green or dull purple. Fruits 6–7 × 6–8 mm, subglobose-trigonous, septicidal, densely lepidote, pale silvery-green or cream-green, sometimes black-speckled. Seeds 4–5.5 × 3 × 2.5 mm, oblong-ovoid, smooth, shiny, grey, sometimes mottled blackish; caruncle 2–2.5 mm wide, somewhat convex, pale yellow.

Caprivi Strip. Mpilila Island, fr. 13.i.1959, *Killick & Leistner* 3357 (K; PRE; SRGH). **Botswana.** N: above Boteti floodplain, fl. 6.xii.1978, *P.A. Smith* 2559 (K; MO; PRE; SRGH). SW: Quarantine Hill, Kuke, galled fr. 22.ii.1970, *Brown* 8687 (K; PRE; SRGH). SE: Orapa, fr. 10.ii.1975, *J. Allen* 274 (MO; PRE). **Zambia**. C: Otto Beit Bridge, fr. 20.iii.1952, *White* 2309 (FHO; K). S: Machili, male fl. 25.xi.1960, *Fanshawe* 5920 (FHO; K; NDO; SRGH). **Zimbabwe**. N: Rukomechi (Rekomitje) R., fr. 3.i.1961, *Goodier* 37 (K; LISC; MO; PRE; SRGH). W: Hwange (Wankie) Nat. Park, o. fr. 28.i.1969, *Rushworth* 1470A (K; LISC; PRE; SRGH). E: Hot Springs, fr. 29.xii.1948, *Chase* 1433 (BM; K; SRGH). **Malawi**. S: Lengwe Game Reserve, st. 7.iv.1970, *Hall-Martin* 295

(SRGH). **Mozambique**. T: Chicoa–Mágoè, fr. 14.ii.1970, *Torre & Correia* 17977 (LISC). MS: Maringa (Marenga), fl. 1890, *Menyharth* 796 (Z). M: Marracuene, Rikatla, fr. 1.xii.1977, *Zunguze* 22A (LMU; MO; SRGH).

Also in Ethiopia, Somalia, Kenya, Tanzania, Angola, Namibia, Swaziland and South Africa (Transvaal, KwaZulu-Natal). Locally common at low to medium altitudes in dense riverine thickets and evergreen forest, also on sandy banks of seasonal watercourses, on wooded rocky outcrops and stony hillsides and in mopane woodland and *Acacia, Combretum, Dichrostachys, Terminalia* thorn scrub on sandy soil; 350–1150 m.

3. **Croton kilwae** Radcl.-Sm. in Kew Bull. **37**: 422 (1982); **42**, 2: 395 (1987); in F.T.E.A., Euphorb. 1: 141 (1987). Type from Tanzania (Southern Province).

A shrub up to 4 m high, monoecious. Twigs greyish-brown. Young growth densely rusty-lepidote. Stipules 1–2 mm long, triangular-subulate. Petioles 3–5 mm long. Leaf blades 2–10 × 1–3.5 cm, elliptic-ovate to ovate-lanceolate, acuminate at the apex, entire, rounded to cordulate at the base, with a pair of shortly-stipitate or sessile basal glands at the petiole apex, membranaceous, evenly minutely stellate-pubescent and green on the upper surface, silvery-lepidote and faintly brown-flecked beneath; lateral nerves in 10–18 pairs, not or scarcely prominent above or beneath. Racemes 3–5 cm long, slender, terminal on short lateral shoots, mostly male with 1–2 female flowers at the base; bracts resembling the stipules. Male flowers: pedicels 7 mm long, slender; sepals 5, 2 × 1.5 mm, ovate, lepidote without, glabrous within, white; petals 5, 2 × 0.5 mm, narrowly oblanceolate, ciliate but otherwise glabrous, white; disk 5-lobed, the lobes truncate; stamens 16, filaments 1 mm long, pubescent, anthers 0.75 mm long. Female flowers: pedicels 3–8 mm long, stouter than in the male, densely lepidote; sepals 5, 2 × 1 mm, elliptic, lepidote without, minutely puberulous within near the apex; petals absent; disk crenellate, thin, flat; ovary 2 mm in diameter, globose, densely lepidote; styles 3–4, 2–2.5 mm long, suberect, deeply 4-partite with filiform segments, lepidote at the base, otherwise glabrous, purplish. Fruits 7 × 7–8 mm, very slightly trilobate or quadrilobate-subglobose, septicidal, evenly lepidote. Seeds 6 × 4 × 3 mm, somewhat compressed-ellipsoid, smooth, slightly shiny, pale greyish-brown; caruncle c. 1.5 mm wide, shallowly convex.

Mozambique. N: Eráti, 12 km Namapa–Alua, Mt. Geovi, fr. 8.i.1964, *Torre & Paiva* 9879 (LISC); Cabo Delgado, 15 km Pemba–Montepuez, fr. 27.i.1984, *Groenendijk, Maite, de Koning & Dungo* 826 (K; LMA; LMU).

Also in southern Tanzania. In grassland with scattered trees and shrubs, and on granite rock outcrops; 400 m.

Since only fruiting material has been seen from Mozambique, the descriptions of the flowers are based on Tanzanian material.

C. kilwae is very close to *C. menyharthii*, but the 4-partite styles are diagnostic.

4. **Croton gratissimus** Burch., Trav. S. Afr. **2**: 263 (1824). —Müller Argoviensis in De Candolle, Prodr. **15**, 2: 516 (1866). —Hutchinson in F.T.A. **6**, 1: 1051 (1913). —Eyles in Trans. Roy. Soc. South Africa **5**: 395 (1916). —Prain in F.C. **5**, 2: 416 (1920). —Engler, Pflanzenw. Afrikas (Veg. Erde 9) **3**, 2: 47 (1921). —Burtt Davy, Fl. Pl. Ferns Transvaal: 301 (1932). —Bremekamp & Obermeyer in Ann. Transvaal Mus. **16**, 3: 421 (1935). —Hutchinson, Botanist in Southern Africa: 667 (1946). —O.B. Miller, Check-list For. Trees Shrubs Bech. Prot.: 32 (1948). —Brenan, Check-list For. Trees Shrubs Tang. Terr.: 206, in adnot. (1949). —White, F.F.N.R.: 196 (1962). —P.G. Meyer in Merxmüller, Prodr. Fl. SW. Afrika, fam. 67: 10 (1967). —Drummond in Kirkia **10**: 251 (1975). —K. Coates Palgrave, Trees Southern Africa, ed. 2, rev.: 416 (1983). —Radcliffe-Smith in F.T.E.A., Euphorb. 1: 139, in adnot. (1987). Type from South Africa (Cape Province).

A shrub or tree up to 12 m tall, monoecious or sometimes dioecious; trunk usually Y-forked with drooping branches; crown sparse, open, rounded or ± pyramidal. Bole up to 40 cm d.b.h.; bark rough and longitudinally or rectangularly fissured on lower trunk, smooth on upper trunk, dark grey to pale brownish-grey, patterned. Branches often in whorls of 3. Young twigs rusty-lepidote to fulvous stellate-lepidote. Stipules 1.5–7 mm long, subulate-filiform. Petioles 0.5–7 cm long. Leaves aromatic when crushed; blades 1.5–18 × 0.5–6 cm, elliptic, elliptic-lanceolate or narrowly oblong-lanceolate, acuminate, obtuse or emarginate at the apex, entire, rounded to shallowly cordulate or cordate at the base, with a pair of sessile to long-stalked basal glands on the lower surface, or less often on the sides or upper surface of the petiole apex, chartaceous to thinly coriaceous, glabrous and deep green to midgreen or stellate-pubescent and olive-green on the upper surface, densely silvery-lepidote and rusty-flecked or fulvous stellate-lepidote beneath; midrib usually impressed above, prominent beneath; lateral nerves in 8–17 pairs, not prominent above, scarcely so to often invisible beneath. Racemes 1–15 cm long, erect at first, later drooping, terminal, androgynous or all male; axis angular; bracts 2–3(5) mm long, subulate, soon falling. Male flowers fragrant; pedicels 1–3(5) mm long; buds 1.5–3.5 mm in diameter, subglobose, yellowish-lepidote; sepals 5, 4 × 2.5 mm, ovate, densely tawny-lepidote or rusty-lepidote without, stellate-pubescent within, greenish-yellow; petals 5, 4 × 1.5 mm, elliptic-oblong, sparingly rusty-lepidote without, puberulous within, ciliate, pale creamy-yellow to whitish; disk glands 5, truncate, glabrous; stamens 17–22, filaments 3–4 mm long, sparingly gland-dotted, otherwise glabrous, pale yellow, anthers 1.5 mm long, with a broad glandular connective, yellow; receptacle pubescent. Female flowers: pedicels 2–3 mm long, extending to 5 mm in fruit, stouter than the male, densely lepidote; sepals and petals ± as in the male; disk scarcely distinguishable; staminodes 4–5, 1.5 mm long, subulate-filiform; ovary 2.5 mm in diameter, globose, densely brown-lepidote; styles 3, 1 mm long, spreading, 2-partite with irregularly multifid segments, abaxially lepidote, adaxially glabrous, dull dark purple. Fruits 8–10 × 8–11 mm, trilobate-subglobose, septicidal, densely greenish- or yellowish- to silvery-lepidote, brown-flecked. Seeds 7 × 5.5 × 3 mm, compressed-ellipsoid, smooth, slightly shiny, light brown; caruncle 1.5 mm wide, pentagonal, flattened.

Leaves glabrous and deep green to mid-green on upper surface; inflorescences 4–15 cm long . . . . var. *gratissimus*
Leaves stellate-pubescent and olive-green on upper surface; inflorescences 1–6(8) cm long . . . . var. *subgratissimus*

Var. **gratissimus**

Croton zambesicus Müll. Arg. in Flora **47**: 483 (1864); in De Candolle, Prodr. **15**, 2: 515 (1866). —Engler, Pflanzenw. Ost-Afrikas **C**: 237 (1895). —Hutchinson in F.T.A. **6**, 1: 758 (1912), excl. spec. *Gillet* 320 & *Seret* 594. —R.E. Fries, Wiss. Ergebn. Schwed. Rhod.-Kongo-Exped. **1**, 1: 122 (1914). —Eyles in Trans. Roy. Soc. South Africa **5**: 395 (1916). —Prain in F.C. **5**, 2: 416 (1920). —Engler, Pflanzenw. Afrikas (Veg. Erde 9) **3**, 2: 47 (1921). —Burtt Davy, Fl. Pl. Ferns Transvaal: 302 (1932). —Bremekamp & Obermeyer in Ann. Transvaal Mus. **16**, 3: 421 (1935). —O.B. Miller, Check-list For. Trees Shrubs Bech. Prot.: 32 (1948). —F.W. Andrews, Fl. Pl. Anglo-Egypt. Sudan **2**: 61 (1952). —Keay in F.W.T.A., ed. 2, **1**, 2: 394 (1958). —Drummond in Kirkia **10**: 251 (1975). —Radcliffe-Smith in F.T.E.A., Euphorb. 1: 138 (1987). —Beentje, Kenya Trees, Shrubs Lianas: 194 (1994). Type: Mozambique, Manica e Sofala, Sena Hill, i.1859, *Kirk* Croton (2) (K, holotype).

Croton amabilis Müll. Arg. in Flora **47**: 537 (1864); in De Candolle, Prodr. **15**, 2: 516 (1866). —N.E. Br. in Bull. Misc. Inform., Kew **1909**: 140 (1909). —Hutchinson in F.T.A. **6**, 1: 757 (1912). —O.B. Miller, Check-list For. Trees Shrubs Bech. Prot.: 32 (1948). Syntypes from Nigeria.

Croton welwitschianus Müll. Arg. in J. Bot. **2**: 338 (1864); in De Candolle, Prodr. **15**, 2: 515 (1866). Type from Angola (Huíla).

Croton microbotryus Pax in Bot. Jahrb. Syst. **10**: 35 (1888). Type from South Africa (Cape Province).

Croton antunesii Pax in Bot. Jahrb. Syst. **23**: 523 (1897). Types from Angola (Huíla).

Leaves glabrous and deep green to mid-green on upper surface; inflorescences 4–15 cm long.

Caprivi Strip. 72 km Katima Mulilo–Singalamwe on Finaughty's Road, y. fl. 30.xii.1958, *Killick & Leistner* 3194 (K; PRE; SRGH). **Botswana**. N: Ngamiland Distr., Tsodilo Hills, y. fl. 2.v.1975, *Müller & Biegel* 2304 (FHO; K; MO; PRE; SRGH). SW: Central Kalahari Game Res., Deception Pan, y. fl. iv.1975, *Owens* 16 (K; PRE; SRGH). SE: 10 km north of Lobatse, fl. 2.x.1977, *O.J. Hansen* 3209 (C; GAB; K; PRE; SRGH). **Zambia**. B: Nangweshi, st. 22.vii.1952, *Codd* 7147 (BM; K; PRE; SRGH). N: L. Kashiba, fr. 22.x.1957, *Fanshawe* 3796 (K; NDO). W: Misaka For. Res., y. fl. 15.viii.1952, *Holmes* 939 (FHO; K). C: Katondwe, fl. 13.xi.1966, *Fanshawe* 9838 (K; NDO). E: Chipata (Ft. Jameson)–Nsefu Road, fl. 26.v.1961, *Angus* 2892 (FHO; K). S: Livingstone, y. fl. 7.i.1953, *Angus* 1116 (BM; FHO; K; MO). **Zimbabwe**. N: Chinhoyi (Sinoia) Cave, y. fl. 21.iv.1948, *Rodin* 4375 (K; MO; PRE; SRGH; UC). W: Hwange (Wankie) Nat. Park, y. fl. 26.ii.1967, *Rushworth* 246 (K; LISC; PRE; SRGH). C: Prince Edward Dam, Harare, y. fl.

31.vii.1934, *Gilliland* Q 600 (BM; K; PRE). E: Mutare Distr., Wengesi R., y. fl. 18.xii.1954, *Chase* 5370 (BM; PRE; SRGH). S: Mnene, y. fl. 26.ii.1931, *Norlindh & Weimarck* 5158 (K; PRE; SRGH; UPS). **Malawi**. N: Rumphi Distr., Livingstonia Road, Rukuru R. Bridge, y. fl. 16.viii.1980, *Patel* 556 (PRE; SRGH). C: 8 km north of Kasungu, y. fl. 7.v.1970, *Brummitt* 10436 (K; PRE; SRGH). S: Mpatamanga Gorge, y. fl. 13.v.1961, *Leach & Rutherford-Smith* 10829 (K; LISC; SRGH). **Mozambique**. T: do Songo para a barragem, y. fl. 5.ii.1972, *Macêdo* 4795 (K; LISC; LMA; SRGH). MS: 29 km Tambara–Serra Lupata, y. fl. 14.v.1971, *Torre & Correia* 18396 (K; LISC; PRE). GI: Mavume, y. fl. vii.1938, *Gomes e Sousa* 2156 (COI; K; LISC). M: Porto Henrique, y. fl. 4.iii.1948, *Gomes e Sousa* 3690 (COI; K; MO; PRE; SRGH).

From the Gambia eastwards to Nigeria; in southern Sudan and Ethiopia and northern Uganda and Kenya; and from Angola and Namibia across to central Mozambique and South Africa (KwaZulu-Natal): a trimodal disjunct distribution. Locally common at medium to low altitudes, in coastal dune forest and riverine fringe vegetation, in mopane woodland on floodplain alluvium, in *Androstachys johnsonii* woodland and thicket, on rocky outcrops and escarpment miombo, in Kalahari Sand woodland and *Baikiaea* mutemwa, and in mixed deciduous woodland and pemba thicket, also on termitaria; 90–1525 m.

Vernacular names as recorded in specimen data include: "mologa" (setstwana); "cassaca" (Tete area); "kanunkila" (Tok.).

The flowers remain in bud all through the dry season and open with the first rains (Mrs Lugard).

The dried flower buds when crushed smell strongly of certain *Labiatae* (e.g. *Thymus, Salvia* spp.).

Var. **subgratissimus** (Prain) Burtt Davy, Fl. Pl. Ferns Transvaal: 301 (1932). Type from South Africa (Transvaal).

Croton subgratissimus Prain in Bull. Misc. Inform., Kew **1913**: 79 (1913); in F.T.A. **6**, 1: 1050 (1913); in F.C. **5**, 2: 415 (1920). —Engler, Pflanzenw. Afrikas (Veg. Erde 9) **3**, 2: 47 (1921). —O.B. Miller, Check-list For. Trees Shrubs Bech. Prot.: 32 (1948). —Brenan, Check-list For. Trees Shrubs Tang. Terr.: 206, in adnot. (1949). —P.G. Meyer in Merxmüller, Prodr. Fl. SW. Afrika, fam. 67: 11 (1967). —Drummond in Kirkia **10**: 251 (1975).

Croton gratissimus sensu Pax in Bot. Jahrb. Syst. **10**: 35 (1888), non Burch.

Leaves stellate-pubescent and olive-green on upper surface; inflorescences 1–6(8) cm long.

Caprivi Strip. 53 km west of Katima Mulilo, y. fl. 17.ii.1969, *de Winter* 9207 (K; PRE). **Botswana**. SW: Olifants Kloof, y. fl. 1889, *Fleck* 453a (K). SE: Dikgomodikae (Dikomu Di Kai), y. fl. 26.ii.1960, *Wild* 5186 (K; MO; PRE; SRGH). **Zimbabwe**: C: Gweru Distr., 4 km north of Lalapansi, subst. 29.i.1973, *Biegel* 4191 (K; SRGH). S: Beitbridge Distr., 6.5 km NE of Pazhi–Limpopo confluence, y. fl. 24.iii.1959, *Drummond* 5975 (K; LISC; PRE; SRGH).

Also in Angola, Namibia and South Africa (Transvaal). Medium to low altitudes, dry sandstone hills, rocky hillsides and serpentine outcrops, also in sandy soil in *Grewia, Commiphora, Kirkia* associations and in wooded grassland; 200–1620 m.

5. **Croton steenkampianus** Gerstner in J. S. African Bot. **12**: 38 (1946). —K. Coates Palgrave, Trees Southern Africa, ed. 2, rev.: 419 (1983). —Radcliffe-Smith in F.T.E.A., Euphorb. 1: 143 (1987). Types from South Africa (KwaZulu-Natal).

A shrub c. 2 m high, or tree to 7 m, dioecious or monoecious. Twigs greyish-brown, lenticellate. Young growth densely fulvous-lepidote. Stipules 3–10 mm long, subulate. Petioles 0.5–6 cm long. Leaf blades 2–15 × 2–10 cm, ovate, acutely shortly acuminate at the apex, sometimes obtuse or emarginate through deformity, entire or subentire on the margin, rounded to shallowly cordate at the base with a pair of sessile discoid basal glands beneath, chartaceous, sparingly scurfily stellate-pubescent and green on the upper surface, silvery-lepidote beneath; 5–9-nerved from the base, lateral nerves in 5–8 pairs, ± impressed above, prominent beneath. Racemes 2–10 cm long, terminal, mostly female with a short apical male portion, or all female; bracts minute. Male flowers: pedicels 3 mm long; sepals 5, 2 × 1 mm, triangular-ovate, yellowish-lepidote without, glabrous within; petals 5, 2 × 0.5 mm, oblong-oblanceolate, pubescent, ciliate, pale yellowish-cream in colour; disk glands 5, rounded or truncate; stamens 17–20, filaments 2–3 mm long, sparingly pubescent, anthers 1 mm long; receptacle pilose. Female flowers: pedicels 3 mm long, stouter than in the male, densely lepidote; sepals 5, 3–4 × 1.5 mm, triangular-lanceolate, subacute, lepidote without, pubescent within, greenish; petals absent; disk shallowly 5-lobed, the lobes rounded; ovary 2 mm in diameter, subglobose, densely fulvous stellate-lepidote; styles 3, 2–3 mm long, erect or spreading, twice bipartite with linear-filiform segments, sparingly stellate-pubescent abaxially, purplish. Fruits 5 × 6–7 mm, trilobate-subglobose, septicidal, evenly stellate-lepidote. Seeds 3.5–4 × 2.5–3.5 × 2–2.5 mm, ovoid, smooth, somewhat shiny, dark brown; caruncle c. 1 mm wide, ± flattened.

Mozambique. GI: Massingir, fl. 23.vii.1982, *Matos* 5069 (LISC). M: Goba, fr. 3.xi.1960, *Balsinhas* 184 (K; LISC; LMA; PRE; SRGH).

Also in eastern Tanzania and South Africa (Transvaal, KwaZulu-Natal). In coastal semi-evergreen woodland, *Androstachys johnsonii* woodland, and wooded grassland with *Entandrophragma, Combretum, Pteleopsis, Vitex* and *Croton gratissimus*; 300–350 m.

6. **Croton megalocarpus** Hutch. in F.T.A. **6**, 1: 760 (1912). —Engler, Pflanzenw. Afrikas (Veg. Erde 9) **3**, 2: 47 (1921). —Robyns & Tournay, Fl. Sperm. Parc Nat. Alb. **1**: 449 (1948). —Brenan, Check-list For. Trees Shrubs Tang. Terr.: 204 (1949). —Dale & Greenway, Kenya Trees & Shrubs: 191 (1961). —J. Léonard in F.C.B. **8**, 1: 54 (1962). —White, F.F.N.R.: 196 (1962). —Biegel, Check List Ornam. Pl. Rhod. Parks & Gard.: 43 (1977). —Troupin, Fl. Pl. Lign. Rwanda: 257, fig. 88/1 (1982); Fl. Rwanda **2**: 217, fig. 65/1 (1983). —Radcliffe-Smith in F.T.E.A., Euphorb. 1: 144 (1987). —Beentje, Kenya Trees, Shrubs Lianas: 192 (1994). Type from Kenya (Nairobi).

A large tree up to 30 m tall with a spreading crown, monoecious or dioecious. Bark and twigs pale grey or brown. Young growth silvery-lepidote or rusty stellate-pubescent. Stipules 7–10 mm long, filiform. Petioles 1.5–11 cm long. Leaf blades 4–16 × 1.5–8 cm, elliptic-ovate to ovate-lanceolate, shortly acuminate, entire or almost so at the apex, cordulate at the base usually with 2(4) sessile or shortly stipitate basal glands on the under surface or near the petiole apex, chartaceous, evenly minutely stellate-pubescent and bright green on the upper surface, silvery-lepidote and brown-flecked beneath; lateral nerves in 10–21 pairs, not or slightly impressed above, prominent beneath. Racemes 2.5–16(30) cm long, terminal, all male or with a few female flowers at the base; axis angular, rusty-lepidote; bracts 1–2 mm long, subulate, soon falling. Male flowers: pedicels 3–10 mm long; buds 4–5 mm in diameter, globose, rusty-lepidote; sepals 5, c. 4 × 2.5 mm, ovate, rusty-lepidote without, sericeous-pubescent within, greenish-yellow; petals 5, 4.5 × 3 mm, obovate, lepidote without, pubescent within, ciliate, pale yellow; disk glands 5(6), free, truncate; stamens c. 30, filaments 5–7 mm long, glabrous above, pilose below, anthers c. 1 mm long; receptacle pubescent. Female flowers: pedicels 4–10 mm long, extending to 2 cm in fruit, stouter than in the male; sepals somewhat smaller than in the male flowers, but otherwise similar; petals 5, 5 × 0.5 mm, linear, lepidote without at the base, pubescent within, ciliate; disk shallowly 5-lobed; ovary 3 mm in diameter, subtrilobate-subglobose, densely lepidote; styles 3, 1.5 mm long, spreading, 2-partite, the segments 2–4-lobulate, lepidote beneath, otherwise subglabrous. Fruits (1.5)3–4.5 × (1.5)2.5–3(4) cm, ellipsoid-ovoid to subglobose, loculicidal, creamy-lepidote; endocarp woody, pitted. Seeds 2 × 1 × 0.7 cm, ellipsoid, shallowly rugulose, slightly shiny, yellowish-grey; caruncle minute.

Zambia. N: Chisau R. Gorge, Saisi Valley, st. 18.xi.1959, *Richards* 11793 (K). C: Lusaka Forest Nursery, y. male fl. 8.viii.1952, *White* 3037 (FHO; K). **Malawi**. N: Musisi (Mussissi) Forest, st. 4.vi.1983, *Dowsett-Lemaire* 766 (FHO; K). C: Dedza, y. male fl. 9.viii.1961, *Chapman* 1436 (FHO; K; SRGH). **Mozambique**. N: Namapa, Eráti, y. male fl. & fr. 17.viii.1948, *Barbosa* 1793 (LISC).

Also in Zaire (Orientale, Kivu and Shaba Provinces), Rwanda, Burundi, Uganda, Kenya and Tanzania. Canopy tree of evergreen rain forest, riverine gully forest and in high rainfall *Brachystegia* woodland; (350)1500–1850 m.

7. **Croton megalocarpoides** Friis & M.G. Gilbert in Nord. J. Bot. **4**: 328, fig. 1 (1984). —Radcliffe-Smith in F.T.E.A., Euphorb. 1: 145 (1987). —Beentje, Kenya Trees, Shrubs Lianas: 192 (1994). Type from Somalia.

Very like *C. steenkampianus* in vegetative features, but with the fruits up to 1.7–2 × 1.7–2 cm, and subindehiscent to tardily loculicidally dehiscent.

Mozambique. N: Eráti, 12 km Namapa–Alua, slopes of Mt. Geovi, fr. 8.i.1964, *Torre & Paiva* 9871 (LISC).

Also in Somalia and Kenya. On granite rock outcrops; 320 m.

8. **Croton longipedicellatus** J. Léonard in Bull. Jard. Bot. État **26**: 387 (1956); in F.C.B. **8**, 1: 66 (1962). —White, F.F.N.R.: 197 (1962). —Radcliffe-Smith in F.T.E.A., Euphorb. 1: 146 (1987). Type from Zaire (Shaba).

A weak, lax, spreading or scrambling shrub or small tree up to 4.5 m tall, monoecious or sometimes dioecious. Bark light grey-brown. Twigs dark grey. Young growth sparingly to evenly stellate-pubescent, scurfily so or not, patent or appressed, or fimbriate-lepidote. Stipules 0.3–1.3 cm long, filiform, soon falling. Petioles 1.5–5.5 cm long, basally geniculate. Leaf blades 2–12.5 × 1–7 cm, ovate to elliptic-ovate, acuminate at the apex, shallowly serrulate to subentire on the margin, usually rounded or cuneate, rarely truncate or shallowly cordate at the base, with a pair of usually long-stipitate discoid glands at or near the base on lower surface, thinly chartaceous, sparingly appressed scurfily stellate-pubescent, or hairs with a patent central ray, to fimbriate-lepidote or subglabrous on upper surface, and sparingly to evenly so beneath; 3–5-nerved from the base, lateral nerves in 3–5 pairs, scarcely to slightly prominent beneath. Racemes 3–17 cm long, terminal or subterminal, androgynous, or male or female on the same or different branches or plants; bracts 1–2 mm long, linear-spathulate to filiform. Male flowers: pedicels 5–8 mm long; sepals 5, 2 × 1.25 mm, elliptic-ovate, sparingly appressed stellate-pubescent without, glabrous within, ciliate, yellowish- or greenish-cream; petals 5, 2 × 1 mm, oblanceolate-oblong, ciliate but otherwise glabrous, yellow; disk glands 5, rounded-subemarginate, fleshy, glabrous; stamens 14–17, filaments 2 mm long, glabrous, anthers c. 1 mm long, yellow; receptacle densely pubescent. Female flowers: pedicels 2–10 mm long, extending to up to 2.5 cm in fruit or not extending; sepals 5, 2.5–3 × 1 mm, accrescent to 7 × 2 mm in fruit, oblong-lanceolate, obtuse, becoming oblong-subspathulate and somewhat foliaceous, sparingly appressed stellate-pubescent without, glabrous within, pale green; petals absent; disk shallowly 5-lobed, the lobes opposite the sepals, rounded, glabrous; ovary 1–1.5 × 1–1.5 mm, subglobose, densely silvery fimbriate-lepidote; styles 3, 2 mm long, connate at the base, divaricate-ascending, deeply 2-partite, the segments filiform, sparingly appressed-stellate-pubescent basally, otherwise glabrous. Fruits 5–7 × 4–7 mm, ovoid-subglobose to trilobate-subglobose, septicidal, evenly to sparingly fimbriate-lepidote, greenish becoming blackish. Seeds 4–4.5 × 2.5–3 × 2 mm, ellipsoid, verruculose, shiny, light greyish-brown with darker warts; caruncle 2 mm wide, 2-partite, yellowish.

Leaves elliptic-ovate, cuneate to rounded at the base; young growth usually appressed-stellate-pubescent to fimbriate-lepidote; fruiting pedicels 1–2.5 cm long · · · · var. *longipedicellatus*
Leaves ovate, truncate to shallowly cordate at the base; young growth patent-stellate-pubescent; fruiting pedicels 2–3 mm long · var. *brevipedicellatus*

Var. **longipedicellatus**

Leaves elliptic-ovate, cuneate to rounded at the base; young growth appressed-stellate-pubescent to fimbriate-lepidote; fruiting pedicels 1–2.5 cm long.

Caprivi Strip. c. 10 km Katima Mulilo–Ngoma, fr. 5.i.1959, *Killick & Leistner* 3305 (K; PRE; SRGH). **Zambia**. B: Masese Forest Station, o. fr. 2.ii.1975, *Brummitt, Chisumpa & Polhill* 14243 (K; MO; SRGH). N: Sumbu, Lake Tanganyika, fl. 16.xi.1957, *Savory* 240 (K; LISC; SRGH). W: Chingola, y. male fl. 2.xi.1957, *Fanshawe* 3826 (K; NDO). S: Machili, fr. 31.xii.1960, *Fanshawe* 6058 (K; NDO). **Zimbabwe**. N: Hurungwe Distr., Zambezi Valley, o. fr. 4.i.1961, *J.H. Goodier* 48 (K; LISC; MO; PRE; SRGH). W: Hwange, y. fl. 24.xi.1974, *Raymond* 293 (K; MO; PRE; SRGH). S: Chiredzi Distr., Pombadzi R., fr. & sd. 5.ii.1971, *Sherry* 128/71 (SRGH). **Mozambique**. T: Songo, fl. & y. fr. 12.xii.1973, *Macêdo* 5432 (LISC).

Also in Zaire (Shaba) and Tanzania. Usually a constituent of dry thicket vegetation, *Baikiaea* mutemwa (understorey thicket of dry deciduous forest on Kalahari Sand), mateshi thicket (dry evergreen thicket in north western Zambia), jesse bush thicket (Zambezi Valley), and in itigi forest/thicket (in northern Zambia), also on granite outcrops and in gully forest; 400–1220 m.

Var. **brevipedicellatus** Radcl.-Sm. in Kew Bull. **45**: 557 (1990). Type: Zambia, Lake Mweru, fr. 13.xi.1957, *Fanshawe* 3967 (K, holotype; NDO).

Leaves ovate, truncate to shallowly cordate at the base; young growth patent-stellate-pubescent; fruiting pedicels 2–3 mm long.

Zambia. N: L. Mweru, fl. 13.xi.1957, *Fanshawe* 3962 (K; NDO). W: R. Lunga, east of Mwinilunga, fl. 27.xi.1937, *Milne-Redhead* 3425 (K; LISC; SRGH).

Also in Angola (Lunda). In evergreen riverine vegetation and lakeshore mushitu (swamp forest).

9. **Croton scheffleri** Pax in Bot. Jahrb. Syst. **43**: 78 (1909). —Hutchinson in F.T.A. **6**, 1: 733 (1912). —Engler, Pflanzenw. Afrikas (Veg. Erde 9) **3**, 2: 48 (1921). —Brenan, Check-list For. Trees Shrubs Tang. Terr.: 206 (1949). —Dale & Greenway, Kenya Trees & Shrubs: 192 (1961). —White, F.F.N.R.: 197 (1962). —Drummond in Kirkia **10**: 252 (1975). —K. Coates Palgrave, Trees Southern Africa, ed. 2, rev.: 419 (1983). —Radcliffe-Smith in F.T.E.A., Euphorb. 1: 147 (1987). —Beentje, Kenya Trees, Shrubs Lianas: 193 (1994). Type from Kenya (Central Province).

Croton madandensis sensu Drummond in Kirkia **10**: 252 (1975), non S. Moore.

A shrub or small tree to 6 m tall, monoecious or sometimes dioecious. Twigs dark purplish-grey. Young growth stellate-hirsute. Stipules up to 7 mm long, filiform, soon falling. Petioles 1–4 cm long. Leaf blades 4–12 × 2–9 cm, broadly ovate to ovate-lanceolate, acuminate at the apex, somewhat irregularly serrulate on the margin, wide-cuneate to rounded or truncate at the base, with a pair of shortly-stipitate or subsessile discoid basal glands on the lower surface, chartaceous, sparingly stellate-pubescent on upper surface, evenly to densely so beneath; 5–7-nerved from the base, lateral nerves in 3–4 pairs, slightly prominent beneath. Racemes 4–7 cm long, terminal, few-flowered, usually androgynous, sometimes male; bracts resembling the stipules. Male flowers: pedicels 5–6 mm long; sepals 5, 2.5 × 2 mm, ovate, stellate-pubescent without, glabrous within, greenish; petals 5, 2.5 × 1.5 mm, obovate-spathulate, ciliate but otherwise glabrous, cream-coloured; disk glands 5, rounded-subtruncate, fleshy, pubescent; stamens 15, filaments 2 mm long, glabrous, anthers 1.25 mm long, orange-yellow; receptacle villous. Female flowers: pedicels 2–3 mm long, extending to 1.5–2 cm in fruit; sepals 5, 3 × 2 mm, accrescent to 10 × 5 mm in fruit, ovate-lanceolate to elliptic-oblong, subacute, becoming foliaceous, with a midrib and 2–3 pairs of lateral nerves, sparingly stellate-pubescent without, subglabrous within, green; petals absent; disk shallowly 5-lobed, the lobes alternating with the sepals, rounded, puberulous; ovary 2 mm in diameter, trilobate-subglobose, densely fulvous stellate-tomentose; styles 3, 3–4 mm long, divaricate, deeply 2-partite, the segments filiform, sparingly stellate-pubescent basally, apically subglabrous. Fruits 5 × 7 mm, trilobate-subglobose, septicidal, evenly scurfily and sparingly patent stellate-pubescent. Seeds 4.5 × 3 mm, ovoid-ellipsoid, smooth, shiny, dark brown; caruncle 1 mm wide.

Zambia. N: Mbala Distr., Vomo Gap, fl. 27.xii.1967, *Richards* 22828 (K). **Malawi**. S: Mangochi Distr., Namaso Bay, fr. 17.iii.1979, *Blackmore & Hargreaves* 659 (K; MAL).

Also in Kenya and Tanzania. On steep rocky slopes in dry deciduous woodland with *Sterculia, Adansonia, Commifera* and *Combretum*, also in riverine thicket; 550–1500 m.

Balaka, Seyani & *Kaunda* 820 (K; MAL), from Malawi, Mulanje Distr., Machemba Hill, 17.xi.1984, differs from typical *C. scheffleri* in having chestnut-brown peltate scales admixed with the whitish stellate hairs on the midribs and main nerves on lower surface of some leaves, and furthermore the female calyx lobes do not appear to be accrescent. The status of this entity remains uncertain at present.

10. **Croton polytrichus** Pax in Bot. Jahrb. Syst. **15**: 533 (1893). —Hutchinson in F.T.A. **6**, 1: 752 (1912). —Engler, Pflanzenw. Afrikas (Veg. Erde 9) **3**, 2: 48 (1921). —Brenan, Check-list For. Trees Shrubs Tang. Terr.: 206 (1949). —F.W. Andrews, Fl. Pl. Anglo-Egypt. Sudan **2**: 60 (1952). —Radcliffe-Smith in F.T.E.A., Euphorb. 1: 148 (1987). —Beentje, Kenya Trees, Shrubs Lianas: 192 (1994). Type from the Sudan (Bahr-el-Ghazal Province).

Subsp. **brachystachyus** Radcl.-Sm. in Kew Bull. **45**: 560 (1990). Type: Zambia, Barotseland, Kataba, fr. 12.xii.1960, *Fanshawe* 5973 (K, holotype; FHO; NDO; SRGH).

Croton sp. cf. *polytrichus* Pax; White, F.F.N.R.: 197 (1962).

A weak, lax shrub or small tree up to 4.5 m tall, but commonly much less, with a rounded, open crown. Twigs dull greyish-purple. Young growth densely stellate-pubescent. Axillary buds becoming pronounced in the late season, obovoid, 4 mm long, scurfily stellate-tomentose. First flush leaves sessile or subsessile, narrowly oblanceolate. Stipules minute and soon falling, or absent. Petioles of mature leaves up to 5.5 cm long. Leaf blades 5–12 × 2–8 cm, ovate to elliptic-obovate, acuminate at the apex, sharply serrulate on the margins, rounded to shallowly cordate at the base, with 2–4 stipitate basal discoid glands on lower surface, membranous to thinly chartaceous, sparingly stellate-pubescent on upper surface, evenly to densely so beneath; 5–7-nerved from the base, lateral nerves in 5–8 pairs, fairly prominent beneath. Racemes 1–5 cm long, terminal, usually androgynous, sometimes male; bracts minute. Male flowers: pedicels 5–7 mm long; sepals 5, c. 2 × 1 mm, ovate, stellate-pubescent without, ± glabrous within, pale yellowish-green; petals 5, 2 × 0.75 mm, oblanceolate-subspathulate, ciliate but otherwise glabrous, creamy-yellow; disk glands 5, rounded-subtruncate, fleshy; stamens 15, filaments 2–2.25 mm long, glabrous, anthers 0.75 mm long; receptacle pilose. Female flowers: pedicels 2–4 mm long, stouter and more densely stellate-pubescent than the male; sepals 5, 2–4 × 0.6–1.3 mm, slightly accrescent in fruit, ovate-lanceolate to linear-lanceolate, sparingly stellate-pubescent without, glabrous within, greenish; petals minute, squamiform; disk rounded-pentagonal, flat, glabrous; ovary 2 mm in diameter, subglobose, densely yellow-brown stellate-pubescent; styles 3, 2.5–3 mm long, spreading, deeply 2-partite with linear segments, sparingly pubescent abaxially, reddish-brown, persistent. Fruits 5 × 7 mm, trilobate-subglobose, septicidal, evenly scurfily and patent stellate-pubescent, green. Seeds 4 × 3 × 2.5 mm, compressed-ovoid, smooth, shiny, chestnut-brown; caruncle 1.5 mm wide, flattened.

Zambia. B: Kataba, fl. 12.xii.1960, *Fanshawe* 5969 (FHO; K; NDO). N: Samfya, fl. 17.xi.1964, *Mutimushi* 1136 (FHO; K; NDO). W: sine loc., male fl. 1931, *Martin* 63/31 (FHO; K). S: Mazabuka Distr., 16 km Choma–Pemba, fr. 30.i.1960, *White* 6634 (FHO; K; SRGH).

Not known elsewhere. Usually a constituent of dry, thicket vegetation; *Baikiaea* mutemwa (understorey thicket of dry deciduous forest on Kalahari Sand), pemba thicket (southern Zambia) itigi forest/thicket (northern Zambia), and ridge thicket under *Pteleopsis anisoptera, Brachystegia spiciformis, Ostryoderris stuhlmannii, Amblygonocarpus androgynus,* also in riparian thicket and lakeshore sand dune scrub; 1000–1280 m.

The typical subspecies ranges from southwest Sudan south through Kenya to southern Tanzania; it differs from the subsp. *brachystachyus* in having longer axillary buds, larger more persistent stipules, leaves of the first flush petiolate and obovate, leaves at maturity subentire or minutely glandular-toothed, basal glands sessile, racemes up to 12 cm long, and shorter male and female pedicels.

Vernacular names as recorded in specimen data include: "muhundaluhunda" (?); "mutunduti" (Tok.).

11. **Croton madandensis** S. Moore in J. Linn. Soc., Bot. **40**: 195 (1911). —Hutchinson in F.T.A. **6**, 1: 756 (1912). —Engler, Pflanzenw. Afrikas (Veg. Erde 9) **3**, 2: 49 (1921). Syntypes: Mozambique, Madanda Forest, fl. 5.xii.1906, *Swynnerton* 1408 (BM; K); fl. 20.xii.1906, *Swynnerton* 1779 (BM).
Croton sp. no. 1. —Drummond in Kirkia **10**: 252 (1975).

A shrub or small tree up to 5 m tall. Twigs greyish-brown, sparingly white-lenticellate. Young growth scurfily patent-stellate-pubescent. Stipules 3–4 mm long, filiform-subulate, soon falling. Petioles 0.5–3.5 cm long. Leaf blades 3.5–8 × 2–4 cm, ovate to elliptic-oblanceolate, shortly acuminate at the apex, entire or subentire, rarely irregularly crenate-serrate on the margin, shallowly cordate at the base, with a pair of long-stipitate or occasionally subsessile discoid glands at the base on the lower surface, membranous, evenly stellate-pubescent on both surfaces when juvenile, later only sparingly so; 3–5-nerved from the base, lateral nerves in 4–5 pairs, scarcely prominent beneath, not so above. Racemes 1–6.5 cm long, subterminal, usually androgynous; bracts minute. Male flowers: pedicels 2 mm long; sepals 5, 1.5 × 1 mm, ovate-oblong, evenly scurfily stellate-pubescent without, glabrous within, yellowish-green; petals 5, 2 × 0.75 mm, oblanceolate-spathulate, ciliate but otherwise ± glabrous, yellow; disk glands minute; stamens 15–16, filaments c. 1 mm long, ± glabrous, anthers 0.5 mm long; receptacle pilose. Female flowers: pedicels 2.5 mm long; sepals 5, 4 × 2 mm, accrescent to 7 × 3 mm, or not accrescent, oblong-lanceolate, evenly scurfily stellate-pubescent without, minutely puberulous within; petals 5, 1 mm long, subulate; disk rounded-pentagonal, flat, glabrous; ovary 2 mm in diameter, subglobose, densely fulvous stellate-tomentose; styles 3, 1.5 mm long, spreading, twice 2-partite or ± 4-partite or twice 3-partite with the segments linear-filiform, abaxially pubescent, adaxially glabrous. Fruits 6 × 7 mm, shallowly trilobate-subglobose, septicidal, sparingly stellate-pubescent, pale green. Seeds 5 × 3.5 × 2.5 mm, compressed-ovoid, ± smooth, slightly shiny, light brown; caruncle 1.5 mm wide, flattened.

Zimbabwe. E: Chipinge Distr., c. 7 km SSW of Chisumbanje, o. fr. & y. fl. 10.iii.1976, *Müller* 2438 (K; PRE; SRGH). S: Buffalo Bend, Mwenezi (Nuanetsi) R., near Malipati, fl. 2.xi.1955, *Wild* 4680 (PRE; SRGH). **Mozambique**. MS: Espungabera (Spungabera) on Maringa Road, fl. & y. fr. 22.xi.1960, *Leach & Chase* 10518 (K; LISC; MO; PRE; SRGH).

Also in South Africa (Transvaal) and Swaziland. At low to medium altitudes often beside seasonal watercourses, in dry thickets and *Androstachys johnsonii* associations, also in *Brachystegia* woodland; 30–400 m.

A fine flowering and fruiting specimen of this species distributed from SRGH appears to bear a wrong label. The SRGH label data is as follows: "Hwange Distr., 900 m, 17.ii.1956, *Wild* 4759, Game Reserve, Ngwashla Road, a trailing perennial herb in *Baikiaea* woodland, flowers and fruit green", and the name *"Plukenetia hastata* Müll. Arg." is written on the label. Neither the distribution nor the description fit *Croton madandensis.*

12. **Croton sylvaticus** Hochst. ex Krauss in Flora **28**: 82 (1845). —Müller Argoviensis in De Candolle, Prodr. **15**, 2: 602 (1866). —Hutchinson in F.T.A. **6**, 1: 771 (1912). —Eyles in Trans. Roy. Soc. South Africa **5**: 395 (1916). —Prain in F.C. **5**, 2: 412 (1920). —Engler, Pflanzenw. Afrikas (Veg. Erde 9) **3**, 2: 48 (1921). —Burtt Davy, Fl. Pl. Ferns Transvaal: 301 (1932). —Brenan, Check-list For. Trees Shrubs Tang. Terr.: 206 (1949). —Dale & Greenway, Kenya Trees & Shrubs: 192 (1961). —J. Léonard in F.C.B. **8**, 1: 72 (1962). —Drummond in Kirkia **10**: 252 (1975). —Biegel, Check List Ornam. Pl. Rhod. Parks & Gard.: 43 (1977). —K. Coates Palgrave, Trees Southern Africa, ed. 2, rev.: 420 (1983). —Radcliffe-Smith in F.T.E.A., Euphorb. 1: 155 (1987). —Beentje, Kenya Trees, Shrubs Lianas: 193 (1994). Type from South Africa (KwaZulu-Natal).

Croton oxypetalus Müll. Arg. in J. Bot. **2**: 339 (1864); in De Candolle, Prodr. **15**, 2: 543 (1866). —Hutchinson in F.T.A. **6**, 1: 774 (1912). —Engler, Pflanzenw. Afrikas (Veg. Erde 9) **3**, 2: 48 (1921). —De Wildeman, Pl. Bequaert. **3**, 4: 361 (1926). —Eggeling & Dale, Indig. Trees Uganda, ed. 2: 122 (1952). Type from Angola (Malanje).

Croton stuhlmannii Pax in Bot. Jahrb. Syst. **19**: 80 (1894); in Engler, Pflanzenw. Ost-Afrikas **C**: 237 (1895). Type from Uganda (Masaka).

Croton verdickii De Wild., Ann. Mus. Congo Belge, Bot. Sér. 5, **2**: 277 (1908). —Hutchinson in F.T.A. **6**, 1: 775 (1912). Type from Zaire (Shaba).

Croton bukobensis Pax in Bot. Jahrb. Syst. **43**: 77 (1909). —Hutchinson in F.T.A. **6**, 1: 765 (1912). —Engler, Pflanzenw. Afrikas (Veg. Erde 9) **3**, 2: 48 (1921). —Brenan, Check-list For. Trees Shrubs Tang. Terr.: 205 (1949). —Eggeling & Dale, Indig. Trees Uganda, ed. 2: 120 (1952). Type from Tanzania (Bukoba).

Croton asperifolius Pax in Bot. Jahrb. Syst. **43**: 79 (1909). —Hutchinson in F.T.A. **6**, 1: 763 (1912). —Engler, Pflanzenw. Afrikas (Veg. Erde 9) **3**, 2: 50 (1921). Type from Zaire (Kasai).

A shrub or tree up to 40 m tall, sometimes deciduous, monoecious; crown spreading. Bole up to 12 m high, and 75 cm d.b.h. Bark smooth or slightly roughened, soft, fibrous, light grey or greyish-brown. Twigs sparingly to densely stellate-pubescent at first, later glabrescent and becoming dark grey-brown. Stipules 0.5–1 cm long, linear, soon falling. Petioles 1.5–7 cm long, bipulvinate. Leaf blades 3–21 × 2–14 cm, ovate, elliptic-ovate or ovate-lanceolate, acuminate at the apex, glandular crenate-serrate on the margins, sometimes shallowly and/or irregularly so and sometimes with small stipitate or sessile discoid glands in the sinuses also, usually cuneate or rounded and with a pair of subsessile or stipitate discoid glands at or near the base, chartaceous, stellate-pubescent on both surfaces when young, later glabrescent, dark green, smelling of walnuts when crushed; 3–5-nerved from the base, lateral nerves in 4–6(8) pairs, sometimes slightly impressed above when dried, prominent beneath. Racemes 6–21 cm long, terminal, male, female or androgynous; axes sparingly to densely stellate-pubescent; bracts smaller than but otherwise resembling the stipules. Male flowers: pedicels 2–6 mm long; sepals 5, 2–3 × 1.5 mm, elliptic-ovate to ovate-lanceolate, sparingly stellate-pubescent without, glabrous within, puberulous at apex, pale yellowish-green; petals 5, 2–3 × 1–1.5 mm, elliptic-lanceolate to oblong-lanceolate, glabrous without, puberulous within, ciliate, greenish-cream in colour; disk glands free, triangular, acutely acuminate; stamens 14–17, filaments 4 mm long, glabrous, anthers c. 1 mm long; receptacle densely villous. Female flowers: pedicels 1–2 mm long, not or slightly extending in fruit, stout, densely stellate-tomentose; sepals 5(7), 3 × 0.5–1 mm, linear-lanceolate, indumented as the male sepals, whitish; petals 0–5, shorter than the sepals, otherwise resembling the male petals; disk 5-lobed, glabrous; ovary 2 mm in diameter, ovoid-subglobose, densely stellate-tomentose; styles 3, 4–5 mm long, deeply 2-partite with the segments filiform. Fruits 9–11 × 7–10 mm when dry, larger when fresh, trilobate-subglobose to ellipsoid, subindehiscent, sparingly to evenly stellate-pubescent, bright pinkish-orange or yellow. Seeds 6 × 4–5 mm, compressed-ovoid, whitish, aril white.

Zambia. N: Lunzua, o. fr. 2.iv.1960, *Fanshawe* 5625 (K; NDO). **Zimbabwe**. C: Avondale, Harare, cultivated, fl. 17.xii.1970, *Biegel* 3426 (K; PRE; SRGH). E: Mutare (Umtali), male fl. 19.xi.1952, *Chase* 4721 (BM; MO; PRE; SRGH). S: Bikita, st. 16.xii.1953, *Wild* 4413 (K; LISC; PRE; SRGH). **Malawi**. N: Mughesse Forest, Misuku Hills, fl. x.1953, *Chapman* 173 (FHO; K). S: Mt. Mulanje foot–Chikomioe Hill, near Likulezi R., fr. 9.iv.1987, *J.D. Chapman, E.J. Chapman & White* 8437 (FHO; K; MAL; MO). **Mozambique**. Z: between Lumba R. and Gueriza (Guerissa), fl. 8.xii.1971, *Müller & Pope* 1960 (K; LISC; SRGH). MS: Bárué Distr., Serra de Choa, near Catandica (Vila Gouveia), fr. 26.ii.1968, *Torre & Correia* 17770 (LISC).

More or less throughout tropical Africa from Guinée eastwards to Ethiopia and Kenya and south to South Africa (KwaZulu-Natal). Mixed evergreen forest, as canopy tree but also a pioneer on forest margins, often on rocky slopes, also in river gully forest and on rocky outcrops; 50–1675 m.

Vernacular name as recorded in specimen data: "musukuta".

Sometimes cultivated as a garden ornamental in Zimbabwe. A useful timber tree with soft easily worked wood.

13. **Croton macrostachyus** Hochst. ex Delile in Ferret & Galinier, Voy. Abyss. **3**: 158 (1848). —A. Richard, Tent. Fl. Abyss. **2**: 251 (1851) sphalm. "*macrostachys*". —Müller Argoviensis in De Candolle, Prodr. **15**, 2: 528 (1866). —Engler, Pflanzenw. Ost-Afrikas **C**: 237 (1895). —Hutchinson in F.T.A. **6**, 1: 772 (1912). —Engler, Pflanzenw. Afrikas (Veg. Erde 9) **3**, 2: 49 (1921). —Robyns & Tournay, Fl. Sperm. Parc Nat. Alb. **1**: 449 (1948). —Brenan, Check-list For. Trees Shrubs Tang. Terr.: 205 (1949). —Eggeling & Dale, Indig. Trees Uganda, ed. 2: 120 (1952). —F.W. Andrews, Fl. Pl. Anglo-Egypt. Sudan **2**: 60 (1952). —Brenan in Mem. N.Y. Bot. Gard. **9**, 1: 71 (1954). —Topham, Check List For. Trees Shrubs Nyasaland Prot.: 50 (1958) sphalm. "*macrostachys*". —Keay in F.W.T.A., ed. 2, **1**, 2: 394 (1958). —Dale & Greenway, Kenya Trees & Shrubs: 191 (1961). —J. Léonard in F.C.B. **8**, 1: 62 (1962). —White, F.F.N.R.: 196 (1962). —Troupin, Fl. Pl. Lign. Rwanda: 256, fig. 88/2 (1982); Fl. Rwanda **2**: 216, fig. 65/2 (1983). —Radcliffe-Smith in F.T.E.A., Euphorb. 1: 149 (1987). —Beentje, Kenya Trees, Shrubs Lianas: 191 (1994). Type from Ethiopia (Tigray Province).

Croton zambesicus sensu De Wild., Ann. Mus. Congo Belge, Bot., Sér. 5, **2**: 278 (1908). —Hutchinson in F.T.A. **6**, 1: 758 (1912), pro parte quoad spec. *Seret* 594; non Müll. Arg.

Croton buluguensis De Wild. in Rev. Zool. Bot. Africaines **9**: 16 (1921); in Pl. Bequaert. **3**: 452 (1926). —Robyns & Tournay, Fl. Sperm. Parc Nat. Alb. **1**: 450 (1948). Type from Zaire (Kivu Province).

A tree up to 30 m tall, dioecious or sometimes monoecious; crown much-branched, spreading, rounded. Bole up to 1.8 m in circumference. Bark smooth to closely reticulate, pale grey or greyish-brown, slightly corky. Young twigs, petioles and inflorescence axes densely, evenly or sparingly greyish appressed or patent stellate-pubescent, later glabrescent and becoming dark grey-brown and lenticellate. Stipules 5–7 mm long, linear to filiform-setaceous, soon falling. Petioles 3–12 cm long. Leaf blades 6–18 × 4–14 cm, ovate, acuminate at the apex, shallowly crenate-serrate to subentire on the margins, cordate or subcordate at the base with a pair of shortly stipitate to subsessile basal discoid glands on the lower surface, chartaceous, sparingly appressed stellate-pubescent to fimbriate-lepidote above, and sparingly to densely so or patent stellate-tomentose beneath; 5–7-nerved from the base, lateral nerves in 4–9 pairs, ± prominent beneath. Racemes up to 35 cm long, terminal, usually androgynous, sometimes male or female; bracts shorter than the stipules but otherwise resembling the stipules. Flowers fragrant. Male flowers: pedicels 0.5–1 cm long; sepals 5, 3 × 2–2.5 mm, ovate, stellate-lepidote without, pubescent within, pale green; petals 5, 3–3.5 × 1.5–2 mm, oblanceolate-oblong, subglabrous without, villous within and on the margin, pale creamy-yellow in colour; disk glands rounded, pilose; stamens 15–20, filaments 4 mm long, villous below, anthers c. 1 mm long; receptacle densely villous. Female flowers: pedicels 2–5 mm long, not or scarcely extending in fruit, stout, densely stellate-pubescent; sepals 5, 3–3.5 × 1.5–2 mm, ovate-lanceolate, evenly stellate-lepidote without, sparingly stellate-pubescent within, grey-green; petals 0 or 5, 0.5–1 mm long, linear to subulate, subglabrous; disk 5-lobed, the lobes truncate, pubescent; ovary 2 mm in diameter, trigonous or quadrangular, densely lepidote; styles 3–4, 3 mm long, spreading, 2-partite with the segments filiform-linear, minutely puberulous. Fruits 0.8–1 × 0.8–1.5 cm, trilobate or quadrilobate, loculicidal or septicidal, densely to evenly scurfily stellate-pubescent, pale greyish-green. Seeds 7 × 4 mm, ellipsoid, rugulose, grey; caruncle 4.5 × 4 mm, waxy.

Zambia. N: Mbala (Abercorn) Agric. Station, fr. 27.iii.1960, *Angus* 2177 (FHO; K; MO; SRGH). E: Chama Distr., Nyika Plateau, Chowo Forest, fr. only 13.viii.1975, *Pawek* 10036 (K; MAL; MO; SRGH; UC). **Zimbabwe**. E: Chimanimani, fl. vi.1958, *Boughey* 4450 (SRGH). **Malawi**. N: Mzuzu, male fl. 15.ii.1976, *Pawek* 10858 (MO; PRE; SRGH; UC). C: Dedza, Mphunzi Hill, fl. & fr. 22.i.1959, *Robson* 1306 (BM; K; LISC; PRE; SRGH). S: Zomba, male fl. 21.xi.1978, *Salubeni* 2398 (MAL; MO; SRGH). **Mozambique**. Z: Mt. Milange, fr. 30.vii.1943, *Torre* 4834a (LISC; PRE). T: Angónia, y. fr. 13.v.1948, *Mendonça* 4223 (LISC).

More or less throughout tropical Africa from Guinée eastwards to Ethiopia and southwards to Angola and Mozambique; also in Madagascar. Evergreen forest, *Brachystegia* woodland and wooded grassland in submontane localities, often on rocky hillsides, in evergreen riverine and gully forest, and mushitu margins (swamp forest), also on termitaria; 825–1830(2165) m.

Vernacular names as recorded in specimen data include: "tensa" (chiNyanja), "mughogha" (Malawi), "nakawalika" (chiYao).

14. **Croton megalobotrys** Müll. Arg. in Flora **47**: 537 (1864); in De Candolle, Prodr. **15**, 2: 598 (1866). —Engler, Pflanzenw. Ost-Afrikas **C**: 237 (1895). —Hutchinson in F.T.A. **6**, 1: 762 (1912). —Engler, Pflanzenw. Afrikas (Veg. Erde 9) **3**, 2: 49 (1921). —Hutchinson, Botanist in Southern Africa: 667 (1946). —O.B. Miller, Check-list For. Trees Shrubs Bech. Prot.: 32 (1948). —Brenan, Check-list For. Trees Shrubs Tang. Terr.: 204, 205 (1949). —White, F.F.N.R.: 197 (1962). —P.G. Meyer in Merxmüller, Prodr. Fl. SW. Afrika, fam. 67: 10 (1967). —Drummond in Kirkia **10**: 252 (1975). —K. Coates Palgrave, Trees Southern Africa, ed. 2, rev.: 417 (1983). —Radcliffe-Smith in F.T.E.A., Euphorb. 1: 151 (1987). Syntypes: Botswana, L. Ngami, Thamalakane (Tamulakane) R., male fl. 1856, *McCabe* in *Herb. Atherstone* 17, 39 (K, syntypes).

Croton gubouga S. Moore in J. Linn. Soc., Bot. **40**: 196 (1911). —Hutchinson in F.T.A. **6**, 1: 766 (1912). —Eyles in Trans. Roy. Soc. South Africa **5**: 395 (1916). —Prain in F.C. **5**, 2: 413 (1920). —Engler, Pflanzenw. Afrikas (Veg. Erde 9) **3**, 2: 48 (1921). —Burtt Davy, Fl. Pl. Ferns Transvaal: 301 (1932). —Topham, Check List For. Trees Shrubs Nyasaland Prot.: 50 (1958). Syntypes: Mozambique, lower Mossurize R. (Umswiriswe), fr. immat. 1.xi.1905, *Swynnerton* 153 (BM; K); Chibabava, lower Búzi R., st. 1.xii.1906, *Swynnerton* 1123 (BM; K). Zimbabwe, lower Save (Sabi) R., fl. 9.xi.1906, *Swynnerton* 1125 (BM; K).

A shrub or tree up to 14 m tall, sometimes sarmentose, often lax branched from the base, monoecious or sometimes dioecious; crown conical to rounded. Trunk up to 60 cm in diameter and 180 cm in circumference, angular. Bark grey, longitudinally lenticellate, smooth at first later becoming fissured. Branches drooping. Twigs grey-green, lenticellate. Stipules 3–7 mm long, linear-lanceolate, often 1–2-lobed near the base, subglabrous, soon falling. Petioles 2–7(11) cm long, densely stellate-pilose at first, soon glabrescent, with 2 stipitate discoid glands at the apex. Leaf blades 3–19 × 2–13 cm, ovate to ovate-lanceolate, usually acutely caudate-acuminate at the apex, somewhat irregularly and coarsely crenate-serrate or dentate on the margins, cuneate to rounded or truncate to cordate at the base, membranous to thinly chartaceous, densely stellate-tomentose on both surfaces at first, later glabrescent above and cottony stellate-pilose beneath, slightly roughened, light or deep green; 3–7-nerved from the base, lateral nerves in 3–5 pairs, slightly prominent. Racemes terminal on the main axis or terminating short axillary or lateral shoots, the male racemes and the androgynous racemes 4.5–17 cm long, the female racemes 2–3 cm long; axes densely stellate-pilose at first, later sparingly so; bracts resembling the stipules, the male bracts several-flowered, the female bracts usually only 1-flowered. Male flowers: pedicels 3–7 mm long, slender; sepals 5, 2.5–3 × 2–2.5 mm, suborbicular, subglabrous without, glabrous within, ciliate, pale green; petals 5, 4 × 1.5 mm, oblanceolate, subacute, pubescent at the apex, pale cream-yellow or whitish; disk glands free, truncate; stamens 15–20(25), filaments 3 mm long, glabrous, anthers 0.5 mm long; receptacle densely villous. Female flowers: pedicels 3–5 mm long, extending to up to 10 mm in fruit, stout; sepals 5, larger than in the male flowers and ovate, but otherwise similar; petals absent; disk shallowly 5-lobed, crenulate; ovary 3 mm in diameter, globose, densely stellate-tomentose; styles 3, 3 mm long, 4-partite, inflexed, glabrous, buff-orange, persistent. Fruits 2–3.5 × 2.2–3.5 cm when dry, up to 4 × 3.8 cm when fresh, trilobate-subglobose, often bilobate or ovoid-ellipsoid by abortion, indehiscent, softly stellate-tomentose at first, later glabrescent, bright green at first, later becoming orange to golden-brown. Seeds 1.7–2 × 1.5–1.8 × 0.9–1.2 cm, broadly compressed-ovoid, subtruncate, dark purplish-brown, dull, ecarunculate.

Caprivi Strip. Bagani (Bagoni) Camp, fr. 19.i.1956, *de Winter & Wiss* 4352a (K; PRE). **Botswana.** N: Zibadianja Lagoon, male fl. 20.x.1972, *Biegel, Pope & Gibbs Russell* 4021 (K; LISC; PRE; SRGH). SE: near Malede, st. 19.xii.1957, *de Beer* 537 (K; PRE; SRGH). **Zambia**. B: Sesheke Boma, fr. 27.xii.1952, *Angus* 1044 (BM; FHO; K; MO). N: Mansa (Fort Rosebery), o. fr. ii.1957, *Fanshawe* 3004 (K; NDO). C: Chingombe, male fl. 26.ix.1957, *Fanshawe* 3734 (K; NDO). E: Machinje Hills, female fl. 12.x.1958, *Robson & Angus* 68 (BM; FHO; K; LISC; PRE). S: 37 km Machili–Mwande, fr. 21.xii.1952, *Angus* 995 (BM; FHO; K; MO). **Zimbabwe**. N: Musukwi (Msukwe) R., y. fr. 18.xi.1953, *Wild* 4166 in *GHS* 44451 (K; LISC; MO; PRE; SRGH). W: Hwange (Wankie) Nat. Park, male fl. 21.xi.1973, *Raymond* 187 (K; MO; PRE; SRGH). C: Munyati (Umniati) R., st. 13.xii.1962, *Müller* 2 (K; MO; SRGH). E: Chimanimani, Umvumvumvu Irrigation Scheme, fr. 16.xi.1952, *Chase* 4699 (BM; MO; SRGH). S: Pazhi R. (Pie R.), male fl. x.1954, *Davies* 811 in *GHS* 48299 (K; MO; PRE; SRGH). **Malawi**. C: Dowa Distr., Chitala, fl. 29.x.1941, *Greenway* 6385 (EA; K; PRE). S: Mangochi (Fort Johnston), fr. 23.xi.1954, *Jackson* 1394 (FHO; K). **Mozambique**. T: Carangache, 19.4 km from Estima, fr. 12.ii.1972, *Macêdo & Esteves* in *Macêdo* 4845 (K; LISC; LMA; PRE; SRGH). MS: Grudja, R. Búzi, fl. & y. fr. 11.xi.1941, *Torre* 3804 (LISC). GI: Estivane–Aldeia da Barragem, male fl. 20.xi.1957, *Barbosa & Lemos* in *Barbosa* 8203 (K; LISC; LUAI). M: Magude, fr. 4.xii.1980, *Jansen, Nuvunga & Petrini* 7618 (K; PRE; WAG).

Also in southwest Tanzania and South Africa (Transvaal). Frequent at medium and low altitudes in riverine fringing vegetation, on sand banks of rivers and seasonal watercourses and on floodplain alluvium forming thickets and as understorey to larger trees, also on seasonal pan margins and island vegetation in swamps (Okavango); 110–1070 m.

Vernacular names as recorded in specimen data include: "chungo-chungo" (Gaza); "mo-tsibi", "phokotsa" (SE Botswana).

The bark and seeds are used as a febrifuge and as a purgative, and also locally in Zimbabwe as a fish poison. Sometimes planted in villages south of Lake Mweru (fide D.B. Fanshawe).

15. **Croton inhambanensis** Radcl.-Sm. in Kew Bull. **45**: 559 (1990). Type: Mozambique, Inhambane, Vilanculos, Muábsa, 27.iii.1973, *Balsinhas* 2496 (K, holotype; LMA).

A shrub or small tree up to 5 m high. Twigs sparingly stellate-pubescent to subglabrous, cinnamon-ochreous at first, later becoming greyish-tawny in colour. Stipules 5 mm long, subulate, sparingly stellate-pubescent, soon falling. Petioles 1–7 cm long, stellate-pubescent at first, later glabrescent. Leaf blades 4–10 × 2–3 cm, lanceolate to elliptic-lanceolate, acuminate at the apex, shallowly and somewhat irregularly glandular-serrulate on the margins, wide-cuneate to rounded at the base, with a pair of small stipitate or subsessile discoid glands on upper surface at the junction with the petiole, chartaceous, minutely glandular-punctate, glabrous or subglabrous on both surfaces, if subglabrous then with very few scattered stellate hairs on the lower midrib and lateral nerves; 5-nerved from the base, lateral nerves in 10–18 pairs, scarcely prominent. Male and female inflorescences and flowers unknown. Fruits 2.5–3 × 2–2.5 cm, obovoid- or ellipsoid-subglobose, scarcely 3-lobed, loculicidal, densely echinate, subglabrous except for the processes; processes up to 7 mm long, subulate, curved, sparingly stellate-pubescent. Seeds 2 × 1.5 × 1 cm, ovoid-ellipsoid, dull, light brown, tawny-mottled, recalling those of the genus *Ricinus*, as indeed also do the fruits.

Mozambique. GI: 30 km Mapinhane–Mavume, fr. ii.1939, *Gomes e Sousa* 2216 (COI; K).

Known only from this locality. Low altitude coastal plain, in dry sandy soil with *Androstachys johnsonii*.

The type and the specimen cited above are the only specimens known of this very distinct species.

16. **Croton leuconeurus** Pax in Bot. Jahrb. Syst. **15**: 533 (1893). —Hutchinson in F.T.A. **6**, 1: 767 (1912). —Engler, Pflanzenw. Afrikas (Veg. Erde 9) **3**, 2: 48 (1921). —F.W. Andrews, Fl. Pl. Anglo-Egypt. Sudan **2**: 61 (1952). —J. Léonard in F.C.B. **8**, 1: 81 (1962). —White, F.F.N.R.: 196 (1962). —Drummond in Kirkia **10**: 252 (1975). —K. Coates Palgrave, Trees Southern Africa, ed. 2, rev.: 417 (1983). —Radcliffe-Smith in F.T.E.A., Euphorb. 1: 156 (1987). Type from Sudan (Equatoria Province).

A shrub, or tree up to 15 m tall, often several-stemmed and with a dense spreading crown, monoecious or dioecious. Bole clean, up to 30 cm in diameter. Bark smooth, grey- and white-blotched. Branches drooping. Twigs grey-green pubescent with dark grey to blackish stellate hairs. Stipules 6–7 mm long, linear to linear-lanceolate, sparingly pubescent, soon falling. Petioles 1.5–6 cm long, with 2 stipitate or ± sessile orange discoid glands at the apex. Leaf blades 3–14 × 2–8.5 cm, elliptic-ovate, gradually acuminate at the apex, glandular crenate-serrate on the margins, the glands conoidal on the teeth, and discoid (when present) in the sinuses, cuneate to rounded or truncate at the base, chartaceous to thinly coriaceous, usually penninerved (palminerved in subsp. *mossambicensis*) with 7–17 pairs of usually conspicuous pale lateral slightly prominent nerves, densely or evenly, rarely sparingly, stellate-pubescent when young, glabrescent on upper surface and sometimes beneath, or remaining softly whitish stellate-pilose beneath, deep dull green to pale yellowish-green on upper surface, turning red prior to falling, often drying blackish (greenish-brown in subsp. *mossambicensis*). Racemes (3)6–20 cm long, terminal on the main axis or terminating lateral shoots, male, female or androgynous; axes usually densely stellate-pubescent; bracts 3–5 mm long, linear-lanceolate, evenly stellate-pubescent to subglabrous, the male bracts several-flowered, the androgynous bracts few-flowered, the female bracts 1-flowered, soon falling. Flowers scented. Male flowers: pedicels 3–5 mm long; sepals 5, 2–3 × 1–1.5 mm, oblong-lanceolate, sparingly stellate-pubescent without, glabrous within, puberulous at apex, green; petals 5, 2–3 × 1–1.5 mm, elliptic, glabrous without, villous within, creamy-yellow in colour; disk glands 5, free, truncate; stamens 15–20, filaments 3 mm long, glabrous, yellow-green, anthers 0.75 mm long; receptacle pilose. Female flowers: pedicels 1–2 mm long, scarcely extending in fruit, stout, densely fulvous stellate-tomentose; sepals 5, 1.5–4 × 0.75–1 mm, slightly accrescent in fruit, lanceolate, otherwise as in male flowers; petals 0–5, 1–2 mm long, linear to filiform, villous on the margin, otherwise glabrous; disk shallowly 5-lobed, pubescent; ovary 2 mm in diameter, subglobose, densely stellate-tomentose; styles 3, 4 mm long, deeply 2-partite with filiform segments, subpersistent, brownish. Infructescences pendent. Fruits 9–14 × 9–12 mm, trilobate-subglobose or obovoid, scurfily stellate-pubescent, yellowish to buff; endocarp thinly crustaceous, breaking up irregularly (frangent). Seeds 8 × 6 × 3.5 mm, compressed ovoid-trigonous, adaxially light brown, abaxially buff and brown-flecked or mottled, dull, ecarunculate.

Leaves broadly elliptic-ovate, mostly penninerved, often drying blackish; racemes (3)6–18 cm long; sepals of female flowers up to 2 mm long · · · · · · · · · · · · · · · · · · · subsp. *leuconeurus*
Leaves narrowly elliptic-ovate, mostly palminerved, often drying greenish-brown; racemes 10–20 cm long; sepals of female flowers up to 4 mm long · · · · · · · · subsp. *mossambicensis*

Subsp. **leuconeurus**

Croton barotsensis Gibbs in J. Linn. Soc., Bot. **37**: 469 (1906). —Hutchinson in F.T.A. **6**, 1: 767 (1912). —Eyles in Trans. Roy. Soc. South Africa **5**: 395 (1916). —Engler, Pflanzenw. Afrikas (Veg. Erde 9) **3**, 2: 48 (1921). —Brenan, Check-list For. Trees Shrubs Tang. Terr.: 205 (1949). Type: Zimbabwe, Victoria Falls, ix.1905, *Miss L.S. Gibbs* 109 (K, holotype; BM).

Croton seineri Pax in Bot. Jahrb. Syst. **43**: 78 (1909). —Hutchinson in F.T.A. **6**, 1: 762 (1912). —Engler, Pflanzenw. Afrikas (Veg. Erde 9) **3**, 2: 49 (1921). —White, F.F.N.R.: 197 (1962). Type: Zambia, Old Livingstone, 15.ix.1906, *Seiner* 18 (B†, holotype; K, illustration of holotype).

Leaves broadly elliptic-ovate, mostly penninerved, often drying blackish; racemes (3)6–18 cm long; female sepals up to 2 mm long.

Botswana. N: Kasane Rapids, Chobe R., fl. 9.x.1984, *P.A. Smith* 4455 (SRGH). **Zambia**. B: Sesheke, y. male fl. 9.viii.1947, *Brenan & Greenway* 7656 (FHO; K; PRE). N: Kasama, male fl. 27.vii.1961, *Lawton* 745 (K; NDO). W: Angolan border west of Kalene Hill Mission, fl. 22.ix.1952, *White* 3325 (BM; FHO; K; MO). S: Victoria Falls, female fl. 5.x.1956, *Gilges* 662 (K; LISC; PRE; SRGH). **Zimbabwe**. W: Victoria Falls, fr. 22.xi.1979, *Ncube* 67 (K; MO; PRE; SRGH).

Also in Sudan, Zaire (Kasai, Shaba), western Tanzania and Angola. Tree of riverine vegetation, in fringing thicket and evergreen forest, often overhanging the water, on sandy banks and in swampy margins of rivers; 880–1650 m.

Fruit edible.

Tab. 57. CROTON LEUCONEURUS subsp. MOSSAMBICENSIS. 1, distal portion of branch with young fruits (× $^{2}/_{3}$); 2, female flower (× 4); 3, male flower (× 4), 1–3 from *Andrada* 1074. Drawn by Christine Grey-Wilson.

Subsp. **mossambicensis** Radcl-Sm. in Kew Bull. **45**: 557 (1990). Tab. **57**. Type: Mozambique, Cheringoma Distr., Inhamitanga, y. fr. 19.ii.1948, *Andrada* 1074 (LISC, holotype).

Leaves narrowly elliptic-ovate, mostly palminerved, often drying greenish-brown; racemes 10–20 cm long; female sepals up to 4 mm long.

Mozambique. Z: 31.2 km Namacurra–Nicuadala, st. 29.viii.1949, *Barbosa & Carvalho* in *Barbosa* 3889 (K; LMA). MS: Without precise locality, y. fr. 1948, *Simão* 266/48 (LISC).
Known only from the specimens cited above. No habitat details available.

17. **Croton gossweileri** Hutch. in F.T.A. **6**, 1: 765 (1912). —Engler, Pflanzenw. Afrikas (Veg. Erde 9) **3**, 2: 48 (1921). Type from Angola (Huíla/Cuando Cubango Provinces).

Very like *C. leuconeurus* Pax, differing chiefly in the leaf blades being more abruptly caudate-acuminate at the apex, and in being only very sparingly stellate-pubescent when young and quite glabrous when mature.

Zambia. B: Kashiji (Kasisi) R., male fl. vii.1933, *Trapnell* 1210 (K).
Also in southwest Angola. Riverine vegetation, in *Syzygium* fringing bush.
In habit resembling a poplar (*Populus* spp.).
Angus 615 (BM; FHO; K) from Zambia, Kabompo R. gorge near Kabompo Boma, fl. 6.x.1952 and *Angus* 1108A (FHO; K) from Zambia, Katombora, Zambezi R., st. 6.i.1953, whilst having the leaf shape of *Croton leuconeurus* subsp. *leuconeurus*, nevertheless have the indumentum character of *Croton gossweileri*.

18. **Croton aceroides** Radcl.-Sm. in Kew Bull. **45**: 555 (1990). Tab. **58**. Type: Mozambique, Homoine Distr., near Inhatechambe (Inhatchambe), 23°36'S, 34°58'E, fl. xi.1973, *Tinley* 3004 (SRGH, holotype).

A many-stemmed tree up to 7 m tall. Young shoots densely fulvous stellate-pubescent and rusty-spotted, later glabrescent and minutely verruculose. Stipules 2–5 mm long, subulate-filiform, sparingly stellate-pubescent. Petioles 1–3 cm long. Leaf blades 3–6 × 3–7 cm, 3(5)-lobed, often with additional smaller lobes alternating with the 3 main lobes, all lobes obtuse at the apex, subtruncate at the base with 2 sessile ochreous disciform glands on the lower surface, chartaceous, fulvous-tomentose when young, later sparingly stellate-pubescent on upper surface and more evenly so beneath; 5–7-nerved from the base, lateral nerves 2–4 in the middle lobe, not prominent above, scarcely so beneath. Racemes 10 cm long, terminal, androgynous; bracts resembling the stipules. Male flowers: pedicels c. 1 mm long; sepals 5, 3 × 2 mm, ovate, acute, stellate-tomentose without, glabrous within; petals 5, 3 × 1 mm, oblanceolate, obtuse, subglabrous but densely ciliate on the margins; disk glands 5, acute, minute; stamens 10–17, filaments 1.5 mm long and glabrous, anthers 0.75 mm long with a broad connective; receptacle densely white-villous. Female flowers: pedicels c. 2 mm long, stouter than in the male; sepals 5, 4 × 1.5 mm, triangular-lanceolate, acute, stellate-pubescent without, glabrous within; petals absent; disk 5-lobed, the lobes triangular-conoidal, fleshy; ovary 3 mm in diameter, bilobate-subglobose, densely stellate-tomentose; styles 2, c. 5 mm long, suberect, deeply bipartite with filiform segments, sparingly appressed stellate-pubescent abaxially, otherwise glabrous. Fruit and seeds unknown.

Mozambique. GI: near Inhatechambe (Inhatchambe) at approx. 23°36'S, 34°58'E, fl. xi.1973, *Tinley* 3004 (SRGH).
Known only from this locality. Locally common, on margins of dry coastal forest, in pallid sands.
Only known from the type.

Doubtfully recorded species

Eyles in Trans. Roy. Soc. South Africa **5**: 395 (1916) records *C. rivularis* Müll. Arg. as being from "south of the Zambesi, *Engler*", but no material of this South African species has been seen from the Flora Zambesiaca area by the author.

Tab. 58. CROTON ACEROIDES. 1, distal portion of flowering branch (× $^{2}/_{3}$); 2, leaf surface (× 12); 3, female flower (× 4); 4, male flower (× 4), 1–4 from *Tinley* 3004. Drawn by Christine Grey-Wilson. From Kew Bull.

47. RICINODENDRON Müll. Arg.

Ricinodendron Müll. Arg. in Flora **47**: 533 (1864); in De Candolle, Prodr. **15**, 2: 1111 (1866).
Barrettia Sim, For. Fl. Port. E. Afr.: 103 (1909).

Dioecious pachycaul trees with a detersible (easily detached) stellate indumentum. Leaves alternate, stipulate, long-petiolate, palmatipartite to palmatisect; petioles sometimes glandular at the apex; leaf segments sometimes pseudopetiolulate, glandular-denticulate, penninerved. Stipules persistent, large, foliaceous, sessile, palmatifid, glandular-dentate. Inflorescences terminal or subterminal, paniculate, the male larger than the female; bracts linear-setaceous. Male flowers shortly pedicellate; buds globose; calyx lobes 4–5, imbricate; petals 5, imbricate, laterally coherent; disk glands 4–6, free, extrastaminal, entire, fleshy; stamens 6–14, filaments united at the base, anthers dorsifixed, introrse, longitudinally dehiscent; pistillode absent. Female flowers pedicellate; calyx lobes 5, imbricate; petals more or less as in the male flowers; hypogynous disk crenellate; ovary 2–3-locular, with 1 ovule per loculus; styles 2–3, bifid. Fruits indehiscent, 2-lobed or 3-lobed, usually wider than long; pericarp coriaceous; mesocarp fleshy; endocarps 2(3), rarely 1 by abortion, distinct, each subglobose, 1-locular, 1-seeded, thinly woody, smooth. Seeds ecarunculate, subglobose; albumen fleshy; cotyledons broad, flat, palminerved.

A monotypic tropical African genus, with a predominantly Guineo-Congolian distribution.

Ricinodendron heudelotii (Baill.) Pierre ex Heckel in Ann. Inst. Bot.-Géol. Colon. Marseille **5**, 2: 40 (1898). —Pax in Engler, Pflanzenr. [IV, fam. 147, iii] **47**: 46, t. 16 (1911). —Engler, Pflanzenw. Afrikas (Veg. Erde 9) **3**, 2: 130 (1921). —Keay in F.W.T.A., ed. 2, **1**, 2: 393 (1958). —J. Léonard in Bull. Jard. Bot. État **31**: 397 (1961). —Radcliffe-Smith in F.T.E.A., Euphorb. 1: 326 (1987). Type from Guinée.
Jatropha heudelotii Baill., Adansonia **1**: 64 (1860).

Subsp. **africanum** (Müll. Arg.) J. Léonard in Bull. Jard. Bot. État **31**: 398 (1961); in F.C.B. **8**, 1: 116 (1962). —Radcliffe-Smith in F.T.E.A., Euphorb. 1: 326 (1987). Type from Bioko (Fernando Po).

A deciduous tree up to 20 m tall. Bark rough, flaking, dark grey. Wood soft, white. Twigs up to 1 cm in diameter, sparingly lenticellate, densely stellate-pubescent when young, soon glabrescent; pith thick. Petiole 5–30 cm long, with 1-more scattered discoid glands c. halfway along. Leaf segments (3)5–7(8), the median segment 10–30 × 5–12 cm, the lateral segments somewhat smaller; all segments broadly elliptic-oblanceolate, acutely long-acuminate to caudate-acuminate at the apex, shallowly glandular-denticulate on the margin, attenuate into the usually scarcely distinct pseudopetiolule, membranous to thinly chartaceous, glabrous or subglabrous on upper surface, subglabrous or densely stellate-tomentose beneath at first, soon glabrescent, or persistently densely fulvous stellate-tomentose beneath; lateral nerves in 12–20 pairs in each segment. Stipules 1–5 × 1.5–4 cm. Male inflorescences up to 40 × 20 cm; axis stellate-pubescent at first, soon glabrescent except in the branch axils; bracts c. 5 mm long. Male flowers: pedicels 1–3 mm long, pubescent; calyx lobes 3–4 × 2–3 mm, obovate, pubescent without and within; petals 5–6 × 2–3 mm, ± oblong, truncate, narrowed at the base, glabrous except at the base, white, greenish-white or pale yellow-green; disk glands ovate-suborbicular, yellow; stamens 6 mm long, anthers less than 1 mm long. Female inflorescences 10–20 × 7–8 cm, more robust than the male; indumentum and bracts as in the male. Female flowers: pedicels 4–6 mm long, pubescent; calyx lobes and petals larger than in the male flowers, but otherwise resembling them; disk yellowish-green; ovary 2(3)-locular, 5 mm in diameter, subglobose, densely stellate-pubescent; styles 4 mm long. Fruit 2(3)-lobed, 2 × 3 cm when dried, 2.5–3.5 × 4–5 cm when fresh, glabrescent, green at first, later blackening. Seeds 1.3–1.7 cm in diameter, reddish-brown; sarcotesta 0.5–2 mm thick, xylotesta 1.5 mm thick.

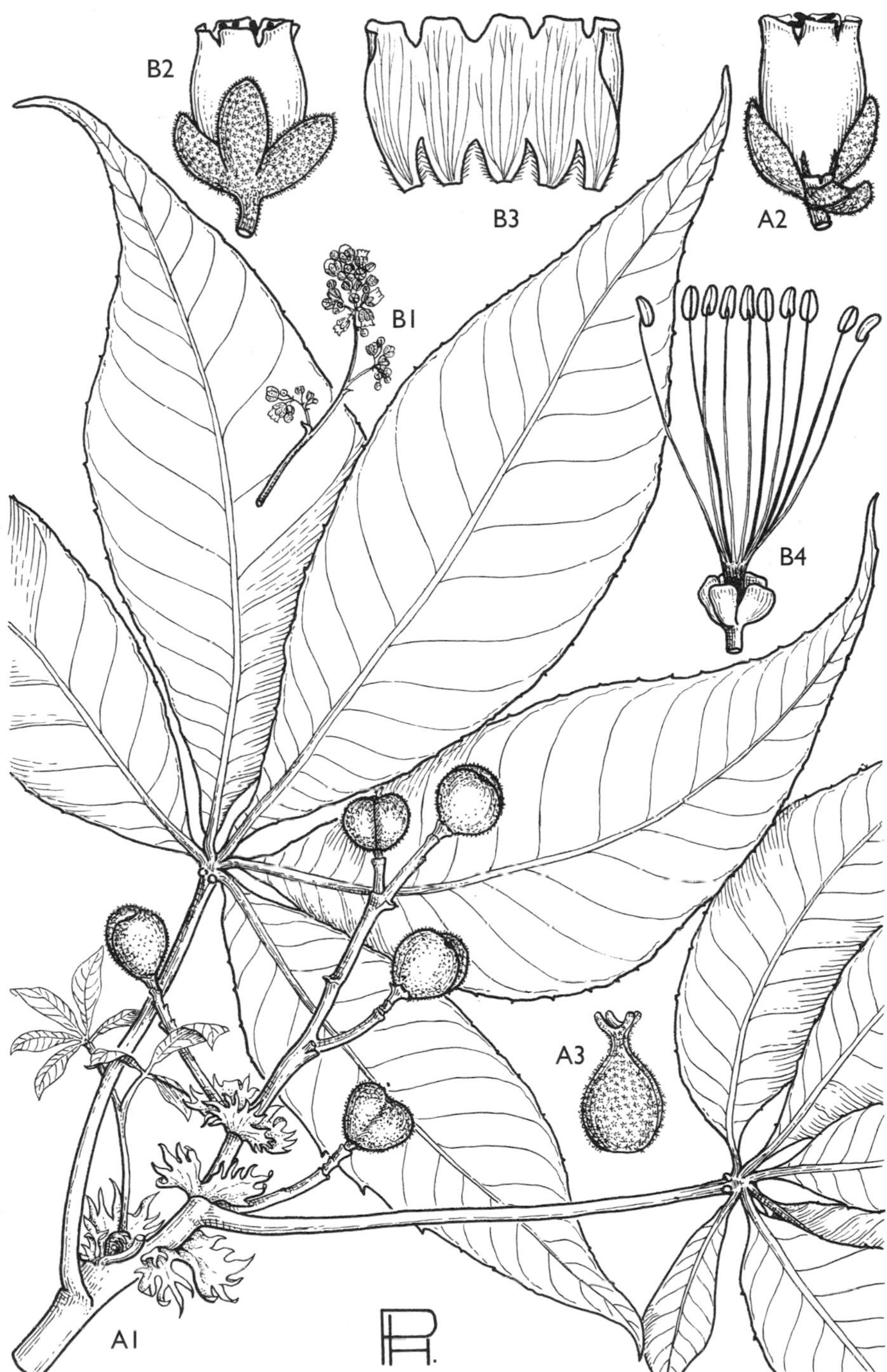

Tab. 59. A —RICINODENDRON HEUDELOTII subsp. AFRICANUM var. AFRICANUM. A1, distal portion of young fruiting branch (× $^{2}/_{3}$), from *Dawkins* 552; A2, female flower (× 4); A3, ovary (× 2), A2 & A3 based on drawing by *Trevithick*. B —RICINODENDRON HEUDELOTII subsp. AFRICANUM var. TOMENTELLUM. B1, male inflorescence (× $^{2}/_{3}$); B2, male flower (× 4); B3, male corolla, opened out (× 4); B4, stamens and disk lobes (× 6), B1–B4 from *Gillman* 873. Drawn by Pat Halliday. From F.T.E.A.

Leaf segments subglabrous, or densely stellate-tomentose on lower surface at first and soon glabrescent . . . . var. *africanum*
Leaf segments persistently densely fulvous stellate-tomentose on lower surface . . . . var. *tomentellum*

Var. **africanum.** Tab. **59**, figs. A1–A3.

Ricinodendron africanum Müll. Arg. in Flora **47**: 533 (1864); in De Candolle, Prodr. **15**, 2: 1111 (1866), pro parte excl. spec. *Mann* 825 ('815'). —Hutchinson in F.T.A. **6**, 1: 745 (1912), pro parte excl. spec. *Mann* 825, *Heudelot* 857, *Johnson* 448, 635. —De Wildeman, Pl. Bequaert. **3**, 4: 506 (1926).

[*Ricinodendron heudelotii* sensu auct. praesertim Pax in Engler, Pflanzenr. [IV, fam. 147, iii] **47**: 46 (1911), pro parte excl. spec. *Heudelot* 857. —Robyns & Tournay, Fl. Sperm. Parc Nat. Alb. **1**: 473 (1948). —Brenan, Check-list For. Trees Shrubs Tang. Terr.: 225 (1949). —Eggeling & Dale, Indig. Trees Uganda, ed. 2: 138 (1952). —F.W. Andrews, Fl. Pl. Anglo-Egypt. Sudan **2**: 95 (1952). —Keay in F.W.T.A., ed. 2, **1**, 2: 393 (1958), pro parte. —Dale & Greenway, Kenya Trees & Shrubs: 218 (1961), non (Baill.) Pierre ex Heckel s. str.]

Ricinodendron gracilius Mildbr. in Notizbl. Bot. Gart. Berlin-Dahlem **12**: 516 (1935) (as "*gracilior*"). —Brenan, Check-list For. Trees Shrubs Tang. Terr.: 225 (1949). Type from Tanzania.

Barrettia umbrosa Sim, For. Fl. Port. E. Afr.: 103, t. 71 (1909). Type: Mozambique, Quelimane Distr., without precise locality, *Sim* 6394 (PRE, holotype).

Leaf segments subglabrous, or densely stellate-tomentose on lower surface at first, and soon glabrescent.

Mozambique. N: 18 km Memba–Mazua, fr. 11.xii.1963, *Torre & Paiva* 9500 (LISC).

From Nigeria to Sudan, Uganda, Tanzania, Zaire (lower Shaba) and Angola. In open woodland, on dark grey sandy clay soils, also in old Cashew plantations; 80–250 m.

The seeds are edible.

Var. **tomentellum** (Hutch. & E.A. Bruce) Radcl.-Sm. in Kew Bull. **27**: 507 (1972). —Beentje, Kenya Trees, Shrubs Lianas: 220 (1994). Tab. **59**, figs. B1–B4. Type from Kenya (Coast Province).

Ricinodendron tomentellum Hutch. & E.A. Bruce in Bull. Misc. Inform., Kew **1931**: 270 (1931). —Battiscombe, Trees & Shrubs of Kenya Colony: 50 (1936).

Ricinodendron schliebenii Mildbr. in Notizbl. Bot. Gart. Berlin-Dahlem **12**: 516 (1935). —Brenan, Check-list For. Trees Shrubs Tang. Terr.: 225 (1949). Type from Tanzania (Southern Province).

Leaf segments persistently densely fulvous stellate-tomentose on lower surface.

Mozambique. N: 3 km Mueda–Nantulo, st. 15.iv.1964, *Torre & Paiva* 12043 (LISC).

Also in Kenya and Tanzania. In open *Parinari* woodland on red clay soil; 800 m.

The seeds are edible.

Ricinodendron heudelotii subsp. *heudelotii* occurs in West Africa from Guinea Bissau to Ghana. It differs from subsp. *africanum* in having fewer leaf segments per leaf (3–5 as opposed to (3)5–7(8)), 3-locular ovaries and (2)3-locular fruits as opposed to 2(3)-locular ones.

48. SCHINZIOPHYTON Hutch. ex Radcl.-Sm.

Schinziophyton Hutch. ex Radcl.-Sm. in Kew Bull. **45**: 157 (1990).
Ricinodendron sensu Schinz in Bull. Herb. Boissier **6**: 744 (1898), pro parte non Müll. Arg. (1864).

Dioecious pachycaul trees with a persistent stellate indumentum. Leaves alternate, stipulate, long-petiolate, digitately compound, (3)5–7-foliolate; petioles glanduliferous; leaflets petiolulate, subentire to minutely and remotely glandular-denticulate, penninerved. Stipules small, fugacious, stipitate, cuneate-flabelliform, glanduliferous. Inflorescences terminal or subterminal, paniculate, the male larger than the female; bracts setaceous or subulate. Male flowers shortly pedicellate; buds subglobose; calyx lobes 5, imbricate; petals 5, imbricate, laterally coherent; disk glands 5, free, extrastaminal, usually 3-fid or 3-lobed; stamens 13–21, united at the base, filaments glabrous above, hairy below, anthers dorsifixed, introrse, longitudinally dehiscent; pistillode absent. Female flowers pedicellate; calyx lobes 5, strongly imbricate; petals more or less as in the male flowers, but more lightly coherent; hypogynous disk shallowly cupular, 5-lobed; ovary 1(2)-locular, laterally compressed, with 1 ovule per loculus; styles 1(2), bifid or bipartite, stigmas verruculose. Fruits indehiscent, ovoid-ellipsoid, usually longer than wide; pericarp coriaceous; mesocarp fleshy; endocarp single, woody, massive, 1(2)-locular, 1(2)-seeded, punctate-foveolate. Seeds ecarunculate, ridged; albumen copious, fleshy, oily; cotyledons broad, flat, palminerved.

A monotypic tropical African genus, largely confined to the Cunene-Cubango-Cuando-Zambezi watersheds.

Schinziophyton rautanenii (Schinz) Radcl.-Sm. in Kew Bull. **45**: 157 (1990). Tab. **60**. Syntypes from Namibia (Ovamboland).

Ricinodendron rautanenii Schinz in Bull. Herb. Boissier **6**: 744 (1898). —Pax in Engler, Pflanzenr. IV. 147, iii] **47**: 48, t. 17 (1911). —Hutchinson in F.T.A. **6**, 1: 746 (1912). —Eyles in Trans. Roy. Soc. South Africa **5**: 398 (1916). —Engler, Pflanzenw. Afrikas (Veg. Erde 9) **3**, 2: 132 (1921). —De Wildeman, Pl. Bequaert. **3**, 4: 505 (1926). —O.B. Miller, Check-list For. Trees Shrubs Bech. Prot.: 33 (1948). —Topham, Check List For. Trees Shrubs Nyasaland Prot.: 53 (1958). —J. Léonard in F.C.B. **8**, 1: 119 (1962). —White, F.F.N.R.: 203 (1962). —P.G. Meyer in Merxmüller, Prodr. Fl. SW. Afrika, fam. 67: 42 (1967). —Drummond in Kirkia **10**: 253 (1975). —K. Coates Palgrave, Trees Southern Africa, ed. 2, rev.: 432 (1983). —Peters in Econ. Bot. **41**, 4: 494–502 (1987). —Radcliffe-Smith in F.T.E.A., Euphorb. 1: 328 (1987). Syntypes as above.

Ricinodendron viticoides Mildbr. in Notizbl. Bot. Gart. Berlin-Dahlem **12**: 517 (1935). —Brenan, Check-list For. Trees Shrubs Tang. Terr.: 225 (1949). Type from Tanzania (Southern Province).

A shrub, or tree up to 20 m tall, with a rounded or spreading crown, and with a trunk up to c. 1 m d.b.h. Bark pale grey, whitish or light brown, smooth at first, later becoming reticulate and flaking, and grey-green beneath. Wood white, very soft. Twigs thick, exuding white gum; twigs, petioles and inflorescence axes ferrugineous stellate-pubescent at first, later scurfily stellate-pulverulent. Leaves digitately compound, (3)5–7-foliolate. Stipules 3–5 × 2–3 mm. Petioles 6–25 cm long; glands 2–4, usually at the petiole apex, prominent, green; petiolules 0.5–1.5 cm long. Leaflets elliptic-ovate to oblanceolate, rarely 3-lobed; median leaflet 5–18 × 2–9 cm, the lateral leaflets slightly smaller; all leaflets obtuse or acute and shortly acuminate at the apex, usually somewhat asymmetrically rounded-cuneate at the base, entire or subentire, edged with marginal dark green gland dots on upper or lower surfaces, densely ferrugineous or fulvous stellate-pubescent on upper surface, later glabrescent, paler or whitish stellate-tomentose beneath; lateral nerves in 6–16 pairs. Male inflorescences 10–22 × 4–8 cm; bracts 3–10 mm long. Male flowers: pedicels 2–5 mm long; calyx lobes 5 × 2.5–3 mm, elliptic-oblong, obtuse, stellate-pubescent without and within; petals 6–7 × 2–3 mm, elliptic-oblong, emarginate-subtruncate at the apex, rounded at the base, glabrous except within at the base, pale lemon-yellow to whitish, drying dark brown; disk glands c. 1 × 1 mm; stamens c. 7 mm long, anthers 0.75 × 0.5 mm. Female inflorescences 5–6 × 2–3 cm; bracts as in the male. Female flowers: pedicels 7–10 mm, stouter than in the male flowers; calyx lobes 8–9 × 5–6 mm, broadly ovate, subacute or obtuse, indumented as in the male; petals 9 × 4 mm, glabrous, otherwise more or less as in the male; disk 4 mm in diameter; ovary 7 × 5 × 3 mm, densely stellate-pubescent; styles 5 mm long. Fruit 3–5 × 2–3.5 cm when dried, up to 7 × 5 cm when fresh, densely stellate-tomentose at first, later glabrescent, green, becoming pale yellow; exopcarp thin, mesocarp 3–4 mm thick, endocarp 4–5 mm thick. Seeds 1.8–2.5 × 1.6–2 cm, compressed-ellipsoid.

Caprivi Strip. 13 km east of Nyangana Mission Station, fr. 9.i.1956, *de Winter* 4198 (K; PRE). **Botswana**. N: 7 km SW of Hyaena Camp on Linyanti River, female fl. 26.x.1972, *Biegel, Pope & Gibbs Russell* 4077 (K; LISC; MO; PRE; SRGH). SE: Orapa–Francistown, st. 30.iv.1976, *Allen* 391 (PRE). **Zambia**. B: Mongu, Kande Lake, male fl. 11.xi.1959, *Drummond & Cookson* 6358 (K; LISC; MO; PRE; SRGH). N: Mpika Distr., South Luangwa Game Reserve, near Katete-Luangwa confluence, st. 5.v.1965, *B.L. Mitchell* 2842 (K; SRGH). C: Kabwe (Broken Hill), male fl. xi.1909, *Rogers* 8638 (K). E: Petauke, fr. immat. 3.xii.1958, *Robson* 816 (BM; K; LISC; PRE; SRGH). S: Mapanza Mission, male fl. 11.xi.1954, *Robinson* 946 (K). **Zimbabwe**. N: Sebungwe Distr., Chicomba Vlei, st. 17.xii.1951, *Lovemore* 223 in *GHS* 35394 (SRGH). W: Hwange Distr., Victoria Falls, male fl. 24.xi.1949, *Wild* 3206 in *GHS* 26359 (K; LISC; PRE; SRGH). C: Lower Gweru (Gwelo) Reserve, st. vi.1962, *Cleghorn* in *GHS* 147206 (K; SRGH). **Malawi**. N: Rumphi Distr.,

Tab. 60. SCHINZIOPHYTON RAUTANENII. 1, distal portion of flowering branch (× $^2/_3$); 2, part of male inflorescence (× 3); 3, male flower, with 2 petals removed (× 3), 1–3 from *Langman* 12; 4, female flower (× 3); 5, female flower, with 3 petals removed (× 2), 4 & 5 from *Biegel et al.* 4077; 6, fruit, with part of pericarp removed (× $^2/_3$), from *Michelmore* 6584. Drawn by Christine Grey-Wilson.

south of Njakwa, st. 4.v.1952, *White* 2844 (FHO; K). S: 20 m from the main Blantyre Road between Namikango Mission and Tondwe, c. 7 m from Zomba, fr. 11.i.1982, *J.D. Chapman* 6087 (FHO; K; MAL; SRGH). **Mozambique**. N: Monapo Distr., 5 km Itaculo-Régulo Chihir, male fl. 2.xii.1963, *Torre & Paiva* 9375 (LISC). Z: 27.2 km Derre–Morrumbala, st. 12.vi.1949, *Barbosa & Carvalho* in *Barbosa* 3047 (K; SRGH). MS: between Serracão Braunstein and Rio Nhamouare, fr. 22.i.1948, *Mendonça* 3679 (LISC).

Also in Zaire (Shaba Province), Tanzania (Eastern and Southern Provinces), Angola (Huíla Province), Namibia (Ovamboland and Hereroland) and South Africa (NW Transvaal). At low to medium altitudes in sandy soil, in well developed deciduous woodland on Kalahari Sand with *Baikiaea, Guibourtia Afzelia* and *Brachystegia,* in short grassland with scattered trees of *Combretum, Terminalia, Burkea* and *Pterocarpus,* and in mopane woodland on Kalahari Sand, also in wooded hills and amongst sand dunes and sandy alluvium beside rivers, sometimes forming pure stands; 50–1220 m.

The Manketti or Mongongo Nut. The seed is hard shelled, and the kernel has a high oil content and is edible.

Vernacular names as recorded in specimen data include: "mgongo", "mugongo", "mungongo" (Barotse area); "mkanga ula" (chiYao); "mkomwa" (chiNyanja); "muhuwi" (chiTumbuka); "mukusu" (chiBwile, chiWemba); "ngoma", "umganumapoba" (siNdebele).

49. ALEURITES J.R. & G. Forst.

Aleurites J.R. & G. Forst., Char. Gen. Pl.: 111, t. 56 (1775).
Camirium Rumph. ex Gaertn., Fruct. Sem. Pl. **2**: 194, t. 125, ii (1791).

Monoecious trees. Indumentum stellate. Leaves alternate, stipulate, long-petiolate, simple, entire or palmately lobed, palminerved. Stipules minute, fugacious. Inflorescences terminal, thyrsiform, androgynous, protogynous, some cymules with terminal female flowers, others all male; bracts subulate, readily caducous. Male flowers: calyx closed in bud, valvately rupturing into 2–3 lobes; petals 5, free, imbricate; disk glands 5, free, alternating with the petals; stamens c. 20, 4-verticillate, the outer free, the inner united into a column, anthers basifixed, introrse, longitudinally dehiscent; receptacle conical; pistillode absent. Female flowers: calyx calyptrate, 2-lobed, splitting down one side; petals resembling those of the male flowers; disk shallowly 5-lobed; ovary 2-locular, with 1 ovule per loculus; styles 2, bipartite, erect. Fruit usually 2-lobed, drupaceous, indehiscent; exocarp fleshy; endocarp thinly woody, 2-locular or 1-locular by abortion. Seeds broadly ovoid, ecarunculate; testa thick, woody; albumen hard; embryo straight; cotyledons broad, flat.

A genus of two species: one a widespread Indopacific species, cultivated in tropical Africa; the other endemic to Hawaii.

Aleurites moluccana (L.) Willd., Sp. Pl. **4**: 590 (1805). —Müller Argoviensis in De Candolle, Prodr. **15**, 2: 723 (1866). —Pax in Engler, Pflanzenr. [IV, fam. 147, i] **42**: 129, t. 45 (1910); Pflanzenw. Afrikas (Veg. Erde 9) **3**, 2: 60, t. 23 (1921). —Brenan, Check-list For. Trees Shrubs Tang. Terr.: 199 (1949). —Keay in F.W.T.A., ed. 2, **1**, 2: 368 (1958), in adnot. —J. Léonard in F.C.B. **8**, 1: 3 (1962). —White, F.F.N.R.: 193 (1962). —Airy Shaw in Kew Bull. **20**: 393 (1967). —Biegel, Check List Ornam. Pl. Rhod. Parks & Gard.: 20 (1977). —Troupin, Fl. Rwanda **2**: 204, fig. 60/1 (1983). —Radcliffe-Smith in F.T.E.A., Euphorb. 1: 176 (1987). Tab. **61**. Type from Sri Lanka.

Jatropha moluccana L., Sp. Pl.: 1006 (1753).

Aleurites triloba J.R. & G. Forst., Char. Gen. Pl.: 112, t. 56 (1775). —Hutchinson in F.T.A. **6**, 1: 814 (1912). Type from Tonga.

Camirium cordifolium Gaertn., Fruct. Sem. Pl. **2** : 195, t. 125, ii (1791). Type from SW Pacific.

A tree up to 10 m tall with a rounded crown. Bark smooth, grey. Young shoots, petioles and inflorescence axes densely scurfily fulvous to ferrugineous stellate-tomentose. Stipules minute, subulate. Petioles 6–22 cm long. Leaf blades 7–24 × 4–20 cm, ovate-lanceolate to elliptic-lanceolate or ovate-trullate, entire or up to 5-lobed, apex and lobes subacute to acutely acuminate, entire or shallowly repand-dentate on the margins, cuneate to truncate or shallowly cordate at the base, with 2 discoid, sessile, contiguous, shiny glands adaxially at the base, 3–5(7)-nerved from the base, densely cinereous-, fulvous- or ferrugineous-stellate-tomentose on both surfaces at first, later glabrescent. Inflorescences 10–16 × 10–14 cm, broadly conical, branching from the

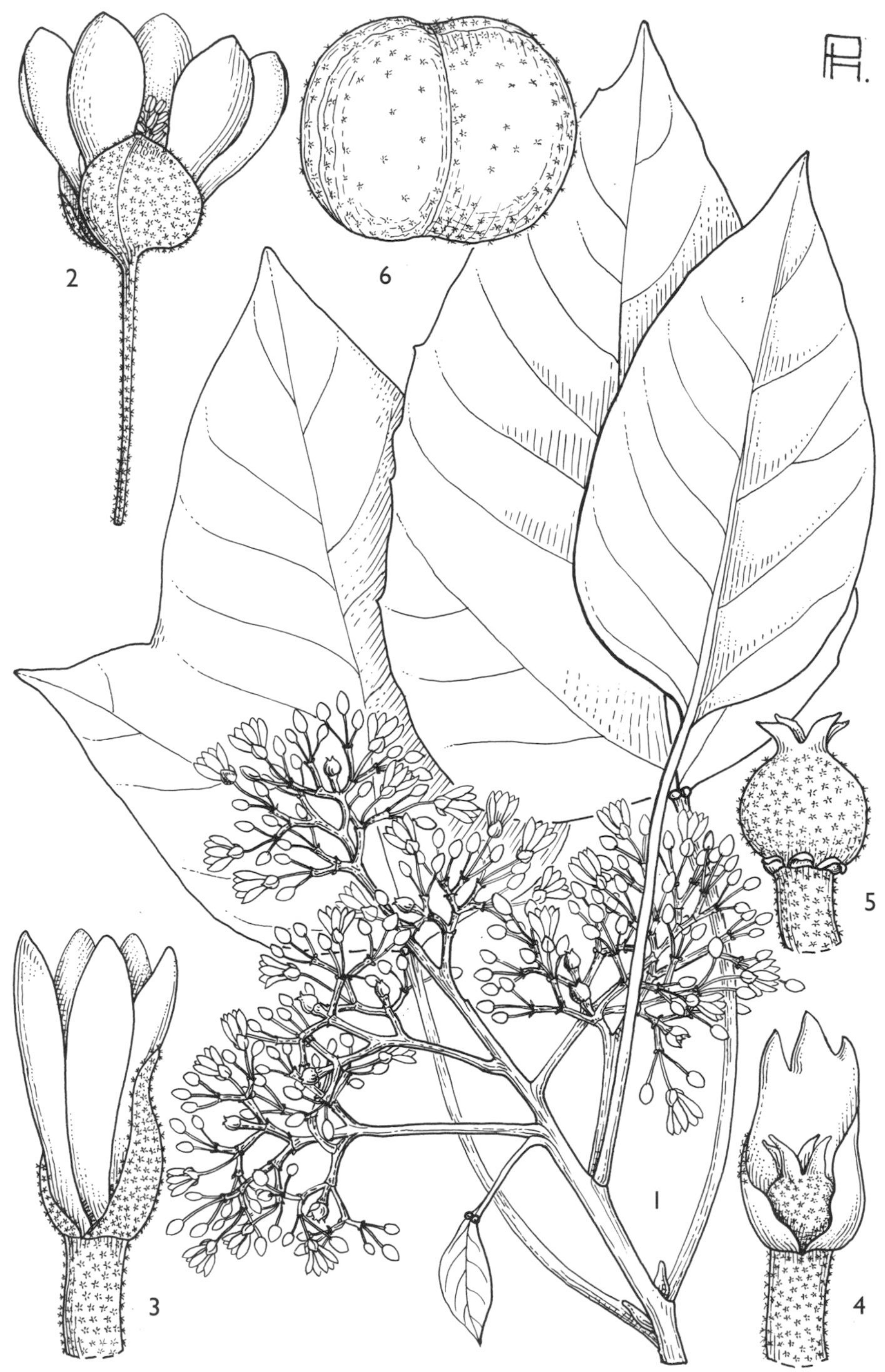

Tab. 61. ALEURITES MOLUCCANA. 1, distal part of flowering branch (× $^{2}/_{3}$), from *Semsei* 1929 and *Thairu* 3; 2, male flower (× 6); 3, female flower (× 6); 4, young fruit (× 6); 5, developing fruit (× 6), 2–5 from *Semsei* 1929; 6, mature fruit (× $^{2}/_{3}$), from *Greenway* 2276. Drawn by Pat Halliday. From F.T.E.A.

base; bracts 2–3 mm long. Male flowers: pedicels 7–10 mm long, slender; buds 2 mm long, ovoid; calyx lobes 3–3.5 × 2–3 mm, ovate, acute, stellate-tomentose without, glabrous within; petals 6–8 × 2 mm, narrowly oblong-spathulate, obtuse, glabrous, yellowish-white; disk glands 0.5 × 0.5 mm, shallowly 3-lobed, flattened, whitish; staminal column 4 mm high, filaments c. 1 mm long, green, puberulous with simple hairs, anthers 0.5 mm long and wide, yellowish; receptacle puberulous. Female flowers: pedicels 3–4.5 mm long, stout, dilated upwards; buds 4 mm long, conical; petals 1.5 mm wide, otherwise as in male flowers; disk c. 2 mm across, flattened; ovary c. 1.5 × 2 mm, subglobose, densely appressed-stellate-tomentose; styles c. 1 mm long, glabrous except at the base without. Fruit 4 × 4–5.5 cm, ovoid-subglobose or transversely-ovoid, shallowly 2-lobed and with 4 low longitudinal ridges, evenly to sparingly appressed- or scurfily stellate-pubescent, green. Seeds 2.5 × 2.75 × 2.25 cm, broadly ovoid, shallowly rugulose, brown mottled cream-coloured or whitish.

Zimbabwe. W: Bulawayo, (cultivated), fl. xi.1955, *Hodgson* 13/55 (K; PRE; SRGH). E: Mutare (Umtali), (cultivated), fl. & fr. x.1948, *Chase* 1312 (K; SRGH). **Malawi**. S: Zomba, Chancellor College (cultivated), fl. & fr. 2.i.1981, *Chapman* 5509 (K; MAL). **Mozambique**. Z: Morrumbala, Massingir, fl. 26.x.1945, *Pedro* 462 (LMA; PRE). GI: Mambone, (cultivated), fl. 24.x.1906, *Johnson* 14 (K). M: Maputo (Lourenço Marques), (cultivated), fl. ii.1946, *Pimenta* 6004 (LISC).

Native of tropical Asia and Oceania from India and China to Polynesia and New Zealand; widely cultivated in the tropics generally. Frequently planted as a street tree in towns, and by roadsides; sea level to 1130 m.

The Candlenut Tree.

50. VERNICIA Lour.

Vernicia Lour., Fl. Cochinch.: 586 (1790).
Aleurites sect. *Dryandra* (Thunb.) Müll. Arg. in DC., Prodr. **15**, 2: 723 (1866).
Dryandra Thunb., Fl. Jap.: 13, t. 27 (1784), non R. Br., (1810), nom. conserv.
Ambinux Comm. ex Juss., Gen. Pl.: 389 (1789).
Elaeococca A. Juss., Euph. Gen.: 38, t. 11 (1824).

Monoecious or subdioecious trees. Indumentum simple or subsimple. Leaves alternate, stipulate, long-petiolate, simple, entire or palmately lobed, palminerved. Stipules soon falling. Inflorescences terminal, thyrsiform, androgynous and protogynous with a single female flower and the rest male, or unisexual; bracts lanceolate, the lower bracts sometimes persistent, the upper soon falling. Flowers large. Male flowers: calyx closed in bud, valvately rupturing into 2–3 lobes; petals 5, free, contorted; disk glands 5, free, alternating with the petals; stamens 8–14, 2(3)-verticillate, the outer free or fused at the base, the inner fused for over half their length into a column, anthers subdorsifixed, introrse, longitudinally dehiscent, with a broad connective; pistillode absent. Female flowers: calyx, petals and disk more or less as in the male flowers; ovary 3(5)-locular, with 1 ovule per loculus; styles erect, 3(5), united at the base, bifid or bipartite. Fruits shallowly 3(5)-lobed, tardily dehiscent; exocarp fibrous; endocarp thinly woody or crustaceous. Seeds ovoid-trigonous, ecarunculate; testa thick, woody; albumen hard; embryo straight; cotyledons broad, flat.

A genus of 3 very closely related species from E & SE Asia, two of which are widely cultivated in Africa.

Indumentum crisped-ferrugineous; glands at base of leaf blade turbinate, divaricately-spreading; inflorescences usually unisexual; flowers appearing with the leaves (coaetaneous); petals 1.5–2.5 cm long; ovary densely tomentose; fruits ovoid-subglobose, ridged and reticulate · 1. *montana*
Indumentum appressed golden-sericeous; glands at base of leaf blade discoid, sessile; inflorescences usually bisexual; flowers precocious; petals c. 3 cm long; ovary sparingly pubescent; fruits ± spherical, smooth or faintly lineate · · · · · · · · · · 2. *fordii*

1. **Vernicia montana** Lour., Fl. Cochinch.: 586 (1790). —Airy Shaw in Kew Bull. **20**: 394 (1966). —Radcliffe-Smith in F.T.E.A., Euphorb. 1: 178 (1987). Tab. **62**, figs. A1–A4. Type from S Vietnam.

Aleurites montana (Lour.) Wilson in Bull. Imp. Inst. **11**: 460 (1913). —Pax & K. Hoffmann in Engler, Pflanzenr. [IV, fam. 147, xiv, Addit. vi] **68**: 8 (1919). —Brenan, Check-list For.

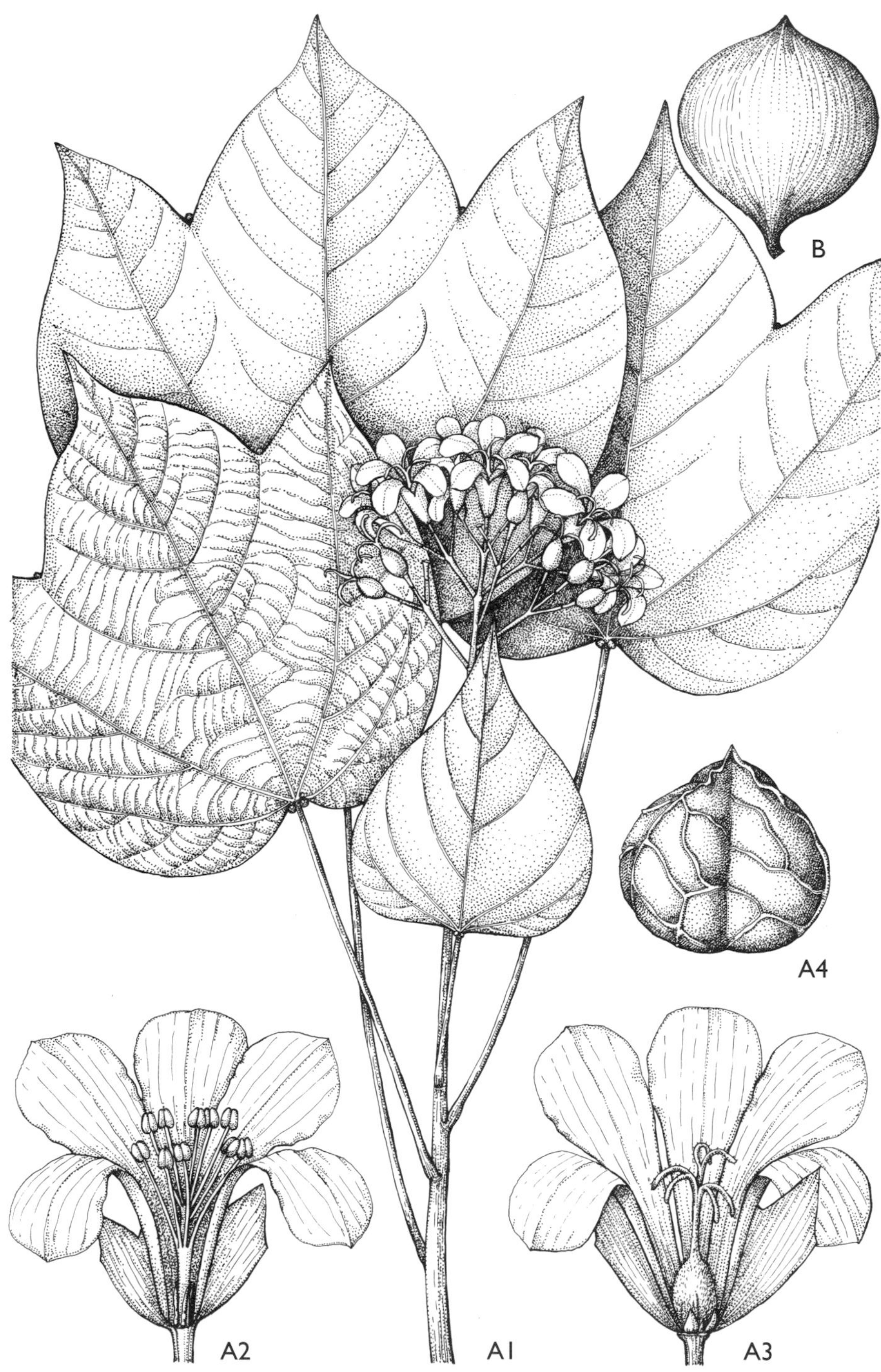

Tab. 62. A —VERNICIA MONTANA. A1, distal part of flowering branch (× 2/3); A2, male flower, partly opened to show glands (× 1½), A1 & A2 from *Semsei* 2367; A3, female flower, partly opened to show glands (× 1½), from *Greenway* 3673; A4, fruit (× 2/3), from *Richards* s.n. B —VERNICIA FORDII, fruit (× 2/3), from *Wilson* 2031. Drawn by Judy Dunkley. From F.T.E.A.

Trees Shrubs Tang. Terr.: 199 (1949). —White, F.F.N.R.: 193 (1962). —Biegel, Check List Ornam. Pl. Rhod. Parks & Gard.: 20 (1977).

Aleurites cordata sensu Müll. Arg. in De Candolle, Prodr. **15**, 2: 724 (1866) pro parte excl. specim. Jap. —Pax in Engler, Pflanzenr. [IV, fam. 147, i] **42**: 132 (1910). —Hutchinson in F.T.A. **6**, 1: 813 (1912). —Engler, Pflanzenw. Afrikas (Veg. Erde 9) **3**, 2: 61 (1921). —De Wildeman, Pl. Bequaert. **3**, 4: 502 (1926), non Thunb.

A tree up to 15 m tall with a large crown, subdioecious. Bark smooth or roughened, grey. Wood soft. Young shoots, base of leaves, inflorescence branches and young fruits crisped ferrugineous-pubescent. Stipules 2–4 × 1 mm, lanceolate, ciliate. Petioles 1–25 cm long, ± glabrous. Leaf blades 4–20 × 4–18 cm, broadly ovate, entire or 3-lobed, leaf blade and lobes acutely acuminate at the apex, entire on the margins, truncate to cordate at the base, usually with 2 turbinate glands 1–3 mm long adaxially at the base, and with one gland in each sinus of the lobed leaves, 3–5-nerved from the base, glabrescent except at the base. Inflorescences usually unisexual. Male inflorescences 15 × 15–20 cm, broadly corymbiform, branching from the base; lower bracts 5–6 × 2–3 cm, foliaceous, elliptic-ovate to linear-lanceolate, persistent, progressively smaller and less persistent upwards. Flowers appearing with the leaves (coaetaneous). Male flowers: pedicels up to 1 cm long, subglabrous; buds 1–1.3 × 0.5 cm, ovoid-ellipsoid, apiculate, rupturing for half to two-thirds of their length into 2(3) subequal or unequal lobes, glabrous; petals 1.5–2.5 × 0.7 cm, oblanceolate-spathulate, unguiculate, rounded, pubescent within on the claw, otherwise glabrous, white; disk glands 3–4 mm long, erect, subulate, sometimes bifid, fleshy, whitish; staminal column 20 mm high, the outer stamens c. 15 mm long, filaments pubescent at the base, otherwise glabrous, anthers c. 2 × 1 mm. Female inflorescences resembling the male, but often smaller. Female flowers: pedicels extending to 15 mm long in fruit; calyx and petals as in the male flowers; disk glands 1–2 mm long, triangular-lanceolate, sometimes bifid; ovary 5 × 3 mm, ovoid-ellipsoid, trigonous, narrowing into the styles, densely ferrugineous-tomentose; styles c. 8 mm long, pubescent at the base, stigmas glabrous. Fruit c. 3.5 × 4 cm, ovoid-subglobose, apiculate, 3(5)-ridged, laxly reticulate, glabrescent. Seeds 2–2.5 × 2–2.5 cm, broadly compressed-ovoid, ± smooth, brownish.

Zambia. N: near Mbala, (cultivated), female fl. v.1969, *Richards* s.n. (K). W: Cherry's Farm, near Ndola, (cultivated), female fl. 27.ix.1963, *Angus* 3769 (FHO; K). C: Lusaka For. Nursery, (cultivated), fl. 5.iii.1952, *White* 2202 (FHO). E: Chadiza, (cultivated), fr. 11.ii.1957, *Angus* 1511 (FHO). **Zimbabwe**. N: Goromonzi, c. 30 km NE of Harare, (cultivated), fr. 22.xi.1970, *Biegel* 3422 (K; SRGH). **Malawi**. N: Mzuzu Boma, (cultivated), fl. 4.ii.1968, *Salubeni* 1193 (MAL). S: Thyolo Distr., Mswika Estate, (cultivated), male fl. 20.x.1941, *Greenway* 6342 (K; PRE). **Mozambique**. Z: Milange, (cultivated), male fl. 22.i.1966, *Torre & Correia* 14049 (LISC). MS: Chimoio, (cultivated), male fl. & y. fr. 20.xi.1906, *W.H. Johnson* 36 (K). M: Namaacha, (cultivated), st. 10.i.1948, *Torre* 7092 (LISC).

Native of south China, Myanmar (Burma), Indochina and Thailand. Widely cultivated in tropical Africa, originally as an oil tree, but also as an ornamental, sometimes becoming naturalized; 700–1650 m.

The Chinese Wood-Oil Tree.

2. **Vernicia fordii** (Hemsl.) Airy Shaw in Kew Bull. **20**: 394 (1966). —Radcliffe-Smith in F.T.E.A., Euphorb. 1: 179 (1987). Tab. **62**, fig. B. Type from China.

Aleurites fordii Hemsl. in Hooker's Icon. Pl. 29, tt. 2801–2 (1906); in Bull. Misc. Inform., Kew **1906**: 120 (1906); **1914**: 3 (1914). —Pax in Engler, Pflanzenr. [IV, fam. 147, i] **42**: 132 (1910). —Brenan, Check-list For. Trees Shrubs Tang. Terr.: 199 (1949). —Biegel, Check List Ornam. Pl. Rhod. Parks & Gard.: 20 (1977).

Very like *V. montana*, but differing in having an appressed golden-sericeous indumentum, sessile discoid glands at the base of the leaf blade, usually bisexual inflorescences, precocious flowers, larger petals (c. 3 cm long), a sparingly-pubescent ovary and smooth or faintly lineate ± spherical fruits.

Zimbabwe. E: Mutare (Umtali), (cultivated), fl. 15.ix.1964, *Loveridge* 1099 (FHO; K; LISC; SRGH). **Malawi**. S: Thyolo (Cholo) Road, 1–4 m beyond Chigumula, fl. 10.ix.1966, *Binns* 356 (MAL). **Mozambique**. M: Namaacha, (cultivated), st. 10.i.1948, *Torre* 7093 (LISC).

Native of south China, south Myanmar (Burma) and north Vietnam. Widely cultivated in the tropics generally. Originally in plantations as an oil tree, but also as an ornamental, sometimes becoming naturalized; 1060 m.

The Tung Oil Tree.

51. CAVACOA J. Léonard

Cavacoa J. Léonard in Bull. Jard. Bot. État **25**, 4: 320 (1955).
Grossera sect. *Racemiformes* Pax & K. Hoffm. in Engler, Pflanzenr. [IV, 147, vi] **57**: 108 (1912).
Grossera subgen. *Eugrossera* Cavaco in Bull. Mus. Natl. Hist. Nat. (Paris), 2ᵉ Sér. **21**: 274 (1949), pro parte quoad sect. *Racemiformes.*
Grossera subgen. *Quadriloculastrum* Cavaco in Bull. Mus. Natl. Hist. Nat. (Paris), 2ᵉ Sér. **21**: 274 (1949).

Dioecious shrubs or trees. Indumentum simple. Leaves alternate, petiolate, stipulate, simple, entire, penninerved; nerves brochidodromous. Stipules soon falling, leaving conspicuous subannular scars. Inflorescences terminal, racemose, strobiliform when young; bracts broad, soon falling. Male flowers: pedicels jointed; calyx closed in bud, later splitting into 2(4) unequal valvate lobes; petals 4–5, free, imbricate; disk glands 4–5(6), free, alternating with the petals, fleshy; stamens 15–35, erect in bud, borne on a convex to columnar receptacle, anthers basifixed, connective broad, thick, thecae soleiform, longitudinally dehiscent; pistillode absent. Female flowers: pedicels stout, jointed; sepals (4)5, free, imbricate; petals 5, free, imbricate; disk hypogynous, cupuliform, fleshy; ovary 3–5-locular, with 1 ovule per loculus; styles 3–5, connate at the base, bifid or trifid. Fruits 3-lobed to 5-lobed, dehiscing loculicidally or septicidally into 3–5 bivalved cocci; pericarp thin, crustaceous; endocarp thick, woody; columella large, persistent. Seeds ecarunculate.

A genus of 3 species in tropical and extratropical Africa.

Cavacoa aurea (Cavaco) J. Léonard in Bull. Jard. Bot. État **25**, 4: 323 (1955). —K. Coates Palgrave, Trees Southern Africa, ed. 2, rev.: 420 (1983). —Radcliffe-Smith in F.T.E.A., Euphorb. 1: 174 (1987). —Beentje, Kenya Trees, Shrubs Lianas: 188 (1994). Tab. **63**. Type: Mozambique, near Salamanga, xii.1947, *Cavaco* 123 (P, holotype).
Grossera aurea Cavaco in Bull. Mus. Natl. Hist. Nat. Paris, 2ᵉ Sér., **21**: 274, 2 (1949).

A shrub or small tree up to 13 m tall, glabrous or subglabrous with a dense crown and regular branching. Trunk irregularly fluted below. Bark thin, scaly, brown. Branches long. Twigs greenish at first, later becoming greyish. Petioles 1–5 cm long. Leaf blades 3.5–18 × 1.5–8 cm, elliptic to elliptic-obovate or elliptic-oblanceolate, obtuse to acutely acuminate at the apex, cuneate or rounded-cuneate at the base, thinly coriaceous; lateral nerves in 9–12 pairs, slightly prominent, tertiary nerves reticulate. Stipules 5 mm long, oblong, subacute. Male inflorescences up to 10 cm long, and up to c. 10-flowered; bracts 4–6 × 2–4 mm; bracteoles 1.5 mm long. Male flowers fragrant; pedicels 1–2.5 cm long; buds subglobose; calyx lobes 3.5–6 × 1–4 mm, lanceolate to broadly ovate, acute or subacute; petals 7–9 × 4–5 mm, elliptic-ovate, obtuse, cream-coloured, bright lemon-yellow or greenish-yellow; disk glands 1 mm across, transversely ovate to reniform; receptacle 4 mm high; stamens 5–6 mm long, anthers 0.5 mm long. Female inflorescences up to 5 cm long, and up to 6-flowered; bracts and bracteoles ± as in the male inflorescence. Female flowers aromatic; pedicels 1–1.7 cm long, extending slightly in fruit; buds conical; sepals 6–8 × 3–4 mm, oblong-lanceolate, subacute; petals 10 × 3–4 mm, elliptic-oblong, otherwise as in the male flowers; disk 4 mm in diameter; ovary 2–3 × 3–4 mm, 3–5-lobed; styles 3–5 mm long. Fruit 1.2–1.5 × 2.3–3 cm. Columella 1.3 cm long. Seeds 10 × 8–9 mm, ovoid-subglobose, smooth, light brown, streaked and mottled darker brown.

Malawi. S: Malawi Hills, fr. 4.i.1964, *Chapman* 2185 (FHO). **Mozambique**. Z: Serra Morrumbala, st. 12.xii.1971, *Müller & Pope* 2015 (SRGH). MS: Missão Catolica de Amatongas, male fl. & fr. immat. 6.xi.1941, *Torre* 3783 (LISC). M: Matutuíne Distr., Salamanga, R. Futi, male fl. 4.xii.1968, *Balsinhas* 1432 (LISC).

Also in Kenya and South Africa (KwaZulu-Natal). In dense, low altitude mixed evergreen forest and coastal forest, often beside streams, and in sandy soil; 500–850 m.

The Mozambique material seen thus far lacks mature fruits. The descriptions of fruit and seed have therefore been made with the help of KwaZulu-Natal material.

The comment made by E.J. Mendes on the LISC material of *Mendonça* 3543 that this number was the source of *Cavaco* 123 is somewhat misleading. Possibly the latter gathering was made from the same tree.

Tab. 63. CAVACOA AUREA. 1, distal part of flowering branch (× $\frac{2}{3}$); 2, male flower (× $2\frac{2}{3}$); 3, sepals (× 6), 1–3 from *Wakefield* s.n.; 4, female flower (× $2\frac{2}{3}$); 5, petals (× 6); 6, ovary (× 6), 4–6 from *Ross & Moll* 2194; 7, 5-lobed fruit (× $\frac{2}{3}$); 8, 4-lobed fruit (× $\frac{2}{3}$), 7 & 8 from *Ross & Moll* 2293; 9, 3-lobed fruit (× $\frac{2}{3}$); 10, mericarp (× $\frac{2}{3}$); 11, seed (× $2\frac{1}{3}$), 9–11 from *Watmough* 442. Drawn by Mary Millar Watt. From F.T.E.A.

52. TANNODIA Baill.

Tannodia Baill., Adansonia **1**: 251, t. 7, f. 1–2 (1861).

Dioecious or polygamo-dioecious (with an occasional male flower on a female inflorescence) trees or shrubs. Indumentum simple. Leaves alternate, petiolate, stipulate, simple, entire or subentire, palminerved. Inflorescences terminal, interruptedly racemose; bracts 1–several-flowered. Male flowers: pedicels jointed; calyx closed in bud, later splitting into 2–5 valvate lobes; petals 4–5, free, imbricate; disk glands 4–5, free, alternating with the petals; stamens 6–14, connate at the base into a short column, 2-seriate with the outer series short and opposite the petals and the inner longer and opposite the sepals, anthers dorsifixed, the outer introrse, the inner extrorse, longitudinally dehiscent, connective broad; pistillode absent. Female flowers: pedicels jointed; sepals 4–5, ± free, ± equal, imbricate; petals 4–5, larger than the sepals, free, imbricate; disk shallowly cupular; ovary 3-locular, with 1 ovule per loculus; styles 3, connate at the base, bifid. Fruits trilobate-subglobose, dehiscing septicidally into 3 bivalved cocci; endocarp thinly woody. Seeds ecarunculate; testa crustaceous; albumen fleshy; cotyledons broad, flat.

A small genus of 4 species, 2 African and 2 from Madagascar and the Comoro Islands.

Tannodia swynnertonii (S. Moore) Prain in J. Bot. **50**: 127 (1912); in F.T.A. **6**, 1: 827 (1912). —Pax in Engler, Pflanzenr. [IV, fam. 147, vi] **57**: 110 (1912). —Eyles in Trans. Roy. Soc. South Africa **5**: 395 (1916). —Engler, Pflanzenw. Afrikas (Veg. Erde 9) **3**, 2: 55 (1921). —Drummond in Kirkia **10**: 252 (1975). —K. Coates Palgrave, Trees Southern Africa, ed. 2, rev.: 421 (1983). —Radcliffe-Smith in F.T.E.A., Euphorb. 1: 172 (1987). Tab. **64**. Syntypes: Zimbabwe, Chirinda Forest, male fl. 24.x.1906, *Swynnerton* 109 (BM; K, chosen here as lectotype; SRGH); fl. 10.xi.1906, *Swynnerton* 109a (BM, female; K, male); fr. 20.i.1907, *Swynnerton* 109b (BM; K; SRGH); fl. & fr. ix.1904, *Swynnerton* 109c (BM, male; K, male; SRGH, female fl. & fr.); female fl. 6.xii.1908, *Swynnerton* 6519 (BM).
Croton swynnertonii S. Moore in J. Linn. Soc., Bot. **50**: 194 (1911).

A tree up to 30 m in height. Trunk buttressed, or slightly fluted at the base. Bark smooth, dark brown. Wood hard. Twigs, petioles and inflorescence axes glabrous. Petioles 0.5–3 cm long, pulvinate. Leaf blades 4–13 × 2–7 cm, elliptic-ovate, acuminate at the apex, entire or subentire on the margins, cuneate to rounded or ± truncate at the base, firmly chartaceous to thinly coriaceous, glabrous except for the domatial hair tufts on lower surface; main nerves 3–5 from the base, lateral nerves in 5 pairs. Stipules 1.5 mm long, lanceolate, soon falling. Male inflorescences up to 23 cm long; bracts resembling the stipules, but slightly larger. Male flowers: pedicels up to 4 mm long, glabrous; calyx lobes 2–3 × 2 mm, ovate, obtuse, pubescent without, glabrous within; petals 3–3.5 × 2–2.5 mm, obovate-suborbicular, rounded, glabrous without, pubescent within at the base, creamy-white; disk glands subglobose; stamens 4–5 mm long, anthers 1 mm long, white. Female inflorescences up to 19 cm long; bracts 1-flowered. Female flowers: pedicels 3–4 mm long, pubescent; sepals 1 mm long, triangular, subacute, pubescent without, glabrous within, green; petals 3–4 × 2.5 mm, suborbicular, rounded, glabrous, pale yellow or creamy-green; disk 1.5 mm in diameter; ovary 1.5–2 mm in diameter, densely fulvous-tomentose; styles 1–1.5 mm long. Fruit 1–1.5 × 0.9–1.4 cm, ± smooth or shallowly muricate, evenly pubescent. Seeds 9 × 6.5 mm, ovoid-ellipsoid, smooth, ± shiny, dark brown.

Zimbabwe. E: Lusitu R. Valley, below Glencoe For. Res., female fl. x.1966, *Goldsmith* 101/66 (FHO; K; LISC; MO; PRE; SRGH). **Mozambique**. MS: Mt. Gorongosa, st. 23.vii.1970, *Müller & Gordon* 1389 (K; LISC; PRE; SRGH).

Also in in Tanzania. Frequent subcanopy tree of mixed evergreen rainforest, and low altitude evergreen forest; 400–1220 m.

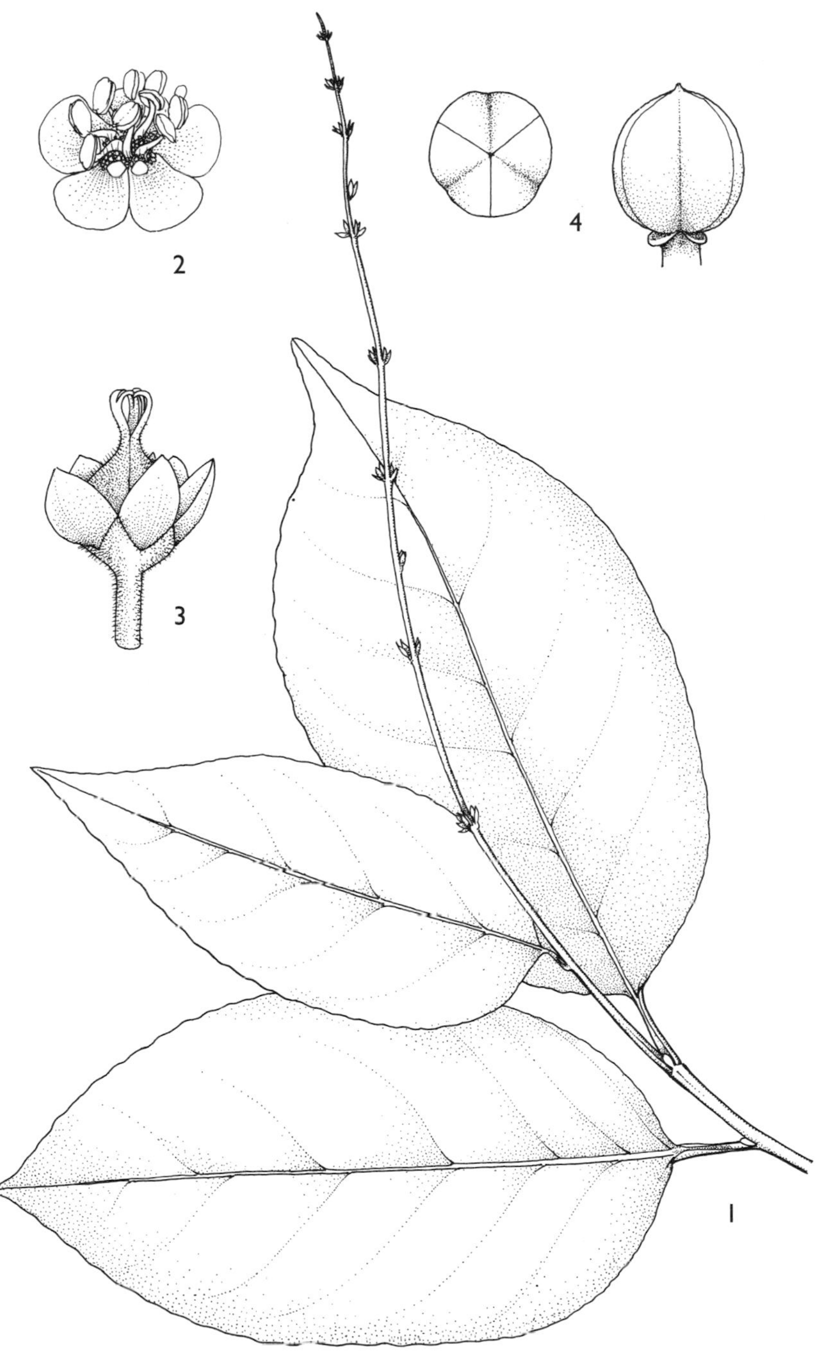

Tab. 64. TANNODIA SWYNNERTONII. 1, distal part of flowering branchlet (× $^{2}/_{3}$), from *Carmichael* 125; 2, male flower (× 4), from *Swynnerton* 109; 3, female flower (× 4), from *Wild* 2164; 4, fruit, two views (× 1), from *Wild* 2094. Drawn by G. Papadopoulos. From F.T.E.A.

53. NEOHOLSTIA Rauschert

Neoholstia Rauschert in Taxon **31**, 3: 559 (1982).
Holstia Pax in Bot. Jahrb. Syst. **43**: 220 (1909), non Hagstr. (1906).
Tannodia sect. *Holstia* (Pax) Prain in J. Bot. **50**: 127 (1912).

Dioecious, or rarely monoecious, shrubs or small trees. Indumentum simple. Buds perulate (furnished with protective scales). Leaves alternate, petiolate, stipulate, simple, subentire to dentate or lobed, palminerved. Inflorescences terminal, interruptedly racemose or subpaniculate; bracts small, 1–several-flowered. Male flowers: pedicels jointed; calyx closed in bud, later splitting into 2–5 valvate lobes; petals (4)5, short, free, imbricate; disk glands 5, free, alternating with the petals; stamens 8–13, connate at the base, anthers basifixed, introrse, thecae parallel, longitudinally dehiscent; pistillode absent. Female flowers: pedicels jointed; sepals 5, free, subequal, imbricate; petals 5, short, free, imbricate; disk shallowly cupular, lobulate; ovary 3-locular, with 1 ovule per loculus; styles 3, ± free, bipartite. Fruits 3-lobed, dehiscing septicidally into 3 bivalved cocci; endocarp thinly woody; columella more or less absent or not persistent. Seeds subecarunculate; testa crustaceous; albumen fleshy; embryo straight; cotyledons broad, flat.

A monotypic genus confined to south east tropical Africa.

Neoholstia tenuifolia (Pax) Rauschert in Taxon **31**: 559 (1982). —Radcliffe-Smith in F.T.E.A., Euphorb. 1: 169 (1987). —Beentje, Kenya Trees, Shrubs Lianas: 216 (1994). Tab. **65**. Type from Tanzania (Southern Highlands).

Holstia tenuifolia Pax in Bot. Jahrb. Syst. **43**: 220 (1909); in Engler, Pflanzenr. [IV, fam. 147, vi] **57**: 108 (1912). —Engler, Pflanzenw. Afrikas (Veg. Erde 9) **3**, 2: 55 (1921). —Brenan, Check-list For. Trees Shrubs Tang. Terr.: 215 (1949). —White, F.F.N.R.: 199 (1962). —Drummond in Kirkia **10**: 252 (1975).

Tannodia tenuifolia (Pax) Prain in J. Bot. **50**: 128 (1912); in F.T.A. **6**, 1: 827 (1912).

Var. **tenuifolia**

A many-stemmed scrambling shrub or liane up to 4 m in height, with pendent branches. Bark grey-brown, shallowly longitudinally fissured. Twigs brown, smooth. Young shoots pubescent. Perulae 2–4 × 1.5–3 mm, broadly ovate, chaffy, puberulous without, glabrous within, chestnut-brown. Petioles 0.5–6 cm long, pulvinate. Leaf blades 3–13 × 1.5–8 cm, broadly ovate to oblong-lanceolate, obtusely acuminate at the apex, entire or sometimes (in shade) coarsely 2–4(5)-lobed, cuneate to rounded or truncate to shallowly cordate at the base, sometimes obliquely so, membranous or thinly chartaceous, sparingly pubescent on both surfaces, and more densely so along the midrib and main nerves beneath, pale green; main nerves 3–5 from the base, lateral nerves in 4–7 pairs. Stipules 1 mm long, subulate, soon falling. Male inflorescences up to 20 cm long, but more usually 5–10 cm long; bracts c. 1–2 mm long, narrowly lanceolate. Male flowers: pedicels 3–10 mm long, capillary; calyx lobes 1.5–2 mm long, ovate, acute, pale greenish-cream coloured or white; petals 1 mm long, obovate, cream-coloured, yellow or reddish; disk glands cylindric-conic; stamens 2–3 mm long, anthers 0.75 mm long, cream-coloured. Female inflorescences up to 11 cm long, but more usually 2–7 cm long, fewer flowered than the male. Female flowers: pedicels 2–4 mm long, sometimes extending to 1 cm in fruit; sepals 2 × 0.5 mm, accrescent to 5.5 × 2 mm in fruit, oblanceolate-oblong, subacute, greenish; petals 0.8 mm long, ± obovate, whitish; disk 1.3 mm in diameter; ovary 1 mm in diameter, densely pubescent; styles 2 mm long, whitish. Fruit 4–5 × 5–8 mm, minutely tuberculate, sparingly setose, greenish. Seeds 3–4 × 2.5–3 mm, ovoid-subglobose, smooth, somewhat shiny, grey, mottled brown; caruncle (when developed) 1 mm across, transversely ovate.

Zambia. N: Kawambwa Distr., near Kafulwe Mission, male fl. 6.xi.1952, *Angus* 730 (BM; FHO; K). W: Hippo Pool, female fl. 9.xii.1951, *Holmes* 594 (FHO). **Zimbabwe**. N: Hurungwe Distr., Musukwi (Msukwe) River, Kanyanga Stream, male & female fl. 18.xi.1953, *Wild* 4176 in *GHS* 44446 (K; LISC; MO; PRE; SRGH). C: Chegutu Distr., Umfuli R., st. 4.vii.1974, *Müller* 2169 (SRGH). E: Chimanimani Distr., Umvumvumvu Gorge, female fl. 19.xi.1956, *Chase* 6246 (K;

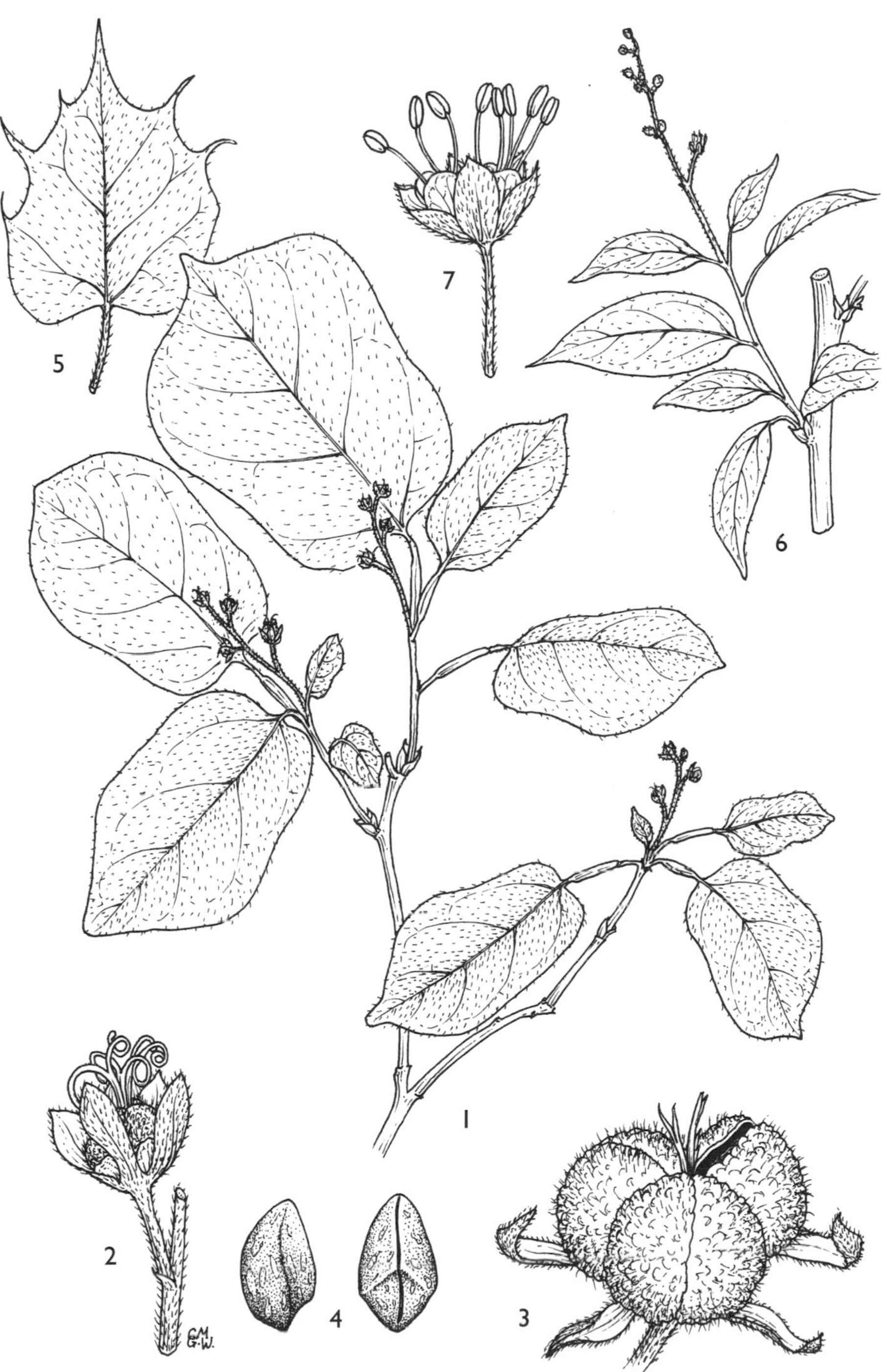

Tab. 65. NEOHOLSTIA TENUIFOLIA. 1, distal part of flowering branch (× $^2/_3$); 2, female flower (× 6), 1 & 2 from *Masterson* 291; 3, fruit (× 6), from *Brummitt & Eccles* 8831; 4, seeds (× 4), from *Patel & Tawakali* 1006; 5, leaf (× $^2/_3$); 6, male inflorescence (× $^2/_3$); 7, male flower (× 6), 5–7 from *Angus* 736. Drawn by Christine Grey-Wilson.

PRE; SRGH). **Malawi**. S: Tumbi Island East, fr. 1.iii.1970, *Brummitt & Eccles* 8831 (K; SRGH). **Mozambique**. N: Nampula, Monapo, fr. & sd. 11.ii.1984, *de Koning, Groenendijk & Dungo* 9568 (K; LMU). Z: Pebane Distr., 33 km Gilé–Nabúri, fr. 20.xii.1967, *Torre & Correia* 16670 (LISC). T: Boruma, "Sambesi-Mittellauf", male fl. xii.1891, *Menyharth* 794 (K). MS: Gorongosa, Parque Nacional de Caça, male fl. 4.xi.1963, *Torre & Paiva* 9041 (LISC).

Also in Tanzania. Valley floor and lake side vegetation, often in riverine thicket (jesse bush) and riverine mixed deciduous woodland with *Combretum, Xylia, Entandrophragma* and *Pteleopsis spp.*, on sandy soil and black clay floodplain alluvium, and on rocky slopes above lake and river side with *Adansonia, Sterculia, Acacia spp.*; 40–1460 m.

Var. *glabrata* (Prain) Radcl.-Sm., contrary to the treatment of this genus in F.T.E.A., Euphorb. 1 (1987), is confined to East Africa, and is not found in Mozambique.

54. NEOBOUTONIA Müll. Arg.

Neoboutonia Müll. Arg. in J. Bot. **1**: 336 (1864); in De Candolle, Prodr. **15**, 2: 892 (1866).

Dioecious, or rarely monoecious, trees or shrubs with a stellate and sometimes also simple indumentum. Leaves alternate, long-petiolate, stipulate, simple, large, more or less entire, minutely gland-dotted, palminerved. Inflorescences paniculate, terminal; male inflorescences larger and more branched than the female. Male flowers: calyx closed in bud, later splitting into 2–3 valvate lobes; petals absent; disk glands extrastaminal, minute; stamens 15–40, filaments free, short, anthers basifixed, introrse, longitudinally dehiscent; pistillode absent. Female flowers: calyx 5–6-lobed, the lobes imbricate; petals absent; disk annular; ovary usually 3-locular, with 1 ovule per locule; styles usually 3, connate at the base, deeply bipartite, rigid, recurved-patent. Fruit usually 3-lobed, septicidally dehiscent into 3 bivalved cocci; endocarp woody. Seeds ovoid or ellipsoid, carunculate; testa crustaceous; albumen fleshy; cotyledons broad, flat.

A small genus of 3 species in tropical Africa.

1. Calyx lobes of female flowers very short, shorter than the calyx tube; leaves sparingly pilose along the midrib and nerves beneath; fruit hirsute · 3. *mannii*
– Calyx lobes of female flowers much longer than the calyx tube · · · · · · · · · · · · · · · · · 2
2. Leaves scurfily stellate-pubescent along the midrib and main nerves beneath; buds of male flowers stellate-pubescent; calyx lobes of female flowers accrescent to 9 mm in fruit; fruit tomentose · 1. *macrocalyx*
– Leaves more or less uniformly stellate-tomentose beneath; buds of male flowers more or less glabrous except at the apex; calyx lobes of female flowers not exceeding 3 mm in fruit; fruit hirsute · 2. *melleri*

1. **Neoboutonia macrocalyx** Pax in Bot. Jahrb. Syst. **30**: 339 (1901). —Prain in F.T.A. **6**, 1: 919 (1912). —Pax in Engler, Pflanzenr. [IV, fam. 147, vii] **63**: 72, t. 10 (1914). —Engler, Pflanzenw. Afrikas (Veg. Erde 9) **3**, 2: 67, t. 26 (1921). —Robyns & Tournay, Fl. Sperm. Parc Nat. Alb. **1**: 451 (1948). —Brenan, Check-list For. Trees Shrubs Tang. Terr.: 221 (1949). —Eggeling & Dale, Indig. Trees Uganda, ed. 2: 135 (1952). —F.W. Andrews, Fl. Pl. Anglo-Egypt. Sudan **2**: 88 (1982). —Dale & Greenway, Kenya Trees & Shrubs: 210 (1961). —Troupin, Fl. Pl. Lign. Rwanda. 266, fig. 91/2 (1982); in Fl. Rwanda **2**: 233, fig. 69/2 (1983). —Radcliffe-Smith in F.T.E.A., Euphorb. 1: 232 (1987). —Beentje, Kenya Trees, Shrubs Lianas: 215 (1994). Type from Tanzania (Southern Highlands Province).

An open-crowned tree 10–20(40) m in height, with a straight clear bole. Bark smooth, pale. Branches brittle. Twigs and petioles scurfily stellate-pubescent. Petioles 5–15 cm long. Leaf blade 10–30 × 10–25 cm, broadly ovate-suborbicular, shortly acutely acuminate at the apex, deeply cordate at the base, usually entire, occasionally shallowly denticulate, sparingly scurfily stellate-pubescent on the nerves on the upper and lower surfaces, otherwise ± glabrous and minutely gland-dotted beneath; main nerves 7–9(11) from the base, lateral nerves in 5–7 pairs. Stipules 0.7–1 × 0.7–1.5 cm, ovate, densely pubescent. Male inflorescences 20–30 × 20 cm, often leafy at the base; axes pubescent; bracts 1–2 mm long, ovate, 1–4-flowered. Male flowers: pedicels 2–3 mm long, ± glabrous; buds ovoid, pubescent; calyx lobes 2.5–3 mm long, broadly ovate, cream-coloured; disk glands 10, 0.3 mm long, cylindric; stamens 30–40, 2–3 mm long, anthers 1.3 mm long, pale yellow. Female inflorescences few-branched, up to 20 cm long and wide; bracts 4 mm long, 1-flowered; otherwise as in the male. Female flowers: pedicels stout, up to 8 mm long, pubescent; calyx lobes 3 mm long, accrescent to 5(9) mm in fruit, linear-lanceolate, 3-nerved, thick, reflexed, velutinous; disk shallowly 5-lobed; ovary 2 mm in diameter, 3-lobed, densely stellate-pubescent; styles 2 mm long, pubescent abaxially. Fruits 1–1.2 × 1.2–1.3 cm evenly scurfily stellate-tomentose. Seeds 7 × 5 mm, ellipsoid, dark brown-mottled; caruncle appressed, 2-lobed.

Zambia. E: Nyika Plateau, st. 30.x.1958, *Robson & Angus* 477 (BM; K; LISC). **Zimbabwe**. E: Stapleford, Mt. Nuza, fr. i.1950, *Goldsmith* 8/50 in *GHS* 26755 (FHO; K; SRGH). **Malawi**. N: Rumphi Distr., Nyika Nat. Park, Kaziyula Forest, male fl. 6.iv.1981, *Salubeni & Tawakali* 2986 (MAL; MO; SRGH). C: Chirobwe Mt., st. 28.xi.1983, *F. Dowsett-Lemaire* 314 (FHO). S: Ntcheu Distr., Chirobwe Mt., male fl. 22.v.1961, *Chapman* 1326 (FHO; K; LISC; MAL). **Mozambique**. N: Ribáuè, Serra de Chinga, male fl. & fr. immat. 12.xii.1967, *Torre & Correia* 16490 (LISC). MS: Serra Zuira, Tsetserra, road to Chimoio (Vila Pery), fr. 2.iv.1966, *Torre & Correia* 15607 (LISC).

Also in Uganda, Kenya, Tanzania, Zaire (Oriental, Kivu), Rwanda and Burundi. Locally common subcanopy tree in evergreen forest; 600–2150 m.

2. **Neoboutonia melleri** (Müll. Arg.) Prain in Bull. Misc. Inform., Kew **1911**: 266 (1911); in F.T.A. **6**, 1: 922 (1912). —Pax in Engler, Pflanzenr. [IV, fam. 147, vii] **63**: 73 (1914). —Eyles in Trans. Roy. Soc. South Africa **5**: 395 (1916). —Engler, Pflanzenw. Afrikas (Veg. Erde 9) **3**, 2: 68 (1921). —Robyns & Tournay, Fl. Sperm. Parc Nat. Alb. **1**: 454 (1948). —Brenan, Check-list For. Trees Shrubs Tang. Terr.: 221 (1949). —Eggeling & Dale, Indig. Trees Uganda, ed. 2: 135 (1952). —Topham, Check List For. Trees Shrubs Nyasaland Prot.: 52 (1958). —Dale & Greenway, Kenya Trees & Shrubs: 211 (1961). —Drummond in Kirkia **10**: 252 (1975). —K. Coates Palgrave, Trees Southern Africa, ed. 2, rev.: 427 (1983). —Radcliffe-Smith in F.T.E.A., Euphorb. 1: 234 (1987). —Beentje, Kenya Trees, Shrubs Lianas: 216 (1994). Tab. **66**. Type: Malawi, Manganja Hills, Sept.–Nov. 1861, *Meller* s.n. (K, holotype).

Mallotus melleri Müll. Arg. in Flora **47**, 30: 468 (1864); in De Candolle, Prodr. **15**, 2: 959 (1866). —Sim, For. Fl. Port. E. Afr.: 107 (1909). Type as above.

Neoboutonia africana Müll. Arg. in J. Bot. **2**: 336 (1864); in De Candolle, Prodr. **15**, 2: 892 (1866). —Bentham in Hooker's Icon. Pl. 13, tt. 1298, 1299 (1879). —Prain in F.T.A. **6**, 1: 921 (1912). —White, F.F.N.R.: 201 (1962), non Pax. Type from Angola.

Croton niloticus Müll. Arg. in Flora **47**: 537 (1864); in De Candolle, Prodr. **15**, 2: 598 (1866). —Engler, Pflanzenw. Ost-Afrikas **C**: 237 (1895). —Hutchinson in F.T.A. **6**, 1: 756 (1912). —Engler, Pflanzenw. Afrikas (Veg. Erde 9) **3**, 2: 49 (1921). —De Wildeman, Pl. Bequaert. **3**, 4: 460 (1926). Type from Uganda.

Neoboutonia velutina Prain in Bull. Misc. Inform., Kew **1911**: 266 (1911); in F.T.A. **6**, 1: 921 (1912). —Keay in F.W.T.A., ed. 2, **1**, 2: 404 (1958). Types from Cameroon.

A several-stemmed tree up to 6–12(15) m in height. Bark smooth, grey-brown. Wood pithy, soft. Twigs and petioles scurfily stellate-pubescent. Petioles 2–10 cm long. Leaf blade (5)10–20 cm long and wide, broadly ovate-suborbicular, shortly acutely or subacutely acuminate to obtuse at the apex, subtruncate to deeply cordate at the base, entire, scabrid and dull green on upper surface, ± uniformly greyish-stellate-tomentose beneath, and minutely gland-dotted beneath; main nerves 7–9 from the base, lateral nerves in 5–8 pairs. Stipules 4 mm long, ovate-lanceolate, pubescent. Male inflorescences 30–40 × 30 cm, with leaf-like bracts at the base; axes pubescent; upper bracts 1 mm long, ovate, 4–6-flowered. Male flowers: pedicels 0.5–1 mm long, glabrous; buds globose, ± glabrous except at the apex; calyx lobes 1.5 mm long, broadly ovate to suborbicular, yellowish; disk glands 8–10, conical; stamens 15–30, 1.5 mm long, anthers 0.75 mm long, pale yellow. Female inflorescences 20–30 × 10–15 cm; bracts 2 mm long, 1–3-flowered; otherwise as in the male. Female flowers: pedicels stout, 2 mm long, pubescent; calyx lobes 2–3 mm long, longer than the calyx tube, not or scarcely accrescent, oblong, 3-nerved, thick, reflexed, velutinous; disk crenulate; ovary 2 mm in diameter, 3-lobed (rarely 4-lobed), densely pubescent; styles 2 mm long, pubescent abaxially. Fruits 6–10 × 7–12 mm, 3-lobed (rarely 4-lobed), scurfily stellate-pubescent and hirsute. Seeds 5–7 × 5 mm, ovoid, brownish, with a small caruncle.

Zambia. N: Mpika, male fl. 1.ii.1955, *Fanshawe* 1914 (FHO; K; NDO; SRGH). W: Mwinilunga Distr., Luakera R., y. fr. 12.xi.1952, *Holmes* 996 (FHO; K; SRGH). E: Makutu Hills, male fl. 28.x.1972, *Fanshawe* 11624 (K; NDO). **Malawi**. N: Chisenga Distr., Mafinga Mts., male fl. 8.xi.1958, *Robson & Fanshawe* 517 (BM; K; LISC; PRE; SRGH). S: Chikala Mt., Liwonde For.

Tab. 66. NEOBOUTONIA MELLERI. 1, distal part of fruiting branch (× ⅔); 2, fruit (× 3); 3, portion of stellate indumentum (× 18), 1–3 from *Fanshawe* 1913; 4, male flower (× 8), from *Fanshawe* 1914. Drawn by Christine Grey-Wilson.

Res., female fl. & y. fr. 10.xii.1981, *Chapman et al.* 6033 (FHO; K; MAL). **Mozambique**. N: Litunde, 15.vi.1934, *Torre* 543 (LISC). MS: Mt. Maruma, male fl. 12.ix.1906, *Swynnerton* 686 (BM; K; SRGH).

From Nigeria eastwards to southern Sudan and south to Angola and Mozambique. Riverine forest margins and riverine woodland, usually in swampy ground beside streams, also in evergreen swamp forest (mushitu); 600–1850 m.

The above description of mature seeds is drawn from F.T.E.A. area material.

The specimen cited by Drummond under *N. melleri* is referable to *Neoboutonia macrocalyx.*

3. **Neoboutonia mannii** Benth. in Hooker's Icon. Pl. 13: sub tt. 1298, 1299 (1879), in adnot. —Prain in F.T.A. **6**, 1: 920 (1912). —Keay in F.W.T.A., ed. 2, **1**, 2: 404 (1958). Type from Principe.

Conceveiba africana Müll. Arg. in Flora **47**, 34: 530 (1864); in De Candolle, Prodr. **15**, 2: 897 (1866). Type as above.

Neoboutonia africana (Müll. Arg.) Pax in Engler and Prantl., Pflanzenfam. **3**, 5: 57 (1890); in Engler, Pflanzenr. [IV, fam. 147, vii] **63**: 75 (1914). —Engler, Pflanzenw. Afrikas (Veg. Erde 9) **3**, 2: 68 (1921). —Brenan, Check-list For. Trees Shrubs Tang. Terr.: 221 (1949). —Topham, Check List For. Trees Shrubs Nyasaland Prot.: 52 (1958), non Müll. Arg. Type as above.

A tree up to 20 m in height, very like *N. melleri* but differing in the following ways: leaves often denticulate on the margins, hirsute-pubescent or pilose on both surfaces; male flower buds pubescent; female calyx lobes 1 mm long, shorter than the calyx tube, broadly triangular-ovate.

Zambia. N: 35 km north of Mpika, Muchinga Escarpment, st. 28.xi.1952, *White* 3761 (FHO; K). W: Mwinilunga, y. fr. 20.x.1955, *Holmes* 1286 (K; NDO). **Malawi**. S: Zomba Mountain, male fl. x.1929, *Clements* 40 (MAL). **Mozambique**. MS: Tsetserra, st. 7.vi.1971, *Müller & Gordon* 1829 (SRGH).

Scattered throughout tropical Africa from Guinea to Uganda and south to Mozambique. Submontane and escarpment evergreen riverine forest beside perennial streams, and plateau or watershed swamp forest (mushitu); 1750 m.

This species may prove to be conspecific with *N. melleri* when more material is examined.

55. SPIROSTACHYS Sond.

Spirostachys Sond. in Linnaea **23**: 106 (1850).

Dioecious or monoecious glabrous trees or shrubs. Leaves alternate, petiolate, stipulate, simple, crenate, penninerved, glanduliferous at the petiole apex. Inflorescences axillary, sessile, spicate and amentiform, or racemose, unisexual or bisexual with the inflorescence mostly male and with 1–3 female flowers at the base; bracts glandular or not at the base, 1-flowered. Male flowers small, sessile; sepals (2)5, imbricate; petals and disk absent; stamens 3, free or united, exserted; anthers extrorse, basifixed, bithecate, longitudinally dehiscent; pistillode absent. Female flowers small, shortly pedicellate; sepals 3–5, open in bud; petals and disk absent; ovary (2)3-locular, with 1 ovule per locule; styles (2)3, united at the base, entire, not strongly coiled. Fruit (2)3-lobed, dehiscing into bivalved cocci; valves not readily separating; exocarp smooth; endocarp thinly crustaceous; columella persistent. Seeds globose or ovoid; testa crustaceous; caruncle minute, separating from the seed and remaining adnate to the columella; albumen fleshy; cotyledons broad, flat.

An African genus of 2 species.

Spirostachys africana Sond. in Linnaea **23**: 106 (1850). —Pax in Engler, Pflanzenr. [IV, fam. 147, v] **52**: 155 (1912). —Prain in F.T.A. **6**, 1: 1006 (1913). —Prain in F.C. **5**, 2: 512 (1920). —Engler, Pflanzenw. Afrikas (Veg. Erde 9) **3**, 2: 139 (1921). —Burtt Davy, Fl. Pl. Ferns Transvaal: 309, t. 45 (1932). —Hutchinson, Botanist in Southern Africa: 667 (1946). —O.B. Miller, Check-list For. Trees Shrubs Bech. Prot.: 33 (1948). —Brenan, Check-list For. Trees Shrubs Tang. Terr.: 226 (1949). —Dale & Greenway, Kenya Trees & Shrubs: 220 (1961). —P.G. Meyer in Merxmüller, Prodr. Fl. SW. Afrika, fam. 67: 44 (1967). —Drummond in Kirkia **10**: 253 (1975). —K. Coates Palgrave, Trees Southern Africa, ed. 2, rev.: 435 (1983). —Radcliffe-Smith in F.T.E.A., Euphorb. 1: 386 (1987). —Beentje, Kenya Trees, Shrubs Lianas: 222 (1994). Tab. **67**. Types from South Africa.

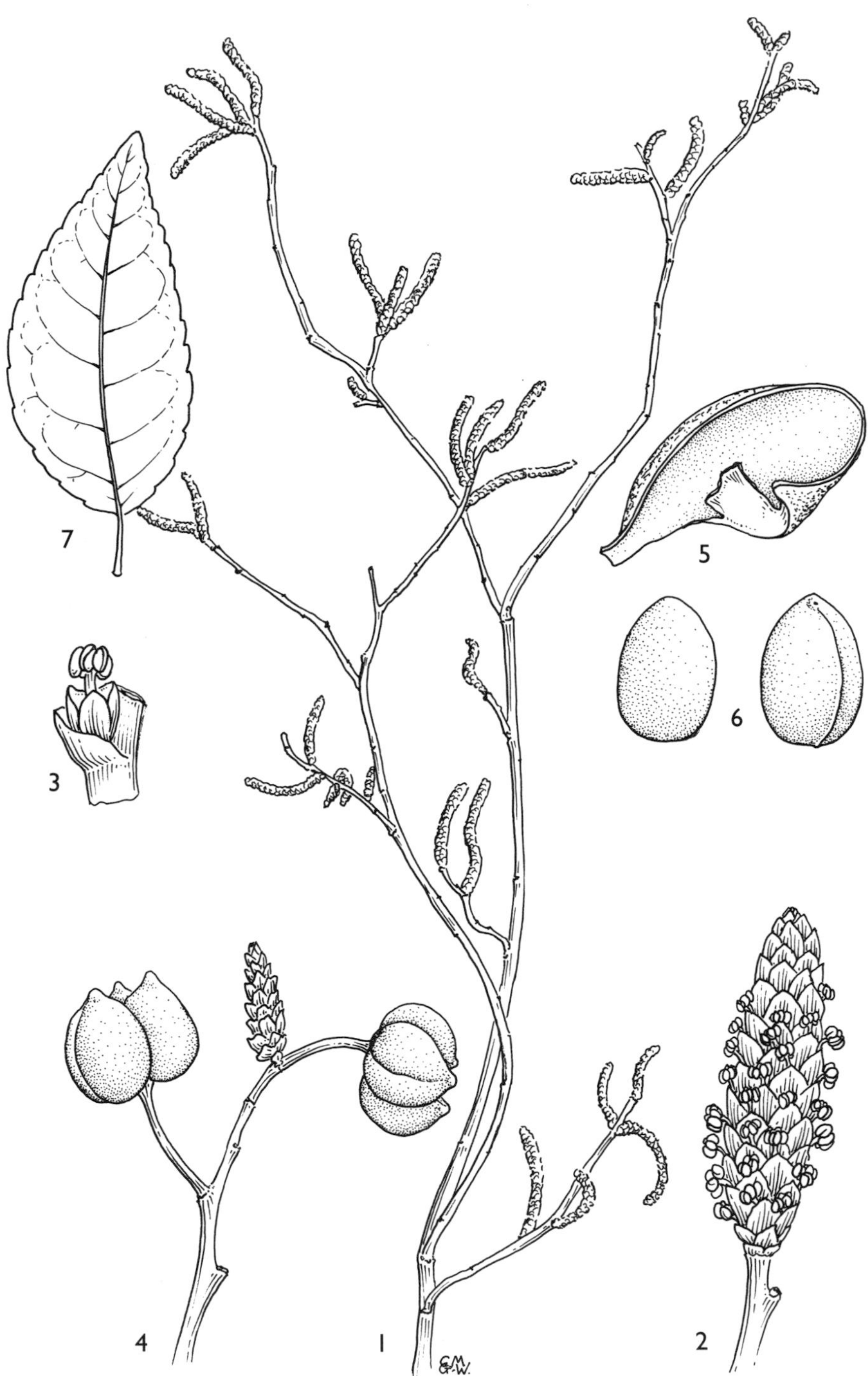

Tab. 67. SPIROSTACHYS AFRICANA. 1, distal portion of flowering branch (× $^{2}/_{3}$); 2, male inflorescence (× 4); 3, male flower (× 6), 1–3 from *Gomes e Sousa* 3342; 4, young fruits (× 2), from *Barbosa* 7762; 5, fruit valve (× 4); 6, seeds (× 4), 5 & 6 from *Gomes e Sousa* 2019; 7, leaf (× 1), from *Barbosa* 7762. Drawn by Christine Grey-Wilson.

Stillingia africana (Sond.) Baill., Étud. Gén. Euphorb.: 522 (1858).
Excoecaria africana (Sond.) Müll. Arg. in Linnaea **32**: 123 (1863); in De Candolle, Prodr. **15**, 2: 1215 (1866). —Sim, For. Fl. Port. E. Afr.: 104, t. C(B) (1909).
Maprounea africana Müll. Arg. in De Candolle, Prodr. **15**, 2: 1191 (1866), quoad spec. *Welwitsch* 401b.
Excoecaria synandra Pax in Bot. Jahrb. Syst. **43**: 223 (1909). Type from Tanzania.
Spirostachys synandra (Pax) Pax in Engler, Pflanzenr. [IV, fam. 147, v] **52**: 155 (1912). —Engler, Pflanzenw. Afrikas (Veg. Erde 9) **3**, 2: 138, 139 (1921).

A deciduous tree usually 12 m tall, sometimes up to 16 m in height, with a copious caustic milky latex, dioecious or monoecious. Bark dark greyish-brown and rough on the bole, smoother and lighter above. Wood hard, dark, smelling of sandalwood. Young twigs brownish, older ones greyish. Stipules minute, fimbriate, soon falling. Petioles 2–10 mm long, adaxially canaliculate, with 2 small hemispherical glands at the apex. Leaf blades 2–7 × 1–3.5 cm, elliptic-ovate, obtusely acuminate or obtuse at the apex, rounded at the base, shallowly crenate, chartaceous to thinly coriaceous, green on upper surface, paler beneath, often becoming bright red or plum-coloured before falling; lateral nerves in 6–14 pairs. Inflorescences often appearing before the leaves, up to 4 cm long, narrowly cylindric, dense flowered, brownish; bracts spirally arranged, 1.5 mm across, transversely ovate, eglandular. Male flowers: sepals (2)3, minute, obovate-suborbicular, concave, brownish-green; staminal column 1 mm high, anthers 0.75 mm long. Female flowers: pedicels very short at first, but extending to 10 mm long in fruit; sepals 5, 1 mm long, triangular-ovate, acute, flat; ovary 1 mm in diameter, 3-lobed, smooth; styles 3, 1.5 mm long, stigmas red. Fruit 5–7 × 12–13 mm, 3-lobed, smooth, green at first, becoming yellowish-brown. Seeds 4 × 3 mm, broadly ovoid-subglobose, smooth, light brown streaked with darker brown.

Botswana. SE: 3 km north of Kanye, fr. 27.x.1978, *O.J. Hansen* 3508 (C; GAB; K; PRE; SRGH). **Zimbabwe**. E: Chirinda, fl. & y. fr. 13.ix.1944, *Cooke-Yarborough* in *GHS* 12773 (SRGH). S: Chitsa's Kraal, y. fl. 5.vi.1950, *Chase* 2314 (BM; COI; SRGH). **Mozambique**. N: Quissanga, y. male fl. 11.ix.1948, *Campos Andrada* 1331 (COI; LISC). Z: Morrumbala, male fl. 3.x.1949, *Campos Andrada* 1940 (COI; LISC). T: between Ancuaze e Dôa, y. fl. 21.vi.1949, *Barbosa & Carvalho* in *Barbosa* 3193 (K; LISC; LMA). MS: Inhamitanga, fr. 31.x.1945, *Simão* 612 (LISC). GI: Mocoduene–Mavume, fr. ix.1937, *Gomes e Sousa* 2019 (COI; K; LISC). M: 10 km Maputo–Matutuíne (Bela Vista), fl. & y. fr. 6.viii.1957, *Barbosa & Lemos* in *Barbosa* 7762 (COI; K; LISC, LMA, SRGH).

Also in Kenya, Tanzania, Angola, Namibia, Swaziland and South Africa (Transvaal, KwaZulu-Natal, Cape Province). Locally frequent and usually gregarious at low altitudes in *Colophospermum mopane* or mixed (*Combretum, Acacia, Brachystegia*) woodlands, sometimes in thickets, the largest trees occur near streams and seasonal watercourses; often also on termitaria and stony slopes; 30–1350 m.

Vernacular names as recorded in specimen data include: "muconite" (north Mozambique); "sandaleen" (Maputo); "umtamboti" (Botswana).

The hard durable ornamental wood is much sought after for furniture making (Miller). The quality of the heartwood and its scent are like those of the Indian sandalwood *(Santalum album)*. It is easily worked, stable and contains an oil which helps to give a beautiful finish. It has been suggested that this may have been the "algum" wood from "Ophir" used in King Solomon's Palace and the Hebrew Temple (Old Testament).

The sawdust of fresh wood and the sap are highly irritant.

The seeds often become infested with the larvae of a small grey moth which cause the seeds to spring several centimeters into the air similar to the "jumping beans" of some South American *Hippomaneae* (*Sebastiania spp., Colliguaya spp.*).

56. EXCOECARIA L.

Excoecaria L., Syst. Nat., ed. 10: 1288 (1759).

Dioecious or monoecious glabrous trees, shrubs or subshrubs. Leaves alternate or opposite, petiolate, stipulate, simple, entire, crenulate or serrulate, penninerved, eglandular. Inflorescences terminal or axillary, spicate or racemose, unisexual or bisexual with most of the inflorescence male with 1–5 female flowers at the base; bracts small, 2-glandular at the base, 1–∞-flowered. Male flowers small, ± sessile; sepals (2)3(4), ± free, imbricate; petals and disk absent; stamens 3 (African species), free, anthers extrorse, basifixed, 2-thecous, the thecae free at the base, longitudinally dehiscent; pistillode absent. Female flowers small, shortly pedicellate; sepals ± as in the male flowers; petals and disk absent; ovary (2)3-locular, with 1 ovule per locule; styles (2)3, free or ± so, entire, strongly coiled. Fruit (2)3-lobed, dehiscing into bivalved cocci; valves readily separating; exocarp smooth; endocarp thinly crustaceous; columella persistent. Seeds globose or ovoid; testa crustaceous; caruncle separating from the seed and remaining adnate to the columella; albumen fleshy; cotyledons broad, flat.

A palaeotropical genus of some 40 species.
Excoecaria bicolor (Hassk.) is cultivated in Zimbabwe, fide Biegel as "*E. cochinchinensis*".

Leaves opposite, entire; inflorescences axillary, densely flowered; fruits 1 cm in diameter · · · · · · · · · · · · 1. *madagascariensis*
Leaves alternate, crenulate; inflorescences terminal, laxly flowered; fruits 5–7 cm in diameter · · · · · · · · · · · · 2. *bussei*

1. **Excoecaria madagascariensis** (Baill.) Müll. Arg. in De Candolle, Prodr. **15**, 2: 1219 (1866). —Pax in Engler, Pflanzenr. [IV, fam. 147, v] **52**: 160 (1912). —Engler, Pflanzenw. Afrikas (Veg. Erde 9) **3**, 2: 140 (1921). —Dale & Greenway, Kenya Trees & Shrubs: 203 (1961). —Drummond in Kirkia **10**: 253 (1975). —K. Coates Palgrave, Trees Southern Africa, ed. 2, rev.: 435 (1983). —Radcliffe-Smith in F.T.E.A., Euphorb. 1: 383 (1987). —Beentje, Kenya Trees, Shrubs Lianas: 207 (1994). Type from Madagascar.

Spirostachys madagascariensis Baill., Étud. Gén. Euphorb., Atlas 17, t. 8, f. 19, 21 (1858). Type as above.

Excoecaria sylvestris S. Moore in J. Linn. Soc., Bot. **40**: 204 (1911). Syntypes: Zimbabwe, Chirinda Forest, 1130–1220 m, 31.i.1906, *Swynnerton* 72 (BM; K; SRGH); x.1908, *Swynnerton* 72a (BM).

Sapium madagascariensis (Baill.) Prain in F.T.A. **6**, 1: 1010 (1913). —Eyles in Trans. Roy. Soc. South Africa **5**: 398 (1916). —Brenan, Check-list For. Trees Shrubs Tang. Terr.: 226 (1949), non Müll. Arg. (1863), nec Pax (1890).

A shrub or small lax tree 1–4 m tall with a white poisonous latex, monoecious. Bark dark brown or blackish, ± smooth or finely reticulate. Leaves opposite. Stipules 2.5 mm long, triangular-ovate, acute, soon falling; scars overlapping. Petioles 5–8 mm long, adaxially canaliculate. Leaf blades 5–18 × 2–6 cm, elliptic to elliptic-oblanceolate, obtusely acuminate at the apex, cuneate at the base, entire, thinly coriaceous, dark green on upper surface, paler beneath, but deep coppery-red when young; lateral nerves in c. 10–20 pairs, subperpendicular. Inflorescences in axils of the upper leaves, 1.5–2 cm long, shortly pedunculate, densely flowered; bracts c. 1 mm across, broadly ovate, imbricate, obtuse. Male flowers: sepals 3, minute, lanceolate, acute, greenish-white to yellowish; stamens 1 mm long, anthers 0.5 mm long. Female flowers: sepals suborbicular-ovate, obtuse or subacute, otherwise ± as in the male flowers; ovary 3-lobed, smooth; styles 3, connate at the base. Fruit 10 mm in diameter, 3-lobed, smooth, green at first, later reddish. Seeds 4 mm in diameter, globose, smooth, greyish-brown, mottled darker brown.

Zimbabwe. E: Chipinge Distr., Chirinda Forest, fl. ix.1962, *Goldsmith* 226/62 (BM; COI; FHO; K; LISC; SRGH); Mt. Selinda Forest near Swynnerton Memorial, fl. & fr. 15.xii.1972, *Müller & Goldsmith* 2070 (K; LISC; SRGH).

Also in southeast Kenya, eastern Tanzania and northwestern Madagascar. Locally common shrub or understorey tree of medium altitude evergreen rainforest; 900–1830 m.

2. **Excoecaria bussei** (Pax) Pax in Engler, Pflanzenr., [IV, fam. 147, v] **52**: 169, t. 31 (1912). —Engler, Pflanzenw. Afrikas (Veg. Erde 9) **3**, 2: 140 (1921). —Dale & Greenway, Kenya Trees & Shrubs: 203 (1961). —Drummond in Kirkia **10**: 253 (1975). —Biegel, Check List Ornam. Pl. Rhod. Parks & Gard.: 53 (1977). —K. Coates Palgrave, Trees Southern Africa, ed. 2, rev.: 434 (1983). —Radcliffe-Smith in F.T.E.A., Euphorb. 1: 383 (1987). —Beentje, Kenya Trees, Shrubs Lianas: 207 (1994). Tab. **68**. Type from Tanzania (Ugogo).

Sapium bussei Pax in Bot. Jahrb. Syst. **33**: 284 (1903). —Prain in F.T.A. **6**, 1: 1010 (1913). —Brenan, Check-list For. Trees Shrubs Tang. Terr.: 226 (1949). —White, F.F.N.R.: 203 (1962).

Excoecaria sambesica Pax & K. Hoffm. in Engler, Pflanzenr. [IV, fam. 147, v] **52**: 170 (1912). Type: Mozambique, Tete Province, Boroma (Boruma), *Menyharth* 746 (Z, holotype).

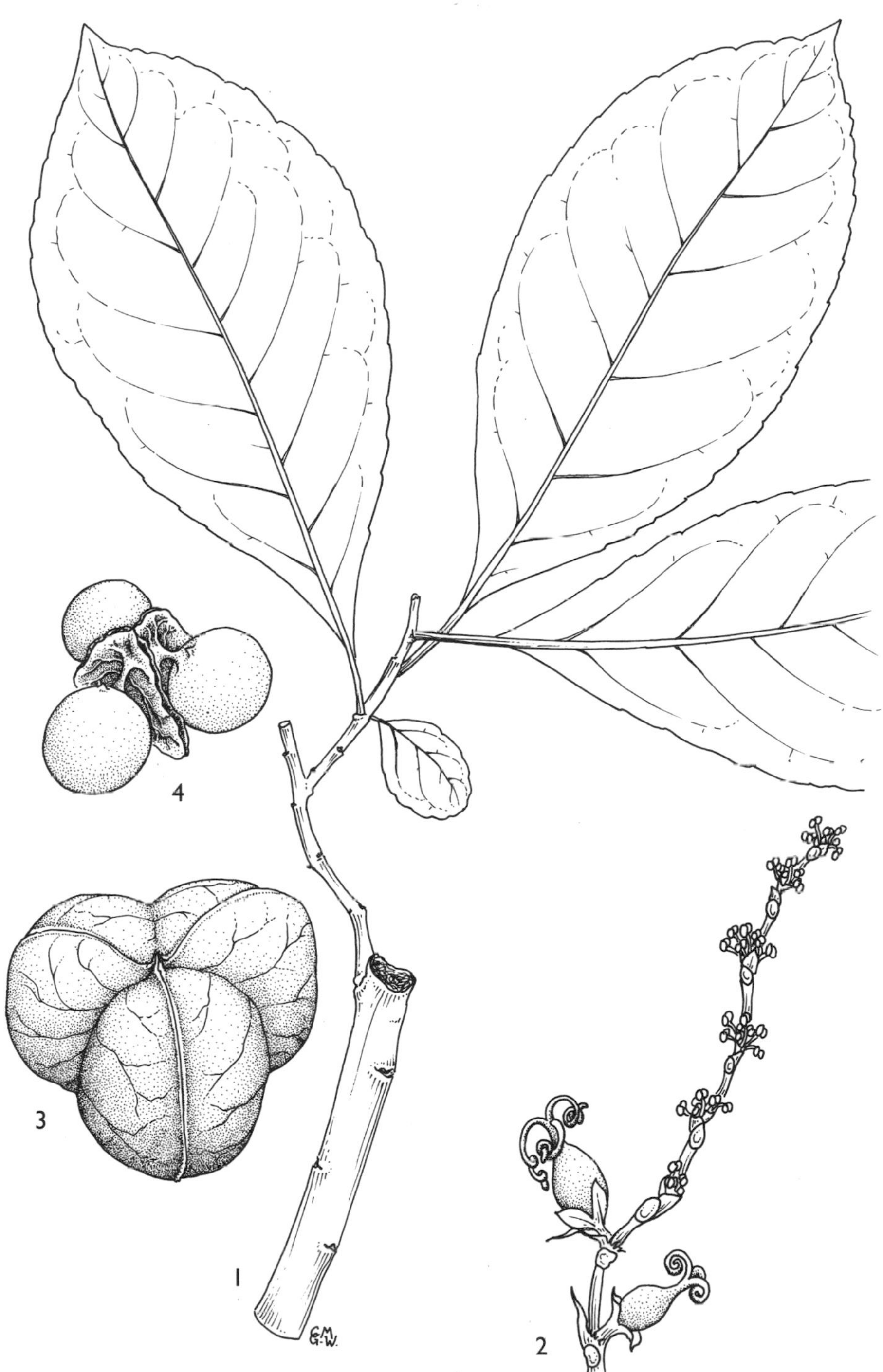

Tab. 68. EXCOECARIA BUSSEI. 1, portion of leafy branch ($\times \frac{2}{3}$), from *Macêdo & Esteves* 4841; 2, inflorescence (× 3), from *Goodier* 439; 3, fruit (× 1); 4, seed cluster (× 1), 3 & 4 from *Macêdo & Esteves* 4841. Drawn by Christine Grey-Wilson.

A deciduous shrub or small tree, sometimes multistemmed, up to 13 m in height, with a dense, rounded crown and white poisonous latex, monoecious. Bark light brown, dark brown, chestnut or purple-grey and fissured on the bole, pale grey and smooth on the branches. Twigs fairly stout, light brown, lenticellate. Buds perulate (furnished with protective scales). Leaves alternate. Stipules c. 1 mm long, ovate-lanceolate, acuminate, soon falling. Petioles 0.5–2 cm long, adaxially canaliculate. Leaf blades 7–25 × 3–10 cm, obovate or elliptic-obovate, acuminate at the apex, cuneate at the base, crenulate, firmly chartaceous, dark green and glossy at first, later becoming bright yellowish-green, sometimes reddish-tinged; lateral nerves in 10–15 pairs. Inflorescences terminal, 2–13 cm long, lax, shortly pedunculate; male bracts 1.5 mm across, orbicular-ovate, apiculate; female bracts 2 mm long, ovate-lanceolate, aristate. Male flowers: sepals 3, c. 1 mm long, oblong-lanceolate, subacute, yellow or yellowish-green; stamens 2–3 mm long, anthers c. 1 mm long, yellow. Female flowers: sepals 3 × 1 mm, elliptic-oblong, mucronulate, otherwise ± as in male flowers; ovary 2 mm in diameter, ellipsoid-subglobose, smooth; styles 3, connate at the base. Fruit 4–5 × 5–7 cm, 3-lobed, inflated, crustaceous, ridged on each valve, otherwise ± smooth, green at first, later yellowish. Seeds c. 12 mm in diameter, ovoid-subglobose, smooth, fawn, mottled brown.

Caprivi Strip. Katima Mulilo, fr. 21.ii.1969, *de Winter* 9271 (PRE). **Botswana**. N: Liversidge's Camp, Okavango River, fl. 21.ii.1979, *P.A. Smith* 2708 (K; PRE; SRGH). **Zambia**. C: Luangwa Game Reserve, Luangwa Wafwa, st. 7.v.1965, *B.L. Mitchell* 2875 (K; SRGH). S: Kalomo, fr. 30.i.1963, *Bainbridge* 700 (FHO; LISC; NDO; SRGH). **Zimbabwe**. N: Hurungwe Distr., 15 miles ESE of Chirundu bridge, fr. 4.ii.1958, *Drummond* 5472 (K; LISC; SRGH). W: Hwange Distr., Gwayi R., fr. 26.ii.1963, *Wild* 6011 (K; LISC; SRGH). C: Chegutu Distr., Munyati (Umniati) River, near Sanyati Reserve Rest House, fl. 13.xii.1962, *Müller* 7 (BM; FHO; K; LISC; PRE; SRGH). E: Chipinge Distr., c. 14 km east of Sabi Experimental Station, fr. 6.viii.1973, *Mavi* 1460 (K; SRGH). S: Mwenezi Distr., Runde (Lundi) River, st. 31.viii.1968, *West* 7612 (SRGH). **Malawi**. S: Chiradzulu, Magomero, fr. 12.iii.1962, *Adlard* 423 (SRGH). **Mozambique**. N: Mueda–Negomano, fr. 2.iv.1960, *Gomes e Sousa* 4556 (COI; K; PRE; SRGH). T: Estima–Candodo, fr. 12.ii.1972, *Macêdo & Baião Esteves* in *Macêdo* 4841 (K; LISC; LMA; SRGH). MS: Chemba, fr. 14.v.1971, *Torre & Correia* 18415 (COI; LISC). GI: Chongoene, fr. i.1969, *Burrows* 263 (SRGH).

Also in S Kenya and Tanzania. Hot dry areas, often in thicket or riverine thicket, also on banks of seasonal watercourses in *Baikiaea* mutemwa and in coastal dune forest and thicket; 300–960 m.

Vernacular name as recorded in specimen data: "mumpobompobo" (Tok.).

57. **SAPIUM** P. Browne

Sapium P. Browne, Civ. Nat. Hist. Jamaica: 338 (1756).

Usually monoecious, glabrous trees, shrubs, subshrubs or rarely perennial herbs, often with a milky latex. Leaves alternate or subopposite, petiolate, stipulate, simple, entire or dentate, penninerved, usually with glands at the base and/or on the margins beneath. Stipules small. Inflorescences usually terminal or leaf-opposed, spicate or racemiform, solitary, bisexual and mostly male with 1 or more female flowers at the base, rarely the female flowers solitary; male bracts 1–many-flowered, usually 2-glandular at the base; bracteoles minute. Male flowers: pedicels jointed or absent; calyx ± open in bud, 2–3(4)-lobed; petals absent; disk absent; stamens 2–3(4), free or slightly connate, usually exserted, anthers extrorse, basifixed, 2-thecous, longitudinally dehiscent; pistillode absent. Female flowers: pedicels usually long; sepals 2–3(4), free or united, imbricate or open; petals absent; disk absent; ovary (2)3(4)-locular, ornamented or not, with 1 ovule per locule; styles (2)3(4), ± free or united, entire, often coiled, usually soon falling. Fruit (2)3(4)-lobed, dehiscing septicidally into (2)3(4) bivalved cocci, or the cocci indehiscent, or the fruit loculicidally or irregularly dehiscent, or indehiscent and drupaceous or bacciform; exocarp smooth or horned, separating or not from the woody or crustaceous endocarp; columella persistent. Seeds ecarunculate or rarely carunculate; sarcotesta thin or fleshy; endotesta crustaceous; albumen fleshy; cotyledons broad, flat.

A pantropical but predominantly neotropical genus of c. 125 species, of which 13 are African. *Sapium sebiferum* (L.) Roxb. is cultivated in Zimbabwe, fide Biegel.

1. A rhizomatous perennial herb; leaves not more than 3 cm long · · · · · · · · · · · 1. *acetosella*
– Shrubs or trees; leaves usually exceeding 3 cm in length · 2
2. Ovary and fruit usually 2-lobed, the lobes smooth; fruit ± indehiscent · · · · · · 7. *ellipticum*
– Ovary and fruit 3-lobed, each lobe provided with 2 ridges or horn-like projections; fruit dehiscent · 3
3. Leaves usually oblong, subacute, obtuse or retuse at the apex, often ± parallel-sided · 2. *oblongifolium*
– Leaves elliptic or lanceolate, acute or subacute at the apex, not parallel-sided · · · · · · · · 4
4. Exocarp thin, not separating from the thick woody endocarp · · · · · · · · · · · · · · · · · · · 5
– Exocarp thick, separating from the woody endocarp · 6
5. Bark of twigs brownish, longitudinally furrowed; glands on the leaf lower surface mostly near the margin; male bracts with 1(2) discoid glands on each side towards the base · 3. *cornutum*
– Bark of twigs breaking up into blackish scales covering a reddish-brown powdery tissue; glands on the leaf lower surface scattered, with up to 3 at the base; male bracts with 2–3(4) poculiform (cup-shaped) glands on each side at the base · · · · · · · · · · · · · · · · 4. *schmitzii*
6. Leaves entire, subentire or only remotely crenate-serrate; glands on the leaf lower surface usually only 2 at or near the base · 5. *integerrimum*
– Leaves crenate-serrate; glands on the leaf lower surface submarginal, not usually at the base · 6. *armatum*

1. **Sapium acetosella** Milne-Redh. in Hooker's Icon. Pl. **32**: t. 3199 (1933). —J. Léonard in Bull. Jard. Bot. État **29**, 2: 140 (1959); in F.C.B. **8**, 1: 163 (1962). —White, F.F.N.R.: 203 (1962). Type: Zambia, c. 25 km west of R. Kabompo, 11.ix.1930, *Milne-Redhead* 1105 (K, holotype).

A glabrous perennial herb, monoecious, forming patches 30 cm in diameter; stems several, erect, up to 15 cm tall, simple or basally-branched, arising seasonally from a thick, apically much-branched woody rootstock. Leaves shortly petiolate, the petioles 1(2) mm long; leaf blades (0.5)1–3 cm × 0.5–9 mm, obovate, elliptic, lanceolate or linear, acute to subacute or occasionally obtuse and mucronate at the apex, attenuate or cuneate at the base, entire to minutely serrulate, eglandular on the surface but often with 1–2 glands on the thickened hyaline margin towards the base, thinly coriaceous, somewhat glaucous, sometimes slightly purplish-tinged; lateral nerves in c. 5–6 pairs, usually scarcely visible. Stipules 0.5 mm long, ovate-lanceolate, acuminate, subentire, soon deciduous. Inflorescences terminal on the main axes, overtopped by lateral leafy shoots, 1.5–3.5 cm long, mostly male with 1–4 female flowers at the base; peduncle up to 1.5 cm long; male bracts 1 mm long, elliptic-lanceolate in outline, with 2 fleshy stipitate reniform concave glands, one on each side, situated halfway along and a terminal triangular, acute, minutely denticulate laminula, 1-flowered, reddish; female bracts resembling the male, except that the glands are often basal; bracteoles minute. Male flowers: pedicels 1 mm long; calyx lobes 3(4), 0.5–1 mm long, triangular, irregularly denticulate, reddish-tinged; stamens 3(4), free, 1 mm long. Female flowers subsessile, becoming shortly stipitate in fruit; calyx lobes resembling those of the male flowers, but larger, and often alternating with minute glands; ovary 2 mm in diameter, 3-lobed, each lobe bearing 2(or more) ± conical appendages; styles 3, 2 mm long, connate to halfway. Fruit 6–8 × 7–8 mm, 3-lobed, septicidally dehiscent into 3 bivalved cocci, each valve bearing a small appendage 0.5–1 mm long; exocarp purplish-tinged, not separating from the thin crustaceous endocarp. Seeds 5 × 3.5 mm, subcylindrical, smooth, slightly shiny, pale grey, with a fleshy depressed-conic caruncle c. 2 mm across.

Leaf blades obovate, elliptic or lanceolate, 2–9 mm wide · · · · · · · · · · · · · · · · · · var. *acetosella*
Leaf blades linear, 0.5–2 mm wide · var. *lineare*

Var. **acetosella**

Leaf blades obovate, elliptic or lanceolate, acute to subacute or obtuse and mucronulate at the apex, subentire or minutely serrulate, 2–9 mm wide.

Zambia. W: Mwinilunga Distr., Cha Mwana (Chibara's) Plain, fl. & fr. 14.x.1937, *Milne-Redhead* 2773 (BM; K; PRE); Solwezi Distr., c. 106 km Solwezi–Nchanga, fl. & fr. 1.x.1947, *Brenan*

& Greenway 8004 (FHO; K); Kitwe, fl. & fr. 20.xii.1972, *Fanshawe* 11714 (K; NDO; SRGH).
Also in Angola and Zaire (Shaba Province). Pyrophyte of sandy watershed plains, usually in dambos or dambo margins in short grassland, often seasonally flooded; 1000–1500 m.

Var. **lineare** J. Léonard in Bull. Jard. Bot. État **31**, 3: 405 (1961). Type: Zambia, Kawambwa, 16.xi.1957, *Fanshawe* 4051 (K, holotype; BR; NDO).

Leaf blades linear, sharply acute, entire, 0.5–2 mm wide.

Zambia. N: c. 10 km from Kawambwa on Mansa (Fort Rosebery) road, fl. & fr. 27.xi.1961, *Richards* 15392 (K); Ntenke, fl. & fr. 2.ix.1963, *Mutimushi* 477 (K; NDO; SRGH).
Known only from northern Zambia. Sandy dambos or grassland with scattered trees of *Protea* and *Terminalia*; 1290 m.
On account of its carunculate seeds, *S. acetosella* might perhaps be considered to have closer links with *Sebastiania* than with *Sapium*.

2. **Sapium oblongifolium** (Müll. Arg.) Pax in Bot. Jahrb. Syst. **19**: 114 (1894), in obs.; in Engler, Pflanzenr. [IV, fam. 147, v] **52**: 247 (1912) —Prain in F.T.A. **6**, 1: 1015 (1913). —Engler, Pflanzenw. Afrikas (Veg. Erde 9) **3**, 2: 142, t. 74 (1921). —J. Léonard in Bull. Jard. Bot. État **29**, 2: 142 (1959); in F.C.B. **8**, 1: 160 (1962). —White, F.F.N.R.: 204 (1962). Type from Angola (Malanje).
Excoecaria oblongifolia Müll. Arg. in J. Bot. **2**: 337 (1864); in De Candolle, Prodr. **15**, 2: 1214 (1866).
Sapium suffruticosum Pax in Warburg, Kunene-Samb.-Exped. Baum.: 284 (1903); in Engler, Pflanzenr. [IV, fam. 147, v] **52**: 247 (1912). Type from Angola (Benguela).

A glabrous subshrub, monoecious; stems one or more, erect, 20–30 cm tall, sometimes up to 60 cm tall, simple or sparingly-branched, arising from a fairly slender horizontal woody rootstock. Leaves petiolate, the petioles 2–9 mm long, adaxially canaliculate; leaf blades (1.5)3–6(8) × (0.5)1–3.5 cm, usually oblong and parallel-sided, sometimes oblong-lanceolate with the sides not parallel, subacute to obtuse or retuse at the apex, rounded or truncate at the base, subentire or shallowly serrate with forwardly directed teeth, with 1–2 glands towards the base, and sometimes also glandular near the margin towards the apex and elsewhere, thinly coriaceous, somewhat glaucous on lower surface; lateral nerves in 8–15 pairs, slightly prominent beneath, scarcely so above, tertiary nerves reticulate. Stipules 2–7 mm long, linear-setaceous, acute, laciniate at the base, otherwise entire, ciliate, deciduous. Inflorescences terminal on the main axes, often overtopped by a succession of lateral shoots, sometimes leaf-opposed, (1)2–4(7) cm long, with a peduncle up to 1.5 cm long or sessile, all male or with 1(2) female flowers at the base; male bracts broad, encircling half the axis, often somewhat 3-lobed with broadly triangular lobes, acute and irregularly denticulate, ciliolate, fleshy at the base, glandular within, 3(5)-flowered; female bracts rather more deeply 3-lobed than the males; bracteoles minute. Male flowers: pedicels 1 mm long; calyx lobes 3, 0.5–1 mm long, ovate, irregularly denticulate-ciliolate, yellowish-green; stamens (2)3, free, 0.7 mm long. Female flowers: pedicels 2–4 mm long, extending to 7–10 mm in fruit; calyx lobes resembling those of the male flowers, but larger, and alternating with sessile glands; ovary 2 mm in diameter, 3-lobed, each lobe bearing 2 triangular-lanceolate appendages 1–1.5 mm long; styles 3, 4 mm long, connate at the base. Fruit 1.2 × 1.6 cm, 3-lobed, septicidally dehiscent into 3 bivalved cocci, each valve bearing a triangular appendage 1–2 mm long; exocarp reddish- or purplish-tinged to dark purplish-brown, not separating from the 2–3 mm thick woody endocarp. Seeds 6–7 × 4–5 mm, ovoid-ellipsoid, smooth, shiny, pale grey mottled blackish, ecarunculate.

Zambia. B: Zambezi (Balovale), fr. 18.iv.1954, *Gilges* 334 (K; SRGH); 3 miles west of Kaoma (Mankoya), fl. 8.xi.1959, *Drummond & Cookson* 6259 (COI; K; LISC; SRGH). W: Mwinilunga Distr., source of Zambezi R., fl. 27.ix.1952, *White* 3393 (BM; FHO; K). S: Kalomo Distr., Nkoya Native Reserve, o. fr. 23.viii.1963, *Bainbridge* 899 (FHO; K; SRGH).
Also in Angola and Zaire (Kasai, Shaba). On Kalahari Sand in ground layer of miombo and mixed deciduous woodlands, also on dambo margins under *Parinari, Isoberlinia, Julbernardia* and *Brachystegia*; 1000–1060 m.
The record for Malawi in Topham's Check List (1958), p. 53, is evidently based on a misidentification.

3. **Sapium cornutum** Pax in Bot. Jahrb. Syst. **19**: 114 (1894); in Engler, Pflanzenr. [IV, fam. 147, v] **52**: 246, t. 46A (1912). —Prain in F.T.A. **6**, 1: 1013 (1913). —Engler, Pflanzenw. Afrikas (Veg. Erde 9) **3**, 2: 142, t. 73A (1921). —De Wildeman, Pl. Bequaert. **3**, 4: 507 (1926). —J. Léonard in Bull. Jard. Bot. État **29**, 2: 141 (1959); in F.C.B. **8**, 1: 156 (1962). —White, F.F.N.R.: 203 (1962). Type from Angola (Lunda).

Sapium cornutum var. *coriaceum* Pax in Bot. Jahrb. Syst. **19**: 115 (1894); in Engler, Pflanzenr. [IV, fam. 147, v] **52**: 246 (1912). Type from Zaire (Kasai).

Sapium poggei Pax in Bot. Jahrb. Syst. **19**: 115 (1894). Type from Zaire (Kasai).

Sapium cornutum var. *poggei* (Pax) Pax in Engler, Pflanzenr. [IV, fam. 147, v] **52**: 246 (1912).

Sapium xylocarpum Pax in Bot. Jahrb. Syst. **19**: 115 (1894); in Engler, Pflanzenr. [IV, fam. 147, v] **52**: 246 (1912). Type from Zaire (Kasai).

Sapium xylocarpum var. *lineolatum* Pax, in Engler, Pflanzenr. [IV, fam. 147, v] **52**: 247 (1912). Type from Zaire (Kasai).

A shrub or small tree 1–4 m tall, sometimes taller, usually glabrous, monoecious. Twigs brownish, longitudinally furrowed. Leaves petiolate, the petioles 2–10 mm long, adaxially canaliculate; leaf blades 2.5–10(15) × 1–4(5) cm, elliptic, elliptic-lanceolate, elliptic-oblong or elliptic-ovate, subacutely or obtusely shortly acuminate at the apex, cuneate to rounded or truncate to shallowly cordulate at the base, serrate or serrulate with forwardly directed teeth, with 1–2 glands towards the base, and sometimes also with other glands not far from the margin, firmly chartaceous to thinly coriaceous; lateral nerves in 10–15(20) pairs, scarcely prominent above or beneath. Stipules 3–5 mm long, narrowly linear, auriculate and laciniate on one side at the base, ciliate, soon deciduous. Inflorescences terminal or leaf-opposed, 3–10(16) cm long, scarcely pedunculate, all male or with 1(2) female flowers at the base; male bracts broad, somewhat 3-lobed, the median lobe narrowly acuminate, denticulate-ciliolate, with 1 large discoid gland on each side towards the base, sometimes accompanied by a smaller one, many-flowered; female bracts 3-partite, larger than the males; bracteoles minute. Male flowers: pedicels 1.5 mm long; calyx lobes 3(4), 0.7–1 mm long, broadly ovate, irregularly denticulate, greenish-yellow; stamens 2–3, free, very short. Female flowers: pedicels 3–5(10) mm long, extending to 1.5–2 cm in fruit; calyx lobes 3(4), triangular, alternating with sessile discoid glands, otherwise as in the males; ovary 2 mm in diameter, 3-lobed, each lobe bearing 2 falcate appendages 2 mm long; styles 3, 7–8 mm long, connate in the lower quarter. Fruit 1–1.5 × 1.5–2 cm, 3-lobed, septicidally dehiscent into 3 bivalved cocci, each valve bearing an obliquely triangular appendage (2)4–5 mm long; exocarp reddish or blackish, not separating from the 3–4 mm thick woody endocarp. Seeds 6–7 × 4–5 mm, ovoid-ellipsoid, smooth, shiny, pale grey mottled blackish, ecarunculate.

Zambia. B: c. 19 km northeast of Mongu, fl. 10.xi.1959, *Drummond & Cookson* 6300 (K; LISC; PRE; SRGH); W: Mwinilunga Distr., Kalene Hill, fl. 26.ix.1952, *Holmes* 912 (FHO; K; PRE; SRGH).

Also in Cameroon, Central African Republic, Equatorial Guinea, Gabon, Congo-Brazzaville, Zaire and Angola. On Kalahari Sand in thickets in open woodland and Kalahari Sand mixed woodland with *Brachystegia, Colophospermum, Pterocarpus* and *Cryptosepalum*, also on rocky hill slopes.

4. **Sapium schmitzii** J. Léonard in Bull. Jard. Bot. État **29**, 2: 142 (1959); in F.C.B. **8**, 1: 159 (1962). —White, F.F.N.R.: 204 (1962). Tab. **69**. Type from Zaire (Shaba).

Very similar to *S. cornutum*, differing chiefly in the bark of the twigs breaking up into small blackish scales overlying a reddish-brown powdery tissue, in having more scattered glands on the leaf lower surface, and in having 2–4 poculiform (cup-like) glands on each side of the male bracts at the base.

Zambia. N: Kawambwa, fr. 23.viii.1957, *Fanshawe* 3554 (K; NDO). W: Mwinilunga, fl. 25.ix.1955, *Holmes* 1204 (K; NDO); c. 96 km south of Mwinilunga on Kabompo road, fr. 3.vi.1963, *Loveridge* 741 (K; LISC; SRGH).

Also in Zaire (Shaba). On Kalahari Sands in dry evergreen thicket dominated by *Cryptosepalum pseudotaxus* with *Isoberlinia paniculata*, and in escarpment and plateau woodlands, and evergreen mist forest margins; 1500–1800 m.

Tab. 69. SAPIUM SCHMITZII. 1, part of flowering branch (× ⅔), from *Holmes* 1204; 2, bisexual inflorescence (× 6), from *Holmes* 1279; 3a, fruit (× 2); 3b, fruit valve (× 2); 4, seeds (× 2), 3 & 4 from *Fanshawe* 3554. Drawn by Christine Grey-Wilson.

5. **Sapium integerrimum** (Hochst. ex Krauss) J. Léonard in Bull. Jard. Bot. État **29**, 2: 142 (1959); in Kew Bull. **14**: 62 (1960). —K. Coates Palgrave, Trees Southern Africa, ed. 2, rev.: 437 (1983). Type from South Africa (KwaZulu-Natal).

Sclerocroton integerrimus Hochst. ex Krauss in Flora **28**, 6: 85 (1845).

Tragia integerrima Hochst. ex Krauss in Flora **28**, 6: 85 (1845), nomen tantum, pro syn.

Sclerocroton reticulatus Hochst. ex Krauss in Flora **28**, 6: 85 (1845). Type from South Africa (KwaZulu-Natal).

Tragia natalensis Hochst. ex Krauss in Flora **28**, 6: 85 (1845), nomen tantum, pro syn.

Stillingia integerrima (Hochst. ex Krauss) Baill., Adansonia **3**: 162 (1862/3).

Excoecaria hochstetteriana Müll. Arg. in Linnaea **32**: 122 (1863). Types as for *Sclerocroton integerrimus* and *S. reticulatus.*

Excoecaria integerrima (Hochst. ex Krauss) Müll. Arg. in De Candolle, Prodr. **15**, 2: 948 (1866).

Excoecaria reticulata (Hochst. ex Krauss) Müll. Arg. in De Candolle, Prodr. **15**, 2: 1213 (1866). —Shinz & Junod in Mém. Herb. Boissier, No. 10: 47 (1900). —Sim, For. Fl. Port. E. Afr.: 103 (1909), pro parte.

Sapium reticulatum (Hochst. ex Krauss) Pax in Engler, Pflanzenr. [IV, fam. 147, v] **52**: 245, t. 46C, D (1912). —Prain in F.C. **5**, 2: 514 (1920). —Engler, Pflanzenw. Afrikas (Veg. Erde 9) **3**, 2: 142, t. 73 C, D (1921). —Mogg in Macnae & Kalk, Nat. Hist. Inhaca Isl., Moçamb., rev. ed.: 148 (1969).

A shrub or small tree 2–8 m tall, glabrous, monoecious. Twigs reddish-brown and flaking at first, later becoming grey-brown and lenticellate. Leaves petiolate, the petioles 3–5 mm long, adaxially ± flattened; leaf blades (2)3–8(10) × 1–3(5) cm, ovate-lanceolate to ovate-oblong, shortly obtusely or subacutely acuminate at the apex, cuneate to rounded or subtruncate and minutely cordulate at the base, entire or shallowly serrate, usually with 1–2 glands near the base, occasionally also with 1 or 2 near the margin away from the base, but not approaching the apex, firmly chartaceous to thinly coriaceous, dark green and shiny on upper surface, paler and glaucous beneath; lateral nerves in 9–12 pairs, scarcely prominent above or beneath, tertiary nerves coarsely reticulate. Stipules 5–6 mm long, linear-lanceolate, broadened and laciniate at the base, acute, ciliate, soon deciduous. Inflorescences terminal on lateral shoots, leaf-opposed or occasionally axillary, 2–7 cm long, very shortly pedunculate to sessile, all male or with 1 female flower at the base; male bracts 1 mm long, broadly ovate, denticulate-ciliolate, with a discoid gland on each side at the base, many-flowered; female bracts 3-partite, larger that the males and somewhat laciniate; bracteoles irregular. Male flowers: pedicels 1 mm long; calyx lobes 3, triangular, irregularly denticulate, yellowish; stamens 3, free, 0.5 mm long. Female flowers: pedicels 3–6 mm long, extending to 10 mm in fruit; calyx lobes 3, broadly triangular, alternating either with sessile discoid glands or lanceolate lacinulae, otherwise as in the males; ovary 2 mm in diameter, 3-lobed, each lobe bearing 2 broadly triangular appendages 1 mm long; styles 3, 7 mm long, connate at the base. Fruit 1–1.5 × 1.5–2 cm, 3-lobed, septicidally dehiscent into 3 bivalved cocci each valve bearing a shortly-conical appendage 2 mm long; exocarp becoming brownish and separating from the 2 mm thick woody endocarp. Seeds 7 × 5 mm, ovoid-ellipsoid, smooth, dull, pale greyish-brown flecked and spotted with darker brown, ecarunculate.

Mozambique. Z: between the beach and Olinga (Maganja da Costa), fr. 27.ix.1949, *Barbosa & Carvalho* in *Barbosa* 4222 (K; LMA). GI: Zavala, fl. x.1937, *Gomes e Sousa* 2047 (COI; K; LISC; MO). M: between Costa do Sol and Marracuene, fr. 13.xi.1960, *Balsinhas* 259 (COI; K; LISC; LUAI; SRGH).

Also in South Africa (KwaZulu-Natal). In coastal thickets and forest margins, and in wooded grassland on coastal plain with *Afzelia, Sclerocarya, Terminalia, Albizia, Strychnos, Garcinia* and *Friesodielsia* on sandy soil; sea level–1500 m.

Drummond in Kirkia **10**: 253 (1975) has erroneously cited *Goldsmith* 5/63 as this species; this gathering is in fact referable to *S. armatum.*

6. **Sapium armatum** Pax & K. Hoffm. in Engler, Planzenr. [IV, fam. 147, v] **52**: 244, t. 46E (1912). —Prain in F.T.A. **6**, 1: 1013 (1913). —Engler, Pflanzenw. Afrikas (Veg. Erde 9) **3**, 2: 142, t. 73E (1921). —J. Léonard in Bull. Jard. Bot. État **29**, 2: 141 (1959). Syntypes from Tanzania (Eastern and Southern Provinces).

Very similar to *S. integerrimum,* differing chiefly in the somewhat larger leaves (up to 14 × 6 cm) which are crenate-serrate and which have the glands on the blade sparingly and irregularly arranged towards the margin, and not necessarily at the base. The fruits too may be somewhat larger (up to 2.5 cm long).

Zimbabwe. E: Chirinda Distr., Ngungunyana (Gungunyana), fr. 22.xi.1943, *Williams* in *GHS* 1159 (SRGH); near Chirinda, fr. iii.1963, *Goldsmith* 4/63, 5/63 (BM; K; LISC; PRE; SRGH). **Mozambique**. N: 21 km Meconta–Corrane, fr. 18.i.1964, *Torre & Paiva* 10057 (LISC).

Also in Tanzania (east and south). In submontane open *Uapaca* woodland, and wooded grassland with *Brachystegia, Sterculia* and *Adansonia*; 220–910 m.

The bimodal distribution of this species recalls that of *Necepsia castaneifolia. Torre & Correia* 14587 (LISC) from Olinga (Maganja da Costa), Mozambique (Z), fr. 14.ii.1966, is intermediate between this and the previous species.

7. **Sapium ellipticum** (Hochst. ex Krauss) Pax in Engler, Pflanzenr. [IV, fam. 147, v] **52**: 253, t. 49 (1912). —R.E. Fries, Wiss. Ergebn. Schwed. Rhod.-Kongo-Exped. **1**: 126 (1914). —Engler, Pflanzenw. Afrikas (Veg. Erde 9) **3**, 2: 143 (1921). —Robyns, Fl. Sperm. Parc Nat. Alb. **1**: 474 (1948). —Brenan, Check-list For. Trees Shrubs Tang. Terr.: 226 (1949); in Mem. N.Y. Bot. Gard. **9**, 1: 75 (1954). —Keay in F.W.T.A., ed. 2, **1**, 2: 415 (1958). —Topham, Check List For. Trees Shrubs Nyasaland Prot.: 53 (1958). —J. Léonard in Bull. Jard. Bot. État **29**, 2: 141 (1959); in F.C.B. **8**, 1: 153 (1962). —White in F.F.N.R.: 204 (1962). —Drummond in Kirkia **10**: 253 (1975). —Troupin, Fl. Pl. Lign. Rwanda 268, fig. 92/1 (1982); Fl. Rwanda **2**: 240, fig. 73/1 (1983). —K. Coates Palgrave, Trees Southern Africa, ed. 2, rev.: 437 (1983). —Radcliffe-Smith in F.T.E.A., Euphorb. 1: 390 (1987). —Beentje, Kenya Trees, Shrubs Lianas: 221 (1994). Type from South Africa (KwaZulu-Natal).

Sclerocroton ellipticus Hochst. ex Krauss in Flora **28**, 6: 85 (1845).

Stillingia elliptica (Hochst. ex Krauss) Baill., Adansonia **3**: 162 (1862/3).

Excoecaria indica Müll. Arg. in Linnaea **32**: 123 (1863); in De Candolle, Prodr. **15**, 2: 1216 (1866) pro parte quoad spec. Krauss.

Excoecaria manniana Müll. Arg. in Flora **47**, 28: 433 (1864); in De Candolle, Prodr. **15**, 2: 1217 (1866). Type from São Tomé.

Excoecaria abyssinica Müll. Arg. in Linnaea **34**: 217 (1865/6); in De Candolle, Prodr. **15**, 2: 1214 (1866). Type from Ethiopia.

Sapium mannianum (Müll. Arg.) Benth. in Bentham & Hooker, Gen. Pl. **3**, 1: 335 (1880). —Prain in F.T.A. **6**, 1: 1016 (1913); in F.C. **5**, 2: 515 (1920). —Eyles in Trans. Roy. Soc. South Africa **5**: 398 (1916). —De Wildeman, Pl. Bequaert. **3**, 4: 509 (1926).

Sapium abyssinicum (Müll. Arg.) Benth. in Bentham & Hooker, Gen. Pl. **3**, 1: 335 (1880). —Engler, Pflanzenw. Ost-Afrikas **C**: 241 (1895).

An evergreen spreading shrub, or tree up to 50 m tall with a clear bole up to 12 m and a thin, open rounded crown with drooping branches, monoecious. Bark smooth and grey at first, later becoming brown or black, rough and fibrous. Twigs pale grey or brownish. Young shoots glabrous or sparingly puberulous and soon glabrescent. Leaves petiolate, the petioles 0.5–1.5 cm long, adaxially shallowly channelled; leaf blades 5–14(18) × 2–7 cm, elliptic-ovate, elliptic, elliptic-lanceolate or elliptic-oblong, subacute or obtusely to acutely shortly acuminate at the apex, cuneate or rounded and often minutely cordulate at the base, shallowly crenate-serrate, rarely subentire, with 2–4 glands on the margin at the base, firmly chartaceous to thinly coriaceous, usually completely glabrous, rarely sparingly puberulous on the midrib at first and later glabrescent, deep green on upper surface apart from the veins, paler dull green beneath; lateral nerves in 10–20 pairs, slightly prominent above and beneath. Stipules 2–2.5 mm long, triangular-ovate, acute, subentire, puberulous or glabrous, soon deciduous. Inflorescences usually terminal or subterminal on lateral shoots, erect, 3–12 cm long, very shortly pedunculate to almost sessile, all male or with 1–3(4) female flowers at the base; male bracts 1 mm long, ovate-suborbicular, denticulate, sparingly puberulous, with a glandular mass on each side at the base, 3–7-flowered; female bracts 2 mm long, 3-lobed, otherwise as in males. Male flowers: pedicels 1.5 mm long, puberulous; calyx lobes 2–3, unequal, 0.75 mm long, broadly ovate, glabrous or puberulous, ciliolate, yellowish-green; stamens 2–3, free, 1 mm long. Female flowers: pedicels 2–5 mm long, extending to 15 mm in fruit; calyx lobes 2–3, subequal, connate only at the base, 1.5 mm long, triangular, ciliolate, green; ovary 1.5 mm in diameter, usually 2-lobed, rarely 3-lobed or 4-lobed, smooth, glabrous; styles 2(4), 3–4 mm long, connate at the base, subpersistent. Fruit 6–9 × 8–12 mm, usually 2-lobed and laterally compressed, rarely 3–4-lobed or by abortion monococcous, indehiscent or subindehiscent, ± smooth, shiny, greenish; mesocarp fleshy; endocarp thinly crustaceous. Seeds c. 6 mm in diameter, subglobose, smooth, dull, pale greyish-brown, ecarunculate.

Zambia. N: Lake Chila, Mbala, fl. 28.i.1971, *Sanane* 1505 (K). W: Ndola, fr. 3.ix.1954, *Fanshawe* 1528 (FHO; K; NDO). N/C: between Mpika and Serenje, fr. 16.vii.1930, *Pole Evans* 291B(23) (K; PRE). **Zimbabwe**. E: Mutare, fl. xii.1944, *Chase* 126 in *GHS* 13099 (BM; COI; LISC; SRGH). **Malawi**. N: Mzuzu, fl. & fr. 24.ii.1977, *Pawek* 12357 (K; MAL; MO; PRE; SRGH; UC). C: Dowa, Ntchisi (Nchisi), st. 7.v.1961, *Chapman* 1209 (SRGH). S: Thyolo (Cholo) Mt. fl. 21.ix.1946, *Brass* 1770 (BM; K; NY; SRGH). **Mozambique**. N: Lichinga (Vila Cabral), fr. 5.iii.1964, *Torre & Paiva* 11027 (LISC). Z: Milange, fl. 11.ix.1949, *Barbosa & Carvalho* in *Barbosa* 4037 (K; LMA). T: between Casula and Furancungo, fr. 14.x.1943, *Torre* 6026 (LISC). MS: Báruè, Serra de Choa, fr. 25.v.1971, *Torre & Correia* 18642 (COI; LISC; LMU; PRE; WAG). M: Salamanga, fl. & fr. 19.ix.1947, *Gomes e Sousa* 3613 (COI; K; LISC; MO; PRE; SRGH).

From Guinea eastwards to Ethiopia and south to Angola and South Africa (KwaZulu-Natal). Locally common at medium and low altitudes in evergreen forest (mixed *Newtonia* forest) and forest margins, and in kloofs and grassy hillsides, and as a canopy tree in coastal forests, also in riverine forest and swamp forest (mushitu); 500–1800 m.

Vernacular names as recorded in specimen data include: "mchenjede", "mchenjele", "mchuwa", "mlanyandi", "mtamamjuwa", "mtepela" (chiYao); "muhlekela" (?); "munyeredze/i" (Manyika); "munyesetsi" (?); "pikapwapwo" (H).

58. MAPROUNEA Aubl.

Maprounea Aubl., Hist. Pl. Guiane **2**: 895 (1775).

Monoecious or sometimes dioecious glabrous trees or shrubs. Leaves alternate, petiolate, stipulate, simple, entire, penninerved, sparingly glandular beneath. Inflorescences terminal on short lateral shoots, spicate or racemose, usually bisexual with 1–5 female flowers at the base and a dense head of male flowers at the apex; bracts glandular at the base, the male bracts small, 3–5-flowered, the female bracts 1-flowered. Male flowers subsessile or shortly pedicellate; calyx lobes 2–3(4), imbricate; petals and disk absent; stamens (1)2–3, filaments united into a column, exserted, anthers basifixed, bithecous, longitudinally dehiscent; pistillode absent. Female flowers pedicellate, the pedicels elongating in fruit; calyx lobes 3(6), ± imbricate; petals and disk absent; ovary (2)3(5)-locular, with 1 ovule per locule; styles (2)3(5), united at the base, entire, straight, becoming reflexed, stigmas canaliculate, papillulose. Fruits ± globose or slightly (2)3(5)-lobed, dehiscing into bivalved cocci; exocarp thin, smooth; endocarp ligneous; columella persistent. Seeds obovoid; sarcotesta smooth, shiny; endotesta crustaceous, smooth, rugulose or foveolate; caruncle large, conical, 2-lobed, fleshy; albumen fleshy; cotyledons broad, flat.

A tropical genus of 4 species, 2 South American and 2 African.

Maprounea africana Müll. Arg. in De Candolle, Prodr. **15**, 2: 1191 (1866). —Pax in Engler, Pflanzenr. [IV, fam. 147, v] **52**: 178 (1912). —Prain in F.T.A. **6**, 1: 1004 (1913). —R.E. Fries, Wiss. Ergebn. Schwed. Rhod.-Kongo-Exped. **1**: 125 (1914–16). —Prain in F.C. **5**, 2: 511 (1920). —Engler, Pflanzenw. Afrikas (Veg. Erde 9) **3**, 2: 140 (1921). —De Wildeman, Pl. Bequaert. **3**, 4: 510 (1926). —Brenan, Check-list For. Trees Shrubs Tang. Terr.: 219 (1949); in Mem. N.Y. Bot. Gard. **9**, 1: 75 (1954). —Topham, Check List For. Trees Shrubs Nyasaland Prot.: 52 (1958). —Keay in F.W.T.A., ed. 2, **1**, 2: 416 (1958). —White, F.F.N.R.: 201 (1962). —J. Léonard in F.C.B. **8**, 1: 145 (1962). —Drummond in Kirkia **10**: 253 (1975). —K. Coates Palgrave, Trees Southern Africa, ed. 2, rev.: 438 (1983). —Radcliffe-Smith in F.T.E.A., Euphorb. 1: 395 (1987). Tab. **70**. Type from Angola (Moçâmedes).

Maprounea obtusa Pax in Bot. Jahrb. Syst. **19**: 116 (1894); in Engler, Pflanzenw. Ost-Afrikas **C**: 241 (1895). Type from Tanzania.

Maprounea vaccinioides Pax in Bot. Jahrb. Syst. **19**: 116 (1894). Type from Angola.

Excoecaria magenjensis Sim, For. Fl. Port. E. Afr.: 104, t. C(A) (1909). Type: Mozambique, Olinga (Maganja da Costa), *Sim* 5590 (PRE, holotype).

A sparingly-branched shrub or small tree up to 8 m tall, with pendent branches, deciduous, monoecious. Bark corky, deeply-fissured, grey or light brown. Twigs reddish-brown. Stipules 1 mm long, ovate, reddish-brown. Petioles 5–30 mm long, wine-red. Leaf blades 2–8 × 1–4 cm, ovate to ovate-lanceolate, obtuse or rounded at the apex, somewhat asymmetrically cuneate to truncate or shallowly cordate at the base, chartaceous to thinly coriaceous, bronze when young later bright green on upper surface with a yellow midrib and nerves, and glaucous beneath; lateral nerves in 6–10 pairs, usually with a few glands on them beneath, quaternary nerve network

Tab. 70. MAPROUNEA AFRICANA. 1, flowering branch (× $^{2}/_{3}$), from *Semsei* 1362; 2, inflorescence (× 2); 3, male flowers (× 8), 2 & 3 from *Shabani* 103; 4, fruits (× 2), from *Mgaza* 346; 5, seeds (× 3), from *Greenway* 5342. Drawn by Christine Grey-Wilson. From F.T.E.A.

becoming verruculose beneath with age. Inflorescences up to 2 cm long; male head 3–8 mm in diameter, subglobose or ovoid, yellowish or reddish; male bracts 1 mm long, triangular, with 1–3 obconical glands on each side near the base; female flowers 0–3 per inflorescence, green; female bracts resembling the stipules. Male flowers: pedicels to 1 mm long; calyx lobes c. 1 mm long, lanceolate, unequal, acute; staminal column 2 mm high, anthers 0.3 mm long, yellow. Female flowers: pedicels extending to 3 cm long in fruit; calyx lobes c. 1 mm long, broadly triangular-ovate, acute; ovary 1 × 1 mm, subtriquetrous-ovoid or lenticular-ovoid, smooth; styles 1–1.5 mm long, subpersistent. Fruits 8–10 × 8–12 mm, smooth, dull green, reddish-tinged, red or crimson. Seeds 7–9 × 6–7 mm, ± smooth or shallowly malleate, greenish-black; caruncle 6–7 mm long, orange or scarlet.

Caprivi Strip. Caprivi, fl. 30.viii.1967, *von Breitenbach* 1229 (PRE). **Zambia**. B: Senanga, fl. 30.vii.1952, *Codd* 7233 (BM; COI; K; PRE; SRGH). N: Samfya, fl. 19.ix.1952, *Angus* 242 (BM; FHO; K; PRE). W: Lisombo R., Mwinilunga, fl. & y. fr. 14.vi.1963, *Loveridge* 999 (FHO; K; LISC; SRGH). C: Mkushi, fr. 27.iii.1961, *Angus* 2508 (FHO; K). E: Chadiza, fr. 27.xi.1958, *Robson* 733 (BM; K; LISC; SRGH). S: Machili, fl. & fr. 16.x.1960, *Fanshawe* 5843 (K; NDO). **Zimbabwe**. E: Makurupini Forest, fr. 25.xi.1967, *Simon & Ngoni* 1300 (K; LISC; SRGH). **Malawi**. N: Vintukhutu Forest, Karonga, fr. 26.iv.1975, *Pawek* 9574 (K; MAL; MO; PRE; SRGH; UC). C: Nkhota-Kota, Chia, fl. 5.ix.1946, *Brass* 17536 (BM; K; NY; SRGH). **Mozambique**. N: 25 km Palma–Nangade, fr. & galls 5.xi.1960, *Gomes e Sousa* 4604 (COI; K; PRE; SRGH). Z: Mocuba to Olinga (Maganja da Costa), 40 km from Mocuba, fl. & y. fr. 26.ix.1949, *Barbosa & Carvalho* in *Barbosa* 4176 (K; LUAI; SRGH). MS: Chiniziua, Beira, fr. & galls 14.x.1957, *Gomes e Sousa* 4401 (COI; K; MO; PRE; SRGH). GI: Inhambane Distr., estrada para Jangamo, fl. 15.ix.1948, *Myre & Carvalho* 205 (K; LMA; SRGH).

From Nigeria eastwards to Tanzania and south to Angola and Namibia. Locally frequent in deciduous woodlands (including miombo, *Baikiaea*, *Burkea* and *Uapaca*), usually where soils are well drained, on dry sandy soil, on Kalahari Sand and on lakeshore and coastal dunes, also on escarpments, rocky hillsides and outcrops, and in wooded grassland and floodplain grassland; 15–1600 m.

Vernacular names as recorded in specimen data include: “kamwaya” (I.); “mdiadothe” (?N); “mumwaa” (chiLozi); “mutimbwa” (chiWisa).

Biegel, Check List Ornam. Pl. Rhod. Parks & Gard.. 54 (1977) reports *Fahrenheitia zeylanica* (Thw.) Airy Shaw [= *Paracroton pendulus* (Hassk.) Miq. subsp. *zeylanicus* (Thw.) Balakr. & Chakrab.], and also *Homalanthus populifolius* Grah. and *Hura crepitans* L., op. cit.: 62, as cultivated in Zimbabwe.

INDEX TO BOTANICAL NAMES